THE STEREO RUBBERS

Edited by

WILLIAM M. SALTMAN
The Goodyear Tire & Rubber Company

A WILEY INTERSCIENCE PUBLICATION

JOHN WILEY & SONS, New York • London • Sydney • Toronto

Copyright © 1977 by John Wiley & Sons, Inc.

All rights reserved. Published simultaneously in Canada.

No part of this book may be reproduced by any means, nor transmitted, nor translated into a machine language without the written permission of the publisher.

Library of Congress Cataloging in Publication Data

Main entry under title:

The Stereo rubbers.

   "A Wiley-Interscience publication.'
   Includes bibliographical references and index.
   1. Elastomers.   2. Polymers and polymerization.
I. Saltman, William M., 1917–

TS1925.S84   678'.72   76-45612
ISBN 0-471-74993-1

Printed in the United States of America

10 9 8 7 6 5 4 3 2 1

Contributors

W. Cooper, Dunlop Research Center, Birmingham, England

Giovanni Crespi, Centro Ricerche e Sviluppo, Montedison, S.p.A., Milan, Italy

G. Dall'Asta, Centro Sperimentale, Milan, Italy

F. Dawans, Institut Français du Pétrole, Rueil-Malmaison, France

E. W. Duck, The International Synthetic Rubber Company Ltd, Southampton, England

R. Dunn, Polysar Ltd., Research & Development, Sarnia, Ontario, Canada

R F. Fedors, Jet Propulsion Laboratory, California Institute of Technology, Pasadena, California

Lewis J. Fetters, Institute of Polymer Science, The University of Akron, Akron, Ohio

Umberto Flisi, Centro Ricerche e Sviluppo, Montedison, S.p.A., Milan, Italy

A. Grosch, European Tire Development Center, Deutsche Uniroyal Englebert AG, Germany

Gerard Kraus, Phillips Petroleum Company, Bartlesville, Oklahoma

J. M. Locke, The International Synthetic Rubber Company Ltd., Southampton, England

Maurice Morton, Institute of Polymer Science, The University of Akron, Akron, Ohio

Willaim M. Saltman, Research Division, Goodyear Tire & Rubber Company, Akron, Ohio

Derek A. Smith, Queen Mary College, University of London, London, England

Ph. Teyssié, Laboratoire de Chimie Macromoleculaire, Université de Liège au Sart-Tilman, Liege, Belgium

G. S. Trick, Manitoba Research Council, Winnipeg, Manitoba, Canada

Alberto Valvassori, Centro Ricerche e Sviluppo, Montedison, S.p.A., Milan, Italy

Preface

This book is intended to be a scientific treatise on the preparation, processing, curing, basic physical, mechanical, and technological properties, and major uses of the elastomers made with solution and stereospecific catalysts. The same Ziegler-Natta revolution that produced many new plastic materials resulted in a host of new and commercially valuable, structurally regular elastomers which became widely known as the "stereo rubbers." The variety of structures arising from the polymerization of dienes, which are the major monomer source for elastomers, is far more diverse than that arising from the polymerization of α-olefins. Thus the determination and understanding of mechanisms and properties are correspondingly more complex.

The goals of the book are twofold. First, they are to treat those subjects of interest to the polymer scientist who wants to know something of the theory, synthesis, structure, and basic physical properties of the stereoregular elastomers. Second, it is to treat problems of reinforcement, vulcanization, processing, and tire behavior. These are of special significance and concern to scientists within the elastomer field but are often quite strange to others.

The authors of each of the chapters in this book have tried to stress areas of importance and of lasting value. Even so, limitations of space have made it necessary to select from the larger literature and to stress those polymers that are in or near production or are of unusual structural interest.

I should like to thank my contributors for their efforts in doing the necessary research and putting together their chapters. I believe the excellence of their efforts speaks for itself.

I wish to thank my colleagues, Drs. N. Calderon, D. I. Livingston, M. C. Morris, C. R. Parks, and G. S. Trick who read various portions of the manuscripts and offered valuable suggestions. I am indebted to Dr. Leora Straka and members of the Goodyear Technical Information Center for help in searching and checking literature references and to Mr R. J. Koch for many of the data in Chapter 1. I also wish to thank Mrs. Barbara Andrus and members of her staff for manuscript typing and other services and to The Goodyear Tire & Rubber Company for permission to publish Chapter 1 and for the use of many of its facilities.

WILLIAM M. SALTMAN

Akron, Ohio
December 1976

Contents

The Stereo Rubbers

William M. Saltman

Research Division,
Goodyear Tire & Rubber Company,
Akron, Ohio

Chapter 1   Introduction

1 INTRODUCTION

The history of the discoveries that led to the current widespread scientific and commercial interest in stereoregular polymerization and its elastomeric products has been reviewed a number of times, notably by Whitby (1), who covered the period through World War II, and then by Tornqvist (2), through about 1968. It is not our intention to repeat these reviews but simply to set the stage for subsequent portions of this book.

The earlier historical period was characterized by grindingly slow research progress because of the lack of both high purity reagents and instrumental means (now available) for identifying microstructures and characterizing the qualities of the high molecular weight polymeric products. Nevertheless, two products, sodium catalyzed polybutadiene in the Soviet Union and free radical initiated emulsion styrene-butadiene rubber (SBR) in the United States and Germany, had been developed to the point where they were suitable for tires and thus the bases of enormous industries. Earlier synthetic rubbers had been poorly suited for tires.

1

The more recent period from about 1952 onward was initiated by the work of Ziegler and Natta on anionic coordination catalysts. In his Nobel Prize acceptance speech, Natta (3) reviewed some of the earliest work on the preparation of polyethylene and subsequent work with other olefinic and nonolefinic monomers. The speech shows the enormous range of Natta and co-workers' activities but also reveals that in their interest in crystalline polymers they allowed others priority in the elastomeric field.

During the same period, in more technologically oriented work, Stavely (4) and Korotkov (5) and their co-workers in the United States and the Soviet Union, respectively, were bringing the value of the organolithium catalysts to fruition. It was during this period from 1952 on that most of the stereospecific catalysts and the elastomers discussed in this book were first prepared, investigated, improved, and commercialized.

The first successful coordination catalyst was a combination of an aluminum trialkyl and titanium tetrachloride which, Ziegler found, produced, at low pressure and at room temperature, a high molecular weight, high density, linear polyethylene (6). The catalyst family swiftly expanded to include, in general terms, an organometallic compound plus a transition metal compound (and perhaps additional components). Many such combinations were found to be capable of polymerizing, under mild conditions, a great variety of hydrocarbon and nonhydrocarbon monomers to polymers with regular and specific types of microstructure. The search for new catalyst combinations continues to this day.

The first elastomer of significance to be synthesized was *cis*-1,4-polyisoprene. It was prepared from isoprene independently and almost simultaneously with lithium (4, 5) and with AlR_3–$TiCl_4$ catalysts (7). These syntheses fulfilled the long sought for goal of duplicating the structure of natural rubber. It is likely that much of the initial research activity in the United States was motivated by fears of further world conflict or unrest which could again cut off natural rubber supplies from the Far East, as had happened during World War II. This event and the recently ended Korean war led to deep concern that the United States not be totally dependent on foreign sources of natural rubber. Subsequently, the economic and technological advantages to be gained from using elastomers not available from nature led to the investigation of many other monomers.

Many of the research workers responsible for these splendors have received due credit for their efforts, but in some instances patent complexities have obscured the identities of the earliest investigators. It is well known that patent offices have almost universally favored, in some perfectly legal way, the native over the foreign inventor.

It is by no means true that any particular polymer structure can be made by any catalyst combination. It is this factor, of course, that has made catalyst research so important a component of industrial research. A tailor who tailored clothes as blindly as we "tailor" polymers would soon smother in a mountain of faulty cuttings. Nevertheless, the variety of polymer structures already prepared is truly amazing and new treasures are still being unearthed. Polymers, copolymers, and terpolymers have been prepared, linear and branched, alternating and block, hydrocarbon and nonhydrocarbon, high and low molecular weight, with narrow and broad molecular weight distributions. As more polymers with well-defined structures are prepared, our evaluation of them becomes more precise and the correlation of structure and physical properties becomes better understood.

The chapters immediately following this one describe the preparation, properties, and theory involved in preparation of many stereoregular elastomers. The types of products are too numerous to describe fully even in a volume of this size. Emphasis has had to be on polymers of fundamental structural interest and/or commercial value.

Subsequent chapters deal with reactions of the polymers with oxygen, which are of concern because of the desirability for stability during the useful life of the product, and with sulfur, which are related to the peculiar property of elastomers that they develop their best physical properties when linked into giant three-dimensional networks.

Later chapters show the considerable effort that has been made to correlate polymer structure with physical properties. Tremendous progress is now possible with the wide variety of new polymers at hand. There are still many unsolved problems, and undoubtedly this will continue to be one of the most fruitful research areas in coming years. Wherever data are available attempts have been made to compare or to correlate properties of the stereoregular elastomers with those of the older elastomers. Such data are, all too often, missing. Often a major portion of our current knowledge is, in fact, based on the many investigations of natural rubber which for such purposes may be regarded as the oldest of the stereoregular elastomers.

It is fitting that following the discussion of the basic viscoelastic and technological properties, the book ends with a discussion of some of the factors involved in tire quality, since tires comprise the major end use for rubber. Tire quality depends greatly on engineering and design elements that are beyond the scope of this book, but there is a basic dependence on the rubber and compounds used. Good tire construction presents a complex of technological problems that will tax the abilities of polymer chemists, physicists, and engineers for years to come.

2 ECONOMICS

The extensive and sometimes feverish research that followed the discovery of the lithium and Ziegler-Natta catalysts soon enlarged on the original polyisoprene investigations to include other elastomeric products. In less than a decade more than a dozen commercial elastomer producing plants had sprung up to exploit the results.

By 1970 there were 5 plants in the United States and about 22 throughout the world (8, 9) producing about 940,000 long tons/year of polybutadiene using at least five different catalyst systems. There were 10 plants (3 in the United States) producing 310,000 tons/year of *cis*-polyisoprene with three catalyst systems. About 40,000 annual long tons of solution styrene-butadiene copolymers were being produced in 7 plants with lithium based catalysts and 179,000 long tons of ethylene-propylene copolymers (EPM or EPR) or terpolymers (EPDM) in 10 plants (5 in United States) with coordination catalysts based on vanadium salts. Except for about 1000 long tons of *trans*-polyisoprene production and a small but swiftly burgeoning production of triblock isoprene-styrene and butadiene-styrene copolymers (over 100,000 long tons in 1976) made with lithium catalysts (10) no other elastomeric solution polymers are known to be in commercial production, although several, based on new catalyst systems or monomers, are in advanced development stages.

The production levels indicated above are still considerably less than levels for natural rubber and emulsion SBR. All indications are that, as population and technology grow, consumption levels of all polymers will increase. However, the property improvements obtainable with the solution elastomers, sometimes only marginal but more often substantial, make it seem inevitable that their rates of growth will continue to exceed those for natural rubber or emulsion SBR. There will be considerable jostling for position in the marketplace among the various solution polymers and also in competition with modified plastics such as the thermoplastic elastomers, plasticized polyvinyl chloride, cross-linked polyethylene, and others.

In 1970 world production of natural rubber was 2.96 million long tons; this was expected (11, 12) to increase to about 6.9 million tons by 1985. However, percentagewise, its use declined from 49% of total rubber in 1960 to about 31% in 1974. In the United States the decline was even more dramatic; during the years 1969–1975 natural rubber consumption was only about 22.5% of total rubber consumed (13). Petrochemical shortages may lead to a resurgence in its use. Because of the various and recurrent economic, energy, and petroleum crises, by 1980 natural rubber may be taking 40% of the world rubber market with 5 million tons/year of tree rubber (12). Some synthetic rubber spokesmen believe it would be wiser to limit new rubber tree acreage and devote more of the world's limited supply of arable land to raising more food crops. Since the entire petrochemical industry of the world consumes less than 5% of total crude oil output, feedstocks should be ample for synthetic rubber, plastics, and other petrochemical requirements (14).

Worldwide, synthetic rubber production in 1970 was 5.60 million long tons, probably growing to about 12 million tons by 1985. About half of this has been emulsion SBR. In the United States the use of synthetics is considerably higher than that in the rest of the world. As indicated by the data in Table 1, about 78% of all rubber now used in the United States is synthetic and 60% of the synthetic is emulsion SBR (12, 15).

The proportion of synthetic rubber consumed which is emulsion SBR is slowly decreasing, although total tonnage continues to increase. Some industrial leaders have been willing to predict a world demand by 1980 for 11.4 million tons of rubber per year of which 25% will be natural rubber, 30% emulsion SBR, and 45% solution rubbers (16). Although not all predictions are so sanguine (13), in the near future solution polymerization and stereospecific catalysts may be the most important in the world.

The data in Table 1 show the remarkable increases in production of the solution rubbers over the last 15 years, from zero production in 1960 to where polybutadiene is now the second most widely used synthetic rubber.

Table 1   Rubber Consumption in the United States (10,15, 22) (Thousands of Long Tons)

Type	1960	1965	1969	1973	1974	1975	1976[a]
Synthetic							
SBR	904	1096	1300	1478	1400	1179	1325
Polybutadiene	0	133	270	314	330	290	335
Neoprene	81	100	120	143	130	144	161
Butyl	62	75	92	140	141	80	140
Polyisoprene	0	42	85	119	105	61	100
Nitrile	32	49	64	63	71	55	60
EPDM	0	13	43	96	106	84	100
Other[a]	—	33	33	35	36	47	54
Total synthetic	1079	1541	2007	2108	2319	1940	2275
Natural	479	515	588	685	660	665	740
Total, new	1558	2056	2595	2793	2979	2605	3015

[a] Estimated.

Data from other countries show the same trends as in the United States. Extensive studies have been made (9) and considerable statistical data are published (17). For example, currently in Japan plant capacity exists for over 300,000 annual tons of SBR production and 95,000 tons of polybutadiene (18). Several new polyisoprene plants are in the planning or building stages; one is already in operation. By 1978 demand may reach 200,000 tons of polybutadiene and 115,000 tons of polyisoprene (19, 20).

The Soviet Union consumed about 1 million tons of synthetic rubber in 1970 and expects to use three times as much in 1975. Of 1968 production, 70,000 tons were polyisoprene, 40,000 were polybutadiene, and about 30,000 were ethylene-propylene rubber (18, 21). Demand for synthetic polyisoprene is expected to reach 350,000 tons by 1978 (19).

3   RAW MATERIALS

Commercial success for a polymer requires a number of qualities. Ultimately, the polymer must have technological qualities such that in one or more large-scale applications its overall suitability is superior to that of any competing polymer. It must also be inexpensive for these applications, which implies that monomers and catalysts be available and inexpensive and that manufacturing procedures be within the capabilities of modern technology.

All the monomers used for commercial polymers are derived from petroleum sources. If oil wells begin to run dry, the chemical and rubber industries (and all of us) will have to cope with entirely new economic patterns. Recent world events foreshadow just such changes. Currently used monomers may become prohibitively expensive and new monomers may begin to look attractive. At present production rates the world may have only about 35 years of known petroleum reserves left. Although our years of grace may be extended substantially by increases in exploratory drilling and by greater recycling of used products it is quite possible that some readers of this book will live to grapple with the problems of preparing monomers from shale oil, coal, carbohydrates, or old tires (23).

Isoprene is the only major new monomer brought into large-scale commercial production because stereospecific catalysts became available to make a desirable elastomer. Small quantities of isoprene had been available from C_5 refinery streams for use in providing unsaturation in butyl rubbers, but this material had neither the quality nor quantity needed for *cis*-polyisoprene production. However, increasing amounts are becoming available as refinery mixes change, and may, in the future, again be a main source for isoprene.

Current supplies of isoprene are prepared chemically (24) in several ways: in a three-step synthesis from propylene (Goodyear-Scientific Design); by dehydrogenating isoamylene (Shell) or other C_5 refinery stream (Enjay); by a two-step process from formaldehyde and isobutylene (Institut Francais du Petrole) (25); or from acetylene and acetone (SNAM) (26). Since most isoprene is used captively, indicated prices of 20–25¢/lb are formal rather than real.

Synthetic *cis*-polyisoprenes must compete in both quality and price with natural rubber. Although basically identical there are several property differences among the natural and the two synthetics (lithium or Ziegler catalyzed) so that one may be preferred over the other two. Often, but not always, price is not the least of these differences. There are almost no other uses for isoprene except for polyisoprene so that consumption of isoprene is reliably reflected by figures for polyisoprene. United States data on consumption of polyisoprene are shown in Table 1. Worldwide plant capacity for polyisoprene was about 300,000 annual long tons in 1970, of which about one-third was within the Soviet nations. In 1973 it was 487,000 long tons. By 1976 it is estimated that worldwide plant capacity will be over 1 million long tons, promising a remarkably high growth rate (19, 21).

Butadiene is the starting monomer for the most widely used stereo rubber as well as for emulsion SBR which is still the most widely used synthetic rubber in the world (Table 1). As a result large butadiene production and purification facilities have been in existence for many years. United States production in 1972 was 1.8 million long tons; world production (excluding the Communist countries) was 3.7 million. Demand for butadiene in the United States decreased in the following years to below the 1.5 million tons expected in 1975, and may not again exceed 1.8 million tons before 1977 because of the business recession and expected slow recovery in the automobile and housing industries (27). Worldwide, demand is expected to reach about 5 million tons by 1980. Shifts to smaller, lighter cars, slower driving speeds, and higher gasoline costs could reduce forecasts for butadiene growth from 4% annually to about 2%.

Almost 85% of the butadiene produced in the United States is used in the manufacture of synthetic elastomers and perhaps 90% of what is produced in Europe. Demand for butadiene is expected to increase more rapidly in Europe than in the United States, since Europe's share of world synthetic rubber production is growing while that of the United States is falling.

Future supplies of butadiene will be involved in very intricate ways with the demand for ethylene. The new naphtha (steam) cracking capacity for ethylene that is being developed will also yield increasing quantities of coproducts butadiene and propylene. The yields of products depend on whether gas oils, naphthas, or lighter feedstocks are used, and thus the proportions of each product will be attuned to achieve maximum profit for each feedstock (28). Much of the new steam cracking capacity is being built in Europe. The access to feedstocks by United States refiners is beset with political and economic restrictions involving the oil producing and exporting nations; these are too well known and too complex to discuss here.

Currently dehydrogenation processes using butenes and butanes are the main sources for butadiene. The price of butadiene is now (1976) about 20¢/lb but no more than three years ago was 9–11¢/lb. The product has been in short supply in the United States since 1969, and as a result the United States imports substantial quantities from both Europe and Japan. In 1974 imports amounted to about 12% of United States production (27). By 1980 butadiene is expected to be in substantial surplus throughout the world.

The advent of solution SBR created no new supply problem for monomeric styrene since styrene, like butadiene, has been used for many years in preparing emulsion SBR. It is even more widely used for polystyrene (crystal and reinforced), acrylonitrile-butadiene-styrene (ABS) terpolymers, and styrene-acrylonitrile (SAN) copolymers.

United States production of styrene in 1974 was about 2.7 million long tons. Of this about 0.3 million is used in elastomers, the remainder going into plastics and resins. Styrene is prepared from ethylbenzene which in turn is made from ethylene and benzene under Friedel-Crafts conditions. The 1974 shortages of styrene, benzene, and other aromatics have become sufficiencies in 1975; shortages are not foreseen for the next few years. The complex interplay of social forces and economics is exemplified by the connection between the reduction and possible banning of lead compounds in gasolines because of air pollution problems and the scarcity of benzene and the price of styrene. The banning (29) of tetraethyl- or tetramethyllead compounds would change the product mix desired from refinery operation to the point that the entire raw material balance of the chemical industry might be changed. Propylene and isobutane needed for alkylation become more valuable to petroleum refiners as lead contents in gasoline drop since relatively small quantities of lead alkyls raise the octane rating of propylene alkylates much more than they raise the rating of other, less sensitive hydrocarbons (30). Aromatics also help raise octane ratings and become more valuable to gasoline makers (28).

Ethylene and propylene are used in many reactions other than polymerization; their total production is enormous. Of over 23 billion lb/year of United States ethylene production, about 40% goes into polyethylene (both high and low density) and most of the remainder into nonpolymer products. Only about 170 million lb are used for elastomeric ethylene-propylene copolymers (EPM) and ethylene-propylene-diene ter-polymers (EPDM). Polypropylene production at about 1.7 billion lb/year (4.5 billion lb worldwide) consumes only about 10% of all propylene production, but yet far more propylene than EPM and EPDM. Monomer sources have therefore been no problem for the ethylene-propylene elastomers. Costs have been as high as 9–11¢/lb for polymerization grade ethylene and propylene in 1976, up from about 3¢ in 1973. The third monomers (dienes), that incorporate unsaturation in EPDM polymers may cost as much as 75¢/lb but are present in only small percentages.

Production projections and estimates of future growth of EPDM have, over the past decade, been the subject of some of the most fanciful predictions in the industry. Many estimates were based on the assumption that EPDM would find widespread acceptance for tire tread and carcass uses, an assumption that has not yet materialized in 1976. For example, in 1962 market researchers predicted 40,000 long tons usage by 1965 (actual 13,000). In 1965 one predicted 50,000 tons by 1970; in 1967 several predicted 57,000, 145,000, and 175,000 tons by 1970 (actual 75,000). For 1975, predictions have ranged between 100,000 and 325,000 tons. Consumption in 1972 was about 67,000 long tons, of which 75% was for nontire uses. The tire uses are primarily for white sidewalls. United States consumption in 1973 reached 92,000 long tons, for a 35% increase over 1972. In 1975, consumption for 1974 is expected to be 122,000 tons and is predicted at 248,000 tons by 1985 (13).

4 TECHNOLOGY AND USES

About two-thirds of all rubber is used (31) in the manufacture of tires and tire products (Table 2). However distribution is quite uneven, as shown in the most recent tabulations (Tables 2–4). Because of its high abrasion resistance, over 95% of polybutadiene production finds its way into tires and only minor amounts into other uses. In 1969 only about 10% of EPDM (5000 long tons) went into tires, and this almost exclusively for sidewall uses; the remainder was used for a wide variety of automotive parts, hose, footwear, and so on. By 1973, 17,000 long tons went into tires (32). Table 3 shows some recent and projected distributions. *cis*-Polyisoprene is a general purpose rubber, as is natural rubber. Lithium catalyzed polyisoprene is less suitable for tire uses whereas the coordination catalyzed material may be used more broadly as a straight replacement for natural rubber. Solution SBR is a general purpose rubber very

Table 2 End Uses for All Rubber (31) (United States Consumption, Thousands of Long Tons)

	1963	1969	1975[a]
Tires and tire products	1081	1685	2050
Nontire products			
Footwear	35	44	44
Belts, belting	23	32	44
Hose, tubing	34	52	67
Sponge rubber products	30	40	45
Foam rubber products	56	86	110
Floor and wall covering	28	20	12
O-rings, packing, gaskets	27	42	55
Pressure-sensitive tape	11	14	18
Industrial rolls	8	14	20
Automotive molded goods	100	125	162
Other molded goods	85	140	190
Military goods	10	13	15
Shoe products	53	53	50
Drugs, medical sundries	7	11	17
Coated fabrics	20	30	40
Thread (bare)	9	12	15
Solvent and latex cement	28	36	48
Toys, balloons	8	10	14
Athletic goods	10	14	21
Wire, cable	36	32	35
Other[b]	65	107	136
Total nontire products	683	915	1115
Total consumption	1764	2600	3165

[a] Estimated.
[b] Includes sealants, sheeting and lining, coatings, latex applications, plus other miscellaneous uses.

similar to its emulsion counterpart (Table 4) but with a number of small differences, such as lower nonrubber content and lower glass transition temperature, which has led to its increasing acceptance in commerce and accounts for optimistic projections for its future growth.

The relative positions of the competing polymers are ever shifting as new or improved versions are brought into production. Because of the enormous market for tires commercial success for a new polymer is at least an order of magnitude easier if it is useful in tires. As implied in Table 2, nontire uses require considerable versatility in rubber properties

**Table 3   EPDM Consumption in the United States (31, 33)
(Thousands of Long Tons)**

Use	1970	1975[a]	1980[a]
Tire products (sidewalls)	5[a]	14	27
Automotive parts	20[a]	48	88
Wire, cable	2	4.5	5
Appliance parts (except hose)	4.5	9	11
Hose	4	8	9
Belting	1	3.5	4
O-rings, seals, gaskets	2.5	4.5	5.5
Rolls	1	2.5	3.5
Proofed goods	2	4	4.5
Footwear	2	4	4.5
Rug underlay	0.5	2.5	3.5
Nonautomotive tires	0.5	2.5	3.5
Other	1	2	3
Total	46	109	172

[a] Estimated.

Table 4   SBR[a] End Uses—United States Consumption, 1969 (31)

Use	Long tons (thousands)
Tires, tire products	890
Foam rubber	50
Footwear, shoe products	62
Belting	8
Hose, tubing	17
Sponge	10
Proofed goods	10
Wire, cable	11
Adhesive	10
Miscellaneous mechanical goods	50
Miscellaneous latex	60
Miscellaneous solid products	132
Total	1310

[a] Almost entirely emulsion SBR.

to satisfy a wide variety of uses. EPDM has been an example of a product with many excellent properties but still not able to satisfy the requirements for use in tires except in sidewalls where ozone resistance is overriding in importance. Early versions of EPDM lacked building tack and were too slow curing even with ultraaccelerators to be compatible with diene elastomers. Faster curing EPDMs are now available but further improvements are still needed before the EPDMs will find wide acceptance in tires. When this happens the optimistic forecasts for EPDM may appear conservative.

Changes in the engineering aspects of tire design also influence the choice of rubber used. Dramatic changes are occurring both in tire construction and in tire cord materials.

There are three basic types of tires in use today. The bias ply (or bias angle) has reinforcing cords extending diagonally across the tire from bead to bead (edge to edge). Alternating layers or plies are angled in opposite directions. The second type, the bias belted tire, has similar reinforcing plies but in addition has a restricting belt running around the circumference of the tire. In the third type, the radial ply tire, the reinforcing cords extend from bead to bead at (almost) right angles to the plane of the tire. On top of the plies (under the tread) there is also an inextensible belt (34).

Bias belted tires, which have been standard equipment on almost all new American cars, were accepted very quickly because of their improved performance over the older bias ply types. Their introduction was accomplished quickly by the tire industry because only relatively minor changes in tire building machinery were required. More radical changes in plant equipment are required to make radial ply tires. The switch from bias ply to belted tires occurred at tremendous speed. In 1968 over 154 million bias ply passenger tires (90%) were made plus 15 million bias belted and 3 million radial tires. In 1970 about 49% were bias belted, 49% bias ply, and 2% radial. Worldwide usage was estimated as 64% bias ply, 25% bias belted, and 12% radial (24). Since then, the changeover in tire constructions is proceeding about as shown in Table 5. The radial tire is expected to be the major construction throughout the world in 1977 by European experts and in 1978–1979 by United States experts (35). It would appear from these estimates that the switch to radial tires will be swifter than earlier projections and almost as swift as the earlier switch to bias belted tires (36). The vast number of tires involved in these changeovers may be seen in Table 6 which indicates United States shipments of passenger, truck, and farm tires during 1972–1974 (37). Although 1974 sales are down about 10% from 1973 some recovery is projected for 1975 (38).

Table 5 Division of Tire Market by Construction Type (36, 38)

Year	Construction	Original Equipment (%)	Replacement (%)
1973	Bias ply	18	49
	Bias belted	64	39
	Radial	18	12
1974	Bias ply	14.6	40.4
	Bias belted	45.4	36.2
	Radial	40	23.4
1975	Bias ply	7	37
	Bias belted	25	33
	Radial	68	30

The success of a particular design may depend on the suitability of the tire cord material. Rayon, nylon, polyester, glass, and steel are all in use. Between 1900 and 1972 almost 5 billion tires were made in the United States. Of these about 33% were made with cotton tire cord, 40% with rayon, 22% with nylon, and 5% with polyester. About 1% contained fiberglass and less than 1% contained steel wire. Cotton is no longer in use and rayon is long past its prime. Nylon and polyester are now most widely used. In recent years steel wire and glass fibers for use in belts

Table 6 United States Shipments of Tires (37, 38)
(Millions of Units)

	1972	1973	1974
Passenger			
Original equipment	51.2	55.4	44
Replacement	141.3	147.0	133
Truck, bus			
Original equipment	12.6	13.6	9.6
Replacement	20.1	21.2	22.8
Farm			
Original equipment	1.6	1.7	2.3
Replacement	3.0	3.1	3.7
Total	229.8	242.0	215.4

Table 7   Tire Cord Usage (23) (Millions of Pounds)

Type	1973	1974[a]	1975[a]
Rayon	92	75	65
Nylon	292	283	278
Polyester	238	270	290
Glass	38	40	45
Steel	80	120	150
Total	740	788	828

[a] Estimated.

have created interest and optimism with regard to their future. Throughout the world well over a billion pounds of tire cord are used each year. About 740 million are used in the United States alone (36). Changeovers in tire cord use are projected to occur as shown in Table 7. In addition to those shown, small amounts of an aramid fiber (DuPont Kevlar, Fiber B) will be used. Threatening all of this, the potential exists for a commercial tire with neither cord nor belts (39) for the somewhat more distant future.

Radial ply tires are reported (40) to be better than bias belted tires in giving better tread wear, less rolling resistance, better ride at higher speeds, and lower running temperatures. The belted tire has better ride characteristics at lower speeds, better cut and bruise resistance, less shake, more uniformity, and costs less to build. The building of radial tires requires rubbers with better uncured (green) strength and better building tack. As radial ply tires become more popular, larger proportions of natural rubber and polyisoprene will have to be used in carcasses and sidewalls of tires in place of SBR and polybutadiene (41) unless improvements are forthcoming in their green strengths.

5   FUTURE TRENDS

By making and selling elastomers and other chemical products in virtually every more or less free enterprise country, the United States rubber companies have pioneered tire business around the world. Outside of Europe, Japan, and Australia, the United States companies are often the dominant or only factor in the tire market. The foreign markets (for United States companies) are most eagerly sought since they are fastest growing. For example, between 1960 and 1970, United States automobile registrations increased 46% to about 90 million. In Europe, Latin America, and other parts of the world registrations more than doubled. In Japan, auto registrations increased over 2000%, from 400,000 to 8.6 million during the same period. The potential for still further growth is plainly apparent. In 1971 the United Kingdom had 186 cars per 1000 inhabitants, West Germany 199, France 240, and Italy 151. The United States had 410 (42). United States shipments were about 190 million units in 1971 for both original and replacement automobile tires. For the rest of the free world they were over 219 million. If truck and farm tires were included, the spread would be even greater (43).

There can be little doubt that the dominant United States position will be seriously contested in the future by Japanese and European producers. Considering the ferment and the multiple crises over oil, energy, and other resources it is likely that vast petrochemical and other industrial complexes will be built in Africa and the Near East in the near future. Only the boldest forecaster should try to go beyond the middle of the next week. Nevertheless, ...

Future research and development trends being perhaps somewhat less difficult to predict than economic trends, we are not lacking in attempts to do so. One very interesting attempt has been made by Kovac (44). In addition to the usual procedures of correlating and extrapolating already existing trends and starting with estimated future needs or goals and working backwards, he assembled a consensus of intuitive forecasts appropriately enough called a Delphi probe as a collective consultation of the oracles. A group of experts was polled to list the most important developments it anticipated between now and the year 2000, and then was asked to estimate a date for realization of each development. The results are shown in Fig. 1.

The estimates seem to fall in no particular order, although it would appear that new fibers and new elastomers will be quickly realized innovations and that developments connected with very high speed tires and/or improved public transportation are most distant. Such estimates are of interest not only because they reflect the then current (1971) expert opinion but also because if authorities believe certain goals to be attainable and others not, they direct their research activities toward the "attainable" ones and thereby create self-fulfilled prophecies. It makes one slightly uneasy, therefore, to contemplate the apparent unrealizability of a biodegradable tire tread. Fortunately, other avenues for recovery of worn tires or other rubber articles are in use or contemplated.

Another similar probe (45) made by a DuPont group found that many experts believed a stepwise trend toward automation and continuous processing would soon change the face of rubber manufacture. The

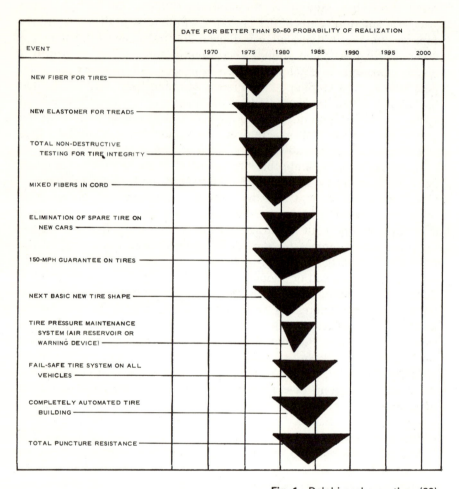

Fig. 1 Delphi probe on tires (29).

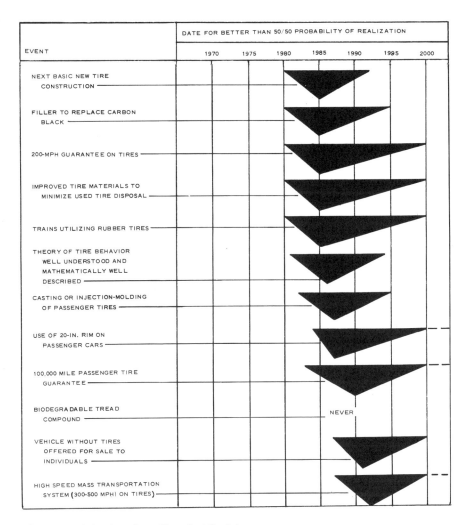

EVENT	DATE FOR BETTER THAN 50/50 PROBABILITY OF REALIZATION							
	1970	1975	1980	1985	1990	1995	2000
NEXT BASIC NEW TIRE CONSTRUCTION								
FILLER TO REPLACE CARBON BLACK								
200-MPH GUARANTEE ON TIRES								
IMPROVED TIRE MATERIALS TO MINIMIZE USED TIRE DISPOSAL								
TRAINS UTILIZING RUBBER TIRES								
THEORY OF TIRE BEHAVIOR WELL UNDERSTOOD AND MATHEMATICALLY WELL DESCRIBED								
CASTING OR INJECTION-MOLDING OF PASSENGER TIRES								
USE OF 20-IN. RIM ON PASSENGER CARS								
100,000 MILE PASSENGER TIRE GUARANTEE								
BIODEGRADABLE TREAD COMPOUND					NEVER			
VEHICLE WITHOUT TIRES OFFERED FOR SALE TO INDIVIDUALS								
HIGH SPEED MASS TRANSPORTATION SYSTEM (300-500 MPH) ON TIRES								

Courtesy of the American Chemical Society.

consensus appeared to be for highly automated factories by the mid-1980s. A tire without fabric in the carcass or belt is not expected to make its mark until the year 2000. Neither probe directed attention to the imminent trends toward smaller cars, smaller tires, slower speeds, and raw material shortages. The most recent Delphi probe (35) again emphasized the trends toward continuous automated processing using pre-compounded powdered or pelletized stocks. Many of the changes predicted to become major have already appeared above the horizon.

REFERENCES

1. G. S. Whitby, ed., *Synthetic Rubber*, Wiley, New York, 1954.

2. E. Tornqvist, in J. P. Kennedy and E. Tornqvist, eds., *Polymer Chemistry of Synthetic Elastomers*, Wiley, New York, 1968, Ch. 2.

3. G. Natta, *Science*, **147**, 261 (1965).

4. F. W. Stavely et al., *Ind. Eng. Chem.*, **48**, 778 (1956).

5. A. A. Korotkov et al., *Dokl. Nauk SSSR*, **110**, 89 (1956); **115**, 545 (1957); R. Ulbrich, *Gummi Asbest.*, **11**, (5), 303 (1958); I. V. Garmonov, International Symposium on Isoprene Rubber, November 20–24, 1972, Moscow.

6. K. Ziegler et al., Angew. Chem., **67**, 541 (1955).

7. S. E. Horne, Jr., U.S. Pat. 3,144,743 (to Goodrich-Gulf Chemicals, December 17, 1963); S. E. Horne, Jr. et al., *Ind. Eng. Chem.*, **48**, 785 (1956).

8. C. F. Ruebensaal, *Rubber World*, **162** (5), 56 (August 1970).

9. C. F. Ruebensaal, *Rubber Plast. Age* (*London*), **49**, 913 (October 1968); **50**, 764 (October 1969); *Rubber World*, **162** (4), 80 (July 1970).

10. *Chem. Eng. News*, **54**, 15 (May 10, 1976); *Rubber Plast. News*, **5**, 12 (March 8, 1976).

11. *Chem. Eng. News*, **48**, 18 (January 12, 1970); S. T. Semegen, *Rubber Plast. News*, **2**, 8 (December 3, 1973).

12. L. Mullins, *Chem. Eng. News.* **53**, 23 (June 16, 1975); *Chem. Mkt. Abstr.*, **68**, 174 (April 1976).

13. *Rubber Plast. News*, **4**, 25 (February 24, 1975).

14. *Chem. Eng. News*, **52**, 10 (September 30, 1974).

15. *Chem. Eng. News*, **47**, 46 (July 14, 1969); **48**, 77a (September 7, 1970); **48**, 106 (October 19, 1970); **54**, 37 (June 7, 1976).

16. *Chem. Eng. News*, **47**, 21 (May 26, 1969).

17. *Rubber Statistical Bulletin*, issued monthly by the Secretariat of the International Rubber Study Group, Brettenham House, 5–6 Lancaster Place, London.

18. *Chem. Eng. News*, **48**, 35 (April 20, 1970).

19. *Chem. Mkt. Abstr.*, **67**, (3), 176 (March 1975).

20. *Chem. Mkt. Abstr.*, **67** (2), 178 (February 1975).

21. *Rubber World*, **162** (4), 80 (July 1970); *Chem. Eng. News*, **51**, 7 (July 9, 1973).

22. *Rubber Age*, **107** (1), 38 (January 1975); **108**, 26 (January 1976).

13521111

23. *Chem. Eng. News*, **48,** 42 (August 24, 1970); **48,** 68 (May 17, 1970).

24. M. J. Rhoad, *Rubber Ind.*, **9** (2), 68 (April 1975).

25. *Chem. Eng. News*, **48,** 60 (July 7, 1970); *Chem. Eng.*, **77** (27), 90 (1970); *Chem. Week*, **108** (12), 39 (March 24, 1970).

26. *Chem. Eng. News*, **43,** 51 (May 17, 1965).

27. B. F. Greek, *Chem. Eng. News*, **53,** 9 (June 16, 1975).

28. B. F. Greek, *Chem. Eng. News*, **53,** 8 (March 31, 1975).

29. *Chem. Eng. News*, **48,** 61A (September 7, 1970).

30. *Chem. Eng. News*, **53,** 10 (March 10, 1975).

31. *Chem. Week*, **106** (11), 56 (March 18, 1970).

32. B. K. Lyckberg, *Rubber Plast. News*, **2,** 6 (December 3, 1973).

33. Noble Keck, *Rubber Age*, **105** (9), 43 (September 1973).

34. F. J. Kovac, *Tire Technology*, 3rd ed., Goodyear Tire & Rubber Company, Akron, Ohio. 1970.

35. J. L. Von Wald, Paper #22, ACS Rubber Division Meeting, Cleveland, Ohio, May 1975.

36. *Rubber World*, **169** (4), 37 (January 1974).

37. *Chem. Mkt. Abstr.*, **65** (8), 967 (August 1973).

38. *Rubber World*, **171** (4), 27 (January 1975); *Modern Tire Dealer*, **56** (6), 52 (January 1975); *Chem. Week*, **118,** 13 (February 25, 1976).

39. *Chem. Eng. News*, **48,** 11 (February 2, 1970).

40. *Chem. Eng. News*, **48,** 36 (April 27, 1970).

41. *Chem. Eng. News*, **43,** 16 (March 11, 1965).

42. *Wall Street Journal*, p. 1 (May 17, 1971).

43. R. DeYoung, Goodyear Tire & Rubber Company Shareholders Meeting, April 5, 1971.

44. F. J. Kovac, *Chem. Technol.*, **1,** 18 (January 1971).

45. *Chem. Eng. News*, **50,** 10 (April 17, 1972).

W. Cooper
Dunlop Research Centre
Birmingham, England

Chapter 2 Polydienes by Coordination Catalysts

1 INTRODUCTION

Polymerization of dienes by coordination catalysts was discovered about 1955 (1); the first major disclosure was the synthesis of *cis*-polyisoprene using essentially the same catalyst as had been found to be effective for the polymerization of ethylene, namely titanium tetrachloride and tri(isobutyl)aluminum (2). Following intensive work, which has continued up to the present, numerous catalyst systems have been discovered which give stereoregular polydienes. Five of these and variations are now in large-scale commercial use for the manufacture of *cis*-polybutadiene and *cis*-polyisoprene. In Japan *ca* 90% syndiotactic 1,2-polybutadiene containing 15 or 25% crystallinity is now available as a thermoplastic elastomer.

Many compounds or complexes of transition metals will polymerize dienes, particularly conjugated dienes, but many of the polymers obtained are of mixed microstructure. Others are low molecular weight oligomers or cross-linked, or insoluble and partly cyclized. There is little merit in examining such work in detail; this chapter is concerned mainly with those systems that give stereoregular polymers, and particularly those that are of actual or potential economic value. For some of the more important, the reaction variables have been examined fully and it is possible to define the optimum conditions, but in many others experimental detail is scanty. Reported variations in polymer yield or microstructure may be significant but more often they are simply the result of poorly optimized conditions or the presence of undisclosed amounts of impurities in the monomers or catalyst ingredients. In general, well-established information will be presented, but, in spite of the volume of scientific and patent literature there are still many gaps in our knowledge and, not infrequently, conflict in the experimental results.

The laboratory synthesis of polymers presents little difficulty, notwithstanding the generally deleterious effects at the parts per million level of moisture, air and other impurities. If pure solvents and monomers are employed and precautions are taken to manipulate and transfer catalyst components under an inert atmosphere, usually nitrogen or argon, most of the polymers described in the literature are fairly readily accessible. (Kinetic studies, of course, present far greater difficulty, since reproducibility of both rate and molecular weight from experiment to experiment is not readily achieved.) It is usual to employ crown or screw-capped glass bottles, stirred multineck glass flasks, or, more rarely, conventional vacuum line equipment. Laboratory processes can be found in the literature and in patent specifications; four typical procedures are given in *Macromolecular Syntheses* (3). Commercial production of the polymers, which began about 1960 and is now very large scale (4) (see also Chapter 1), posed considerable engineering problems, both in control of the polymerization and in the isolation of the polymers. A flow diagram for a manufacturing process is shown in Fig. 1 (5).

2 BUTADIENE POLYMERS

2.1 cis-1,4-Polybutadiene

The synthesis of *cis*-polybutadiene was a long sought goal attained by the use of coordination catalysts. The interesting properties of the polymer when finally obtained, its low glass transition temperature and high resilience, its ability to be blended with advantage with SBR and natural rubber, and the good abrasion resistance of its carbon black reinforced vulcanizates (6) encouraged extensive investigations. As a result several catalyst systems have been found to give polymers of 90% cis structure or above. Three of these are used in manufacture, based on organoaluminum compounds with compounds of titanium, cobalt, or nickel. Consumption in 1969 in the United States alone was over 260,000 tons (7).

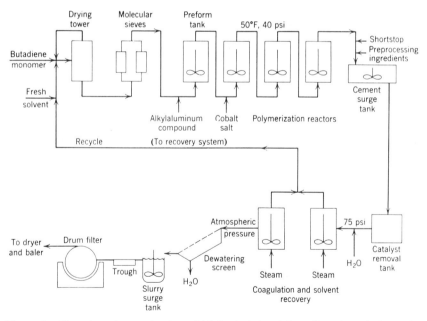

Figure 1. Flowsheet for production of high *cis*-1,4–polybutadiene in solution using an aluminum-cobalt catalyst system (5). Courtesy of *Chemical Engineering*, McGraw-Hill Book Company.

2.1.1 *Titanium Catalysts*

Titanium compounds and metal alkyls were first described for the polymerization of butadiene in an early patent of wide scope. Polymers of predominantly 1,4 structure are disclosed but the patent does not include the specific systems later found to give high cis polymer. Many of the examples relate to the use of catalysts based on titanium tetrachloride and, although the polymers obtained are of moderate cis content (60–70%) (8), they have not proved to be very satisfactory. In addition to considerable variation in structure with relatively minor changes in experimental conditions, there is a marked tendency to form low molecular weight oligomers and gelled polymer. Titanium tetrabromide and aluminum trialkyls give higher cis contents (80–90%) (9) but here also there is a tendency to produce gelled polymer; all satisfactory catalysts for high *cis*-polybutadiene with titanium as the transition metal also contain iodine. The first of these were based on titanium tetraiodide and aluminum trialkyls, which give 90–95% cis-1,4 polymer of high molecular weight that is essentially gel free (10). The commonly used aluminum alkyls are the triethyl and the triisobutyl but the reaction does not seem to be particularly sensitive to the nature of the alkyl group The optimum Al/Ti mole ratio is 3–6:1, and at higher ratios the yield falls slightly; the cis content is relatively insensitive to the Al/Ti ratio. Polymerization rate and polymer yield increase and the molecular weight decreases with increase in catalyst concentration. Catalyst efficiencies range from about 50 to 120 g polymer/mmole titanium tetraiodide. Table 1 gives some data on the polymerizations with this catalyst. Although good yields of polymer can be obtained with about 0.4–0.5 mmhm (millimoles per hundred grams of monomer) of titanium tetraiodide with an Al/Ti ratio of 5, the polymer is of very high molecular weight and for processable

Table 1 Polymerization of Butadiene by TiI$_4$/Al(i-Bu)$_3$ (10)

Butadiene 100 g Benzene 400 g TiI$_4$ (mmoles)	(wt %)	Al/Ti	% Conversion (2 hours at 30°C)	$[\eta]$
0.35	0.195	20	28	5.3
0.70	0.39	10	73	3.87
0.70	0.39	5	92	3.71
2.33	1.29	3	92	2.66
3.5	1.94	2	95	2.10

materials ($[\eta]$ 2–3) the amount of iodide required appears to be in the region of 1–2 mmhm. Under these conditions, polymer contains 92–93% cis-1,4 structure; 2–3% trans-1,4 structure; and 4–6% 1,2 structure.

An aromatic solvent seems to be necessary for fast polymerizations. Toluene and benzene are equivalent and mixtures of aromatic and aliphatic solvents can be employed. However, if the solvent is entirely aliphatic, there is a marked fall in polymerization rate. The polymer is mainly linear (11), of moderate molecular weight distribution (12, 13), and the average molecular weight changes only slightly with conversion. It has been claimed (14) that a small amount of isoprene (up to 5 parts/100 of butadiene) reduces the molecular weight of the polymer without affecting the cis content and permits some degree of breakdown on mastication. In the example quoted there is indeed a fall in solution viscosity on mastication at 300°F but since ultimately a small amount of gel is produced branching is evidently occurring concurrently.

There are some disadvantages to the TiI_4/AlR_3 catalyst in that titanium tetraiodide is of high density (4.3 g/cm^3), is insoluble in hydrocarbon solvents and, therefore, must be used in finely divided form—without contamination from the atmosphere. This can be done either by dry grinding under an inert atmosphere or by ball-milling with the solvent, but clearly the elimination of this step would be advantageous. The catalyst reaction product is heterogeneous at all Al/Ti ratios, but neither the solid portion nor the supernatant solution alone is effective for polymerization. The insoluble portion is a complex of titanium triiodide with an aluminum alkyl iodide, and possibly contains some alkyltitanium iodides. The solution contains free triisobutylaluminum, which is necessary for high catalytic activity. The valence state of the titanium is reduced to an average of about 2.8 as the Al/Ti ratio is increased. The approximately constant composition of the insoluble portion of the iodide catalyst with increase in Al/Ti ratio may be partly responsible for the relative insensitivity of the catalyst compared with those from titanium tetrachloride in regard to polymerization rate and polymer structure.

It would seem reasonable that titanium triiodide could be obtained in other ways, and there are now many examples to prove that the iodine need not be attached in the first instance to the titanium atom. If the correct molecular ratios are chosen, good yields of high cis, high molecular weight polymer can be obtained from systems such as $TiCl_4/I_2/Al(i\text{-}Bu)_3$ (15) and $TiCl_4/AlI_3/AlHCl_2 \cdot D$ (16) (D is a donor such as Et_2O or NMe_3 which helps to stabilize the aluminum hydrides and makes them soluble in hydrocarbon solvents, but does not, in 1:1 complexes, influence the reaction adversely).

Iodine in the presence of an excess of trialkylaluminum gives the dialkylaluminum iodide, which can reduce titanium tetrachloride to the trivalent state, and, since iodine-chlorine interchange occurs, iodine becomes attached to the titanium. Saltman and Link (17) have shown that as the Al/Ti ratio is increased there is a fall in the Cl/Ti ratio in the solid precipitate to below 0.5 and a rise in the I/Ti ratio to above 3. The exchange reaction is concurrent with liberation into the supernatant solution of diisobutylaluminum chloride. High polymer yields, cis contents above 90% and dilute solution viscosities above 2 are obtained at Al/I$_2$/Ti catalyst ratios of 8:1.5:1, a total catalyst concentration of 3.2 mmhm, monomer concentration of 12.5% w/w in benzene, and polymerization for 1 hour at 50°C. Yield, cis content, and molecular weight all fall with an increase in the proportion of titanium tetrachloride, whereas a reduction reduces the yield of polymer without affecting the cis content.

Other catalyst variants employ mixtures of titanium tetrachloride with the tetraiodide, in the ratio 0.5–5:1 (18). Mixed halides such as titanium trichloroiodide or titanium dibromodiiodide can also be used, but there is no special merit in their use; there is a lack of flexibility in that the ratio of the halogen atoms is fixed—whereas, as shown in the I$_2$/TiCl$_4$ system, the optimum is sharply defined with an I/Cl ratio of 3:4.

The catalysts from mixtures of aluminum hydride derivatives, aluminum iodide, and titanium tetrachloride will obviously be of comparable but not identical composition. They are dark and are partly insoluble unless prepared in the presence of monomer, where they appear to be homogeneous. The composition of the catalysts can be varied widely and yields of polymer pass through maxima as the ratios are varied. Cis contents of polymer are rather less sensitive to catalyst composition and reach values of 90–95%. A substantial excess of the aluminum hydride appears to give the best results but is not always essential, and the AlI$_3$/TiCl$_4$ ratio is in the range 1–2.5:1. Examples of typical systems are given in Table 2. The molecular weight is also seen to be controlled by catalyst concentration.

The organoaluminum compound can be replaced by other organometal compounds, including the alkyls of lithium, sodium, zinc, and lead. The BuLi/TiI$_4$ system has been studied in detail (19). It exhibits rather complex behavior and the ratio of butyllithium to titanium tetraiodide must be carefully controlled for high cis content polymers. The yield passes through a maximum at about Li/Ti = 2 and then falls almost to zero. At still higher ratios the yield increases but gives polymer of mixed microstructure. Butyllithium differs in its reducing power compared with aluminum alkyls and may not fully reduce titanium tetraiodide. It may produce a relatively stable ionic Ti(IV) complex quite different in catalytic

Table 2 Polymerization of Butadiene by Catalysts Containing Aluminum Iodide (16)

Polymerizations: 2 hours at 5°C; $TiCl_4$ 1.8–2 mmhm
Monomer concentration: ca. 15% w/v in toluene

(a) Conversion and Polymer Structure
1. $AlHCl_2Et_2O/AlI_3/TiCl_4 = 6:1.5:1$
   90% conversion; 93% cis-1.4
2. $AlH_3NMe_3/AlI_3/TiCl_4 = 1.2:1:1$
   99% conversion; 93.7% cis-1,4
   4.5% 1,2
3. $AlH_2ClNMe_3/AlI_3/TiCl_4 = 3:2:1$
   92% conversion; 90% cis-1,4
   4.3% 1,2

(b) Effect of Catalyst Concentration on Molecular Weight
$AlHCl_2 \cdot Et_2O/AlI_3/TiCl_4 = 5.5:1.3:1$

$TiCl_4$ (mmhm)	$[\eta]$
0.34	7.2
0.75	4.46
1.14	3.43
1.70	1.62
2.22	0.88

activity from the aluminum complexes. At high ratios (Li/Ti 5–10) there will be a considerable amount of free lithium butyl which can initiate polymerization in its own right, and indeed, the polymers so obtained have structures similar to those from the lithium alkyl alone. A recipe for high cis polymer using this system is the following (20):

Butadiene	30 ml	Temperature 30°C; time 12 hours
Benzene	100 ml	Conversion ca. 100%
TiI_4	0.33 mmole	Polymer structure:
LiBu	0.66 mmole	93% cis; 3.5% 1,2

With lead alkyls better yields of higher cis content polymer are obtained if the titanium tetraiodide is replaced by a 1:1 mixture of $TiCl_4/TiI_4$ (21). A possible explanation for the increase in catalyst activity is that since titanium tetrachloride is soluble, it gives an initiator of high surface area which would help remove impurities that behave as catalyst poisons.

Sodium alkyls behave like the lithium alkyls, and high cis, high molecular weight, low gel content polymers have been reported (22). Optimum yields are obtained with Na/Ti = 2.5; at higher ratios (4–8) there is a rapid fall in polymer yield and cis content.

Donors such as ethers, thioethers, and tertiary amines may be present in titanium halide catalyzed polymerizations. Polymers of 90–93% cis content were obtained from $TiI_4/Al(i$-$Bu)_3$ i-Pr_2O (13), except that polymers of relatively low molecular weight ($[\eta]$ below 1) had cis contents falling to below 80%. The heptane solvent allowed fast initial polymerization (titanium tetraiodide concentration in the range ca. 0.7–2 mmhm) at 2°C, but polymerization died out at 30–50% conversion. Turov et al. (23) have shown ethers to affect the structure of polybutadiene obtained with $Al(i$-$Bu)_3/TiCl_4$, the lower ethers having a greater effect in reducing the cis content. Additional diethyl ether in the $AlHCl_2 \cdot D$ systems likewise caused a fall in yield and in cis content (16). Trimethylamine and other aliphatic tertiary amines have a more marked effect, stopping the polymerization at relatively low concentrations; triphenylamine had less effect on the polymerization.

A donor may thus be beneficial in small amounts (e.g., as 1 : 1 complex with the organometal compound) by giving more finely divided or "solubilized" catalysts or by moderating the reducing power of the organometal compound, but in large amounts or if it is a strong donor, it may compete with monomer for coordination at the organometal complex and adversely affect both polymerization rate and polymer structure.

2.1.2 Cobalt Catalysts

These catalysts are also of great commercial importance and although developed somewhat later than the titanium catalysts there is a reference in a patent filed in 1956 (24) to an all cis-polybutadiene produced from cobalt chloride and triisobutylaluminum. Combinations using the trialkylaluminum have not achieved importance although there are references in the patent literature (25, 26). The more important systems use aluminum alkyl halides and cobalt salts (27–29). Those containing cobalt halides, the sulfate, or oxides are heterogeneous, whereas those containing soluble salts (e.g., the acetylacetonates, octanoate, or halide-pyridine complexes) (30) remain homogeneous. The distinction is somewhat artificial since the heterogeneous catalysts probably produce a small amount of soluble catalyst which is responsible for the polymerization. Much higher concentrations of cobalt are thus required for fast polymerizations, and they are of minor interest.

The presence of halide groups, preferably chloride or bromide, appears necessary in all these catalysts in high molar ratio compared with the cobalt. Also, although cis-polybutadiene has been reported to be produced by polymerization on cobalt chloride alone (31), all effective catalyst systems contain aluminum. The aluminum cocatalysts show important differences in their behavior. When aluminum diethyl chloride is the co-catalyst an additional activator is required. The most commonly employed activator is water. This has its maximum effect at a concentration of about 10 mole % on the aluminum diethyl chloride; in its absence polymerization does not occur, and the yield falls off rapidly above about 30 mole %. The water converts a proportion of the aluminum diethyl chloride into a stronger Lewis acid $[O(AlEtCl)_2]$ which appears to be essential for high activity (32). If water is added in the form of $CoCl_2 \cdot 6H_2O$ the catalyst is inactive—presumably here the water ligands protect the cobalt salt against reduction to an active form. Many other compounds have been found to be activators; among these are oxygen, alcohols, organic hydroperoxides, halogens and hydrogen halides, aluminum chloride, and reactive organic halides such as allyl chloride. All these substances result in an increase in Lewis acidity (33); the molar concentrations required for high activity differ somewhat but in all cases are in the range 10–20 mole % on the aluminum diethyl chloride.

Catalysts employing aluminum ethyl sesquichloride are highly active and are stated not to require water as activator, which, in contrast, has a retarding effect on the polymerization rate. However, Van de Kamp has found (34) that whereas it is not necessary when cobalt octanoate or acetylacetonate is used, a small amount of water is necessary in the solvent when cobalt chloride is the salt employed; in addition, it must be added to the organometal compound before formation of the catalyst.

Lewis acids stronger than the alkylaluminum sesquihalides, such as are produced by $Al_2Et_3Cl_3/H_2O$ or $AlEtCl_2$, with cobalt salts are also active catalysts but have a tendency to form gel (35); this is particularly true of the $AlCl_3/CoCl_2$ catalyst (36). Soluble gel-free polymers are obtained from these with the addition of Lewis bases, e.g., ethers and thioethers (35, 37). The effect of the donors is not completely explainable but they might be expected to scavenge cationic species which could cause cross-linking to occur.

The heterogeneous $CoCl_2/AlEt_2Cl$ catalysts require about 10 mmhm of cobalt for high conversion to polymer. The yield is usually in the range 150–200 g/g cobalt irrespective of the Al/Co ratio, which is normally in the range 1–5; further excess has no effect on the polymerization. The soluble catalysts normally require cobalt concentrations in the region 0.01–0.1 mmhm with Al/Co ratios in the range 100–1000. Lower ratios have been employed but the highest cis contents and rates are obtained with ratios of above 200. Almost any soluble Co(II) or Co(III) compound or complex may be used; those most frequently encountered in the literature are the carboxylates (e.g., octanoate or naphthenate), the acetylacetonates Co(acac)$_2$ and Co(acac)$_3$, and donor complexes such as CoCl$_2$ · pyridine. Other complexes in which the cobalt is chelated to some other function in the molecule, e.g., salicylaldehyde, 1-nitroso-2-naphthol, 8-hydroxyquinoline, and mercaptobenzthiazole, can be used (38) but they do not appear to have any advantage over the simple salts. A yield of about 80 g of polymer could be expected from a catalyst prepared with 0.01 mmhm of cobalt, which indicates their very high efficiency (ca. 10^5 g polymer/g Co). The catalysts are decomposed by excess of oxygen or moisture but when correctly formulated they can be relatively stable on prolonged storage. Cobalt alkyls which might be expected to be formed are very unstable (although they are stabilized by powerful donors to some extent), decomposing to give the metal. The complexes obtained from aluminum alkyl halides, which may be presumed to have organic groupings attached to the cobalt, are much more stable, the stability increasing with increase in the halogen/alkyl ratio. Thus, cobalt chloride, alone or as its pyridine complex, has been found not to be reduced by aluminum diethyl chloride (39), and the mixture CoCl$_2$/AlCl$_3$/AlEt$_3$ = 1 : 1 : 1 in benzene solvent has been found to retain its catalytic activity almost unimpaired for many months (40). In this case, a dark brown lower layer was produced which was satisfactory for the polymerization to high cis polymer. A reduction in the proportion of chlorine, as in the replacement of cobalt chloride by cobalt tris(acetylacetonate), gives less stable intermediates, and reduction to cobalt metal occurs in the absence of monomer (41). Complexes containing π-allylic groupings or π complexed olefins are more stable than σ bonded organocobalt compounds, and in accord with this it is found that the catalysts are more stable when prepared in the presence of diene monomer or aromatic solvents. The solvent also affects the rate of polymerization and the molecular weight of the polymer; the advantage of an aromatic solvent is shown by the data in Table 3 (32).

Control of molecular weight is a major problem in polymerizations with most coordination catalysts. Transfer reactions, which are desirable for molecular weight control, are unpredictable and imperfectly understood and the addition of compounds that will intercept the propagation step may not only reduce the rate of polymerization but may also change the polymer structure. Changes in temperature or in catalyst concentration may be effective but both of these procedures have the serious disadvantage in that the rate of polymerization is also changed. As shown in Table 3, limited variation of the solvent composition may be effective without

Table 3 Effect of Solvent in Polymerization of Butadiene by AlEt$_2$Cl/CoCl$_2$-py/H$_2$O^a (32)

Benzene/Hexane	Yield (%)	Cis-1,4 (%)	$[\eta]$
100:0	95.6	98	6.2
80:20	99.6	98	5.9
60:40	90.2	98	3.6
20:80	82.4	98	2.1
0:100	37.8	96	1.0

^{a} Al/Co/H$_2$O = 20:0.02:2 mmhm at 5°C.

too great an effect on polymer structure or yield. Amine complexes of the cobalt salt are effective in reducing molecular weight, as shown in Table 4.

The rate of polymerization is reduced at higher concentrations of pyridine although this may not be very apparent from Table 4, where a lengthy polymerization time was taken. Generally, a small amount of the donor increases the rate, presumably by increasing the concentration of active catalyst, but further amounts reduce the rate significantly or may even stop the polymerization. There is also an effect on polymer structure. The cis content falls (the trend can be seen in Table 4), and at sufficiently high donor concentrations products of mixed microstructure and low molecular weight are produced (42). Hydrogen and olefins have a

Table 4 Effect of Pyridine Concentration on the Polymerization of Butadiene Using AlEt$_2$Cl/CoCl$_2$/py^a (30)

Pyridine/Co (mole ratio)	Conversion	$[\eta]$	Gel	Cis1,4 (%)
0	10	4.85	0	—
1:1	23	7.18	0	98.6
3:1	86	4.13	0	—
4:1	94	3.24	0	96.4
5:1	94	2.67	0	—
6:1	97	1.97	0	—
7:1	92	1.84	0	—
8:1	91	1.37	0	95.1

No polymer formed at Py/Co 9–10

^{a} Butadiene 100 g, toluene 1200 g, AlEt$_2$Cl 10 mmoles (Al/Co = 10). Polymerization 16 hours at 5°C.

positive effect in reducing the molecular weight of the polymer. With a partial pressure of 20 atm of hydrogen, a reduction of dilute solution viscosity from 5.5 to 1.4 is reported (43). Ethylene and propylene up to 6 mole % can be used, and copolymerization does not occur, but most effective are allene and methylallene which reduce the dilute solution viscosity (DSV) to about 2 with concentrations on the monomer of about 0.2 mole %. The commercially desired molecular weights of polybutadienes depend on the use to which the polymer is put. The medium molecular weight polymer usually has a Mooney viscosity (ML_4) of 45–50 (DSV $\sim$ 2.0), while that for use in tires is usually an oil extended polymer of much higher molecular weight (ML_4 up to 100). From a production standpoint it may be poor economics to carry out the polymerizations to produce high molecular weight polymer directly since the high viscosity of the solution will reduce reactor output. It is possible to achieve the necessary high molecular weight after polymerization is complete by the addition of a reactive halide such as t-butyl chloride (44). A controlled interlinking of polymer molecules occurs probably by a cationic mechanism to give the desired rubber plasticity without the formation of gel. The cis content is unaffected but the molecular weight distribution and solution properties of the polymer are changed (45). The oil can then be added prior to isolation of the polymer. The process is in commercial use in Germany (46) but does not seem to have caught on elsewhere.

The alkyl-free initiators for the preparation of high cis-polybutadiene in which the active complexes have the structure $Co(AlX_4)_2$ where $X = Cl$, Br, I, are of considerable theoretical interest, although whether they are suitable for large-scale polymer production is obscure. The complexes are simply prepared by stirring together the metal halides in cyclohexane or benzene (36). The cobalt halides are insoluble in hydrocarbons but the aluminum halides and the complexes are appreciably soluble. Cobalt concentrations using the chlorides ranged from about 100 to 400 ppm (1–5 mmoles/liter) in cyclohexane and up to five times this concentration in benzene. An excess of aluminum halide is present—particularly in those catalysts prepared from the more soluble bromide and iodide. Al/Co ratios quoted vary from 2.5 to 12.5 for chloride complexes and up to 70 for those containing bromide ions. From a $CoCl_2/AlCl_3$ complex containing 15 ppm of cobalt a 45% yield of polymer of 96% cis structure and a dilute solution viscosity of 2.85 was obtained in 33 minutes at 25°C (butadiene 8% w/v in cyclohexane). The polymers are apparently prone to rapid gelation, possibly by a cationic reaction with the aluminum chloride. The complexes prepared from the bromide gave polymer of somewhat lower cis content (ca. 80%) and generally contained gel, which was attributed to the high concentration of aluminum bromide. The oily complexes from the mixed metal halides have been found to produce low molecular weight polymers similar to those obtained with cationic catalysts, but this can be prevented by the addition of aluminum powder (ca. 1 wt %) in the preparation of the catalyst (37). Alternatively, high cis content is claimed to be produced by the addition of excess thiophene to the catalyst mixture (thiophene/Al/Co 8–12:4:1). The thiophene is considered to participate in the polymerization and not simply to act as an acid scavenger.

The rate of polymerization is increased with increase in concentration of the aluminum chloride/thiophene complex (at constant cobalt concentration) and the molecular weight increases greatly with increase in thiophene/cobalt ratio (110).

Catalysts from cobalt salts and aluminum trialkyls have been reported to give high cis-polybutadiene, crystalline syndiotactic 1,2-polybutadiene, or polymers of mixed microstructure. In most cases the conditions that give rise to a particular polymer can be distinguished, but in some cases catalysts of the same or similar composition have apparently given different polymers, and it must be presumed that some essential facts have not been disclosed. Catalysts with a low Al/Co ratio, in which the cobalt is present as the chloride or bromide, made up in the absence of butadiene, and aged before use, produce high cis-polybutadiene (25).

Catalysts prepared in this way contain a solid precipitate, but the solution also contains cobalt and is an active catalyst (26). Thus cobalt chloride (95 mmoles) and triisobutylaluminum (55 mmoles) in benzene (180 ml) were heated at 70–80°C for 12 hours and allowed to stand for 34 days at room temperature. The liquid solution after removal of the solid contained chlorine (24.9 g/liter) and cobalt (0.35 g/liter) and polymerized butadiene to a polymer of high molecular weight and a cis content of 99%. Under these conditions aluminum alkyl halides are produced by reduction of some of the cobalt halide, which would then be available to form the catalytic complex with additional cobalt chloride. The following is an idealized reaction, which represents the formation of the dialkyl chloride:

$$2AlEt_3 + CoCl_2 \rightarrow 2AlEt_2Cl + Co + 2Et\cdot$$

The products actually formed depend on the stoichiometry (39). With excess cobalt chloride approximately equivalent amounts of aluminum diethyl chloride and cobalt are obtained but even with excess alkyl the cobalt does not appear to be fully reduced. It is apparent that this is merely a relatively inefficient way of producing the aluminum alkyl halide catalysts discussed previously.

The observations on catalysts with higher Al/Co ratios (above 100) are more difficult to explain. In the system Al(i-Bu)$_3$/CoCl$_2$-py, Gippin (32) obtained no polymer in the absence of water, but good yields of polymer were obtained (and, when prepared at 3°C, of high cis content) provided 1 mole of water per mole of triisobutylaluminum had been previously added to the metal alkyl. The water would hydrolyze the trialkyl in this stoichiometry to give either a polymerized aloxane or a mixture of more highly hydrolyzed products and O(Al i-Bu$_2$)$_2$. However, O(AlEt$_2$)$_2$, prepared from triethylaluminum and water with the same solvent and cobalt complex (benzene, CoCl$_2$ · 2py, Al/Co = 200 : 1) has been reported to give a high yield of crystalline syndiotactic 1,2 polymer (47). It is conceivable that the explanation here is concerned with catalyst aging, since Susa, in a comparable study of the AlR$_3$/cobalt salt system (48), states that for syndiotactic 1,2-polybutadiene the catalyst should be made up in the presence of the monomer (49). The initial reaction product from AlEt$_3$/CoCl$_2$-py at high Al/Co ratio is unstable and loses activity rapidly (48). It seems likely, therefore, that the monomer stabilizes the intermediate—possibly as a π complex which then gives the 1,2 polymer. However, these observations do not explain completely the results obtained from the use of Al(i-Bu)$_3$/H$_2$O, and this aspect is worthy of further clarification.

There are an increasing number of low valency complexes of cobalt becoming available, including π hydrocarbon cobalt carbonyls; trisallylcobalt has also been prepared. It would be expected that such compounds would give catalysts for diene polymerization. Aluminum triethyl and dicobalt octacarbonyl give oligomers of butadiene, once again indicating the importance of the electronegative halogen for the formation of high polymer, since in contrast Co$_2$(CO)$_8$/AlEt$_2$Cl, like other cobalt systems with aluminum alkyl halides, gives high cis polymer (50). Dicobalt octacarbonyl with other metal halides, e.g., the chlorides of molybdenum and tungsten, gives polymers of high 1,2 (atactic) or mixed microstructure (Table 5) (51).

Table 5 Polymerization of Butadiene with Catalysts from Dicobalt Octacarbonyl^a (51)

Catalyst	Conversion (%)	[η]	Gel (%)	Cis (%)	Trans (%)	1,2 (%)
Co$_2$(CO)$_8$/MoCl$_4$	31	3.36	48	1.5	1.9	96.6
Co$_2$(CO)$_8$/MoCl$_5$	93	5.23	0.5	—	2	98
Co$_2$(CO)$_8$/WCl$_6$	44	4.20	—	56.7	34.9	8.4

^{a} Concentration of Co$_2$(CO)$_8$ 0.46 mmhm, Co/Mo = 1 : 1. Monomer concentration in benzene ca. 10% w/v. Polymerization 18 hours at 50°C.

2.1.3 *Nickel Catalysts*

Many compounds of nickel have been found to give high molecular weight stereoregular polybutadiene either alone or more commonly with some other metal compound. The most active nickel catalysts are of comparable efficiency to the cobalt catalysts and have the necessary features for a viable industrial process; since 1965 *cis*-polybutadiene has been manufactured in Japan by Japan Synthetic Rubber Company using a soluble nickel catalyst (52).

As with the cobalt systems heterogeneous and homogeneous catalysts exist and it is necessary for halogen atoms or some other highly polar grouping to be present. The structure of the polymer is influenced more by the type of halogen atom than the corresponding cobalt systems; in general high cis producing catalysts should contain fluorine, chlorine, or bromine, since iodine gives high trans structure polymers.

According to the patent literature, nickel salts can be used as replacements for cobalt salts to prepare high cis content polymers, but there is little advantage in their use as such. The data in Table 6 show that with the soluble salts high cis contents are obtained at low temperatures and low concentrations of catalyst. The cis contents are somewhat lower than

Table 6 Polymerization of Butadiene with Nickel Catalysts

(a) $NiX_2/Al(i\text{-Bu})_3$ (25)

        X = Cl, Al/Ni = 1   Conversion 80%; cis content 95.5%

        X = Br, Al/Ni = 1.5 Conversion 76%; 1,4 structure 98%

        NiX_2 3.23 mmhm; butadiene in benzene 1:6

        Polymerization 18 hours at 50°C

(b) $Ni(acac)_2/AlEt_2Cl$ (27)

Ni (mmhm)	Al/Ni	Polymerization Temperature (°C)	Cis(%)	Trans(%)
1.4	11	0	84	10
0.63	190	−10	95	3
0.32	50	−10	95	2
0.07	225	−10	95	2.5

    Butadiene/benzene 1:2 w/v

    Polymerization 1 hour

    Yield not stated

for the corresponding polymers prepared with the cobalt catalysts. However, variants containing fluorine are of high activity and give polymers of 98% cis structure. Three-component catalyst systems containing a nickel complex or salt, an organometal compound, and a Lewis acid are the subject of patents (53) and are the basis of a commercial process in Japan (52) and in the United States. The components are open to wide variation in structure and concentration but the following is a typical example:

Ni naphthenate/AlEt$_3$/BF$_3$ ·Et$_2$O
Ni = 0.8 mmhm; Ni/Al/B = 1 : 7.7 : 7.7
Butadiene in benzene: 2 : 5 w/v
Polymerization: 1 hour at 45°C
93% conversion; 97% cis-1,4

The literature shows great variations in the yields of polymer based on the nickel salt employed but in the systems of the highest efficiency yields are in the region 10^3–10^4 g polymer/g Ni. The presence of fluoride seems necessary to produce high molecular weight polymers. The catalysts are very susceptible to impurities in the monomer, particularly to allene, 1,2-butadiene, and acetylenes (54, 55).

The structure of the active nickel catalysts, as with those from cobalt, is by no means clear, but it appears to be generally accepted that a hydrocarbon derivative is produced by an exchange reaction between groups attached to the metal and the organometal compound. The intermediate so produced, which may be stabilized by the solvent or other ligands, then reacts with monomer to give a propagating chain. It would therefore be expected that nickel compounds containing a metal-carbon bond would form the basis for active catalysts.

Nickel forms many hydrocarbon derivatives of reasonable stability which can be prepared independently, such as (π-C$_3$H$_5$)$_2$Ni, π-C$_8$H$_{12}$Ni, and π-C$_{12}$H$_{18}$Ni. These do not convert butadiene to high molecular weight polymers, although they do give dimers and trimers, but by replacement of some of the hydrocarbon-metal attachments with strongly electronegative groupings, complexes that give higher molecular weight polymer are formed. π-Allylnickel chloride gives cis polymer, and the iodide gives *trans*-polybutadiene (56). The bromide produces polymer of mixed or high trans microstructure (56, 57). The rates of polymerization and molecular weights are very low (56, 58) and do not compare with the Ni/Al catalysts illustrated in Table 6. Much more active catalysts are produced from these nickel complexes by the addition of Lewis acids (56, 59) such as titanium or aluminum halides. Essentially similar behavior is exhibited by other soluble complexes of nickel, such as nickel carbonyl and bis(π-C_5H_5)$_2$Ni. Presumably displacement of the ligand by monomer gives a growing chain of comparable structure in all these systems. The following are two recipes illustrating the use of these soluble nickel complexes with metal halides:

1. (π-C_4H_7)$_2$Ni/NiCl$_2$ = 1 : 10; nickel: 41.7 mmhm (59)
   Butadiene in benzene: 1 : 2 w/v
   Polymerization 8 hours at 50°C
   78% conversion; 95% cis-1,4

2. Ni(CO)$_4$/AlBr$_3$; Al/Ni = 2 : 1; nickel: 1.85 mmhm (60)
   Butadiene in benzene: 2.5 moles/liter
   Polymerization 15 hours at 50°C
   100% conversion; 90% cis-1,4

Protonic acids are also effective in conjunction with nickel complexes. Trifluoroacetic acid, picric acid, and hydrochloric and hydrobromic acids give 70–93% cis polymer, and hydrogen iodide 100% trans polymer (61). The precipitate from the reaction of the first of these with cyclooctadiene nickel gives catalysts of high activity (62):

Ni(C_8H_{12})/CF_3COOH = 1 : 1; nickel: 7.7 mmhm
Butadiene: 4.8 moles/liter
Conversion: 65% after 1 hour at 0°C
98–99% cis-1,4; [η] = 2.0–2.4

The formation of high cis polymer requires the use of an aliphatic hydrocarbon as solvent. In benzene the polymer contains alternate cis and trans units. The addition of triphenyl phosphite or alcohols gives pure *trans*-1,4-polybutadiene (63).

Other variants of the nickel catalysts are π-C_4H_7NiX/benzoyl peroxide or oxygen; from the latter at Ni/O$_2$ = 2 : 1, 91% cis polymer was obtained (64). An effective system is from Ni(PCl$_3$)$_4$ in conjunction with titanium tetrachloride and aluminum chloride or bromide. Ratios are typically in the range Ni/Ti/Al = 1 : 1–2 : 0.05–1.0 and high conversions (in heptane solution) are obtained at Ni concentrations in the range 1–2 mmoles/liter (65).

It is possible to polymerize butadiene to high cis polymer with supported metallic nickel alone (66). The colloidal metal itself is inactive, as is Raney nickel deposited on carbon or alumina, but when an acidic carrier such as silica or silica-alumina is used, polymerization occurs.

Lewis acids (BF$_3$ etherate, TiCl$_4$) and alkyl or silyl halides will also activate colloidal nickel (67). In general, yields based only on nickel are low with these catalysts. Nickel chloride that has lost chlorine by photolysis or by heating under high vacuum gives moderate yields of 92–95% cis polymer (31). The polymers are of rather low molecular weight ([η] = 0.7) and contain about 50% gel. The effect here is not due to metallic nickel, since if heating is sufficiently intense to produce the metal, the salt is ineffective; catalytic activity is ascribed to NiCl.

2.1.4 *Cerium Catalysts*

That compounds of rare earth elements would give catalysts for high *cis*-polybutadiene was disclosed in 1963–1964 (68); the performance of soluble cerium salts was examined in detail by Throckmorton (69). Good yields of polymer were obtained from cerium octanoate with a mixture of a trialkylaluminum and an aluminum alkyl halide (Et$_2$AlCl, EtAlCl$_2$). Typically 0.25 mmole of cerium salt, 10 mmoles of triisobutylaluminum, and 0.25 mmole of aluminum diethyl chloride per 100 g of butadiene were used. The solvent was benzene or hexane, reaction velocities being significantly faster in the latter solvent. Ratios of Cl/Ce = 2 and Al/Ce = 40 gave the best yields of polymer. The cis content declined from 99 to 91.5% as the ratio of Al/Ce was increased from 18 to 200, and the dilute solution viscosity fell over the same range from above 4 to below 1. Higher cis contents were obtained with branched chain compared with linear aluminum alkyls (97–98% compared with 93–95%). Aluminum diethyl bromide and iodide gave higher yields of polymer compared with the chloride over wider ranges of halogen/Ce ratio, all three organometal halides optimizing at X/Ce = 2:1. With AlEt$_2$F maximum activity was observed at a F/Ce ratio of 40:1. The yield of polymer increased to a maximum (at Ce/Cl = 2) with increase in cerium content at constant triisobutylaluminum concentration and then declined. The dilute solution viscosity of the polymer fell with increasing cerium content.

From this it is clear that the cerium catalyst is of considerable interest. The polymers are of high cis content, not critically dependent on reagent concentrations. They are gel free and a good control of molecular weight is possible. Yields of polymer are approximately 400 g/mmole (3000 g/g) of cerium, less than the yields from optimized cobalt or nickel catalysts but more than is generally obtained with titanium catalysts.

The insensitivity of the cerium system to the nature of the halogen atom in the catalyst compared with the others is shown in the data in Table 7.

Table 7 Dependence of Cis Content of Polybutadienes on Halogen Atom in Catalyst (69)

Transition Metal	Halogen			
	F	Cl	Br	I
		cis-1,4 (%)		
Ti	35	75	87	93
Co	93	98	91	50
Ni	98	85	80	10
Ce	97	98	98	98

2.1.5 Uranium Catalysts

Tris(allyl)uranium halides have recently been found to give high cis polymers of exceptional steric purity (up to 99% cis content) with melting points substantially higher than those reported previously for polybutadiene (70). Uranium alkoxides are more convenient to handle than the allylic complexes and give catalysts of comparable efficiency (71). The following values are quoted for uranium based polymers in comparison with other polybutadienes:

Catalyst	T_m (°C)
Uranium	2.2–3.0
Nickel	−1.2
Cobalt	−2.3
Titanium	−8.0

Other uranium catalyzed samples gave considerably higher melting points, one being 12.5°C. Other titanium catalyzed polybutadienes have been reported with melting points of −15 to −4°C (98).

The high cis structure of the polymer is unaffected by the nature of the halogen attached to uranium but catalytic activity is greatly improved by the addition of a Lewis acid such as $AlEtCl_2$ or $TiCl_4$ (72). Low levels of uranium are then required and polymer yields are of the order of 200–500 g/mmole of uranium. The polymer yield is maximum at Al/U = 0.8 and improves on aging the catalyst. The following is a typical recipe:

Butadiene	100 g
Hexane	660 ml
$U(\pi\text{-allyl})_3Cl$	0.47 mmole
$AlEtCl_2$	0.14 mmole

Time 1 hour, temperature 20°C
83% yield; 98.9% cis-1,4; 4–5 DSV

Molecular weights are high and distributions are broad ($\bar{M}_v/\bar{M}_n = 4$–7). The cis content falls slightly (from 99 to 98.5%) with increase in molecular weight. Owing to the broad molecular weight distributions, the polymers have good processability; the greater tendency for crystallization is exhibited in gum vulcanizates, with superior physical properties. The cerium and uranium catalysts are among the few known to give both high cis-polybutadiene and high cis-polyisoprene (68, 70).

2.2 trans-1,4-Polybutadiene

The polymer can be prepared with a variety of catalysts but, possibly owing to its minor importance, few of them have been studied in depth. Crystalline polymers have been obtained using catalysts from titanium, vanadium, chromium, rhodium, iridium, cobalt, and nickel. The only product available commercially melts at about 85°C compared with 145°C for the pure isomer, and has a trans content in the range 80–90% (73).

2.2.1 Vanadium Catalysts

Vanadium halides in conjunction with aluminum alkyls are remarkably specific in producing polymer of high trans content. The highest trans contents (up to 99%) are obtained from the heterogeneous $VCl_3/AlEt_3$ catalyst (74). Vanadium tetrachloride and vanadium oxychloride are reported to give trans contents in the range 94–98% (75). The concentration of vanadium halide required is relatively high (1–2% w/w) on the monomer, and the Al/V ratio is 2–10. Polymerization temperatures are in the range 20–50°C and polymer yields fall at higher temperatures. Triethylaluminum can be replaced by aluminum diethyl chloride without significantly affecting the polymer structure. With the soluble vanadium halides a fraction of ether soluble, low molecular weight, amorphous polymer is formed which can be minimized by adding the organoaluminum compound slowly to the vanadium halide rather than the reverse mode of addition (75). Vanadium trichloride gives high molecular weight, highly crystalline polymer irrespective of the order of catalyst makeup.

The formation of highly crystalline polymer that precipitates during polymerization has a marked effect on the polymer yield which depends markedly on the degree of subdivision of the catalyst and the manner in which the catalyst becomes immobilized in a mass of solid polymer. The polymerization, which commences rapidly, falls to a slow rate, and conversions may be limited to about 25–50%. In such systems the yields of polymer are low (50–100 g/g V). Higher yields can be obtained when the catalyst is prepared in a highly dispersed form on a carrier. Coprecipitation of vanadium trichloride with titanium trichloride (either the β or the γ form) increases the rate of polymerization of butadiene by a factor of up to about 30 (with triethylaluminum at Al/Ti/V = 15:4:1) without reducing the trans content of the polymer (76).

2.2.2 *Titanium Catalysts*

Several titanium-containing catalysts have been disclosed for the preparation of high *trans*-polybutadiene. Titanium tetrachloride on reaction in hydrocarbon solution with aluminum trialkyl gives a finely divided precipitate containing mixed subhalides of titanium. Under relatively mild reducing conditions, i.e., with the trialkyls or aluminum diethyl chloride at Al/Ti in the range of 1–2, the subhalide is predominantly the β form of $TiCl_3$, complexed with the organometallic compound. These catalysts are of a rather variable character and probably contain more than one complexed organometallic compound. The crude mixture gives polymer with cis contents varying from 60 to 80%. If the titanium trichloride is obtained via the reaction at ambient temperatures of 1 mole of triethylaluminum with 3 moles of titanium tetrachloride, the precipitate is β-$TiCl_3$ admixed with aluminum trichloride (76). When activated with triethylaluminum or aluminum diethyl chloride this produces polymers with a substantial cis content, but with aluminum ethyl dichloride (Al/Ti = 3:1) the high trans polymer is obtained (76). Pure β-$TiCl_3$, obtained from the thermal decomposition of methyl titanium trichloride, when complexed with the alkyls or dialkyl halides of aluminum gives either polymers of mixed microstructure or mixtures of high *cis*- and high *trans*-polybutadiene (77); it does not appear to have been tried with aluminum ethyl dichloride (however, see the discussion near the end of Section 3.1 and ref. 78).

Two other forms of titanium trichloride exist—α-$TiCl_3$, obtained by reducing titanium tetrachloride with hydrogen, and γ-$TiCl_3$, obtained by heating the β form above 140°C. A convenient method of preparing γ-$TiCl_3$ is to heat the reaction product from $AlEt_3$/$TiCl_4$ (Al/Ti = 1:3) in a high boiling hydrocarbon at 170°C for 1 hour. Both α- and γ-$TiCl_3$, although they differ in crystalline lattice structures, give high *trans*-polybutadiene when activated with triethylaluminum and both have low polymerization activity as judged by polymer yields (77).

Polymers of high trans content are obtained from lithium aluminum hydride and titanium tetraiodide (79), and this catalyst may be the one

Table 8 Polymerization of Butadiene by LiAlH₄/TiI₄^a (79)

LiAlH₄ (mmoles)	TiI₄ (mmoles)	Al/Ti	Conversion (%)	Microstructure (%)			[η]
				trans-1,4	cis-1,4	1,2	
8.8	10.3	0.86	81	82	3	15	1.5
13.2	10.1	1.31	91	86.5	4.5	9	1.03
12.0	10.0	1.2	86	78.5	9.5	12	0.54

^{a} Butadiene 100, benzene 440. Polymerization 17–24 hours at 50°C. Yields of polymer based on the TiI₄ tend to be low; the best reported yields are in the region 400 g/g Ti.

used for the manufacture of the commercial polymer (73). Polymerization is carried out in aromatic or aliphatic solvents at 50°C with 2–5% of catalyst (total) on the monomer, and high conversions are obtained. Suitable Al/Ti ratios are in the range 0.9–3.0, and typical data are given in Table 8.

2.2.3 Rhodium Catalysts

trans-1,4-Polybutadiene has been obtained with simple salts such as rhodium nitrate or chloride (80). These produce polymer without the addition of an alkylating cocatalyst and are normally used in aqueous or alcoholic solution. These characteristics almost certainly result from the ability of rhodium to form π complexes with diolefins and π-allylic complexes that are sufficiently stable to resist rapid hydrolysis by water or alcohol. Similar behavior by other metals has been reported but not confirmed (81). The polymer obtained from rhodium is highly stereoregular (99–100% trans) and has a melting point of 145°C. The polymerization rate is dependent on pH, the nature of the rhodium salt, and the other materials present, particularly the emulsifier. The nitrate is effective in water or alcohol but rhodium chloride requires an emulsifier. The type of surfactant is important, best results being obtained with sodium dodecyl sulfate or sodium dodecylbenzene sulfonate; normal anionic soaps and cationic emulsifiers cannot be used (82). Polymerization rates are linear to high conversion and there is no induction period. The rate levels off above 1% w/w of rhodium salt on the monomer and is higher at all catalyst concentrations in the presence of the surfactant (80). The following is illustrative of polymerization conditions:

Water (100 ml), rhodium trichloride trihydrate (0.25 g), sodium dodecylbenzene sulfonate (2.5 g), and butadiene (50 g) are shaken for 20 hours at 50°C. Polymer yield 25–30%, [η] = 0.4; 99% trans-1,4.

Molecular weights are generally low. In aqueous systems $[\eta]$ is 0.37–0.50, whereas in methanol or ethanol the molecular weight is even lower ($[\eta] = 0.14$–0.16). The valence state of the metal required for polymerization by rhodium salts is not known with certainty; it is possibly a lower valency form [e.g., Rh(I)] produced by reaction with the solvent or the butadiene. Additives that would reduce the rhodium salt to a low valency state (or in the absence of monomer to the metal) give greatly increased rates. Thus cyclohexadiene (CHD) at a CHD/Rh molar ratio of 16–20:1 markedly increases the rate (83); formic acid is also effective (84).

Butadiene is polymerized in benzene solution by catalysts from rhodium salts and organometallic compounds (85). Rhodium chloride suspended in benzene and reacted with triisobutylaluminum gives very low yields of polymer, and homogeneous catalysts from Rh(acac)$_3$/AlEt$_2$Cl, although effective, require high concentrations of catalyst for high conversion:

Rh(acac)$_3$ 13 mmhm; Al/Rh = 14.4
Butadiene: 2.3 moles/liter in benzene
Polymerization 10 hours at 32°C
Conversion ca. 80%; $[\eta] = 0.10$

Molecular weights are low in all cases and fall with increase in polymerization temperature; in one series of experiments $[\eta]$ went from 0.2 at −20°C to 0.14 at 40°C.

2.2.4 Cobalt Catalysts

The halides or soluble salts of cobalt, aluminum diethyl choride, and trimethylamine or triethylamine give *trans*-polybutadiene (42). The reactions have been carried out in benzene or aliphatic hydrocarbon solution; although the reactions were much slower than those giving the high cis polymers, good yields of polymer were obtained with ratios in the range Al/Co/NR$_3$ (R = Me, Et) = 10:1:7.5–11.0. Methanol (ca. 2 mole % on the aluminum diethyl chloride) was used as the activator. Fairly high concentrations of cobalt chloride (ca. 10 mmhm) were nesessary but lower concentrations (ca. 2 mmhm) of soluble cobalt salts were effective. Sigmoidal time-yield curves were observed, and the rate increased with rise in temperature over the range 20–50°C without change in polymer structure. Typically, with 20 mmhm CoCl$_2$ and Al/Co/NEt$_3$ = 10:1:8, a polymer yield of 80% with 94% trans, 6% 1,2 structure was obtained at 20°C in 30 hours, after an induction period of about 6 hours. The slow initial rate probably results from slow formation of active catalyst from the insoluble halide. The catalyst solution has the bright blue color characteristic of tetrahedral cobalt complexes and is readily distinguished from the yellow cis or mixed-microstructure-producing catalysts. Essentially similar results have been obtained with cobalt acetylacetonate, octanoate, or oleate. Variation of catalyst composition and concentration results in some change in structure (72–95% trans) but there is no cis content, the residue being wholly 1,2 structure. As the donor concentration is further increased the rate decreases to zero at $NR_3/Al = 2$.

2.2.5 *Nickel Catalysts*

These are characterized by the presence of iodide ligands (56). It is usual to employ a hydrocarbon derivative of the metal [e.g., $(\pi\text{-}C_4H_7)_2Ni$, $(\pi\text{-}C_5H_5)_2Ni$] in conjunction with a metal iodide, e.g., nickel iodide or titanium tetraiodide. The polymers contain from 95 to 97% trans structure, the residue being vinyl. Reactions can be carried out either in benzene or a polar solvent such as tetrahydrofuran. Yields are very variable but are better when two-metal systems, e.g., Ti–Ni, are employed. Most other ligands produce high cis or mixed microstructure polybutadiene but $\pi\text{-}C_3H_5NiO(2,3,6)C_6H_2Br_3$ gives 96% trans polymer (86).

2.2.6 *Other Catalysts*

Chromium catalysts do not show high activity and have not been extensively studied. Their tendency is to form 1,2 structures. Trisallylchromium and oxygen ($Cr/O_2 = 2:1$) have been reported to give 92.5% trans polymer (86), and chromic oxide deposited on silica-alumina and activated at 500–550°C will polymerize butadiene to the trans form (87). Polymerization with the latter at 80°C using a 2:1 v/v solution of monomer in benzene or toluene gave a product of 135°C melting point and density 1.015 but no details of yields and rates of polymerization were given. Encapsulation of the supported catalyst and a sharp fall in polymerization rate are generally observed with these catalysts when used with butadiene and isoprene. Iridium trichloride has been reported to give high *trans*-polybutadiene (81) but there is little other information available.

2.3 1,2-Polybutadiene

This polymer exists in crystalline isotactic and syndiotactic forms and as an amorphous elastomeric atactic polymer. The latter high vinyl polybutadienes are readily prepared using alkali metal catalysts in the presence of donors such as ethers (diglyme, tetrahydrofuran) or tertiary amines, and the molecular weight and molecular weight distribution are readily controlled.

Certain coordination catalysts do give high (up to 90%) vinyl structure polymers that are amorphous elastomers. In the absence of crystallinity no information can be gained on the distribution of the stereoisomers; although one form may be predominant, the polymers are regarded as simply atactic. The high vinyl elastomers can be reinforced by carbon black and vulcanized by sulfur and may become of industrial importance provided they can be made gel free and of high molecular weight. One coordination catalyst that appears to meet these requirements is $VO(naphthenate)_3/MoCl_5/AlEt_3$:

Triethylaluminum (2 mmhm), vanadyl naphthenate (1 mmhm), and molybdenum pentachloride (3 mmhm). Butadiene in cyclohexane: 1:7 w/v, polymerization 20 hours at 65°C. Conversion 31%, $[\eta] = 5.1$, gel free. Cis-1,4, 8.4%; trans-1,4, 10.1%; vinyl, 81.5%.

Dicobalt octacarbonyl/$MoCl_4$ or $MoCl_5$, Co/Mo = 1:1–2, also gives atactic 1,2-polybutadiene (51).

The tactic polymers have achieved no commercial significance, and consequently it is unlikely that the preparative conditions have been optimized. There is little information on polymerization rates or molecular weight control. Several catalyst systems have been found for the syndiotactic polymer but only one has given the isotactic structure.

2.3.1 Syndiotactic 1,2-Polybutadiene

Vanadium Tris(acetylacetonate)/Triethylaluminum (88). Halogen-free vanadium catalysts give this polymer, but it is not of homogeneous composition; the vinyl contents of the crude material range from 78 to 86%. The best yields of highly crystalline, 90–95% vinyl structure polymer (mp 156°C) are obtained with higher Al/V ratios, in the range from 3 to 30. The amorphous fraction of lower vinyl content was removed by extraction with ether or low boiling aliphatic hydrocarbons.

Cobalt Catalysts. Syndiotactic 1,2-polybutadiene is obtained from cobalt salts with halogen-free aluminum alkyls such as $AlEt_3$, $(AlEt_2)_2O$, and $(Et_2Al)_2SO_4$. The significance of the halogen is demonstrated by $CoCl_2$-py/(AlEtCl)NPh(AlEt_2), which gives a mixture of high cis and high 1,2 polymer, whereas corresponding systems containing $(AlEtCl)_2NPh$ and $(AlEt_2)_2NPh$ give exclusively the cis-1,4 and the syndiotactic 1,2 polymers. Moreover, with mixtures of triethylaluminum and diethylaluminum chloride there is a sharp changeover from 1,2 to cis as the proportion of the latter increases (47). The significance of aging in determining catalyst activity has been referred to in the discussion of the cis polymer.

The literature gives clear indications as to what the experimental conditions for high cis or 1,2 polymer should be, but the active species present have not been identified. Therefore, the following rationalization is most probably incomplete and oversimplified. Organocobalt compounds alone are ineffective for the formation of high molecular weight polymer and the active catalyst is presumably an organocobalt-aluminum complex with the two metal atoms linked by hydrocarbon radical or halogen bridges; when there is no halogen only the former will be present. Complexes of this type are relatively unstable and decompose to give metallic cobalt except when monomer is present, in which case polymerization to give the high 1,2 polymer occurs. In mixtures of trialkylaluminum and dialkylaluminum chloride, the former, being the stronger reducing agent, would be expected to react preferentially to give the complex which, again provided butadiene is present, would produce the high vinyl polymer. However, halogen bridges are more stable and in the absence of butadiene these should be formed by displacement or exchange reactions to give the same cis producing catalyst as is formed from aluminum alkyl halide alone. When cobalt chloride is used some aluminum diethyl chloride will be produced and this on aging in the absence of monomer will give the same stable cis producing catalyst. Mixtures of the two polymers no doubt result from the presence of two distinct catalytic entities propagating independently.

Other than this, the conditions that have been disclosed are similar to those employed for the high cis polymer. Benzene is the preferred solvent; monomer concentration 10% v/v, triethylaluminum (15–30 mmhm) and soluble cobalt salts or complexes (0.1–0.2 mmhm) are used. Polymerization is carried out at 15–20°C. Reported yields (20–50%) tend to be lower than with the high cis recipes.

This cannot be the complete explanation for the dependence of polymer structure on catalyst composition since cobalt catalysts have been found which contain no halogen but which give cis enchainment of monomer. Thus $Co(acac)_3/AlEt_3/H_2O$ gives a mixed 1:1 *cis*-1,4/1,2-butadiene copolymer with a random distribution of monomer units (89). The Al/Co ratio is not critical, $Al/H_2O = 1$, and polymerization is conducted at low temperatures (below 0°C).

$Co(SCN)_2(Ph_3P)_2/(Et_2Al)_2SO_4$ differs from other cobalt catalysts containing donors in that instead of the trans polymer it produces a 95% syndiotactic 1,2-polybutadiene (90). The reactions are slow (16 hours at 80°C) and require relatively high catalyst concentrations (3 mmhm, Al/Co = 0.5–2).

Titanium Catalysts. The catalysts that have been used consist of a trialkylaluminum with halogen-free titanium compounds, $Ti(OR)_4$ or $Ti(NR_2)_4$, in homogeneous media (91). The polymerizations are carried out in hydrocarbon solution at fairly low temperatures (0–15°C) using Al/Ti ratios of 7–10; the crude polymers contain from 88 to 96% of 1,2 enchainment. As with the cobalt catalysts the important feature is the absence of halogen since triscrotyltitanium without an aluminum cocatalyst also gives 1,2-polybutadiene (59).

The polymers obtained from $Ti(OR)_4/AlEt_3$ are mainly amorphous in character and from 40 to 60% is soluble in ether. The ether insoluble fraction is of greater than 95% 1,2 structure, but exhibits only weak crystallinity and is evidently not highly stereoregular. There is a marked tendency for this catalyst to give gelled polymers, possibly by polymerization initiated from pendant vinyl groups.

Miscellaneous Catalysts. There are a number of other transition metal complexes that have produced high vinyl polybutadiene. Molybdenum compounds [$MoCl_5$, $MoO_2(acac)$, and $MO_2(OR)_2$] combined with the alkyls of aluminum or gallium or lithium aluminum hydride give polymers of either random or syndiotactic structure (92–94). The reaction variables do not appear to have been examined in detail for these systems. Trisallylniobium has been reported to give polybutadiene of exclusively 1,2 structure (86). Palladium salts ($PdCl_2$, K_2PdCl_4), in aqueous solution with sodium hexadecylbenzene sulfonate as emulsifier, are reported to polymerize butadiene to high vinyl polymer (91–98%) (95). For moderate yields of polymer (23–30%), long reaction times (99 hours at 50°C) and high catalyst concentrations (5 mole % on monomer) are required. Solid polymer was obtained with the halides, whereas palladous sulfate or nitrate gave only low molecular weight polymer.

2.3.2 *Isotactic* **1,2-***Polybutadiene*

The crystalline polymer (mp 126°C) is the least accessible of the isomers, and no satisfactory procedure for its synthesis in good yield has been described. It has been obtained using soluble complexes of chromium with aluminum alkyls. The nature of the chromium complex does not appear to be critical, and provided it is soluble [e.g., chromium tris(acetylacetonate) and tris(bipyridyl)chromium halides], high 1,2 polymer is produced. Trisallylchromium itself gives crystalline polymer of comparable structure (whether isotactic or syndiotactic is not stated) (59). The addition of chromium trichloride to this does not greatly reduce the 1,2 structure but addition of nickel chloride results in high cis polymer, presumably ligand transfer forming the more active cis catalyst from nickel.

The crude product (1,2 structure 70–90%) is generally a mixture of amorphous and crystalline polymer, the former predominating (96). The crystalline polymer is either the isotactic form or a mixture of isotactic and syndiotactic forms. The amount of crystalline polymer increases on aging the catalyst solution before adding the monomer. About 45% of isotactic polymer (97% 1,2 structure) has been obtained using an aged catalyst, and $AlEt_3/Cr(acac)_3 = 5:1$.

A study of the kinetics of the polymerization with $Cr(acac)_3/AlEt_3$ shows polymerization to be slow, and for reasonable rates (20%/hr) chromium concentrations of about 7 mmhm are required. The rate falls continuously with increase in Al/Cr ratio from 4 to 20. Molecular weights reported are low (200–3000), which could account for much of the amorphous polymer from unaged catalysts. The molecular weights increase with monomer concentration. In contrast to the soluble cobalt system the molecular weight is independent of chromium concentration and falls with increase in triethylaluminum concentration owing to transfer of growing chains (97).

3   ISOPRENE POLYMERS

3.1   *cis*-1,4-Polyisoprene

This important polymer is manufactured on a substantial scale (production capacity in the United States in 1969, 152,000 tons) (7) by two processes, one using lithium alkyls and the other trialkylaluminum/ titanium tetrachloride. Both materials are partial or complete replacements for natural rubber in many applications and have specific advantageous features. The former, although of lower cis content (92–94%), can be obtained as linear, very high molecular weight polymer with a narrow molecular weight distribution, and is free from traces of transition metals which could be troublesome in certain applications. The latter polyisoprene is of higher cis content (96–98% and virtually indistinguishable from natural rubber by infrared spectroscopy), ensuring more rapid crystallization and hence higher tear strength in vulcanizates, which are close in mechanical properties to those from the natural product (98). Unlike butadiene, only one coordination catalyst type is in commercial use for the preparation of high cis-polyisoprene. In these β-TiCl$_3$ is complexed with an organoaluminum compound such as the alane, the triisobutyl, diisobutyl chloride, or hydride. The simplest method for preparing this catalyst and the first to be used (2) is obtained by mixing titanium tetrachloride and triisobutylaluminum.

This catalyst tends to produce some low molecular weight liquid polymers and gel. The oligomers are detrimental to the physical properties of the vulcanized rubber, whereas high cis content, linearity of polymer structure, and high molecular weight are desirable characteristics. The catalyst composition, the order of addition of reagents, and the structure of the aluminum alkyl all have significant effects on polymerization. The Al/Ti ratio should lie between 0.9 and 1.2; much outside this range polymerization rates become very low (99). The order of addition of reagents is important in order not to overreduce the titanium. The alkylaluminum is added to the titanium tetrachloride as a dilute solution in a hydrocarbon solvent with good agitation. Low yields of mixed microstructure polymer are obtained by the reverse procedure (100). The reaction is normally carried out at room temperature but it is also possible to form a Ti(IV) complex at low temperatures which decomposes to give β-TiCl$_3$ on raising the temperature. It is equally important to avoid the occlusion of titanium tetrachloride in the precipitate since this could give rise to gelled polymer of low cis content or liquid polymers. Moreover, in the reaction of titanium tetrachloride with triethylaluminum it may be presumed that both triethylaluminum and aluminum diethyl chloride are complexed onto the β-TiCl$_3$ in variable proportions dependent on the conditions. These problems can be avoided if the precipitate is separated and washed with hydrocarbon solvent, whereby it loses its catalytic activity (101), but may then be reactivated by the addition of alkylaluminum. The additional metal alkyl required is about 0.025 mole/mole of titanium in the precipitate (102). Aluminum diisobutyl choride also gives an active catalyst but aluminum isobutyl dichloride does not.

The structure of the alkyl group in the catalyst has an effect, the best yields being obtained with higher alkyl groups, and branching in the alkyl group is an advantage. Aluminum triisobutyl is generally favored; the results obtained with a number of reducing agents are given in Table 9. Aluminum trimethyl and triethyl are the least satisfactory and give poor yields of solid high molecular weight polymer. Aging the catalyst before adding the monomer permits the decomposition and disappearance of undesirable Ti(IV) compounds, but its effect on yield or polymer properties is not very great. Aromatic ethers (103) and other donors such as triphenylamine (104) or diphenyl sulfide (105) have a beneficial effect, increasing the polymerization rate and the quality of the product. The effect is very dependent on the structure of the donor. The most effective are those of low basicity, and diphenyl ether is particularly useful (106). The optimum rate of polymerization is at an $Al/Ph_2O/Ti$ ratio of $0.9:0.9:1$ and the fractions of oligomers and gelled polymer, at ca. 3 and 10%, respectively, are low. The catalyst is still made up by adding the etherate to the titanium tetrachloride, and although the triisobutylaluminum gives the best results compared with other metal alkyls, aluminum trimethyl and triethyl catalysts are upgraded, the latter being better than an unmodified $Al(i\text{-}Bu)_3/TiCl_4$ system. The catalysts change little with time on aging at room temperature but slowly lose activity when stored at 60°C. Aliphatic ethers are much less effective than aromatic compounds and triisobutylaluminum-ethyl etherate gives a less active catalyst than the alkyl itself (106).

Table 9   Comparison of Aluminum Alkyls in the Polymerization of Isoprene by AlR₃/TiCl₄ (102)

R_3 in AlR₃	Al/Ti	Yield (%)	$[\eta]$	Gel (%)
Me₃	1.1	12	4.4	16
Et₃	1.0	29	2.2	13
n-Bu₃	1.0	45 (49)[a]	3.6 (3.3)	12 (13)
i-Bu₃	1.0	33	3.7	10
n-Octyl₃	1.0	59 (77)	2.7 (3.4)	8 (25)
(i-Bu)₂H	1.0	53 (55)	2.8 (2.5)	20 (7)

[a] Figures in parentheses refer to aged catalysts.

Aliphatic hydrocarbons are the most suitable solvents for polymerization; the temperature is normally in the range 20–50°C. A recipe for the polymerization of isoprene using $Al(i\text{-}Bu)_3/TiCl_4$ is the following (3):

To isoprene (55 g, 15% w/w in pentane) is added triisobutylaluminum (0.2 mmole) and titanium tetrachloride (0.185 mmole). The catalyst components are used as 0.2–0.5 M solutions in heptane. Polymerization is carried out at 50°C for 16 hours. The polymer on isolation (protected with antioxidant) has $[\eta] = 2.0$–3.5, the cis content is >94%, and the yield is 70–80%. An essentially similar reaction occurs replacing the triisobutylaluminum by its diphenyl etherate (103). For the preparation of *cis*-polyisoprene the monomer must be of high purity. Minute traces of acetylenes, cyclopentadiene, and, to a lesser extent, pentadienes have an unfavorable effect on the polymerization. The solvents and monomer must be rigorously dried and there should be no sulfur or oxygen-containing poisons.

There are several variants of the basic $AlR_3/TiCl_4$ system which can be used. Thus β-TiCl$_3$ can be prepared from methyltitanium trichloride, and with aluminum triethyl or diethyl chloride gives high cis-polyisoprene; clearly this route is only of theoretical interest. The aluminum alkyls can be replaced by derivatives of aluminum hydride such as $AlHCl_2 \cdot NMe_3$, $AlHCl_2 \cdot OEt_2$, $AlH_2Cl \cdot NMe_3$, $AlH_3 \cdot NMe_3$, AlH_2NMe_2, and AlH_2OEt_2 (107). The polymerization rates are maximum at Al/Ti in the region 1.1–1.6, depending on the structure of the complex, and high conversions are obtained with ca. 9 mmhm of titanium tetrachloride. Molecular weights rise to a maximum with an increase in the Al/Ti ratio to 1.5 and fall with an increase in catalyst concentration. Molecular weight rises with conversion in the early stages of polymerization and then remains constant. In contrast to $TiCl_4/Al(i\text{-Bu})_3$, benzene is the preferred solvent. Generally, the cocatalysts containing Al–N or Al–O bonds are less active than the ether or amine complexes. Recently poly(iminoalanes), $(AlHNR)_n$, have been found to be effective cocatalysts for the polymerization of isoprene to gel-free high cis polymer (108). The following illustrates the experimental conditions:

Isoprene, 20.4 g; n-heptane 100 ml.; $TiCl_4$ 1.82 mmoles
$(AlHNn\text{-Bu})_n$ 2.92 mmole; Al/N = 1.6
Polymerization 2 hours at 15°C
73% yield; $[\eta] = 6.14$; 96.5% cis-1,4; 3.0% 3,4

Another variant of the titanium catalyst is a ternary system consisting of a titanium tetraalkoxide, an aluminum alkyl sesquichloride, and a silicon compound (triethylsilane, trimethylmethoxysilicon) (109). The effect of the silicon compound which is added to the prereacted $Ti(OR)_4/Al_2Et_3Cl_3$ is to increase the cis content of the polymer to 97–98% compared with about 90% in its absence, and to eliminate the formation of gel. Presumably by halogen-alkoxyl interchange and reduction some type of β-TiCl$_3$ complex is produced. A recipe containing tetra(n-butoxy)titanium (1.5% v/v), aluminum ethylsesquichloride (4.2% v/v), and trimethylmethoxy-silicon (1.5% v/v) (catalyst components based on monomer) with iso-prene in n-heptane (20% v/v) gave 56% conversion to 98% cis polymer with low gel content (1%) and $[\eta] = 2.4$. The very high concentration of catalyst would seem to be a major disadvantage to its use on a large scale.

The preparation of high cis-polyisoprene of high molecular weight from $AlRCl_2/Ti(OR)_4$ has also been reported recently (96% cis-1,4 at

Al/Ti = 4–6) (78). The active catalyst is shown to be a complex between β-TiCl$_3$ and an aluminum compound such as Al(OR)RCl. With higher Al/Ti ratios the polymers are of mixed cis-trans structure. [With butadiene the polymer obtained at Al/Ti = 6 is of 60–70% cis structure, not greatly different from that obtained with β-TiCl$_3$/AlEt$_2$Cl; but at high Al/Ti ratios the high trans (88–94%) polymer, similar to that from β-TiCl$_3$/AlEtCl$_2$ (76), is obtained.]

A recipe for the polymerization has been given (111):

Isoprene	45 g
Pentane	500 ml
(i-PrO)$_4$Ti	0.672 mmole
Al-i-BuCl$_2$	3.60 mmoles
n-Bu$_3$N	0.378 mmole

Time 48 hours; conversion >90%
Gel: 8–10%; DSV 2.4–3

The nature of the alkyl radical in the titanate is important since it influences the form of titanium trichloride that is produced. Primary titanates give the cis polymer, whereas tertiary titanates give the trans polymer. Secondary titanates give the cis polymer in aliphatic solvent, as indicated in the recipe and trans polymer in aromatic solvent (111).

High (76–93%) *cis*-polyisoprenes are obtained using catalysts from Zr(π-allyl)$_4$–TiCl$_4$ at Zr/Ti ratios of 0.8–1.2. Reduction of TiCl$_4$ occurs with the formation of a precipitate, and the active species is considered to be a complex of TiCl$_3$ with Zr(π-allyl)$_3$Cl (112). π-Allyl nickel trifluoroacetate catalysts and variants will give up to 94% 1,4 addition, but the maximum cis content reported is only 79% (113).

Aluminum alkyls occupy a unique position among the cocatalysts in that they are used in all the important commercial processes, and there are very few effective catalysts that do not contain them. One reported example replaces the aluminum alkyl with a magnesium alkyl (114). The products are inferior in cis content (70–55%), falling with increase in Mg/Ti ratio over the range 0.75:1 to 1.25:1, and they contain substantial quantities of gel. The recipe below is included purely for interest as one not based on aluminum:

Di-n-butyl magnesium: 1.4 mmhm
Titanium tetrachloride: 1.77 mmhm; Mg/Ti = 0.79:1
Isoprene in n-hexane: 13.2% w/w
Polymerization 16 hours at 50°C
Yield of polymer: 65%; gel content 23%; inherent viscosity of sol: 3.92
Structure: 75% cis-1,4; 6.2% 3,4

With all elastomers a higher molecular weight improves the final properties of the vulcanizate, but with most, processing difficulties place a practical upper limit on the bulk viscosity of the polymer. Since polyisoprene can be readily broken down by mastication, this is not a severe limiting factor and an increase in the viscosity of the polymer is advantageous, except insofar as it influences output and the handling of the viscous polymer solutions during manufacture. The dependence of viscosity on polyisoprene concentration has been discussed by Litvin et al. (115). The polymer molecular weight can be increased by reducing the catalyst concentration or by lowering the temperature—both of which decrease the polymerization rate. The concurrent formation of low molecular weight polymer and gel makes it difficult to generalize on the relationship between molecular weight and reaction variables. It has been reported that the molecular weight distribution obtained using $TiCl_4/Al(i-Bu)_3$ is independent of conversion (116). Although this appears to be broadly true, a detailed examination of the changes shows them to be complex (117). There is an initial rise in the early stages of reaction followed by a leveling off and then a decrease. Changes in gel content are somewhat unpredictable; rises and falls have been recorded but a full explanation of these effects has not been given. A useful index for assessing polymer quality is given by $[\eta] (100 - G)/100$, where G is the percentage of gel content.

3.2 *trans*-1,4-Polyisoprene

trans-Polyisoprene is of particular interest in that it is the synthetic counterpart of gutta-percha or balata. The former, once produced in the East Indies, is no longer available and balata, obtained from the forests of tropical South America, is expensive and in limited supply. The polymerization of isoprene with coordination catalysts quickly led to the synthesis of the pure 1,4 isomers, the trans form being patented in 1955 (118). The limited outlets for it and its consequent high price have restricted its commercial development, but it has been manufactured in the United Kingdom and in Canada since 1963–1964 (119).

The synthetic polymer of high trans content (>98%) is very similar in its properties to the natural polymer (120). Melting transitions for the α and β forms are very similar in the natural and synthetic polymers, as is the rate of crystallization, particularly when nucleating agents are added to the latter. The molecular weights of the synthetic polymers are usually higher than that of the natural ($[\eta]$ about 1.0), and mechanical or other means of breakdown is often necessary for their use.

Several systems have been found to give trans polymer, but the best, from the point of view of isomer purity, are those based on vanadium halides. In the early literature there are many reports to the effect that polyisoprenes of high trans content are obtained using $Al(i\text{-Bu})_3/TiCl_4$ catalysts in which the Al/Ti ratio is below 1.0 (1). It is true that with $Al/Ti \simeq 0.5$ hard resinous polymers are obtained which have infrared absorptions corresponding to predominantly trans-1,4 structure, but the products are mechanically very dissimilar to balata. They are generally highly branched and contain gel. A considerable amount of cyclization occurs since the catalyst appears to be cationic and will usually contain some free titanium tetrachloride. For these reasons this system will not be considered further.

3.2.1 *Vanadium Catalysts*

All vanadium halides give high *trans*-polyisoprene with aluminum trialkyls; the highest stereoregularity is obtained using vanadium trichloride. The main problem with the unmodified $VCl_3/AlEt_3$ catalyst is the relatively poor yield of polymer obtained, usually about 50–100 g polymer/g vanadium trichloride. Conditions for the polymerization are not critical. The yield of polymer passes through a broad maximum at Al/V in the range 5–7 and diminishes slowly as the ratio is increased above this (74). The solvent is preferably an aliphatic hydrocarbon; no advantage results from substitution of benzene or toluene. The yield of polymer is markedly increased either by the addition of an ether (e.g., diisopropyl ether) or by the deposition of the vanadium chloride on a finely divided support.

The effect of diisopropyl ether has been studied in detail (121). In large amounts it decreases the polymerization rate, although the amount that can be tolerated depends on the Al/V ratio. Thus, at Al/V 2:1 the optimum ratio of $i\text{-Pr}_2O/V$ is 5:1, giving a fourfold increase in yield of polymer compared with the unmodified catalyst. The ether has a negligible effect on the microstructure of the polymer and appears to accelerate the rate by increasing the amount of alkylvanadium chloride produced (122), which is the probable active species, and there is a marked increase in the number of growing centers in its presence. It is presumed that ethers or other electron donors help to cleave and partly solubilize the laminar vanadium trichloride crystals at lattice defects. Vanadium dichloride is inactive for the polymerization of isoprene, although mixed with the trichloride it does not appear to be disadvantageous. It would be expected, therefore, that the polymerization would be sensitive to temperature, since decomposition of the active propagating chains might be expected to lead to the formation of vanadium dichloride. In fact, it is observed that at temperatures above 50°C the polymerization dies out before complete conversion, while below 30–40°C the polymer is obtained in a crystalline form at a slow polymerization rate. The optimum temperature is in the region 45–50°C.

Vanadium trichloride deposited on an inactive support such as clay or titanium dioxide is an equally or more effective system than donors in increasing the yield of polymer (123). The method of deposition is to decompose vanadium tetrachloride in solution in benzene or heptane with ultraviolet radiation or by heating in the presence of the dried finely divided support. (Inferior results are observed if the catalyst is merely prepared from vanadium tetrachloride and the trialkylaluminum in the presence of the support.)

Thus, kaolin (20 g, surface area $10 \, m^2/g$) in benzene (200 ml) and vanadium tetrachloride (9 g) was refluxed under argon for 4 hours. The filtered solid contained 24.6% of the trichloride. A sample of this (0.29 g/0.07 g VCl_3) and triisobutylaluminum (0.444 g as a 10% solution in benzene) in benzene (350 ml) and isoprene (60 g) were reacted at 50°C for 72 hours. Conversion was 40%. (Polymer yields are rather variable but are, at maximum, above 400 g polymer/g vanadium trichloride, i.e., four to eight times better than the unsupported catalyst.)

Vanadium tetrachloride behaves in a broadly similar manner to the trichloride, but the polymers are inferior (124). The precipitate formed is mainly the trichloride and the aluminum diethyl chloride produced does not interfere seriously. The more finely divided form of the catalyst gives a somewhat faster initial rate of polymerization but it quickly becomes encapsulated with polymer; yields as low as 20 g/g vanadium trichloride have been found. The lower melting point and crystallinity are possibly due to the presence of cationic substances, since the polymer contains gel not broken down on milling.

Vanadium oxytrichloride with triethylaluminum gives polymer of lower molecular weight and lower stereoregularity than the other vanadium halides, and although the product is usually over 90% trans content, it is not satisfactory for the preparation of highly crystalline polymer.

3.2.2 *Mixed Vanadium and Titanium Catalysts*

These catalysts give high trans polymer and are the most active both in regard to speed of polymerization and yield of polymer. Two types are described—one heterogeneous, consisting of vanadium trichloride mixed with or deposited on titanium trichloride with triethylaluminum, and the other a soluble system from titanium alkoxide, trialkylaluminum, and vanadium trichloride on an inert support.

VCl₃/β-TiCl₃/AlEt₃. This catalyst may be prepared by reduction of mixtures of titanium and vanadium tetrachloride (typically Ti/V = 4 : 1 and Al/V = 1 : 3), followed by activation of the reduced metal halides with an excess of triethylaluminum to give an Al/(V + Ti) ratio of 5. The preparation of the mixed halides can be carried out at temperatures in the range 80–170°C. These catalysts were found to give average polymerization rates up to 50 times faster than those from vanadium trichloride alone (76).

Particularly reactive catalysts are obtained by first reducing titanium tetrachloride to β-TiCl₃, conversion of this to the γ form by heating, followed by the addition of vanadium tetrachloride which is decomposed to the trichloride by heating. The mixed trichlorides are then separated from the mother liquor and activated by triethylaluminum (125). In these systems the optimum V/Ti ratio is 1 : 2 with Al/(V + Ti) = 5–10. Catalyst activity is not very reproducible but is very high compared with the individual halides, and is in the range (V + Ti)/V/Ti = 1000–3000 : 50 (200) : 1 (10) g polymer yield/g VCl₃ (or TiCl₃) for a fixed polymerization time. The values in parentheses for the latter two refer to supported catalysts (10% transition metal halide on TiO₂). The following gives details for the polymerization using the mixed halides (125).

Titanium tetrachloride (20 mmoles) and triethylaluminum (8 mmoles) were reacted in liquid paraffin (20 ml) for 30 minutes at room temperature and then at 170°C for 1 hour. On cooling, vanadium tetrachloride (10 mmoles) and triethylaluminum (5 mmoles) were added and the mixture heated at 120–134°C for 1 hour. The solid precipitate was washed with benzene to remove liquid paraffin. To a portion of the deep purple-red precipitate containing 0.3 g VCl₃ (1.9 mmoles) was added purified dry petrol (1000 ml; bp 90–105°C), triethylaluminum (15 mmoles), isopropyl ether (0.75 ml), and isoprene (250 ml). Conversion to polymer after 1 hour at room temperature was 80%; trans content 98%.

Molecular weights of the polymers obtained by this method are high ($[\eta]$ = 4–5).

Ti(OR)₄/VCl₃–TiO₂/AlR₃ (124). This catalyst is prepared by adding the titanium alkoxide [or PhTi(OR)₃] and alkylaluminum to vanadium trichloride supported on clay. The optimum V/Ti ratios for different Al/V ratios are Al/V = 5 : 1, V/Ti = 2–4 : 1; Al/V = 10 : 1, V/Ti = 1–2 : 1; Al/V = 20, V/Ti = 0.5–1 : 1. Typically, vanadium trichloride (0.45 mmole) is deposited on kaolin (5–6% VCl₃ on the clay), and reacted with tetra(2-ethylbutoxy)titanate (0.22 mmole) and triisobutylaluminum (4.5 mmoles) in benzene (170 ml), followed by isoprene (150 ml). Conversion to polymer after 6 hours at 50°C is 90%; trans content 98%; $[\eta] = 3$–5. Catalyst activity is about 50 times greater than the supported vanadium trichloride catalyst and about twice that of the VCl_3/γ-$TiCl_3/AlEt_3$ catalyst. There is a marked fall in activity without the carrier for the vanadium trichloride, but it is still significantly more active (by a factor of about 2) compared with the supported vanadium catalyst without the titanium alkoxide.

3.2.3 *Other Catalysts*

The α and γ forms of titanium trichloride with an alkylaluminum give crystalline *trans*-polyisoprene (77). The rates of polymerization are slow, possibly due to the deposition of insoluble polymer on the catalyst (126), and although the polymerizations have a fairly high activation energy (127), conversions are low at temperatures from 60 to 100°C. Yields of the order of 0.03–0.05 g/g $TiCl_3$/hr have been reported (128). Deposition of γ-titanium trichloride on titanium dioxide or kaolin and the addition of an ether increases the rate by a factor of from 10 to 30 (128), although compared with the vanadium or mixed metal catalysts reactions are still very slow. Little advantage results from an excess of triethylaluminum and the Al/Ti ratio can range from 2 to 12. Petrol or toluene are suitable solvents and the best reported yield (1.26 g polymer/g $TiCl_3$/hr) was with Al/Ti/i-$Pr_2O = 12:1:1$ in toluene at 50°C.

Chromium trioxide on silica-alumina gives crystalline *trans*-polyisoprene (87) but the catalyst is of low efficiency and of little practical importance.

3.3 1,2- and 3,4-Polyisoprenes

In general those catalysts which give 1,2-polybutadiene give the 3,4 isomer from isoprene; the polymers are amorphous and not of high steric purity. So far no predominantly 1,2-polyisoprene has been synthesized, although small proportions of this structure are found in mixed microstructure polymers. Alkali metal based catalysts have been used to prepare such polymers (see Chapters 4 and 5).

The only coordination catalyst that has been studied in detail is $Cr(acac)_3/AlEt_3$ (129). The structure of the polymer is not completely specified but is apparently mixed 1,2 and 3,4, and there may be some cyclization. Molecular weights are very low ($[\eta] = 0.07$–0.16) and follow the same pattern as with butadiene (97). Reactions are likewise slow [8% conversion per hour at 40°C, using 16.7 mmhm $Cr(acac)_3$ and a monomer concentration of 5 moles/liter.] Several more interesting catalysts giving polymer containing high 1,2 or 3,4 structures have been found, but as these are of regular structure they have been considered separately as "equibinary" polymers.

4   1,3-PENTADIENE (PIPERYLENE)

This monomer exists in the trans and cis forms, **1** and **2**:

$$CH_2= CH$$
$$\backslash$$
$$CH = CH$$
$$\backslash$$
$$CH_3$$

$$CH_2 = CH \qquad CH_3$$
$$\backslash \qquad /$$
$$CH = CH$$

1   **2**

Both isomers are polymerized by some catalysts but with others only the trans isomer will polymerize. It is apparent that the boat form of the cis monomer (**3**) will be sterically unfavorable (see also Chapter 4, Section 3.5):

$$CH = CH$$
$$/ \qquad \backslash$$
$$CH_3 \qquad CH$$
$$\qquad //$$
$$\qquad CH_2$$

3

Thus 1,4 polymerization will give the trans polymer unless there is rotation in the transition state to give the cis polymer. If two-point S-cis coordination is essential at the catalyst site, polymerization will not occur. In addition to the isomerism about the double bond the configuration of the substituted carbon atom leads to isotactic and syndiotactic forms.

1,3-Pentadiene (or any other α- or ω-substituted butadiene) has two reaction sites which may lead to stereoisomers; i.e., its polymers are ditactic. There are four possible 1,4 addition stereoisomeric polypentadienes: the isotactic and syndiotactic cis-1,4 and a similar trans-1,4 pair. There are also cis-1,2 polymer (from cis monomer) and trans-1,2 polymer (from trans monomer). Both the 1,2 and the analogous 3,4 addition polymers permit isotactic and syndiotactic forms. Only isotactic and syndiotactic cis-1,4 and isotactic trans-1,4 polymers and the syndiotactic 1,2-polypentadiene have been prepared. 1,3-Pentadiene is a potentially abundant hydrocarbon and although polymer is not produced from it on a commercial scale the elastomeric cis polymer gives strong resilient vulcanizates (130), comparable with other hydrocarbon polymers.

4.1   Syndiotactic *cis*-1,4-Polypentadiene

The catalysts for this isomer are obtained from soluble cobalt compounds and organoaluminum chlorides. The nature of the product obtained

Table 10 Polymerization of 1,3-Pentadiene by Cobalt Catalysts (131)

Organo-aluminum Compound	Al/Co	Water Co-catalyst	Solvent	Polymer Structure
AlEt$_2$Cl	50	–	C$_6$H$_6$	1,2
AlEt$_2$Cl	50	–	C$_7$H$_{16}$	~55% 1,4
		+		Up to 85% 1,4
O(AlEtCl)$_2$	50	–	C$_6$H$_6$	~90% 1,4
AlEtCl$_2$	50	–	C$_6$H$_6$	~85% 1,4 (thiophene added to suppress cationic activity)
AlEt$_2$Cl	90–3000	+ or –	C$_7$H$_{16}$	>90% 1,2
O(AlEtCl)$_2$	85–1000	–	C$_7$H$_{16}$	Up to 94% 1,4
Co(acac)$_2$ or Co(acac)$_3$ ca. 10^{-4} mmhm				

depends on the conditions, being either predominantly syndiotactic cis-1,4 polymer or 1,2 polymer, or a mixture of the two. The conditions have been studied in detail by Porri and co-workers (131), and are summarized in Table 10.

It is apparent that the highest purity syndiotactic 1,4 isomer is obtained with the aloxane. Experimental conditions which give good yields of highly stereoregular polymer are the following:

Co(acac)$_2$: 1.6 × 10^{-3} mmole; O(AlEtCl)$_2$: 0.96–4 mmoles
Al/Co = 600; thiophene: 2 mmoles
Benzene: 10 ml; 1,3-pentadiene: 2 ml
Polymerization 20 hours at 18°C
86% conversion; 92% syn-cis-1,4; [η] = 1.15

4.2 Isotactic *cis*-1,4-Polypentadiene

This isomer has a slightly lower melting point (43–46°C) than the syndiotactic polymer (50–53°C) and a shorter x-ray identity period (8.1 Å compared with 8.4 Å). It has been obtained from both the cis and trans forms of the monomer using aluminum alkyl/titanium alkoxide catalysts. Both forms of the monomer give high cis-1,4 structure (>85%) but the polymer from the trans monomer is evidently sterically purer and has been obtained in crystalline form.

The preparative conditions can be varied widely but the optimum in terms of rate, cis content, and molecular weight is obtained with Al/Ti ratios in the region of 6:1 (132):

Monomer (98–99% trans isomer)		16 g
Benzene		100 ml
AlEt₃	} Al/Ti = 5	8 mmoles
	} Aged at room temperature	
Ti(O-n-Bu)₄	} for 20 minutes	1.6 mmoles

Polymerization 35 hours at 0°C. Crude yield: 40%. Cis-1,4 content: 80%. Extraction of the crude polymer with butanone left a crystalline high molecular weight material: 85% cis; 10% trans; 5% 3,4; $[\eta] = 5–7$.

The trans double bonds in this polymer are predominantly internal following 1,4 addition but a small proportion of pendant trans double bonds may also be present.

By use of an optically active titanium alkoxide [titanium tetra(*l*-menthoxide)] the crystalline polymer obtained possesses optical activity (133).

4.3 Isotactic *trans*-1,4-Polypentadiene

The polymerization of either *cis*- or *trans*-1,3-pentadiene with heterogeneous catalysts from vanadium trichloride and aluminum trialkyls has given the crystalline isotactic trans-1,4 polymer (mp 95°C) (134). The polymer obtained from the pure isomers was mainly crystalline, but when mixed monomer isomers were used the product contained a substantial fraction (ca. 40%) of ether soluble polymer. The conditions employed for the preparation of the crystalline polymer were as follows:

Monomer (100% trans isomer) 21% v/v in *n*-heptane
Catalyst: vanadium trichloride 70 mmhm; triethylaluminum 175 mmhm (Al/V = 2.5)
Polymerization 70 hours at room temperature
85% conversion; crystalline, ether insoluble fraction, 85%

As with diene polymerizations employing vanadium trichloride high catalyst concentrations are required but the amount required here, about 30% w/w on the monomer, seems excessive.

4.4 Syndiotactic 1,2-Polypentadiene (131, 135)

The catalyst for this polymer is a soluble cobalt salt and an organoaluminum chloride. The conditions for its formation are shown in

Table 10, namely an aliphatic hydrocarbon as solvent, or if aromatic solvents are used, a low Al/Co ratio. Experimental conditions are the following:

Heptane: 20 ml, Co(acac)$_3$: 2.2×10^{-3} mmole, 0.063 mmhm AlEt$_2$Cl: 2 mmoles   (Al/Co = 910)
1,3-Pentadiene (99% trans): 5 ml
Polymerization 10 hours at 20°C
68% conversion; 98% 1,2 structure

The polymer is a noncrystalline elastomer in the unstretched state, but on stretching an x-ray diffraction pattern has been obtained from which the identity period was calculated to be 8.1 ± 0.5 Å.

5 OTHER CONJUGATED DIENES

Very few dienes other than butadiene, isoprene, and pentadiene have been the subject of detailed study. A number of stereoregular polymers have been described, usually by the application of a single catalyst under arbitrarily chosen experimental conditions. None has received serious consideration as a commerical rubber or plastic, which, no doubt, explains the mainly superficial and scanty information in the literature.

In general, it appears that 2-substituted butadienes give predominantly cis-1,4 polymers with TiCl$_4$ based catalysts, although there is a tendency for the cis content to decrease as the size of the alkyl grouping is increased (136). As in the case of isoprene, high cis polymers are almost unique to titanium compounds. A variety of π-allylic complexes of nickel, cobalt, chromium, and zirconium, with or without an organoaluminum cocatalyst, have been tested in the polymerization of isoprene and other 2-alkylbutadienes. Cis contents were low and polymers were generally of mixed microstructure although in a few instances 1,4 contents were as high as 90% (137).

5.1 2,3-Dimethylbutadiene

This monomer has been polymerized using titanium tetrachloride and aluminum triisobutyl, and two crystalline polymers have been described. One obtained with equimolar proportions of the catalyst components has a melting point of 189–198°C and an x-ray identity period of 7.0 Å; it is identified as the cis-1,4 isomer (138). The other, obtained using Al/Ti = 1:4 in heptane, of melting point 253–259°C, $[\eta] = 0.5$–1.3, is the trans isomer (139). No information is available on rates of polymerization and polymer structure as a function of reaction variables. The TiCl$_4$/AlEt$_2$Cl catalyst has also been used with this monomer (140). Low conversions to partly crystalline polymer were reported.

5.2   2-Ethylbutadiene

Only the cis-1,4 isomer has been reported (141), prepared using TiCl$_4$ (2 mmhm)/Al(i-Bu)$_3$Ph$_2$O, at an Al/Ti ratio of 0.9:1. Yield (15% solution of monomer in pentane, 215 hours at 50°C) was 47%; cis content 97%; [η] = 3.2.

5.3   2-Propylbutadiene

Polymers of high cis-1,4, trans-1,4, and 3,4 structure have been obtained using catalysts of the type found to polymerize isoprene to polymers of corresponding structure (142). The catalysts employed (benzene or heptane solvent) were TiCl$_4$/Al(i-Bu)$_3$ (Al/Ti = 1.6), VCl$_3$/AlEt$_3$ (Al/V = 3.5), and Ti(OR)$_4$/AlEt$_3$ (Al/Ti = 5), respectively. Conversions reported for relatively high concentrations of transition metal compound were low, from below 1 to about 8 g polymer/mmole titanium or vanadium compound. The titanium catalyzed polymers were amorphous rubbers and the trans-1,4 polymer a crystalline material, (mp 42°C).

5.4   2-t-Butylbutadiene

This monomer, as a result of the bulky substituent, adopts the S-cis conformation of the double bonds and gives the cis-1,4 polymer (mp 103–106°C, identity period 15.3 Å) with titanium tetrachloride based catalyst (143). Aluminum alkyls are suitable cocatalysts (Al/Ti = 0.8–1.0) but higher molecular weight and more crystalline polymers are obtained with complexes of the aluminum hydrides, AlH$_3$NMe$_3$, AlH$_2$NMe$_2$, and AlHCl$_2$ · D where D is a donor such as NMe$_3$ or Et$_2$O. Yields of polymer were of the order of 5 g/mmole of titanium tetrachloride when polymerized in benzene or heptane at 15°C.

5.5   2-Phenylbutadiene

With conventional TiCl$_4$/Al(i-Bu)$_3$ catalysts (Al/Ti = 1.3) this monomer gives low molecular weight solid polymers. They are generally amorphous and at the higher Al/Ti ratios the cis-1,4 content maximum is estimated to be in the range 80–95%. The x-ray diffraction pattern indicates slight crystallinity in the polymer (144).

5.6 2-Methyl-1,3-pentadiene

Three catalysts have been described for this monomer (145). Using Ti(O-n-Bu)$_4$/VCl$_3$/Al(i-Bu)$_3$ in a Ti/V/Al ratio of 1:2:20, a slow polymerization occurred (26 days at room temperature to ca. 50% conversion) to give a crystalline polymer (mp, two peaks, 182 and 192°C) of ca. 100% trans-1,4 structure. The CoCl$_2$-py/AlEt$_2$Cl/thiophene catalyst (thiophene/Al/Co = 200:100:1 at 0°C in benzene) gave an amorphous high (94%) trans-1,4 polymer. Triethylaluminum in conjunction with vanadium trichloride or oxytrichloride also produced the amorphous trans polymer. In contrast TiCl$_4$/Al(i-Bu)$_3$ (Al/Ti = 1.0, aged for 30 minutes) gave a crude product containing 90% cis-1,4 units. Extraction of this with butanone left a residue (ca. 65%) that was pure crystalline cis isomer (mp 164.5°C).

5.7 Myrcene, $Me_2C{=}CH(CH_2)_2\underset{\underset{\displaystyle CH{=}CH_2}{|}}{C}{=}CH_2$

This monomer when polymerized with TiCl$_4$/Al(i-Bu)$_3$ (Al/Ti = 1.2–3:1), behaves as a 2-substituted butadiene and gives 1,4 polymers with pendant 4-methylpent-3-enyl groupings (146). Catalyst concentrations used were in the region of 1–2 mole % of triisobutylaluminum, with conversions in the range 50–80% after 48 hours at 25°C. Polymers of relatively low molecular weight and low gel content of unknown steric structure were obtained. Tough rubbery polymers of high molecular weight and containing some gel were obtained with VCl$_3$/Al(i-Bu)$_3$ (Al/V = 5.8).

5.8 Cyclic Dienes

There is not a great deal of information on the polymerization of cyclic dienes by coordination catalysts. Polymerization occurs readily with 1,4 and 1,2 addition, to give autooxidizable, resinous polymers of relatively low molecular weight with cationic and anionic initiators as well as with coordination type catalysts. The active catalytic species participating in the latter type of reaction have not been identified but there seems little advantage in their use. Cyclopentadiene (147) and 1,3-cyclohexadiene (148–150) have been polymerized. TiCl$_4$/Al(i-Bu)$_3$ has been found to give polymers with both monomers; α-TiCl$_3$/Al(Et)$_3$ and NiCl$_2$/AlEt$_2$Cl have proved effective for 1,3-cyclohexadiene. 1,4-Cyclohexadiene gives polymers of the same structure as the 1,3 isomer, double bond migration occurring during polymerization (151).

The semicyclic diene 1-methylene cyclohexene-2 has been polymerized in high yield by $AlEt_3/TiCl_4$ or VCl_4 to give polymers mainly of 1,4 structure (152).

6 NONCONJUGATED DIENES

6.1 Aliphatic Diolefins

Only those monomers in which there can be interaction between the two double bonds require consideration. Where one double bond is more reactive than the other (153), e.g., in 1,4-hexadiene and 5,6-dimethyl-1,6-octadiene, a linear chain polymer would be expected with pendant double bonds and with only a minor proportion of cyclic structures. Possible exceptions arise when a favored five- or six-membered ring can form by participation of the internal double bond, e.g., in 1,5-heptadiene (154). α-ω-Diolefins present two possible alternatives (**4** and **5**) for cooperative polymerization.

The more usual structure is **5** and this is the one present in 1,5-hexadiene ($n = 2$). It is a highly crystalline, high melting plastic (mp 119–146°C; identity period 4.8 Å; $[\eta] = 1.2$–2.3) (155).

The polymer with the best physical properties was obtained from $TiCl_4/AlEt_3$ (Al/Ti $\simeq 2$) but faster polymerization and higher catalyst efficiencies (up to 80 g/g catalyst) were obtained with modified systems, $AlEt_3/TiCl_3 \cdot 0.22\,AlCl_3$ and $AlEt_3/TiCl_2 \cdot 0.5\,AlCl_3$. The polymer contains unsaturation resulting from independent polymerization of some of the double bonds.

6.2 Allenes

Allene polymerizes readily with conventional coordination catalysts to give high molecular weight, crystalline (mp 115–125°C), and amorphous polymers which have been considered to contain one or more of the groupings **6–8** (156).

$$-\underset{\underset{CH_2}{\overset{\|}{C}}}{C}-CH_2- \qquad -\underset{\underset{CH=CH_2}{\overset{|}{}}}{CH}- \qquad -CH=CH-CH_2-$$

$$\text{6} \qquad\qquad\qquad \text{7} \qquad\qquad\qquad \underset{cis}{\text{8}}$$

Polymers have been prepared with saturated solutions of monomer in a variety of solvents (aliphatic, aromatic, and halogenated hydrocarbons, ethers, and thioethers), or in high pressure reactions (900 psi) at temperatures in the range 30–180°C. Catalysts were from aluminum alkyls and transition metal halides, particularly vanadium oxytrichloride. The Al/V ratio was generally in the region of 20:1. Yields were variable but on occasion exceeded 40 g/mmole $VOCl_3$. Aluminum alkyls are more efficient than the aluminum alkyl chlorides but aluminum trimethyl has a lower activity than the other trialkyls (157).

A less sterically pure isomer (mp 60–61°C) has been obtained in high yield using hydrocarbon π complexes of nickel [$(\pi$-$C_8H_{12})_2$Ni, $(\pi$-$C_3H_5)_2$Ni, or π-C_3H_5NiBr] (158). These form more stable complexes by displacement of the hydrocarbon by monomer prior to polymerization. Allene as a 2.6 M solution in toluene with π-allylnickel bromide (8.4 mmoles) gave an 85% yield of polymer ($\bar{M}_n = 1.38 \times 10^5$) after 20 hours at 0°C. Tris(π-allyl)cobalt is also a catalyst at low temperatures for allene. The structure of the polymers has been studied in detail (159) and although not all aspects are completely known, it seems that only polymer with structure **6** has been obtained in a pure form. Polymer of the highest melting point has been obtained either with $VOCl_3$/Al(i-Bu)$_3$ or noble metal catalysts. The structure, as judged from infrared measurements, depends on the solvent and temperature of polymerization.

Substituted allenes polymerize under conditions similar to those of allene with $VOCl_3$/Al(i-Bu)$_3$ (160) or π-C_3H_5NiBr (158). The rates of polymerization with the former decreased in the order $CH_2{=}C{=}CH_2 >$ $R_1CH{=}C{=}CH_2 > R_1R_2C{=}C{=}CH_2 > R_1CH{=}C{=}CHR_2 > R_1R_2C{=}C{=}$ $CHR_3 \simeq R_1R_2C{=}C{=}CR_3R_4$. These polymers are mainly amorphous except for unsymmetrical dimethylallene (3-methyl-1,2-butadiene) which is a crystalline material. Substituted allenes could polymerize at either double bond (**9** and **10**). With methylallene the ratio of 1,2 to 2,3 structure is in the range 80–90% : 10–20%. Crystalline poly(3-methyl-1,2-butadiene) is exclusively the 1,2 polymer (158). Rhodium(I) complexes such as cis-Rh(CO)$_2$PPh$_3$Cl and Rh(CO)$_3$Cl will polymerize propadiene to 1,2 polymer (161).

$$-CH_2-\underset{\underset{CR_1R_2}{\overset{\|}{C}}}{C}- \qquad\qquad -CR_1R_2-\underset{\underset{CH_2}{\overset{\|}{C}}}{C}-$$

$$\text{(1,2)} \qquad\qquad\qquad\qquad \text{(2,3)}$$

$$R_1 = H. \quad R_2 = Me ; \qquad\qquad R_1 = R_2 = Me$$

$$\text{9} \qquad\qquad\qquad\qquad\qquad \text{10}$$

7   COPOLYMERIZATION OF DIENES

A substantial number of copolymerizations of dienes have been reported and in many cases, as judged from the reactivity ratios, the copolymers are of random structure. However, the application of copolymerization theory for assessing randomness requires caution in polymerizations with coordination catalysts. Reliance cannot be placed on the reactivity ratios unless they have been shown to hold for a substantial extent of polymerization and are in accord with the fractional composition of the polymer. The existence of independent simultaneously propagating reactions from more than one catalyst entity is clearly a problem, particularly with heterogeneous catalysts, and specificity at either the monomer coordination or propagation steps may be a hindrance to copolymerization between monomers of different structure. Copolymerization in an ideally random manner is likely to occur only between dienes of comparable reactivity and will vary from catalyst to catalyst. Thus, with rhodium catalysts only 1,3-pentadiene appears to copolymerize with butadiene (83), whereas with titanium and cobalt based catalysts other monomers such as isoprene or 2,3-dimethylbutadiene will copolymerize fairly readily. The addition of other substances such as donors can influence copolymerization as well as polymer microstructure. In general, full details of composition, distribution, and microstructure are not available for the systems examined. In some cases the microstructures of the comonomers are the same as in the homopolymers, whereas in others they differ. Since there may be several different but not dissimilar stereochemical forms present in the copolymer, analysis can be a major problem. Of the various monomer pairs butadiene-isoprene has been studied in greatest detail. The quoted reactivity ratios are given in Table 11.

With cobalt catalysts the cis content of the butadiene fraction falls and the 1,4 content of the isoprene portion increases (165). Results on the very similar $Co(acac)_3/Al_2Cl_3Et_3$ catalyst (Al/Co = 17; Co = 0.6 mm/liter) showed a fall in cis content of the butadiene from 97% in the homopolymer to 60% in the copolymer obtained from a 47:53 molar mixture of butadiene and isoprene. The isoprene was present in the polymer as mixed 1,4–3,4 structure, the 1,4/3,4 ratio decreasing with increase in isoprene content (166). In contrast to the cobalt systems the microstructures of the units in the copolymer prepared by $AlEt_3/TiCl_4$ do not differ from the corresponding homopolymers (162, 167). Other diene copolymerizations for which reactivity ratios have been reported are listed in Table 12. Butadiene–2-phenylbutadiene copolymers are nonideal and except for a restricted range of monomer compositions are block structures, the microstructures of the copolymer units being cis-1,4 (168).

Table 11   Reactivity Ratios for Butadiene (M_1)/Isoprene (M_2) Copolymerization

Catalyst	r_1	r_2	Ref.
$AlEt_3/TiCl_4$	1.6	1.1	162
$AlEt_3/TiCl_4$ (washed ppt)	1.49	1.03	162
$Al(i\text{-}Bu)_3/TiCl_4$	1.0 ± 0.05	1.0 ± 0.05	163
$Al(i\text{-}Bu)_2Cl/TiCl_4$	2.3	1.15	163
$AlEt_3/TiI_4$	1.88	0.55	162
$Al(i\text{-}Bu)_3/TiI_4$ (Al/Ti = 8)	2.8 ± 0.27	0.53 ± 0.05	164
$Al(i\text{-}Bu)_3/TiI_4/Bu_2S$	4.63 ± 3.7	0.76 ± 0.08	164
$\quad$ (Al/Ti/Bu_2S = 8:1:16)			
$CoCl_2\text{-}py/AlEt_2Cl$	0.99	1.37	165
$CoCl_2/AlEt_2Cl/EtOH$	0.92	1.25	165
$CoCl_2/Al(i\text{-}Bu)_2Cl$	1.0 ± 0.1	0.9 ± 0.1	164
$CoCl_2/Al(i\text{-}Bu)_2Cl/Bu_2S$			
$\quad$ (Al/Bu_2S = 1:1)	1.0 ± 0.2	1.0 ± 0.1	164
$\quad$ (Al/Bu_2S = 1:3)	2.0	0.62	164
$NiCl_2\text{-}py/AlEt_2Cl$	1.15	0.59	165
$\pi\text{-}C_4H_7NiCl/TiCl_4$	6.15 ± 0.35	1.15 ± 0.08	164
$\quad$ (Ni/Ti = 1)			

2,3-Dimethylbutadiene and pentadiene will both copolymerize with butadiene with cobalt catalysts (164), and copolymers have been obtained from the latter with cobalt (164) and vanadium based catalysts (169, 170). With the soluble vanadium catalysts polymerizations were carried out at low temperatures (-20 to $0°C$) since the catalysts lose activity rapidly on aging. Polymers of variable composition have been obtained by varying the monomer feed, but pentadiene copolymerizes preferentially (166, 169). The cis isomer of pentadiene does not polymerize. The heterogeneous $VCl_3/AlEt_3$ catalyst gives very poor yields of copolymer. The following is a recipe for the soluble catalyst:

Monomer: 45 g (17.5 mole % pentadiene); toluene 350 ml; V(acac)$_3$ 0.14 mmole (Al/V = 65). Conversion: 85% after 20 hours at $-20°C$. The pentadiene in the copolymer was 19.3 mole %.

The butadiene units in the copolymers are trans-1,4 and the pentadiene is mixed trans-1,4 and 1,2 structure. The polymers are crystalline, and pentadiene enters the crystalline lattice of *trans*-polybutadiene isomorphously. The polymers of low melting point crystallize on stretching and vulcanize to strong elastomers (169).

Table 12 Monomer Reactivity Ratios for Diene Copolymerization

Catalyst	Monomer$_1$	Monomer$_2$	r_1	r_2	Ref.
CoCl$_2$/Al(i-Bu)$_2$Cl (Al/Co = 1.4)	Butadiene	2,3-Dimethyl-butadiene	0.74 ± 0.02	0.67 ± 0.05	164
CoCl$_2$/Al(i-Bu)$_2$Cl/Bu$_2$S (Al/Co = 1.4, Bu$_2$S/Al = 3)	Butadiene	2,3-Dimethyl-butadiene	0.35 ± 0.02	0.63 ± 0.04	164
CoCl$_2$/Al(i-Bu)$_2$Cl (Al/Cl = 1.4)	Butadiene	1,3-Pentadiene	0.81 ± 0.14	0.12 ± 0.02	164
Al(i-Bu)$_3$/TiI$_4$ (Al/Ti = 8)	Butadiene	2,3-Dimethyl-butadiene	6.28 ± 0.21	0.15 ± 0.06	164
Al(i-Bu)$_3$/TiI$_4$/Bu$_2$S (Al/Ti/Bu$_2$S = 8 : 1 : 16)	Butadiene	2,3-Dimethyl-butadiene	5.04 ± 0.31	0.28 ± 0.13	164
AlHCl$_2$Et$_2$O/TiCl$_4$/AlI$_3$ = 5.5 : 1 : 1	Butadiene	2-Phenyl-butadiene	1.90 ± 0.086 1.85 ± 0.15	2.58 ± 0.036 2.6 ± 0.6	168
TiCl$_4$/Al(i-Bu)$_3$	Isoprene	Styrene	0.63 ± 0.39 11 ± 14	1.8 ± 0.9 0.08 ± 0.84	171

Monomer pairs that have been studied, but for which there are no details of reactivity ratios, are given in Table 13. Of particular interest are the alternating copolymers of butadiene and acrylonitrile. Quite apart from their potential interest as oil resistant elastomers (178) these are unique in that the acrylonitrile is first complexed with the catalyst (the order of addition of components is important) before adding the butadiene.

Other analogous combinations made with similar catalysts have also been reported but not yet in great detail (179, 180 and references therein) covering butadiene or isoprene and a variety of acrylic compounds. Furukawa and his co-workers propose a ternary complex intermediate such as Lewis acid-acrylic monomer–butadiene and consider the mechanism to have an intermediate species similar to that of the Diels-Alder reaction (181).

The mechanical properties of alternating poly(butadiene-acrylonitrile) were examined (182). The alternating copolymer was found to have a lower glass transition temperature, higher tensile strength, and greater chain flexibility than the comparable emulsion copolymer.

Using solely Lewis acids, such as zinc chloride, Gaylord (183) obtained a wide variety of alternating copolymers. He proposed that the co-monomers form a charge transfer complex which then polymerizes spontaneously or under the influence of free radicals. Polymer formation is thus envisaged as a homopolymerization of the complex intermediate.

The most interesting and potentially important butadiene copolymers obtained with transition metal catalysts are alternating copolymers with α-olefins, particularly propylene. These have been obtained with vanadium catalysts, e.g., VCl_4 or $VOCl_3$, with aluminum alkyls or alkyl halides prepared and used at low temperatures (ca. $-70°C$) (184); additives such as carbonyl-containing compounds increase the polymer molecular weight. The vanadium catalyst can also produce *trans*-polybutadiene or the olefin homopolymer; the conditions for formation of copolymer are considered to involve a divalent vanadium complex with three sites available for monomer coordination and active polymer chain, which predisposes to copolymerization rather than homopolymerization. Soluble vanadium catalysts from $VOCl_3$, VCl_4, $VO(OEt)_2$ give polymer of low molecular weight ($[\eta] \sim 0.1$) although with catalyst modification, e.g., by addition of benzoyl peroxide or CrO_2Cl_2, rather higher values are obtained (185).

Organotitanium compounds, e.g., $Ti(OCOPh)Cl_3$ and $O[Ti(OCOEt)_3]_2$, with aluminum alkyl cocatalysts, are also effective (186, 187). One titanium catalyst ($TiCl_4/AlEt_3/COCl_2$), however, has been claimed to give a random copolymer (188), $r_1 = 6.36$, $r_2 = 0.42$ (butadiene, M_1), which

Table 13 Copolymers from Dienes

Catalyst	Comonomer	Comments	Ref.
Butadiene			
$AlEt_2Cl/V(acac)_3$	1,3-Pentadiene	Pentadiene units,	169
$AlEt_3/VCl_3$		trans-1,4 and	
$AlEt_2Cl/VCl_3 \cdot 3THF$		1,2; butadiene,	
$Al/V = 50–100:1$		trans-1,4	170
$Al(i\text{-}Bu)_3/$	Ethylene	Polymerized at $-78°C$;	172
$Al(i\text{-}Bu)_2Cl/$		small yield of	
$VCl_4/PhOMe$		crystalline alter-	
$=2:2:1:2$		nating copolymer	
$AlEtCl_2/VOCl_3$	Acrylonitrile	Alternating copolymer	173
$AlEt_3/CrO_2Cl_2$	Acrylonitrile	irrespective of	
		monomer composition	174
$AlH_{0.5}Cl_{2.5}OEt_2/$	Styrene	Preferential polymer-	166
$Co(acac)_3/AlCl_3$		ization of butadiene.	
$60–120:1:22–45$		Cis content of butadiene	
		portion decreases with	
		increase in styrene	
		content	
Isoprene			
$AlEt_3/TiCl_4$	Styrene	Rate of polymerization	175
$Al/Ti = 1.08$		and random character	
		of copolymer increased	
		by the presence of high	
		molecular weight poly-	
		propylene oxide	
$AlEt_3/TiCl_4$	Propylene	Propylene content of	176
		copolymer increases if	
		catalyst is prepared	
		in the presence of	
		propylene and with an	
		increase in Al/Ti ratio	
$Al(i\text{-}Bu)_3/TiCl_4$	Cyclopentadiene	Strong inhibition of	177
		isoprene polymerization	
		by cyclopentadiene. No	
		information on polymer	
		structure	

emphasizes the specific conditions required for the production of alternating copolymer. Organotitanium catalysts have been shown to give high molecular weight copolymer, and although commercially practical conditions have not been specified, the following recipe illustrates the preparation in reasonable yield of high molecular weight copolymer (186).

Butadiene	2.0 ml
Propylene	2.5 ml
Ti(OCOPh)Cl$_3$	0.44 mmole
Al(i-Bu)$_3$	1.00 mmole
Reaction time:	16 hours at −20°C
Copolymer yield:	1.67 g; $[\eta] = 4.00$
Structure of butadiene units:	13% cis-1,4; 85% trans-1,4; 2% vinyl

In a larger scale experiment at −40°C without solvent (6 kg polymer, 34% conversion) the polymer yield was ca. 11 g/mmole of TiCl$_3$(OCOPh).

The copolymer appears to be of considerable technological interest, but it is not yet available commerically (189).

Copolymers of butadiene with other olefins and copolymers of isoprene can also be prepared with these catalysts but have received less attention. One patent disclosed for the preparation of alternating ethylene-butadiene copolymers, in which the butadiene is largely in the 1,4 configuration, employs TiCl$_4$/AlR$_3$ in conjunction with ketones, carboxylic acids, and anhydrides (190). The following is one formulation:

Butadiene	13,000 g
Ethylene	1380 g
Toluene	50 liters
Al(i-Bu)$_3$	246 g
TiCl$_4$	94.5 g
Benzoic acid	60.8 g
Reaction time:	5 hours at −30°C
Copolymer yield:	3800 g
$[\eta]$ (dl/g):	2.0

Black reinforced vulcanizates prepared from this copolymer had good tensile strength and elongation. The copolymers are crystalline, which presumably reduces their value as elastomers.

The noble metal catalysts are generally ineffective for copolymerization. The addition of cyclohexadiene or pentadiene accelerates the polymerization of butadiene by rhodium chloride whereas relatively small amounts of isoprene and 2,3-dimethylbutadiene markedly reduce the yield of polymer. In all cases, with the exception of 1,3-pentadiene, only *trans*-polybutadiene was obtained (83). Styrene, methyl methacrylate, acrolein, and acrylonitrile likewise affect the rate, but no copolymer is formed when using rhodium or palladous chloride catalysts (191).

Polybutadienes of mixed microstructure with some molecular order are not unknown and in such cases the explanation is that the polymers contain block structures. Quite different are the regular polymers prepared very recently which, although obtained from a single diene monomer, possess equal numbers of units of different structure in the polymer chain. Such polymers appear to be the result of an alternating copolymerization where the monomer is coordinating or reacting in two different ways. The polymers have been termed "equibinary." The catalysts used contain a high proportion of polar groupings (e.g., acid, halogen, alkoxide), and specific ratios are required.

With π-Ni$(C_8H_{12})_2$/CF$_3$COOH (Ni = 2×10^{-2} mole/liter; Ni/CF$_3$COOH = 5–50:1) butadiene has given good yields of high molecular weight polymer containing equimolar cis-1,4 and trans-1,4 structures (192). The polymer structure was independent of polymerization temperature over the range 0–50°C. Styrene also has the effect of changing the polymer structure from high cis to equibinary without copolymerizing.

With isoprene both cis-1,4/3,4 and 1,2/3,4 polymers have been obtained. The former was prepared using cobalt halides (Co = 6×10^{-2} mole/liter) with aluminum diethyl ethoxide (Co/Al = 1:3.4) or phenylmagnesium bromide plus methanol (Co/Mg/MeOH = 1:2.2:2) (193). Complexes of phenylmagnesium bromide with tributylphosphine and triethylamine were also effective in the latter system. The reactions were carried out in toluene solution (2.3 moles/liter of isoprene) for 7 hours at 55°C. Conversions to polymer were in the range 10–70% and with the donor-free catalysts the intrinsic viscosities of the polymers were greater than unity. Cobalt fluoride and phenylmagnesium bromide alone (Mg/Co = 2.4) in highly polar solvent mixtures containing 45–60% of hexamethylphosphoramide gave polymers with approximately equal proportions of 1,2 and 3,4 structures (194).

REFERENCES

1. Belgian Pat. 543,292 (to Goodrich-Gulf, December 2, 1955).
2. S. E. Horne, Jr. et al., *Ind. Eng. Chem.*, **48,** 785 (1956).
3. *Macromolecular Syntheses*, Vol. 2, Wiley, New York, 1966, pp. 39–53.
4. C. F. Ruebensaal, *Rubber Plast. Age*, **50,** 764 (1969).
5. F. C. Price, *Chem. Eng.*, **70** (2), 84 (1963).

6. J. N. Short, G. Kraus, R. P. Zelinski, and F. E. Naylor, *S.P.E. Tech. Papers*, **5,** 8 (1959).

7. *Chem. Eng. News*, **July 14, 1969,** 55; *Chem. Week*, **106** (11), 56 (*March* 18, 1970).

8. W. Cooper, in *Progress in High Polymers I*, Heywood, London, 1961, p. 302.

9. British Pat. 824,201 (to Chem. Werke Hüls, December 31, 1956).

10. British Pat. 848,065 (to Phillips Petroleum Co., April 16, 1956).

11. W. Cooper et al., *J. Polym. Sci.*, **50,** 159 (1961).

12. I. Ya. Poddubnyi et al, *Makromol. Chem.*, **94,** 268 (1966).

13. J. F. Henderson, *J. Polym. Sci.*, **C4,** 233 (1963).

14. British Pat. 931,313 (to Phillips Petroleum Co., February 23, 1961).

15. British Pat. 938,089 (to Phillips Petroleum Co., December 2, 1960).

16. W. Marconi et al., *J. Polym. Sci.*, **A3,** 735 (1965).

17. W. M. Saltman and T. H. Link, *Ind. Eng. Chem., Prod. Res. Dev.*, **3,** 199 (1964).

18. British Pat. 865,337 to Phillips Petroleum Co., August 11, 1958).

19. M. H. Lehr and P. H. Moyer, *J. Polym. Sci.*, **A3,** 231 (1965).

20. British Pat. 931,579 (to Polymer Corp., January 31, 1961)

21. British Pat. 931,440 (to Phillips Petroleum Co., April 7, 1961).

22. British Pat. 920,244 (to Phillips Petroleum Co., October 3, 1960).

23. B. S. Turov et al., *Dokl. Akad. Nauk SSSR*, **146,** 1141 (1962).

24. U.S. Pat. 2,977,349 (to Goodrich Gulf Chem. Inc., November 7, 1956).

25. British Pat. 859,698 (to Goodyear Tire & Rubber Co., December 4, 1958).

26. British Pat. 906,056 (to Bataaf. Petroleum Maat., February 28, 1958).

27. Belgian Pat. 573,680 (December 6 and 24, 1957); Brit. Pat. 849,589 (to Montecatini S.p.A., June 27, 1958).

28. Belgian Pat. 575,671 (to Goodrich-Gulf Chem. Inc., February 13, 1958).

29. Belgian Pat. 579,689 (to Shell Int. Res., June 16 and September 9, 1958).

30. British Pat. 948,288 (to Phillips Petroleum Co, December 5, 1960).

31. W. S. Anderson, *J. Polym. Sci.*, **A1,** 429 (1967).

32. M. Gippin, *Ind. Eng. Chem., Prod. Res. Dev.*, **1,** 32 (1962).

33. M. Gippin, *Ind. Eng. Chem., Prod. Res. Dev.*, **4,** 160 (1965).

34. F. P. Van de Kamp, *Makromol. Chem.*, **93,** 203 (1966).

35. W. M. Saltman, in *Encyclopedia of Polymer Science and Technology*, Vol. 2, Wiley, New York, 1965, pp. 717–718.

36. J. G. Balas et al., *J. Polym. Sci.*, **A3,** 2243 (1965).

37. H. Scott et al., *J. Polym. Sci.*, **A2,** 3233 (1964).

38. A. Takahasi and S. Kambara, *J. Polym. Sci.*, **B3,** 279 (1965).

39. R. N. Kovalevskaya et al., *Polym. Sci. USSR*, **4,** 414 (1963).

40. C. R. McIntosh et al., *J. Polym. Sci.*, **A1,** 2003 (1963).

41. C. E. H. Bawn, *Rubber Plast. Age*, **46,** 510 (1965).

42. W. Cooper et al., *Adv. Chem. Ser.*, **52,** 46 (1966).

43. C. Longiave et al., *Chim. Ind.* (*Milan*), **44,** 725 (1962).

44. F. Engel et al., *Rubber Plast. Age*, **45,** 1499 (1964).

45. W. Ring and H. J. Cantow, *Macromol. Chem.*, **89,** 138 (1965).

46. H. Höfermann et al., Paper to U.N. Industrial Development Organization Petroleum Symposium, Baku, U.S.S.R., October, 1969.

47. C. Longiave and R. Castelli, *J. Polym. Sci.*, **C4,** 387 (1963).

48. E. Susa, *J. Polym. Sci.*, **C4,** 399 (1963).

49. Belgian Pat. 597,165 (to Montecatini S.p.A., November 18, 1959).

50. E. I. Tinyakova, *Vysokomol. Soedin.* **4,** 828 (1962).

51. S. Otsuka and N. Kawakami, ACS Meeting, Chicago, 1964. Preprints, Division of Petroleum Chemistry A-33.

52. S. Kitagawa and Z. Harada, *Jap. Chem. Q.*, **IV-1,** 41 (1968).

53. British Pats. 905,099 and 906,334 (to Bridgestone Tire K.K., December 31, 1959).

54. R. Sakata et al., *Makromol. Chem.*, **139,** 73 (1970).

55. E. W. Duck et al., *Eur. Polym. J.*, **7,** 55 (1971).

56. L. Porri et al., *J. Polym. Sci.*, **C16,** 2525 (1967).

57. V. A. Kormer et al., *J. Polym. Sci.*, **C16,** 4351 (1969).

58. J. F. Harrod and L. R. Wallace, *Macromolecules*, **2,** 449 (1969).

59. E. I. Tinyakova et al., *J. Polym. Sci.*, **C16,** 2625 (1967).

60. B. D. Babitskii et al., *J. Polym. Sci.*, **C16,** 3219 (1968).

61. J. P. Durand et al., *J. Polym. Sci.*, **B5,** 785 (1967).

62. J. P. Durand et al., *J. Polym. Sci.*, **B6,** 757 (1968).

63. J. Marechal et al., *J. Polym. Sci.*, *A-1*, **9,** 1993 (1970).

64. T. Matsumoto et al., *J. Polym. Sci.*, **B6,** 869 (1968).

65. D. K. Jenkins et al., *Polymer*, **7,** 419 (1966).

66. British Pats. 905,097 (December 31, 1959) and 913,520 (January 26, 1959) (to Bridgestone Tire K.K.).

67. T. Otsu and M. Yamaguchi, *J. Polym. Sci.*, *A-1*, **7,** 387 (1969).

68. Belgian Pat. 644,291 (to Union Carbide Corp., June 15, 1964).

69. M. C. Throckmorton, *Kaut. Gummi Kunstst.*, **22,** 293 (1969).

70. A. deChirico et al., *Makromol. Chem.*, **175,** 2029 (1974).

71. M. Bruzzone et al., *Rubber Chem. Technol.*, **47,** 1175 (1974).

72. G. Lugli et al., *Makromol. Chem.*, **175,** 2021 (1974).

73. H. E. Railsback et al., Phillips Petroleum Co., Research Division Report 2283-59R, January 26, 1959.

74. G. Natta et al., *Chim. Ind. (Milan)*, **40,** 362 (1958).

75. G. Natta et al., *Chim. Ind. (Milan)*, **41,** 116 (1959).

76. G. J. van Amerongen, *Adv. Chem. Ser.*, **52,** 136 (1966).

77. G. Natta et al., *Gazz. Chim. Ital.*, **89,** 761 (1959).

78. S. Cucinella et al., I.U.P.A.C., Symposium on Macromolecular Chemistry, Budapest, 1969, Paper 4/26.

79. British Pat. 848,067 (to Phillips Petroleum Co., April 20, 1956).

80. R. E. Rinehart et al., *J. Am. Chem. Soc.*, **83,** 4864 (1961).

81. A. J. Canale et al., *Chem. Ind. (London)*, **1962,** 1054.

82. M. Morton et al., *Rubber Plast. Age*, **46,** 404 (1967).

83. R. Dauby et al., *J. Polym. Sci.*, **B2,** 413 (1964); **C16,** 1989 (1967).

84. T. M. Shryne, U.S. Pat. 3,168,507 (to Shell Oil, February 2, 1965).

85. J. Zachoval and B. Veruovic, *J. Polym. Sci.*, **B4**, 965 (1966).

86. E. I. Tinyakova et al., I.U.P.A.C. Symposium, Budapest, 1969, Preprint 4/03.

87. E. I. Tinyakova et al., *J. Polym. Sci.*, **52**, 159 (1961).

88. G. Natta et al., *Chim. Ind. (Milan)*, **41**, 526 (1959).

89. J. Furukawa et al., *Polym. J. (Japan)*, **2**, 371 (1971).

90. M. Iwamoto and S. Yuguchi, *J. Polym. Sci.*, **B5**, 1007 (1967).

91. G. Natta, *J. Polym. Sci.*, **48**, 219 (1960); G. Natta, L. Porri, and A. Carbanaro, *Makromol. Chem.*, **77**, 126 (1964); A. Mazzei, D. Cucinella, W. Marconi, and M. DeMalde, *Chim. Ind. (Milan)*, **45**, 528 (1963).

92. G. Natta, *Nucleus (Paris)*, **4**, 97, 211 (1963).

93. German Pat. 1,144,925 (to Phillips Petroleum Co., March 6, 1958).

94. German Pat. 1,124,699 (to Phillips Petroeum Co., March 6, 1958).

95. A. J. Canale and W. A. Hewitt, *J. Polym. Sci.*, **B2**, 1041 (1964).

96. G. Natta et al., *Chim. Ind. (Milan)*, **41,** 12 (1959).

97. C. E. H. Bawn et al., *Polymer*, **5**, 419 (1964).

98. K. W. Scott et al., *Rubber Plast. Age*, **42**, 175 (1961).

99. W. M. Saltman et al., *J. Am. Chem. Soc.*, **80,** 5615 (1958).

100. British Pat. 992,189 (to Goodyear Tire & Rubber Co., August 3, 1962).

101. W. M. Saltman, *J. Polym. Sci.*, **A1**, 373 (1963).

102. E. Schoenberg et al., *Adv. Chem. Ser.*, **52,** 7 (1966).

103. French Pat. 1,393,714 (to Goodyear Tire & Rubber Co., April 11, 1963); *CA*, **65,** 13442 (1965).

104. British Pat. 1,017,889 (to Goodyear Tire & Rubber Co., November 26, 1963).

105. British Pat. 1,043,439 (to Goodyear Tire & Rubber Co., November 26, 1963).

106. E. Schoenberg et al., Paper 45, Division of Industrial and Engineering Chemistry, 145th ACS Meeting, New York, September 1963.

107. W. Marconi et al., *Makromol. Chem.*, **71,** 118, 134 (1964).

108. A. Mazzei et al., *Makromol. Chem.*, **12,** 168 (1969).

109. British Pat. 1,168,397 (to Montecatini S.p.A., September 19, 1967).

110. V. S. Bresler and S. S. Medvedev, I.U.P.A.C. Symposium, Budapest, 1969, Paper 4/25.

111. S. E. Horne, Jr., and C. J. Carman, *J. Polym. Sci.*, *A-1*, **9**, 3039 (1971).

112. V. A. Kormer et al., *J. Polym. Sci. (Chem.)*, **11**, 2557 (1973).

113. J. P. Durand and F. Dawans, *J. Polym. Sci.*, **B8**, 743 (1970).

114. British Pat. 946,906 (to Phillips Petroleum Co., August 14, 1961).

115. Y. A. Litvin et al., *Sov. Rubber Technol.*, **21** (2), 2 (1962).

116. S. E. Bresler et al., *Rubber Chem. Technol.*, **33**, 689 (1960).

117. W. M. Saltman et al., *Rubber Plast. Age*, **46**, 502 (1965).

118. British Pat. 834,554 (to Montecatini S.p.A., March 12 and December 22, 1955).

119. Dunlop Rubber Co. Ltd., *Rubber Plast. Age*, **43**, 1115 (1962); Trans-Pip, *Polymer Product Bulletin*, Polymer Corp., Sarnia, Canada.

120. W. Cooper, Atti. XV, Congress Int. Mat. Plast. Elast. (Mat. Plast. Elast. Ind., Milan), Turin, September 1963, p. 309.

121. W. Cooper et al., *J. Polym. Sci.*, **C4,** 211 (1964).

122. W. Cooper et al., *J. Polym. Sci.*, **B4,** 309 (1966).

123. British Pat. 877,371 (to U.S. Rubber Co., March 16, 1959).

124. J. Lasky et al., *Ind. Eng. Chem., Prod. Res. Dev.*, **1,** 82 (1962).

125. British Pat. 1,024,179 (to Dunlop Rubber Co. Ltd., March 2, 1962).

126. A. M. Vladimirov et al., *Sov. Rubber Technol.*, **18,** (12) 4 (1959).

127. Y. V. Zabolotskaya et al., *Polym. Sci., USSR*, **6,** 88 (1965).

128. British Pat. 1,007,646 (to Dunlop Rubber Co. Ltd., December 28, 1962).

129. A. B. Deshpande et al., *J. Polym. Sci.*, *A-1*, **5,** 761 (1967).

130. Belgian Pat. 617,545 (to Montecatini S.p.A., May 13 and October 16, 1961).

131. L. Porri et al., *J. Polym. Sci.*, **B5,** 321 (1967); *Eur. Polym. J.*, **5,** 1 (1969).

132. G. Natta et al., *Makromol. Chem.*, **77,** 114 (1964).

133. G. Natta et al., *Makromol. Chem.*, **67,** 225 (1963).

134. G. Natta et al., *J. Polym. Sci.*, **51,** 463 (1961).

135. G. Natta et al., *Eur. Polym. J.*, **1,** 81 (1965).

136. R. Ohno et al., *Polym. J. (Japan)*, **4,** 56 (1973).

137. V. A. Vasiliev et al., *J. Polym. Sci. (Chem.)*, **11,** 2489 (1973).

138. T. F. Yen, *J. Polym. Sci.*, **35,** 533 (1959).

139. T. F. Yen, *J. Polym. Sci.*, **38,** 272 (1959).

140. O. Solomon and C. A. Stoicesu, *Rev. Chim. (Bucharest)*, **9,** 507 (1958); *CA*, **55,** 20906 (1961).

141. British Pat. 1,004,665 (to Goodyear Tire & Rubber Co., October 12, 1961).

142. W. Marconi et al *J. Polym. Sci.*, **A3,** 123 (1965).

143. W. Marconi et al., *J. Polym. Sci.*, **A2,** 4261 (1964).

144. W. Marconi et al., *J. Polym. Sci.*, **C16,** 805 (1965); J. K. Stille and E. D. Vessel, *ibid.*, **49,** 419 (1961).

145. D. Cuzin et al., *Eur. Polym. J.*, **5,** 283 (1969).

146. C. S. Marvel and C. C. L. Hwa, *J. Polym. Sci.*, **45,** 25 (1960).

147. S. P. S. Yen, *Polym. Prepr.*, **4** (2), 82 (1963); ACS Meeting, New York, 1963.

148. A. Martinato et al., *Bull. Soc. Chim. France*, **10,** 2800 (1965).

149. D. A. Frey et al., *J. Polym. Sci.*, **A1,** 2057 (1963).

150. G. Lefebvre and F. Dawans, *J. Polym. Sci.*, **A2,** 3277 (1964).

151. B. Reichel and C. S. Marvel, *J. Polym. Sci.*, **A1,** 2935 (1963).

152. K. Mabuchi et al., *Makromol. Chem.*, **81,** 112 (1965).

153. J. M. Wilbur and C. S. Marvel, *J. Polym. Sci.*, **A2,** 4415 (1964).

154. L. M. Romanov et al., *Polym. Sci., USSR*, **4,** 1424 (1963).

155. H. S. Makowski et al., *J. Polym. Sci.*, **A2,** 1549 (1964).

156. W. P. Baker, Jr., *J. Polym. Sci.*, **A1,** 655 (1963).

157. R. Havinga and A. Schors, *J. Macromol. Sci.*, **A2,** 1 (1968).

158. S. Otsuka et al., *Eur. Polym. J.*, **3,** 73 (1967).

159. S. Otsuka et al., *J. Am. Chem. Soc.*, **87,** 3017, (1965); H. Tadokaro and Y. Takahashi, *J. Polym. Sci.,* **B3,** 697 (1965).

160. R. Havinga and A. Schors, *J. Macromol. Sci.*, **A2,** 31 (1968).

161. J. P. Scholten and H. J. van Ploeg, *J. Polymer Sci. (Chem.)*, **10**, 3067 (1972).

162. J. Furukawa et al., *J. Chem. Soc. Japan (Ind. Chem. Sect.)*, **65**, 2074 (1962).

163. L. S. Bresler et al., *Polym. Sci., USSR*, **5**, 1012 (1964).

164. I. N. Smirnova et al., I.U.P.A.C. Meeting, Budapest, 1969, Paper 4/23.

165. T. Saegusa et al., *J. Chem. Soc. Japan (Ind. Chem. Sect.)*, **65**, 2082 (1962).

166. H. Weber et al., *Makromol. Chem.*, **101**, 320 (1967).

167. T. Suminoe et al., *Chem. High Polym. (Japan)*, **21**, 9 (1964).

168. W. Marconi et al., *J. Polym. Sci.*, **C16**, 805 (1965).

169. G. Natta et al., *Makromol. Chem.*, **53**, 52 (1962).

170. L. Porri et al., *Makromol. Chem.*, **61**, 90 (1963).

171. N. Yamazaki et al., *J. Chem. Soc. Japan (Ind. Chem. Sect.)*, **64**, 103 (1961).

172. G. Natta et al., *Makromol. Chem.*, **79**, 161 (1964).

173. J. Furukawa and Y. Iseda, *J. Polym. Sci.*, **B7**, 47 (1969); J. Furukawa et al., *Polym. J.*, **1**, 155 (1970); **2**, 337 (1971).

174. M. Taniguchi et al., *J. Polym. Sci.*, **B7**, 411 (1969).

175. T. Tsuruta and Y. Ishizuka, *Kobunshi Kagaku*, **26**, 311 (1969); *J. Macromol. Sci.*, **A3**, 319 (1969).

176. T. Suminoe et al., *Chem. High Polym. (Japan)*, **20**, 262 (1963).

177. N. V. Shcherbakova, *Sov. Rubber Technol.*, **26**, (3), 5 (1967).

178. J. Furukawa et al., *J. Polym. Sci.*, **B7**, 561 (1969).

179. J. Furukawa et al., *J. Polym. Sci.*, *A-1*, **8**, 1147 (1971).

180. J. Furukawa, *Polym. J.*, **1**, 442 (1970).

181. J. Furukawa et al., *Polym. J.*, **2**, 475 (1971).

182. J. Furukawa and A. Nishioka, *J. Polym. Sci.*, **B9**, 199 (1971); *J. Appl. Polym. Sci.*, **15**, 1407 (1971).

183. N. G. Gaylord and A. Takahasi, *J. Polym. Sci.*, **B7**, 443 (1969); *Adv. Chem. Ser.*, **91**, 94 (1970).

184. J. Furukawa, *Angew. Makromol. Chem.*, **23**, 189 (1972).

185. British Pat. 1,297,165 (to Maruzen Petrochemicals, December 1969).

186. K. Hayashi, A. Kawasaki, and I. Maruyama, U.S. Pat. 3,737,416 (to Maruzen Petrochemicals, May 1973).

187. K. Hayashi, A. Kawasaki, and I. Maruyama, U.S. Pat. 3,737,417 (to Maruzen Petrochemicals, May 1973).

188. J. Furukawa et al., *Polym. Sci. (Chem.)*, **11**, 629 (1973).

189. J. Furukawa, *J. Polym. Sci. Symposia*, **48**, 19 (1974).

190. British Pat. 1,340,554 (to Maruzen Petrochemicals, April 1971).

191. R. S. Berger and E. A. Youngman, *J. Polym. Sci.*, **A2**, 357 (1964).

192. J. P. Durand and Ph. Teyssie, *J. Polym. Sci.*, **B6**, 299 (1968).

193. G. Delheye and F. Dawans, *Makromol. Chem.*, **98**, 164 (1966).

194. Ph. Teyssie et al., *J. Polym. Sci.*, **C22**, 221 (1968).

Ph. Teyssié

Institut de Chimie
Université de Liège
au Sart Tilman
4000 Liège, Belgium

F. Dawans

Institut Français du Pétrole
92 Rueil-Malmaison, France

Chapter 3 Theory of Coordination Catalysis

Manuscript received December 1970, revised March 1975.

1 INTRODUCTION

1.1 Scope of the Chapter

A great deal has been achieved since the early discovery by Ziegler of the "Aufbau" reaction (1) and its striking modification in the presence of transition metal derivatives. The tremendous amount of research work in polymer chemistry stimulated by these discoveries provided a very strong support for the rapid development of coordination catalysis and coordination chemistry. In turn, the advances realized from the study of definite complexes in simple organic reactions provided powerful tools for a better understanding of coordination polymerizations. In particular the determinant role of the transition metal complexes is now well documented. The first part of this chapter is devoted to surveying the most important recent advances using monometallic catalysts.

This situation has also allowed a fruitful attack on the unsolved and complex problems regarding the oxidation state of the transition metal; the mono- versus bimetallic mechanism; and the heterogeneous versus homogeneous control of the stereospecificity.

A much better insight into the detailed mechanism of stereospecific coordination polymerization has been obtained. Rather direct and clear correlations have been established between the structure of the active species and its catalytic properties, i.e., activity and stereospecificity (2). In addition, it has been realized that the reaction schemes used to describe α-olefin polymerization by Ziegler type catalysts $M_T X_n + AlR_m X_{3-m} + L$ are very general and applicable to many reactions, and to very different types of substrates and bonding (e.g., to epoxide polymerization; see Section 5).

Other fields of specific polymerization contain one or another aspect of the basic features of coordination complex control; stereocontrol by organolithium derivatives (3, 4) and the alternating copolymerization of hydrocarbon and polar monomers in the presence of metal salts (5) are particularly interesting examples.

1.2 Importance of Coordination Catalysis in Polymerization

The breadth of the field may be indicated by a few examples (6).

Both the efficiency and the versatility of the coordination catalysts can be controlled closely by systematic modification of the catalytic structure, with monometallic and with bimetallic complexes. Very high activities have been obtained, allowing use of minute amounts of catalyst even on an industrial scale, in the parts per million range. Steric, structural, and geometric isomerism have been controlled in practice in most hydrocarbon polymers, leading, for instance, to the preparation of high performance isotactic polyolefins and synthetic equivalents of natural rubber. Still more refined types of stereoregulation (which could be considered as second-order control) have also been realized, for instance by preparing ditactic polymers (erythro- and threotactic polyolefins) or equibinary polydienes, and by performing stereoelective polymerizations where a particular optical isomer of a racemic monomer is preferentially incorporated. Polymerization of different types of monomers has also been achieved, for example by ring opening of cycloolefins or heterocyclic compounds (see Chapter 6). Finally, strong indications have been obtained that direct methods to prepare block copolymers, involving coordination catalysis at least for one type of sequence, could be designed with a sufficient degree of specificity to prepare interesting new products in one operation.

Considering both the quality and the volume of the products resulting from processes based on coordination catalysis, mainly of the Ziegler-Natta type, one must admire the tremendous development that has taken place in less than 25 years.

2 THE DETERMINANT ROLE OF TRANSITION METAL COMPLEXES IN STEREOSPECIFIC POLYMERIZATION OF UNSATURATED HYDROCARBONS

2.1 Present Conceptions

Numerous papers have been concerned with the nature of the active sites and the origin of stereospecific control in the Ziegler-Natta catalysts. Various and sometimes contradictory schemes have been offered to explain the principles underlying their action.

These multicomponent active systems (involving a transition metal derivative $M_T X_n$, a metal alkyl, e.g., $AlR_m X_{3-m}$, and perhaps a ligand L) might be grouped into catalysts acting in a homogeneous phase and those acting in a heterogeneous phase, the solubility or insolubility being determined, of course, by the structure and the ratio of the different components which make up the catalyst. In the case of homogeneous catalysis, it is clear that the factors controlling the stereospecificity must be linked to the specific interaction of the monomer and/or the growing chain with the active complex (including metal, counterion, ligands, and solvent). In the case of catalysts acting in a heterogeneous phase, it was speculated that the determinant factor might be the structure of the catalytic surface. Comparison of homogeneous and analogous heterogeneous systems allows one to forecast that there should be no essential difference in their basic mechanism of action. This point is discussed in more detail in Section 3.4.

Although the exact mechanism by which stereospecific catalysts operate is still a matter of controversy, two conclusions have been reached which have adequate experimental foundation and are widely accepted at the present time:

1. Interaction of the components of the bimetallic catalytic systems results in the formation of alkyl derivatives of transition metals which are capable of coordinating unsaturated hydrocarbon molecules.

2. Growth of the polymer chain takes place by repeated insertion of the monomer into a bond between transition metal and one carbon atom belonging to the alkyl group or later to the growing polymer chain.

It is now clear that these polymerization catalysts, in mechanism, represent a particular although very important example of a broad class of complex catalysts for organic reactions, including hydrogenation, carbonylation, oligomerization, isomerization, etc., of unsaturated compounds.

The exact role of the organometallic compound is still debated. Some scientists propose that its function is limited to alkylating the transition metal, the propagation involving two or more coordination sites on this central transition metal (Fig. 1). Others believe the organometallic compound not only alkylates the transition metal but also participates in the formation of the active center (Fig. 2). There are still some who propose

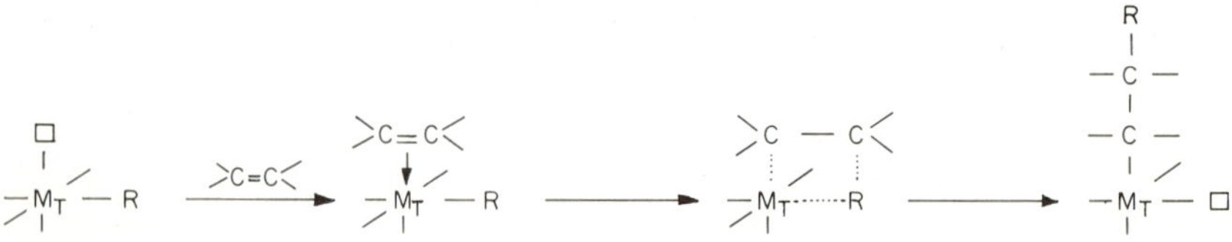

Figure 1

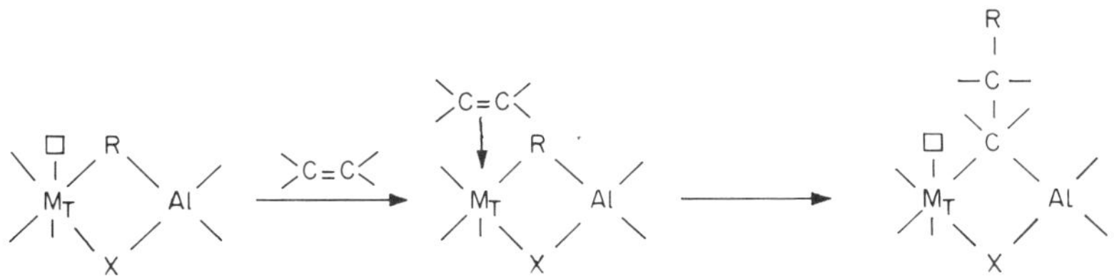

Figure 2

that the growing chain is attached to the aluminum atom (8). It seems more accurate to admit that there may well be internal or external transfer reactions to the aluminum alkyl during polymerization, thus leading at some stage to the eventual formation of different metal-polymer bonds. However, the dependence or reactivity ratios in α-olefin copolymerization on the structure of the transition metal derivative and not of the organometallic cocatalyst, has now been accepted for several years to indicate growth on the transition metal atom. This hypothesis, according to which a transition metal-carbon is involved in the propagation step, is further supported by a number of reported experiments, such as the electrodialytic investigation of the soluble catalyst $(CH_3)_2AlCl +$ $(C_5H_5)_2TiCl_2$ (9–12), the dependence of the propagation rate constants on the transition metal component and not on the metal alkyl structure (13), and the kinetic study of ethylene polymerization combined with ESR and magnetic susceptibility measurements on soluble catalysts (14–17).

If the role of the alkyl derivative of a metal of the first three groups is limited to alkylating the transition metal compound, proportionality should be expected between the alkylating power of the organometallic compound used for the catalyst preparation and the activity of the catalyst itself. Different studies (18, 19) have shown that this is not the case. Hence in addition to the alkylating power of the organometallic compound, its stronger or weaker tendency to form complexes with the transition metal compound is also important and might even influence the stereospecificity of the entire catalytic system. Thus even admitting that the growth of the polymer chain takes place through repeated insertion of the monomer into a transition metal-carbon bond, it is now evident that active centers are formed whose activity and eventual stereospecificity depend on the nature of other groups bound to this metal, as is the case of many other reactions. In other words, in the presence of metal alkyl compounds and particularly of aluminum alkyls with a high ability to form complexes with transition metal compounds, active centers also containing aluminum may be formed. This observation is further corroborated by the fact that the value of the propagation rate constants for ethylene polymerization by catalysts obtained by the interaction of tetrakis-π-allyl zirconium with an oxide support, changes markedly as the nature of the support is changed (7). This seems to be related to the difference in the composition of the active centers of these catalysts, and is further evidence that an alkylating organometallic cocatalyst is not absolutely required for the formation of highly active catalytic species.

Notwithstanding the large amount of work accomplished in recent years and reported in a number of excellent reviews (6, 20–23), this interesting problem has not yet been completely solved. This is owing mostly to its complexity but above all to the various natures of the numerous combinations which belong to the vast class of Ziegler-Natta catalysts, which are still difficult to represent by a unitary model. Even where studies have been carried out on homogeneous systems, at least two organometallic compounds are involved, making accurate determination of the structure of very small amounts of the propagating species difficult.

These problems are discussed in more detail in the following sections. However, the ability to polymerize, stereospecifically, unsaturated hydrocarbons in the presence of monometallic complexes is now well documented; it has been definitely proved that a second metal is not essential for catalyst formation. Accordingly, the present discussion will center on simple catalytic transition metal complexes having well-defined structures and on the factors, reported mostly during the last 10 years, controlling both the activity and the stereospecificity of monometallic catalysts. The ability to modify the catalyst structure systematically is to be considered as particularly important, since an understanding of the mechanism of action might reveal more of the important features of the chemistry of the active site involved.

These active monometallic catalysts reported for the polymerization of unsaturated monomers (principally butadiene and ethylene) are essentially transition metal salts and complexes (e.g., hydride, alkyl, carbene, cyclopentadienyl, carbonyl, benzyl, and allyl complexes), used as such or more often in the presence of some added electron donor or acceptor. Many catalytic derivatives used appear to be originally associated as binuclear complexes, involving, in some cases, two different transition metals; but even in these cases the catalytic site may be regarded as being formally monometallic, the other metal derivative functioning as a ligand.

2.2 Types of Catalytic Complexes

2.2.1 Transition Metal Salts

Noble Metal Salts. Considerable interest was aroused by the discovery that rhodium salts catalyze the stereospecific polymerization of 1,3-butadiene to a high trans-1,4 polymer in protonic media (24); indeed, noble metal salts were the first example of stable and formally simple catalysts inducing a highly stereospecific polymerization. Although this polymerization can be carried out in an aqueous emulsion system, it has been convincingly demonstrated that these systems do not operate by means of a conventional free radical mechanism (25), but they do imply the coordination of the monomer on the metal atom. Consequently, these catalysts enjoy a high versatility and many factors influence the course of the reaction; in particular, the nature of the metal used appears to be the determinant of the stereospecificity. For example, the microstructure of the polybutadiene obtained in the presence of salts and complexes of palladium is predominantly 1,2, whereas high trans-1,4 contents are observed in the polymers produced with rhodium derivatives and, in spite of some controversy, the polybutadiene produced by means of complex cobalt fluorides is reported to be essentially cis-1,4 (220).

Various coordinating compounds such as those containing nitrogen atoms can markedly affect the activity of the rhodium in the polymerization systems, confirming that its $4d$ orbitals are involved in the catalytic process (28). Addition of certain additional diolefins was shown to give superior catalysts (29). The so-called emulsifiers play an important role as ligands that take an active part in polymerization. Only anionic emulsifiers give active catalysts, and at emulsifier/rhodium molar ratios greater than 2 (30). The emulsifier is consumed during the reaction, the polymer produced containing approximately 1 mole of sulfur per chain (31). Studies conducted in homogeneous solution have confirmed that the effective surfactants are of the sulfate or sulfonate types (sodium lauryl sulfate and sodium alkylbenzene sulfonates having alkyl chains greater than C_5) (32).

It has thus become obvious that the rhodium based catalytic systems involve the formation of a complex between the metal, butadiene, and various ligands, including the emulsifier itself. In fact, the close analogies between these reactions and Ziegler-Natta catalysis—the only major difference being the stability of the noble metal catalysts toward protonic media—suggest some similarity in the mechanism of both types of polymerization, most probably implying the formation of an allyl type species, indicative of a coordination propagation proceeding by cis rearrangement. One of the possible paths is shown in Fig. 3, although initiation by a π-crotyl complex arising from an intermediary hydrido-Rh(I) species (33) might also be considered, as well as the incorporation of some chlorine at chain ends.

The observations made during a study of butadiene polymerization in homogeneous solution, in the presence of dichloro-2,6,10-dodecatriene-1,12-di-yl-ruthenium, $RuCl_2(C_{12}H_{18})$, and a tertiary phosphine, also suggest a close similarity between the propagation mechanism in this

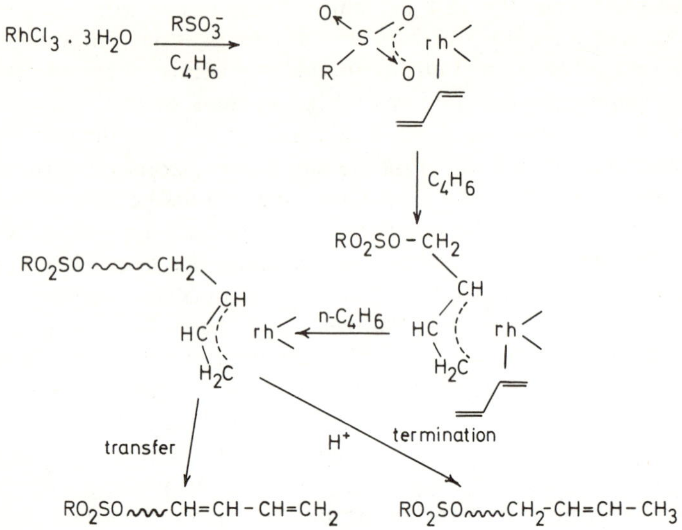

$$RhCl_3 \cdot 3H_2O \xrightarrow[C_4H_6]{RSO_3^-}$$

(reaction scheme; rh designates rhodium)

$$\downarrow C_4H_6$$

$RO_2SO - CH_2$

transfer / H^+ termination

$$RO_2SO \sim\sim CH=CH-CH=CH_2 \qquad RO_2SO\sim\sim CH_2-CH=CH-CH_3$$

Figure 3 Proposed mechanism for rhodium catalyst coordination propagation proceeding by cis rearrangement. (rh designates a rhodium atom in an octahedral environment but whose ligands are not all determined. RSO_3^- may be replaced by Cl^- or H^-.)

homogeneous solution polymerization and that in the heterogeneous emulsion polymerization with a ruthenium trichloride-triphenylphosphine catalyst. The NMR spectrum of the homogeneous system indicates the coexistence of both π- and σ-allylic structures, considered to be active intermediates for the polymerization of butadiene (34).

Additional strong evidence in favor of coordinated mechanisms is the selective behavior of metal derivatives toward specific monomers; for example, rhodium salts do not catalyze ring opening polymerization of norbornene but they yield addition polymers from cyclobutene (35). On the other hand, ruthenium salts which as such are not good catalysts for the polymerization of butadiene promote the ring opening polymerization of both cyclobutene and norbornene (36–38). A comparative structural study of polynorbornadienes produced by complexes of three different metals (e.g., rhodium, iridium, and palladium) emphasizes the dependence of the polymer structure upon the nature of the metal. Indeed the polymer produced exhibits a unique structure depending on the specific nature of the metal used as catalyst (39). In Fig. 4 rhodium gives a saturated polymer with a nortricyclene repeating unit from a 1,5 polyaddition (scheme *A*); with palladium, a 1,2 addition leads to a polymer containing one unsaturation per repeating unit (scheme *B*), while with

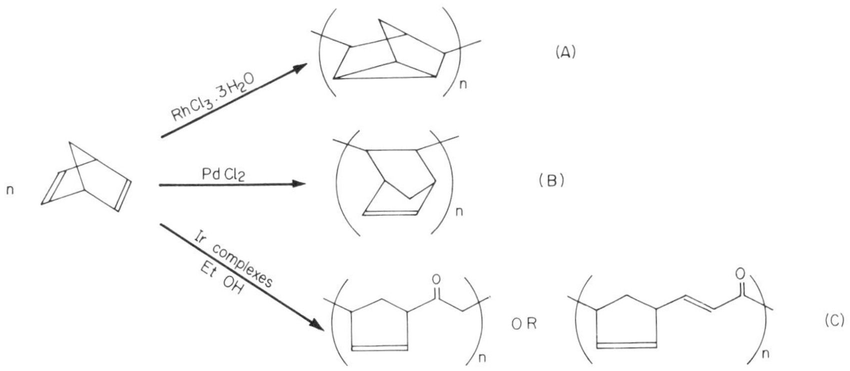

Figure 4

iridium oxygenated polymers containing one oxygen atom per repeating unit are obtained on performing the reaction in an oxygenated solvent (scheme C). Some noble metal complexes are thus specific for addition polymerization whereas others are effective only for ring opening polymerization.

Another significant example indicating close interaction between the monomer and the noble metal atom during the propagation reaction is the polymerization of propylene in the presence of palladium cyanide, yielding a copolymer containing 93% 1,3 units and 7% 1,2 units (40). Indeed, since palladium cyanide is completely insoluble in the medium, the reaction probably takes place at the crystal surface involving two adjacent palladium atoms. The peculiar structure of the polymer produced might be due to the intermediary formation of a π-allyl type complex (Fig. 5).

First Row Transition Metal Salts.   A propagation reaction mechanism involving coordination of the monomer at the crystal surface may be compared with the catalytic activity promoted through γ-irradiation [see, e.g., Pino (41) and Allegra et al. (42)] or through mechanical activation (ball-milling) (43, 44) of crystalline titanium halides which are, in the absence of organometallic cocatalysts, otherwise very poor catalysts for the polymerization of ethylene. Indeed, when γ-irradiated or ball-milled,

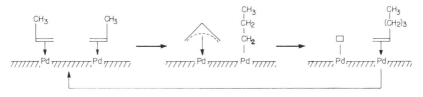

Figure 5

they are converted into active catalysts (e.g., for ethylene) in the absence of organometallic cocatalysts or of metals and their derivatives (such as Al or $AlCl_3$). The polymerization rate was shown to be proportional to surface area, and a direct function of Ti^{2+} content of the catalyst. The proposed reaction scheme postulates an active catalytic alkyl complex, where both initiation and propagation could take place at a single metal atom or alternatively on two adjacent Ti^{2+} ions.

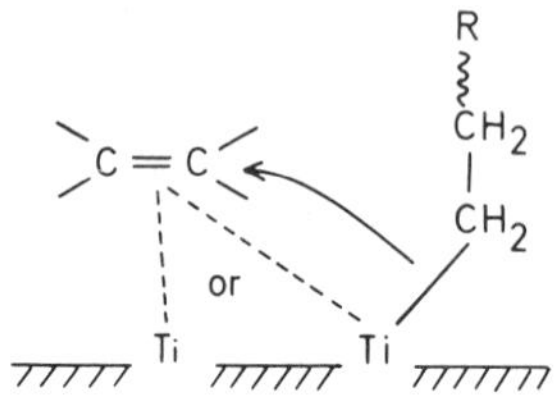

Termination probably proceeds through the formation of hydride species, as suggested by the presence of one terminal double bond per polyethylene chain.

Just as titanium subchlorides were shown to be catalysts for the polymerization of monoolefins, 1,3-butadiene was reported to be polymerized to cis-1,4 polymer on irregularly stacked, halogen deficient crystals of cobalt(II) or nickel(II) halides (45, 46). In this case, halogen is removed from the halides by heating the salts under high vacuum or by photolyzing them in the presence of butadiene, transferring halogen to butadiene. The species responsible for the initiation of butadiene polymerization are probably nickel or cobalt monohalide complexes.

Catalytic subhalides of nickel inducing cis-1,4 polymerization of butadiene may also be obtained by reacting nickel hydride (46, 47) or reduced nickel (48–50) with some Lewis acids and various organic halides. Once again, the results suggest that the cis-1,4 polymerization proceeds by a coordinated mechanism, involving the formation of π-crotyl complexes with the monomer:

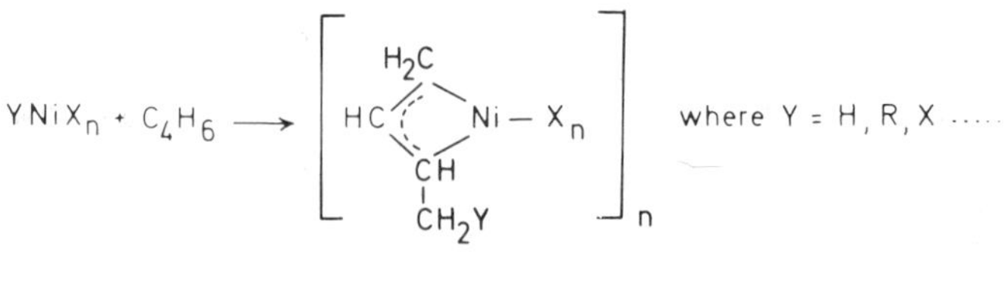

More recently, halide derivatives of transition metal haloacetates were found to be efficient catalysts for the stereospecific polymerization of unsaturated monomers; for example, $CF_3COONiCl$ yields a high cis-1,4-polybutadiene in hydrocarbon solutions (51, 52). In this case, the transitory formation of a π-chlorobutenyl complex was confirmed, using a ^{36}Cl-labeled nickel salt (208).

Finally, stereospecific polymerization reactions can also be promoted by transition metal subhalides complexed with strong electron donating ligands. For example, $TiCl_3 + N(C_2H_5)_3$ or $P(n-C_4H_9)_3$ as well as $[(C_6H_5)_3P]_3NiX$ or $(Cl_3P)_4Ni$ complexes, when reacted with Lewis acids, induce the coordinated polymerization of various monomers such as 1,3-butadiene, isoprene, ethylene, allene, or isobutyl vinyl ether (53–56). In the presence of $(Cl_3P)_4Ni + WCl_6$, the ring opening polymerization of cyclopentene to a linear polymer is also promoted (57).

2.2.2 Alkyl Complexes of Transition Metals

As summarized in Table 1, numerous catalytic systems containing transition metal alkyl derivatives are known which polymerize α-olefins, cyclooolefins, conjugated diolefins, and vinyl or acetylenic monomers, in the absence of any alkyl derivatives of metals of groups IA to IIIA.

For α-olefin polymerization these systems usually contain two compounds of the same metal, namely titanium, which suggests that the active centers might involve two metal atoms. Although it is well known that organoaluminum can impart a high catalytic activity to organometallic compounds of a transition metal, which otherwise would be practically inactive [e.g., $(C_5H_5)_2Ti(C_2H_5)Cl + (C_2H_5)_2AlCl$] or at least poorly active $[(C_6H_5CH_2)_4Ti + (C_6H_5CH_2)_3Al]$, the fact that some of the titanium catalysts containing no aluminum [such as $RTiCl_3$, $(C_6H_5CH_2)_3TiCl$, or $(CH_3)_2Ti(C_5H_5)_2 + TiCl_3$ or $TiCl_4$] give crystalline polypropylene may be considered support for the idea that the *origin* of the stereoregulation is solely in the structure of the transition metal complex.

Another interesting observation concerns the ability of some alkyl complexes of transition metals to induce the polymerization of vinyl polar monomers, avoiding the marked tendency of polar groups to react with the metal and to inhibit completely the polymerization process. Such a specific interaction with alkyl complexes of transition metals was suggested, for example, in the course of the polymerization of acrylonitrile in the presence of diethyldipyridylnickel (81); infrared absorption data indicate that under suitable conditions acrylonitrile may be coordinated to the nickel atom through the vinyl bond and not by the nitrile groups; this interaction causes the weakening of the ethyl-nickel bonds, leading to their scission and the initiation of polymerization by a coordinated mechanism (Fig. 6).

Table 1   Catalysts from Alkyl Complexes of Transition Metals

Monomers	Alkyl Complexes	Cocatalysts	Ref.
α-Olefins[a]	$RTiCl_3$	Cr, V, Ti halides	58–64
	$C_2H_5TiCl_3$	3–25 Megarads	66
	R_2TiCl_2		58–65
	$R_2Ti(C_5H_5)_2$	$TiCl_3$, $TiCl_4$	64, 67–71
	$(CH_3)_2Ti(OR)_2$	$TiCl_3$, $TiCl_4$, VCl_4	72, 73
	$(C_6H_5CH_2)_{4-n}TiCl_n$		74
	$(C_6H_5CH_2)_{4-n}ZrCl_n$		75
	$(C_3H_5)_4Zr$	SiO_2	7, 209–211
	$(C_7H_7)_4Ti$	Al_2O_3	7, 209–211
Vinyl ethers	$RTiCl_3$		76
Vinyl compounds	$(C_5H_5)_2TiCl_2$	Fatty acids	77
Diarylvinylamine	$(CH_3CO_2)_2TiCl_2$		78
Alkoxystyrenes	$(CH_3CO_2)_2TiCl_2$		79
Styrene	$C_6H_5Ti(i\text{-}C_3H_7O)_3$		80
Acrylonitrile	$(C_2H_5)_2Ni$ dipyridyl		81
Alkyl isocyanates	$(C_2H_5)_2Ni$ dipyridyl or $(PPh_3)_4Ni$		205
Allene	$(C_8H_{12})_2Ni$		82, 83
1,2-Butadiene	$(C_8H_{12})_2Ni$		84
1,3-Butadiene	$(C_8H_{12})_2Ni$	Metal halides	85, 86
	$(C_8H_{12})_2Ni$	Protonic acids	87, 88
	$C_8H_{12}NiX$		89
	$(C_6H_5)_3CNiCl$		90
	$(C_5H_5)_2Ni$	Metal halides	47, 91–93
	$(C_5H_5)_2Ni$	F_3CCOOH	94
	$(C_5H_5)_2Ni$	p-Chloranil	47

[a] Most of the reported catalytic systems were applied to ethylene polymerization. Consequently, valuable data on the stereoregularity control by monometallic species are difficult to sort out. However, in the case of propylene polymerization, partially crystalline polymers are obtained only in the presence of VCl_4, $TiCl_4$, or $TiCl_3$ as cocatalysts.

Finally, it may be interesting to point out that some carbene complexes of tungsten were recently shown to enjoy a very high catalytic activity for the ring opening polymerization of cyclopentene in the presence of Lewis acids as cocatalysts (212) (see also Chapter 6).

In butadiene polymerization performed in the presence of hydrocarbon complexes of nickel, the insertion of monomer molecules occurs between

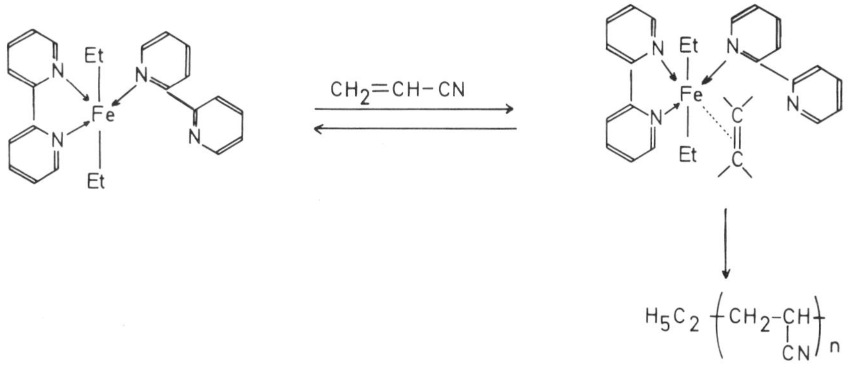

Figure 6

the nickel atom and a π-allylic unit formed by direct interaction of 1,3-butadiene with the nickel complex as shown by Wilke and co-workers (95) (Fig. 7). This π-allyl group might also be formed by the reaction of a cocatalyst with the nickel complex, followed by interaction with the monomer (Fig. 8), as proposed by Dolgoplosk and co-workers (47).

2.2.3 Carbonyl Complexes of Transition Metals

Manganese carbonyl, $Mn_2(CO)_{10}$, induces the ring opening polymerization of propylene oxide to yield a tactic polymer (107, 108). In addition to this reaction, an important series of catalysts involving carbonyl complexes of transition metals was discovered by Otsuka and Kawakami (96) for the stereospecific polymerization of 1,3-butadiene. They found that the reaction product of $Co_2(CO)_8$ with $MoCl_5$ polymerizes butadiene in benzene to an amorphous 1,2 polymer, while the system $Ni(CO)_4 + MoCl_5$ gives a high (more than 85%) cis-1,4-polybutadiene. Since that time, other nickel and cobalt carbonyl complexes, which as such do not exhibit any catalytic activity in butadiene polymerization, have been tested further together with numerous Lewis acids as cocatalysts (see Table 2). The interaction between carbonyls and Lewis acids results in carbon monoxide evolution and precipitation of products insoluble in hydrocarbons. The rate and stoichiometry of the reaction are influenced by the nature of the Lewis acid used, vanadium and tungsten derivatives

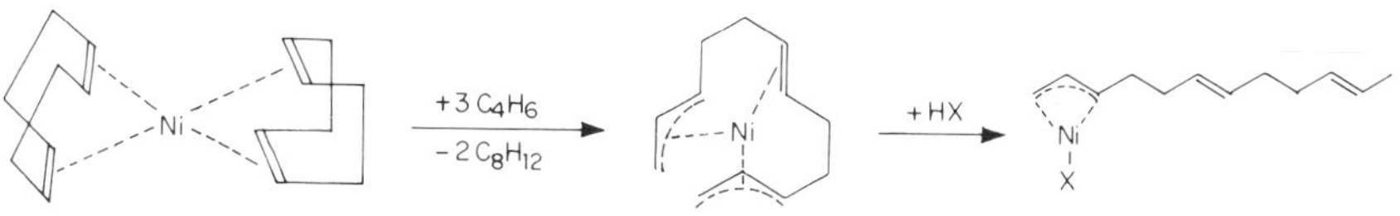

Figure 7

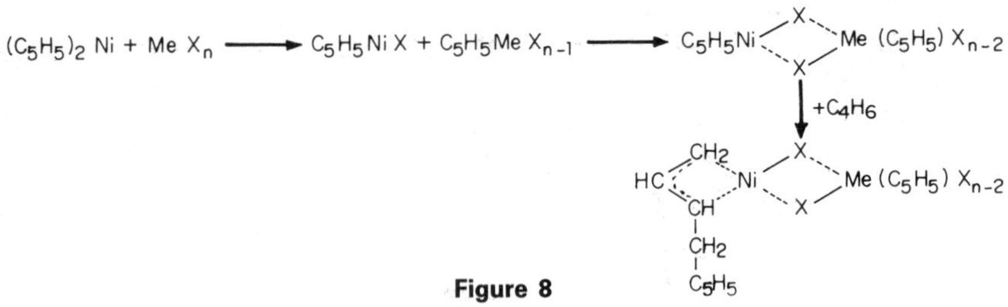

$$(C_5H_5)_2\ Ni + Me\ X_n \longrightarrow C_5H_5Ni\ X + C_5H_5Me\ X_{n-1} \longrightarrow C_5H_5Ni \overset{X}{\underset{X}{\diamondsuit}} Me\ (C_5H_5)\ X_{n-2}$$

$$\Big\downarrow +C_4H_6$$

$$HC \overset{CH_2}{\underset{CH}{\diamondsuit}} Ni \overset{X}{\underset{X}{\diamondsuit}} Me\ (C_5H_5)\ X_{n-2}$$
$$\underset{\underset{C_5H_5}{CH_2}}{|}$$

Figure 8

being the most active. The polymers produced all exhibit high cis-1,4 contents (around 90%). Cyclopentadienylnickel carbonyl behaves as $Ni(CO)_4$ but substitution of one carbonyl group by triphenylphosphine in $Ni(CO)_4$ results in a complete loss of activity of the products formed by reaction with Lewis acids (47). Catalysts produced by reaction of nickel carbonyl with Lewis acids in aromatic solvents were shown to be arene-nickel complexes:

$$Ni(CO)_4 + MeX_n + ArH \longrightarrow Ni(ArH)_2 \cdot (MeX_n)_m$$

The activity of these catalysts is considered to be associated with their π complex nature, and therefore dependent on the ability of aromatic ligands to be substituted by butadiene, i.e., on the lability of the arene-metal bond; in fact, butadiene polymerization is completely suppressed in the presence of mesitylene or hexamethylbenzene.

Table 2 Catalysts from Carbonyl Complexes of Transition Metals for 1,3-Butadiene Polymerization

Carbonyl Complex	Cocatalyst	Polybutadiene Structure (%)			Ref.
		Cis-1,4	Trans-1,4	1,2	
$Ni(CO)_4$	Various metal halides	75–95	3–20	1–6	47, 96–105
$Ni(CO)_4$	$AlCl_3 + KCN$	59	37	4	103
$Ni(CO)_4$	$AlCl_3 + KI$	3	95	2	103
$(C_5H_5NiCO)_2$	$TiCl_4$, $VOCl_3$	91–94	4–6	2–3	47, 97
$[Co(CO)_4]_2$	$MoCl_4$, $MoCl_5$	2	3	95	47, 96 97, 102 106
$[Co(CO)_4]_2$	WCl_6	56	35	9	102
$[Co(CO)_4]_2$	$AlCl_3 + thiophene$	96	2	2	105
$[Rh(CO)_2Br]_2$ $[RhNH_3(CO)_2Cl]$ $[Rh(CO)Br_2]_2 \cdot [N(C_4H_9)_4]_2$ $[Rh(CO)_2Cl_2] \cdot [N(C_4H_9)_4]$			78–84		46

On the other hand, halide derivatives of rhodium carbonyl are catalysts for the stereospecific polymerization of butadiene in hydrocarbon media, to yield 1,4-trans polymers (46).

Butadiene polymerizations induced by carbonyl complexes on the one hand, and by other transition metal π complexes, especially alkyl, arene, and cyclopentadienyl complexes, on the other, present striking similarities. This similar behavior can be ascribed to the intermediate formation in all cases, of π-allylic complexes resulting from the interaction of the monomer with the transition metal compound and its cocatalyst. Reactions of this type are well documented in organometallic chemistry. In other words, the catalytic activity does not necessarily depend on the existence of a preformed transition metal-carbon bond. This bond may be generated by the action of the monomer on the complex, as is also suggested in the case of the metal salts.

2.2.4 π-Allylic Complexes of Transition Metals

It has been shown experimentally that dienes react with a number of group VIII metal derivatives to give complexes having a π-allyl type of structure. NMR spectroscopy indicates that the metal–ligand bond involves three carbon atoms with delocalized π electrons; the π-allylic group is thus considered to be a bidentate ligand that can be converted, under the influence of another suitable ligand, to a σ-allyl group able to behave like an alkylated initiating center.

Since many recent results indicate the formation of intermediate π-allylic types of structures, the polymerization of conjugated diolefins by simple π-allyl derivatives of transition metals is of great interest. This subject has been recently reviewed (109). Such π-allyl groups represent good models of the active site structure. Since the first examples of stereospecific polymerization of 1,3-butadiene by π-allylic complexes of nickel and cobalt were reported independently by Natta and Wilke, numerous studies have been concerned with polymerizations in the presence of π-allyl derivatives of transition metals, as summarized in Tables 3 and 4.

Without a cocatalyst, only very few π-allyl type catalysts achieve the activity of the bimetallic Ziegler-Natta types, nor do they produce the same quality polymers (discussed in Section 4). As is evident from part A

Table 3 Catalysts from π-Allyl Transition Metal Complexes for 1,3-Butadiene Polymerization in Hydrocarbon Solutions

π-Allyl Complex	Cocatalysts	Polybutadiene Structure (%)			Ref.
		cis-1,4	trans-1,4	1,2	
A. $(\pi\text{-Allyl})_x M_T$					
$(C_3H_5)_2Ni$, $(C_4H_7)_2Ni$		1,5,9-Cyclododecatriene 95			
$(C_3H_5)_3Co$, $(C_4H_7)_3Co$		Linear oligomers			46, 95, 110
$(C_3H_5)_3Cr$, $(C_4H_7)_3Cr$			19–10	81–90	46, 47, 110–112
$(C_4H_7)_3Nb$				up to 97	46, 113
$(C_4H_7)_3Ti$		0	17	83	46
$(C_4H_7)_3Rh$		0	94	6	46
B. $(\pi\text{-Allyl})_x M_T + MeX_n$					
$(C_3H_5)_2Ni$	$SnCl_4$	30	68	2	91, 92
	SnI_4	0	95	5	47, 91
	$SnCl_2$	52	46	2	91
$(C_4H_7)_2Ni$	$AlCl_3$, $MoCl_5$, $TiCl_4$	81–95	3–18	1–3	46, 47, 90
	$NiCl_2$, $SnCl_4$, $SnCl_2$				
	$AgClO_4$, $Mg(ClO_4)_2$				
	NiF_2	28	66	6	47
	$NiBr_2$	14	82	4	47
	NiI_2	0	95	5	47
	$NiCl_2 + 10THF$	71	24	5	47
	$NiCl_2 + 1P(C_6H_5)_3$				
	$NiCl_2 + 100H_2O$	0	95	5	47
	$NiCl_2 + 1Bu_2S$	87	11	2	47
$(C_4H_7)_3Cr$	$CrCl_3$	10	15	75	47
	$NiCl_2$	92	2	6	46, 47
	$NiBr_2$	0	95	5	47
	TiI_4	90	6	4	46, 47, 114
	$TiCl_4$	45	50	5	114
	$Cr(AcAc)_3$	35	0	65	114
	$Zn(OOCCCl_3)_2$	82	5	13	46
	$Mn(OOCCCl_3)_2$	64	11	25	46
$(C_4H_7)_3Co$	$CoCl_2$	40	13	47	46
$(C_4H_7)_3Fe$	$NiCl_2$	92	6	2	46
$[(C_4H_7)_2Mo]_2$	$MoCl_5$	15	4	81	47
	$TiCl_4$	61	15	24	47
$(C_4H_7)_3Ti$	$NiCl_2$, TiI_4	85–88	5–8	4–10	46
$(C_3H_5)_4Zr$	$TiCl_4$	45	45	10	114
$(C_4H_7)_3Rh$	$AlCl_3$, $SnCl_4$	0	90–98	2–10	46

Table 3 (Continued)

C. $(\pi\text{-allyl})_x M_T + HX$ or X_2

$(C_3H_5)_2Ni$	$1CF_3COOH$	71–77	22–28	1	115
	$2CF_3COOH$	48–53	46–50	1–2	115
	$C_7H_7SO_3H$	48–50	48	2–4	115
	$CH_{3-n}Cl_nCOOH$	92–95	4–6	1–2	47
	$C_6H_3(NO_2)_2OH$	93	5	2	46
$(C_4H_7)Ni(C_5H_5)$	$C_6H_2(NO_2)_3OH$	90	8	2	116
$C_{12}H_{18}Ni$	HCl	84	13	3	87, 88, 117,
	HBr	72	25	3	87, 88, 117
					118
	HI	0	100	0	87, 88, 117,
					118
	Cl_2	74	23	3	117, 118
	I_2	12	86	2	117, 118
	$CH_{3-n}Cl_nCOOH$	89–91	4–6	3–5	87, 88, 117,
					118
	$1CF_3COOH$	92	4	4	87, 88, 117,
					118
	Up to $3CF_3COOH$	50	50	0	121, 122
	or $CF_2ClCOOH$				
	$C_6H_2(NO_2)_3OH$	93	4	3	87, 88, 117,
					118
$(C_4H_7)_3Cr$	1HCl	14	19	67	46
	2HCl	90	5	5	46, 113, 114
	CCl_3COOH	93	4	3	46, 47, 113
$(C_3H_5)_3Co$, $(C_4H_7)_3Co$	1HCl	36	14	50	46
	2HCl	88	2	10	46, 113
	CCl_3COOH	78	8	14	47
$(C_4H_7)_3Co$	I_2	45	12	43	46, 111, 112
$(C_3H_5)_3Co$	I_2	90	2	8	110
$[(C_3H_5)_2Mo]_2$	HCl or HI	0–4	6–7	90–93	47
$(C_4H_7)_3Rh$	HCl	10	88	2	46
$(C_4H_7)_3Nb$	HCl	91	5	4	46, 113

D. $(\pi\text{-Allyl})_x M_T + $ quinones

(π-Allyl, methallyl, crotyl)$_2$Ni	p-Chloranil or p-bromanil	88–94	3–9	3	119, 120
$[(C_3H_5)_2Mo]_2$	p-Chloranil or benzoquinone	0	1–5	95–99	47, 116

E. $(\pi\text{-Allyl } M_T X)_n$

$C_3H_5^-$ or C_4H_7NiCl		92	6	2	46, 47, 125–
					127
$C_3H_5^-$ or C_4H_7NiBr		45–72	25–53	2–3	46, 47, 125–
					127

Table 3 (Continued)

π-Allyl Complex	Cocatalysts	Polybutadiene Structure (%)			Ref.
		cis-1,4	trans-1,4	1,2	
π-C_4H_7NiI		4	93	3	46, 47, 125–127
$C_3H_5NiOOCCH_{3-n}Cl_n$		92–97	2–6	1	113, 123, 124
π-Allyl, methallyl, or crotyl $NiOOCCF_3$		91–98	1–8	1	52, 113, 123, 124, 170
$C_3H_5NiOC_6H_2(NO_2)_3$		97	3	0	52
$C_3H_5NiOC_6H_2Br_3$		0	96	4	47, 113
$C_3H_5NiO_3SC_6H_4CH_3$		48	48	4	47, 52, 113
$C_{12}H_{19}NiOOCCF_3$		98	2	0	94, 128
F. $(\pi$-Allyl $M_TX)_n + MeX_m$					
C_3H_5 or C_4H_7NiX $X = Cl, Br, I$	Ti, V, Mo, W, Sn, Mg, Co, Ni, Zn, B, Al halides	80–95	4–20	1–6	46, 47, 91, 113, 115, 116, 126, 127
	Mg, K sulfates, KCNS, K_2CO_3, $AgNO_3$ $Mg(ClO_4)_2$ Ni, Co, Mn, Mg, Zn trichloro- or trifluoroacetates				
C_4H_7NiCl	$SnCl_2$	43	56	1	91
C_4H_7NiCNS	$TiCl_4$	79	19	2	129
$C_4H_7NiOOCCH_3$	$TiCl_4$	91	6	3	129
C_4H_7NiI	SnI_4	85	13	2	46
G. $(\pi$-Allyl $M_TX)_n +$ Halogen or Organic Electron Acceptors					
C_4H_7NiCl	I_2	94	4	2	47
C_4H_7NiI	I_2	84	15	1	46
C_4H_7NiX $X = Cl$ or I	$CHCl_2COOH,$ $CCl_3COOH,$ $C_6H_2(NO_2)_3OH,$ CF_3COOH $CCl_3COH,$ $(CCl_3)_2CO$ CCl_3COCl $C_2H_5OOCCCl_3$	87–95	1–7	2–12	46, 47, 113 115, 116
C_4H_7NiCl	C_6H_5COOH	82	13	5	127
$C_4H_7NiOOCCl_3$	CF_3COOH	94	3	3	115
$C_4H_7NiOOCCF_3$	CCl_3COOH	90	7	3	115

Table 3 (Continued)

C$_3$H$_5$ or C$_4$H$_7$NiOOCCF$_3$	CF$_3$COOH	50	49	1	115, 130
C$_4$H$_7$NiCl	p-Benzoquinone, dichloro- or bromoquinones p-chloranil, p-iodanil, p-fluoranil	95–98	1–3	1–2	47, 113, 116 119, 120
C$_4$H$_7$NiBr	p-Chloranil	92	6	2	47, 116, 119, 120
C$_4$H$_7$NiI or –CNS	p-Chloranil	49–51	46–48	3	116, 119, 120
C$_4$H$_7$NiOOCCH$_3$ or OOCCF$_3$	p-Chloranil	83–94	3–14	3	115, 116, 129
C$_4$H$_7$CoI	p-Chloranil	96	3	1	116
π-Methallyl CoCl	p-Chloranil	70	7	23	119

H. (π-Allyl M$_T$X)$_n$ + Electron Donors

C$_4$H$_7$NiCl	O$_2$, benzoyl or t-butyl peroxides, acetylacetone, thiophene, (C$_2$H$_5$)$_2$O	71–90	7–24	3–5	47, 127, 131
	Na$_2$O$_2$	52	44	4	127
	AIBN	45	43	12	127
	Tetrahydrofuran	54	41	5	47
	C$_2$H$_5$OH	30	70	0	130
	H$_2$O + KI	14	82	4	130
C$_3$H$_5$NiBr	O$_2$, benzoyl peroxide	84–88	7–11	5	131
C$_4$H$_7$NiBr	(C$_2$H$_5$)$_2$O	60	36	4	131
	C$_2$H$_5$OH	32	66	2	131
	H$_2$O or H$_2$O + KI	4–6	92–94	2	131
C$_4$H$_7$NiI	Benzoyl peroxide	0	96	4	132
	p-Nitrobenzoyl peroxide	84	11	5	132
	O$_2$	57	38	5	132
	(C$_2$H$_5$)$_2$O, C$_2$H$_5$OH, H$_2$O	4–18	80–94	2–3	131
	H$_2$O + KI, HCOOH, CH$_3$COOH				
C$_3$H$_5$NiOOCCF$_3$	Aromatic derivatives	49	49	2	130
C$_{12}$H$_{19}$NiOOCCF$_3$	C$_2$H$_5$OH, (C$_6$H$_5$O)$_3$P	0	96	4	133

Table 4   Catalysts from π-Allyl Transition Metal Complexes for the Polymerization of Monomers Other Than 1,3-Butadiene

Monomers	π-Allyl Complex	Cocatalysts	Polymer Structure	Ref.
Isoprene	$(C_4H_7)_3Cr$		35% 1,4; 28% 1,2; 37% 3,4	46
	C_4H_7NiCl	$TiCl_4$	50% cis-1,4; 38% trans-1,4; 12% 3,4	46
		$ZnCl_2$, $Ni(OOCCCl_3)_2$ Cl_3CCOOH	23–30% cis-1,4; 52–59% trans-1,4; 10–20% 3,4; 0–2% 1,2	46
		p-Chloranil	46% cis-1,4; 39% trans-1,4; 15% 3,4	46
	$C_{12}H_{19}NiOOCCF_3$		68–79% cis-1,4; 15–26% trans-1,4; 6% 3,4	135
	$C_3H_5NiOOCCF_3$	o-$Cl_2C_6H_4$	51–55% cis-1,4; 45–49% trans-1,4	135
Ethylene	$(C_3H_5)_3Cr$,			110, 114
	$(C_3H_5)_2CrI$		Linear, high density	136, 137
	$[(C_3H_5)_2Cr]_2$	$TiCl_4$	Linear, high density	
	$(C_3H_5)_4Zr$		Linear, high density	136
	$(C_4H_7)_4Ti$,		Linear, high density	136, 137
	$(C_3H_5)_2Ti(OC_2H_5)_2$			
	$(C_3H_5)_4Ti$,			
	$(C_3H_5)_4Nb$	$TiCl_4$	Linear, high density	139
Propylene	$(C_3H_5)_3Cr$,			
	$(C_3H_5)_2CrCl$	$TiCl_3$, $TiCl_4$	Partially crystalline	114
	$(C_3H_5)_4Mo$,			
	$(C_3H_5)_4Zr$	$TiCl_3$, $TiCl_4$	Partially crystalline	114
	$[(C_3H_5)_2Cr]_2$	$TiCl_3$	85% crystalline	139
Chloroprene	(Allyl or methallyl)$_3$Cr			138
Acetylene	C_3H_5NiCl	Phosphine, isonitrile		140
Allene	C_3H_5NiBr		1,2 polymer	82, 83, 84
1,2-Butadiene				
Styrene	C_3H_5NiCl		Low molecular weight	127, 141, 142
	$C_3H_5NiOOCCF_3$		atactic polymers	52
	$C_{12}H_{19}NiOOCCF_3$			135
	$(C_3H_5)_3Cr$	$TiCl_4$		114
Methyl meth-acrylate	$(C_3H_5)_3Cr$,			
	$(C_4H_7)_3Cr$		About 20% isotactic,	138, 143
	$(C_3H_5)_2CrAcAc$		35% heterotactic,	138
	$(C_3H_5)_3Rh$,		and 45% syndiotactic	138, 143
	$(C_4H_7)_4Ti$			
	$(C_3H_5)_4Mo$			138
Methacrylo-nitrile	$(C_3H_5)_3Cr$,			138, 143
	$(C_4H_7)_3Cr$			
	$(C_3H_5)_4Zr$			138, 143
Butyl vinyl ether	C_4H_7NiCl	Benzoyl peroxide	Atactic	127
Cyclobutene	$(C_4H_7)_2Ni$, C_4H_7NiX	$AlBr_3$, $TiCl_4$	Polycyclobutene	144
	$[(C_4H_7)_2Mo]_2$, $(C_4H_7)_4W$	$AlBr_3$, $TiCl_4$, $MoCl_5$, WCl_6	80–90% cis-1,4-, 10–20% trans-1,4-poly-butadiene	144
Cyclopentene	$(C_4H_7)_2Ni$, C_4H_7NiX	$AlBr_3$, $TiCl_4$ $MoCl_5$	Polycyclopentene	144
	$[(C_4H_7)_2Mo]_2$	$AlBr_3$, WCl_6	35–40% cis-1,4-, 60–65% trans-1,4-poly-pentenamer	144

of Table 3, only a few π-allylic complexes of transition metals are able to promote polymerization; most of them first induce oligomerization reactions and become active polymerization catalysts only when halides or other anions are bound in the coordination sphere, most often through addition of various cocatalysts to the reaction medium (as in parts B, C, and D of Table 3).

This influence of the counterion in the catalytic π-allyl complexes is a determinant not only of the overall activity but also of the stereospecificity (see part E, Table 3). The addition of various metallic or organic electron acceptors to π-allyl $M_T X$ complexes, resulting in some cases in the formation of charge transfer complexes, results in enhanced catalytic activity and, for most polybutadienes, a further increase in the cis-1,4 content (parts F and G, Table 3). The addition of electron donating ligands usually results in decreased catalytic activities with a simultaneous modification of the stereospecificity (part H, Table 3). The significance of these effects is considered in more detail in Section 4, when discussing the factors influencing the stereoregulation process in diolefin polymerization.

As shown in Table 4, π-allylic complexes of transition metals are able to initiate the polymerization not only of other conjugated diolefins such as isoprene and chloroprene, but also of monoolefinic, acetylenic, and polar vinyl monomers. Partially crystalline polypropylene was also obtained by means of these complexes, with titanium tetrachloride or trichloride as cocatalysts. It is possible that, in the latter case, polymerization actually takes place on titanium catalytic species formed by reaction of the cocatalysts with the π-allylic complexes. Polymerization of cyclobutene and cyclopentene in the presence of π-crotyl complexes of molybdenum, tungsten, or nickel (eventually combined with Lewis acids) has also been reported (198). For example, catalytic systems based on bis(π-crotyl)nickel and π-crotylnickel halides polymerize cyclobutene and cyclopentene exclusively through the double bond, whereas ring opening is observed in the presence of catalysts containing π-crotyl complexes of molybdenum or tungsten.

3 THE EVOLUTION OF IDEAS ON THE MECHANISM OF OLEFIN COORDINATION POLYMERIZATION

The evolution of ideas on reaction mechanism has undoubtedly been dominated by the mechanism proposed by Cossee a few years ago. By then a general consensus had been reached on several essential points, and these key points were included in the Cossee hypotheses. These were determinant role of the transition metal; formation of an alkyl derivative of this metal foreshadowing the polymeric chain on the catalyst; accessibility of at least one free coordination position on the complex to allow a proper positioning of the monomer; and existence in this complex of steric factors responsible for the stereospecificity of the propagation reaction. Cossee's proposal left several questions unanswered. In particular, it did not detail the role of the alkyl metal derivative of groups I–III (especially aluminum) in controlling the stereospecificity and activity of the catalyst, nor the exact importance of the heterogeneous or homogeneous state of this catalyst. With our current knowledge, these questions are no longer as crucially important as they once were, but these aspects will be discussed later.

3.1 The Cossee-Arlman Mechanism: Cis Rearrangement

In the Ziegler-Natta bimetallic catalysts, a dominant role is played by the transition metal derivative in controlling the stereospecific polymerization reaction. The influence of the aluminum compound is much less essential than had been frequently proposed for historical reasons [Aufbau reaction (1) of ethylene to low molecular weight polymers by aluminum alkyls]. The assumptions as to the greater importance of the transition metal have now been experimentally proved by performing the polymerization of α-olefins in the presence of monometallic catalysts (containing only one type of metal, although sometimes binuclear) displaying good activities and stereospecificity.

Emphasizing the prominent role of the transition metal, and more specifically of its d orbitals, in 1964 Cossee and Arlman presented in a very lucid series of papers a mechanism for the stereospecific polymerization of unsaturated hydrocarbons on a previously alkylated monometallic complex, for instance, of titanium (145–148) as shown in Fig. 9. Their reaction scheme specifically involves the following important considerations.

3.1.1 Formation of the Active Center (Steps 1 and 2 of Fig. 9)

The essential role of the alkylaluminum compound is to alkylate the titanium by substituting one of the chloride ions of a pentacoordinated metal atom exposed at the surface and displaying a chloride vacancy. Of these five chloride ions, three are completely embedded in the interior of the crystal; of the two remaining, the first is still attached to two metal ions, and only the other is considered to be loose enough to undergo the alkyl exchange easily. The alkyl group introduced by this reaction foreshadows the growing polymer chain. From crystallographic models

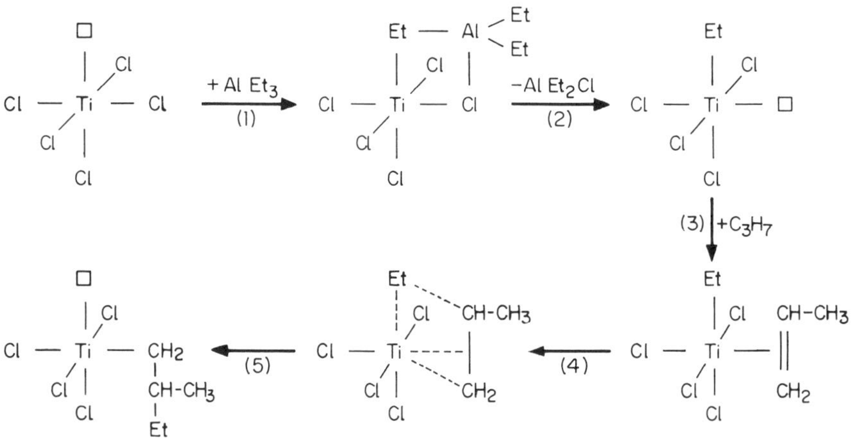

Figure 9

and calculations of the energies needed to remove a chlorine atom from the surface of α-TiCl$_3$ crystals, Arlman showed that the vacancies should be found on the edges of the elementary planes in the crystal. The number of such vacancies, potentially giving active sites, was shown to correspond reasonably to the number of the centers experimentally determined by kinetic estimates and by use of radiotracers. These deductions have been substantiated (see Section 3.2) by electron micrographs taken by Rodriguez et al. (12, 149), which indeed show dramatically that the growing chains are not in a 001 face, but are located along a spiral, most probably the edge of a crystal growth spiral.

The alkylation reaction (steps 1 and 2) itself has been put in evidence first on a model system devoid of side reactions, ScCl$_3$–Zn(Et)$_2$, where both zinc and ethyl groups could be labeled (150); a further detailed study of the TiCl$_3$–Al(CH$_3$)$_3$ system was also performed by Rodriguez et al. (151).

3.1.2   *Reaction of Monomer at the Active Center (Steps 3–5)*

This reaction implies the preliminary bonding of the olefinic monomer to the free coordination position of the alkylated octahedral titanium complex, TiCl$_4$R $\square$. Step 3 occurs through π bonding (152), a type of bonding that is now well documented; the knowledge gained from studies of different olefin-transition metal complexes is readily applied to polymerization catalysts. The heats of formation of the complexes are usually small and almost the same as their heats of solution (see, e.g., ref. 157); also, small complex formation constants have been indicated by several thermodynamic and kinetic studies. Molecular models reveal that there is no special steric hindrance to a very close approach of the monomer to this sixth coordination position on the $TiCl_3$ crystal.

The greatest merit of this scheme is most certainly the avoidance of any important nuclear displacement during the reaction of the coordinated monomer with the alkyl group; only the first CH_2 group of the growing chain attached to the transition metal has to undergo a limited translation of about 1.9 Å. This displacement starts with the angular vibration of the Ti–CH_2 bond. In transition metal complexes, owing to an additional overlap of the CH_2 group with the metal d_{yz} orbital, the amplitude of this vibration in the nonequilibrium position may be larger than in a nontransition metal complex. This situation also allows a much greater overlap of the two potential wells corresponding to the equilibrium positions of the R groups and the coordinated olefin's orbitals. The energy of activation for the rearrangement (step 5) is consequently lowered, and the migrating group finds simultaneously a combination of overlapping orbitals all along its reaction path (Fig. 10).

In this reaction sequence, it is claimed that the overall measured activation energy is essentially representative of the energy involved in the rearrangement, since the heats (ΔH) of complex formation are usually small.

The cis migration mechanism presented here explains very well many essential features of the olefin's specific polymerization by transition metal complexes. An elegant illustration of its versatility came from observations made in the course of vinylcyclopropane polymerization (158); indeed, the simultaneous occurrence of 1,2 and 1,5 additions implies a mechanism involving both normal and abnormal alkylation of the coordinated monomer during cis migration (Fig. 11). In fact, it is of much

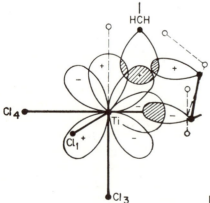

Figure 10

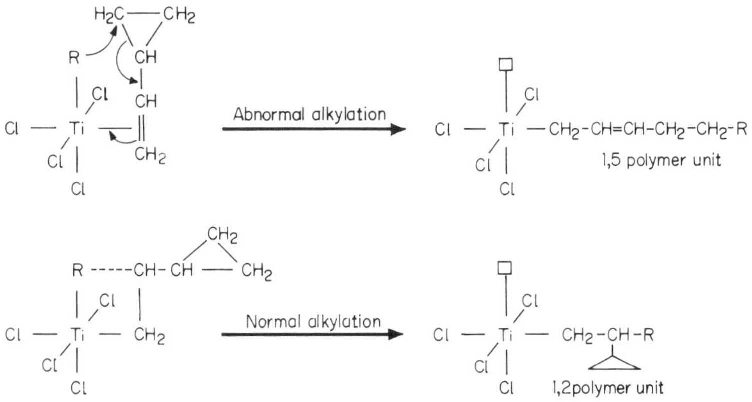

Figure 11

broader scope than for polymerization only, and has been applied to a number of reactions catalyzed by transition metal complexes, for example in the dimerization of ethylene in the presence of rhodium salts, which has been shown to involve a $[C_2H_5-RhCl_3(C_2H_4)(solvent)]^-$ complex. Some catalytic hydrogenations are claimed to proceed by a similar migration of the metal bonded hydrogen to the coordinated olefinic substrate. Several carbonylation reactions (e.g., hydroformylation) have also been explained by this type of rearrangement.

The general nature of this reaction, corroborated by new examples discovered every year, was also demonstrated by the specific activation and reaction of other interesting substrates including, e.g., molecular oxygen or nitrogen.

3.1.3 Quantitative Aspects of the Rearrangements

In an attempt to get a more precise picture of the reaction (148), it has been proposed that in the nonequilibrium position the important factor is the mixing of ϕ with ψ_2 orbitals; orbital d_{yz} (originally in the common $3d$ level) is the one that connects π^* and σ_R (or performs the mixing). This will occur to an appreciable extent only when the energy spacings (π^*, σ_R and d_{yz}) are not too large; i.e., the best catalytic activity will be attained when the metal $3d$ level is somewhere between π^* and σ_R (Fig. 12). It is also clear that these levels, and consequently the relative activity, will be influenced not only by the nature of the metal but also by that of the more or less electron attracting surrounding ligands.

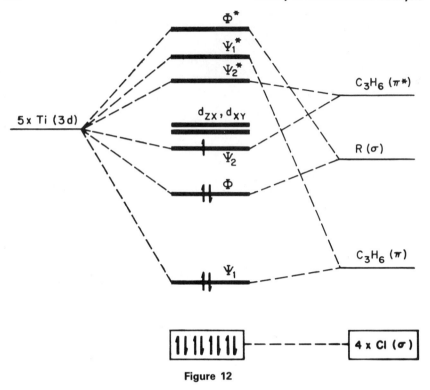

Figure 12

Quantum mechanical calculations allowed a more quantitative evaluation of this cis rearrangement (in a complex carrying a methyl group to simplify the problems) through an "iterative extended Hückel" approach (153). Results were obtained for several situations along the reaction coordinate; the energy diagrams of a number of relevant orbitals, their composition, and the evolution of charges and bond orders give a fairly detailed picture of this catalytic reaction. The calculations confirm the qualitative picture that the d level is between π and π^* of C_2H_4, and the C–Ti bond is located 16,000 cm^{-1} below that of the nonbonding d_{xy} and d_{yz} of the metal; they also show, in the initial titanium alkyl, the small electronegativity difference in the Ti–C bond, and the reason for the low kinetic stability of this bond (short distance between the M_T–C bond orbital and the empty d orbitals). Moreover, the total splitting of the d levels is of a very reasonable order of magnitude (28,700 cm^{-1}), and the chloride ions also show some π bonding to the metal.

In the complex with ethylene, d_{xy} is depressed by the olefin antibonding orbital and the distance σ-$CH_3 \leftrightarrow d_{xy}$ is thereby reduced from 14,900 to 11,100 cm^{-1}. If this is indeed a measure of the M_T–C bond lability, we have a quantitative indication of the path of action of the transition metal ion in catalysis. Simultaneously the π-π^* distance in ethylene increases from 48,500 to 53,200 cm^{-1}. An interesting conclusion is the unsymmetrical bonding of the two carbon atoms of the olefin, the one closer to the methyl group being somewhat more weakly bonded.

During the reaction (moving the CH_3 group), which is accompanied by delocalization, the charge on the Ti atom (as well as on the four Cl ions) remains practically constant despite the temporary storage of some negative charge. The important conclusion is that the alkyl group *does not move as an anion*, as was often suggested in the early studies on the Ziegler-Natta catalysts.

A comparison has been made among Ti, V, and Cr indicating that the lability of the methyl group increases in this series (from Ti to Cr), while there is a corresponding decrease in the M_T–ethylene interaction. This lability could be a favorable factor in the propagation, which fits Pasquon and Zambelli's determination of the absolute activity of active sites in Ti and V catalysts; however, it also becomes a limiting factor for the formation of a sufficient number of M_T–R bonds, which may explain the very poor performance of Cr in Ziegler-Natta catalysts as well as the improved performances in better ligand environment as in oxide type catalysts.

3.1.4 *Stereospecificity*

The isotacticity of many polyolefins obtained with the Ziegler-Natta $TiCl_3$–AlR_3 catalyst has also been explained by Cossee and Arlman on the basis of a monometallic complex.

Considering the octahedral $[(RTiCl_4 \,\square\,]$ in the surface of the crystalline lattice, they came to two important conclusions: (*a*) This octahedron is placed unsymmetrically, and does not contain a plane of symmetry; and (*b*) the positions of the alkyl group and the coordination vacancy in the lattice are not equivalent, either sterically or ionically.

This situation might completely predetermine the configuration of the new asymmetric carbon being formed (i.e., only one orientation of the coordinated olefin is allowed), explaining the tacticity of the chain (ref. 148, pp. 154 and 162).

There are still two possible paths for the reaction:

1. After the rearrangement, the alkyl group stays on its new position, the monomer is coordinated on the newly formed vacancy, and the same process is repeated, giving an alternating type propagation and a syndiotactic polymer. This eventuality will be favored by a substantial

lowering of the temperature (decreasing the rate of both alkyl back-shift and alkyl migration, but increasing complex formation on the new vacancy).

2. In most cases, however, the extended alkyl group moves back to its original position before a second monomer is incorporated. This ensures that the insertion of every monomer unit in the chain is repeated in a stereospecifically identical manner and an isotactic polyolefin is produced. This scheme fits in with the nonequivalence of the two positions involved, and also with the kinetic analysis of the reaction. This back-shift will proceed by the same mechanism as the migration in the insertion step, with the help of metal t_{2g} orbitals. Consequently, the preference for an isotactic or syndiotactic placement will be governed essentially by the ratio of the rates of the alkyl back-shift and the insertion in the growing chain.

When using a monomer with two potentially asymmetric carbon atoms, e.g., a 1-deuterated α-olefin, a ditactic polymer is obtained whose formation can be exactly explained by the spatial control of the monomer coordination.

Finally, the formation of stereoblocks, i.e., isotactic homopolymers containing sequences of, say, d configurations followed by other sequences of the mirror image l configuration, can also be easily visualized. The growing chain may exchange with a metal alkyl molecule in the solution and exchange again with another active titanium site from which position growth resumes but possibly with the opposite configuration. It would be interesting to see if these stereoblock structures are absent when using a monometallic catalyst that was prepared using no metal alkyl.

3.2 The Modified Model for Titanium Trichloride

Despite its outstanding merits, the Cossee-Arlman scheme still leaves a series of questions unanswered. For instance, (a) why does the aluminum alkyl, in some cases, influence the activity and even the stereospecificity of the catalyst (e.g., AlR_2I); (b) why do the organozinc derivatives, which are good alkylating agents, yield less active and stereospecific catalysts; (c) why do electron donor ligands (e.g., amines) significantly improve the stereospecificity of a given catalyst; (d) why do only a small proportion of the alkylated transition metal sites give rise to polymer chains, as shown by a comparison of the degree of alkylation with the dispersion of growing chains along crystal growth spirals or faults, as indicated by electron microscopic observations; and (e) why does β-$TiCl_3$ still promote the formation of a significant proportion of isotactic polypropylene, despite its great number of symmetrical active sites with two vacancies?

These points were considered in a modified scheme presented by Rodriguez and Van Looy (149). Keeping the basic features of Cossee's proposal (in particular the cis rearrangement reaction), these authors suggest that the active site is a tetracoordinated alkylated titanium atom carrying two coordination vacancies. One vacancy is occupied by the halogenated aluminum alkyl (or eventually another ligand) and forms a bridged bimetallic complex with the titanium atom; the other is filled by the monomer. The resulting catalytic sequence is schematized in Fig. 13.

After the rearrangement (breaking the initial bimetallic bridge), the formation of a new bridge under the influence of the aluminum atom is the driving force that will restore the active site to its initial configuration. This will give rise to an isotactic polymerization. Such a mechanism may furnish a satisfactory answer to the questions raised above (at least for β-TiCl$_3$). It explains in particular the influence of external ligands (amines) acting in place of the aluminum alkyl; the inhibitory effect of strongly coordinating entities such as ZnCl$_2$; the stereospecificity of TiCl$_2$–Al(CH$_3$)$_3$; and the influence of aluminum alkyls. For the α form of TiCl$_3$, one has to admit that although the pentacoordinated form is predominant, there are still a few tetracoordinated sites arising from lack of stoichiometry in the crystal and from a double chlorine vacancy permitted by the relatively small differences of retention energies within the edges of this crystal. The latter observation would explain the small number of active sites indicated by electron micrographs.

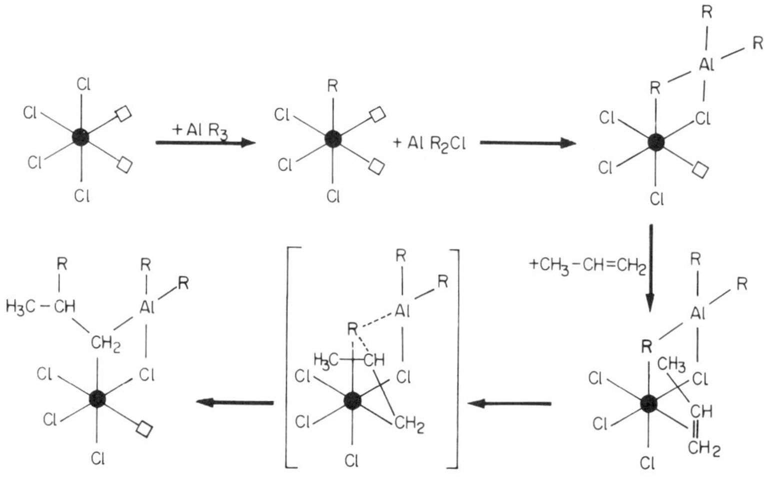

Figure 13

The proposal does not imply that Cossee's pentacoordinated center must be inactive, since the aluminum alkyl coordination is reversible, but its stereospecificity is not guaranteed. It reconciles the antagonistic views of the supporters of the mono- or bimetallic mechanisms (see the following section), since the aluminum alkyl is playing the role of a ligand to the determinant transition metal complex. It must be admitted that the essential role of the titanium-carbon-aluminum bridge is questionable in view of the relative instability of such structures (154) compared to the corresponding chloride bridges. It represents a very good working hypothesis to use to compare the results described below for the homogeneous polymerization of ethylene and propylene.

3.3 Soluble Bimetallic Catalysts

The polymerization of olefins by two families of *soluble* Ziegler-Natta catalysts has been investigated in detail. These give a better insight into the polymerization mechanism and support the main ideas arising from studies of heterogeneous catalysts. These investigations include the polymerization of ethylene by biscyclopentadienyl titanium dichloride catalysts, and the syndiotactic polymerization of propylene by vanadium catalysts in the presence of an electron donor molecule.

3.3.1 *Cp_2TiCl_2—$EtAlCl_2$ for Ethylene Polymerization*

This interesting system, very suitable for a systematic investigation because of its solubility and its rather well-defined chemistry, has been extensively studied by Henrici-Olivé and Olivé (14). Combining EPR and kinetic measurements, they came to the conclusion that the active species was again an alkylated octahedral titanium complex on which polymerization took place through a scheme very similar to the one proposed by Cossee:

The studies also emphasized a number of interesting features:

1. *Oxidation state and geometry.* The active center contained Ti(IV) exclusively in an octahedral arrangement, which is especially interesting in view of the time-honored discussions on the importance of the oxidation state of the transition metal, and particularly since most heterogeneous catalysts involve Ti(III).

Hence it becomes clear that the key factor is the *geometry* of the complex. For *soluble* species, only the Ti(IV) oxidation state ensures the octahedral geometry with a vacancy for the monomer coordination in the suitable cis position relative to the R group (growing chain), while the corresponding Ti(III) gives an inactive tetrahedral complex. In heterogeneous catalysts, the crystalline $TiCl_3$ lattice itself ensures this same octahedral structure.

2. *Stability of the* Ti–R *bond and reactivity of the complex.* Taking as a criterion either the rate of reduction from Ti(IV) to Ti(III) or the rate of ethylene polymerization, it was also proved experimentally that the Ti–R bond was destabilized both by the alkyl groups of the aluminum ligand (acting presumably through its bridge in the position trans with respect to the R group) and by the coordinated monomer. This particular point, detailed by simplified molecular orbital calculations, was also confirmed by experiments involving nonpolymerizable olefins and strongly supports the predictions of Cossee.

The stability of this Ti–R bond, which is directly related to the catalytic activity of the complex, can accordingly be "tailored" to a certain extent by modifying the donor-acceptor properties of the surrounding ligands, e.g., the aluminum compound. In similar fashion the π-allylnickel complexes exhibit tremendous changes in the butadiene polymerization rate when the electronegativity of the counteranion is modified (123). This explains how the alkyl metal part of the catalyst can influence the polymerization course, and *closely fits the ideas of Rodriguez and Van Looy summarized above.*

In summary, catalysis results from three cooperative influences: destabilization of the Ti–R bond by coordination of the monomer and by the bridged aluminum alkyl, enhancement of the olefin double bond reactivity by π coordination to titanium, and a suitable spatial arrangement in the complex for the "cis migration" of the alkyl group (or growing chain) by only a small "in plane" vibrational displacement as stressed by Cossee. All these features are related to an octahedral geometry of the complex.

Chain termination is also satisfactorily accounted for as owing to the bimolecular mutual deactivation of two catalyst molecules with loss of the R group (growing chain).

3.3.2 *VCl₄–Al(C₂H₅)₂Cl for Syndiotactic Propylene Polymerization*

Boor and Youngman (155) as well as Zambelli and Natta (156) and their co-workers have studied these systems, and although suggesting slightly different active centers, both groups came to the conclusion that again, polymerization took place on an octahedral complex, on two vicinal positions occupied, respectively, by the chain and the monomer.

Zambelli used a Lewis base L (particularly anisole) as an additional ligand replacing one of the organometallic molecules, and obtained a complex to which the following structure was ascribed:

This species is more stable than in the absence of anisole since polypropylene yield and molecular weight both increase, indicating a considerable decrease in the termination process.

The stereospecificity of the syndiotactic polymerization of this monomer is favored both by low reaction temperatures ($-78°C$) and the presence of the Lewis base. In the proposed catalyst structure, the vanadium atom is asymmetrical, and the syndiotactic type of stereospecificity is explained by an inversion of configuration every time a previously coordinated monomer molecule enters the vanadium-alkyl bond. From a steric point of view this asymmetry does not seem important enough to justify a preferential presentation of the monomer. The explanation can probably be found in Boor's proposal that the free rotation of the last added unit is sterically hindered. The addition of a new monomer molecule would then be faster if it approaches the growing chain end from a direction imposed by the presence of the counterions in the most favorable position, i.e., presenting the side of the molecule having a configuration opposite to that of the last unit to minimize the interaction of the methyl groups. A rise of the reaction temperature will obviously lower this rotation barrier and consequently reduce the stereospecificity of the process.

It was also shown that the exchange of ligands (when favored by the reaction conditions), or their dissociation leading to two coordination vacancies (in particular when using a metal alkyl having a lower tendency to coordinate than the aluminum compound), introduces steric irregularities in the growing chain, e.g., isotactic diads or even sterically disordered sequences.

3.4   Homogeneous and Heterogeneous Coordination Catalysis

The importance of the insolubility or solubility of the catalyst in the polymerization medium has been a disputed point for years.

The fact that until recently, it has been impossible to prepare *isotactic* unhindered polyolefins with a soluble catalytic system has been considered an indication that π bonding of the monomer in itself is not sufficient to ensure this type of stereoregulation even on a sterically crowded complex. Moreover, this is in agreement with what we know about the chemistry and mechanism of such reactions. It corroborates the hypotheses of Cossee and Rodriguez that an additional factor, such as the severe steric requirements of the crystalline matrix or the control exerted by the bridged aluminum alkyl, is essential to ensure at every step the insertion of the olefin in an identical stereoconfiguration.

One apparent exception to this is the report by Mazzanti (44) of a soluble catalyst able to perform the polymerization of 4-methylpentene into an isotactic high polymer. It is not yet certain, however, that the catalytic species is a truly isolated complex in solution, or an aggregate that reproduces locally, under apparently homogeneous conditions, on a microscopic scale, the same steric requirements as the matrix of the insoluble systems. In another apparent exception Ballard (211) has recently reported that soluble $Zr(allyl)_4$ and $Zr(allyl)_3Cl$ can produce soluble isotactic polypropylene. It is reasonable to assume that this is a homogeneous process. Thus, a single transition metal atom appears to be able to form isotactic polymers and does not require the environment of a solid surface. Even in this case, the environmental changes of the metal atom afforded by adding insoluble cocatalysts lead to increased rates of polymerization.

The beneficial effects of a second factor for steric control in the orientation of the monomer, besides coordination of the double bond to produce isotactic polymers, have also been demonstrated in anionic coordination polymerization. It is worthwhile to remember that lithium alkyls polymerize styrene to atactic macromolecules unless the lithium compound is incorporated in a crystalline lattice such as LiBr, yielding then a highly isotactic polystyrene (159). When the monomer itself contains a second suitably located group, its orientation may be predetermined enough to yield isotactic products even in completely homogeneous solution; this is the case for the polymerization of *o*-methoxystyrene by butyllithium (160), and of methyl methacrylate by 9-fluorenyllithium in toluene (161).

In summary, one can consider the influence of the homogeneous or heterogeneous state of the catalyst as purely incidental; it does not at all affect the basic coordination polymerization mechanism, as confirmed by the fact that individual transition metal atoms can generate isotactic placements. However, if not essential, it is important in controlling the steric environment of the active center (and sometimes the electronic environment too, when the matrix acts as a delocalized ligand); in this respect, the presence of a crystalline lattice or any kind of rigid hindered environment contributes to ensure the isotacticity of a chain composed of monomeric units that do not have more than one binding point to the catalyst.

Interesting new developments (162, 164) indicate that the solid catalytic phase can govern the morphology of the polymer formed; in such cases, we have a direct and specific action of the solid surface. Some systematic control of this effect could represent a new and valuable procedure for modifying polymer properties.

3.5 Conclusion

The cis rearrangement reaction in an alkylated octahedral transition metal complex offering a coordination vacancy to the monomer, affords a clear and useful picture of the actual active center in both heterogeneous and homogeneous stereospecific polymerization of α-olefins. Taking into account recent additional considerations, one obtains a better idea of the relative importance of factors such as the oxidation state of the metal atom, the solubility of the catalyst, and principally the role of the alkylaluminum derivative. This last problem has been the subject of many controversies, but can be rationalized by considering that, besides its alkylating role, the *aluminum alkyl can also act as a ligand* in the coordination sphere. In this manner it exerts a control over the electronic distribution responsible for the reaction rate, and sometimes to a lesser extent steric influence over the course of the polymerization.

These descriptions have been of great help in tailoring the catalyst structure for an ever more precise and elaborate control of both activity and structure, at the molecular and hopefully supermolecular level. Elastomer synthesis represents a special aspect of monoolefin coordination polymerization, since one usually tries to avoid too much crystallinity and starts with different monomers, usually dienes. With dienes stereoregularity involves not only isotactic and syndiotactic placements but also cis, trans, vinyl, and related isomerism. In the preparation of elastomeric polyenamers from cycloolefins some new data recently gathered about the control of olefin metathesis suggest, once again, a process involving a cis rearrangement through the intermediate formation of carbene species (165). Although recently these species were successfully isolated (221), the results can also be explained by hypothesizing transient metallocyclopentanes instead of metallocyclobutanes.

4   ADVANCES IN DIOLEFIN COORDINATION POLYMERIZATION

The ability to initiate stereospecific polymerizations by means of well-defined monometallic catalysts that can be modified systematically (particularly the π-allyl complexes; see Table 3) and to characterize the microstructure of the corresponding polymers (by x-ray diffraction, infrared, and NMR spectroscopy) has helped to establish correlations between the composition of the catalytic species and both its overall activity and stereospecificity. The recent advances in stereospecific polymerization of conjugated diolefins appear to be most enlightening as to the propagation control mechanism, the more so as several geometric isomers are possible. Subtleties in behavior are observable which are absent in olefin polymerization.

4.1   Factors Influencing Stereospecificity

The structure of polydienes is determined not only by the transition metal itself but also by the nature of the active center as a whole, influenced by the ligands bound into the coordination sphere. These include monomer, the counterion, the solvent, and any electron donating or withdrawing additives used. Three major functions of a complexing agent or ligand have to be considered:

   1.   The ligand can help to stabilize the transition metal in a lower oxidation state where d-p_π bonding is highly favored.
   2.   The ligand is very important in directing traffic in the coordination sphere. This can be done by physically blocking one or more coordination sites and/or by changing the electron density in the d orbitals which are available for bonding to an incoming substrate.
   3.   Some ligands are better leaving groups than others and can facilitate the substitution of olefin or diolefin reactants on the transition metal, consequently influencing the overall kinetics of the process, as will be shown later.

A significant illustration of ligand influence on the structural course of polymerization comes from appropriate modifications of π-allylic catalysts. It is possible to prepare by such modifications not only the different isomeric structures of polybutadiene (46) but also three different isomeric compositions of 1,4-polybutadiene (130):

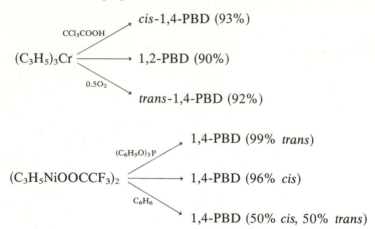

4.1.1 *The Transition Metal*

As already mentioned, when transition metal compounds having unfilled *d* orbitals form complexes with unsaturated hydrocarbons, there is an overlapping of the filled π orbitals of the olefin with the free *d* orbitals of the metal, and an opposite overlap of the filled *d* orbitals of the metal with the vacant antibonding orbitals of the olefin. Depending on the nature of the orbitals involved, the degree of overlapping of these orbitals, and the atomic radius of the metal, monomer-transition metal complexes will be formed which differ in their structure and consequently exhibit different stereospecificities. On this basis, it should be expected that the structural course of the reaction will be determined primarily by the nature of the transition metal, as widely confirmed in organometallic chemistry. For example, investigations of butadiene trimerization by simple π-allylic complexes indicated the specific formation of either cyclic or linear trimers, implying the same intermediate complex, when bis-π-allylnickel or palladium, respectively, is used as catalyst (95) (Fig. 14).

$$\pi-(C_3H_5)_2 \ M_T \ + \ 3 \ C_4H_6 \longrightarrow$$

Figure 14

The nature of the transition metal is also a determining factor in stereo-specific polymerizations with homogeneous Ziegler-Natta catalysts, as detailed with the aid of molecular orbital theory (see ref. 166). As appears from Table 3, the results obtained with monometallic catalysts of the π-allyl type confirm that metals belonging to the right side of the transition series in the periodic system (e.g., cobalt, nickel, iron, or rhodium) are the most appropriate for the preparation of 1,4-polybutadienes, whereas metals from the center left yield predominantly 1,2-polybutadiene and promote the polymerization of monoolefins (e.g., chromium, molybdenum, niobium). However, the original specificity due to the metal may be completely modified by counteranions; e.g., the addition of hydrogen chloride to chromium or niobium π-allylic com-plexes results in the formation of *cis*-1,4-polybutadiene instead of a vinylic polymer (46, 113). This indicates that the counterion present in the coordination sphere of the catalyst may be another determinant in controlling the stereospecificity.

4.1.2 *The Counterion*

The microstructure of polybutadienes produced by π-allylic catalysts depends closely on the nature of the anion in the complexes, i.e., chloro complexes give the cis-1,4 isomer whereas iodo complexes favor the formation of the trans-1,4, and bromo complexes produce polymers of mixed intermediate structure. All these products are essentially free from vinyl structure (see Table 3, part E). In fact, since the decrease of the electron density on the transition metal atom becomes more pronounced as the halogen electron affinity increases, we can ascribe the relatively high cis-1,4 contents of polybutadienes obtained with the chloro catalysts to the lower electron density of transition metal orbitals, relative to that in the bromo or iodo systems.

The cis content is still higher when a more electron withdrawing anion, such as a trihaloacetate, is used. Even slight modifications in the counter-ion used can result in change of microstructure in the polymers obtained, e.g., high *cis*-1,4-polybutadiene is produced in the presence of π-allylnickel picrate whereas the tribromophenate derivative yields a high trans-1,4 polymer (47, 52).

In fact, the observations made with different counterions support the concept that the change of stereospecificity is probably not due exclu-sively to a steric influence, but mostly to the inductive effect of the counterion present in the coordination sphere of the catalytic complex. Moreover, the role of the counterion appears to be a determinant in controlling not only the stereospecificity but also the overall rate of polymerization, as is shown in Section 4.2.1.

4.1.3 Ligands

The ligands bound to the transition metal may strongly influence the steric course of the polymerization, depending on their electron donating or accepting properties; for example, stereoregular 1,4-polybutadienes with high cis-1,4 contents are obtained in the presence of nickel complexes stabilized by carbonyl, cyclopentadiene, cyclooctadiene, or π-allylic ligands with various acids as cocatalysts, whereas when starting from tetrakis triphenyl phosphite nickel, a crystalline trans-1,4-polybutadiene is formed (133).

The addition of electron donating ligands to a complex giving the cis-1,4 polymer originally completely modifies its stereospecificity and usually leads to an increase in the amount of trans-1,4 (or even, in a few cases, of 1,2) units as has also been demonstrated with several Ziegler-Natta catalytic systems. This is clearly illustrated by the addition of stoichiometric amounts of ethanol or triphenyl phosphite to π-allylnickel trifluoroacetate, leading to a high trans-1,4-polybutadiene instead of a high cis-1,4 polymer (130). Cryoscopic determinations performed on the catalytic species indicate these changes in stereospecificity are due to the occupation of vacant coordination positions on the metal atom by these additional polar ligands. These observations strongly support Arlman's mechanistic assumptions in the case of diolefin stereospecific propagation on α- and β-TiCl$_3$ catalysts (146, 178).

In contrast to the electron donors, the majority of electron accepting molecules used as additives have practically no effect on the microstructure of the polymer chain when starting from catalysts originally yielding a cis-1,4 structure, such as π-allylnickel chloride or the haloacetates. But when starting from π-allylnickel bromide or iodide, electron accepting compounds change the stereospecificity, favoring the cis-1,4 structure formation (see Table 3, part G). In this case, it is obvious that the change in stereospecificity cannot be attributed to a halogen exchange reaction between catalyst components, since systems containing only iodine [e.g., (C$_4$H$_7$NiI)$_2$ + SnI$_4$ or I$_2$] give predominantly cis-1,4-polybutadiene while the corresponding original π-allylnickel iodide brings about the formation of the trans-1,4 polymer. Therefore variations in stereospecificity in the presence of additional electron acceptors appear to be due essentially to the formation of charge transfer complexes in which the electropositive charge on the transition metal atom is increased.

Other ligands have been found with a specific behavior that is different from those discussed above, and not yet observed with the usual bimetallic catalytic systems; indeed, a new kind of stereoregular polymer, namely the "equibinary polydienes" containing equimolecular amounts of two different structural units, were synthesized by modifying, with suitable additional ligands, catalytic species yielding polymers containing only one type of structural unit. Such an elaborate control of the propagation reaction may be very helpful in getting a more intimate view of the stereoregulation processes, since it appears to be a general phenomenon. Various equibinary polydienes have been synthesized by similarly modifying different catalysts; for example, equibinary poly(cis-1,4–3,4)- or (1,2-3,4)-isoprenes have been prepared with modified cobalt catalysts (19, 167, 168), and equibinary poly(cis-1,4-trans-1,4)-butadiene (121, 130, 133) or isoprene (135) in the presence of modified nickel catalysts. An interesting characteristic of the catalysts promoting the formation of equibinary polydienes is their reversible character. The removal of the ligand or the addition of electron accepting or donating compounds promotes the formation of polymers having an all cis-1,4 or crystalline trans-1,4 structure, respectively; consequently, the binary selectivity appears to enjoy some of the versatility usually observed for monoisomeric stereospecificity. A study devoted to the catalytic species involved in the formation of equibinary poly(cis, trans)-butadiene, strongly suggests a competitive modification of the initial symmetry of the catalytic complexes when coordinating the additional ligand (130). A tentative scheme for the formation of these equibinary polydienes is proposed (213) on the basis of a binuclear complex in which the insertion of cis and trans units in the two bound growing chains is coupled and complementary.

A most interesting mechanistic indication is the possibility to control to a large extent the distribution of the two isomeric units in the polymeric chains, without disturbing their equimolecular composition; for instance, equibinary 1,4-polybutadienes (containing practically 50% cis and 50% trans) have been obtained in which the isomer placement switches, depending on the nature of the polymerization solvent, from a purely random one (in benzene) to a highly alternate one (in CH_2Cl_2) (214). A tentative interpretation of this behavior might involve kinetic control of the lifetime of the σ form (allowing butadiene insertion) versus the π form of the bound chain.

4.2 Factors Controlling the Kinetics of Coordinated Polymerization

The nature of the ligands bound in the coordination sphere of the transition metal may influence not only the stereospecificity but also the catalytic activity; i.e., changes in the nature of the counterions as well as of the ligands (including monomers) in the catalytic complexes, result in drastic modifications of the overall polymerization rate.

4.2.1 *The Counterion*

An increase of the counterion electron withdrawing character in monometallic catalytic complexes, for instance, the substitution of the halogen in π-allyl derivatives of transition metals by anions of halogenocarboxylic acids, leads to a noticeable increase in activity. Moreover, the behavior of a series of π-allylnickel haloacetates, all of which are highly stereospecific catalysts for the cis-1,4 polymerization of butadiene, indicates an interesting correlation between the over-all polymerization rate and this electron withdrawing character of the coordinated counter anion. The catalytic activity increases sharply in the sequence $CH_3COO^- < CH_2ClCOO^- < CHCl_2COO^- < CCl_3COO^- < CF_3COO^-$, and the overall polymerization rate with the trifluoroacetate complex is about 150 times as great as that in the presence of the corresponding monochloroacetate (124). In addition to their high catalytic activity, the trifluoroacetate complexes produce high molecular weight *cis*-1,4-polybutadiene displaying characteristics similar to those of elastomeric polybutadiene obtained with Ziegler-Natta catalysts. These data confirm the interesting possibility of preparing, by the use of appropriate counterions, stable monometallic transition metal complexes that enjoy catalytic properties similar to those of bimetallic systems without the addition of any cocatalyst.

4.2.2 *Ligands*

Ligands Originally Bound to the Transition Metal.   Since the first step in coordinated polymerization is the formation of a π complex between the monomer and the transition metal, it should be expected that the ligands bound originally to the catalytic complexes may strongly influence the initiation step and also the overall polymerization rate. Many monometallic catalysts (i.e., π-allylic derivatives) are normally associated as binuclear complexes. Often, the active centers for chain growth are not the dimeric molecules of the original complexes but a mononuclear species, and the first step will be a dissociative equilibrium involving monomer (Section 4.2.3) which might also determine the overall rate of the polymerization process depending on the types of ligands primarily bound to the transition metal. Only a few accurate data have been reported on these points, partly because of the variable degree of purity of the complexes studied.

In a series of hydrocarbon-nickel π complexes, the catalytic activity for the polymerization of butadiene was found to decrease in the sequence (133)   2,6,10-dodecatriene-1,12-diyl $\geq$ 1,5,9-cyclododecatriene > 1,5-cyclooctadiene $\gg$ cyclopentadiene > cyclooctatetraene.

Furthermore, in a homogeneous series of π-allylic nickel halides (169) or haloacetates (170), the overall rate constants appear to vary with the type of substitution of the π-allylic ligands in the order allyl > methallyl > crotyl. This parallels the effective activation energies of the hydrogenolysis of these π-allylic complexes (171), indicative of the respective heats of dissociation of the corresponding binuclear complexes.

Monomers. The influence of π-allylic ligands bound to the transition metal may be related to the specificity of such catalysts toward the polymerizing monomer and also to the practical impossibility to obtain a true copolymerization with most of them; this is the case, for example, with butadiene, since the coordination or the insertion into the chain of a different monomer molecule, e.g., of vinyl type, will modify the structure of the catalytic complex and thereby its relative stability. For instance, as observed long ago with cobalt based Ziegler-Natta catalysts, by adding suitable amounts of monoolefins to butadiene, it is possible to control closely the mean molecular weight of the polybutadiene produced. Indeed, insertion of the monoolefin yields a growing chain that must be bonded to the transition metal by a σ type bond, and the σ-cobalt (or nickel) to carbon bonds are known to be much more unstable at room temperature than π-allylic bonds. In some cases copolymers were obtained, e.g., from styrene with butadiene, using π-allylic nickel derivatives as catalysts. The determination of the monomer reactivity ratios indicated that butadiene is much more reactive than styrene; moreover, the overall polymerization rate, the copolymer intrinsic viscosity, and the cis-1,4-polybutadiene content decreased sharply with an increase of the styrene content in the starting mixture.

Homopolymerization of styrene by π-allylic nickel complexes yields oligomers having an average degree of polymerization ranging from 8 to 10 and, in some cases, only diphenylbutene-1 (216). These results again confirm that the bond formed by nickel with styryl radical is less stable than the π-allylic bond formed from a butadiene molecule (52, 126, 141, 142, 172). This also seems the reason why nickel and cobalt complexes polymerize diolefins other than 1,3-butadiene (e.g., isoprene, 1,2-butadiene, or allene, as shown in Tables 1 and 4), but not α-olefins. This coordinated polymerization of monoolefins is promoted by the presence of chromium, zirconium, and titanium complexes, supporting the aforementioned concept that the reaction is controlled essentially by the transition metal. Monometallic transition metal compounds were even found to be active catalysts for the polymerization of vinyl polar monomers, i.e., methyl methacrylate or acrylonitrile (144). However, these compounds show considerable variations in activity, and not all polar monomers can be polymerized because of their marked tendency to react with the metal complex to form a new organometallic compound.

Additional Ligands. The coordination of additional electron accepting or donating molecules in the coordination sphere of the transition metal may influence not only the stereospecificity, as mentioned above, but also the overall activity.

For instance, higher catalytic activities are achieved by adding various electron acceptors (e.g., metallic salts, iodine, derivatives of haloacetic acids, or quinones) to the π-allyl complexes of transition metals. In either case, this interaction leads to an increase of the effective charge on the transition metal atom and to an increase in its coordinating ability for monomer, together with the enhanced activity in polymerization.

In the presence of metal halides polymerization proceeds in a heterogeneous manner which implies a change not only in catalyst molecular structure but also in phase constitution. A similar initial situation is encountered when adding organic electron accepting molecules but, in this case, the polymerization proceeds essentially in a homogeneous medium, since the catalytic precipitate usually redissolves in the presence of butadiene when the reaction reaches a few percent conversion. For a series of trichloroacetic derivatives added to π-crotyl-nickel chloride, the activity increases as one goes from trichloroacetic acid to chloral and still further to the trichloroacetic acid chloride and nickel salt; this corresponds qualitatively to the increasing electron accepting ability of these ligands. Similar observations were made on substituted quinones; their half-wave electron addition potentials, which are a measure of the electron affinity of the acceptor, are $+0.01$, -0.18, and -0.51 V for p-chloranil, dichloro- or dibromo-p-benzoquinone, and unsubstituted benzoquinone, respectively. This sequence indicates their decreasing complexing ability as well as their decreasing efficiency in enhancing polymerization rates.

On the other hand, addition of electron donor materials usually leads to decreased catalytic activities and cis-1,4 contents, as shown previously. In fact, when bases coordinate with transition metal ions, they donate electrons to hybrid σ orbitals of the metal ions, but at the same time, part of the resultant negative charge on the metal is back-donated to the ligands in order to keep the metal atom nearly neutral. In other words, the d electron density of the metals is expected to be more reduced through back-donation when the basicity of the coordinated ligand is higher. If, as generally admitted, the d orbitals play an important role in the coordination of the monomer and the chain to transition metal atoms, the overall activity as well as the cis content of polybutadiene should depend on the coordination of these bases (e.g., ethers, phosphines, phosphites) to the transition metal atom in the active species, as indicated in Table 3, part H.

4.2.3 *Kinetic Data on Polymerization by* π-*Allylic Complexes of Transition Metals*

The rates of butadiene polymerization by various π-allylic complexes of transition metals are given by

$$-\frac{dM}{dt} = kC^{0.5}M^{1}$$

for bis(π-allylic nickel halides (47, 169, 173) or bis(π-crotylpalladium chloride) (174); by

$$-\frac{dM}{dt} = kC^{1}M^{1}$$

for bis(π-allylic nickel haloacetates) (47, 130) or bis(π-allylic nickel halides) plus electron acceptors (46, 112, 175); and

$$-\frac{dM}{dt} = kC^{2}M^{0}$$

for tris(π-allylic chromium) (46).

The overall activation energy (ranging from 16 to 14.5 kcal/mole) is surprisingly constant for all the bis(π-allylnickel halides) independently of the nature of both halogen and allylic groups; it ranges from 13 to 10 kcal/mole in the case of bis(π-allylnickel haloacetates), and is decreased from 13 to 6 kcal/mole, depending on the nature of the electron acceptor added.

Cryoscopic determinations have shown that π-allylnickel trifluoracetate is normally associated as a binuclear complex, but in the experimental conditions yielding *cis*-1,4-polybutadiene, the coordination of stoichiometric amounts of butadiene converts this binuclear species into a mononuclear one (130, 176). These data indicate that, at least in this case, the active centers for chain growth are not the dimeric molecules of the original complexes but that, as already suggested, the first step in the polymerization reaction is a dissociative equilibrium involving the monomer:

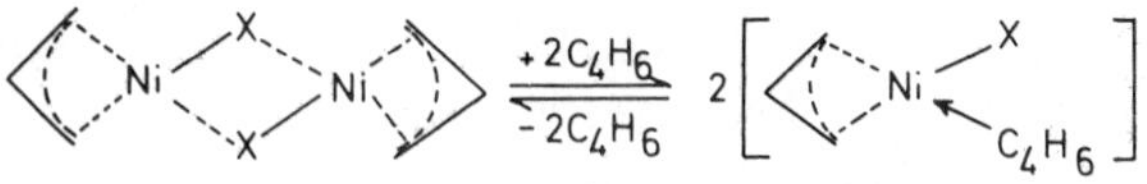

It has also been considered (13) that in some coordination polymerizations, the rate determining step could be the coordination of the olefin to the transition metal. However, as mentioned above, some complementary data have recently shown that the polymerization can also be carried out on a binuclear species, e.g., to obtain the equibinary *cis,trans*-1,4-polybutadiene (213).

In fact, most of the observations made in the course of kinetic studies of coordinated polymerizations by monometallic catalytic complexes fit the scheme of dissociation of the initial complex, followed by propagation through an iterative cis rearrangement, and termination by a variety of reactions. Using labeled atoms, some details of this termination mechanism in the coordinated polymerization with π-allylic complexes have been elucidated. In principle, the following routes may be proposed for the termination of growing chains: thermal deactivation by disproportionation, coupling in the deactivation stage, termination onto monomer with chain transfer, or addition of the growing chain to compounds from the allylic substrate (e.g., hexadiene) which might be present in the reaction mixture. Analysis of the experimental data obtained with ^{14}C-labeled π-crotyl complexes shows that the last route is not realized under the usual polymerization conditions. The polymers formed contained one or two crotyl radicals from the catalyst in every macromolecule, supporting the two first routes considered (174, 177). Chain transfer reactions that determine the molecular weight of the polymer produced apparently occur as follows:

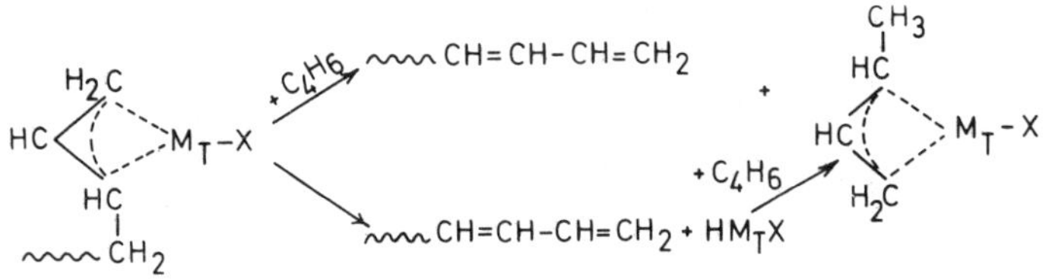

and the formation of two conjugated double bonds at the end of the polymer chains has been confirmed by reaction with maleic anhydride (47).

4.3 Possible Stereospecific Polymerization Mechanisms

The enormous amount of sometimes contradictory experimental data pertaining to the same polymerization process, increasingly indicates the difficulty of describing a universal mechanism for stereospecific synthesis. The study of the intimate behavior of transition metal π complexes is of interest because most of these complexes are relatively stable and well-defined species; in addition, these structures, mainly of the π-allyl type, are probably general intermediates in the stereospecific polymerization of conjugated dienes.

In fact, the stereoregularity of polybutadiene should be determined by the conformation that the butadiene molecule or the end of the growing polymeric chain assumes when coordinated to the catalyst metal atom in the transition state of the polymerization reaction. Many reaction mechanisms find it difficult to explain cis-trans isomerism.

One of the most comprehensive hypotheses was proposed several years ago by Arlman (178) as an extension of Cossee's scheme for α-olefin polymerization on heterogeneous catalysts. Its salient features may be summarized as follows. When the transition metal atom carries two vacancies, the monomer is coordinated as a bidentate ligand by its two double bonds, and is incorporated in the growing chain R (by the cis rearrangement reaction) in the same configuration, i.e., as a cis-1,4 unit. If only one vacancy is available, the coordination of the same monomer by one double bond allows it to take its more probable transoid conformation, giving rise to a trans-1,4 unit. These views have been substantiated by the preparation of *trans*-1,4-polybutadiene on α-TiCl$_3$, and of a resolvable mixture of both *trans*-1,4- and *cis*-1,4-polybutadiene on β-TiCl$_3$ where both types of sites coexist. The formation of 1,2 units would be dependent on the relative distance from the C_2 or the C_4 carbon of the monocoordinated monomer to the first carbon of the R group carried on the same transition metal (growing chain). A shorter C_2–R distance (depending on metal radius and complex geometry) would favor this vinyl structure.

It should be stressed that the whole mechanism involves a σ bond between chain and metal in the transition state, whether or not this chain is stabilized through a π-allyl bond (involving its first double bond) between every insertion step. Indeed, the recently reviewed (206) σ-π rearrangements of organotransition metal compounds are well known to play an important role in many processes, including polymerization, isomerization, and hydroformylation of unsaturated hydrocarbons.

Moreover, in elucidating the mechanism of 1,3-diene insertion into the allylic palladium bonds of the complexes (π-allyl-PdX) by means of NMR spectra, Hughes and Powell (217) have confirmed that the 1,3-diene, acting as a monodentate ligand, coordinates to the palladium through the least substituted double bond to give a σ-allylic intermediate [(σ-allyl) (diene)PdX]. The key step of the successive electrocyclic insertions of

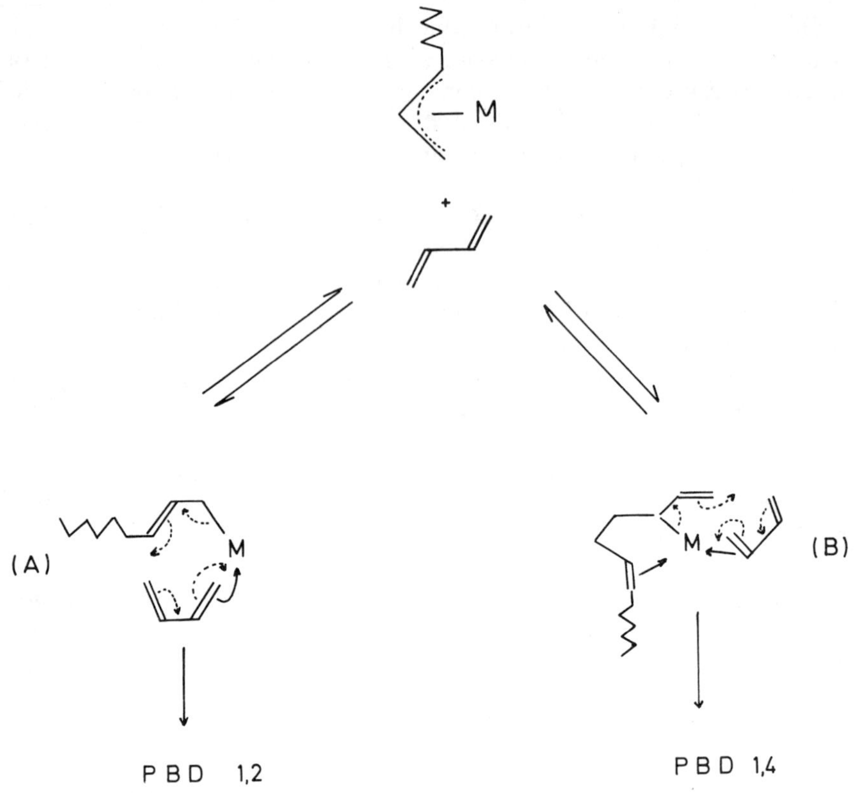

Figure 15

1,3-dienes into allyl-transition metal bonds is a ready $\pi \to \sigma \to \pi$ process via an intermediate such as A in Fig. 15, providing a plausible mechanism for 1,2 polymerization of 1,3-dienes. It should be noted, however, that this type of mechanism differs essentially from a cis rearrangement process; it is more like an allylic transposition reaction. Changes in catalytic systems which lead to an increased number of available coordination sites were shown to generally change the mode of polymerization from 1,2 to 1,4. The electrocyclic insertion mechanism may thus conceivably lead to a 1,4 polymeric chain, according to the Hughes-Powell report, via anchimerically assisted formation of an intermediate (B). However, this interpretation raises several difficulties since it implies less probable types of substitution and coordination for the allylic group and the monomer molecule.

In the case of polymerization by π-allyl complexes, another mechanism frequently proposed in the past (see, e.g., refs. 47, 101, 173, 179) was associated essentially with the isomerism of the growing chain end in the form of a π-allyl complex; it was speculated that the "syn" form would favor trans polymerization and the "anti" form cis polymerization, the anti form being favored by coordination to the metal of the double bond nearest the π-allylic polymer end. However, NMR data obtained in

" syn" form → trans - 1,4   "anti " form → cis - 1,4

solution seem to exclude the coordination of the penultimate double bond, and to indicate that only the syn form exists as a stable complex in both substituted π-allylnickel trifluoroacetate, chloride and iodide which yield high *cis*-1,4- and high *trans*-1,4-polybutadienes, respectively (46,

133, 180, 215). These observations suggest the lack of correlation between the isomerism of the π-allylic ligand and that of the resulting polymer, although as an unstable transition state, an anti π-allyl complex that would undergo polymerization prior to rapid isomerization to the stable syn form is not necessarily ruled out.

Another tentative interpretation was based on the dimeric or monomeric form of the catalytic species in solution (46, 126). At least in the case of pure cis-1,4 or trans-1,4 polymerization in the presence of π-allylnickel haloacetates, the kinetic and cryometric data exclude that a polymer chain grows on a binuclear catalytic complex, so the changes of microstructure cannot be explained on this basis.

In summary, the data currently available on the homogeneous polymerization of conjugated diolefins by π complexes, increasingly support Arlman's and Hughes and Powell's concepts that the polymerization proceeds through a σ type of transition complex; this mechanism does not exclude the coordination of the growing chain to the metal by a π-allyl bond for most of the time, but it assumes that the allylic structure passes into an alkyl one under the influence of the coordinated monomer, prior to its insertion into the chain. This mechanism implies also that the σ-π isomerization does not modify the potentially cis structure of the last incorporated monomer unit, a point that is still open to discussion. The formation of cis or trans units will be determined essentially, as in the case of propagations on α- or β-TiCl$_3$, by the capacity of a transition metal to coordinate both or only one of the two double bonds of the diene molecule. The formation of a predominantly cis structure of polybutadiene in the presence of cobalt or nickel catalysts would be rather easy, especially so since the distance between the first and fourth carbon atoms in the cis configuration of free butadiene is practically the same, i.e., 2.87 Å, as in the molecule coordinated by both double bonds:

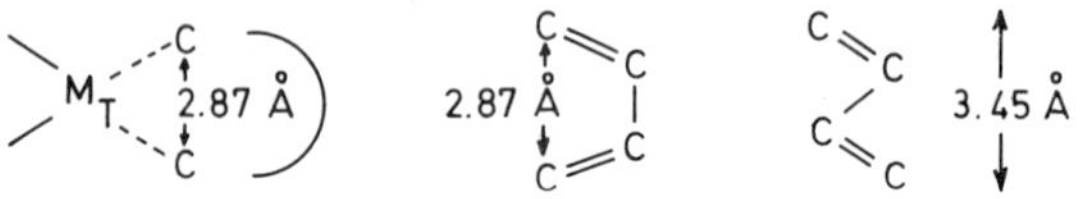

Moreover, it seems quite probable that, as already suggested (181) the coordination of the monomer through both its double bonds occurs in fact in two consecutive steps. Chain propagation results, as indicated earlier, from a sequence of iterative events, involving formation of a π complex between the catalyst and the diene, isomerization of the π-allyl (if any) to a σ complex, and insertion of the coordinated monomer into the σ-alkyl to metal bond to give a complex similar to the initial one, with further coordination of another monomer molecule. Within the framework of this mechanism, implying six- to four-membered transition states, the structure of the initial complex between the catalyst and the diene determines the microstructure of the polymer chain.

The presence of strong electron accepting ligands (anions or organic molecules) should favor this chelation, increasing the cis-1,4 content as experimentally observed. The formation of 1,2 units might take place when monomer coordination involves only one double bond and when simultaneously the atomic distances favor the 1,2 binding, or when the type of conversion of the π-allylic complex into the σ complex determines 1,2 (instead of 1,4) stereoregulation:

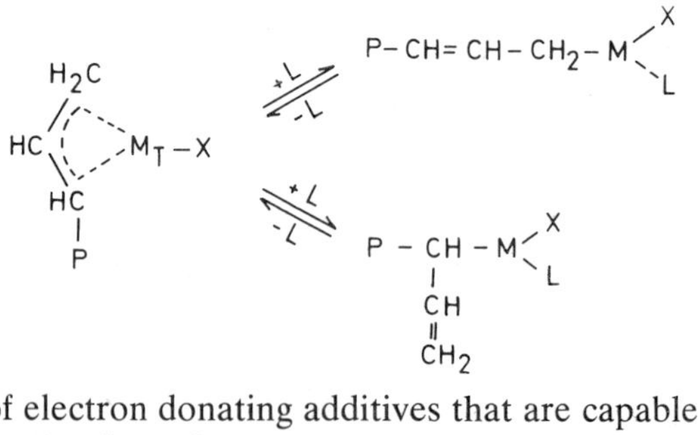

The action of electron donating additives that are capable of competing with the bidentate diene for a site in the coordination sphere of the catalyst becomes easily understandable. Only monodentate complexes with monomer may be formed under those conditions and this leads to trans-1,4 or 1,2 polymers. The inductive influence of the counteranion on the cis-trans isomerism can also be accounted for by admitting that only the sufficiently electronegative anions are able to promote quantitatively, perhaps through a "trans" effect, the coordination of both butadiene double bonds.

This propagation reaction proceeds alternatively on two different sites of the coordination complex. This is most probably the key to the interpretation of the peculiar effect of specific π ligands promoting the formation of equibinary polydienes (see Section 4.1.3). A different isomer is alternatively inserted in the chain on each side, or otherwise the whole complex switches from one geometry to the other after every insertion step.

In conclusion, and accounting for most of the experimental data currently available, the mechanism of formation of stereoregular polybutadienes in solution can be tentatively summarized as follows:

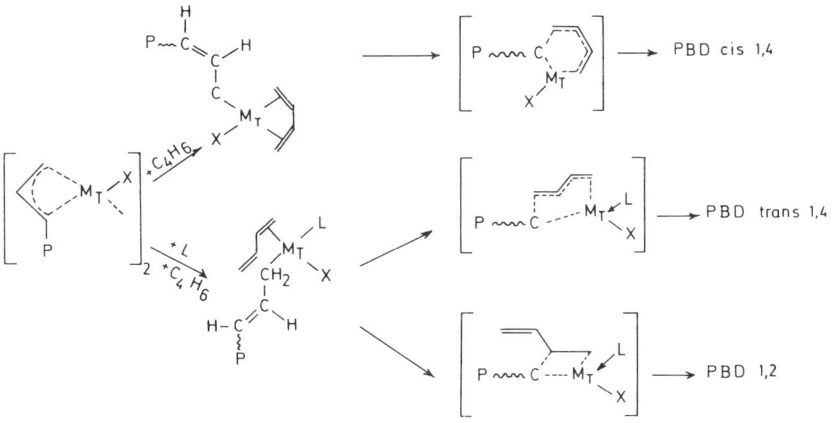

5 CURRENT PROGRESS IN RELATED AREAS

5.1 Basic Catalytic Structures

The relationships between the catalyst's structure and its activity and stereospecificity have been further elucidated by investigations of the coordination polymerization of heterocyclic monomers, such as the oxiranes and thiiranes, to high molecular weight polymers.

The first efficient catalysts were obtained either by reaction of the monomer with ferric chloride (182) followed eventually by subsequent hydrolysis or by controlled hydrolysis of organometallic derivatives (183, 184) (mostly zinc and aluminum alkyls).

These results strengthened the idea that the responsible catalytic species were essentially composed of –M–O–M– groupings. Indeed, such species have been directly synthesized (185) by condensing a metal alcoholate with the acetate of a different metal, in a 2:1 ratio, e.g.,

$$2Al(OR)_3 + Zn(OAc)_2 \rightarrow (RO)_2Al—O—Zn—O—Al(OR)_2 + 2ROAc \uparrow$$

The new bimetallic μ-oxoalkoxides obtained are highly active for the homogeneous ring opening polymerization of oxiranes and thiiranes, which confirms that the presence of a metal to carbon bond is not essential to this type of polymerization. These results were confirmed by preparing similar species (186) with comparable activities through hydrolysis of a double metal alcoholate (Meerwein complex), e.g.,

$$[Al(OR)_3]_2 \cdot Zn(OR)_2 + 4H_2O \rightarrow (RO)_2Al—O—Zn—O—Al(OR)_2$$
$$+ 4ROH\uparrow$$

Such catalysts are very active and rank among the best for preparation of polypropylene oxide elastomers (189). Both the activity and the stereospecificity of these compounds depend in a very specific manner on the nature of M_1, M_2, X, and Y in the general structure

$$X_nM_1—O—M_2—O—M_1X_n$$
$$\underset{Y_m}{|}$$

as well as on the nature of the solvent and the degree of association of the catalyst.

A comparison of the polymerization rates of different types of oxiranes, together with the competitive nature of the polymerization process, led (187) to a coordination mechanism that can be expressed as shown in Fig. 16. This scheme reveals striking similarities to the one proposed by Cossee for olefin polymerization: foreshadowing of the polymer chain by the OR group carried on the metal, and an alternating flip-flop mechanism governed by an electronic rearrangement where both chain and monomer exchange places on two different coordination positions without any important nuclear displacement. The similarity persists despite the different nature of the bonds involved, and the fact that at least two metal atoms seem necessary (188) to account for the reaction characteristics.

5.2 Catalysts with Bifunctional Behavior

A number of attempts have been made to impart to stereospecific catalyst systems a still more elaborate behavior.

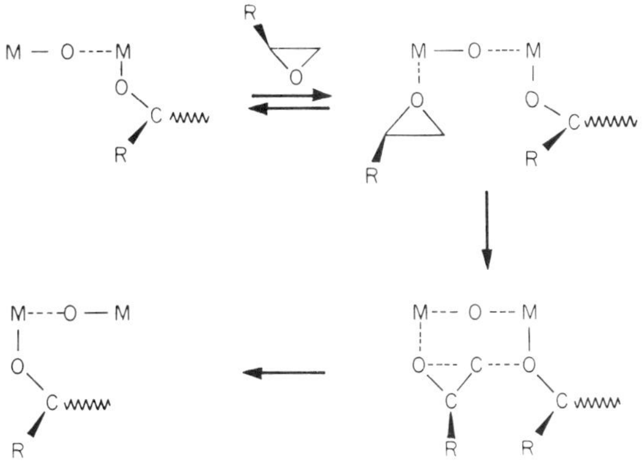

Figure 16

One of the earliest and most interesting realizations was to promote "stereoelectivity," i.e., the ability of a complex catalyst to favor the preferential incorporation in the growing chain of one of the optical isomers of a racemic monomer. This asymmetric synthesis is difficult to perform with high specificity in the polymerization of α-olefins (190), but has been more successfully achieved for oxirane and thiirane polymerization (191, 192), by substituting asymmetric groups on the catalyst, as close as possible to the determinant metal atom. A similar result has been obtained in organic synthesis, e.g., in homogeneous hydrogenation by rhodium complexes carrying an asymmetric phosphine (193).

Several research groups have tried to control the sequential composition of stereoregular polymers, particularly in copolymerization reactions. It would be of the utmost interest to be able to prepare block copolymers with these coordination catalysts since anionic systems, despite their outstanding efficiency in this respect, do not polymerize certain monomers (α-olefins in particular) and cannot give rise to given types of stereoregularity (e.g., pure cis-1,4-polybutadiene). In the case of polyolefins, although the products have not been thoroughly characterized in terms of sequential (194) molecular weight, preparation of block copolymers of ethylene and propylene has been claimed as well as an absence of termination in the syndiotactic polymerization of propylene (156). Since polar vinyl monomers are sometimes difficult to polymerize with coordination catalysts, a two-step procedure has also been devised (195–197) involving chemical transformation of the "living" polyolefin end into a reactive group (usually a peroxide) capable of initiating a radical type of formation of another (polar) sequence; this procedure usually yields mixtures of different products. Branched polymers have been prepared either by a simple "jumping" reaction (198) or by grafting techniques (207) where, for example, double bonds in a polydiene are reacted by metathesis with cyclopentene, thus promoting the formation of grafted polypentenamer side chains.

Even if the formation of block polymers by coordination catalysis with transition metals is not yet fully mastered, this is a very challenging and interesting area which could give rise to important developments in the near future, particularly in view of possible supermolecular organizations (199).

Finally, a third and exciting avenue was opened a few years ago, when it was shown that functions other than polymerization could be imparted simultaneously to the same catalytic system. For instance (200), a trimetallic catalyst, involving an aluminum alkyl, titanium chloride, and a nickel salt, has been used to polymerize internal olefins into poly-α-olefins, the nickel component ensuring a rapid isomerization of the internal olefin into the terminal α isomer which could be polymerized by the titanium moiety of the system. The polymers obtained have properties similar to those of the regular products arising from classical Ziegler type polymerization of the corresponding α-olefins.

5.3 Other Polymerization Mechanisms Involving Coordination Complexes

A long time ago, it was shown by Bier (201) that the coordination of an olefin such as ethylene to a metal salt such as silver nitrate could enhance its reactivity and allow radical type polymerization under rather mild conditions to a practically linear product. Other examples of this behavior have been described recently, such as the low activation energy polymerization of butadiene into a crystalline trans-1,4 polymer in the presence of rhodium chloride, and into a 1,4-trans-1,2 equibinary polymer in the presence of silver nitrate, in aqueous emulsion under γ-irradiation (202).

The thermal or radical induced polymerization of monomers coordinated to metal salts has also given rise to a very interesting research area. Gaylord (5, 203), among others, has attempted to rationalize several results indicating that in the presence of metal salts displaying an acceptor character, different monomers could copolymerize, either spontaneously (thermal activation) or in the presence of radicals (peroxides, radiations), to yield 1:1 alternating copolymers; this is the case for the copolymerization of butadiene and acrylonitrile in the presence of zinc chloride, as well as propylene-acrylonitrile with dichloroethylaluminum. Gaylord has claimed that the polar monomer coordinates to the Lewis acid with a resultant increase in its electron accepting ability. This complexed monomer participates in an electron transfer process with the more

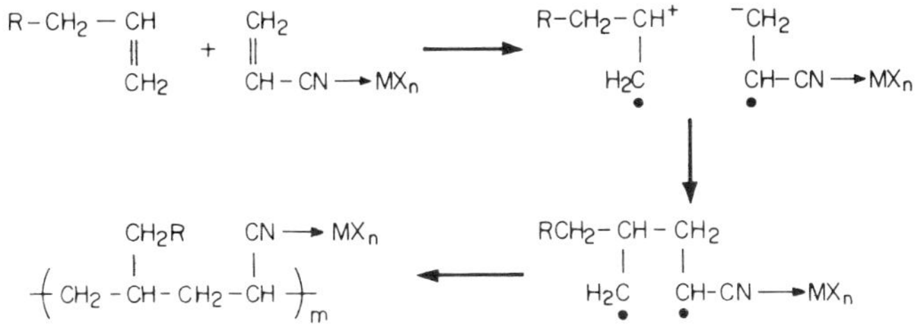

Figure 17

electron donor monomer, such as an olefin or conjugated diene, to form a charge transfer complex that could be considered a diradical species. Chain growth involves spontaneous or radical initiated homopolymerization of this Lewis acid activated diradical charge transfer complex, as illustrated in Fig. 17.

The intermediate complex explains the constancy of the polymeric composition irrespective of the monomeric ratio, mode of initiation, and reaction conditions, when the metal salt is present. (The metal is no longer required when the two monomers have sufficiently different donor-acceptor properties, for instance, the styrene-maleic anhydride pair). This reaction mechanism, which may have some formal similarities to the Diels-Alder reaction, has not been convincingly substantiated. In a more detailed analysis of these reactions, Zubov (218) has concluded that they can be described in terms of common kinetic schemes of addition radical polymerization, despite "anomalous" features for radical processes. Complex formation simply changes the relative reactivities by modifying resonance and polar characteristics of the monomers and radicals, or by changing the structure of the transition complex.

This type of structural control opens the way to preparing a broad class of new products, whose properties can be evaluated in terms of a regular arrangement of the two different monomeric units. Some of them, e.g., the alternating equimolar butadiene-acrylonitrile and butadiene-propylene copolymers, have been studied as specialty rubbers in view of their good overall set of properties (204, 222). The fact that under certain experimental conditions, the chains might grow without termination is still an additional attractive feature of these reactions.

6 CONCLUSION: TRENDS AND DIRECTIONS OF RESEARCH

We now have a much clearer view of the active center structure and the stereospecific propagation mechanism in coordination polymerization.

Very plausible reaction mechanisms have been elaborated which involve complexes of definite structure, and which are able to explain in detail at the atomic level the stereospecific insertion of the monomer into the growing chain. All these descriptions, often based on direct experimental evidence indicating not only the structure of the complex but also its behavior in the polymerization medium, stress the same conclusion: The active center is a metal complex, carrying different ligands and the growing polymer chain (attached on two coordination positions) and offering to the monomer to be incorporated, one or more suitable free coordination positions. Both the activity and the stereospecificity of this complex are determined by its geometry and by the electronic distribution dictated by the presence of different ligands, such as another monomer and/or a second metallic entity.

Despite the tremendous variety of catalytic ingredients used in these reactions, all of them apparently proceed through the common general type of structure outlined above. This means that the responsible species may be formed in situ by numerous and very different reaction paths. In other words, the coordination polymerization of very different monomers by very different catalysts may proceed through the same basic mechanism, which in fact explains a much broader class of catalytic processes involving much of organic chemistry. We have here a rewardingly unifying concept which is of great help in interpreting and forecasting catalytic behavior. This is the case, for example, for the epoxide/oxoalkoxide as well as the diolefin/π-allyl-metal-X systems. Some unity between the behavior of transition metal complexes and lithium derived catalysts might also be expected (3, 4).

New directions include the exploratory search for catalysts able to polymerize new monomers stereospecifically or known monomers into polymers of yet unknown stereoregular structure such as 1,3 polymerization of propene, controlled ring opening polymerization, new alternating copolymers, and equibinary polydienes. There is no doubt that unexpected and interesting results will be obtained.

The use of already known coordination catalysts to perform more involved polymerization reactions, in particular to produce di- or multi-sequential block copolymers, will certainly yield new products of great interest, in particular for the study of materials organized on a super-molecular scale, access to which has been restricted until now to the use of anionic catalysts. Other related interesting fields in rapid development are the studies of the mechanism by which heterogeneous Ziegler-Natta catalysts control the direct growth of fibrillar structures, and also by which some inorganic or polymer-supported coordination catalysts are able to increase both the activity (parts per million range) or lifetime of the catalytic species, and its specificity or electivity (219).

The last step in the elucidation of polymerization control by catalytic structure has yet to be accomplished in many cases. We have a rather good "chemical" picture of the situation, in terms of the well-defined composition of the active complexes as well as of the gross electronic influence of the ligands and metal involved. However, it appears that a significant part of this control is due to the stereogeometry and/or the fine electronic balance in the complex *in the polymerizing mixture*. It is highly probable that this knowledge will be gained only through the simultaneous application of several elaborate physical methods.

REFERENCES

1. K. Ziegler, in H. Zeiss, ed., *Organometallic Chemistry*, Reinhold, New York, 1960.
2. P. Teyssié, *Rev. Gén. Caoutch. Plast.*, **46**, 851 (1969).
3. M. Morton and A. D. Sanderson, Symposium on Stereoregular Polymerization, CIC/ACS Meeting, Toronto, May 1970.
4. S. Bywater, Symposium on Stereoregular Polymerization, CIC/ACS Meeting, Toronto, May 1970.
5. N. G. Gaylord and A. Takahashi, *J. Polymer. Sci.*, **B6**, 749 (1968).
6. D. Ketley, *Stereochemistry of Macromolecules*, Vols. I and II, Dekker, New York, 1969, and references cited therein.
7. V. A. Zakharov, et al., *Makromol. Chem.*, **175**, 3035 (1974).
8. K. Vesely et al., *J. Polym. Sci.*, **C16**, 417 (1967).
9. E. S. D'yachkovskii et al., *Polym. Sci., USSR*, **8**, 336 (1966).
10. E. S. D'yachkovskii et al., *J. Polym. Sci.*, **C16**, 2333 (1967).
11. A. K. Shilova et al., *Polym. Sci., USSR*, **9**, 903 (1967).
12. E. A. Grigoryan et al., *Polym. Sci. USSR*, **9**, 1372 (1967).
13. I. Pasquon et al., *J. Polym. Sci.*, **C16**, 2501 (1967).
14. G. Henrici-Olivé and S. Olivé, *Adv. Polym. Sci.*, **6**, 421 (1969).
15. G. Henrici-Olivé and S. Olivé, *Angew. Chem., Int. Ed.*, **7**, 821 (1968).
16. G. Henrici-Olivé and S. Olivé, *Makromol. Chem.*, **121**, 70 (1969).
17. G. Henrici-Olivé and S. Olivé, *J. Polym. Sci.*, **C22**, 965 (1969).
18. A. Zambelli et al., *Chim. Ind. (Milan)*, **46**, 1464 (1964).
19. F. Dawans and P. Teyssié, *Eur. Poly. J.*, **5**, 541 (1969).
20. W. Cooper and G. Vaughan, *Progress in Polymer Science*, Vol. I, Pergamon Press, Oxford, 1967.
21. J. Boor, Jr., *Ind. Eng. Chem., Res. Dev.*, **9**, 437 (1970).
22. J. P. Kennedy and E. G. M. Tornqvist, *Polymer Chemistry of Synthetic Elastomers*, Wiley-Interscience, New York, 1969.
23. A. Rawe, *Organic Chemistry of Macromolecules*, Dekker, New York, 1967, p. 111.
24. R. E. Rinehart et al., *J. Am. Chem. Soc.*, **83**, 4864 (1961).
25. R. E. Rinehart et al., *J. Am. Chem. Soc.*, **84**, 4145 (1962).
26. A. J. Canale et al., *Chem. Ind.*, **1962**, 1054.

27. R. S. Berger and E. A. Youngman, *J. Polym. Sci.*, **A2**, 357 (1964).
28. P. Teyssié, *C. R. Acad. Sci. (Paris)*, **256**, 2846 (1963).
29. P. Teyssie and R. Dauby. *Bull. Soc. Chim. France*, **1965**, 2842.
30. M. Morton et al., *Rubber Plast. Age*, **46**, 404 (1965).
31. R. Dauby et al., *J. Polym. Sci.*, **C16**, 1989 (1967).
32. M. Morton and B. Das, *J. Polym. Sci.*, **C27**, 1 (1969), and references cited therein.
33. R. Cramer, *J. Am. Chem. Soc.*, **89**, 1633 (1967), and references cited therein.
34. K. Hiraki and H. Hiroi, *J. Polym. Sci.*, **B7**, 449 (1969).
35. G. Natta et al., *J. Polym. Sci.*, **B2**, 349 (1964).
36. G. Natta et al., *Makromol. Chem.*, **81**, 253 (1965).
37. F. W. Michelotti and W. P. Keaveney, *J. Polym. Sci.*, **A3**, 895 (1965).
38. R. E. Rinehart and H. P. Smith, *J. Polym. Sci.*, **B3**, 1045 (1965).
39. R. E. Rinehart, *J. Polym. Sci.*, **C27**, 7 (1969).
40. A. D. Ketley and J. A. Broatz, *J. Polym. Sci.*, **B6**, 341 (1968).
41. P. Pino, *Adv. Polym. Sci.*, **4**, 393 (1966), and references cited therein.
42. G. Allegra et al., *Makromol. Chem.*, **90**, 60 (1966).
43. F. X. Werber et al., *J. Polym. Sci.*, **A6**, 743, 755 (1968).
44. G. Mazzanti, *Chem. Ind. (London)*, **1969**, 1204, and references cited therein.
45. W. S. Anderson, *J. Polym. Sci.*, **A1**, 5, 429 (1967).
46. B. A. Dolgoplosk and E. I. Tinyakova, *Izv. Akad. Nauk SSSR, Ser. Khim.*, 1970, 344, and references cited therein.
47. V. A. Yakovlev et al., *Vysokomol. Soedin.* **A11**, 1645 (1969), and references cited therein.
48. K. I. Ueda et al., *Kogyo Kagaku Zasshi*, **66**, 1103 (1963).
49. T. Otsu et al., *J. Polym. Sci.*, **B5**, 835 (1967).
50. T. Otsu and M. Yamaguchi, *J. Polym. Sci.*, *A-1*, **7**, 387 (1969).
51. U.S. Pat. 3,739,003 (to Institut Français du Pétrole, 1973).
52. F. Dawans et al., *J. Polym. Sci.*, **B10**, 493 (1972).
53. S. Kunichika et al., *Kogyo Kagaku Zasshi*, **72**, 1814 (1969).
54. J. Boor, Jr., *J. Polym. Sci.*, **3**, 995 (1965).
55. French Pat. 1,361,677 (to United States Rubber Co., 1964).
56. U.S. Pat. 3,414,555 (to Int. Synthetic Rubber Co., 1964).
57. U. A. Kormer et al., Neth. Appl. Pat. 69, 13 450 (1969).
58. U.S. Pat. 3,021,319 (to Hercules Powder Co., 1962).
59. Belgian Pat. 595,685 (to Union Carbide Corp., 1961).
60. K. Kuhlein and K. Clauss, *Makromol. Kolloq.*, Freiburg, 1969.
61. Canadian Pat. 620,209 (to Farbwerke Hoechst, A. G., 1961).
62. U.S. Pat. 3,037, 971 (to Hercules Powder Co., 1962).
63. C. Beerman and H. Bestian, *Angew. Chem.*, **71**, 618 (1959).
64. G. L. Karapinka et al., *J. Polym. Sci.*, **50**, 143 (1961).
65. British Pat. 856,434 (to National Lead Co., 1960).
66. U.S. Pat. 2,951,796 (to Union Carbide Corp., 1960).
67. U.S. Pat. 3,000,870 (to Montecatini, 1960).

68. British Pat. 881,004 (to Montecatini, 1958).
69. G. Natta et al., *Chim. Ind. (Milan)*, **39**, 1032 (1957).
70. U.S. Pat. 2,992,212 (to Hercules Powder Co., 1961).
71. K. Yomo et al., *Kogyo Zasshi*, **68**, 352 (1965).
72. French Pat. 1,437,892 (to Farbwerke Hoechst, A. G., 1965).
73. French Pat. 1,540,898 (to Idemitsu Kosan Co., 1968).
74. U. Giannini and U. Zucchini, *Chem. Commun.*, **1968**, 940.
75. U. Zucchini et al., *Chem. Commun.*, **1969**, 1174.
76. S.A. Pat. 59/2269 (to Montecatini, 1959).
77. French Pat. 1,249,666 (to National Lead Co., 1961).
78. Belgian Pat. 608,604 (to Montecatini, 1962).
79. British Pat. 889,591 (to Montecatini, 1962).
80. D. F. Herman and W. K. Nelson, *J. Am. Chem. Soc.*, **75**, 3877, 3882 (1953).
81. A. Yamamoto and S. Ikeda, *J. Am. Chem. Soc.*, **89**, 5989 (1967).
82. U.S. Pat 3,405,112 (to Japan Synthetic Rubber Co., Ltd., 1968).
83. S. Otsuka et al., *J. Am. Chem. Soc.*, **87**, 3017 (1965).
84. S. Otsuka et al., *Eur. Polym. J.*, **3**, 73 (1967).
85. F. Dawans and P. Teyssié, *C. R. Acad. Sci. (Paris)*, **261**, 4097 (1965).
86. F. Dawans and P. Teyssié, *J. Polym. Sci.*, **B3**, 1045 (1965).
87. F. Dawans and P. Teyssié, *C. R. Acad. Sci. (Paris)*, **263C**, 1512 (1966).
88. U.S. Pat. 3,400,115 (to Institut Français du Pétrole, 1968).
89. L. Porri et al., *J. Polym. Sci.*, **B5**, 629 (1967).
90. T. Arakawa, *Kogyo Kagaku Zasshi*, **70**, 1738 (1967).
91. T. G. Golenko et al., *Izv. Akad. Nauk SSSR, Ser. Khim.*, **1968**, 2271.
92. B. D. Babitskii et al., *Dokl. Akad. Nauk SSSR*, **161**, 836 (1965).
93. Jap. Pat. 25,007 (1965) and 19,833 (1966) (to Japan Synthetic Rubber Co., Ltd.)
94. J. P. Durand et al., *J. Polym. Sci.*, **B6**, 757 (1968).
95. G. Wilke et al., *Angew. Chem., Int. Ed.*, **5**, 151 (1966), and references cited therein.
96. S. Otsuka and M. Kawakami, *Angew. Chem.* **75**, 858 (1963).
97. B. D. Babitskii et al., *Dokl. Akad. Nauk SSSR*, **165**, 95 (1965).
98. German Pat. 1,209,752 (to Badische Anilin und Soda Fabrik A. G., 1966).
99. German Pat. 1,269,811 (to Badische Anilin und Soda Fabrik A. G., 1968).
100. British Pat. 1,138,817 (to Stamicarbon N. V., 1969).
101. Jap. Pat. 17,687 (to Bridgestone Tire Co., Ltd., 1962).
102. S. Otsuka and M. Kawakami, *Kogyo Kagaku Zasshi*, **68**, 874 (1965).
103. U.S. Pat. 3,255,170 (to United States Rubber Co., 1966).
104. French Pat. 1,416,412 (to Badische Anilin und Soda Fabrik A. G., 1964).
105. British Pat. 1,027,867 (to Badische Anilin und Soda Fabrik A. G., 1966).
106. Jap. Pat. 6960 (to Japan Synthetic Rubber Co., Ltd., 1968).
107. W. Strohmeier and P. Hartmann, *Z. Naturforsch.*, **21b**, 1119 (1966).
108. W. Strohmeier and P. Hartmann, *Z. Naturforsch.*, **24b**, 777 (1969).
109. F. Dawans and P. Teyssié, *Ind. Eng. Chem. Res. Dev.*, **10**, 261 (1971).
110. French Pat 1,410,430 (to Studiengesellschaft Kohle, 1964).

111. G. Wilke, *Angew. Chem., Int. Ed.*, **2,** 105 (1963).

112. G. Wilke, 8th International Conference on Coordination Chemistry, Vienna, September 1964.

113. E. I. Tinyakova et al., International Symposium on Macromolecular Chemistry, Budapest, 1969.

114. French Pat. 1,470,794 (to Farbwerke Hoechst A. G., 1966).

115. V. A. Yakovlev et al., *Dokl. Akad. Nauk SSSR*, **187,** 354 (1969).

116. S. V. Lebedev, French Pat. 1,550,097 (to All-Union Scientific Research Institute of Synthetic Rubber, 1967).

117. U.S. Pat. 3,497,488 (to Institut Français du Pétrole, 1970).

118. J. P. Durand et al., *J. Polym. Sci.*, **B5,** 785 (1967).

119. French Pat. 1,551,887 (to SNAM Progetti S.p.a., 1967).

120. G. Lugli et al., *Inorg. Chim. Acta*, **3,** 151 (1969).

121. J. P. Durand and P. Teyssié, *J. Polym. Sci.*, **B6,** 299 (1968).

122. French Pat. 1,590,083 (to Institut Français du Pétrole, 1970).

123. U.S. Pat. 3,542,695 (to Institut Français du Pétrole, 1970).

124. F. Dawans and P. Teyssié, *J. Polym. Sci.*, **B7,** 111 (1969).

125. L. Porri et al., *Chim. Ind. (Milan)*, **46,** 426 (1964).

126. L. Porri et al., *J. Polym. Sci.*, **C16,** 2525 (1967).

127. T. Matsumoto et al., *J. Polym. Sci.*, **B6,** 869 (1968).

128. French Pat. 1,556,962 (to Institut Français du Pétrole, 1969).

129. V. I. Skoblikova at al., *Vysokomol. Soedin.*, **B10,** 590 (1968).

130. J. C. Maréchal et al., *J. Polym. Sci.*, **A8,** 1993 (1970).

131. V. A. Kormer et al., *Dokl. Akad. Nauk SSSR*, **180,** 665 (1968).

132. T. Matsumoto et al., *J. Polym. Sci.*, **B7,** 541 (1969).

133. J. P. Durand et al., *J. Polym. Sci.*, **A8,** 979 (1970).

134. E. V. Kristalnyi et al., *Vysokomol. Soedin.*, **A12,** 836 (1970).

135. J. P. Durand and F. Dawans, *J. Polym. Sci.*, **B8,** 743 (1970).

136. French Pat. 1,561,515 (to Imperial Chemical Industries, Ltd., 1969).

137. Jap. Pat. 13,091 (to Idemitsu Kosan Co, Ltd., 1967).

138. British Pat. 1,091,296 (to Imperial Chemical Industries, Ltd., 1968).

139. British Pat. 1,184,592 and 3 (to Idemitsu Kosan Co. Ltd., 1970).

140. Jap. Pat. 9925 (to Nippon Gosei Gomu K.K., 1970).

141. L. I. Redkina et al., *Dokl. Akad. Nauk SSSR*, **186,** 397 (1969).

142. L. I. Redkina et al., International Symposium on Macromolecular Chemistry, Budapest, 1969.

143. D. G. H. Ballard et al., *J. Chem. Soc. (B)*, **1968,** 1168.

144. V. A. Kormer et al., *Dokl. Akad. Nauk SSSR*, **185,** 873 (1969).

145. P. Cossee, *J. Catal.*, **3,** 80 (1964).

146. E. J. Arlman, *J. Catal.*, **3,** 89 (1964).

147. E. J. Arlman and P. Cossee, *J. Catal.*, **3,** 99 (1964).

148. P. Cossee, in D. Ketley, ed., *Stereochemistry of Macromolecules*, Dekker, New York, 1967, p. 145.

149. L. A. M. Rodriguez and H. M. Van Looy, *J. Polym. Sci.*, **4**, 1971 (1966).

150. P. Cossee, *Tetrahedron Lett.*, **17**, 12 (1960).

151. L. A. M. Rodriguez, et al., *J. Polym. Sci.*, **4**, 1905 (1966).

152. J. Chatt and L. A. Duncanson, *J. Chem. Soc.*, **1953**, 2939.

153. P. Cossee et al., 4th International Congress on Catalysis, Moscow, 1968.

154. D. Hoeg, in A. Ketley, ed., *Stereochemistry of Macromolecules*, Dekker, New York, 1967, p. 95.

155. J. Boor, Jr., and E. A. Youngman, *J. Polym. Sci.*, *A-1*, **4**, 1861 (1966).

156. A. Zambelli et al., *Makromol. Chem.*, **112**, 160 (1968).

157. B. Hessett and P. G. Perkins, *Chem. Ind. (London)*, **1970**, 747.

158. C. G. Overberger and G. W. Halek, *J. Polym. Sci.*, *A-1*, **8**, 359 (1970).

159. V. Desreux, personal communication.

160. H. Yuki et al., *J. Polym. Sci.*, *A-1*, **7**, 1933 (1969).

161. T. G. Fox et al., *J. Am. Chem. Soc.*, **80**, 1768 (1958).

162. R. St John Manley, *J. Macromol. Sci.-Phys.*, **B2**, 501 (1968), and references cited therein; Symposium on Stereoregular Polymerization, CIC/ACS Meeting, Toronto, May 1970.

163. J. Y. Gutman and J. E. Gillet, Symposium on Stereoregular Polymerization, CIC/ACS Meeting, Toronto, May 1970.

164. R. H. Marchessault et al., Symposium on Stereoregular Polymerization, CIC/ACS Meeting, Toronto, May 1970.

165. J. L. Herisson and Y. Chauvin, *Makromol. Chem.*, **141**, 161 (1971).

166. K. Matsuzaki and T. Yosukawa, *J. Polym. Sci.*, **A1**, 511 (1967).

167. F. Dawans and P. Teyssié, *Makromol. Chem.*, **109**, 68 (1967).

168. P. Teyssié et al., *J. Polym. Sci.*, **C22**, 221 (1968).

169. S. S. Medvedev, *Russ. Chem. Rev.*, **37**, 842 (1968).

170. B. P. Bourdauducq and F. Dawans, *J. Polym. Sci.*, *A-1*, **10**, 2527 (1972).

171. O. P. Parenago et al., *Kinet. Katal.* **10**, 273 (1969).

172. I. V. Ostrovskaya et al., *Dokl. Akad. Nauk SSSR*, **181**, 892 (1968).

173. J. F. Harrod and L. R. Wallace, *Macromolecules*, **2**, 449 (1969); Symposium on Stereoregular Polymerization, CIC/ACS Meeting, Toronto, May 1970.

174. G. M. Khvostik et al., *Dokl. Akad. Nauk SSSR*, **186**, 894 (1969).

175. N. I. Pakaro et al., *Dokl. Akad. Nauk SSSR*, **185**, 1323 (1969).

176. F. Dawans et al., *J. Organomet. Chem.*, **21**, 259 (1970).

177. B. D. Babitskii et al., *Dokl. Akad. Nauk SSSR*, **180**, 420 (1968).

178. E. J. Arlman, *J. Catal.*, **5**, 178 (1966).

179. B. A. Dolgoplosk et al., *J. Polym. Sci.*, **C16**, 3685 (1968).

180. T. Matsumoto and J. Furukawa, *J. Polym. Sci.*, **B5**, 935 (1967).

181. L. Porri et al., *Eur. Polym. J.*, **5**, 1 (1969).

182. M. E. Pruitt and J. M. Baggett, U.S. Pat. 2,706,189 (to Dow Chem. Co., 1955).

183. E. J. Vandenberg, U.S. Pat. 3,135,706 (to Hercules Powder Co., 1964); *J. Polym. Sci.*, **47**, 149 (1960).

184. H. Tani et al., *J. Am. Chem. Soc.*, **89**, 173 (1967).

185. M. Osgan and P. Teyssié, *J. Polym. Sci.*, **B5**, 789 (1967).

186. J. P. Pasero, Thesis, University of Paris, 1971.

187. N. Kohler et al., *J. Polym. Sci.*, **B6**, 559 (1968).

188. E. J. Vandenberg, *J. Polym. Sci.*, *A-1*, **7**, 525 (1969).

189. M. Osgan et al., ACS 157th Symposium Division of Petroleum Chemistry, A89, 1968.

190. P. Pino, *Adv. Polym. Sci.*, **4**, 393 (1965).

191. S. Inone et al., *Makromol. Chem.*, **90**, 131 (1966).

192. P. Sigwalt and N. Spassky, personal communication.

193. W. S. Knowles et al., *Chem. Tech.*, **1972**, 590.

194. I. Pasquon et al., in A. Ketley, ed., *Stereochemistry of Macromolecules*, Dekker, New York, 1967, p. 220.

195. French Pat. 1,531,409 (to Avisun Corp., 1967).

196. French Pat. 1,563,716 (to Rexall Drug and Chem. Co., 1967).

197. Neth. Appl. Pat. 69, 12354 (to Société Nationale des Pétroles d'Aquitaine, 1969).

198. W. Ring and H. J. Cantow, *Makromol. Chem.*, **89**, 138 (1965).

199. C. Sadron, *Chim. Ind.*, **96**, 507 (1966).

200. Y. Chauvin et al., *Bull. Soc. Chim. France*, **1966**, 3223.

201. Belgian Pat. 602,153 (to Farbwerke Hoechst, A. G., 1961).

202. J. Duchemin et al., French Pat. 1,603,773 (1971).

203. A. Takahashi and N. G. Gaylord, *J. Macromol. Sci.*, **A4**, 127 (1970).

204. J. Furukawa et al., *J. Polym. Sci.*, **B7**, 561 (1969).

205. T. Koshiwagi et al., *J. Polym. Sci.*, **B8**, 173 (1970).

206. M. Tsutsui et al., *Angew. Chem.*, *Int. Ed.*, **8**, 410 (1969).

207. F. Dawans, *Inf. Chim.*, No. 122, 105 (1973).

208. V. M. Frolov et al., *Kinet. Katal.*, **15**, 1475 (1974).

209. Y. I. Yermakov et al., *Kinet. Katal.*, **13**, 1422 (1972).

210. F. I. Karol et al., *J. Polym. Sci.*, *A-1*, **10**, 2621 (1972).

211. B. G. H. Ballard, *Adv. Catal.*, **23**, 263 (1973).

212. Y. Chauvin, private communication.

213. M. Julemont et al., Ziegler Symposium on Coordination Polymerization, 167th ACS Meeting, Los Angeles, 1974.

214. M. Julemont et al., *Makromol. Chem.*, **175**, 1673 (1974).

215. R. Warin et al., *J. Polym. Sci.*, **B11**, 177 (1973).

216. F. Dawans, *Tetrahedron Lett.*, No. 22, 1943 (1971).

217. R. P. Hughes and J. Powell, *J. Am. Chem. Soc.*, **94**, 7723 (1972).

218. V. P. Zubov, 23rd International Congress of Pure and Applied Chemistry, Boston, *Special Lectures*, **8**, 69 (1971).

219. N. Kohler and F. Dawans, *Rev. IFP*, **27**, 105 (1972).

220. W. A. Hewett, private communication.

221. C. P. Casey and T. J. Burkhardt, *J. Am. Chem. Soc.*, **96**, 7808 (1974).

222. J. Furukawa, *Angew. Makromol. Chem.*, **23**, 189 (1972); *J. Polym. Sci.*, **C48**, 19 (1974).

E. W. Duck and J. M. Locke

International Synthetic Rubber Co., Ltd.
Southampton, England

Chapter 4 Polydienes by Anionic Catalysts

1 INTRODUCTION

Synthetic rubbers produced by anionic catalysts such as alkali metals were the first to become commercially important. It is only within the last 15 years, however, that significant quantities of widely used general purpose synthetic rubbers have been manufactured by anionic processes. Today, lithium alkyls and their complexes are among the most favored of the anionic initiators for polydienes because they offer several advantages, and it is with these catalysts that commercial growth has been so rapid. Compared with the other alkali metal compounds the lithium alkyls show

139

(*a*) high activity in very small amounts; (*b*) excellent rates of polymerization to 100% conversion; (*c*) good solubility in hydrocarbons; and (*d*) a widely adjustable charge separation depending on the solvent used or the complexing ligand which leads to control over structure in the polymer so that a wide range of different polymers can be made. For these reasons, this chapter deals mainly with polydienes and copolymers resulting from anionic polymerization of dienes using lithium based catalysts.

Polymerization with alkali metals and their compounds as catalysts is, of course, not new. The early German workers who created methyl rubber in World War I (1914–1918) using 2,3-dimethylbutadiene, sodium metal, and carbon dioxide knew only that they had produced a rubberlike polymer but had very little idea of its structure. They had not yet established the importance of cis-1,4, trans-1,4, or 1,2 isomerism, molecular weight, molecular weight distribution, block sequences, branching, cross-linked gel, and so on, although they did have a crude understanding of toughness and processability. One heard of methyl rubber W (W = weich = soft) and methyl rubber H (H = hart = hard). The soft or hard masses obtained by contacting sodium with the undiluted monomer were of uncontrolled molecular weight. Nevertheless, after excessive mastication these were usable synthetic rubbers that were commercialized under the difficult supply and demand conditions of wartime.

The name BUNA (*bu*tadiene-*na*trium) used to describe these rubbers persisted and became almost synonymous with the term "synthetic rubber"—so much so that even the present day German emulsion polymerized styrene-butadiene rubbers are referred to as Bunas, though a sodium (natrium) catalyst is no longer employed in the process.

The sodium rubbers were displaced by those produced using emulsion processes, particularly during World War II (1939–1945), as a matter of convenience and of rapid plant output of a good synthetic material. Emulsion polymerization was an easier reaction to control and yielded a greater output of a most useful synthetic rubbery styrene-butadiene copolymer.

In a significant exception the Russians alone persisted in following the development of alkali metal catalysis while every other country decided to pursue emulsion processes.

2 POLYDIENES PREPARED USING CATALYSTS BASED ON ALKALI METALS OTHER THAN LITHIUM

The earliest polymerization of a diene with an alkali metal was that of isoprene polymerized using sodium and credited to the British workers,

F. E. Matthews and E. H. Strange, in a patent filed in 1910 (1). Between 1910 and the outbreak of the first world war in 1914 several patents and papers appeared which made it evident that a number of people had been working independently in the area of alkali metal catalyzed synthetic rubbers. Harries for the Bayer Company (2) was the most notable. He disclosed his work in 1911 and claimed that the first sodium initiated polymerizations were observed with butadiene. The convenience of obtaining and working with liquid monomers in those early times led researchers to devote most of their efforts to isoprene and dimethyl-butadiene. A recounting of all these early publications and the controversy which surrounded the claims to priority has been presented by Törnqvist (3).

In the light of present-day experience and the facets of anionic polymerization described later in this chapter, three points from these early publications have special significance. First, Kyriakides (4) noted the effect of monomer purity on rate of polymerization. He observed that the isoprene prepared by reacting ethylmagnesium chloride with chloroacetone and catalytically reducing the methyl-2-butylene oxide, polymerized more rapidly with sodium than the isoprene obtained by Harries. Harries himself stressed that the purity of isoprene was important when using a sodium catalyst. The second point was the observation made by Holt (5) that sodium initiation of dienes in the presence of carbon dioxide gave products differing from those prepared in its absence. This was surely the earliest reference to some change in structure and/or molecular weight resulting from a modification of the anionic catalyst. Third, Labhardt's neglected work (6) on alkali metal alkyls in 1912 might conceivably have led to the present generation of the so-called solution rubbers at a much earlier time had it been followed up. A review of the early work on sodium polymerization of dienes has been given by Taft and Tiger (7).

2.1 Polybutadienes

Butadiene, the lowest homolog of the conjugated diene series, has received more detailed attention than any other diene. As a monomer it is readily available in large quantities economically, either using the modern petrochemical processes of cracking and dehydrogenation or as a by-product of ethylene production.

It has been polymerized by every alkali metal or a suitable organic derivative from lithium to cesium, as well as with the related alkaline earth metal derivatives (e.g., calcium). In examining the differences in structure of the polybutadienes when using the alkali metals in a nonsolvating

Table 1   Polybutadiene Prepared with Alkali Metal Initiators in Pentane at 0°C (8)

Alkali Metal	Polybutadiene Structure (%)		
	Cis-1,4	Trans-1,4	1,2
Li	35	52	13
Na	10	25	65
K	15	40	45
Rb	7	31	62
Cs	6	35	59

medium such as a straight chain saturated hydrocarbon, it is immediately apparent that lithium differs from the other alkali metals (Table 1).

Lithium gives a much lower figure for the 1,2 structure and hence a much higher overall 1,4 content than the other alkali metals. It is particularly for this reason that lithium and its organic derivatives are considered in a separate section. The remaining alkali metals result in relatively high levels of 1,2 structure. This structure is favored when there is a more definite separation into ionic species in the catalyst or in the succeeding polymerization stages. The lithium–carbon bond formed by the reaction of lithium metal with the monomer, or present in a lithium alkyl, does not ionize to the same degree as the metal–carbon bonds in the other alkali metals. In solvating media such as tetrahydrofuran very high 1,2 contents result in the polybutadiene in all cases, as shown in Table 2 (9). With the alkali metal naphthalene derivatives in THF the cis-1,4 configuration is noticeably absent.

Tables 1 and 2 show that the early German workers on BUNA and their Russian successors on SK-B, who polymerized butadiene with

Table 2   Polybutadiene Prepared with Alkali Metal Naphthalene Complexes in THF at 0°C (9)

Alkali Metal Naphthalene	Polybutadiene Structure (%)		
	Cis-1,4	Trans-1,4	1,2
Li	0	3.6	96.4
Na	0	9.2	90.8
K	0	17.5	82.5
Rb	0	24.7	75.3
Cs	0	25.5	74.5

sodium or potassium, had synthesized rubbers rich in 1,2 structure which no doubt had a predominating influence on the properties of these rubbers. The specific properties of controlled 1,2-polybutadienes are examined in conjunction with other structural effects (Section 3) in lithium polymerized polybutadienes. The significant point to note is that the sodium and potassium rubbers, by virtue of a high content of randomly distributed amorphous 1,2 polymerized butadiene, had relatively high glass transition temperatures compared to emulsion SBR (GR-S).

Crouch and Kahle (10) remark on a shipment of 18,000 lb of Russian polybutadiene made by the sodium process and received by the U.S. Office of Rubber Reserve in 1946. The material showed a wide variation of plasticity, even within a single bale. Compared with the GR-S styrene-butadiene emulsion polymer of that time it was a relatively soft sticky material that exhibited a high degree of cold flow in storage; its properties were regarded as only fair. The tensile strength of a standard vulcanizate was of the order of 2000 psi (140–145 kg/cm^2) with elongation 350–500%. Flex life and hysteresis properties were equivalent to GR-S but the products stiffened at -20 to $-25°F$ ($-30°C$) and were thus poorer than those of GR-S.

Potassium as catalyst for producing BUNA 85 and BUNA 115 was used on a relatively small scale during and after World War I by the Germans, who developed a preference for the emulsion polymers as mentioned above. These BUNAs prepared with potassium were apparently used mainly in hard rubber applications rather than in tires.

A more recent and complete examination of a sodium polymerized polybutadiene has been carried out by Blümel (11) who compared it with other polybutadienes. The sample examined was prepared by a bulk polymerization technique which is reported to be still employed by Eastern European countries. The catalyst was finely divided sodium metal suspended in solvent-free butadiene. Polymerization occurred in a horizontal autoclave in which a rotating screw discharged the finished rubber. No additives were used to inactivate the catalyst during working-up operations.

The polybutadiene produced in this manner had a very high gel content ($>40\%$) and a high ash content (1–2%). Investigation of the sol portion separated from the gel showed it to have a high 1,2 content, low cis-1,4, and extremely nonuniform molecular weight and configuration distributions. The steric composition of the gel portion was similar.

The soluble portion showed an inhomogeneity factor u [where $u = (M_w - M_n)/M_n$, M_w = weight average molecular weight, M_n = number average molecular weight] between 3 and 10 times greater than that of lithium alkyl polymerized polybutadienes and almost twice that of emulsion polybutadiene.

Summarizing the basic nature of the total polymer, it exhibited a high degree of precross-linking, branching, and gel combined with a great inhomogeneity in composition and distribution. This all sounds quite detrimental when one considers how much effort has been put into obtaining greater steric purity and control of molecular weight distribution in recent years. Shatteringly, it was observed that this material had excellent processing characteristics. It was the only polybutadiene that gave a milling behavior as good as cold SBR and approaching natural rubber. The cold flow was greater than cold SBR or natural rubber but less than several other polybutadienes. Compounding in an internal mixer showed mixing characteristics not too dissimilar from those of cold SBR and natural rubber. Extrusion profiles with compounds based on natural rubber–polybutadiene blends were among the best of the series compared and the precross-linking seemed to have exerted no unfavorable effect on the extrusion appearance; moreover, a particularly low swelling was observed. One significant point arising from the assessment of this polybutadiene appears to be that the right shape of unsymmetrical molecular weight distribution can overcome many disadvantages associated with high gel contents as far as processing is concerned. It is also fairly obvious that we do not yet know all the factors which relate microstructure and fundamental rheological properties to the factors important for factory processing.

Table 3 shows the static vulcanizate properties of this polybutadiene compared with a high *cis*-1,4-polybutadiene of similar Mooney viscosity and in the same compound formulation.

Generally speaking, these values indicate a rubber that has properties inferior to those of both high *cis*-1,4-polybutadiene and cold emulsion SBR. In this respect this evaluation would seem to be in complete accord with the 1946 comparison on Russian polybutadiene.

The dynamic properties of the vulcanizates (Table 4) indicated that the precross-linking, high inhomogeneity, and different configuration composition had a deleterious influence on the sodium polymerized polybutadiene. The abrasion resistance was significantly worse, as was crack growth on flexing. It was also stated that the sodium polybutadiene showed high heat development under dynamic stress.

Comparable and corresponding evaluations for polybutadienes prepared with potassium and the other alkali metals (except lithium) have not been located in the literature. However, one can assume that results would be roughly analogous to those observed with the sodium polybutadiene insofar as the properties tie in with the 1,2 content.

Table 3 Static Properties of Vulcanizates

Rubber	Configuration Composition			Raw Mooney ML$_4$	Compound Mooney	Tensile Strength (kg/cm^2)	Elongation at Break (%)	Modulus 300% (kg/cm^2)	Permanent Set (%)	Shore Hardness	Rebound (%)		Tear Resistance (kg)
	Cis-1,4	Trans-1,4	1,2								22°C	75°C	
Sodium PBD	20	40	40	35	38	117	365	92	9	58	38	47	14
High cis-PBD (cobalt cat.)	98	1	1	31	58	165	505	69	9	58	52	56	14
Cold SBR (23% styrene 77% butadiene)	15	69	16	53	58	227	505	110	10	61	40	55	17

Table 4   Dynamic Properties of Vulcanizates

Rubber	Abrasion[a]	Crack Growth[b]		
		10 mm	15 mm	20 mm
Sodium PBD	110	2,000	3,500	5,300
High cis-PBD	20	2,300	5,000	10,000
Cold SBR	90	13,000	30,000	53,000

[a] Laboratory method using 3 kg load. Quoted as "AP method," presumably "abrasive paper."
[b] De Mattia, DIN 53522.

Precross-linking, gel, and molecular weight distribution must, however, be accounted for.

2.2 Polyisoprenes

As in the case of butadiene polymerization the alkali metal and solvent used influence the proportions of the various isomeric structures produced in polyisoprene.

Lithium catalysts in a nonsolvating hydrocarbon solvent are unique in producing with isoprene a polymer of 92–94% cis-1,4 structure. With the other alkali metals and their alkyl derivatives very low cis contents are in evidence and the trans-1,4 and 3,4 structures predominate. Together these two structures account for about 90% of the contents; the amount of trans-1,4 increases (and 3,4 decreases) in the order $Na < K \le Rb < Cs$; see Table 5 (12).

Table 5   Polyisoprenes Prepared with Alkali Metal Catalysts in Nonpolarizing Media

Catalyst	Solvent	Polymer Structure (%)			
		Cis-1,4	Trans-1,4	1,2	3,4
Li	Pentane	94	—	—	6
LiBu	Pentane	93	0	0	7
Na	Pentane	—	43	6	51
NaEt	Pentane	6	42	7	45
K	Bulk	—	52	8	40
Rb	Bulk	(3)	50	8	39
Cs	Bulk	(8)	67	8	17

Figures in parenthesis indicate values assumed by difference.

Table 6 Effect of Donor Solvent on Polyisoprene Structure

Catalyst (Alkali Metal)	Solvent	Polymer Structure (%)			
		Cis-1,4	Trans-1,4	1,2	3,4
Li	Pentane	94	—	—	6
Li	Et$_2$O	—	49	5	46
Na	Pentane	—	43	6	51
Na	THF	—	33	13	54

As in the case of butadiene the presence of a polarizing solvent or donor compound has a marked effect on the polymer structure as exemplified by lithium and sodium (13) in Table 6. This comparison shows how the 3,4 structure is favored as the polymerization system becomes more extremely anionic.

We can now readily appreciate why the early workers with isoprene failed to obtain a polymer that could adequately replace the 100% cis-1,4-polyisoprene in natural rubber. Because of the easier access of butadiene and the concentration of commerical developments around its polymers, the amount of relevant test data on anionically prepared polyisoprenes (other than the lithium higher cis types) is scanty. Taft and Tiger (7) give some data for a polymer prepared from pure isoprene (99.8 mole %) using sodium at 30°C which indicate a rubber rather inferior to natural rubber in stress-strain properties but similar to polybutadiene. Flex hysteresis relationship and low temperature properties were inferior.

3 POLYDIENES PREPARED USING LITHIUM CATALYSTS

As already observed, lithium metal or a lithium alkyl gives a predominantly 1,4 polymerization of both butadiene and isoprene in a nonpolar medium. Lithium differs markedly from the other alkali metals in its small ionic size and in that its hydrocarbyl compounds, particularly the normal alkyl derivatives, show a high degree of association in nonpolar solvents to give aggregates (14, 15). The degree of association depends on the alkyl group, the normal alkyls showing higher degrees of association than those having a branch at the first or second carbon atom from the lithium. The associated species exhibit equilibria with the monomeric species:

$$(LiR)_n \rightleftharpoons (LiR)_{n-1} + LiR$$

In the case of n-butyl, $n = 6$. The iso-, secondary, and tertiary butyl lithiums are less associated.

Diene polymerization is initiated by the monomeric alkyllithium and the rate of initiation depends on the structure of the alkyl group (16, 17), decreasing in the order

$$s\text{-BuLi} > i\text{-PrLi} > t\text{-BuLi} > i\text{-BuLi} > n\text{-BuLi}$$

From a practical point of view, only n-butyllithium is widely used as initiator in commercial polymerizations. Compared to the other alkyllithiums, n-butyllithium is stable in hydrocarbon solution indefinitely at room temperature. Because of the higher rate of initiation s-butyllithium would be advantageous and has been used commercially; however, it is less stable and hence storage and transport of this material are less convenient on a large scale.

Strangely, the initiation reaction with n-butyllithium of diene polymerization in hydrocarbon solution shows a higher order than the $\frac{1}{6}$ order expected from the n-BuLi being present as the hexamer. Polymeric ion pairs and also the effect of catalytic trace impurities such as lithium butoxide have been suggested to account for these observations. Whatever the explanation, it has not hindered commercial development. A more detailed interpretation of the kinetics and reaction mechanisms involved in polymerization with lithium alkyls is given in Chapter 5.

3.1 Polybutadiene

The first manufacturer to produce and market large quantities of polybutadienes using butyllithium was The Firestone Tire & Rubber Company in the United States. Current producers of this type of polybutadiene and the trade names used are given in Table 7. In all these

Table 7 Manufacturers of Low *cis*-Polybutadienes Using Lithium Catalysts

Company	Country	Trade Name
A. A. Chemical Co.	Japan	Solprene
Asahi Chemical Industry Co.	Japan	Asadene
Calatrava, Emperesa, etc.	Spain	Solprene
Coperbo	Brazil	Coperflex
Firestone	United States	Diene
Firestone	France	Diene
International Synthetic Rubber Co. (ISR) Ltd.	United Kingdom	Intene
N.V. Petrochim. S.A.	Belgium	Solprene
Phillips Imperial Chemicals Ltd.	Australia	Solprene
Phillips Petroleum Co.	United States	Solprene

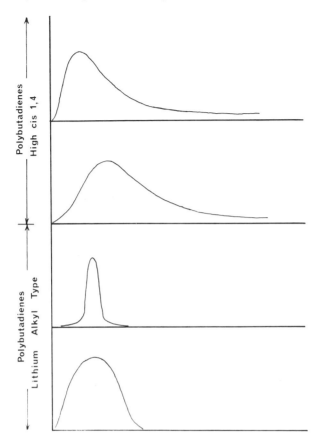

Figure 1 Typical molecular weight distribution curves for polybutadienes.

cases analysis of structure gives, depending on source and method, 36–44% cis-1,4, 48–50% trans-1,4, and 8–10% 1,2.

In the early days of "low cis" polybutadiene, many users expressed the opinion that its processing characteristics and physical properties were inferior to the "high cis" polybutadiene obtained using Ziegler catalyst combinations such as an aluminum alkyl with cobalt salts. However, it is now generally acknowledged that in optimally compounded and vulcanized black stocks the physical properties show very little difference. Although there is some isomerization from cis to trans-1,4 during the vulcanization of the high cis material (18), the essential requirement seems to be that the polymers have a high overall 1,4 content, i.e., over 90%. Whether this is made up of all cis material or approximately half cis and half trans is of little consequence as far as vulcanized rubber properties are concerned.

The high cis materials prepared with Ziegler catalysts containing cobalt, nickel, or titanium do exhibit slightly more tack and process a little more easily than the polybutadiene obtained from lithium alkyl polymerizations. This is caused mainly by the narrower molecular weight distribution of the latter material and is a difference that can be largely eliminated by the use of variations in compounding techniques and the judicious use of extender oils and plasticizers.

Typical curves for molecular weight distributions are given in Fig. 1 (11). A number of more detailed curves for alkyllithium initiated polymers are shown in Chapter 5.

3.1.1 Cold Flow and Gel

Another significant difference between high *cis*-1,4-polybutadiene and "low cis" polybutadiene has been the tendency of the latter to exhibit cold flow (Fig. 2), i.e., to behave like a very viscous liquid at low shear

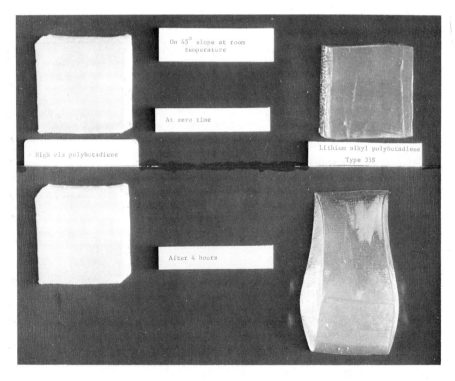

Figure 2 Cold flow characteristics of polybutadiene.

rates. As can be imagined, for purposes of packing and storage this property is a serious disadvantage.

Polybutadiene produced using lithium alkyls, e.g., Intene, Diene, and Asadene rubbers, are now offered in two forms—a flow grade (or standard grade) and a nonflow grade. In the latter a small amount of an incipient cross-linking monomer may be introduced during polymerization; the cross-linking induced is sufficient to inhibit cold flow but does not produce gel in the polymer or affect its processability significantly. In the patent literature (19) divinylbenzene has been mentioned as an ingredient for inducing nonflow characteristics. In modern operations the introduction of a few long chain branches achieves similar results.

Lithium alkyl catalyzed polybutadiene is far less prone to contain gel and never contains transition metal catalyst residues which may be present in polybutadienes prepared with cobalt- or nickel-containing catalysts. The presence of these transition metal residues requires the addition of stabilizers to attain good oxidation stability.

3.1.2 *Processing and Compounding of "Low Cis" Polybutadiene Blends*

It is not usual to use polybutadiene in general purpose applications without blending. This is because properties such as abrasion resistance and high resilience are not improved pro rata above the levels that are obtained at a 60% concentration in the blend. Processing difficulties are also encountered at high polybutadiene contents.

Blends of polybutadiene with natural rubber or with styrene-butadiene rubber result in compounds with good processing characteristics and permit full advantage to be taken of the improved abrasion resistance, resilience, tensile strength, and tear resistance. To obtain the maximum physical properties from a blend it is essential that the viscosities of the polymers prior to blending should be as close as possible and at the most not more than 5–10 Mooney points apart. The lithium polybutadiene is normally obtainable at three Mooney viscosity levels (35, 45, and 55), as standard (S) or nonflow (NF) grades. Thus it is possible to choose the correct grade to give the optimum conditions for successful blending. Oil extended grades have also been introduced, e.g., which contain 20 or 37.5 parts of aromatic oil per 100 parts of rubber.

Blends process satisfactorily on conventional rubber processing equipment, though some changes in compounding technique may have to be employed, the most important of which concerns the correct choice and amount of plasticizer to be used. While high plasticizer levels may facilitate good mixing, even higher levels may be required for satisfactory subsequent processing, that is, calendering and extruding. In the majority of cases, an increase in the proportion of polybutadiene present in the blend should be accompanied by a corresponding increase in plasticizer level. Such increases in plasticizer level allow for, and in some cases necessitate, an increase in filler content.

As a general rule, blends with styrene-butadiene copolymers require higher plasticizer levels than corresponding formulations based on natural rubber. For easier incorporation of the high plasticizer levels required in blends with styrene–butadiene copolymers, use can be made of oil extended types.

Polybutadiene is more resistant to mastication and peptization than natural rubber but, because of its availability at various viscosity values, the grade with correct plasticity for blending can be chosen. This obviates any need for mastication prior to mixing. The Mooney viscosity of a blend of natural rubber can, however, be reduced either by constant milling or by the use of peptizing agents. This mechanical or chemical method of reducing the molecular weight is accompanied by a loss in physical properties, e.g., tensile strength, tear, and elongation at break.

3.1.3 *Compounding*

In general no polymer or blend can be compounded or processed exactly like another. Thus the basic compounding rules that apply when polybutadiene is used differ somewhat from those for other general purpose polymers. These differences are mainly due to the very narrow molecular weight distribution of the polymer which enables it to accept much larger quantities of oil than have hitherto been used. By using these larger quantities of oil, it is also possible to incorporate larger volumes of fillers which, taken together, give compounds which offer economies. This ability to accept large quantities of oil and filler is not limited to the cheaper ranges of compounds; it is usual practice to use higher quantities of black and oil in such products as tire treads and conveyor belt cover stocks. By using larger amounts of oil in these premium grade compounds, the best mechanical properties are obtained from the polymer. These high black and oil containing compounds also give the best processing characteristics with reduced die swell and improved surface finish on extrusions.

To achieve the best processing for compounds having low oil loadings, the lower viscosity raw polymer should be used; with increased black and oil loadings the higher viscosity raw polymers should be used. To achieve maximum tensile strength, tear resistance, and resilience it is advisable to use the highest Mooney grade available.

The vulcanization characteristics of polybutadiene fall between those of styrene-butadiene and natural rubber. The accelerator and sulfur requirements are, therefore, slightly different and have to be modified when formulating polybutadiene compounds.

3.1.4 *Use of Oil*

The oils normally used in rubber compounding are of three types: (*a*) paraffinic, (*b*) naphthenic, and (*c*) aromatic. In the following tables the effects of these oils have been evaluated in a compound based on the polybutadiene/natural rubber blend and using HAF black as the main filler.

	parts by weight
Polybutadiene	75
RSS1	25
HAF black	50
Zinc oxide	5
Stearic acid	2
Sulfur	2
N-Oxydiethylenebenzothiazole-2-sulfenamide	0.7
Oil	Variable

Mooney Viscosity. This is reduced as the quantity of oil included in the mix is increased. The paraffinic types yield the maximum reduction in Mooney viscosity. The naphthenic and aromatic oils give similar viscosities. This is shown in Table 8.

Scorch Characteristics. The inclusion of oil increases the scorch time because plasticity is lowered, which reduces the amount of heat generated. All grades of oil have the same effect.

Tensile Strength. Increasing the quantity of oil in a stock without alteration in filler loading automatically reduces the tensile strength. The

Table 8 Compound Mooney Viscosity[a]

Loading (phr)	0	5	15	25
Paraffinic	85	63	45	35
Naphthenic	85	63	48	43
Aromatic	85	64	50	43

[a] ML_4 at 100°C.

most drastic reduction is brought about by using a paraffinic type. This is shown in Table 9.

Table 9   Tensile Strength[a]

Loading (phr)	5	15	25
Paraffinic	2600 (182–8)	2300 (161–7)	1300 (91–4)
Naphthenic	2600 (182–8)	2500 (175–8)	1900 (133–6)
Aromatic	2600 (182–8)	2500 (175–8)	2000 (140–6)

[a] In units of psi (kg/cm^2).

Modulus.   The softening effect of the oils is generally the same and modulus results at 300% elongation do not differentiate among the grades of oil.

Elongation at Break.   The elongations at break of the vulcanized compounds are increased progressively, the aromatic oil giving the highest figures. This is shown in Table 10.

Table 10   Elongation at Break (%)

Loading (phr)	0	5	15	25
Paraffinic	400	500	650	800
Naphthenic	400	500	650	850
Aromatic	400	600	700	950

Hardness.   The hardness of the vulcanizates is reduced by the addition of oil. The paraffinic types tend to reduce the hardness more than the naphthenic or aromatic oils. This is shown in Table 11.

Table 11   Hardness, IRHD

Loading (phr)	0	5	15	25
Paraffinic	68	57	53	44
Naphthenic	68	62	55	47
Aromatic	68	61	57	48

Tear Strength.   For the highest tear values, aromatic oils should be used. The paraffinics give the lowest results with naphthenics at the intermediary level.

Rebound Resilience.   If it is necessary to use oils when compounding for rebound resilience, paraffinic types give the highest resilience values followed by naphthenic and aromatic.

Heat Buildup. This is complementary to rebound resilience and, therefore, aromatic oils give the greatest heat buildup.

3.1.5 Influence of Carbon Blacks

To obtain maximum physical properties, HAF, ISAF, and SAF blacks should be used. In addition, as already outlined, the use of higher than normal amounts of oil in compounding makes the use of these finer particle size blacks much more practical because the mixing temperatures can be more easily contained within practical limits. The maximum temperature that can be tolerated for mixing black masterbatches is of the order of 160°C. Temperatures above this can lead to processing troubles in the final stock. In certain instances, it has become common practice to judge the efficiency of masterbatch mixing by temperature measurements alone rather than by timing the cycle itself. In this way the undesirable side effects that occur when the carbon/rubber matrix is exposed to excessive temperature can be prevented.

In general terms, the highest degree of reinforcement is obtained by using SAF, ISAF, or HAF in blends of polybutadiene with SBR or NR. The softer, coarser particle size blacks confer only moderate degrees of reinforcement, although the combined effect of such blacks in blends with harder blacks can be employed to obtain an overall balance of physical properties.

The final physical properties that can be achieved by the use of conventional blacks alone is shown in Table 12.

High Structure Blacks. High structure blacks are now being employed extensively in compounds containing polybutadiene. They modify certain properties in both the vulcanized and unvulcanized state.

In compounds based solely on NR or SBR, the changes generally observed in processing characteristics when a high structure HAF black compound is used in place of an HAF black are (a) more rapid dispersion of black during mixing, (b) quicker buildup of temperature during mixing, (c) improved surface finish to extrusions, (d) reduced die swell and extrudate shrinkage, (e) increased compound viscosity, and (f) possible lowering of scorch safety times.

In the vulcanized state the changes that occur in NR and SBR stocks reinforced with high structure blacks are mainly (a) higher modulus, (b) increased hardness, (c) possible better abrasion resistance, and (d) improved tread crack resistance in tire tread compounds.

Some of these modifications in vulcanized and unvulcanized properties are disadvantageous to the rubber technologist and, in particular, to the

Table 12 Physical Properties Conferred by Conventional Carbon Blacks

Type	TS	M 300	E at B	Hardness	Resilience
A. 100 Parts Polybutadiene; 55 Parts Carbon Black;					
15 Parts Oil					
SAF	2500	1300	490	61	58.9
ISAF	2400	1270	460	60	61.0
HAF	2500	1500	430	60	69.0
FEF	2130	1600	390	59	69.7
GPF	1870	1180	450	55	72.3
SRF	1400	1000	440	53	73.6
FT	830	400	600	47	75.8
MT	650	360	480	47	77.8
EPC	2250	870	410	58	66.5
B. 75 Parts Polybutadiene; 25 Parts SBR 1712;					
70 Parts carbon Black; 25 Parts Oil					
SAF	2240	780	610	60	50.0
ISAF	2320	780	600	61	51.4
HAF	2300	780	620	56	56.8
FEF	2160	1190	540	58	63.8
GPF	1460	770	490	51	65.1
SRF	1200	610	400	51	66.6
FT	450	190	610	42	68.6
MT	380	220	440	44	71.7
EPC	1980	510	730	56	54.9
C. 50 Parts Polybutadiene; 50 Parts SBR 1712;					
70 Parts Carbon Black; 15 Parts Oil					
SAF	3070	1070	630	61	49.0
ISAF	2870	1040	600	60	50.6
HAF	2770	1060	610	57	60.9
FEF	2100	1340	460	57	60.9
GPF	1940	850	630	51	64.5
SRF	1950	750	650	51	65.6
FT	1000	300	720	44	68.6
MT	720	280	670	45	70.4
EPC	2650	720	720	59	52.9

Note: TS, tensile strength (psi); M 300, modulus at 300% elongation (psi); E at B, elongation at break (%).

tire compounder because they are accompanied by higher stock viscosity, higher modulus and hardness, and possible reduction in processing safety.

In all NR and SBR based stocks, these effects can be compensated for by the addition of higher volumes of process oil which gives lower compound costs, but is limited by the decrease in physical properties that can be tolerated in the final vulcanized stock.

In similar compounds based on blends of polybutadiene with either NR or SBR, the desirable effects of high structure HAF blacks can be used to particular advantage because the addition of oil, necessary to offset the increased hardness and compound viscosity, can be made without seriously reducing the level of reinforcement and resistance to abrasion.

Table 13 shows the physical properties and improvement in extrusion characteristics obtained by using a high structure HAF black (HS.HAF). The physical property changes can be summarized as (*a*) slightly higher tensile strength, (*b*) increased hardness, (*c*) higher modulus values, (*d*) increased heat buildup characteristics, and (*e*) lower rebound resilience.

The addition of extra oil will modify hardness and modulus and will in turn cause heat buildup to be increased. The lower resilience value on its own suggests the possibility of improved road holding properties. The use of high structure blacks is likely to be of the greatest value in passenger tire formulations where the resultant improved processing and tread wear can be used to advantage and the effects of reduced resilience can be of benefit in improved road holding. In truck tire formulations the acceptance of these types of blacks depends on whether the advantages obtained from improvements in processing and tread wear outweigh the effects of higher hysteresis.

The compounds containing HS.HAF exhibit increased rates of extrusion, improved smoothness of extruded surfaces and a considerable reduction in die swell. These are shown in Fig. 3.

Variations in Black/Oil Ratios.   If the carbon black loading of a range of compounds is based on the combined total of rubber plus oil, only small alterations in black loadings are required as the proportion of oil is increased in order to maintain the physical properties at substantially the same level. This is illustrated in Table 14.

Figure 3 Extrusion profiles with HAF and HS.HAF carbon blacks. From left to right: HS.HAF, HAF, HS.HAF, HAF.

Table 13 Physical Properties of Compound (HAF vs. HS.HAF)

	Comparison of HS.HAF and HAF Carbon Black in Polybutadiene/SBR Tread Stocks			
	A	B	C	D
SBR (Intol 1712)	68.75	68.75	35	35
PBD (Intene 35NF)	50	50	65	65
HS.HAF black	—	60	—	50
HAF black	60	—	50	—
Aromatic oil	18	18	15	15
Zinc oxide	5	5	5	5
Stearic acid	2	2	2	2
4-Isopropylaminodiphenyl- amine	1.2	1.2	1.5	1.5
Sulfur	2.0	2.0	1.7	1.7
CBS (cyclohexyl- benzthiazylsulfenamide)	1.0	1.0	0.85	0.85
Mooney viscosity (ML$_4$ at 100°C)	36	42	37.5	45.5
Mooney scorch (minutes to 10 pt. rise at 144°C)	9'12"	9'55"	8'32"	8'44"
Vulcanization time at 144°C	60	60	60	60
100% modulus (psi)	225	350	350	250
(kg/cm^2)	(15.8)	(24.6)	(24.6)	(17.6)
300% modulus (psi)	1075	1600	1500	1575
(kg/cm^2)	(75.6)	(112.5)	(105.5)	(110.7)
Tensile strength (psi)	2650	2750	2675	2675
(kg/cm^2)	(186.3)	(193.3)	(188.1)	(188.1)
Elongation at break (%)	570	475	470	490
Tear strength at 20° (lb)	26	22	21	21
(kg)	(11.7)	(9.9)	(9.5)	(9.5)
Resilience at 20°C (%)	54.4	50.7	60.4	55.2
Resilience at 70°C (%)	62.4	55.8	67.6	61.0
Hardness, IRHD	59	64	59	64
De Mattia cut growth resistance (kcs to failure)	24	23	19	74
Compression set (25% strain 22 hours at 70°C)	18.60	18.08	15.61	15.25
Goodrich heat buildup (°C)	22	26	22.5	30
Extrusion characteristics: Extruded through standard Garvey die (g/min)	535	614	436	452
% Die swell	83.1	36.8	88.4	47.3

Table 14   The Influence of Black and Oil Variations on Physical Properties

SBR (Intol 1717)	87.5	87.5	87.5	87.5	87.5
PBD (Intene 55NF)	50.0	50.0	50.0	50.0	50.0
HAF Black	67.5	75.0	81.5	95.0	103.0
Naphthenic oil	—	12.5	22.5	32.5	45.5
Stearic acid	1.5	1.5	1.5	1.5	1.5
Zinc oxide	3.0	3.0	3.0	3.0	3.0
CBS (cyclohexylbenzthiazyl- sulfenamide)	1.25	1.25	1.5	1.75	1.75
Sulfur	2.0	2.25	2.25	2.25	2.5
Oil (phr)	37.5	50	60	70	80
HAF black (phr)	49	50	51	56	57
Vulcanization time at 153°C (minutes)	15	20	15	17	16
Physical Properties:					
Specific gravity	1.12	1.13	1.13	1.14	1.14
Mooney viscosity (ML$_4$ at 100°C)	56	45	37	34	33
200% modulus (psi)	675	900	700	700	725
(kg/cm^2)	(47.5)	(63.3)	(49.2)	(49.2)	(50.9)
Tensile strength (psi)	2050	2200	3000	2075	1800
(kg/cm^2)	(144.1)	(154.7)	(210.9)	(145.9)	(126.6)
Elongation at break (%)	380	350	400	400	370
Tear strength (lb)	38.0	38.0	32.7	30.2	35.5
(kg)	(17.4)	(17.4)	(14.9)	(13.7)	(16.3)
Hardness, IRHD	59	60	56	59	58
Resilience at 20°C (%)	59.8	54.4	52.0	44.9	42.0
After Aging 7 days at 70°C:					
Tensile strength (psi)	1725	2175	2200	1875	1575
(kg/cm^2)	(121.3)	(152.9)	(154.7)	(131.8)	(110.7)
Elongation at break (%)	220	250	250	260	200

The vulcanization system requires modification in order to allow for the increase in oil content. These adjustments also apply to polymers containing other kinds of oil or less oil than Intol 1717. As stated previously, if the oil is aromatic, the level of physical properties will be higher. This maintenance of physical properties only applies within certain limits of black and oil, but the limits can be easily established by examining ranges similar to those just shown. As the volumes of oil and black are increased there is a corresponding decrease in both weight and volume cost; savings of up to 20% of the volume cost can be achieved without any drastic loss in physical properties.

3.1.6 *Nonblack Fillers*

The most effective nonblack reinforcing agent for compounds containing
polybutadiene is silica. Silicates produce a moderate degree of reinforcement but this is considerably lower than that obtained from silica fillers.

For this reason, it is necessary to use a proportion of silica with other
nonreinforcing fillers such as whiting or china clay when light-colored
compounds are required. It must also be appreciated that higher than
average filler and process oil levels are necessary to obtain satisfactory
physical and processing properties.

3.1.7 *Vulcanization*

The vulcanization characteristics of polybutadiene with respect to accelerator and sulfur requirements lie intermediate between those of
natural rubber and styrene-butadiene copolymers. Greater accelerator
and lower sulfur concentrations are required than with natural rubber
(see Figs. 4–6). However, some reduction in the rate of cure can also

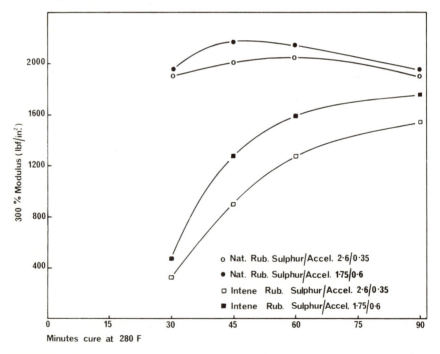

Figure 4 Cure curves for natural rubber and polybutadiene (Intene) rubber (50 phr
HAF black).

result from increasing the oil and filler loadings and, therefore, further increases in both sulfur and accelerator levels may be necessary.

Accelerator systems already widely used for styrene-butadiene compounds and natural rubber compounds exhibit the same characteristics when used for compounds containing polybutadiene. Activator levels are similar to those used in the compounding of natural rubber because solution polymerized rubbers do not contain fatty acid.

High Temperature Vulcanization of Polybutadiene/Natural Rubber Blends. Manufacturers, especially in the tire, injection molding, and direct vulcanized footwear industries, show increasing tendencies to use higher vulcanization temperatures ($\geq 160°C$) as a convenient means of increasing production capacity. Where natural rubber forms the basis of compounds to be cured at high temperatures, as in heavy truck tires, loss of physical properties due to reversion can be expected.

Blends of natural rubber and low *cis*-polybutadiene show a remarkable degree of retention of properties at high vulcanization temperatures.

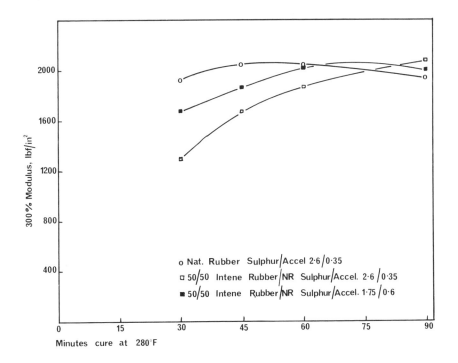

Figure 5 Cure curves for natural rubber and 50–50 Intene polybutadiene rubber/natural rubber (50 phr HAF black).

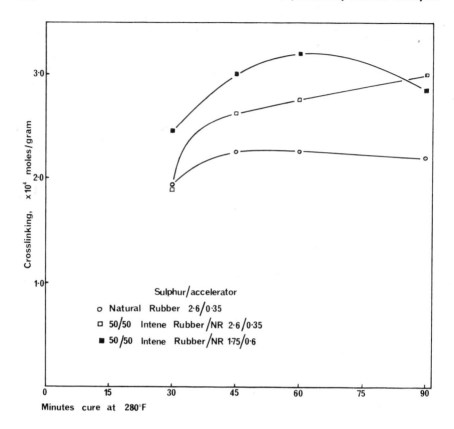

Figure 6 Cure rate as measured by cross-linking (50 phr HAF).

Figure 7 shows the falloff in the tensile strength of a natural rubber tread vulcanizate with increasing temperature of cure compared with a vulcanizate based on a 50:50 blend of polybutadiene and natural rubber. From the graphs shown it can be seen that the reduction in tensile strength with increasing temperature of cure is proportionately much less for the polybutadiene/NR vulcanizate until, at a temperature of 180°C and cure times in excess of approximately 16 minutes, the tensile strength of the polybutadiene/NR vulcanizate is superior to that of the natural rubber vulcanizate.

For more detailed information on the processing, compounding, and applications of low *cis*-polybutadiene, the reader is referred to the Technical Service booklets issued by the manufacturing companies.

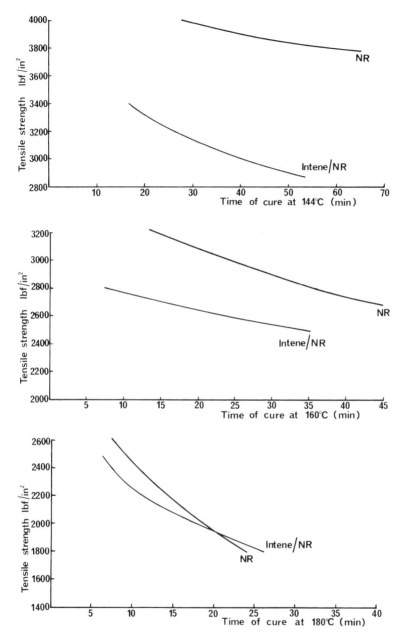

Figure 7 Effect of cure temperature on cure time and ultimate tensile strength.

3.1.8 *Use of Polybutadiene in Plastics*

High Impact Polystyrene. One of the most interesting, well-established, and fastest growing applications of polybutadiene is as an impact improver in polystyrene. It is well known that it is necessary to modify polystyrene with a rubbery polymer to achieve a degree of flexibility in the product.

In the past SBR fulfilled this need but now the most widely used and most effective rubber offering polystyrene good color, flexibility, and impact strength, especially under low temperature conditions, is low *cis*-polybutadiene.

It is well worth examining the mechanism of the reinforcement of polystyrene. When a material is subjected to tension it can either stretch or break, depending on whether the yield stress is lower or higher than the tensile stress at rupture. Consider a piece of a brittle material such as polystyrene which is under impact and in which a small crack has started (Fig. 8*a*). If the material is not able to stretch, the crack will continue to travel quickly through the material under the energy of the impact. However, if the material is able to stretch without breaking (Fig. 8*b*), the shape at the leading edge of the crack will change, and instead of being a sharp wedge where large forces are concentrated, it becomes bridged by the stretched material which tends to pull the crack together as well as allowing the forces to be spread over a larger area. Furthermore, the stretching of the material absorbs some of the energy of the impact and the result is that the progression of the crack is arrested. Reinforcement results from the use of plasticizers in PVC, but this method is not useful for polystyrene because of the deterioration of other properties.

Reinforced polystyrenes are prepared commercially by dissolving 5–10% of rubber in styrene monomer and then polymerizing the styrene in bulk, in suspension, in emulsion, thermally, or some combination of these. When polystyrene is reinforced with rubber, the rubber is not actually dissolved in the polystyrene, but is present as a separate phase (which may contain polystyrene inclusions) in the form of small globules between 1 and 10 μ in diameter and which can be seen under a phase contrast microscope. A considerable fraction of the styrene is grafted to the rubber. Thus, in a successfully reinforced polystyrene a crack traveling through the plastic will meet one of these globules of rubber which will stretch and hold the crack together, absorbing some of the energy of the impact as it does so (Figs. 8*c* and 8*d*). To get satisfactory reinforcement of the plastic the following conditions must exist:

1. The adhesion between the polystyrene and the rubber must be high to prevent the crack from simply traveling around the edge of the rubber

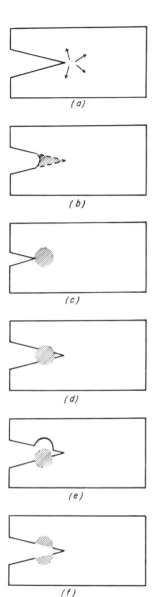

Figure 8 Crack propagation in high impact polystyrene. (*a*) A crack traveling through a brittle plastic has a sharp leading edge where the forces concentrate. Thus the fracture travels quickly through the material. (*b*) A plasticized material can yield without breaking. The tension in the stretched material (shaded) holds the edges together, preventing the crack from traveling further. (*c*) Rubber reinforced polystyrene is not homogeneous but contains globules of rubber present as a separate phase. (*d*) The rubber particles stretch and the tension holds the edges together, preventing further travel of the fracture. Part of the energy of the impact is absorbed at the same time. (*e*) If compatibility is poor (no grafting), there is no adhesion between the rubber and the plastic and the crack travels around the interface. (*f*) The rubber particle will rupture if it is below its glass transition temperature or if it has low tensile strength.

particle and on through the plastic (Fig. 8*e*). This is achieved during graft polymerization by the formation of molecules containing both the rubber backbone and grafted polystyrene side chains. These molecules probably bridge the polystyrene/rubber interface and because of entanglement of the ends of the molecule in both phases, molecular rupture would have to occur before the rubber could separate from the polystyrene.

2. The rubber particles must have as high a tensile strength as possible to avoid being ruptured (Fig. 8*f*). This condition is satisfied after graft polymerization by the rubber being part of a cross-linked network.

3. The rubber particles must be above their glass transition temperature or they will be as brittle as the polystyrene and will offer no reinforcing action. The lower glass transition temperature of polybutadiene is reflected in the better low temperature properties of the polystyrene made from it.

4. The rubber particles must be of optimum size. If they are too large, the surface finish of the polystyrene is impaired; if they are too small, a crack will be able to travel around them, so that they will have no chance of arresting its progress.

Simple mixtures of polystyrene and polybutadiene have the same low impact strength as the unreinforced material partly because the compatibility between the two polymers is poor and there is no adhesion between the two phases and partly because raw uncross-linked polybutadiene has very low tensile strength (conditions 1 and 2 above). Simple mixtures of SBR and polystyrene are somewhat better and can be improved still more if some degree of reaction or grafting can be achieved during the actual mixing procedure. However, it is only when a high degree of grafting and gelation of the rubber is obtained, as in the graft copolymerization process, that optimum properties are achieved.

The effect of the glass transition temperature of the rubber can be seen in the impact strengths of SBR reinforced polystyrene compared with polybutadiene reinforced polystyrene at low temperatures (Fig. 9). The two polystyrenes chosen have the same impact strength at 20°C, but that of the SBR type falls off below 0°C whereas the polybutadiene type retains its impact strength down to −60°C. The temperature at which the impact strength falls to that of the unreinforced polystyrene is somewhat higher than the glass transition temperature of the pure rubber, because it has been considerably modified by grafting. In SBR, the rubber has about twice its weight of grafted polystyrene.

The size of the rubber particles in the final product is one of the most important features of manufacture and one of the most difficult to control

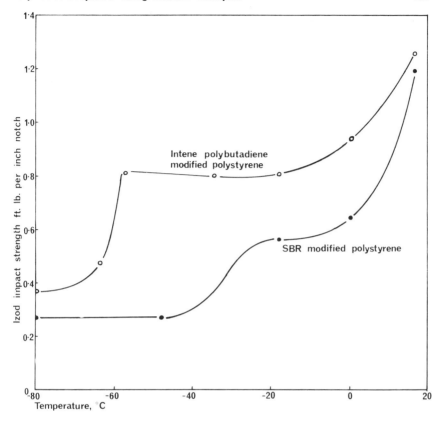

Figure 9 Low temperature impact properties of rubber reinforced polystyrene.

reproducibly. It is in this facet that the various manufacturing processes differ and where the know-how of the different companies lies. The rubber, whether it be SBR or stereopolybutadiene, is dissolved in the styrene monomer and remains so during the early part of the polymerization. At around 25–30% conversion, the amount of polystyrene in the system becomes so great that the rubber becomes incompatible with the styrene/polystyrene phase and is thrown out of solution. Shortly after this the rubber is gelled by the free radicals present in the system and the dimensions of the particles become fixed, controlled by the stirring action or amount of shear at the critical time between the rubber coming out of solution and its gelling. It is this stirring action which largely controls the impact strength and flow properties of the final product.

Table 15   Properties of Polyethylene-Polybutadiene Blends

Properties	Unmodified Polyethylene	Polyethylene + 10% Polybutadiene
Tensile (psi)	2875	2700
Elongation at break	600	670
Tear strength (lb)	5.13	4.98
Impact strength (ft-lb) (falling dart method)	0.55	1.45

ABS.   Attempts to use solution polybutadiene in the preparation of ABS (acrylonitrile–butadiene–styrene) in the same manner as high impact polystyrene have been less successful. Impact strengths, flow characteristics, hardness, and gloss in the final product do not compare favorably with those obtainable from latex graft techniques of the more conventional methods.

PVC.   Low *cis*-polybutadiene has been grafted in solution with a styrene–methyl methacrylate mixture to produce a modified rubber which is suitably compatible with PVC (20). The modified rubber disperses in suitable form, and if it is of the correct refractive index, a reinforced, transparent PVC is obtained. The application of electron microscopy in the field of such composite structures containing rubber has been described in the literature (21).

Blends with Polyethylene.   Polybutadiene has been used to improve the impact strength of polyethylene, particularly of the sheet grade used for containers and sacks. The improvement in this property allows for a polyethylene sack to compare favorably in strength and price with the five-ply paper sacks used widely for such materials as cement and fertilizers.

The changes in properties of a polyethylene containing 10% of a low *cis*-polybutadiene are shown in Table 15 (22).

Polypropylene Blends.   Polypropylene in commercial use is usually of high crystallinity and hence exhibits poor impact strength and is brittle at low temperature. By incorporating 15% of low *cis*-polybutadiene into polypropylene the impact strength can be increased from 2 to 7 ft-lb (22).

Ethylene-Vinyl Acetate Copolymer (EVA) Blends.   Low *cis*-polybutadiene has been blended with EVA to improve its low tempera-

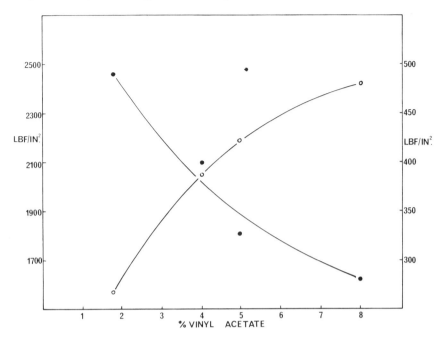

Figure 10 The tensile strength values for black loaded (○) (left scale), and gum (●) (right scale). EVA/polybutadiene (30/70) vulcanizates, showing the effect of varying vinyl acetate content.

ture flexibility (22). In other studies where the EVA was added as the minor component in a blend with polybutadiene to give vulcanizable rubbers it was found that the tensile strength of black loaded compounds increased with increasing vinyl acetate content of the EVA used. In gum stocks without black present a completely opposite effect was observed (22) (Fig. 10).

3.1.9 *Use of Lithium Alkyl in Producing Other Polybutadiene Stereostructures*

High 1,2-Polybutadiene. It was shown earlier that polar solvents and electron donor compounds such as ethers lead to an increase in the 1,2 content of polybutadiene when prepared using lithium catalysts (see Table 2). By introducing small quantities of diethylene glycol dimethyl ether (diglyme) or tetramethylethylenediamine (TMEDA) to the normal polymerization of butadiene in a nonpolar hydrocarbon with butyllithium, almost any desired level of 1,2 content can be obtained (23) (Fig. 11).

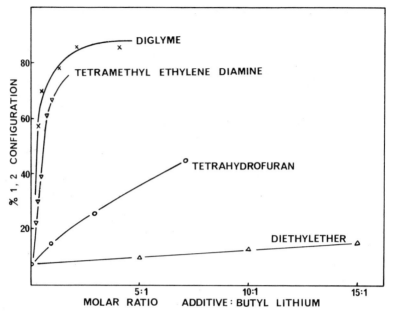

Figure 11 Effect of donor additives on 1,2 content of polybutadiene.

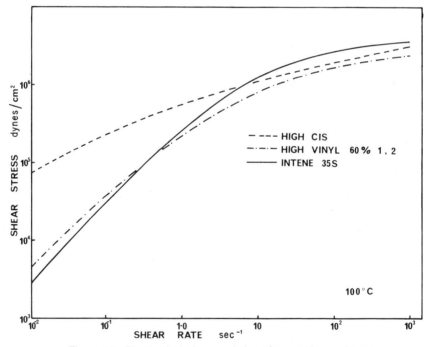

Figure 12 Rheological characteristics of raw polybutadienes.

The higher 1,2 materials show interesting differences in processability and low temperature properties compared with the high 1,4 materials (23) (Figs. 12–16).

Examination of Fig. 15 is especially enlightening. SBRs were found commercially attractive because they were elastomers with convenient processability characteristics combined with good cured physical properties. They especially had good road holding and wet skid properties combined with acceptable abrasion resistance. It is known that hysteresis and skid resistance improve as the second-order glass transition temperature T_g increases. Good abrasion resistance and good tread wear are found in low T_g rubbers such as the high 1,4-polybutadienes, but these have not been satisfactory tread rubbers by themselves because of inadequate road holding and wet skid deficiences. Thus commercial tire rubbers have always had compositions which represented best balance compromises of wear and traction properties. This is reflected in the T_g, and, indeed, if one examines the T_g of the rubber *blends* used in tires, they will be found to fall generally in the range -50 to $-75°C$.

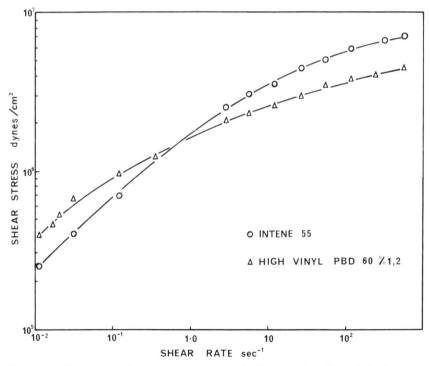

Figure 13 Processing of tread stock compounds showing the effect of vinyl content of polybutadiene used.

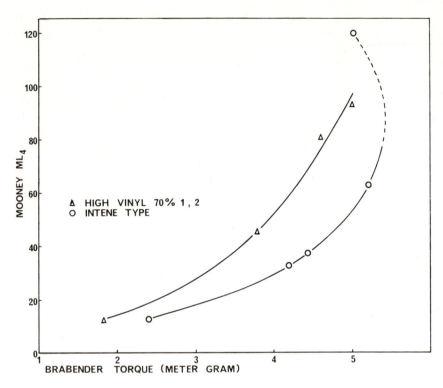

Figure 14 Brabender processing characteristics of high and low vinyl, linear polybutadienes.

SBR achieves this T_g range by incorporating styrene along with butadiene in the polymer chain. However, T_g also increases with the vinyl content in a butadiene homopolymer as shown in Fig. 15. If one obtains polybutadienes with between 35 and 55% vinyl configuration, this should include elastomers with properties spanning the range of the SBRs commonly in use. Three manufacturers have been active in large-scale development of medium vinyl content polybutadienes as a general purpose rubber, but primarily in tires, in blends with SBR and, perhaps, high 1,4-polybutadiene. Shortages in styrene supplies have accelerated this development.

These vinyl polybutadienes are described in patents as being made with alkyllithium catalysts by adding a donor, such as tetramethylethylenediamine (TMEDA), at some intermediate point in the reaction (24); or by using a donor modified lithium initiator and allowing the reaction to

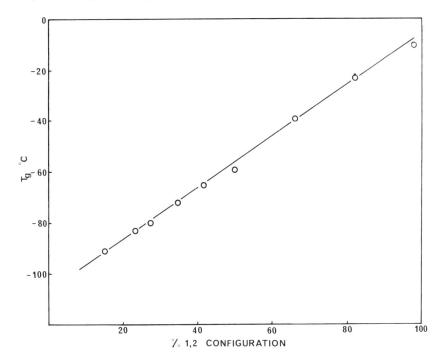

Figure 15 Relationship of glass transition temperature to percent 1,2 configuration in polybutadiene.

exotherm without restraint (25). Since lower vinyl contents occur at higher reaction temperatures, both of these procedures produce polybutadienes with an irregular, tapered distribution of vinyl structures. At the same time there is probably some broadening of the normally narrow molecular weight distribution. It is very likely that to improve processability further, techniques such as those used to broaden the distribution for solution SBR or the block polymers via branching, coupling, or end linking are used for the medium vinyl polybutadienes as well (e.g., see Section 3.4).

Some results (26) comparing the properties of SBR with several polybutadiene homopolymers of varying vinyl contents are shown in Table 16. The vinyl polybutadienes have higher Mooneys, lower die swell, lower tensile strengths, and poorer handling qualities. The abrasion resistance and traction depend on vinyl content and on oil and black

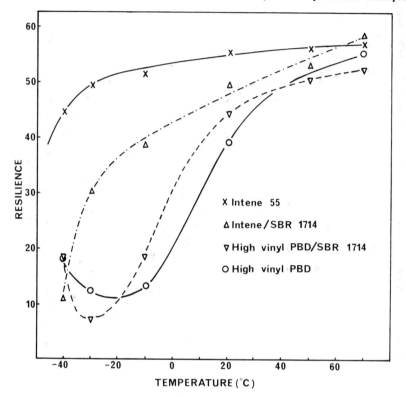

Figure 16 Comparison of temperature-resilience curves for different polybutadienes and blends with SBR.

levels, but beneficial balances of properties comparable to those obtained in current commercial blends are possible.

Of course, it is also obvious that in block copolymers high 1,2 structures can be combined with the high 1,4 structures to give further modifications of physical properties.

Hydrogenated Polybutadiene.   The versatility of lithium based polymerizations of butadiene can be further extended by an in situ hydrogenation reaction of the polymers while still in solution. Such hydrogenation can bring about profound changes in the physical properties of the raw polymer (23). A combination of butyllithium and nickel diisopropylsalicylate enables controlled hydrogenation of the polymer to be carried out

Table 16 Medium Vinyl Polybutadienes in Passenger Tire Treads[a]

		Polybutadiene			
	SBR 1712	35% Vinyl	45% Vinyl	55% Vinyl	65% Vinyl
Rubber hydrocarbon	100	100	100	100	100
Total oil (phr)	45	45	45	45	40
N339 black (phr)	75	75	75	75	70 (N220)
Compounded ML$_4$	58	65–80	70–80	72	76
Extrusion rate (g/min)	79	83	86	88	—
Extrusion appearance (3–12, 12 best)	10+	12	12	12−	12−
Die swell (%)	96	72	68	58	—
Scorch time (minutes)	14	12	12	13	12
Dispersion (0–10, 10 best)	8	7	7	7	7
300% modulus (psi)	1260	1280	1320	1290	1120
Tensile (psi)	3080	2540	2670	2600	2460
Elongation (%)	600	520	530	530	530
Shore A hardness	58	60	59	59	58
Heat buildup (°C)	45	42	41	41	41
Resilience (%)	52	60	59	58	55
Blowout time (minutes)	11	>60	>60	35->60	>60
Gehman freeze point (°C)	−46	−59	−50	−46	−38
Abrasion index[b]	100	(145)	120	100	75–85
Skid and traction index (wet)[b]	100	(90)	90–100	95	95–100

[a] Tread formulations contained 2.1 phr sulfur and from 0.9 to 1.2 phr accelerator (Santocure NS) plus typical amounts of zinc oxide, stearic acid, and antioxidants.

[b] Determined on retread passenger tires. Results for 35% vinyl PBD estimated from results at higher black and oil levels.

under very mild conditions, even at room temperature and ambient pressure if necessary.

Normal low *cis*-polybutadiene, which tends to be a soft cheeselike polymer in the raw state, is converted by hydrogenation into either a tough thermoplastic elastomer or a very tough thermoplastic, depending on the degree of hydrogenation. Figures 17 and 18 illustrate the enormous influence on properties occurring as the degree of hydrogenation exceeds 50% of the double bonds present.

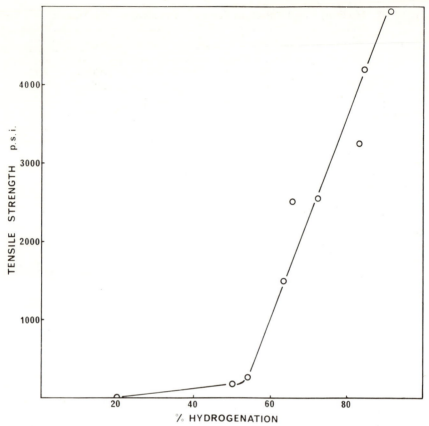

Figure 17 Effect of hydrogenation on tensile strength of raw polybutadiene.

3.2 Polyisoprene

3.2.1 *Preparation*

Earlier it was seen that with lithium catalysts in the absence of polar solvents or electron donor additives, isoprene polymerizes to a high cis-1,4 content, i.e., about 90% cis-1,4. This is somewhat less than the 96% obtainable with Ziegler catalyst combinations containing an aluminum alkyl and titanium tetrachloride. The synthetic polyisoprenes are primarily desired as direct replacements for natural rubber, which is 97–100% *cis*-1,4-polyisoprene if one disregards the nonrubber and oxygenated constituents (27). Since, in the case of polyisoprenes, the processing characteristics, crystallinity under stress, and green strength depend mainly on the steric purity and also on the molecular weight distribution, the differences between these two types of synthetic polyisoprenes are

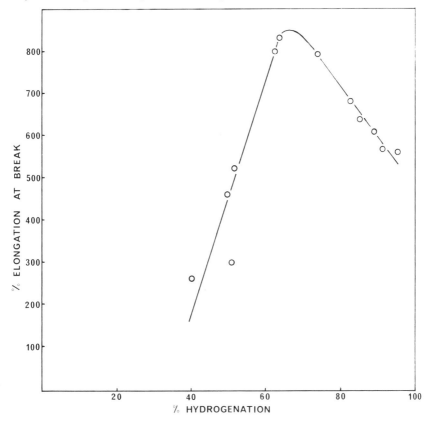

Figure 18 Effect of hydrogenation on elongation at break.

more significant in relation to their applications than was the analogous case for the butadiene rubbers. Although the very high *cis*-polyisoprene prepared with the Ziegler catalysts can be used as a total replacement for natural rubber, the lithium alkyl type should not be treated as if it were natural rubber (28).

At present only the Shell group in Europe manufactures polyisoprene using a lithium alkyl catalyst; all other manufacturers of polyisoprene use Ziegler catalysts. In the Shell method (*a*) the polymerization is simple and straightforward in initiation, propagation, and reaction control; (*b*) polymer can be prepared directly from the dehydrogenation product of mixed isoamylenes; (*c*) corrosive and oxidation promoting catalyst remnants are absent; and (*d*) the product is pure, colorless, and consistent in quality with no gel present even at the highest molecular weights.

Shell has utilized its access to isoamylene/isopentane fractions from refinery streams to reduce one of the major cost items in the production of synthetic polyisoprene, namely, the monomer. Dehydrogenation of this fraction yields isoprene in isoamylenes (or isoamylene/isopentane mixture) since high selectivity does not occur at high conversion. This isoprene can be polymerized using the residual isoamylenes as the solvent medium for both the catalyst and the rubber. After separation of the finished rubber the isoamylenes can be recycled for further dehydrogenation. Neither lithium nor Ziegler types of catalysts require high purity isoprene, per se, but both require the systems to be free of catalyst poisons, the most common of which are air, moisture, acetylenes, cyclopentadiene, and most polar molecules.

The Shell process is believed to employ preferably secondary butyl lithium as the initiator in order to control the molecular weight spread. If this is used as the initiator, the usually slow initiation reaction for isoprene is completed before 10% conversion has been reached and no free lithium alkyl is present after this stage (29). n-Butyllithium in isoprene is not completely initiated even at the end of the polymerization (17). This contrasts with n-butyllithium in the polymerization of butadiene where no unreacted n-butyllithium remains after 10% conversion has been reached (29). More recent work (30) which reexamined this phenomenon, allowing for impurities such as lithium butoxides and lithium hydroxide, has reaffirmed the differences between n- and s-butyllithium.

The active catalyst concentration has, of course, a profound effect on the rate of reaction, intrinsic viscosity, and cis content of the polymer obtained. A decrease in the catalyst concentration favors higher cis contents and higher molecular weights but slows the rate of polymerization. Minimum requirements are 89% cis-1,4 content (as measured by infrared techniques) (31) with an intrinsic viscosity (in toluene or isooctane at 30°C) (32) of between 6 and 10. In order to achieve this with the mixed feed of isoprene/isoamylenes some monomer purification step such as treatment with a sodium dispersion is necessary. Because of the higher cement viscosities involved, polyisoprene is run to about 15% concentration in solution compared with nearer 25% in the case of polybutadiene.

In contrast to the Ziegler process using transition metal halides, where gel problems can be encountered (at intrinsic viscosities of 3–4, as much as 25% gel may be formed) (33), the lithium alkyl process produces practically no gel even at intrinsic viscosities up to 10.

3.2.2 Processing: Green Strength and Milling

A very important property in processing a raw rubber is its ability to pull itself over the mill rolls. In the case of polyisoprene this is related to its ability to crystallize under stress and to the rubber's green strength. A

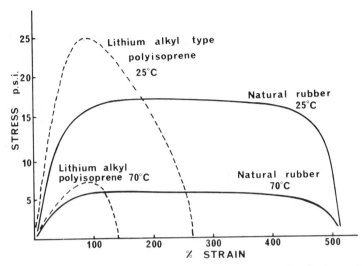

Figure 19 Green strength: stress-strain properties of unvulcanized tread stocks (extension values 0.2 in./min).

small amount of loose gel can assist in improving the green strength, though it is generally agreed that both green strength and tack improve significantly as the cis content increases toward 96%. The green strength can be determined by examining the stress-strain relationship of uncured compounded stocks as shown in Fig. 19.

Isomerized natural rubber in which the cis content has been reduced

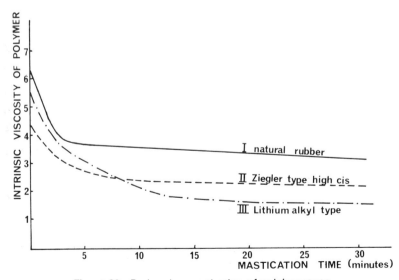

Figure 20 Brabender mastication of polyisoprenes.

exhibits a marked reduction in green strength. However, trans-1,4 struc-ture has a much less deleterious effect on crystallization than 1,2 or 3,4. For this reason the butadiene-isoprene copolymers prepared using the Alfin catalysts (see later) are of particular interest. For polyisoprenes with about 90% cis content, which tend to tear before crystallizing, a blend with about 25% natural rubber has adequate green strength and good tack. Blending with butadiene rubber or SBR presents no problems.

Figure 20 shows the effect of masticating in a Brabender natural rubber (I), Ziegler type, very high cis-polyisoprene (II), and the polyisoprene prepared using alkyllithium catalyst (III).

Table 17 Typical Formulations and Properties of Vulcanized Pure Gum Polyisoprenes

	Cis-1,4 Content (%)		
	Synthetic Polyisoprene		Natural Rubber
	96.5	92	100
Component	Formulation, Parts by Weight		
Polymer	100.0	100.0	100.0
Zinc oxide	6.0	3.0	5.0
Stearic acid	4.0	3.0	2.0
Antioxidant	1.0	1.0	1.0
Sulfur	3.0	1.5	2.75
Zinc salt of 2-mercapto- benzothiazole	1.0		
N-Cyclohexylbenzothiazole- 2-sulfenamide		0.3	
Hepteen base		0.1	
Ridacto		0.3	
Benzothiazyl disulfide			1.0
Tetramethyl thiuram disulfide			0.1
Total	115.0	109.2	111.85
Cure	60 minutes at 250°F	15 minutes at 293°F	15 minutes 287°F
Tensile strength (psi)	3900	3710	4300
Elongation (%)	735	910	730
Modulus at 300% (psi)		160	300
at 500% (psi)	750		
Hardness (Shore A)	42	35	41

As in the case of the two types of linear polybutadienes described earlier, it would appear once again that with optimum processing and compounding, both synthetic polyisoprenes give similar results in the main applications of radial tires and truck tires.

Tables 17 and 18 compare the gum vulcanizates and HAF black reinforced vulcanizates, respectively (34). The polyisoprene in Table 18 is a lithium polymer. The manufacturer's literature should be consulted for up-to-date recipes and properties.

Table 18 Typical Properties of Vulcanized Black Reinforced Polyisoprene Compounds (50 phr HAF Black)

	cis-1,4-Polyiso-prene	cis-1,4-Polyiso-prene, Oil Extended	Natural Rubber	SBR	cis-1,4-Poly-butadiene
Polymer	100	100	100	100	100
HAF Black	50	50	50	50	50
Processing oil	5		5	8	8
Sulfur	2.75	2.75	2.75	1.75	1.75
Tensile strength (psi)	3570	3080	3960	3760	2880
Modulus at 300% (psi)	1630	1330	1990	1690	1300
Elongation (%)	520	530	550	550	500
Hardness (Shore A)	56		63	58	56
Angle tear (lb/in.)	380	430	600	320	300
Yerzley resilience (%)	71	73	72	60	74
Heat buildup Goodrich flexom- eter, ΔT ($^\circ C$)	19	22	18.5	24	18

3.2.3 Solution Masterbatching

As yet no solution carbon black masterbatches of synthetic polyisoprenes have appeared on the commercial market, though much has been made of this idea in the patent literature (35). One can foresee the potential of solution blending polybutadienes and polyisoprenes followed by solution masterbatching with carbon black in order to maximize efficiency and reduce the need for Banbury mixing.

3.2.4 *Latex Manufacture*

The direct manufacture of a high *cis*-1,4-polyisoprene by emulsion polymerization to yield a latex has not yet been achieved. Shell has produced a latex from their solution polyisoprene (36). This has resulted from emulsification of the hydrocarbon solution in water and subsequently removing the hydrocarbon phase by flash distillation. Such lattices have been used to make surgical gloves and other dipped goods. The lower modulus of the alkyllithium polyisoprene gum vulcanizate (Table 17) gives a softer, more pleasant glove. Tear strength and other properties are excellent (37).

3.25 *Nonrubber Applications*

Synthetic polyisoprenes that are free from gel are particularly satisfactory for chlorination, hydrochlorination, and cyclization in solution. The chlorinated product and the cyclized rubber find ready application in the surface coatings industries. The hydrochlorinated material has been used as a transparent plastic wrapping for foodstuffs. All these materials were prepared previously from natural rubber.

3.3 Solution Copolymers of Butadiene with Styrene

3.3.1 *Random Copolymers*

"General purpose" applications of a synthetic rubber usually means use in various aspects of tires. Linear butadiene polymers cannot be used successfully in general purpose applications unless blended with a suitable SBR or natural rubber. Although polybutadienes are excellent in improving abrasion resistance and reducing heat buildup in tires, in the unblended state they lack the easy processability, oxidation stability, tear strength, and road holding properties associated with emulsion styrene-butadiene copolymers. The attempt to copolymerize these two monomers in solution by simple addition of an alkyllithium initiator to a mixture of styrene and butadiene in a hydrocarbon solvent results in a copolymer containing long sequences of polybutadiene blocks and polystyrene blocks.

Although in the homopolymerization of these monomers with alkyllithiums in hydrocarbon solution styrene polymerizes more rapidly than butadiene, it does not polymerize first in the mixture. Butadiene is incorporated (with a small amount of styrene) into the copolymer first and only then is the residual styrene free to form blocks by homopolymerization onto the polybutadienyllithium (38-40).

Such solution styrene-butadiene block copolymers are very useful elastomers for specific applications such as molded shoe soles, but they give inferior properties as tire rubber. Abrasion resistance under dynamic stress conditions is poor compared with that of normal emulsion SBR.

It was found that random distribution of the styrene units could be achieved by either incremental addition of the butadiene monomer (41) or by the use of "randomizing" agents such as potassium tertiary butoxide (42) added to the lithium alkyl catalyst system. The polar additive used as a randomizing agent must not significantly alter the steric form of the butadiene unit and hence materials such as ethers and amines cannot be used since they result in relatively high levels of 1,2 configuration being formed, which raises the glass transition temperature and is an important factor in determining the properties of the copolymer.

Several companies now produce random solution styrene-butadiene copolymer rubbers commercially (Firestone, Phillips, Asahi, and ISR).

The first to appear on the market was Firestone's Duradene (now called Stereon). The Duradene styrene-butadiene copolymers were claimed to offer dimensional stability during storage, good processing characteristics in factory operations at 100% level, and better abrasion resistance and equal wet pavement traction compared with emulsion SBR (43). Table 19 summarizes the basic differences between the raw random solution styrene-butadiene copolymer and emulsion SBR.

In the work on tire tread stocks from random solution styrene-butadiene copolymers Firestone did the main pioneering (43, 44), entailing exhaustive examination of the many factors and parameters involved in the synthesis, structure, and macrostructure of these materials in addition to compounding, vulcanization studies, and tire testing. From the mass of observations the following points can be made.

Stereostructure. An increase in the 1,2 content of the butadiene linkages beyond 10%, increases the glass transition temperature (T_g) of the raw polymer and increases the Young's modulus index (YMI) of the tread compound. The YMI is defined as the temperature at which the rubber compound has stiffened to the point of having a modulus value of 10,000 psi. Rubber is considered to be too stiff to be usable at or below the YMI. The YMI is considered a more relevant measure than T_g since it takes into account the effect of other constituents in the compound, e.g., oil and black. As the YMI increases, the coefficient of friction and traction increase, tread wear rating and rebound decrease.

Table 19 Comparison of Solution SBR and Emulsion SBR

Property	Random Solution SB Copolymer	Emulsion SBR (Cold)
Stereostructure of butadiene units	10% 1,2 35–40% cis-1,4 50–55% trans-1,4	20% 1,2 10% cis-1,4 70–73% trans-1,4
Molecular weight distribution	Narrow (or of controlled moderate broadness)	Broad to very broad
Presence of very low and very high molecular weight fractions and gel	Absent	Present
Chain branching	Absent	Present
Distribution of styrene units	Blocks should be <2%; styrene units are substantially equally randomly distributed	No blocks present; tendency for styrene percentage to increase in chain at higher conversions
Nonrubber constituents	Only trace of lithium compounds present	7% of fatty acid/or rosin acid present

Styrene Units.   An increase in the styrene content increases both the T_g and YMI. Processability (thermoplasticity) and tensile strength increase while rebound decreases as the styrene content is increased. In the Firestone Stereons the optimum bound styrene level compatible with good processing, improved wear resistance, and no loss of wet road traction appears to be between 18 and 21%. The Stereons cover styrene contents of 10% (Stereon 720), 18% (Stereon 750, which is oil extended), and 20% (Stereon 700 and 702) according to the IISRP *Elastomers Manual* (45). Above 18% styrene an appreciable loss in abrasion resistance is noticed (43). The distribution of styrene in the form of blocks has a detrimental effect on wear rating and increases thermoplasticity.

Molecular Weight Distribution.   A very narrow molecular weight distribution leads to poor processing characteristics. A controlled, broader distribution gives greatly improved processing characteristics and Garvey die extrusions. The elimination of very low molecular weight fractions improves physical properties, abrasion resistance, rebound (heat buildup), and oil extension (43). The presence of very high molecular weight polymer fractions and gel leads to poor processing. The gel may lead to weak vulcanizates if not broken down during compounding.

Table 20 Effect of Butadiene Microstructure and Copolymer Composition on Tread Properties (44)

Polymer Type	SBR	Copolymer with High Vinyl	Stereon Rubber	Diene Rubber
Catalyst	Free radical	Alkyllithium plus ether	Alkyllithium	Alkyllithium
Medium	Emulsion	Solution	Solution	Solution
Molecular weight distribution	Very broad	Moderate	Moderate	Moderate
% styrene	25	25	21	0
Microstructure butadiene portion				
% cis-1,4	10	23	33	36
% trans-1,4	70	49	58	55
% vinyl (1,2)	20	28	9	9
% steel ball rebound				
73°F	31	33	42	52
212°F	54	61	62	63
Young's modulus index (°C)	-39	-42	-55	-70
Coefficient of friction on wet concrete	100	100	98	85
Wear rating	100	100	135	145

Chain Branching. The presence of branching increases processability since linear polymers show a greater degree of Newtonian rheological behavior. Branched polymers exhibit non-Newtonian characteristics (44). Tri- and tetrachain branching can be produced intentionally in both polybutadiene and solution styrene butadiene copolymers by coupling reactions on the live chains, e.g., by the addition of $SiCl_4$ to the polymer cement (46). The introduction of chain branching reduces the tendency to cold flow.

Some of these effects are illustrated for tread stocks in a comparison given by Weissert and Johnson (44) in Table 20.

3.3.2 Oil Extension of Random Solution Styrene-Butadiene Copolymers

In the same way that the polybutadienes lend themselves to high oil extension, the solution copolymers can take good loadings of oil and black. Firestone's Stereon 750 contains 37.5 parts of an aromatic oil, as does Asahi Tufdene 1530. These rubbers are particularly suited to tire manufacture i.e., treads and camelback (47). Tables 21 and 22 show the results with a passenger tread and a truck tread compound, respectively, using this oil extended solution styrene-butadiene copolymer.

Table 21 Properties-Passenger Tread Compound

Recipe	Parts by Weight
Solution SBR (Tufdene 1530)	137.5
Sulfur	1.65
Accelerator CZ	1.7
Zinc oxide	5
Stearic acid	2
Antioxidant HP	1
Sonic X-140[a]	17.5
Seast 6[b]	80
Total	245.9

Properties[c]	
Compound Mooney viscosity (ML$_4$, 100°C)	55
Mooney scorch time (minutes, t_5)	49
Hardness (Hs, JIS)	60
300% modulus (kg/cm^2)	82
Tensile strength (kg/cm^2)	194
Elongation (%)	550
Tear strength (kg/cm)	79
Akron abrasion (cc/1000 rads)	0.044
De Mattia (cut growth, cycle)	3700
De Mattia (cut initiation, 100°C, cycle)	8200
Resilience (%)	37
Goodrich heat buildup (°C, ΔT)	20
Heat aging resistance (100°C, 96 hours):	
Hardness change (%)	+13
300% modulus change (%)	+66
Tensile strength change (%)	−18
Elongation change (%)	−36
Tear strength change (%)	−20

[a] Aromatic oil.
[b] ISAF carbon black.
[c] Note: Press cure, 141°C, 60 minutes.

Table 22 Properties-Truck Tread Compound

Recipe	Parts by Weight
Solution SBR (Tufdene 1530)	55
NR RSS 4	60
Sulfur	1.5
Accelerator NOBS	1.4
Zinc oxide	3.9
Stearic acid	3.4
Antioxidant HP	1.3
Antioxidant B	1.3
Suconox	0.4
Sonic X-140[a]	10
Seast 3[b]	20
Seast 6[c]	25
Seagal 300[d]	20
Total	203.5

Properties[e]	
Compound Mooney viscosity (ML_4, 100°C)	37
Mooney scorch time (minutes, t_5)	60
Hardness (Hs, JIS)	60
300% modulus (kg/cm^2)	94
Tensile strength (kg/cm^2)	243
Elongation (%)	560
Tear strength (kg/cm)	108
Akron abrasion (cc/1000 revs.)	0.046
De Mattia (cut growth, cycle)	14,000
De Mattia (cut initiation, 100°C, cycle)	49,000
Resilience (%)	38
Goodrich heat buildup (°C, ΔT)	15
Heat aging resistance (100°C, 96 hours):	
Hardness change (%)	+13
300% modulus change (%)	+49
Tensile strength change (%)	−38
Elongation change (%)	−43
Tear strength change (%)	−41

[a] Aromatic oil.
[b] HAF carbon black.
[c] ISAF carbon black.
[d] CRF carbon black.
[e] Note: Press cure, 141°C, 60 minutes.

3.3.3 *Severity of Tire Testing*

Polybutadiene tires wear much better than emulsion SBR tires under conditions of high severity, i.e., high speed and low temperature. At low severity the reverse was first observed (48) but it was shown that the relative wear rating improved with decreasing YMI of the tread compound. At high severity, random solution styrene-butadiene copolymers gave significantly better results than emulsion SBR (44).

3.3.4 *Hydrogenation of Random Solution Styrene-Butadiene Copolymers*

In an analogous manner to the hydrogenation of linear polybutadiene random solution styrene-butadiene copolymers hydrogenate to give interesting strong thermoplastic elastomers (49). The optimum bound styrene content for this purpose was 8%.

3.4 Block Copolymers

3.4.1 *Preparation*

The versatility and uniqueness of living anionic systems can be demonstrated in the synthesis of block copolymers. Here the still living anionic end of a polymer chain A can initiate the polymerization of another monomer B present in the polymerization medium to result in the formation of an A–B block copolymer (50). If termination and chain transfer reactions have been eliminated, the A–B block copolymer will be free of A or B homopolymer. With mastery of the techniques of anionic polymerization, it is possible to synthesize a wide range of block polymers. With the use of difunctional initiators A–B–A block terpolymers can be prepared by the initiation of polymerization of monomer A at both growing ends of polymer B. Alternatively, A–B–A block terpolymers can be synthesized by sequential addition of B to living A followed by A to the growing A–B, or by chemical coupling of two growing A–B moieties.

This method allows the strict control of the molecular weight, chemical composition, and order of the blocks in the polymerization. The anionic method is the only good method now known which allows the production of block copolymers with such a controlled composition. Chapter 5 describes the theory and techniques for preparation and many properties of the block polymers.

Other methods used for the preparation of block copolymers are by the polymerization of mixtures of monomers or even by the sequence polymerization of one monomer resulting in a stereo block copolymer; for example, an A block of predominantly 1,4-polybutadiene units can be copolymerized with a B block consisting of mainly 1,2 units.

There are limitations on the versatility of this technique. For example, an A–B block copolymer where A is styrene and B is ethylene glycol can be synthesized by anionic methods, whereas the reverse is not true since living polyethylene glycol will not initiate polymerization of styrene (51).

Finally, mention must be made of a class of block copolymers known as tapered polymers. These are A–B block copolymers linked through a copolymeric block which is initially richer in A than B but becomes richer in B. The symbol A→B has been proposed for such a unit (52) and thus the total block copolymer just described can be illustrated by A–A→B–B.

The study of block copolymers gained impetus in 1963 when workers at Shell Development Company showed that potentially commercial products having novel and desirable properties could be obtained by the living anionic polymerization technique (53). The block copolymers with the interesting properties were of the A–B–A type, where A was a polystyrene block and B a polybutadiene block. These materials, now commonly referred to as thermoplastic elastomers, can be processed *without vulcanization* as thermoplastics using conventional machinery to give products having high resilience, high tensile strength, reversible elongation, and good abrasion resistance. The unusual physical properties can be ascribed to the incompatibility of the elastomeric polybutadiene portion and the polystyrene clusters which arise from the aggregation of the polystyrene ends. At ambient temperatures these domains act as labile cross-linking sites (54). This is shown schematically in Fig. 21 and more quantitatively in Fig. 8 of Chapter 5. The polystyrene phase can act as both a cross-link and a filler.

The variations in a three-block system such as this are numerous but, generally speaking, the terminal nonelastomeric blocks should preferably have an average molecular weight of 5000–50,000 and a glass transition temperature at least above 30°C and preferably above 50°C, whereas the elastomeric middle block should have a preferred molecular weight of 50,000–500,000 and a glass transition temperature at least below 10°C and preferably below −25°C (53). Thus the elastomeric middle block can be prepared from almost any conjugated diene. A host of possibilities exists for the nonelastomeric end block, but vinyl aromatic polymers are preferred, styrene being the usual choice. The alkyllithium and especially secondary alkyllithium compounds are usually employed as catalysts in this process. The microstructure of the middle elastomeric block should be high 1,4 in order to achieve elastic behavior, and consequently the

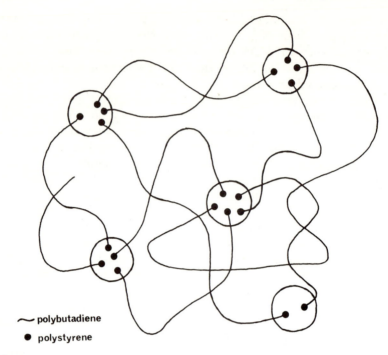

~ polybutadiene

● polystyrene

Figure 21 Schematic representation of domain formation of A–B–A block copolymers.

polymerization is usually carried out in a hydrocarbon solvent. This criterion prevents the use of most dilithio initiators which are preparable or stable only in ether solution since the presence of an ether in an anionic polymerization is known to lead to a high 1,2 and/or 3,4 content in the polydiene (24). Recently, however, an A–B–A block copolymer having the structure poly-α-methylstyrene-polyisoprene-poly-α-methylstyrene has been prepared (55) using 1,4-dilithio-1,1,4,4-tetraphenylbutane in cyclohexane containing a small amount of anisole which has little or no effect on the 1,4 structure of anionically prepared polydienes (8). Tensile strength measurements showed that a higher tensile strength was obtained if α-methylstyrene replaced styrene in an A–B–A block copolymer. Dilithiostilbene has also been used (56) as a diinitiating catalyst for the preparation of a three-block thermoplastic elastomer.

One of the most suitable methods for the preparation of an A–B–A block copolymer is by the coupling of two A–$B_{1/2}$ moieties. This method has the advantage that only two polymerization steps are involved instead of three, and that a doubling of molecular weight occurs in one step. A wide variety of reagents have been used to couple two living block copolymers, including halogens (57), dihaloalkanes (58), and diethylenic aryl hydrocarbons (59). However, the reaction of living polymers with halogen containing compounds to eliminate lithium bromide is somewhat inefficient due to side reactions. The use of gases to chain extend living polymers is difficult because of dispersion problems although both carbon monoxide (60) and carbon dioxide (61) have been claimed as coupling agents. A more efficient system would appear to be the use of dicarboxylic acid esters (62).

The polybutadiene portion of an A–B–A block copolymer is subject to oxidative attack but this inherent weakness can be reduced by hydrogenation of the central diene portion (63). This can be accomplished conveniently by using a homogeneous hydrogenation catalyst which will operate at ambient temperatures in hydrocarbon solvent at atmospheric pressure. Another disadvantage of the A–B–A block copolymers is that since they are not vulcanized they possess low solvent resistance. However, if the polystyrene portion of the A–B–A block is replaced by vinyl pyridine and the resultant block copolymer is treated with hydrogen chloride (64), a solvent resistant thermoplastic elastomer can be obtained. Recently thermoplastic elastomers of the A–B–C type (65) have been claimed where A is polystyrene, B polyisoprene, and C polyethylene sulfide. Novel monomers such as octamethylcyclotetrasiloxane have been incorporated into A–B–A block copolymers (66).

Block copolymers of the A-B type prepared from styrene and butadiene have very different properties from those having the A–B–A structure. Unless they are vulcanized in the conventional manner A–B block polymers containing up to 33% styrene have very poor tensile strength and elongation (67).

Other types of stereo block copolymers may also be produced by varying the ionic nature of the organolithium initiator. A case in point (68) is the synthesis of a polybutadiene stereo block copolymer. The polymerization of butadiene with butyllithium to a certain block size followed by the addition of tetrahydrofuran and further butadiene gave a stereo block copolymer. The microstructure of the first block was 46% trans-1,4, 45% cis-1,4, and 9% 1,2, while that of the second was 16% trans-1,4, 11% cis-1,4, and 73% 1,2. Stereo block polyisoprenes have been prepared in a similar manner.

In a similar way (69) butadiene and isoprene can be copolymerized in blocks to give A–B–A copolymers. It has been shown that treatment of such block copolymers with hydrogen chloride results in a self-curing elastomer. Thus, a polyisoprene-polybutadiene-polyisoprene block when treated with hydrogen chloride gives a thermoplastic elastomer (70). It

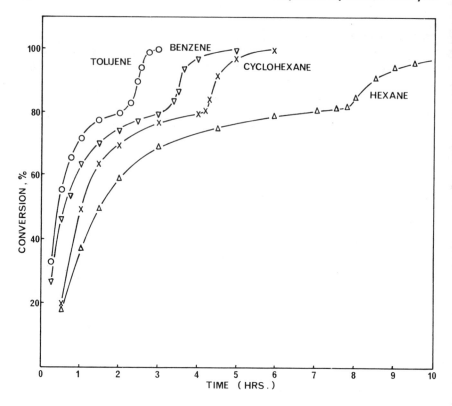

Figure 22 Polymerization of butadiene-styrene (75:25) in different hydrocarbons at 50°C.

would appear that in this treatment only the polyisoprene moiety of the block copolymer is affected by hydrochlorination.

If a mixture of styrene and butadiene is copolymerized in a hydrocarbon solvent using an anionic catalyst, a tapered block copolymer will result (see Section 3.3). The formation of the essentially A–B→B–B block copolymer can be followed visually in that when all the butadiene is used up the reaction mixture assumes the deep red color of the polystyrene anion. The point at which the color appears corresponds to an inflection in the conversion curve at 80%, as shown in Fig. 22. The fact that polystyrene blocks occur only after the inflection point has been reached has been shown by taking samples at various conversions and determining the polystyrene content after oxidative degradation (71) using osmium tetroxide and t-butyl hydroperoxide (72).

3.4.2 Branches and Grafts

A recent advance in the chemistry of living anionic polymers and the synthesis of new elastomers by anionic methods has been the synthesis of tailor-made star and comb shaped polymers (Fig. 23).

To form a star shaped polymer, the living anionic end of the polymer chain is made to react with a polyfunctional reagent. The ultimate number of arms of the final polymer then depends on the functionality of the reagent. Although this novel approach to new polymers is theoretically correct, in practice it is difficult to obtain a monodisperse star polymer, mainly because of steric effects. Other than theoretical interest in the preparation and properties of this type of molecule, the production of star shaped polymers has great industrial significance since a method exists here for trebling or quadrupling the molecular weight of a polymer in a one-step reaction.

The production of a high Mooney polymer from a low one has several obvious advantages in that the need to polymerize to a high molecular weight is eliminated and the resultant nonlinear polymers do not exhibit "cold flow" properties. Some results obtained (46) for the reaction of polybutadienyllithium with silicon chlorides are given in Table 23.

A polybutadiene with the greatest molecular weight is obtained when stoichiometric equivalents are used. This is readily observed by viscosity measurements since the viscous flow is drastically changed by branching. As can be seen in Fig. 24, a peak in the measurement of intrinsic viscosity or Mooney was observed in the reaction of polybutadienyllithium and silicon tetrachloride when a slight excess over a PLi : SiCl$_4$ mole ratio of 4 : 1 was used. Gel and very high molecular weight fractions were absent.

In a similar manner, carbon tetrachloride has been used as the coupling agent in the presence (73, 74) and absence (75) of ethers. Other active halogen compounds that have been used to increase the molecular weight

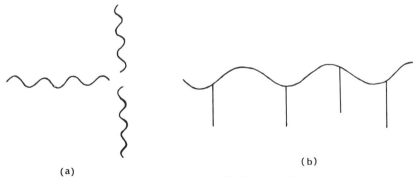

(a)   (b)

Figure 23 Star and comb shaped polymers.

Table 23   Typical Results for Reaction of Polybutadienyllithium with Silicon Chlorides

Silicon Chloride	BuLi (effective) (mmoles/50 g butadiene)	Mole ratio (PLi/silicon)	Intrinsic Viscosity $[\eta]$		Mooney of Product
			Parent	Product	
$(CH_3)_3SiCl$	—	0.5	1.35	1.39	—
$(CH_3)_2SiCl_2$	0.25	1.2			56
		1.8	1.97	2.99	103
		2.0			67
CH_3SiCl_3	0.75	1.9		1.58	11
		2.6		1.68	40
		2.8	0.99	1.71	51
		3.2	1.00	1.63	40
		3.5		1.55	31
		4.4		1.45	19
		9.7	0.99	1.17	4
$SiCl_4$	1.1	3.0		1.41	40
		4.0		1.51	51
		4.3	0.84	1.61	52
		4.8	0.82	1.57	50
		6.1		1.49	30

of a live polymer are based on polyhalo heterocycles (76) such as 2,4,6-trichloro-s-triazine.

The use of divinylbenzene has been investigated for polyisoprene (77) and as a coupling agent for alkyllithium polymerized polymers and copolymers (78).

There seem to be many ways of synthesizing comb shaped polymers (Fig. 23b). One method which is being investigated widely is that of anionic grafting of a monomer onto a metallated backbone polymer. In 1965 Russian workers (79) reported the direct metallation of polystyrene using the complex formed by the reaction of n-butyllithium with tetramethylethylenediamine. The formation of a lithium-containing polymer was proved by reaction with solid carbon dioxide, benzaldehyde, benzophenone, and the anionic grafting of methyl methacrylate and acrylonitrile. No mention was made of the anionic grafting of a diene monomer onto the polystyrene backbone. The reaction was thought to proceed via a 1:1 butyllithium-tetramethylethylenediamine complex.

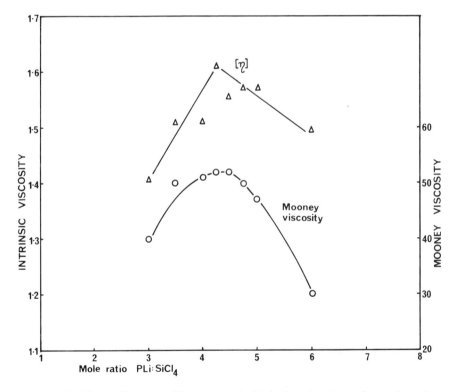

The direct metallation of *cis*-1,4-polybutadiene and *cis*-1,4-polyisoprene using the same catalyst system has also been reported (80). In this case the lithiated polymer was characterized by reaction with various reagents, including carbon dioxide, Michlers ketone, trimethyl- and triphenylchlorosilane, benzaldehyde, and pyridine. The same group later reported (81) the graft polymerization of styrene, methyl methacrylate, and acrylonitrile on the polydiene backbone. In the styrene graft the

Figure 24 The effect on Mooney and intrinsic viscosity of varying the polybutadienyllithium/silicon tetrachloride ratio.

final product could be separated into the true grafted polymer and polystyrene resulting from polymerization of styrene by residual butyllithium.

A copolymer of butadiene (80–90%) and vinyl naphthalene (10–20%) was lithiated by a butyllithium-tetrahydrofuran complex followed by the addition of styrene or vinyl naphthalene to produce a grafted polymer with laterally attached segments of polystyrene (82). Little or no homopolymer is formed since the unreacted butyllithium is selectively destroyed.

In a similar manner poly(2,6-dimethyl-1,4-phenylene ether) was lithiated with butyllithium-tetramethylethylenediamine or tetrahydrofuran and the resultant polymer grafted with isoprene (83). Amines other than tetramethylethylenediamine have been used in conjunction with organolithium compounds (84) and the use of a potassium t-butoxide-butyllithium complex (85) has been suggested for the metallation of unsaturated polymers. It is interesting to speculate on the nature of the metallating agent in this case since it has been found that the reaction between an alkyllithium and an alkali metal t-butoxide results in formation of the alkali metal alkyl (86).

$$t\text{-BuOK} + \text{BuLi} \rightleftharpoons \text{BuK} + t\text{-BuOLi}$$

It has been suggested (82) that the products obtained by grafting styrene onto a butadiene–2-vinyl naphthalene copolymer should exhibit elastomeric properties comparable to the styrene-butadiene A–B–A block copolymers mentioned earlier, provided the number of grafts per backbone is small. Two recent patents from Dunlop have described good physical properties for a polyisoprene (87) and a polybutadiene (88) styrene graft obtained using a butyllithium-tetramethylethylenediamine complex.

3.4.3 *End Group Reactions*

The chemistry of anionic polymerization is primarily concerned with the anionic end of the polymer chain. With monofunctional initiators polymerization occurs at this one living end, and termination of the polymerizing carbanion with various reagents can give rise to a whole host of polymers containing a functional group at this end. By the use of a suitable diinitiator species a polymer can be produced having two "live" ends which can be terminated to give α,ω difunctional polymers. Although α,ω difunctional polymers can be produced by the oxidative degradation of unsaturated elastomers or by the free radical polymerization of an unsaturated monomer, such as butadiene, by an initiator containing a functional group, the use of dianionic species is advantageous in that better control of the degree of polymerization and functionality of the polymer can be exercised than with other methods.

The term *telechelic* has been proposed (89) for polymer molecules possessing two terminal functional groups. Usually the molecular weight of the difunctional polymer is kept below 5000, and consequently the term *liquid polymers* has been used to describe this class of polymer. The molecular weight is purposely kept low in order to produce a polymer that can be poured and used as a true prepolymer. The functional groups are usually hydroxyl, carboxyl, amine, halogen, or thiol and the polymer can be used with a wide range of curing, cross-linking and chain extending agents. The diene that has been most widely investigated is butadiene. These materials have so far been limited to specialty uses such as binders for rocket propellants and in mastics and adhesives applications. However, the curing of telechelic polymers through their end groups results in a three-dimensional network with (theoretically) no free chain ends, whereas in a normal vulcanizate free chain ends exist which do not contribute to the modulus.

In a cured telechelic system the distance between cross-links is constant and contributes to a more efficient three-dimensional network. For a perfect network to be formed it is essential that the functionality of the telechelic polymer be precisely two; experimentally this is most difficult to achieve. In other words, the two living ends of the polymer must both terminate by reaction with the chosen reactant rather than with solvent or impurity or by an isomerization process. The termination reaction itself must also go to completion. In termination by ethylene oxide (90) or carbon dioxide anions are formed which tend to associate with a resultant increase in viscosity. In some cases gels are formed which are broken only on the addition of a proton donor.

The initiators that have been used to polymerize dienes to provide telechelic polymers include the alkali metal-polynuclear aromatic hydrocarbon complex or dimetallic alkyls. In the former, the initiators normally used are a combination of sodium or lithium and naphthalene or biphenyl. In either case a strongly polar solvent must be used for the preparation and stabilization of the complex. Usually an ether such as tetrahydrofuran is employed and the 1,2 or vinyl content in the case of a polybutadiene is then extremely high. With a lithium-naphthalene initiated polymerization in tetrahydrofuran the final 1,2 content of the polymer is over 90% (91). Of the dimetallic alkyls, the most widely used example appears to be the reaction product from lithium and *trans*-stilbene (92). Other initiators that have been reported are the dilithium complexes of aromatic ketones (93).

Many reagents can be used to terminate the anionic ends of a living polymer, but the following end groups are the most important and interesting:

1. *Hydroxyl.* The most widely used terminating agent for the introduction of hydroxyl groups is ethylene oxide (94–96):

$$\sim\!CH_2^{\ominus}Li^{\oplus} + CH_2\!\!-\!\!CH_2 \longrightarrow \sim\!CH_2\!\!-\!\!CH_2\!\!-\!\!CH_2\!\!-\!\!OLi$$
$$\overset{O}{\diagdown\diagup}$$

$$\xrightarrow{H^+} \sim\!CH_2\!\!-\!\!CH_2\!\!-\!\!CH_2\!\!-\!\!OH$$

As mentioned previously, the association of alcoholate terminals leads to an increase in the solution viscosity. Other reagents that have been used are aldehydes and ketones (97).

2. *Carboxyl.* The usual reagent for the introduction of carboxyl groups is carbon dioxide (89, 90):

$$\sim\!CH_2^{\ominus}Li^{\oplus} + CO_2 \longrightarrow \underset{O}{\sim\!\overset{\|}{C}\!-\!OLi} \longrightarrow \underset{O}{\sim\!\overset{\|}{C}\!-\!OH}$$

The carbonation of a living polymer can involve bubbling the gas through the polymer solution or pouring the solution over solid carbon dioxide. Both methods are inefficient due to incipient gel formation. Methods have been described involving turbulent flow at a T junction to overcome this problem (98) as well as for the preparation and purification of a commercially available carboxy telechelic polymer (99).

3. *Thiol terminated polybutadienes.* The thiol group has been introduced by terminating polymerization with sulfur (92, 100), cyclic disulfides (101), or an episulfide (102):

a)　$\sim\!\!CH_2^{\ominus}Li^{\oplus} + S \rightarrow \sim\!\!CH_2\!\!-\!\!SLi$

b)　$\sim\!\!CH_2^- Li^+ + H_2C\!\!\underset{S-S}{\overset{CH_2}{\diagup\diagdown}}\!\!CH_2 \longrightarrow \sim\!\!CH_2\!\!-\!\!S\!\!-\!\!CH_2\!\!-\!\!CH_2\!\!-\!\!CH_2\!\!-\!\!S\,Li$

c)　$\sim\!\!CH_2^- Li^+ + H_2C\!\!\underset{S}{\overset{}{\diagup\diagdown}}\!\!CH_2 \longrightarrow \sim\!\!CH_2\!\!-\!\!CH_2\!\!-\!\!CH_2\!\!-\!\!SLi$

The lithium mercaptide formed can then be destroyed on addition of an acid.

4. *Amine or substituted amine terminated polybutadiene.* The indirect introduction of a substituted amine group has been effected by terminating a living polymer with an aldehyde such as N,N'-p-dimethylamino benzaldehyde (103).

By use of reagents of this type reactive groups of two varieties can be introduced into a polymer chain and these can be reacted with two entirely different coupling agents.

The preparation of liquid aziridinyl telechelic polybutadiene has been reported (104). In this case the living polymer end is reacted with an excess of a substituted aziridinyl phosphine oxide.

The lithium derivative is then hydrolyzed as usual.

The phosphine oxide is added in excess to avoid chain extension and coupling via the second aziridinyl group. The aziridinyl substituted telechelic polybutadienes are extremely reactive and can be readily coupled using di- or polyacids or anhydrides.

An aziridinyl telechelic styrene-butadiene copolymer has been cured with thiomalic acid in an accelerated sulfur cure and compared (92) with a sulfur cured emulsion SBR 1500. The physical properties of the cured telechelic polymer were found to be superior to those of the SBR 1500.

The future large-scale usage ot telechelic polymers lies in their replacing the conventional solid polymers in certain applications. It is possible to visualize a process for the manufacture of rubber articles where an α,ω difunctional polymer is premixed with various fillers and a second reagent and poured or injected into a mold. The advantages of a process of this type are many and include the elimination of expensive power- and time-consuming mixing steps. Firestone has been reported to have made a cordless tire by injecting a heated liquid rubber into a mold. Although no mention has been made of the lqiud rubber used, it has been conjectured (105) that the liquid rubber could be a low molecular weight dilithium initiated polymer in which the terminal groups are carboxyl or hydroxyl.

3.5 Alfin Rubbers

The Alfin catalyst has been known since 1947 when A. A. Morton and his co-workers observed that the sodium salt of a methyl n-alkylcarbinol and the sodium salt of an olefin such as propylene catalyze the polymerization of butadiene and isoprene (106). The name Alfin was coined for this type of catalyst (*al*cohol and ole*fin*). The original polybutadiene obtained was highly cross-linked although the polyisoprene was soluble. The active catalyst system apparently consists of two products, sodium isopropoxide and allyl sodium.

$$C_5H_{11}Na \ + \ (CH_3)_2-CH-O-CH-(CH_3)_2 \ \longrightarrow$$

$$(CH_3)_2-CHONa \ + \ C_5H_{12} \ + \ CH_2=CH-CH_3 \qquad\qquad [1]$$

$$C_5H_{11}Na \ + \ CH_2=CH-CH_3 \ \longrightarrow \ CH_2=CH-CH_2Na \ + \ C_5H_{12} \qquad [2]$$

Morton proposed the following structure as a working model of the catalyst:

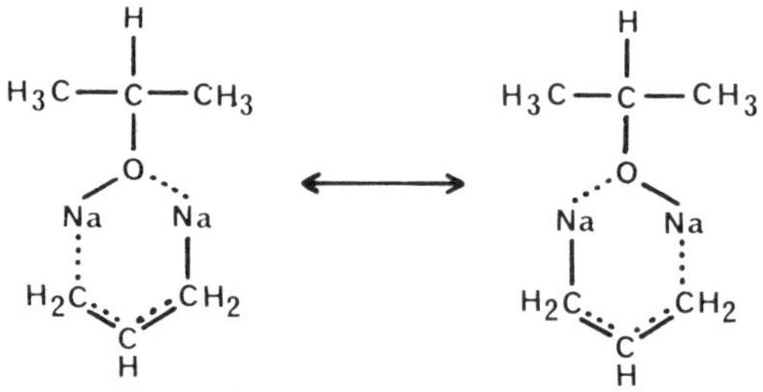

He recognized the importance of the sodium chloride which must be present in the catalyst system from the reaction of sodium with amyl chloride to form amyl sodium. He also recognized that polymerization using an Alfin catalyst was very different from the known alkali metal polymerization of diolefins since reaction with the Alfin catalysts was finished in minutes whereas the conventional sodium catalysts took much longer. Although the Alfin catalysts gave extremely fast reactions the products were very tough unworkable polymers of extremely high molecular weight. With an improved catalyst a polybutadiene was prepared with less than 5% of gel and having intrinsic viscosity of 11 or 13 (107). By 1952 it had been conclusively shown that sodium chloride was essential to the functioning of the Alfin catalyst (108, 109).

The effect of associated salts on the polymerization of butadiene by allyl sodium are shown in Table 24.

Improvements were gradually made in the preparation of the catalyst, including the substitution of the diisopropyl ether by isopropanol (107).

$$C_5H_{11}Na + (CH_3)_2CHOH \rightarrow C_5H_{12} + (CH_3)_2CHONa \qquad [3]$$

$$C_5H_{11}Na + (CH_3)_2CHONa + CH_2{=}CH{-}CH_3 \rightarrow CH_2{=}CH{-}CH_2Na$$
$$+ (CH_3)_2CHONa + C_5H_{12} \quad [4]$$

Table 24[a] Effect of Salts on Allyl Sodium Catalysis

Allyl Sodium with	Yield (%)	Intrinsic Viscosity	% 1,2
NaCl	4	0.5	53
NaOCH(CH$_3$)$_2$	4	0.4	47
NaCl + NaOCH(CH$_3$)$_2$	65	13.0	28

[a] Reprinted from Morton (109) by courtesy of Palmerton Publishing Company.

Recent advances in the preparation of the catalyst with regard to the amounts and the order of addition of the reagents have been made (110). The reaction of a threefold excess of sodium with the isopropanol followed by addition of the butyl chloride results in a sodium alkyl stabilized by the alkoxide. This procedure is quoted as resulting in savings of 50% for the alkyl halide and 25% for sodium; Wurtz type reactions of the sodium alkyl are avoided (110).

The main disadvantage to the use of Alfin rubbers has been their very high molecular weight since early results on gum and tread stocks showed them to have good physical properties (111, 112). Values of 7 million for the molecular weight can be obtained using the Alfin catalyst compared to values of 100,000–200,000 experienced in emulsion polymerization. The Alfin rubbers might have been the workhorse of the synthetic rubber industry instead of SBR during World War II, and indeed even now if the high molecular weight problem could have been solved.

It was not until the early 1960s that the addition of a dihydroaromatic compound to a butadiene polymerization was found to reduce the molecular weight of the final polymer drastically (113). This important discovery (114) was made during an attempt to copolymerize butadiene with monomers obtained by the Birch reduction of aromatic compounds using sodium and alcohol in liquid ammonia.

1,4-dihydrobenzene   1,2-dihydrobenzene

1,4-Dihydrobenzene can be readily isomerized to the conjugated 1,2-dihydrobenzene. Both isomers failed to copolymerize with butadiene but in the presence of 1,4-dihydrobenzene, the viscosity of the final polybutadiene solution was much lower than without it and the yield of polymer was virtually unchanged. Figure 25 shows the effect observed when using dihydrobenzene or dihydronaphthalene as the moderator.

By using 1.5% of dihydronaphthalene as the moderator a final molecular weight of 300,000 can be attained; 1,4-dihydronaphthalene is about 10 times as active as 1,4-dihydrobenzene. Other simple dihydroaromatic compounds are either ineffective or only slightly active. Recently, diallylbenzene or diallylnaphthalene (115), monoalkyl ethers of dihydrobenzene

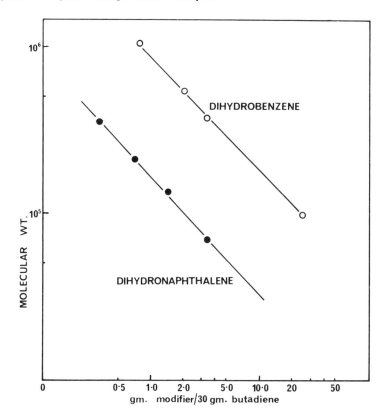

Figure 25 The effect of modifier concentration on polybutadiene molecular weight.

or dihydronaphthalene (116) and aromatic halogen containing com-pounds (117) such as chlorobenzene, bromobenzene, iodobenzene, and 1- and 2-bromonaphthalene have all been claimed as active modifiers.

The use of the modifiers mentioned above does not appear to influence the final microstructure of the polymer (114) which remains close to 70:10:20 trans-1,4 cis-1,4 1,2 and, in fact, the infrared spectra of the modified and unmodified Alfin polybutadienes are virtually superimposable.

A recent patent has indicated (118) considerable advances in both the preparation of the Alfin catalyst and the modification of the resultant molecular weight of the polymer. The Alfin catalyst used in this instance is prepared from disodiooctadiene, isopropyl alcohol, and sodium chloride and results in a slurry in hexane of sodium chloride, sodium isopropoxide, disodiooctadiene, and octadiene in the ratio 3:2:1:1.

Polymerization of butadiene results in a polybutadiene having a molecular weight of 400,000. If the polymerization is carried out in the presence of 1,4-dihydronaphthalene, the molecular weight is reduced to 100,000.

Cyclooctadiene appears to be incorporated into the structure of the polymer. It is interesting to speculate whether the cyclooctadiene is acting as a modifier in a manner similar to 1,4-dihydrobenzene even though 1,5- and 1,3-cyclooctadiene are mentioned in a list of ineffective molecular control agents (114) since it has been noted that the polymerization of butadiene in the presence of ^{14}C-labeled 1,4-dihydrobenzene results in the incorporation of one C_6 ring per polymer chain (114).

3.5.1 *Alfin Copolymers*

The copolymerization of styrene and butadiene has been investigated for some time (111, 112, 114, 119). The polymerization characteristics of Alfin reagents toward butadiene and styrene are unusual (107). With ordinary organosodium reagents such as amyl sodium, styrene polymerizes much faster than butadiene whereas when Alfin catalysts are used butadiene polymerizes five times as quickly as styrene.

In copolymerization of styrene and butadiene the rates are nearly the same (111, 112). The characteristics of a styrene-butadiene copolymer differ from those of polybutadiene in many ways (119). The amount of gel in the copolymer is less than that in the homopolymer. It was found to decrease until a 60–70% conversion was reached.

It has been suggested (114) that as little as 2% of styrene copolymerized with butadiene will reduce the gel content essentially to zero. However, a 15% styrene level polymer has improved resilience and tack, and easier processing than one containing 2% styrene. Dihydroaromatic compounds can also function effectively in the preparation of copolymers. The Alfin catalyst and technique has also been used to copolymerize butadiene with butene-2 (121). A plant for the production of Alfin butadiene-styrene and butadiene-isoprene copolymers has been erected and operated in Japan but has not been a commercial success.

Although the initiation and termination steps of Alfin catalysis are similar to polymerization by organosodium compounds, the different chain growth process yields a polymer with a different microstructure. This difference in chain growth may be due to a surface effect; in this and other respects Alfin catalysts resemble Ziegler catalysts in that the monomer molecules are adsorbed and oriented before polymerization.

A free radical mechanism for Alfin polymerization was suggested by Morton (122), and indeed some pieces of evidence tend to support the involvement of free radicals. The difficulty in interrupting the polymerization at low conversion to give low molecular weight products and the fact that the polymer has a microstructure similar to free radical polybutadiene support this hypothesis.

An anionic mechanism (123) has been suggested. Here the sodium allyl is considered to be associated in a highly polar surrounding with the sodium isopropoxide and sodium chloride. The metal–carbon bond is strongly stabilized and the carbonium ion of the activated monomer is sufficiently stable to allow rearrangement of the double bond.

The presence or absence of sodium chloride in the catalyst and its importance have already been mentioned. It has been suggested that electron defect centers of the "color center" type have a role in the activity of the catalyst. These centers can act in the activation of the monomer leading to polymerization and can arise during preparation of the alkali metal halides as a by-product of the Wurtz reaction.

$$M + RX \rightarrow M^+X^- + M^+R^- \quad (R = alkyl, \quad M = alkali\ metal)$$

In the Wurtz reaction a deeply colored salt is often precipitated and it is possible that such colored compounds may influence the course of polymerization, although Morton has shown that the blue color associated with a prepared Alfin catalyst is unconnected to the activity of the catalyst (124).

3.6 Polymers from Piperylene and Chloroprene

In this section we shall attempt to cover miscellaneous anionic polymerizations that have been reported using catalysts other than those previously mentioned and the polymerization of novel monomers. In many cases a particular monomer can only be polymerized anionically by employing an unconventional catalyst.

The homo- and copolymerization of 1-methylbutadiene, also known as piperylene or 1,3-pentadiene, have been studied extensively, since this compound occurs as a constitutent in refinery C_5 streams. Piperylene

consists of a mixture of cis and trans isomers, both of which can exist in
the cisoid or transoid form (see Chapter 2, Section 4).

cisoid   transoid

trans   cis   cis   trans

In the cisoid conformation there is considerable steric hindrance to the
approach of a reagent; this is not as pronounced in the case of the
transoid conformation. The polymer that can be obtained from piperylene
can have the following microstructures:

cis–1,4   trans–1,4

cis –1,2   trans –1,2

as well as that arising from 3,4 addition.

Polymerization of piperylene containing 95.7% trans and 4.3% cis
isomer with ethyl-, butyl-, or hexyllithium gave a vulcanizable elastomer.

Polymerization of the purified cis and trans isomers showed that the
trans isomer reacted faster than the cis and resulted in a product with a
higher molecular weight (125). Ozonolysis of a polypiperylene rubber

showed (126) the presence of three types of 1,4 units: 1,4-1,4 (24%), 1,4-4,1 (6%), and 4,1-1,4 (8%) in addition to 1,2 units.

The anionic polymerization of 2,3-dimethylbutadiene has also been investigated (127, 128) and as with butyllithium polymerization of butadiene, the 1,2 content of the poly(2,3-dimethylbutadiene) increases from 6 to 60–80% on changing the polymerization solvent from hexane to an ether or amine. The polymers obtained in all instances were of a white glassy mass (128). The polymerization of 2-ethylbutadiene with organolithium initiators has been reported to give elastomers which when vulcanized have good strength and elastic properties (129).

The incorporation of a cyano group into a butadiene unit might be thought to yield an interesting polymer having useful properties. By the anionic polymerization of both *cis*- and *trans*-1-cyanobutadiene a polymer having a predominantly trans-1,4 structure is obtained (130), the complete absence of any 1,2 configuration being shown by infrared spectroscopy. Polymerization of the trans isomer of 1-cyanobutadiene results in formation of an amorphous polymer while the cis isomer gives a polymer having some crystallinity.

A considerable amount of work has been carried out on the polymerization of 2-chlorobutadiene (chloroprene) using anionic systems. With butyllithium as the initiating species the polymerization reaction terminates at a conversion of 7–9% (131). This limiting conversion can be increased (132) to about 90% in 65 hours using a catalyst system based on butyllithium and tributyldimagnesium iodide at a $Bu_3Mg_2I/BuLi$ ratio of 0.5. The system is heterogeneous and contains the complex BuLi–LiI in the precipitate and Bu_2Mg in the solution. The temination step is thought to occur by elimination of the alkali metal halide from the growing end of the chain (133).

$$\text{mCH—CH}_2^{\ominus}\text{Li}^{\oplus} \rightarrow \text{mCH=CH}_2 + \text{LiCl}$$
$$\mid$$
$$\text{Cl}$$

It would appear (134) that the polymers obtained are almost completely insoluble and molecular weights could not be obtained. The microstructure of the polychloroprene is mainly trans-1,4.

Organomagnesium compounds have also been used in the polymerization of butadiene. Phenylmagnesium bromide, when treated with hexamethylphosphoramide, is reported (135) to polymerize butadiene to a high 1,2 polymer. The replacement of the hexamethylphosphoramide by benzene resulted in an inactive catalyst. A polyisoprene having a 3,4 content of 93–99% can be obtained by the polymerization of isoprene using simple Grignard reagents prepared free of ether (136). The nature of the halogen in the Grignard reagent as well as the alkyl groups appears to have no effect on the microstructure of the final polymer. The polymerization rate is reduced considerably by the addition of an electron donating compound such as an ether or amine and the 3,4 content of the polymer is reduced.

A catalyst system consisting of dibutylmagnesium-butylmagnesium iodide in hexane or toluene will polymerize butadiene or 2,3-dimethyl-butadiene (137) resulting in a partially cyclized polymer.

The use of organocalcium compounds as initiators for the polymerization of diolefins has also been investigated, although not to such a great extent as organomagnesium compounds. Butylphenylcalcium has been claimed to polymerize isoprene (138). A thorough study has been made of the system dibutylcalcium/butylcalcium iodide (139, 140) prepared by the reaction of butyl iodide with metallic calcium in diethyl ether. The polymerization of butadiene using this catalyst system gives a product having 90% 1,4 units and 10% 1,2 units. The proportion of cis-1,4 varies but is generally about 20%. The 1,2 content can be substantially increased by the replacement of the hydrocarbon solvent by an ether such as tetrahydrofuran.

The use of the alkaline earth metals as initiators for the polymerization of dienes has been thoroughly investigated by workers at the Institut de Chemie at Strasbourg and reported in a series of papers (141–145). The dienes examined were isoprene, 2,3-dimethylbutadiene, and piperylene and the metals used (strontium and barium) were usually prepared from a solution of the metal in liquid ammonia by evaporation of excess ammonia followed by heating under vacuum to destroy the metal–ammonia complex (141) or by metal grinding (142). The microstructures of the polymers obtained were determined by NMR and infrared spectroscopy (143). The polymerization of isoprene in bulk using barium gave a polymer containing 21% 3,4 and 79% 1,4 structure whereas solution polymerization in heptane using strontium gave a polymer with a 3,4 content of 44% and a 1,4 content of 56%. The kinetics of polyisoprene formation using strontium have been studied (144) and an anionic mechanism has been proposed (145). Similarly, metallic calcium has been used to polymerize isoprene (146). The microstructure of the final polymer depends on the method used for the preparation of the calcium as does the yield of polymer obtained. When butadiene is polymerized using metallic calcium (147) the final polymer has a high 1,2 content whether the polymerization is carried out in bulk, benzene, or tetrahydrofuran.

REFERENCES

1. F. E. Matthews and E. H. Strange, British Pat. 24,790 (1911).

2. C. Harries, *Ann.*, **383**, 184 (1911); U.S. Pat. 1,058,056 (to Bayer & Co., 1913).

3. E. G. M. Törnqvist, in J. P. Kennedy and E. G. M. Törnqvist, eds. *Polymer Chemistry of Synthetic Elastomers*, Part 1, Wiley-Interscience, New York, 1968, pp. 46–50.

4. L. P. Kyriakides, *Chem. Eng. News*, **23**, 531 (1945); *J. Am. Chem. Soc.*, **36**, 531, 657, 663, 980 (1914).

5. A. Holt, *Chem. Ztg.*, **38**, 188 (1914); German Pat. 287,787 (to Badische Anilin & Soda-Fabrik, 1915).

6. H. Labhardt, German Pat. 255,786 (to Badische Anilin & Soda-Fabrik, 1913).

7. W. K. Taft and G. J. Tiger, in G. S. Whitby, ed., *Synthetic Rubber*, Wiley, New York, 1954, Ch. 21, pp. 734–747.

8. A. V. Tobolsky and C. E. Rogers, *J. Polym. Sci.*, **40**, 73 (1959).

9. A. Rembaum et al., *J. Polym. Sci.*, **61**, 155 (1962).

10. W. W. Crouch and G. R. Kahle, *Pet. Refiner*, **37**, 187 (1958).

11. H. Blümel, *Kaut. Gummi Kunstst.* **16**, 571 (1963); reprinted and translated in *Rubber Chem. Technol.*, **37**, 408 (1964).

12. W. Cooper and G. Vaughan, in A. D. Jenkins, ed., *Progress in Polymer Science*, Vol. 1, Pergamon Press, Oxford, 1967, pp. 93–160.

13. J. E. Mulvaney et al., *Fortschr. Hochpolym. Forsch.* **3**, 106 (1961).

14. G. E. Coates, *Organometallic Compounds*, Methuen, London, 1960, p. 20.

15. D. Margerison and J. P. Newport, *Trans. Faraday Soc.*, **59**, 2058 (1963).

16. I. Kuntz, *J. Polym. Sci.*, **A2**, 2827 (1964).

17. H. L. Hsieh, *J. Polym. Sci.*, **A3**, 163 (1965).

18. J. I. Cunneen et al., *J. Polym. Sci.*, **40**, 1 (1959).

19. British Pat. 951,831 (to Farbenfabriken Bayer, 1964); British Pat. 968,756 (to Firestone Tire & Rubber Co., 1964).

20. British Pat. 1,086,888 (to Kanegafuchi Chemical Industry, 1967); B. E. Bailey and J. D. Moore, British Pat. 1,143,872 (to Int. Synthetic Rubber Co., Ltd., 1969).

21. R. J. Seward, *Rubber J.*, **150**, 47 (1968).

22. E. W. Duck and P. W. Milner, *Proceedings of the International Rubber Conference, Brighton*, 1967, Maclaren, London, 1967, pp. 359–376.

23. E. W. Duck and J. M. Locke, *J. Inst. Rubber Ind.*, **2**, 223 (1968); E. W. Duck, *Eur. Rubber J.*, **155** (12), 38 (1973).

24. E. W. Duck and B. J. Ridgewell, British Pat. 1,231,645 (to Int. Synthetic Rubber Co., Ltd., May 12, 1971); R. Zelinski, U.S. Pat. 3,301,840 (to Phillips Petroleum Co., January 31, 1967).

25. K. Nordsiek and N. Sommer, German Pat. 2,158,574 and 2,158,575 (to Chem Werke Hüls, May 30, 1973).

26. J. R. Haws et al., *Rubber Ind.*, **9** (3), 107 (June 1975).

27. D. W. Fraga, *J. Polym. Sci.*, **41**, 522 (1959); B. Ellis, *Chem. Ind.* (*London*), **1962**, 1447.

28. G. W. Atkinson and J. M. Goppel, Third TNO Conference on New Commercial Technical Developments, Rotterdam, February 26–27, 1970, pp. 55–72.

29. H. L. Hsieh, *J. Polym. Sci.*, **A3**, 181 (1965).

30. A. Guyot and J. Vialle, *Polym. Lett.*, **6**, 403 (1968).

31. J. L. Binder, *J. Polym. Sci.*, **A1**, 37 (1963).

32. W. H. Beattie and C. Booth, *J. Appl. Polym. Sci.*, **7**, 507 (1963).

33. C. F. Gibbs et al., *Kaut. Gummi*, **13**, 336 WT (1960).

34. A. R. Bean et al., *Encyclopedia of Polymer Science and Technology*, Vol. 17, Wiley, New York, 1967, pp. 782–855.

35. British Pat. 866,490 (to Shell Internationale Research Mij., 1961).

36. British Pats. 957,967 (1964), 1,016,235 (1966), and 1,016,236 (1966) (to Shell Internationale Research Mij.).

37. A. R. Bean et al., ref. 34, p. 829.

38. A. A. Korotkov and N. N. Chesnokova, *Polym. Sci., USSR*, **2**, 284 (1960).

39. I. Kuntz, *J. Polym. Sci.*, **54**, 569 (1961).

40. A. F. Johnson and D. J. Worsfold, *Makromol. Chem.*, **85**, 273 (1965).

41. U.S. Pat. 3,094,512 (to Phillips Petroleum Co., 1963).

42. C. F. Wofford and H. L. Hsieh, *J. Polym. Sci.*, *A-1* **7**, 461 (1969); C. F. Wofford, U.S. Pats. 3,294,768 (1966), 3,496,154 (1970), and 3,498,960 (1970) (to Phillips Petroleum Co.).

43. J. M. Willis and W. W. Barbin, *Rubber Age*, **100**, 53 (1968).

44. F. C. Weissert and B. L. Johnson, *Rubber Chem. Technol.*, **40**, 590 (1967).

45. *The Elastomers Manual*, International Institute of Synthetic Rubber Producers, Inc., New York, 1970 edition.

46. R. P. Zelinski and C. F. Wofford, *J. Polym. Sci.*, **A3**, 93 (1965).

47. Asahi Chemical Industry Co., Ltd., Technical Literature on Tufdene 1530.

48. D. V. Sarbach, *Rubber Age*, **89**, 283 (1961); D. R. Henson, *Rubber Plast. Age*, **45**, 1331 (1964).

49. E. W. Duck et al., Advances in Polymer Science and Technology, Symposium, London, October 1970, *J. Inst. Rubber Ind.* **6**, 19 (1972).

50. M. Szwarc et al., *J. Am. Chem. Soc.*, **78**, 2656 (1956).

51. D. H. Richards and M. Szwarc, *Trans. Faraday Soc.*, **55**, 1644 (1959).

52. J. F. Henderson and M. Szwarc, in A. Peterlin, M. Goodman et al., eds., *Macromolecular Reviews*, Vol. 3, Wiley-Interscience, New York, 1968, p. 367.

53. Belgian Pat. 627,652 (1963); British Pat. 1,000,090 (1965) (to Shell Internationale Research Mij. N.V.).

54. G. Holden et al., *J. Polym. Sci.*, **C26**, 37 (1969).

55. L. J. Fetters and M. Morton, *Macromolecules*, **2**, 453 (1969).

56. U.S. Pat. 3,251,905 (to Phillips Petroleum Co., 1966).

57. British Pat. 1,121,978 (to Polymer Corp., Ltd., 1968).

58. British Pat. 1,014,999 (to Shell Internationale Research Mij., 1965).

59. British Pat. 1,025,295 (to Shell Internationale Research Mij., 1966).

60. British Pat. 1,074,276 (to Shell Internationale Research Mij., 1967).

61. British Pat. 1,103,939 (to Polymer Corp., Ltd., 1968).

62. German Offen. 1,905,422 (to Shell Internationale Research Mij., 1969); *CA*, **71,** 92470v (1969).

63. British Pat. 1,028,388 (to Shell Internationale Research Mij., 1966).

64. British Pat. 1,032,150 (to Shell Internationale Research Mij., 1966).

65. German Offen. 1,911,241 (to Dunlop Co., Ltd., 1969); *CA*, **72,** 4201a (1970).

66. M. Morton et al., *J. Appl. Polym. Sci.*, **8,** 2707 (1964).

67. R. E. Cunningham and M. R. Treiber, *J. Appl. Polym. Sci.*, **12,** 23 (1968).

68. U.S. Pat. 3,140,278 (to Esso Res. Eng. Co., 1964).

69. British Pat. 1,090,207 (to Shell Internationale Research Mij., 1967).

70. British Pat. 1,163,674 (to Polymer Corp., Ltd., 1969).

71. I. M. Kolthoff et al., *J. Polym. Sci.*, **1,** 429 (1946).

72. H. L. Hsieh, *Rubber Plast. Age*, **46,** 394 (1965).

73. German Offen. 1,915,812 (to International Synthetic Rubber Co., Ltd., 1969).

74. German Offen. 1,909,825 (to Firestone Tire & Rubber Co., 1969); *CA*, **72,** 44830m (1970).

75. Firestone Tire & Rubber Co., S. African Pat. 67/04,573 (1968); *CA*, **70,** 38666e (1970).

76. German Offen. 1,901,900 (to International Synthetic Rubber Co., Ltd., 1969).

77. J. Zilliox et al., *J. Polym. Sci.*, *C*, No. 22, 145 (1968).

78. German Pat. 1,128,666 (to Farbenfabriken Bayer A.G., 1962).

79. M. A. Yampol'skaya et al., *Polym. Sci., USSR*, **8,** 848 (1966).

80. Y. Minoura et al., *J. Polym. Sci.*, *A-1*, **6,** 559 (1968).

81. Y. Minoura and H. Harada, *J. Polym. Sci.*, *A-1*, **7,** 3 (1969).

82. J. Heller and D. B. Miller, *Polym. Lett.*, **7,** 141 (1969).

83. A. J. Chalk and T. J. Hoogeboom, *J. Polym. Sci.*, *A-1*, **7,** 2537 (1969).

84. S. African Pat. 67/07,680 (to Firestone Tire & Rubber Co., 1968); *CA*, **70,** 48458c (1969).

85. German Offen. 1,911,886 (to Firestone Tire & Rubber Co., 1969); *CA*, **72,** 13610k (1970).

86. L. Lochmann et al., *Tetrahedron Lett.*, **1966,** 257.

87. S. African Pat. 68/02,597 (to Dunlop Co., Ltd., 1968); *CA*, **70,** 107324f (1969).

88. S. African Pat. 68/02,474 (to Dunlop Co., Ltd., 1969); *CA*, **71,** 31203p (1970).

89. C. A. Uraneck et al., *J. Polym. Sci.*, **46,** 535 (1960).

90. E. Schoenberg, *J. Polym. Sci.*, **49,** No. 152, S9 (1961).

91. K. Hayashi and C. S. Marvel, *J. Polym. Sci.*, **A2,** 2571 (1964).

92. C. A. Uraneck et al., *J. Appl. Polym. Sci.*, **13,** 149 (1969).

93. U.S. Pat. 3,410,836 (to Phillips Petroleum Co., 1968).

94. U.S. Pat. 3,055,952 (to E. I. duPont de Nemours, 1962).

95. U.S. Pat. 3,175,997 (to Phillips Petroleum Co., 1965).

96. British Pat. 1,029,451 (to National Distillers and Chemical Corp., 1966).

97. British Pat. 946,300 (to Polymer Corp., 1964).

98. U.S. Pat. 3,281,335 (to Phillips Petroleum Co., 1966).

99. C. A. Wentz and E. E. Hopper, *Ind. Eng. Chem., Prod. Res. Dev.*, **6,** 209 (1967).

100. U.S. Pat. 3,135,716 (to Phillips Petroleum Co., 1964).

101. U.S. Pat. 3,048,568 (to Phillips Petroleum Co., 1962).

102. British Pat. 964,259 (to B. F. Goodrich Co., 1964).

103. U.S. Pat. 3,109,871 (to Phillips Petroleum Co., 1963).

104. British Pat. 944,538 (to Phillips Petroleum Co., 1963).

105. R. F. Wolf, *Rubber Age*, **102**, No. 3, 83 (1970).

106. A. A. Morton et al., *J. Am. Chem. Soc.*, **69**, 950 (1947).

107. A. A. Morton, *Ind. Eng. Chem.*, **42**, 1488 (1950).

108. A. A. Morton et al., *Ind. Eng. Chem.*, **44**, 2876 (1952).

109. A. A. Morton, *Rubber Age*, **72**, 473 (1953).

110. Neth. Appl. 6,408,062 (to National Distillers and Chemical Corp., 1966).

111. J. D. D'Ianni et al., *Ind. Eng. Chem.*, **42**, 95 (1950).

112. R. A. Stewart and H. L. Williams, *Ind. Eng. Chem.*, **45**, 173 (1953).

113. British Pat. 943,625 (to National Distillers and Chemical Corp., 1963).

114. V. L. Hansley and H. Greenberg, *Rubber Chem. Technol.*, **38**, 103 (1965).

115. British Pat. 1,144,105 (to National Distillers and Chemical Corp., 1969).

116. British Pat. 1,132,410 (to National Distillers and Chemical Corp., 1968).

117. U.S. Pat. 3,448,093 (to National Distillers and Chemical Corp., 1969).

118. U.S. Pat. 3,380,984 (to National Distillers and Chemical Corp., 1968).

119. R. G. Newberg et al., *Rubber World*, **161**, 67 (1969).

120. British Pat. 1,060,685 (to National Distillers and Chemical Corp., 1967).

121. British Pat. 872,163 (to Richardson Co., 1961).

122. A. A. Morton, in A. Farkas, ed. *Advances in Catalysis*, Vol. 9, Academic Press, New York, 1957, p. 751.

123. H. Uelzmann, *J. Polym. Sci.*, **32**, 457 (1958).

124. A. A. Morton, *Solid Organoalkali Metal Reagents*, Gordon & Breach, New York, 1964.

125. I. A. Livshits et al., *Kauch. Rezina*, **28**, 4 (1969); *CA*, **70**, 116022h (1969).

126. A. I. Yakubchik et al., *Zh. Prikl. Khim.*, **35**, 405 (1962); *CA*, **57**, 4808e (1962).

127. H. Yuki et al., *J. Polym. Sci.*, A-1, **6**, 3333 (1968).

128. I. A. Livshits et al., *Polym. Sci.*, *USSR*, **9**, 2856 (1967).

129. I. A. Livshits and L. M. Korobova, *Vysokomol. Soedin.*, **3**, 891 (1961).

130. U. Giannini et al., *Makromol. Chem.*, **61**, 246 (1963).

131. B. L. Yerusalimskii et al., *Polym. Sci.*, *USSR*, **6**, 1429 (1964).

132. B. L. Erusalimiskii et al., *Dokl. Akad. Nauk. SSSR*, **169**, 114 (1966).

133. I. G. Krasnosel'skaya et al., *Polym. Sci.*, *USSR*, **9**, 2406 (1967).

134. I. G. Krasnosel'skaya and B. L. Yerusalimiskii, *Polym. Sci.*, *USSR*, **9**, 952 (1967).

135. U.S. Pat. 3,347,912 (to Esso Research and Engineering Co., 1967).

136. W. Fo-shung et al., *Vysokomol. Soedin.*, **2**, 541 (1960).

137. B. A. Dolgoplosk et al., *Vysokomol. Soedin.*, **4**, 1333 (1962); *CA*, **59**, 4045d (1963).

138. German Pat. 1,122,708 (to Farbenfabriken Bayer A. G., 1962); *CA*, **56**, 15679c (1962).

139. E. J. Tinyakova and E. Z. Eivazov, *Izv. Akad. Nauk SSSR, Ser. Khim.*, **8,** 1508 (1965); *CA*, **63,** 18267*c* (1965).

140. Ye. I. Tinyakova et al., *Polym. Sci., USSR*, **9,** 2720 (1967).

141. J. P. Kistler et al., *Bull. Soc. Chim. France*, **1964,** 3149.

142. J. P. Kistler, *J. Polym. Sci., C*, **16** (Pt. 5), 2825 (1967).

143. J. P. Kistler et al., *Bull. Soc. Chim. France*, **1967,** 4759.

144. J. P. Kistler et al., *Bull. Soc. Chim. France*, **1968,** 732.

145. J. P. Kistler et al., *Bull. Soc. Chim. France*, **1968,** 735.

146. J. Parrod et al., *J. Polym. Sci., C*, **16,** 4059 (1968).

147. E. Z. Eivazov and E. I. Tinyakova, *Vysokomol. Soedin., Ser. B*, **9,** 764 (1967); *CA*, 13444*b* (1968).

Maurice Morton and Lewis J. Fetters

Institute of Polymer Science
The University of Akron
Akron, Ohio

Chapter 5 Anionic Polymerization of Vinyl Monomers in Hydrocarbon Solvents

Portions of this chapter have appeared in *Rubber Reviews*, **48,** 359 (1975).

1 INTRODUCTION

Addition polymerizations involving soluble organometallic species have received intensive attention in recent years with special reference to the type of counterion and solvent. An anionic mechanism is proposed for those systems in which there is good reason to assume that the metal is strongly electropositive relative to the carbon (or other) atom at the tip of the growing chain. Hence the metal (e.g., lithium) becomes a cation either in the free state or coupled with the growing carbanion. Under appropriate experimental conditions, spontaneous termination is avoidable in many of those systems when one of the metals of group I is used as the counterion.

The alkali metals of sodium and potassium were revealed to be polymerization initiators of isoprene in the disclosures of Matthews and Strange in 1910 (1) and Harries in 1911 (2, 3). The first unambiguous report of the use of lithium in reactions with diolefins appears to be that of Ziegler et al. in 1934 (4–6) who investigated the reaction between alkali metals (lithium, sodium) or alkyllithium species and butadiene, isoprene, 2,3-dimethylbutadiene, or piperylene.

Ziegler was able to show that the reaction of sodium with butadiene involves the formation of a disodium adduct. Proof of the existence of this adduct was obtained by adding methylaniline (6) and quantitatively determining the yield of sodium methylanilide. Two moles of sodium methylanilide were produced for every mole of butadiene consumed. Ziegler also proved the difunctionality of the lithium 2,3-dimethylbutadiene adduct by isolating a family of products of the general formula

$$\underset{\begin{array}{cc} | & | \\ CH_3 & CH_3 \end{array}}{Li(CH_2C = C - CH_2)_n Li}$$

where n ranged from 1 to 6. It was thus demonstrated that polymerization occurred by some process other than a free radical mechanism, but Ziegler (1934) did not specifically suggest carbanionic propagation.

The polymerization process occurring in these reactions was not recognized as anionic until quite recently. It was not realized until about 1950 that basic initiators, such as organometallics, lead to an anionic polymerization mechanism. Since the counterions in these were metallic in nature, the propagating ion was assumed to be a carbanion in analogy with the cationic systems. One of the earliest kinetic studies of an anionic polymerization was the work of Higginson and Wooding (7) on the polymerization of styrene by potassium amide in homogeneous solution in liquid ammonia. Because of the homogeneity of the system, they were able to study the kinetics of this reaction and to propose the following mechanism:

$$KNH_2 \rightleftharpoons K^+ + NH_2^-$$ [1]

$$NH_2^- + M \rightarrow NH_2{-}M^- \rightarrow NH_2{-}M_x^-$$ [2]

$$NH_2{-}M_x^- + NH_3 \rightarrow NH_2{-}M_xH + NH_2^-$$ [3]

where M represents the styrene monomer. It is noteworthy that the nature of the solvent (liquid NH_3) led to a transfer reaction with the active chain (equation [3]), which would thus strongly control the chain length.

A similar transfer step was also observed in the polymerization of butadiene by sodium metal in toluene (8), as evidenced by the short chain products which resulted from the frequent transfer reaction, i.e., the metallation of the toluene:

$$\text{\textasciitilde\textasciitilde\textasciitilde}CH_2{-}CH{=}CH{-}\overset{=}{C}H_2Na^+ + C_6H_5CH_3 \rightarrow$$
$$\text{\textasciitilde\textasciitilde\textasciitilde}CH_2{-}CH{=}CH{-}CH_3 + C_6H_5\overset{=}{C}H_2Na^+ \quad [4]$$

Unfortunately, because of the heterogeneous nature of this system, no kinetic studies were possible here or in other solvents, or in bulk polymerization.

The use of lithium as the counterion in the polymerization of isoprene was shown (9) to cause the formation of predominantly *cis*-1,4-polyisoprene in either bulk or solution polymerizations where the solvent was a hydrocarbon. This capability of lithium metal generated a vast amount of interest and research in the homopolymerization of diolefins, their copolymerizations with monomers such as styrene, and most recently in the area of block copolymerizations, where a single chain is made up of segments formed from monomers such as styrene and a diene. This chapter is thus concerned with the application and description of the various anionic polymerization systems where alkali metal counterions are employed to fashion elastomeric materials of novel and often unique architecture. In the main, the discussion is limited to systems in which both the initiation and the propagation reactions occur under homogeneous conditions.

2 INITIATION METHODS IN ANIONIC POLYMERIZATIONS

2.1 Aromatic Complexes of Alkali Metals

The true nature of the carbanionic mechanism of polymerization only began to become apparent through studies of homogeneous systems involving initiators of the sodium naphthalene type. The latter species, a form of radical anion, had been known for some time, but its exact structure and behavior were elucidated only fairly recently. Thus Schlenk and co-workers (10, 11) showed that alkali metals, such as sodium, were able to react with the higher aromatics, such as anthracene, in the presence of ether solvents, without involving a substitution reaction. However, it was Scott (12) in 1935 who showed that the greenish-blue solution that resulted from the reaction of sodium with naphthalene, or biphenyl, in dimethoxyethane was capable of initiating a very rapid polymerization of styrene, butadiene, and other conjugated monomers. Scott and co-workers also proved that these complexes were carbanionic in nature by reacting them with carbon dioxide to form dicarboxylic acids. For this reason, they considered these species to be dimeric.

Later work, of a more sophisticated type, by Lipkin, Weissman, and co-workers (13, 14) led to the actual elucidation of the structure of these aromatic complexes. Using electron spin resonance measurements, they showed that these species are monomeric radical anions, possessing an extra electron in the lowest unoccupied π orbital. (These complexes are not to be confused with stable compounds such as sodium naphthyl, $NaC_{10}H_7$, which is formed by displacement of one of the ring hydrogen atoms.) The complex formation is due to the nature of both the solvent and the aromatic hydrocarbon. The solvent actually aids in the transfer of the electron from the sodium to the naphthalene and then stabilizes the resultant complex through interorbital exchanges with available electrons, e.g., from the oxygen of the ether solvent. Thus

$$\text{naphthalene} + Na + THF \rightleftharpoons \left[\text{naphthalene}\right]^{-} Na^{+} \ (THF) \qquad [5]$$

where THF represents tetrahydrofuran. Lipkin et al. also showed that the ease of formation of these complexes depends on the electron affinity of the hydrocarbon. They determined the equilibrium constant for electron-transfer reactions between aromatic hydrocarbons, e.g.,

$$\text{naphthalene}^{-} + \text{phenanthrene} \rightleftharpoons \text{naphthalene} + \text{phenanthrene}^{-}$$

and were able to set up an electron affinity scale as follows:

$$\text{biphenyl} < \text{naphthalene} < \text{phenanthrene} < \text{anthracene}$$

Shortly thereafter, Szwarc et al. (15, 16) showed that it was an electron transfer process that enabled these complexes to initiate the polymerization of styrene and other conjugated monomers. By careful exclusion of oxygen, water, and other impurities, they were able to determine the stoichiometry of the homogeneous anionic polymerizations and to demonstrate for the first time the absence of any termination (or transfer) reactions. They proposed the following mechanism for the polymerization of styrene:

Reaction [6] shows the initiation by electron transfer, leading to styrene radical anions which couple very rapidly (reaction [7]) to form styrene dianions capable of further propagation by anionic attack on styrene monomer. Substituting a diene for styrene does not alter the basic mechanism. A diene dianion is also formed by the coupling of the ion radical, and polymerization then proceeds through an anionic mechanism.

Szwarc and colleagues pointed out the unique features of this polymerization, namely (a) the extreme rapidity of the initiation step, as evidenced by the instantaneous color change from greenish-blue to the deep red of styrene anion; (b) the absence of any termination, or transfer, processes, so that all the chains continue growing until complete depletion of the monomer; and (c) the potential for a Poisson distribution of the chain lengths due to simultaneous growth of all the chains. They coined the picturesque and descriptive term "living polymer" for this system and also pointed out its potentialities for block polymerization and for the formation of polymers having functional end groups (e.g., COOH, OH) by termination of the "living" ends with suitable reactants.

Thus the phenomenon of a living polymer, i.e., the absence of any termination or transfer reaction, was discovered in these homogeneous systems becuase it was possible to study their stoichiometry. It is obvious, for example, that the sodium naphthalene system must lead to the stoichiometric relation:

$$\bar{M}_s = \frac{\text{grams of monomer}}{0.5 \text{ mole initiator}}$$

between the stoichiometric number average molecular weight $\bar{M}_s$ at 100% conversion, and the concentrations of monomer and sodium naphthalene used.

2.2 Initiation by Alkali Metals

The discovery of the special characteristics of the sodium naphthalene polymerizations (1956) followed very closely upon the discovery of the stereoregular polymerization systems (1954), including the synthesis of a high cis-1,4-polyisoprene by lithium metal catalysts (9). This remarkable achievement of the long-awaited synthesis of the molecular structure of natural rubber by use of a simple alkali naturally focused increased attention on the mechanism of these polymerizations. It thus became apparent that the so-called anionic mechanism of polymerization induced by the alkali metals must also involve initiation by electron transfer from the metal to the monomer, just as in the case of the soluble sodium naphthalene system. For example, in the lithium initiated polymerization of butadiene, we have

$$Li + CH_2{=}CH{-}CH{=}CH_2 \rightarrow \dot{C}H_2{-}CH{=}CH{-}\overset{\doublebar}{C}H_2 Li^+ \qquad [8]$$

$$2\dot{C}H_2{-}CH{=}CH{-}\overset{\doublebar}{C}H_2 Li^+ \rightarrow$$
$$Li\overset{\doublebar}{C}H_2{-}CH{=}CH{-}CH_2{-}CH_2{-}CH{=}CH{-}\overset{\doublebar}{C}H_2\overset{+}{L}i \qquad [9]$$

and/or

$$\dot{C}H_2{-}CH{=}CH{-}\overset{\doublebar}{C}H_2\overset{+}{L}i + Li \rightarrow \overset{+}{L}i\overset{\doublebar}{C}H_2{-}CH{=}CH{-}\overset{\doublebar}{C}H_2\overset{+}{L}i \qquad [10]$$

Equation [8] involves the initial electron transfer from the metal to the monomer, leading to formation of a radical anion. (For simplicity, only one resonance form of the latter is pictured above.) The latter may then participate in the reactions shown in [9] and [10], i.e., by coupling of two radical ions or by transfer of another electron from the metal, both processes leading to a dianionic species. It is obvious that the choice between these two reactions would depend on the conditions prevalent, e.g., the concentration and reactivity of the alkali metal. Thus, unlike soluble aromatic complexes of the alkali metals, the formation of the radical anions may proceed only relatively slowly at the metal surface, leading to a low concentration of these species. Depending on the reactivity of the particular alkali metal, reaction [10] could occur while the radical anion was still in the vicinity of the metal surface and before it could diffuse away to meet another radical anion.

Actually, the occurrence of two electron transfer steps, as suggested in [9] and [10], is strongly supported by the work of Ziegler (4–6) and later by that of Robertson and Marion (8); both groups were able to show the presence of a monomeric disodium adduct in this type of polymerization. Hence it is reasonable to assume that both reactions [9] and [10] can occur, depending on the particular system used, to a greater or lesser extent. It has even been suggested by O'Driscoll and Tobolsky (17), who studied the copolymerization of styrene and methyl methacrylate by lithium metal, that such a radical anion, under these circumstances, can undergo a certain amount of propagation simultaneously as a radical and an anion before becoming a dianion. They based their conclusions on the markedly different results obtained when these two monomers were copolymerized by lithium metal as compared with soluble organolithium initiators, e.g., butyllithium; the latter, as expected, showed a "pure anionic" behavior. However, other studies (18) of this copolymerization by lithium metal indicate that these results may be due to the heterogeneous nature of the system rather than to the presence of a radical mechanism.

2.3 Initiation by Organolithium Compounds

The third type of anionic initiation mechanism is exemplified by systems involving organolithium initiators. These initiators are the most versatile, since they are soluble in a variety of solvents, including hydrocarbons as well as ethers. Furthermore, since they function by direct anionic attack rather than by electron transfer, they lead to a monofunctional propagating chain, thus avoiding some of the complications of a difunctional chain, especially with regard to the molecular weight distribution. As with initiators activated by the electron transfer process, there is no chain transfer or termination in the absence of compounds having an active hydrogen. Since the chains grow only at one end, the stoichiometric number average chain length, at 100% conversion, is given by the relation

$$\bar{M}_s = \frac{\text{grams of monomer}}{\text{moles initiator}}$$

This absence of a termination step for monomers such as the dienes or styrene has facilitated studies of the kinetics of their initiation and propagation steps. The work has been directed, in the main, toward an understanding of the mechanism involved. The four characteristics of the organolithium induced polymerizations of dienes or styrene which any viable mechanism must explain are as follows:

1. The fractional order rate dependence of the propagation rate on initiator concentration in nonpolar solvents.
2. The pronounced preference for butadiene or isoprene by the growing chain in copolymerization with styrene in nonpolar solvents, and the reversal of these reactivities in polar solvents.

3. The striking effect of polar solvents in favoring side vinyl polydiene microstructures.

4. The effect of different dienes on the cis/trans ratio obtained in nonpolar solvents, e.g., the high (70–90%) cis-1,4 content of polyisoprene and the relatively low ($\sim$40%) cis-1,4 content of polybutadiene.

3 ASPECTS OF HOMOGENEOUS ANIONIC POLYMERIZATION SYSTEMS

3.1 Molecular Weight Distributions

Perhaps the most prominent feature of some homogeneous anionic polymerizations is that spontaneous termination is avoidable through a judicious choice of experimental conditions. The absence of a "built in" termination reaction in many of these anionic systems is the feature which permits the preparation of block copolymers by the predetermined sequential addition of different monomers. This method can lead to the formation of a wide variety of linear or "star" (branched) block copolymers, where the various blocks differ in their properties, e.g., polar or nonpolar, crystalline or amorphous, blocks whose glass transition temperatures differ, or that have different conformations (random coil and rodlike), elastomeric and nonelastomeric, or hydrophilic and hydrophobic blocks.

The absence of a spontaneous termination reaction is also the feature that allows the synthesis of polymers possessing narrow molecular weight distributions when the initiation rate is of the same order of magnitude as the rate of propagation. Under these conditions the resultant polymer will have a distribution of molecular weights approaching the Poisson distribution, as predicted by Flory (19), where the number fraction of j-mers, P_j, is given by

$$P_j = \frac{e^{-x}x^{j-1}}{(j-1)!} \qquad [11]$$

Here x denotes the number of monomers reacted per initiator molecule. The weight fraction distribution is given by

$$W_j = \frac{x}{x+1}\frac{je^{-x}x^{j-2}}{(j-1)!} \qquad [12]$$

Thus, the weight/number average chain length ratio is given by

$$\frac{\bar{x}_w}{\bar{x}_n} \approx 1 + \bar{x}_n^{-1} \qquad [13]$$

When $\bar{x}_n$ nears 100 a virtually monodisperse molecular weight distribution can be achieved.

In order to obtain predictable molecular weights and narrow molecular weight distributions, the following conditions must be met: (*a*) the exclusion of terminating impurities; (*b*) an initiation rate competitive with the propagation rate; and (*c*) homogeneity during both the initiation and propagation steps. Under these conditions, the anionic systems offer, to date, the only method for the preparation of polymers having narrow molecular weight distributions. This is demonstrated in Fig. 1 which shows a gel permeation curve for polyisoprene (20).

Table 1 lists the various attempts to prepare polymers of narrow molecular weight distributions. Although the conventional $\bar{M}_w/\bar{M}_n$ ratio for a material is near unity, these two moments of the overall molecular weight distribution should not be taken alone as authoritative proof that the sample in question is "monodisperse." A sample consisting of equal parts by weight of three monodisperse fractions of molecular weights of 4×10^5, 5×10^5, and 6×10^5 will have a value for $\bar{M}_w/\bar{M}_n$ of 1.03. Thus at

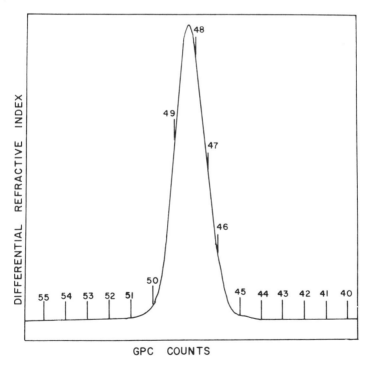

Figure 1 Gel permeation chromatogram of polyisoprene prepared with *s*-butyllithium in cyclohexane at 34°C. $\bar{M}_s = 1.3 \times 10^5$; $\bar{M}_n = 1.35 \times 10^5$; $\bar{M}_w = 1.4 \times 10^5$.

Table 1   Polymers with Narrow Molecular Weight Distributions Prepared via Anionic Polymerization

Initiator	Polymerization Solvent	$\bar{M}_w/\bar{M}_n$	Ref.
	Polystyrene		
Sodium naphthalene	Tetrahydrofuran	1.0_4–1.1_0	21
α-Phenylethylpotassium	Tetrahydrofuran	1.0_5	22
Sodium biphenyl	Tetrahydrofuran	1.0_7	23
n-Butyllithium	Benzene[a]	1.0_3–1.0_9	24
α-Methylstyrene sodium dianion	Tetrahydrofuran	1.0_5–1.0_9	25
n-Butyllithium	Benzene[a]	1.0_1	26
Ethyllithium[b]	Benzene	1.0_5–1.1_2	27
Sodium naphthalene	Tetrahydrofuran	1.1–1.5	28
Styryllithium	Tetrahydrofuran	$1.1_2, 1.1_7$[c]	29, 30
	Poly(α-methylstyrene)		
Sodium naphthalene	Tetrahydrofuran	1.0–1.0_3	31, 32
n-Butyllithium	Tetrahydrofuran	1.0_5	33
α-Methylstyrene sodium dianion	Tetrahydrofuran	1.0_5	34
	Poly(p-$tert$-butylstyrene)		
s-Butyllithium	Benzene	$<1.0_4$	35
	Polybutadiene		
n-,[b] s-, and t-Butyllithium	Benzene, cyclohexane, and n-hexane	$<1.0_5$	36–39
	Polyisoprene		
n-[b] and s-Butyllithium	Benzene, cyclohexane, and n-hexane	1.0_5	20, 40
	Poly(acrylonitrile)		
Sodium triethyltriisopropoxyaluminate	Dimethylformamide	~ 1.1	41

least three moments, namely $\bar{M}_n$, $\bar{M}_w$, and $\bar{M}_z$, of the molecular weight distribution are necessary in order to gain meaningful insight into the molecular weight distribution. However, in the main, the accumulated evidence does indeed indicate that the polymers listed in Table 1 can be prepared, under the appropriate conditions, with narrow molecular weight distributions.

Table 1   (Continued)

Initiator	Polymerization Solvent	$\bar{M}_w/\bar{M}_n$	Ref.
	Poly(t-butyl crotonate)		
2-Methylbutyllithium	Tetrahydrofuran	1.0_1	42
	Poly(n-butyl isocyanate)		
Fluorenyl sodium and n-butyllithium	Toluene	~ 1.1	43, 44
	Poly(dimethylsiloxane)^d		
n-Butyllithium	Tetrahydrofuran	~ 1.1	33
	Poly(methyl methacrylate)		
Sodium biphenyl	Tetrahydrofuran	$< 1.0_8$	45
Lithium biphenyl	Dimethoxyethane	~ 1.1	46
			47–52
	Poly(propylene sulfide)		
n-Butyllithium and sodium naphthalene	Tetrahydrofuran	< 1.1	47–52

^{a} A small amount of tetrahydrofuran (THF) was added to increase the rate of initiation. It should be noted that triethylamine is as effective as THF for this purpose and, unlike the ether, will not cause the slow termination of the growing chains.

^{b} Ethyl- and n-butyllithium were reacted with monomer, before polymerization, under conditions designed to effect complete initiation (27, 36, 40). The low molecular weight ($\bar{M}_n \simeq 5 \times 10^3$) polymer-lithium species was then used as the initiator.

^{c} These are the $\bar{M}_z/\bar{M}_w$ values for polystyrenes with weight average molecular weights of 43.7 $(\pm 0.2) \times 10^6$ and 27.3 $(\pm 0.5) \times 10^6$, respectively.

^{d} The monomer used in this preparation of poly(dimethylsiloxane) was the cyclic trimer hexamethyltrisiloxane.

3.1.1   *Experimental Procedures*

Gel permeation chromatography has become a widely used analytical tool for the determination of molecular weights and molecular weight distributions. However, this particular technique can yield inaccurate information if too few columns are used or if the columns used fail to cover the porosity range necessary to adequately resolve the molecular weight distribution of the examined sample. The gel permeation chromatograms shown in this chapter were run on the Waters Ana-Prep instrument with tetrahydrofuran as the carrier solvent. Solution concentrations were 0.25 (w/v). A differential refractive index detector was used. The following seven Styragel columns were used for the chromatograms shown in Figs. 1–5 and 15–20: 2–5×10^3 Å, two columns with 5–15×10^3 Å, 1.5–5×10^4, 5–15×10^4, 1.5–7×10^5, and 7–50×10^5 Å. The chromatogram in Figs. 9, 10, 12, and 13 was obtained with the first five columns of the column arrangement listed above. The chromatogram shown in Fig. 7 was obtained with the seven column set listed above plus the following five columns with porosity ranges of 50–80, 80–100, 100–350, 350–700, and 700–2000 Å. The flow rate used for the chromatograms shown in Figs. 1–5, 9, 12, 13, 15–18, 19, and 20 was 0.25 ml/min. A flow rate of 1 ml/min was used for the other chromatograms. The Ana-Prep instrument was equipped with a 5 ml syphon. The plate count for the seven column set was 750 ppf at the 1 ml/min flow rate and 950 ppf at the 0.25 ml/min flow rate. The seven column Styragel set yielded a linear calibration, using polystyrene standards, over the molecular weight range of 4×10^3 to 2.1×10^6 g/mole. The calibration for the five column set was linear over the range of 4×10^3 to 8.6×10^5 g/mole. The 12 column set exhibited a linear calibration from styrene monomer to a molecular weight of 2.1×10^6 g/mole.

3.2 Initiation Reactions Involving Soluble Organolithium Compounds

It is a generally recognized fact that the reaction between some organolithium compounds and the dienes or styrene is a slow process, relative to propagation, in hydrocarbon media. For example, in the n-butyllithium initiation of styrene in benzene at 30°C, the following relation (53) was reported to hold for the minimum ratio of monomer to initiator required for the complete consumption of the initiator:

$$\frac{[M]}{[n\text{-}C_4H_9Li]^{4/3}} > 504 \qquad [14]$$

The slow rates of initiation exhibited by some organolithium species are related, at least in part, to the fact that these compounds are strongly associated (hexameric to dimeric) in hydrocarbon solvents. Table 2 displays the reported association states of a variety of organolithium species in hydrocarbon solvents.

Table 2 Association States of Organolithium Compounds in Hydrocarbon Solutions

Compound	Solvent	Concentration Range (RLi Molality)	n Degree of Association	Method[a]	Ref.
C_2H_5Li	Benzene	Not Stated	~6	F	54
	Benzene	0.03–0.4	~6	F	55
	Benzene	0.03–0.4	4.5–6.0	F	56
	Benzene	0.02–0.23	6.07 ± 0.35	F	57
	Cyclohexane	0.02–0.10	5.95 ± 0.3	F	58
	Benzene	0.006–0.19	6.1 ± 0.18	F	59, 60
	Cyclohexane	0.006–0.08	6.0 ± 0.12	F	59, 60
n-C_4H_9Li	Benzene	0.5–3.4	6.25 ± 0.06	I	61
		Not Stated	~7	B	62
	Cyclohexane	0.4–3.3	6.17 ± 0.12	I	61
	Benzene	0.002–0.6	6.0 ± 0.12	F	59, 60
s-C_4H_9Li	Benzene	0.17–0.5	4.13 ± 0.05	F	63
	Cyclohexane	0.11–0.4	4.12 ± 0.09	F	63
i-C_3H_7Li	Cyclohexane	0.004–0.02	4.0 ± 0.08 (>4.0 above 0.02 M)	F	59, 60
t-C_4H_9Li	Benzene	0.05–0.18	3.8 ± 0.2	B	64
	n-Hexane	0.05–0.23	4.0 ± 0.2	B	64
	Benzene	0.26–0.66	4.0 ± 0.04	F	59, 60
	Cyclohexane	0.0005–0.3	4.0 ± 0.05	F	59, 60
n-$C_5H_{11}Li$	Benzene	0.23–2.32	6.00 ± 0.09	V	65
n-$C_8H_{17}Li$	Benzene	0.19–2.32	5.953 ± 0.016	V	65
$(CH_3)_3SiCH_2Li$	Benzene	0.6–2.78	4.0 ± 0.2	B	66
	Benzene	0.03–1.1	4.0 ± 0.03	F	59, 60
	Benzene	0.06–0.49	4.0 ± 0.11	F	67
	2-Methylpentane	0.2–1.2	3.9 ± 0.2	B	66
	Cyclohexane	0.002–0.05	6.0 ± 0.18	F	59, 60
Menthyllithium	Benzene	0.10 and 0.38	2.17, 1.95	F	68
	Cyclohexane	0.23 and 0.29	1.93, 2.04	F	68
$C_6H_5CH_2Li$	Benzene	0.0072–0.036	$2.2 - \pm 0.3$	F	59, 60
3-Neopentyl-allyllithium	Benzene	0.33–0.0495	3.7–2.14[b]	F	69
3-Butenyllithium	Cyclopentane	5.06–0.77	6.1 ± 0.8	V	70

[a] I, isopiestic; F, freezing point depression; B, boiling point elevation; V, vapor pressure depression.

[b] For 3-neopentylallyllithium the degree of association decreased as the concentration decreased.

Attempts have been made by some workers to relate the observed reaction orders of numerous initiation studies to the inverse of the association state of the initiating organolithium, based on the concept that only the unassociated organolithium species have the capacity to react with monomer. Conversely, it was *assumed* that the associated species had no reactivity with monomer. Thus, for example, the following steps were envisaged for the *n*-butyllithium initiation of styrene in benzene:

$$(n\text{-}C_4H_9Li)_6 \; \rightleftharpoons \; 6n\text{-}C_4H_9Li \hspace{3cm} [15]$$

$$n\text{-}C_4H_9Li + CH_2{=}CH \longrightarrow n\text{-}C_4H_9CH_2CHLi \hspace{2cm} [16]$$

This proposed mechanism had as its foundation the observed 1/6 dependence of the initiation rate on initiator concentration *and* the fact that *n*-butyllithium is associated as a hexamer in hydrocarbon solvents. A variety of reaction orders for the reaction of organolithium species with olefinic and diolefinic compounds have been reported and these are shown in Table 3.

It is quickly apparent from an examination of Tables 2 and 3 that the *proposed* correspondence between organolithium association state and initiation reaction order is not generally obeyed, e.g., initiation reaction orders for *n*-butyllithium range between one-sixth and first order. Variations in reaction orders were also found on changing the solvent from an aromatic to an aliphatic one.

Hence it appears that the convenient concept that the dissociated form of an organolithium species represents the sole reactive form is in error. This concept has been questioned, and correctly so, on energetic grounds by Brown (80). He considers the reaction scheme presented in equations [15] and [16] to be an improbable event since the enthalpy change for the dissociation of the hexamer into the monomeric form is too high, i.e., about 100 kcal/mole, to allow the concentration of unassociated initiator to reach a level adequate to account for the observed rates of initiation. This, coupled with an unfavorable entropy effect, would make the free energy change for equation [15] so large (i.e., positive) that no balanced assumption about the kinetic behavior of the various candidate species will suffice to make the concentration of unassociated organolithium rate determining.

As an alternative, Brown (80) has suggested that styrene or a diene can react directly with associated initiator, e.g., hexamer, tetramer, or the

Table 3 Initiation Studies with Organolithium Species in Hydrocarbon Solvents[a]

Monomer	Initiator[a]	Solvent	Reaction Order	T (°C)	Method of Analysis	Ref.
Styrene	s-C_4H_9Li	Cyclohexane	1.4	40	UV	63
Styrene	s-C_4H_9Li	Benzene	0.25	30	UV	63
Styrene	n-C_4H_9Li	Cyclohexane	0.5–1.0	40	UV	71
Styrene	n-C_4H_9Li	Benzene	0.167	30	UV	53
Styrene	n-, s-, t- and i-C_4H_9Li	Toluene	~1	30	Gas chromatography	72
Isoprene		Cyclohexane		and 50		
Butadiene		n-Hexane				
Styrene	n-C_4H_9Li	Benzene	0.33	25	UV	73
Butadiene	n-C_4H_9Li	Cyclohexane	0.5–1.0	40	UV	71
Isoprene	n-C_4H_9Li	Cyclohexane	0.5–1.0	30	UV	74
Isoprene	s-C_4H_9Li	Cyclohexane	0.75	30	UV	63
Isoprene	s-C_4H_9Li	Benzene	0.25	30	UV	63
Isoprene	s-C_4H_9Li	n-Hexane	0.70	30	UV	75
Isoprene	s-C_4H_9Li	Cyclohexane	0.66	25	Gas chromatography	76, 77
Isoprene	t-C_4H_9Li	Cyclohexane	0.2–0.7	25		76, 77
2,4-Dimethylstyrene	n-C_4H_9Li	Benzene	0.19	?	UV	78
Butadiene	s-C_4H_9Li	Benzene	0.9	30	IR	79

[a] Initiator concentrations were between 10^{-4} and 5×10^{-3} M with the exception of ref. 73, where the n-C_4H_9Li concentration was 1.29 M. All of the studies revealed a first-order dependence on monomer.

dimer form. It should also be noted, with regard to the foregoing point, that the initiating organolithium species may well exist, at least in part, in lower states of association at concentrations ($\sim 10^{-3}$ M) appropriate for polymerization, which are generally much lower than those used for association measurements. A dependence of the degree of organolithium association on concentration has been observed for several systems (60, 69).

Brown has instead proposed several plausible mechanisms which adequately account for the low fractional reaction orders observed. His conclusions have been confirmed in a quite satisfactory fashion by the results of other workers (81–85), results that show that the fractional orders of these initiation reactions can be accounted for by involving associated organolithium species as reactive entities. Hence it would appear that the establishment of any reaction order with respect to each associated species is by no means a trivial operation.

The mechanism proposed in equations [15] and [16] also fails to take into account the existence of cross-associated complexes between poly(styryl)lithium and *n*-butyllithium. Viscosity measurements have qualitatively shown (75, 86–88) that poly(isoprenyl)lithium can form cross-associated complexes with *n*-butyllithium, *s*-butyllithium, and *t*-butyllithium while poly(styryl)lithium will do the same with these three initiating organolithium species (33). The cross-association of ethyllithium and poly(isoprenyl)lithium has been studied quantitatively (88) by means of concentrated solution viscosity measurements in *n*-hexane, a method that yielded insight into the relative molecular weights of the polymeric species, i.e., self-associated and cross-associated. The following association equilibrium was proposed (88) for this system:

$$(RM_jLi)_2 + (C_2H_5Li)_6 \rightleftharpoons 2[RM_jLi \cdot (C_2H_5Li)_3] \qquad [17]$$

where RM_jLi denotes the poly(isoprenyl)lithium chain. This proposed equilibrium is in agreement with the known states of association of poly(isoprenyl)lithium and of ethyllithium and yields a consistent value of 6.4 (± 0.2) for the equilibrium constant, indicating preferential cross-association.

Thus the order and rate of these initiation reactions can be influenced by the extent and degree of the association state of the initiator and the initiator-polymer complex. Based on equation [17], as the initiator supply dwindles and the chain end concentration increases, the majority of the initiator would be found in the cross-associated complex. Since nothing is known about the reactivity of the initiator in these complexes, it is currently impossible to define the role that such a complex has in the kinetics of either the initiation or propagation steps. It is evident, though, that the reactions of alkyllithium species with nonpolar monomers in hydrocarbon solvents constitute, kinetically, very complex systems.

Worsfold and Bywater (53) conducted their n-butyllithium-styrene initiation study under conditions that are not encountered in polymerization reactions leading to the formation of high polymer, i.e., the styrene–n-butyllithium ratio they used was about 10 or less. One result of their kinetic study is that the minimum molecular weight necessary to ensure complete utilization of n-butyllithium under conditions appropriate for polymerization is about 3×10^4 at 30°C. However, recent work in these laboratories (111) has shown that complete consumption of initiator is achieved much more rapidly than predicted by their results. Figure 2 shows a gel permeation chromatogram of a polystyrene prepared by n-butyllithium in benzene (89). The molecular weight distribution $(\bar{M}_w/\bar{M}_n)$ of this polymer was narrower than anticipated from the results of Worsfold and Bywater (53). Analysis for residual n-butyllithium revealed that complete utilization of the initiator had taken place, a result that is supported by the good agreement of the predicted molecular

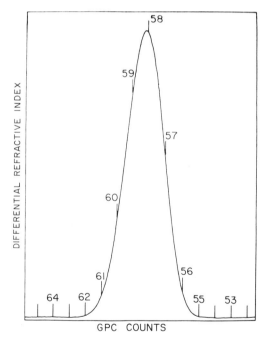

Figure 2 Gel permeation chromatogram of polystyrene prepared with n-butyllithium in benzene at 30°C. $\bar{M}_s = 1.15 \times 10^4$; $\bar{M}_n = 1.17 \times 10^4$; $\bar{M}_w = 1.26 \times 10^4$. The initial monomer concentration was 0.1 M.

Table 4   Relative Reactivities of Organolithium Compounds in Initiation (39, 72)

Dienes

Menthyllithium > s-butyllithium > i-propyllithium > t-butyllithium > n-butyllithium and ethyllithium

Styrene

Menthyllithium > s-butyllithium > i-propyllithium > n-butyllithium and ethyllithium > t-butyllithium

Solvents

Toluene > benzene > n-hexane > cyclohexane

---

weight $\bar{M}_s$ and the measured number average molecular weight (via membrane osmometry). Similar results (89) have been obtained for a series of polystyrenes with molecular weights ranging from 5×10^3 to 2.5×10^4. In all instances complete initiation was achieved at a much more rapid rate than predicted by the results of Worsfold and Bywater. Hence their kinetic analysis of this particular monomer-organolithium pair apparently cannot reflect the initiation behavior of n-butyllithium with styrene under conditions necessary for the preparation of high molecular weight polymer.

Although many of the initiation reactions listed in Table 3 have been only partially elucidated, it has been possible to reach some semiquantitative conclusions as to the relative rates of reaction of various alkyllithium species in the initiation of styrene and the diene monomers in hydrocarbon solvents. These conclusions are presented in Table 4 along with the influence of solvents.

Table 4 shows that, with one exception, the initiating organolithium species exhibit identical relative reactivities toward the dienes and styrene. The observed exception involves t-butyllithium and its change from a moderately fast initiator of the dienes to a very slow initiator of styrene. The low reactivity of t-butyllithium was originally conceived (95) as being due to steric hindrance existing between the initiator and styrene. It has been noted that there is an initial, relatively rapid reaction between styrene and a fraction of the t-butyllithium (20, 89). The rate of initiation then rapidly decreases, leading to a drawn-out initiation step and the formation of a low molecular weight tail in the molecular weight distribution of the resultant polymer. The gel permeation chromatogram in Fig. 3 illustrates this effect and the lack of an analogous high molecular weight tail. This latter feature would be expected to be present if this particular initiation was characterized by a gradual addition of all the t-butyllithium to the styrene. Hence this type of asymmetrical molecular

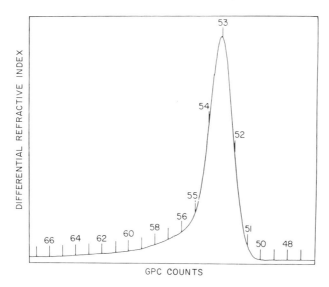

Figure 3 Gel permeation chromatogram of polystyrene prepared with t-butyllithium in benzene at 30°C. $\bar{M}_s = 1.9 \times 10^4$; $\bar{M}_n = 2.9 \times 10^4$; $\bar{M}_w = 3.8 \times 10^4$. The initial monomer concentration was 0.5 M. At the completion of this polymerization, unreacted initiator was present.

weight distribution seemingly demonstrates that a fraction of t-butyllithium reacts relatively rapidly with styrene while the remainder reacts at a markedly decreased rate.

Additional insight (89) into the slow reaction between styrene and t-butyllithium can be gained by inspection of Figs. 4 and 5. Figure 4 is a gel permeation chromatogram of a polystyrene prepared in cyclohexane

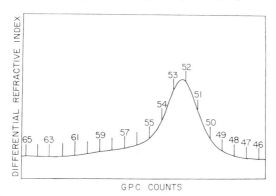

Figure 4 Gel permeation chromatogram of polystyrene (PS–48–I) prepared with t-butyllithium in cyclohexane at 35°C. $\bar{M}_s = 1.11 \times 10^4$; $\bar{M}_n = 4.0 \times 10^4$; $\bar{M}_w = 5.2 \times 10^4$. The initial monomer concentration was 0.5 M.

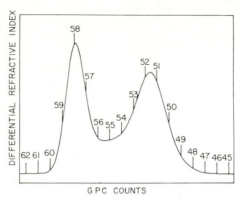

Figure 5 Gel permeation chromatogram of polystyrene prepared by a second monomer addition to sample PS–48–I. $\bar{M}_s = 2.22 \times 10^4$; $\bar{M}_n = 2.3 \times 10^4$; $\bar{M}_w = 4.1 \times 10^4$. The monomer concentration after the second addition was 0.5 M.

at 35°C. Figure 5 is, on the other hand, a chromatogram of polystyrene prepared by the addition of a second increment of styrene, along with 1 vol % tetrahydrofuran, to the polystyrene system shown in Fig. 4. The second peak (count 57.9) in this bimodal distribution is the result of initiation by the residual t-butyllithium.

The absence of any polystyrene above count 62 (Fig. 4) demonstrates the dormant nature of the residual initiator. It is of interest to note that the lowest molecular weight polystyrene found via gel permeation chromatography is about 6×10^3 g/mole. However, the addition of tetrahydrofuran clearly activates the residual initiator (which is nearly 75% of the total amount added), leading to the bimodal molecular weight distribution shown in Fig. 5. It is also pertinent that although the predicted and number average molecular weights are markedly different for the polystyrene shown in Fig. 4, they are in excellent agreement for the "bimodal" polymer.

This retardation of the initiation process signals the formation of a less reactive form of the initiator. It may well be that the t-butyllithium which is cross-associated with the poly(styryl)lithium is less reactive toward styrene than is the self-associated initiator. Although the question has been raised (90) as to whether the low reactivity of t-butyllithium is, in the main, due to an impurity, our analysis (89) (via the 300 MHz spectrometer) of the t-butyllithium used by us does not reveal the presence of any impurity. Obviously any quantitative resolution and explanation of the behavior of this initiator-monomer system will require additional work. Nonetheless the initiation behavior of this particular organolithium does serve to demonstrate the kinetic complexities that can be encountered in the initiation of dienes or styrene by organolithium species in hydrocarbon solvents.

3.3   Kinetics of Alkyllithium Initiated Polymerizations and Association Effects

Propagation kinetics studies on the dienes and styrene have, in the main, shown that the propagation reaction has a first-order dependence on monomer concentration. This was found both from studies of the rates of monomer disappearance and from the dependence of the propagation rate on initial monomer concentration. These observations apply to both hydrocarbon and ether solvents.

For those monomers which offer no opportunity for side reactions, the system should consist of a fixed number of growing chains and a dwindling monomer supply once initiation is completed. Thus, it would be expected that the polymerization rate should be directly proportional to the number of growing chains, i.e., the initiator charge. For organolithium polymerizations of isoprene and butadiene in ethers (principally tetrahydrofuran) or amines (86, 91–93), it was found that the propagation reaction shows a first-order dependence on the concentration of growing chains. This is a convincing indication that the active chain end species is of one form e.g., an ion pair. For the polymerization of styrene in ether solvents, though, the polymerization rate is too rapid to measure by ordinary means. However, a special flow technique has been used (94) to measure propagation rates and it was found that the dependency on growing chain concentration was somewhat greater than one-half order but definitely less than first order. This behavior was rationalized by proposing the simultaneous activity of both ion pairs and free ions. While the free ions represent only a small fraction of the chain ends, they account for most of the propagation because of their much greater reactivity, and a kinetic order of approximately one-half results. The extent of ionization of these species was determined by conductivity measurements. This led to a calculation of the equilibrium constants (ca. 10^{-6}), and the rate constants for the ion pairs (ca. $10^2 \, M^{-1} \, sec^{-1}$) and free ions ($10^4$–$10^5 \, M^{-1} \, sec^{-1}$).

In hydrocarbon solvents, dissociation of ion pairs is unlikely and one would expect only ion pairs and thus a first-order dependency of propagation rate on active chain end concentration. However kinetic orders for styrene, isoprene, and butadiene show a much greater complexity.

The propagation rate of styrene in benzene, toluene, and cyclohexane was found to show a kinetic order of one-half in all three solvents. Since the concept of dissociation of the ion pairs into free ions is untenable in these solvents, the half-order kinetics have been explained by an association-dissociation equilibrium of dimers into ion pairs. The physical evidence for such an association became available only when the degree of association was determined by viscosity measurements on their concentrated solutions (20, 86, 95–97). This technique involves the relation between viscosity η and molecular weight M, given by

$$\eta = KM^{3.4} \qquad [18]$$

where K is a constant that includes the concentration term. Because of the 3.4 exponent, this method yields a sensitive measure of the relative molecular weight of the living polymer-lithium chain compared with that of the terminated chain. Thus poly(styryl)lithium was found to be dimeric. On the assumption that a small equilibrium concentration of unassociated chain ends is responsible for propagation, the half-order kinetics follow:

$$\frac{R_p}{[\text{M}]} = k_p K_d^{1/2}[(\text{RM}_j\text{Li})_2]^{1/2} \simeq k_p K_d^{1/2}\left(\frac{[\text{RM}_j\text{Li}]_0}{2}\right)^{1/2} \qquad [19]$$

where R_p is the propagation rate; $[\text{M}]$ is monomer concentration; K_d is the dissociation constant for ion pair dimers $(\text{RM}_j\text{Li})_2$; and $[\text{RM}_j\text{Li}]_0$ is the total concentration of poly(styryl)lithium. Implicit in this equation is the assumption that the unassociated ion pairs are the sole reactive species and that they represent only a very small fraction of the total ion pair concentration. A value, at 30°C, for k_p of about $17\,\text{M}^{-1}\,\text{sec}^{-1}$ has been estimated (36).

In contrast to styrene, the propagation kinetics of the dienes are far more complex. Measurements on the dienes in aliphatic and aromatic solvents have all indicated kinetic orders between one-fourth and one-sixth. Some representative data for isoprene polymerization (98) are shown in Fig. 6.

By analogy with styrene, this one-fourth to one-sixth order dependency was taken to indicate the existence of association numbers 4–6 values that received some apparent support from the application of light scattering (99). However, these data appear (20) to be erroneous; the overwhelming experimental evidence points to a dimeric state of association for the poly(dienyl)lithium chains as well as for poly(styryl)lithium. The dimerized association state, at chain end concentrations occurring during polymerization, is supported by the results from (a) light scattering measurements done in n-hexane (97); (b) concentrated solution viscosity measurements done on high molecular weight poly(dienyl)lithium chains (20, 86, 95–97); (c) bulk viscosity measurements done on low molecular

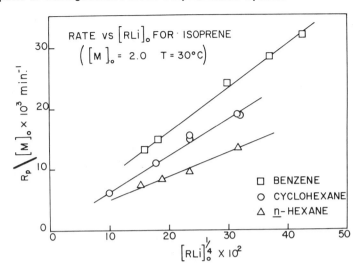

Figure 6 Dependency of the propagation rate of isoprene on chain end concentration in hydrocarbon solvents ($[M]_0 = 2.0$, $T = 30°C$).

weight poly(butadienyl)lithium chains (85); (*d*) cryoscopic measurements on 3-neopentylallyllithium solutions (69); and (*e*) the results obtained from the comparison of the concentrated solution viscosities before and after the linking of poly(dienyl)lithium species into star shaped chains (20).

These results also leave little room for doubt that the kinetics of propagation bear no direct relation to the state of association; a situation which. it will be recalled, was found for the initiation kinetics involving organolithium species (cf. Tables 2 and 3). It is thus clear that the kinetics of the propagation reaction for isoprene and butadiene cannot be adequately expressed by a relation of the type shown in equation [19], although it appears to describe satisfactorily the propagation kinetics for styrene in hydrocarbon solvents.

Table 5 summarizes the available information on propagation rates in the organolithium induced polymerizations of the two dienes and styrene in hydrocarbon solvents. It can readily be seen that many of these findings are discordant with regard to the reaction order with respect to chain end concentration (initiator dependence). As mentioned previously, the polymerization of styrene consistently shows a one-half order whereas the most reliable data on the polymerization of the dienes indicate kinetic orders between one-fourth and one-sixth.

Table 5 Propagation Rate Studies in Hydrocarbon solvents

Monomer	Initiator	Solvent	Initiator Concentration (moles/liter)	Order of Monomer Dependence	Order of Initiator Dependence	Ref.
Styrene	n-C_4H_9Li	Benzene	10^{-3}–10^{-1}	2	0	100
			$<2\times10^{-2}$	1	1	101
			$>2\times10^{-2}$	Variable	0	101
		Toluene and cyclohexane	8×10^{-3}–7×10^{-2}	1	0.5	102
			0.16×10^{-4}–3.9×10^{-2}	1	0.5	53
	n-and t-C_4H_9Li	Toluene and cyclohexane	5×10^{-4}–2.5×10^{-3}	2	0.33–0.5	103
				2	0.33–0.5	103
	s-C_4H_9Li	Toluene and cyclohexane	3×10^{-3}–4×10^{-2}	1	0.5	103
	n-C_4H_9Li	Cyclohexane	10^{-3}–5×10^{-5}	1	0.5	71
Isoprene	n-C_4H_9Li	n-Hexane	2.4×10^{-4}–5.7×10^{-3}	1	0.5	93
	n-C_4H_9Li	n-Hexane	2×10^{-5}–1.4×10^{-3}	1	0.5	104
	t-C_4H_9Li	n-Hexane	10^{-3}–10^{-4}	1	0.5	105
	C_2H_5Li	Cyclohexane	6×10^{-4}–4.1×10^{-3}	1	0.5	106
	n-C_4H_9Li	n-Heptane	10^{-2}–10^{-6}	1	0.17–0.5	107
	n-C_4H_9Li	Benzene	2×10^{-3}–16×10^{-3}	1–2	0	108
		n-Hexane				

n-C_4H_9Li	n-Heptane	$<5\times10^{-3}$	1–2	0.25	109
		5×10^{-3}–2×10^{-2}	1–2	0.25–0	
		$>2\times10^{-2}$	1–2	Negative	
n-, sec-, t-C_4H_9Li	Cyclohexane	2.6×10^{-3}–6.5×10^{-3}	1	0.17	110
n-C_4H_9Li	Benzene	10^{-3}–10^{-2}	1–2	0	111, 112
s-C_4H_9Li	Benzene	7×10^{-4}–10^{-2}	1	0.5	113
s-C_4H_9Li	Cyclohexane	5.8×10^{-5}–1.6×10^{-2}	1	0.25	74
n-C_4H_9Li	Benzene	3×10^{-3}–3.5×10^{-2}	1	0.21–0.4	114
1,1-Diphenyl-hexyllithium	Benzene	10^{-4}–5×10^{-3}	1	0.25	115
Butadiene n-C_4H_9Li	Cyclohexane	4×10^{-4}–10^{-2}	1	0.17	71
n-C_4H_9Li	n-Heptane	7×10^{-5}–6×10^{-3}	1	0.25	109
n-C_4H_9Li	n-Hexane	9×10^{-4}–1.2×10^{-2}	1	0.5	93
s-C_4H_9Li	Cyclohexane	8×10^{-2}–2×10^{-2}	1	0.5	103
		2×10^{-2}–2×10^{-3}	1	0.33	103
n-, s-, t-C_4H_9Li	Toluene and cyclohexane	5×10^{-4}–4×10^{-2}	2	1–0	103

3.4   Chain Transfer

The process of chain transfer in anionic polymerization systems, although recognized, has received virtually no quantitative attention. Thus there is no list of chain transfer constants for carbanion systems as for free radical polymerizations. The work of Higginson and Wooding (7) on the homogeneous polymerization of styrene by potassium amide in liquid ammonia and that of Robertson and Marion (8) on the heterogeneous polymerizations of butadiene by sodium in toluene first demonstrated the important role of solvent in transfer reactions. Parallel results were obtained by Bower and McCormick (116) and Brooks (117) for the organosodium initiated polymerization of styrene in toluene. Both groups detected molecular weights lower than calculated from the monomer/initiator ratio.

Low molecular weight chains have been found (118–120) in the polymerization of ethylene by n-butyllithium in a benzene-tetramethylenediamine (TMEDA) system. TMEDA promotes a chain transfer reaction between the primary carbanion and benzene and oligomeric polyethylene containing phenyl groups can be isolated. Changing the solvent to n-hexane yielded a polymerization free of any chain transfer (121), although only very low molecular weight material was obtained.

The oligomerization of butadiene with sodium counterion in a tetrahydrofuran–toluene solution was also studied (122). The presence of the ether promoted the transfer reaction between the chain ends and toluene as witnessed by the low degree of polymerization (10 or less) found. For this system, where the chain transfer constant is near unity, the weight average degree of polymerization can be given by (123)

$$\bar{x}_w = \frac{2\bar{x}_n - 1 + \beta}{1 + \beta} \qquad [20]$$

where $\beta = \alpha/\bar{x}_n$ and $\alpha = 1.703$ when butadiene and toluene are used. The α factor (the ratio of the molecular weight of butadiene to that of toluene) is necessary since the contribution of the benzyl group to the overall molecular weight of each chain is large at low values of $\bar{x}_n$. Hence, when $\bar{x}_n$ is known, the ratio of the weight average to number average molecular weights can be estimated.

Low molecular weight polydienes of various microstructures are being prepared commercially by an anionic chain transfer polymerization (124–127). These polymerizations involve toluene as both solvent and chain transfer agent, and lithium as the counterion. The transfer reaction is promoted by the use of diamines or potassium t-butoxide. Depending on

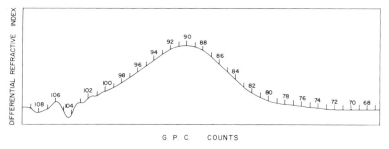

Figure 7 Gel permeation chromatogram of PM polybutadiene (Lithium Corporation of America) prepared by chain transfer anionic polymerization.

reaction conditions low molecular weight polybutadienes are prepared with variations in both microstructure and molecular weights. In addition, the foregoing process has proved to be amenable for the preparation of low molecular weight copolymers of butadiene and α-methylstyrene. Figure 7 is a gel permeation chromatogram of one of these commercial low molecular weight polybutadienes. As anticipated, the polybutadienes obtained from this polymerization system possess broad molecular weight distributions ($\bar{M}_w/\bar{M}_n \simeq 2$). The magnitude of the molecular weight distributions is consistent with the random nature of the transfer step and the relatively rapid nature of the initiation via the metallated toluene.

A chain transfer mechanism has been invoked (128–129) to explain the fact that although 100% conversion is obtained in the n-butyllithium initiated polymerization of 9-vinyl anthracene in tetrahydrofuran, only material with degrees of polymerization from 4 to 12 is obtained. Unfortunately, no experiments were done to elucidate in an unambiguous fashion the mechanism of the chain transfer step.

Gatzke (130) has investigated the chain transfer reaction between poly-(styryl)lithium and toluene at ca. 60°C. A relationship between the number average degree of polymerization and the chain transfer constant was derived as follows:

$$\bar{x}_n = \frac{[M]X}{[SLi]} - C_{RH}[RH]\ln(1-X) \qquad [21]$$

where [M] is monomer concentration, X is degree of conversion, [SLi] is chain end concentration, [RH] is toluene concentration, and C_{RH} is the transfer constant. The value of C_{RH} was found to be 5×10^{-6}.

These results of Gatzke vitiate some conclusions reached by Smets and co-workers (131, 132). The latter polymerized o- and p-methoxystyrene in toluene with n-butyllithium initiator (131) and styrene in toluene with various anisoles (132) as cosolvents. For the p-methoxystyrene and styrene studies a termination reaction was claimed to occur via metallation of either the p-methoxystyrene or the anisoles. It was assumed that the only type of transfer present involved either monomer or the aromatic ether but the known chain transfer reaction involving toluene was not considered. Some polymerization kinetic measurements on o-methoxystyrene were interpreted as indicating the absence of chain termination, although the benzyllithium resulting from chain termination by toluene can initiate new chains, thus leading to a constant concentration of chain ends. Furthermore, the presence of methoxy groups in these polymerizations will serve to increase the frequency of the chain transfer reaction with toluene. A puzzling feature of both papers by Smets was the claim that both number average and viscosity average molecular weights were measured although number average data were reported in only one paper (132). These data are obviously needed to clarify these reactions.

3.5 Copolymerization of Dienes and Styrene

In copolymerizations involving organolithium chain ends unusual solvent effects have been observed. Although it might be anticipated that different counterions would influence the reactivity of a monomer pair, it was surprising to observe that the nature of the solvent can also exhibit a marked influence. This is especially pronounced in solvents that exert a specific solvation effect (such as ethers or amines). This is well demonstrated (133–135) for the monomer pairs, styrene-butadiene and styrene-isoprene. It was found that, in *hydrocarbon solvents*, it is the diene which dominates the initial copolymerization to the virtual exclusion of styrene and exhibits a propagation rate nearly identical to that of the diene alone. Only when the diene supply is nearly depleted does styrene begin to become incorporated into the chain. The faster polymerization rate then observed is that of styrene alone.

The addition of small amounts of ethers or amines can markedly alter the situation, so that the styrene predominates over the dienes. This behavior demonstrates that the ionic species present in the organolithium polymerizations are different in different solvent systems, and that solvents having atoms with unshared electron pairs can exert a pronounced solvating effect on the cations and thus alter the structure of the ion pair. This solvation of the ion pair manifests itself in two ways: incorporation of styrene in the copolymer chain is enhanced, and the vinyl mode of addition of the diene units is increased.

The reversal of reactivity of styrene and the dienes in copolymerization in hydrocarbon solvents also requires explanation. Two possibilities have been suggested. On a kinetic basis, the observed behavior requires a crossover reaction between styryllithium and the diene which is rapid compared to the other three propagation steps. Since there are pronounced differences in absorption in the visible and ultraviolet regions between the poly(styryl)lithium and poly(dienyl)lithium species it is easy to compare the relative rates of the individual homopropagation steps with the corresponding cross-propagation steps of the styrene-butadiene and styrene-isoprene pairs. Thus the individual crossover steps can be illustrated as follows:

$$\text{\small$\wedge\!\!\wedge$}\text{SLi} + \text{D} \xrightarrow{k_{\text{SD}}} \text{\small$\wedge\!\!\wedge$}\text{SDLi} \qquad\qquad\qquad [22]$$
$$\underset{red}{} \qquad\qquad \underset{colorless}{}$$

$$\text{\small$\wedge\!\!\wedge$}\text{DLi} + \text{S} \xrightarrow{k_{\text{DS}}} \text{\small$\wedge\!\!\wedge$}\text{DSLi} \qquad\qquad\qquad [23]$$
$$\underset{colorless}{} \qquad\qquad \underset{red}{}$$

where D and S denote the diene and styrene, respectively. In this way, the crossover rate constants k_{SD} and k_{DS} were measured (136–138), and values were then combined with the apparent propagation rate constants of the monomers in homopolymerization to yield r_1 and r_2 values for the copolymerization (Table 6). The values of k_{SD} were found, as anticipated, to be very high.

It is reasonable to expect a faster reaction rate for poly(styryl)lithium with a diene than with styrene, if not from resonance energy considerations, then from entropy considerations in view of the shielding effect of the bulky phenyl group in styrene. This type of rate reversal is also noted in free radical polymerization although not to as pronounced an extent.

The large differences observed in the reactivity ratios of the styrene-diene pairs have raised some doubts as to the validity of an explanation solely on kinetic grounds, and an alternative explanation has been proposed (133) based on a specific solvent effect exerted by the dienes. This

Table 6 Copolymerization of Styrene and Dienes in Organolithium Systems

Monomer[a] (M$_2$)	Solvent	Temperature (°C)	r_1	r_2	Ref.
Butadiene	Benzene	29	0.08–0.41	4.5	136
	Cyclohexane	40	<0.04	26	137
Isoprene	Benzene	29	0.13–0.46	4.9–23.1	136
	Cyclohexane	40	0.046	16.6	138

[a] Styrene = M$_1$.

involves the dienes "solvating" the ion pair chain end which, it is assumed, styrene cannot do. The presence of a more powerful solvating agent such as an ether would prevent the diene from effective participation in complex formation and thus allow both monomers to polymerize solely on the basis of their reactivities. A complete resolution of the mechanism operating in the styrene-diene hydrocarbon systems will depend on better insight into the interaction of these unsaturated compounds (i.e., the conjugated dienes) with the organometallic chain ends.

Germane to this point is the work of Smart et al. (70, 139) who observed an interaction (apparently of a dipole-dipole nature) between the lithium counterion and the double bond of 3-butenyllithium. This interaction, demonstrated by the application of ^{7}Li NMR, ultraviolet, and infrared spectroscopy, is the *first* unambiguous demonstration that π electrons can, at least in hydrocarbon solvents, interact with the lithium of an organolithium species even when the latter is in the associated state. These results thus suggest that the dienes can interact with the organolithium chain end of a growing polymer chain. The ramifications and potential role of this interaction on the kinetics and mechanisms of these homo- and copolymerizations are considered in a later section.

The copolymerization of styrene and butadiene has achieved commercial importance (140, 141) with an apparent random placement of the two monomer units. In the main, the copolymers are formed from a 75:25 butadiene/styrene charge. One, Solprene 1204 (Phillips Petroleum Company), has a vinyl content of ca. 30%, indicating that a polar cosolvent (e.g., triethylamine or ether) may have been used to randomize the placement of the monomer units. It has also been noted (142, 143) that the presence of lithium, sodium, potassium, rubidium, or cesium *t*-butoxide exerts a profound effect on the organolithium initiated polymerization of styrene in butadiene. By the adjustment of the ratio of butoxide to alkyllithium and variation in the counterion combinations, copolymers with varying degrees of randomness were prepared. As an example, at an organolithium/*t*-C$_4$H$_9$OK mole ratio of 2.0, the styrene content, expressed as a function of the conversion curve for a 75:25 butadiene/styrene charge, was found to be identical to the conversion obtained when phenylpotassium was used as the initiator. These results are markedly different from those obtained when organolithium is used alone. It was suggested (142, 143) that the two cations are in tautomeric equilibrium with the chain end and the butoxide, leading to monomer placements characteristic of each of the counterions. This could be a natural result of a system of cross-associated species, viz., the *t*-butoxide with a chain end, as previously discussed for the organolithium initiators. Each species will undoubtedly differ from the others in regard to rate of monomer addition, choice of monomer, and, for butadiene, in chain microstructure. No information is currently available as to the structure of the proposed cross-associated species.

Of course it is possible to obtain a random copolymer of styrene and butadiene, without a concomitant high vinyl microstructure, even in hydrocarbon media, by maintaining a constant styrene/butadiene monomer ratio throughout the reaction. This can be accomplished either by a programmed replenishment of the butadiene or by a continuous polymerization process (144).

3.6 Block Copolymerization

The quest for new elastomeric materials has led to the discovery and evaluation (145) of "thermolastic" elastomers which are block polymers of the ABA type. These polymers exhibit the properties of a vulcanized rubber at ambient temperature but can be molded and remolded at elevated temperatures (146). They can have tensile strengths in excess of 300 kg/cm with elongations of 1000% or greater at temperatures of ca. 20°C (147).

These properties are a reflection of the chain structure of these triblock polymers in which the interior block (B) is a polydiene ($M_n = 50,000-100,000$) while the exterior block (A) is polystyrene ($M_n = 10,000-15,000$) (147). Upon exceeding the glass transition temperature of the polystyrene, the end blocks soften and diffuse to join other end segments and thus form discrete spherical aggregates when cooled. This phase separation occurs since the entropy gained by mixing is very small—a consequence of the relatively small number of molecules per gram involved. Hence a normally insignificant positive free energy of interaction is sufficient to overwhelm this small combinatorial entropy of mixing.

The formation of ordered domains of polystyrene-rich phases embedded in a polydiene-rich matrix has been corroborated by electron micrographs of the triblock polymers (148–150). Additional insight on the *three*-dimensional structure of these two-phase systems has been obtained through the application of low angle x-ray scattering (151) on annealed SBS block copolymers. The molecular weights of these samples ranged from 4.9×10^4 to 9.0×10^5 while volume percent polystyrene ranged from 23.5 to 26.6. For the highest molecular weight block copolymer simple cubic packing of *spherical* polystyrene domains was found while a face-centered cubic system was found for the remaining three samples. This series of packings is shown to scale in Fig. 8.

These results have been questioned by Pedemonte and Alfonso (152) who claimed a lamellar arrangement for the polystyrene domains in an

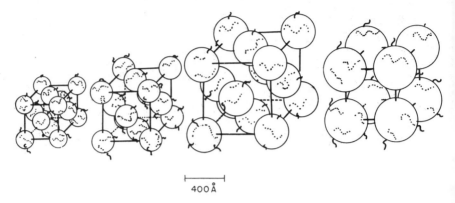

Figure 8 Scale models of lattice packings of polystyrene spheres. From left to right: $\bar{M}_n$ of the block copolymers = 49×10^3, 94×10^3, 142×10^3, and 90×10^4 g/mole, respectively.

SBS (styrene-butadiene-styrene) block copolymer via electron microscopy. Questions of technique and interpretation still remain to be resolved, but misinterpretation may have resulted from the overlap of (spherical) domains in adjacent layers of film which gave rise to apparently striated structures (189).

The synthesis of precisely tailored block polymers is possible through the application of the homogeneous, termination-free anionic polymerizations initiated by organolithium compounds. Such systems thus far offer the only mechanism by which control over molecular weights and purity of composition can be exerted, as well as being the only addition polymerization system to achieve commercial importance for the preparation of high molecular weight block polymers. The preparation of the ABA polymers can follow any one of several procedures. Each is briefly outlined below and then discussed in more detail (153):

1. A three-stage process using a monofunctional initiator.

2. A two-stage process with a difunctional initiator leading to the polymerization of monomer B followed by the polymerization of monomer A.

3. The two-stage process using a monofunctional initiator to synthesize AB polymer with subsequent coupling to form ABBA polymer.

4. The two-stage process with a monofunctional initiator to form an initial A block followed by the copolymerization of A and B during which the latter is preferentially polymerized. Thus a tapered segment made of monomers A and B is formed as the middle block between terminal A blocks.

3.6.1 *Three-Stage Process of Monofunctional Initiators*

When a monofunctional organolithium initiator is used to prepare block polymers of polystyrene-polydiene-polystyrene (SDS), styrene is polymerized first. Usually the polymerization is carried out in a hydrocarbon solvent (e.g., benzene) since a less elastomeric, high 1,2-polydiene microstructure results from the presence of polar solvents.

The second step, initiation of the diene by styryllithium, is quite rapid (see Table 6), as demonstrated by the rate of disappearance of the red color of the polystyryllithium chain end. In the third and final step, the addition of styrene to the SD block polymer, the crossover reaction from dienyllithium to styrene is not fast but can be accelerated by the addition of a small amount of diethyl ether, tetrahydrofuran, or dimethoxyethane. The ether concentration used is usually about 10^2 times greater than the chain end concentration.

The efficiency of the three-stage process is demonstrated by the GPC curve shown in Fig. 9 for an SIS (styrene-isoprene-styrene) block copolymer (89) prepared in benzene and initiated with *sec*-butyllithium.

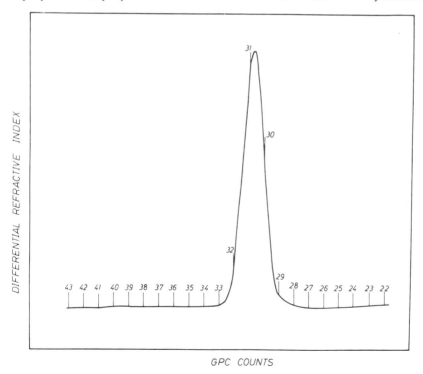

GPC COUNTS

Figure 9 Gel permeation chromatogram of an SIS block copolymer.

3.6.2 *Two-Stage Polymerization Involving Dilithium Initiators*

The difunctional initiators, e.g., sodium naphthalene, are potentially the best for the synthesis of ABA block polymers. They involve only two sequential monomer additions, B followed by A, and thus termination of the chain ends by adventitious impurities is minimized. Second, they can be used in unique "unidirectional" block polymerizations, i.e., where B can initiate A but A cannot initiate the polymerization of B.

A major hindrance in the use of difunctional initiators for SDS block polymers is that they can be prepared only in aliphatic ethers, which leads to a low 1,4 content in the polydiene, i.e., a nonrubbery polymer. However, recent developments (154) in dilithium initiators make it possible to circumvent these difficulties and to use a difunctional initiator to generate ABA block polymers of high purity. The block polymers synthesized were poly-α-methylstyrene-polyisoprene-poly-α-methylstyrene (mSImS) and poly-α-methylstyrene-polybutadiene-poly-α-methylstyrene (mSBmS). α-Methylstyrene was selected as a good candidate for an A block with interesting high temperature potential since it is known (155) to have a glass transition temperature of 173°C.

In order to prepare a block polymer with end blocks of poly-α-methylstyrene, a dilithium initiator is necessary for two reasons. First, α-methylstyrene has a ceiling temperature (156) of 60°C and the active polymer is therefore in equilibrium with a substantial amount of monomer at room temperature. Second, this monomer has very slow rate of polymerization at ambient temperature or below, in hydrocarbon solvents (157), and hence requires the presence of polar solvents to accelerate the polymerization rate. This does not pose a particularly difficult problem if α-methylstyrene is the last monomer polymerized, since polar solvents can safely be present once the diene has been polymerized.

A variety of dilithium initiators have been prepared (158–159) in aliphatic ether solvents, but these are of little interest for the reason cited earlier. Although organic dilithium compounds have been prepared in hydrocarbon media (160) they are known to be insoluble as well as impure. However, a dilithium compound of high purity, namely 1,4-dilithio-1,1,4,4-tetraphenylbutane, has been synthesized in a hydrocarbon medium containing a small amount (~ 15 vol %) of an aromatic ether such as anisole. The success of this synthesis is based on the fact (161) that such aromatic ethers have little or no effect on the 1,4 content of anionically prepared polydienes.

With the use of this initiator, SIS, mSImS, and mSBmS block polymers were synthesized and some of their physical properties evaluated. Table 7

Table 7   Molecular Weights and Composition of Block Polymers Made by Dilithium Initiation

Sample	$M_s \times 10^{-3}$ (g/mole)	$M_n \times 10^{-3}$ (g/mole)	$M_w \times 10^{-3}$ (g/mole)	End Blocks (wt %)
SIS-6	20 – 87–20	131	—	33
mSImS-1	21 – 85–21	132	141	33
mSImS-2	27 – 66–27	117	125	45
mSImS-3	20 – 87–20	127	—	32
mSImS-4	14 – 61–14	89	—	32
mSImS-5	14 – 87–14	115	—	25
mSImS-6	16 – 73–16	105	—	30
mSBmS-1	13.7– 51–13.7	78	—	35
mSBmS-2	12.2– 57–12.2	82	—	30
mSBmS-3	14 –102–14	131	—	20
mSBmS-4	14.6– 42–14.6	72	—	41
mSBmS-5	18.5– 78–18.5	115	—	32

contains the characterization and composition data on these block polymers (154, 160).

In conjunction with the molecular weights, further characterization of these block polymers was provided by the gel permeation chromatograph and the ultracentrifuge. Some of these data are presented in Figs. 10 and 11. Both the GPC curves and the Schlieren pattern illustrate that these block polymers possessed a narrow distribution of molecular weights with the absence of any detectable amount of low molecular weight material caused by either monofunctional initiator or premature termination reactions.

Measurements (154) of the tensile strength of SIS and mSImS block polymers of virtually identical composition and molecular weight revealed that the replacement of polystyrene with poly-α-methylstyrene endowed the mSImS block polymer with higher tensile strengths at comparable temperatures as well as higher temperature capability (Table 8). Thus the presence of glassy microdomains made of a polymer having a higher modulus than polystyrene apparently results in a higher tensile strength at rupture than is found in block polymers containing polystyrene. This, of course, implies that the poly-α-methylstyrene domains are capable of absorbing more energy than the polystyrene domains, presumably because of the higher stress that the poly-α-methylstyrene domains can support at comparable temperatures. This view is in line with that of Smith and Dickie (163) who demonstrated that the plastic characteristics

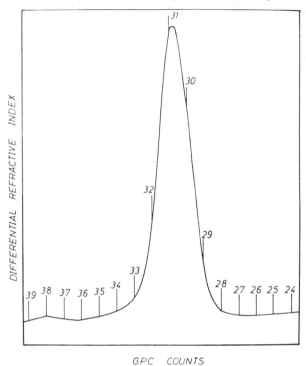

GPC COUNTS

Figure 10 Gel permeation chromatogram of mSImS-1 block copolymer with 33 wt% poly(α-methylstyrene). $M_n = 13.2 \times 10^4$ g/mole.

of the microdomains have a controlling effect on crack growth and thus on the ultimate properties of these block polymers. The higher T_g of the poly-α-methylstyrene blocks manifests itself in the higher temperature capability of the mSImS block polymers. Hence molded goods can be made which possess the ability to undergo sterilization at 100°C without loss of shape.

The synthesis of an mSImS block polymer in toluene was also reported (164) using the tetramer of α-methylstyrene as the initiator with lithium as the counterion. Unfortunately, no characterization data were included nor was there given any information pertaining to the projected molecular weight. The isoprene microstructure was not high cis-1,4 since the dilithium initiator was made in tetrahydrofuran, which was not removed. The polymerization of the α-methylstyrene was accelerated by the addition of a small amount of hexamethylphosphortriamide. This polar compound has been shown (165–168) to be capable of vastly accelerating alkali metal polymerizations. The tensile strength of this mSImS polymer

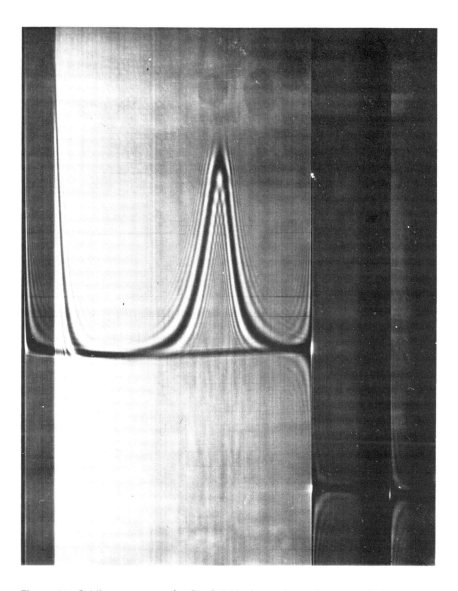

Figure 11 Schlieren pattern of mSlmS-1 block copolymer in tetrahydrofuran.

Table 8   The Tensile Strength and Elongation at Break at Various Temperatures for the mSImS and SIS-6 Block Polymers

Temperature (°C)	mSImS-1		SIS-6	
	σ_b (kg/cm^2)	λ^a	σ_b (kg/cm^2)	λ^a
−60	615	4	440	5
−40	600	6	380	6.5
−20	580	7	360	8.5
0	530	8.5	345	10
10	500	9		
20	455	10	260	11
40	295	12	235	14
60	225	13	120	16
80	110	11.5	30	16
100	65	11		
115	45	12.5		
130	20	11		
145	6	4		

^{a} The values of λ (extension ratio) are only approximate for temperatures other than 20°C.

was reported to be ca. 170 kgcm^2. This relatively lower tensile strength probably reflects the presence of diblock polymer resulting from monofunctional initiator impurities.

3.6.3   Monofunctional Initiation and Coupling

This mode of block polymer preparation is valuable from several standpoints in the preparation of ABA polymers. A two-stage polymerization is followed by a difunctional coupling agent added to unite the diblocks and thus form triblock polymer. The polymerization time is one-half that necessary for the preparation of a triblock polymer, of the same molecular weight and composition, by the three-stage process. Since it involves only two monomer additions, the possibility of termination due to residual impurities in a third monomer addition is avoided.

One difference between this procedure and those systems involving a dilithium initiator is that a monomer pair can be used where the chain end of A will initiate B whereas the chain end of B is incapable of initiating A. Thus it is possible to synthesize a block polymer where B is poly(propylene sulfide) and A is poly-α-methylstyrene (169). However, the offsetting disadvantage of this system is that it demands high precision in the stoichiometry of the coupling reaction as well as a high degree of efficiency. Any deficiency in the coupling reaction will leave diblock material which is quite deleterious, even in small amounts (170), to the physical properties of block polymers.

In Figs. 12 and 13 GPC traces (171) of two of the commercially available Kraton block polymers are shown. These traces were generated with a five-column system and a flow rate of 0.25 ml/min. It is important that the correct columns be used with a slow flow rate in order to achieve maximum resolution. It is readily seen that the Kratons possess a bimodal molecular weight distribution and contain ca. 15–20 wt% of what appears to be diblock polymer, which is one-half the molecular weight of the triblock polymer. The existence of this diblock material can be explained on the basis of inefficiency in the coupling step—if this is the preferred method in the commercial preparation of these SDS polymers.

A new block polymer, Kraton G, has been prepared by the Shell Chemical Company. This material contains an elastomeric center block resulting from the hydrogenation of the center segment. This block was originally polybutadiene containing about 65% 1,4 and 35% 1,2 units as determined by NMR. As with the other Kratons, the end blocks are

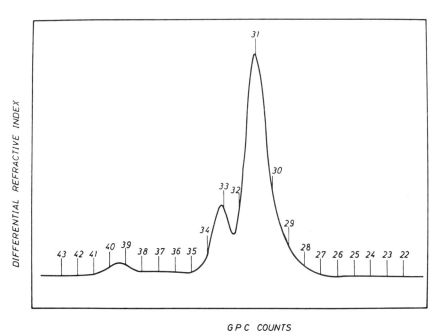

GPC COUNTS

Figure 12 Gel permeation chromatogram of block copolymer Kraton 1101.

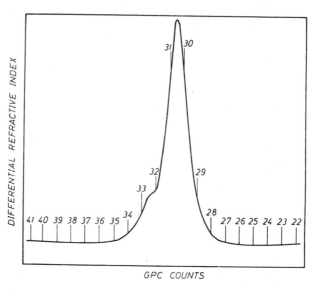

Figure 13 Gel permeation chromatogram of block copolymer Kraton 1107.

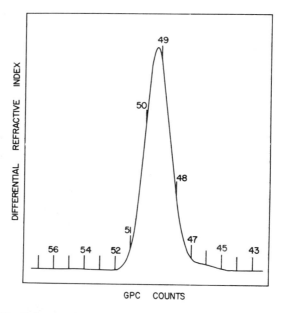

Figure 14 Gel permeation chromatogram of block copolymer Kraton G.

254

polystyrene. A GPC trace of this material is shown in Fig. 14. Two points can be made. The sample is devoid of low molecular weight diblock polymer, results that would be consistent with a three-stage sequential polymerization. In addition, the GPC trace demonstrates the absence of low molecular weight material resulting from degradation during the hydrogenation reaction.

3.6.4 Two-Stage Block Polymerization and Copolymerization of Styrene and Dienes by Organolithium

This procedure is based on the previously discussed low copolymerizability of styrene with dienes in organolithium-hydrocarbon systems. It has the advantage of a two-stage system in that the effects of termination can be minimized. The second step consists of the addition of a styrene-diene mixture, since the diene will be preferentially polymerized. Even if this styrene-diene mixture contains some terminating impurities, the effects may not be particularly troublesome since termination at this stage may produce free polystyrene which is known (170) to be not too important in reducing the properties of triblock polymers. Thus no diblock material should be formed if the termination reaction is rapid in relation to propagation. The marked disadvantage of this system is caused by the fact that a portion of styrene becomes copolymerized with the diene leading to decreased incompatibility of the two phases and hence poorer tensile properties (172). This copolymerization results in a "tapered" segment of random composition which "buffers" the two predominantly homopolymer end blocks. The effect on tensile properties of this copolymerization is especially pronounced in SIS block polymers since isoprene does not exclude styrene to the extent that butadiene does. Thus the polyisoprene block contains substantially more styrene than a comparable polybutadiene block, and this is reflected in the poorer physical properties of these block polymers (172) compared to those made by the three-stage process.

This procedure of prior mixing of the two monomers can also be used with a dilithium initiator. Thus a one-step monomer addition procedure is possible. However, virtually no termination can be tolerated since this would result in the synthesis of diblock polymer. This situation would then be identical to that encountered in a coupling procedure where the reaction between a dihalide and the chain ends fails to completely convert the diblock material to triblock polymer.

The mixing of monomers for one addition can also be used for α-methylstyrene-diene pairs. In one respect α-methylstyrene is better suited for this procedure than is styrene since virtually no α-methylstyrene is incorporated into the polydiene block, irrespective of whether isoprene or butadiene is used. Thus, the maximum possible incompatibility between the two phases can be achieved even though both monomers are added in one step.

3.6.5 *Star Block Copolymers*

The technique of forming "star" polymers has proved to be feasible through the use of polyfunctional linking agents. This procedure was first used by us (173) and later by others as a means of synthesizing tri- and tetrabranched polystyrenes. This synthetic procedure permits the predetermined manipulation of molecular architecture while maintaining control over molecular weight. Thus, a comparison of the solution and bulk properties of branched polymers with their linear counterparts is possible. Table 9 contains the types of branched polymers synthesized and characterized and the linking agents used. It can be seen that star polymers with various numbers of branches up to nine arms per molecule have been synthesized.

It is apparent from Table 9 that the attempts to synthesize well-defined branched polymers by anionic methods have followed two lines: (*a*) the reaction of polyfunctional electrophilic species with monocarbanionic polymers, and (*b*) the application of an anionic sequential copolymerization of a vinyl or diene monomer with a di- or trivinyl aromatic monomer.

Steric considerations can also influence the extent and efficiency of these linking reactions. It was noted (173) that the reaction between polystyryllithium and silicon tetrachloride yields a mixture of tri- and tetrabranched species along with linear chains. Similar results were also obtained when 1,2,4,5-tetrachloromethylbenzene was used (176). It has also been noted (88) that poly(isoprenyllithium) will not react completely with silicon tetrachloride when the latter is present in stoichiometric amounts. Predominantly trichain material was obtained. Figure 15 is a gel permeation chromatogram of a trichain polyisoprene prepared from silicon tetrachloride. The polymer registering at counts 47–50 is unlinked material. On the other hand, it was observed some time ago (185–187) that poly(butadienyllithium) will react in a quantitative fashion with silicon tetrachloride. These observations are fortified by the gel permeation chromatogram shown in Fig. 16. In this case, poly(isoprenyllithium) chains were capped with about six butadiene units before the addition of the linking agent. It would thus seem that the methyl group in the 4,1 chain end (191) of poly(isoprenyllithium) is sufficiently bulky to inhibit a complete reaction with silicon tetrachloride. It should also be noted that a quantitative reaction takes place when poly(isoprenyllithium) is reacted with methyltrichlorosilane.

The only commercially available star polymers are the Solprene and K-Resin block copolymers produced by the Phillips Petroleum Company. The former are four-armed elastomeric block copolymers of polystyrene and polybutadiene. A gel permeation chromatogram typical of these radial block copolymers is shown in Fig. 17. As can be seen the linked four-arm block copolymer is present with very little unlinked material. Table 10 contains the characterization data for the Solprene block copolymers.

Star block copolymers with a degree of linking of approximately 10 have been prepared (189) in benzene by the use of *m*-divinylbenzene as the linking agent. The linking reaction is carried out by adding a pre-determined amount of divinylbenzene to the polystyrene-*b*-polydienyl-lithium solution. As can be seen from Table 11 a variety of star block copolymers can be synthesized with various degrees of branching. Furthermore, the degree of branching can be controlled reasonably well and material nearly free of unlinked diblock can be prepared. This latter point is demonstrated by the gel permeation chromatogram shown in Fig. 18.

A new polystyrene-*b*-polybutadiene thermoplastic material containing about 75% styrene has been developed by the Phillips Petroleum Company (192). These K-Resins are block copolymers (known as KR01 and KR03) and like the Solprenes are the result of a postpolymerization linking reaction. However, as can be seen in Figs. 19 and 20, they possess a tetramodal molecular weight distribution. This unusual distribution is either the result of extensive blending to achieve a uniform commercial product or the result of a two-stage initiator addition during the polymerization. The addition of the second initiator (*sec*-butyllithium) takes place along with the addition of more styrene. Thus, at the close of the polymerization of the styrene, a bimodal molecular weight distribution is obtained. This two-step styrene polymerization is then followed by an addition of butadiene which, of course, yields diblock copolymer. Once linking is accomplished (by the addition of a polyfunctional epoxidized linseed oil) star block copolymers containing SB blocks are obtained. The multimodal nature of the diblock molecular weight distribution and the polyfunctionality of the star block copolymers formed are credited with improving the processibility of the KR01 and KR03 block copolymers.

Table 9 Synthesis of Branched Polymers

Monomer	Initiator	Solvent	Linking Agent	p^a	Ref.
Styrene	n-Butyllithium	Benzene	Silicon tetrachloride	3 and 4	173
	n-Butyllithium	Benzene/THF	1,2,4-Tri(chloromethyl)benzene	3	174
	n-Butyllithium	Benzene/THF	1,2-Bis(trichlorosilyl)ethane	3 and 4	175
			Cyclic trimer of phosphonitrilic chloride	2, 4 and 6	
	n-Butyllithium	Benzene/THF	1,2,4,5-Tetra(chloromethyl)benzene	2–10	176
			Hexa[p-(chloromethyl)phenyl]benzene	6	177
	Amylpotassium	THF, toluene/THF	Tri(allyloxy)-2,4,6-triazine	3	178
	α-Phenylethyl potassium	THF	1,2,4,5-Tetra(chloromethyl)benzene	3 and 4	179
	n-Butyllithium	Benzene/THF	Silicon tetrachloride	3 and 4	180
	sec-Butyllithium	Benzene	1,2-Bis(methyldichlorosilyl)ethane^b	4	181
	sec-Butyllithium	Benzene	p- and m-Divinylbenzene	7–15.5	182
	sec-Butyllithium	Benzene	p- and m-Divinylbenzene	6–13.4	183
	Phenylisopropyl potassium	THF	p-Divinylbenzene	3.4–22.3	184

Butadiene	n-Butyllithium	Cyclohexane	Silicon tetrachloride	4	185
		Cyclohexane	Methyltrichlorosilane	3	185
	n-Butyllithium	Benzene	Silicon tetrachloride	4	186
	n-Butyllithium	Hexane	Silicon tetrachloride	4	38,187
Isoprene	sec-Butyllithium	Cyclohexane	Methyltrichlorosilane	3	188
			Silicon tetrachloride	3	
Styrene butadiene block copolymers	sec-Butyllithium	Benzene	m-Divinylbenzene	4–6	189
Styrene-isoprene block copolymers	sec-Butyllithium	Benzene	Silicon tetrachloride	4	189
			m-Divinylbenzene[b]	4–9	189
	n-Butyllithium	Benzene	Silicon tetrachloride[b]	3	189
			Silicon tetrachloride	3 and 4	190

[a] Number of branches per macromolecule.
[b] The polystyryllithium was reacted with isoprene, about 3 units per active chain end. This served to reduce steric hindrance which retards the linking reaction.

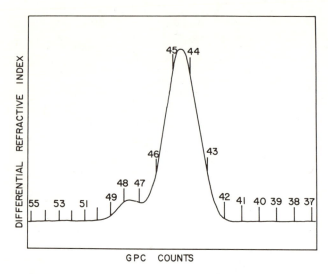

Figure 15 Gel permeation chromatogram of trichain polyisoprene. $\bar{M}_n = 3.6 \times 10^5$ g/mole.

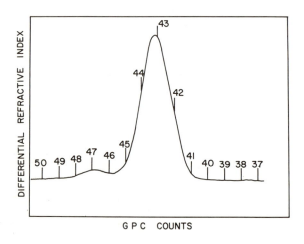

Figure 16 Gel permeation chromatogram of tetrachain polyisoprene. $\bar{M}_n = 6.0 \times 10^5$ g/mole.

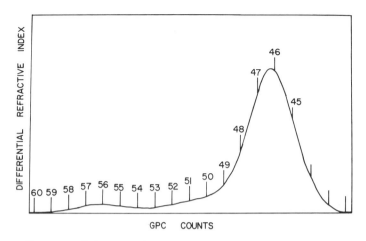

Figure 17 Gel permeation chromatogram of Phillips Solprene 406.

Table 10 Molecular Weights and Characterization of Phillips Radial Block Copolymers (193)

Phillips No.	Polystyrene (wt %)	$\bar{M}_n \times 10^{-5a}$ (g/mole)	$\bar{M}_n \times 10^{-4}$ (Polystyrene Portion)[b] (g/mole)
406	40	2.5	2.5
411	30	3.0	2.2
414	40	1.6	1.6

[a] Membrane osmometry; solutions in toluene at 37°C.
[b] Via gel permeation chromatography.

Table 11 Molecular Parameters of Star Block Copolymers

Star	Diblock $\bar{M}_s \times 10^{-3}$ (g/mole)	$\dfrac{m\text{-DVB}}{Li}$	Star $\bar{M}_n \times 10^{-3}$ (g/mole)	Number of Branches (p)	Unlinked Diblock (%)
SIS-DVB-3	66	3.26	448	7	~3
SIS-DVB-4	70	6.12	626	9	~3
SIS-DVB-5-1	71.6	12.64	326	7	9
SIS-DVB-6	51	3.23	305	6	~2
SIS-DVB-8	78.3	4.4	485	6	~2

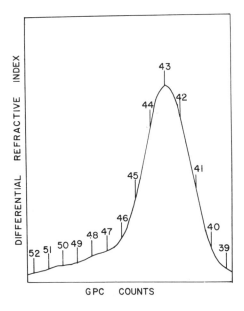

Figure 18 Gel permeation chromatogram of SIS nine-armed star block copolymer with 27 wt% polystyrene. $\bar{M}_n = 62.6 \times 10^4$ g/mole.

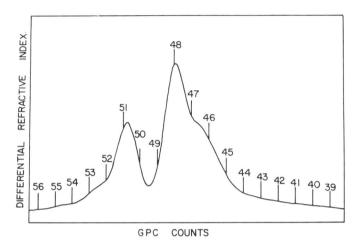

Figure 19 Gel permeation chromatogram of Phillips KR01 block copolymer.

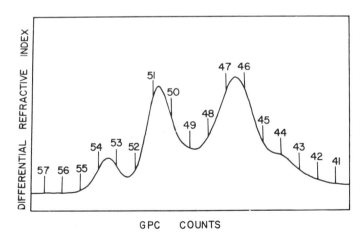

Figure 20 Gel permeation chromatogram of Phillips KR03 block copolymer.

4   STEREOSPECIFICITY IN DIENE POLYMERIZATION

4.1   Chain Microstructure

There are two truly outstanding features of the anionic mechanism of polymerization. The first is the "living" character of the growing chain, which has already been discussed. The second is the control of the stereoisomeric placement of the incoming monomer units. Although the homogeneous anionic polymerization systems appear to exert little or no control over the stereoisomerism of the asymmetric carbon atom (tacticity), the mechanism does appear to regulate the geometric isomerism of polydiene units. This is best exemplified by the discovery (9) that the polymerization of isoprene by lithium metal or derivatives leads to a high cis-1,4 chain structure, close to that of natural rubber.

It had been known for some time before this discovery that the microstructure of polybutadiene rubber was somewhat different when potassium metal was used as catalyst instead of the usual sodium. This was probably the first indication that the "counterion" could influence the mode of placement of the incoming monomer unit. The effect of lithium in isoprene polymerization was, of course, much more striking. The intensive investigations which followed this discovery demonstrated that it was not only the type of alkali metal which affected the chain microstructure, but that the solvent also could exert a profound influence (194–199). Fairly comprehensive reviews of these findings have already been published (200, 201). The older data are based mainly on analysis by infrared spectroscopy, which is not unequivocal in assigning values to the cis/trans ratio in polyisoprene. More recent data from NMR spectroscopy appear to be more definitive in this respect.

In general, all of these investigations have led to the following conclusions:

1.   The proportion of 1,4 chain units is highest for lithium and decreases with increasing electropositive character of the alkali metal.

2.   Polar solvents (e.g., ethers, amines) lead to an increase in side vinyl content of the polymer chain. This is especially marked in the case of lithium.

Thus the formation of side vinyl units (1,2 or 3,4, etc.) has been correlated (194, 196) with the increasing "ionic" character of the carbon–metal bond, polar solvents presumably increasing this ionicity by solvating the metal cation.

This gradation of ionic character of the carbon–metal bond has also been correlated with behavior during copolymerization (202). Thus data

Table 12 Effect of Initiator and Solvent in Isoprene-Styrene Copolymerization (202)[a]

Initiator	Solvent	% Styrene	% 1,4
Lithium metal or n-C₄H₉Li	Tetrahydrofuran	80	30
Na metal	Tetrahydrofuran	80	32
Na metal	Dimethyl ether	75	36
n-C₄H₉Li	Dimethyl sulfide	75	36
Li metal or n-C₄H₉Li	Diethyl ether	68	51
Na metal	Hydrocarbons	66	45
Li metal	Triethylamine	60	52
n-C₄H₉Li	Dibutyl sulfide	50	62
n-C₄H₉Li	Diphenyl ether	30	82
Li metal or n-C₄H₉Li	Hydrocarbons	16	95

[a] Equimolar monomer ratios, 25°C.

for isoprene-styrene copolymers in Table 12 show that there is excellent (inverse) agreement between the styrene content and the amount of 1,4 addition in such copolymers, the more ionic systems apparently favoring a greater styrene inclusion and a lower 1,4 content.

As previously stated, it has been proposed (94) that the propagating chain in ethers and other polar solvents consists of ion pairs in equilibrium with free carbanions, and this has been corroborated both by kinetic data and physical measurements of free ion concentrations. However, the existence of such ionic species has not been proposed in the case of hydrocarbon media, where the carbon–metal bond is presumably not ionic in nature.

From the viewpoint of reaction stereospecificity, the most thoroughly studied systems have been those involving lithium and organolithium initiators. The latter offer the most scope for investigations since they are soluble in a variety of solvents, both polar and nonpolar, and undergo few side reactions. Any mechanistic considerations will therefore be restricted to the organolithium polymerizations.

It is apparent from the foregoing discussion that the 1,4 content of a polydiene chain will be decreased by the presence of polar solvents in any organolithium initiated polymerization. Measurements of polyisoprene microstructure (Table 13) have shown that polar solvents such as tetrahydrofuran will alter the microstructure from the 70–90% cis-1,4

Table 13 Polydiene Microstructure via NMR Analysis

Initiator	Solvent	Cosolvent	$\dfrac{[\text{Cosolvent}]}{[\text{Initiator}]}$	Structure (%)				Polymerization Temperature (°C)	Ref.
				Cis-1,4	Trans-1,4	1,2	3,4		
			Polyisoprene						
t- and s-Butyl- lithium	Benzene	—	—	71[a]	23	—	6	20	203
				66[a]	27	—	7	40	203
				64[a]	27	—	8	60	203
				62[a]	30	—	8	80	203
n-Butyllithium	Cyclohexane	—	—	80[a]	15	—	5	30	204
t-Butyllithium[c]	Benzene[c]	—	—	69[b]	25	—	6	20	205
t-Butyllithium[d]	Benzene[d]	—	—	70[b]	24	—	6	20	205
t-Butyllithium[e]	n-Hexane[e]	—	—	70[b]	24	—	6	20	205
t-Butyllithium[f]	n-Hexane[f]	—	—	80[b]	14	—	6	20	205
t-Butyllithium[g]	n-Hexane[g]	—	—	91[b]	4	—	5	20	205
		THF	2	68[a]	19	—	13	30	203
		THF	15	—	69	—	31	30	203

				Polybutadiene[b]				
Cyclohexane +10% THF	THF	—	—		26	66	9	30	204
		—	—		15	30	55	30	206
n-Butyllithium	n-Hexane	—	35	6.3	57	8	—	30	207
	Ethyl ether	29	1.0	50	21	—	30	208
	THF	25	8.2	40	35	—	30	208
		21	17	31	45	—	30	208
		14	53	28	58	—	30	208
		7		10	85	—	30	208

[a] By means of the HR-100 NMR spectrometer.
[b] By means of the HR-300 NMR spectrometer.
[c] [Initiator] = 9×10^{-3} M, [Monomer] = 0.5 M.
[d] [Initiator] = 4×10^{-5} M, [Monomer] = 0.5 M.
[e] [Initiator] = 1.2×10^{-2} M, [Monomer] = 0.5 M.
[f] [Initiator] = 3.5×10^{-5} M, [Monomer] = 0.5 M.
[g] [Initiator] = 5×10^{-5} M, [Monomer] = 8.5 M.

microstructure obtained in hydrocarbon solvents to that of a chain containing mainly the 3,4 and 1,2 forms of addition. A parallel effect is seen for polybutadiene, which has a high (>90%) 1,4 content when prepared in hydrocarbon solvents while ethers convert the structure to a high vinyl content.

For polybutadiene it has been shown that the cis/trans ratio is 41:59 for material prepared in n-hexane. These NMR measurements compare favorably with infrared results reported by Forman (201) for polybutadiene prepared by n-butyllithium initiation in hydrocarbon media. The cis/trans ratio of ~40:60 is apparently maintained even when the polybutadiene is prepared under conditions that lead to a high vinyl content (207, 208).

In the ^{13}C NMR spectra, Mochel (209) could find no observable cis-trans linkages in polybutadienes polymerized in the n-butyllithium-hydrocarbon system. However, more recent work (207, 210) with the HR-300 NMR spectrometer has shown that the isomeric distribution is random and that the conclusion that no cis-trans linkages exist in butyllithium initiated polybutadienes is in error.

4.2   Chain End Structure Studies of Poly(dienyllithium) Species

The most recent investigations of the nature of the propagation step in the organolithium polymerization of dienes have utilized the techniques of nuclear magnetic resonance to study these carbon-lithium chain ends (69, 191, 211–217). The dienes studied included butadiene (69, 213, 214, 217), isoprene (191, 212, 213), 2,3-dimethylbutadiene (212), several pentadienes (215, 216), and 2,4-hexadiene (215, 216). Proton resonances of the chain end units were determined using approximately 1 M concentration of organolithium initiators in order to obtain precise spectra. Investigations (191, 213, 214) were made of the spectra of the monomer protons both in the presence and absence of the active chain ends. This was done in order to note whether any perturbation of these protons by the carbon–lithium chain ends occurs, an event that might indicate an intermediate complex formation. However, no differences were detected in the monomer spectra, indicating that a monomer-lithium complex, if one existed, was quite fugitive. Any such complex, if present, must have a lifetime too short to be observed in the NMR time scale.

However, as noted previously, interaction (proposed to be of a dipole-dipole nature) was reported (69, 84) between lithium and the double bond in 3-butenyllithium, which exhibits a degree of association of 6. This result suggests that olefinic and diolefinic species may, in general, interact (i.e., complex) with the counterion of associated organolithium compounds.

Typical ^{1}H resonance spectra for the chain end units of the oligomers (213) obtained from butadiene, isoprene, and 2,3-dimethylbutadiene in benzene are shown in Figs. 21 and 22. The term "pseudoterminated" refers to the addition of butadiene-d_6, which when added to the chain end, in effect, removes the lithium from the chain end without the use of a polar terminating agent such as water or an alcohol. The "transparent" butadiene-d_6 is also useful in noting the effect of chain length on chain end structure without "swamping" the NMR spectrum with a high concentration of in-chain units.

The peaks in Figs. 21 and 22 corresponding to the chain end protons are clearly recognizable by comparing the active and pseudoterminated species. These oligomers were prepared by the reaction of ethyllithium-d_5 with monomer. Since the organolithium is a slow initiator, there was a considerable amount of residual initiator (transparent) present with the oligomers of isoprene and 2,3-dimethylbutadiene. This was later remedied, in part, by using s-butyllithium to make the oligomers of

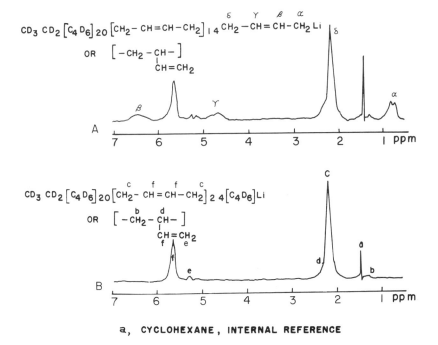

a, CYCLOHEXANE, INTERNAL REFERENCE

Figure 21 100 MHz NMR spectra of poly(butadienyl)lithium in benzene-d_6 at 23°C. (A) Active; (B) pseudoterminated.

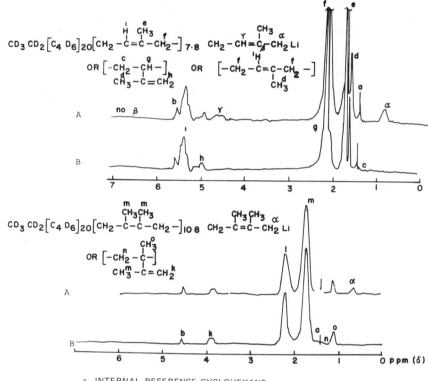

a INTERNAL REFERENCE CYCLOHEXANE
b REMNANT NONDEUTERATION IN PERDEUTEROBUTADIENE UNITS

Figure 22 100 MHz NMR spectra of poly(isoprenyl)lithium and poly(2,3-dimethylbutadienyl)lithium in benzene-d_6 at 23°C. (A) Active; (B) pseudoterminated.

isoprene (191, 211). As for butadiene, the chains were grown to a length of 20 or more units (using transparent butadiene-d_6) in order to avoid the high in-chain vinyl microstructure reported for very low molecular weight polybutadiene (85).

An examination of these NMR spectra (Figs. 21 and 22) led to the following conclusions:

1. The poly(butadienyl)lithium chain end is virtually 100% 1,4 with no 1,2 structures observable although the in-chain units show a 9% 1,2 content. The lithium is σ bonded to the α-carbon, as indicated by the presence of the two equivalent α protons. Hence, there is no evidence of a π-allyl type of delocalized bonding involving the γ-carbon.

2. The poly(isoprenyl)lithium chain ends show an exclusive 4,1 structure with no 4,3 addition despite the 10% 3,4 in-chain units which are present. In this case too, the lithium is σ bonded to the α-carbon. This conclusion is based on the observation that there are two equivalent α protons, no β protons, and no perturbation of the γ protons which would be expected for the π-allyl structures.

3. Poly(dimethylbutadienyl)lithium also shows exclusive 1,4 structures in its chain end units, despite the presence of 15–20% 1,2 in-chain units. There are, of course, no β or γ protons in this case, but the α protons again show only a localized carbon–lithium bond.

Although these spectra showed no noticeable π bonded organolithium chain ends, the presence of vinyl in-chain units was taken as evidence for the presence of an undetectable amount of such delocalized chain ends in equilibrium with the observed σ bonded chain ends, as depicted in Fig. 23. Presumably the π bonded chain ends can account for the observed vinyl in-chain units, by the addition of the incoming monomer at the γ-carbon. However, the type of equilibrium denoted in Fig. 23 should not lead to any isomerization of the β–γ double bond, and no such isomerization was in fact observed (191, 213).

The hypothesis that it is the π bonded chain ends, undetectable in the presence of hydrocarbon media, which lead to in-chain vinyl units, is strengthened by observation of the ^{1}H spectra in polar solvents such as tetrahydrofuran (THF). The spectra of poly(butadienyl)lithium in THF

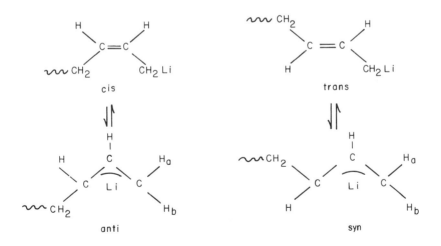

Figure 23 Localized-delocalized equilibrium in polybutadienyllithium chain ends.

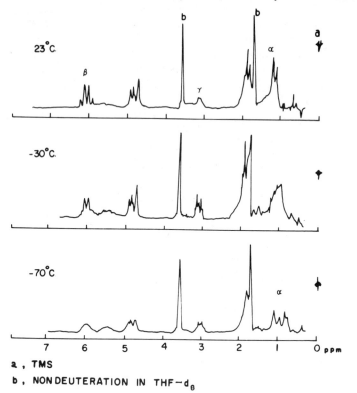

Figure 24   100 MHz NMR spectra of poly(butadienyl)lithium in THF-d_8 at three temperatures.

are shown in Fig. 24. The π character, i.e., delocalized, of the carbon–lithium bond was readily deduced from the following observations:

1. Two distinguishable α protons were observed (two overlapping doublets). This is presumably due to the restricted rotation of the α–β partial double bond.

2. The γ protons appear upfield at 3.3 ppm instead of at 4.7 ppm as in benzene, indicating the effect of the lithium close to the γ position.

Thus the correlation of the π bonded lithium found in polar solvents with the large amount of in-chain vinyl structure obtained in these systems seems to be justified.

The evidence for the existence of σ and π bonded lithium deduced from these NMR studies provides a basis for proposing a mechanism to account for the effect of solvents on the chain structure. This is depicted in Fig. 25 for the case of butadiene. Thus the 1,4 units are considered to

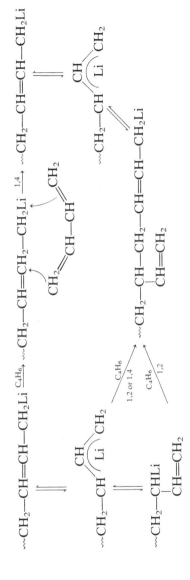

Figure 25 Proposed mechanism for butadiene polymerization

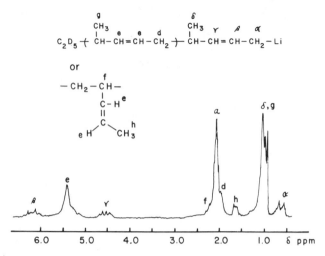

Figure 26 100 MHz NMR spectrum of poly(*trans*-1,3-pentadienyl)lithium in toluene-d_8. (a, nondeuteration in toluene-d_8.)

result from a concerted four-center reaction between the incoming monomer and the σ bonded carbon lithium bond, whereas the 1,2 units result from the attack by the incoming monomer at the γ-carbon, π bonded to the lithium. Figure 25 also indicates the possibility for the existence, in the presence of THF, of a σ bonded 1,2 chain end as one of the equilibrium forms. This was proposed to account for the fact that isomerization of the β–γ double bond was found to occur in THF.

Further corroboration for this proposed mechanism was obtained from the NMR studies of the active chain ends in the polymerization of pentadienes and hexadiene (216). Thus Fig. 26 shows the 100 MHz spectra of poly(1,3-pentadienyl)lithium prepared with the initiator ethyllithium-d_5. The α, β, and γ peaks of the chain end unit protons occur in positions similar to those found for butadiene and isoprene, and indicate a 4,1 chain end unit in this case also. However, the in-chain units had a mixed structure consisting of 80–85% 1,4 and 15–20% 1,2 with no 3,4 structure. Since there is no way to rationalize the occurrence of 1,2 in-chain units from 4,1 chain ends, this must be taken as evidence for the existence of an undetectable amount of 1,4 chain ends. The evidence for such chain ends will be discussed later. However, it is the occurrence of the 1,2 in-chain units which offers convincing evidence to corroborate the mechanism proposed in Fig. 25. Thus, although the monomer used in this work was the trans isomer, the 3,4 double bonds in the 1,2 in-chain units showed about 25–35% cis structure. Since direct isomerization of the monomer was shown to be impossible under these conditions, this change must involve isomerization of the active chain ends, which occurs during the σ-π equilibrium, as illustrated in Fig. 27. Thus, unlike the case of the previous 1,4 chain ends (butadiene, isoprene, etc.), where the σ-π equilibrium could not lead to isomerization of the 1,4 structure, such an equilibrium can lead to the isomerization of the 3,4 bond in pentadiene because its configuration can alter during an equilibrium of this type.

Figure 27 illustrates the previously proposed σ-π equilibrium on the basis of an undetectable amount of π-allyl 1,4 chain ends in the pentadiene case. Fortunately, in this case, direct evidence of the presence of 1,4 chain ends is available. This is demonstrated in Fig. 28, which shows 100 MHz spectra of a poly(pentadienyl)lithium prepared with the more reactive initiator, isopropyllithium-d_7, which leads to shorter chain lengths (e.g., about 2.7 units as shown). These low oligomers demonstrate the actual presence of 1,4 chain ends, which presumably result more frequently from the initiation step, i.e., from the attack of the secondary carbon–lithium species on the diene. The 1,4 chain ends apparently exist, at least to some extent, in the delocalized π-allyl form, as shown by the position of the γ peak at 3.1 ppm. This corresponds quite closely to that of the γ–H of the delocalized chain end of poly(butadienyl)lithium in THF (see Fig. 24). Hence this is the first instance where such π-allyl equilibrium forms have actually been observed in nonpolar solvents. Apparently the relative instability of the secondary carbanion at the α-carbon provides the driving force toward a greater extent of electron delocalization.

As already mentioned, the 1,4 chain ends are no longer detectable at higher chain lengths, due apparently to the preference of the more stable

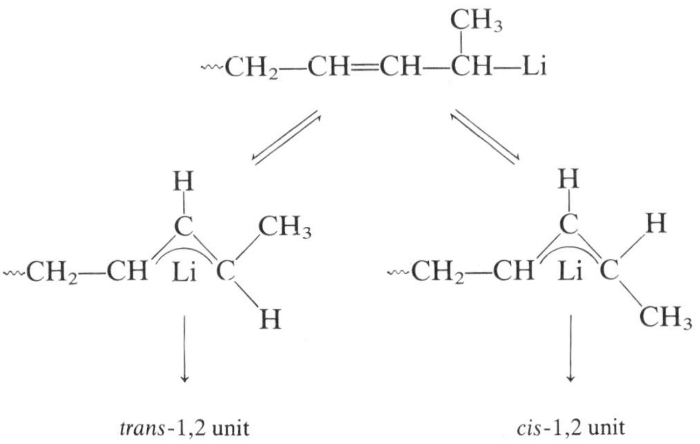

Figure 27 Mechanism for isomerization of the 3,4 bond in 1,3-pentadiene during polymerization.

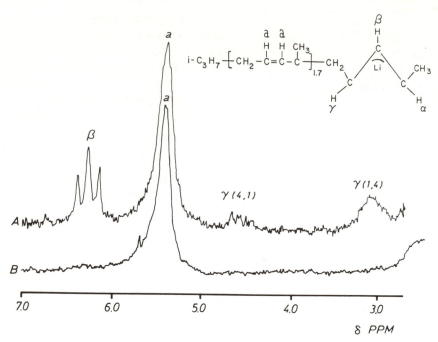

Figure 28   100 MHz NMR spectra of poly(*trans*-1,3-pentadienyl)lithium in benzene-d_6 at 23°C. (*A*) Active; (*B*) pseudoterminated.

4,1 chain ends to have a primary carbanion structure at the α-carbon. No such choice is available in the case of 2,6-hexadiene, where only 1,4 addition is possible, leading to a secondary carbanion. As might be expected, such chain ends show up in the delocalized form (215, 216) as shown in the NMR spectra of Fig. 29. No detectable amount of γ protons at 4.7 ppm, corresponding to the covalent form of the chain end, could be found.

Thus it appears that the chain end structures of these terminally substituted conjugated dienes offer firm evidence for the existence of the delocalized carbon–lithium bond even in nonpolar media, as called for in the proposed mechanism for the propagation step.

The mechanism proposed above can, therefore, account for the effect of polar solvents on the chain structure, at least insofar as the vinyl versus linear chain units are concerned. It can also help to rationalize the effects of nonpolar and polar solvents on the copolymerization of styrene and the dienes. Thus, for example, in hydrocarbon solvents, the σ bonded diene chain ends apparently predominate, making it difficult for styrene to participate in the necessary reaction, as illustrated in Fig. 25. Thus the diene is favored in the copolymerization. On the other hand, as the Lewis base strength of the solvent toward the carbon–lithium increases, a greater amount of styrene is incorporated into the initially formed copolymer (202, 218). This increase in styrene content is paralleled by an increase in the in-chain vinyl groups of the incorporated diene component. This is consistent with the foregoing mechanism (Fig. 25) where polar solvents increase the amount of the π bonded lithium species at the expense of the σ bonded species. The former species, as has been discussed, leads to the formation of in-chain vinyl groups during polymerization. The π bonded lithium species would also be expected to be more reactive toward styrene than the σ bonded lithium chain end, since the four-center reaction involving monomer and the active chain end is no longer necessary with this delocalized species. An additional facet to be considered is the fact that, under polymerization conditions, the σ bonded lithium chain ends are strongly associated in the dimeric state while the π bonded lithium chain ends exist as predominantly unassociated species (95, 96). However, the role of association on the reactivity of these carbon–lithium chain ends (in both homo- and copolymerizations) has yet to be quantitatively elucidated.

This mechanism, though, cannot adequately explain two other phenomena, i.e., the 70–90% cis-1,4 structure of polyisoprene, as compared to the 40% cis-1,4 content of polybutadiene, and the low kinetic orders of the propagation rates, in hydrocarbon media. Since it appears

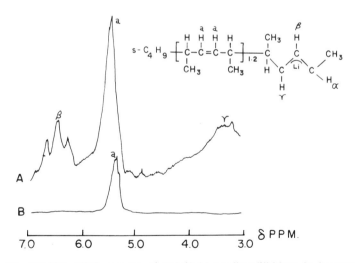

Figure 29 100 MHz NMR spectra of poly(2,4-hexadienyl)lithium in benzene-d_6 at 23°C. (*A*) Active; (*B*) pseudoterminated.

from some NMR spectra, that the 4,1-polyisoprene chain ends have a high cis structure whereas the 1,4-polybutadiene chain ends have a high trans structure, similar to the in-chain units for both polymers, it would be attractive to postulate that the configuration of the chain units is decided at the time of insertion of the monomer by the predominant conformation of the latter. However, the most recent evidence from microwave spectroscopy (219) indicates that both butadiene and isoprene have a predominant transoid conformation. Hence the cis inclusion of isoprene must be occurring during the actual propagation step.

In this connection, some very recent work (206, 220) has attempted an unequivocal resolution of the isomeric structure of the 4,1 chain end of poly(isoprenyl)lithium. Thus the one-to-one adduct of *t*-butyllithium and isoprene, 2,5,5-trimethylhexen-2-yllithium, was prepared in benzene and allowed to isomerize in the presence of THF, leading to a double bond of presumably exclusive cis configuration (220). Hence it was proposed that the predominant (~70%) form of the 4,1-poly(isoprenyl)lithium chain ends in *hydrocarbon* media must be trans, and this was presumably confirmed by a combination of ^{1}H and ^{13}C spectra (206). This particular finding is in sharp contrast to the chain end structure (~70% cis) *inferred* previously (191, 213). This proposal that the predominating form of the poly(isoprenyl)lithium chain ends has a trans structure generates the anomaly of how to accommodate the fact that the polymer produced is predominantly cis. Even if this analysis is reliable, the experiment had to be performed at very high concentrations, where the 2,2,5-trimethyl-hexen-2-yllithium species is probably associated as a tetramer (69), whereas at concentrations appropriate for the preparation of high molecular weight polymer, poly(isoprenyl)lithium has been shown to be associated as a dimer (20, 95–97). The role (if any) of the association state in determining the configuration of the poly(isoprenyl)lithium chain end is as yet unresolved.

As for the elucidation of the propagation kinetics, it would be necessary, among other things, to detect any effect of the chain end *concentrations* on the chain end structure. Since the poly(dienyl)lithium chain ends are strongly associated as dimers, the effect of concentration on reactivity is not unexpected, especially in view of their highly polar nature and the nonpolar character of the solvents. NMR studies have not as yet been able to throw any light on this perplexing problem since ordinary techniques are not sufficiently sensitive to work in this range of concentrations.

An additional complicating factor in these diene polymerizations is the lack of insight into the relative contribution of the associated and unassociated species toward propagation. For example, a total of 14 different reactive species *may* exist in these systems, i.e., 10 types of associated dimers resulting from the association and cross-association of the cis and trans σ bonded lithium chains and the syn and anti delocalized π bonded lithium species, as well as four types of unassociated chain ends. This complexity of structures may indeed account for the complex and baffling kinetics found in studies on organolithium reactions in hydrocarbon solvents. Hence the concept that only unassociated organolithium species are reactive in these systems appears to be an oversimplification, and it is far more likely that the associated species participate directly in polymerization. Future work could well be directed toward the elucidation of the role of the associated chain ends in these polymerizations.

REFERENCES

1. F. E. Matthews and E. H. Strange, British Pat. 24,790 (1910).
2. C. H. Harries, *Ann.*, **383,** 157 (1911).
3. C. H. Harries, U.S. Pat. 1,058,056 (April 8, 1913).
4. K. Ziegler et al., *Ann.*, **511,** 13 (1934).
5. K. Ziegler and L. Jakob, *Ann.*, **511,** 45 (1934).
6. K. Ziegler et al., *Ann.*, **511,** 64 (1934).
7. W. C. E. Higginson and N. S. Wooding, *J. Chem. Soc.*, **1952,** 760.
8. R. E. Robertson and L. Marion, *Can. J. Res.*, **26B,** 657 (1958).
9. F. W. Stavely and co-workers, *Ind. Eng. Chem.*, **48,** 778 (1956).
10. W. Schlenk et al., *Bereichte*, **47,** 479 (1914).
11. W. Schlenk and E. Bergman, *Ann.*, **463,** 91 (1928).
12. N. D. Scott et al., *J. Am. Chem. Soc.*, **58,** 2442 (1936).
13. D. Lipkin et al., *Science*, **117,** 534 (1953).
14. D. E. Paul et al., *J. Am. Chem. Soc.*, **78,** 116 (1956).
15. M. Szwarc et al., *J. Am. Chem. Soc.*, **78,** 2656 (1956)
16. M. Szwarc, *Nature*, **178,** 1168 (1956).
17. K. F. O'Driscoll and A. V. Tobolsky, *J. Polym. Sci.*, **37,** 363 (1959).
18. C. G. Overberger and N. Yamamoto, *J. Polym. Sci.*, **B3,** 569 (1965).
19. P. J. Flory, *J. Am. Chem. Soc.*, **62,** 1561 (1940).
20. L. J. Fetters and M. Morton, *Macromolecules*, **7,** 552 (1974).
21. F. Wenger, *Makromol. Chem.*, **64,** 151 (1963).
22. S. P. S. Yen, *Makromol. Chem.*, **81,** 152 (1965).
23. F. Wenger, *Makromol. Chem.*, **36,** 200 (1960).
24. T. A. Altares et al., *J. Polym. Sci.*, **A2,** 4533 (1964).
25. J. M. G. Cowie et al., *Trans. Faraday Soc.*, **57,** 705 (1961).
26. L. J. Fetters and M. Morton, in W. Bailey, ed., *Macromolecular Synthesis*, Wiley, New York, 1972, p. 77.

27. M. Morton et al., *J. Polym. Sci.*, **A1,** 461 (1963).

28. M. Morton et al., *J. Polym. Sci.*, **A1,** 443 (1963).

29. D. McIntyre et al., *Science*, **176,** 1041 (1972).

30. E. Slagowski et al., *Macromolecules*, **7,** 394 (1974).

31. H. W. McCormick, *J. Polym. Sci.*, **25,** 488 (1957); **41,** 327 (1959).

32. A. F. Sirianni et al., *Trans. Faraday Soc.*, **55,** 2124 (1959).

33. L. J. Fetters and Y. Kesten, unpublished results.

34. F. Wenger, *J. Am. Chem. Soc.*, **82,** 4281 (1960); *Makromol. Chem.*, **37,** 143 (1960).

35. E. Firer, Ph.D. Thesis, University of Akron, Ohio, 1973.

36. J. F. Meier, Ph.D. Thesis, University of Akron, Ohio, 1963.

37. H. E. Adams et al., *Ind. Eng. Chem., Prod. Res. Dev.*, **5,** 126 (1966).

38. B. L. Johnson et al., *Proceedings International Rubber Conference*, MacLaren, London, 1967, p. 29.

39. H. L. Hsieh and O. F. McKinney, *J. Polym. Sci.*, **B4,** 843 (1966).

40. M. Morton et al., *J. Polym. Sci.*, **A1,** 475 (1963).

41. R. Chiang et al., *J. Polym. Sci.*, *A-1*, **4,** 3089 (1966).

42. T. Kitano et al., *Macromolecules*, **7,** 719 (1974).

43. L. J. Fetters and H. Yu, *Macromolecules*, **4,** 385 (1971).

44. A. J. Bur and L. J. Fetters, *Macromolecules*, **6,** 874 (1973).

45. A. Roig et al., *J. Polym. Sci.*, **B3,** 171 (1965).

46. G. M. Guzman and A. Bello, *Makromol. Chem.*, **107,** 46 (1967).

47. R. S. Nevin and E. M. Pearce, *J. Polym. Sci.*, **B3,** 491 (1965).

48. S. Boileau et., *Makromol. Chem.*, **69,** 180 (1963).

49. P. Sigwalt, *Chim. Ind. (Milan)*, **96,** 909 (1966).

50. R. F. Kammereck, Ph.D. Thesis, University of Akron, Ohio, 1969.

51. M. Morton et al., *Macromolecules*, **4,** 11 (1971).

52. M. Morton et al., *Brit. Polym. J.*, **3,** 120 (1971).

53. D. J. Worsfold and S. Bywater, *Can. J. Chem.*, **38,** 1891 (1960).

54. F. Hein and H. Schram, *Z. Phys. Chem.*, **A151,** 234 (1930).

55. T. L. Brown and M. T. Rogers, *J. Am. Chem. Soc.*, **79,** 1859 (1957).

56. T. L. Brown et al., *J. Am. Chem. Soc.*, **84,** 1371 (1962).

57. T. L. Brown et al., *J. Am. Chem. Soc.*, **86,** 2134 (1964).

58. T. L. Brown et al., *J. Organomet. Chem.*, **3,** 1 (1965).

59. T. L. Brown, *Acc. Chem. Res.*, **1,** 23 (1968).

60. H. L. Lewis and T. L. Brown, *J. Am. Chem. Soc.*, **92,** 4664 (1970).

61. D. Margerison and J. P. Newport, *Trans. Faraday Soc.*, **59,** 2058 (1963).

62. G. Wittig et al., *Ann.*, **571,** 167 (1951).

63. S. Bywater and D. J. Worsfold, *J. Organomet. Chem.*, **10,** 1 (1967).

64. M. Weiner et al., *Inorg. Chem.*, **1,** 654 (1962).

65. D. Margerison and J. D. Pont, *Trans. Faraday Soc.*, **67,** 353 (1971).

66. R. H. Baney and R. J. Krager, *Inorg. Chem.*, **3,** 1967 (1964).

67. G. E. Hartwell and T. L. Brown, *Inorg. Chem.*, **3,** 1656 (1964).

68. W. H. Glaze and C. H. Freman, *J. Am. Chem. Soc.*, **91,** 7198.

69. W. H. Glaze et al., *J. Organomet. Chem.*, **44**, 39 (1972).

70. J. B. Smart et al., *J. Organomet. Chem.*, **64**, 1 (1974).

71. A. F. Johnson and D. J. Worsfold, *J. Polym. Sci.*, **A3**, 449 (1965).

72. H. Hsieh, *J. Polym. Sci.*, **A3**, 163 (1965).

73. K. F. O'Driscoll et al., *J. Polym. Sci.*, **A3**, 3241 (1965).

74. D. J. Worsfold and S. Bywater, *Can. J. Chem.*, **42**, 2884 (1964).

75. J. E. L. Roovers and S. Bywater, *Macromolecules*, **1**, 328 (1968).

76. A. Guyot and J. Vialle, *J. Macromol. Sci.-Chem.*, **A4**, 79 (1970).

77. A. Guyot and J. Vialle, *J. Polym. Sci.*, **B6**, 403 (1968).

78. G. M. Burnett and R. N. Young, *Eur. Polym. J.*, **2**, 329 (1966).

79. V. Sgonnik et al., *Makromol. Chem.*, **174**, 81 (1973).

80. T. L. Brown, *J. Organomet. Chem.*, **5**, 191 (1966); in F. G. A. Stone and R. West, eds., *Advances in Organometallic Chemistry*, Vol. 3, Academic Press, New York, 1965, p. 365; *Pure Appl. Chem.*, **23**, 447 (1970); *Trans. N.Y. Acad. Sci.*, **136**, 95 (1966).

81. W. Glaze et al., *Organomet. Chem.*, **44**, 49 (1972).

82. P. D. Partlett et al., *J. Am. Chem. Soc.*, **91**, 7425 (1969).

83. W. H. Glaze et al., *J. Org. Chem.*, **30**, 248 (1966).

84. J. P. Oliver et al., *J. Am. Chem. Soc.*, **88**, 4101 (1966).

85. H. S. Makowski and M. Lynn, *J. Macromol. Chem.*, **1**, 443 (1966).

86. M. Morton et al., *Rubber Plast. Age*, **42**, 397 (1961).

87. F. Schue and S. Bywater, *Macromolecules*, **2**, 458 (1969).

88. M. Morton et al., *Macromolecules*, **3**, 333 (1970).

89. L. J. Fetters et al., unpublished results.

90. J. E. L. Roovers and S. Bywater, *Macromolecules*, **8**, 251 (1975).

91. Yu. L. Spirin et al., *Dokl. Akad. Nauk SSSR*, **139**, 899 (1961); *CA*, **56**, 2551*b* (1961).

92. S. Bywater and D. J. Worsfold, *Can. J. Chem.*, **45**, 1821 (1967).

93. M. Morton et al., *J. Polym. Sci.*, **A1**, 1735 (1963).

94. D. N. Bhattacharyya et al., *J. Phys. Chem.*, **69**, 612 (1965).

95. M. Morton and L. J. Fetters, *J. Polym. Sci.*, **A2**, 3311 (1964).

96. M. Morton et al., *J. Polym. Sci.*, C, No. 1, 311 (1963).

97. M. Morton et al., *Macromolecules*, **3**, 327 (1970).

98. M. Morton et al., *Prepr. I.U.P.A.C. Macromol. Symp.*, Tokyo, **1**, 69 (1966).

99. D. J. Worsfold and S. Bywater, *Macromolecules*, **5**, 393 (1972).

100. K. F. O'Driscoll and A. V. Tobolsky, *J. Polym. Sci.*, **35**, 259 (1959).

101. F. J. Welch, *J. Am. Chem. Soc.*, **81**, 1345 (1959); *J. Polym. Sci.*, **82**, 6000 (1960).

102. R. C. P. Cubbon and D. Magerison, *J. R. Soc.*, **268A**, 260 (1962); *Polymer*, **6**, 102 (1965).

103. H. Hsieh, *J. Polym. Sci.*, **A3**, 153, 173 (1965).

104. L. J. Fetters, *J. Res. Nat. Bur. Stand.*, **69A**, 159 (1965).

105. I. Kuntz, *J. Polym. Sci.*, **A2**, 2827 (1964).

106. B. Francois et al., *J. Polym. Sci.*, C, No. 4, 375 (1964).

107. C. Lundborg and H. Sinn, *Makromol. Chem.*, **41**, 242 (1960); H. Sinn and C. Lundborg, *ibid.*, **47**, 86 (1961); H. Sinn and O. T. Onsager, *ibid.*, **52**, 246 (1962); H. Sinn and W. Hoffman, *ibid.*, **56**, 234 (1962); H. Sinn and F. Bandermann, *ibid.*, **62**, 134 (1963); H. Sinn, C. Lundborg, and O. T. Onsager, *ibid.*, **70**, 222 (1964); W. Gerbert, and J. Hinz, and H. Sinn, *ibid.*, **144**, 97 (1971).

108. A. Korotkov et al., *Vysokomol. Soedin.*, **1**, 46 (1959); *Polym. Sci., USSR*, **1**, 10 (1960); *Rubber Chem. Technol.*, **33**, 610 (1960).

109. Yu. L. Spirin et al., *Dokl. Akad. Nauk SSSR*, **146**, 368 (1962); *CA*, **58**, 1537d (1963); Yu. L. Spirin, D. K. Polykov, A. R. Gantmakker, and S. S. Medvedev, *J. Polym. Sci.*, **53**, 233 (1961).

110. A. Guyot and J. Vialle, *J. Macromol. Sci.-Chem.*, **A4**, 107 (1970).

111. J. Minoux et al., *Makromol. Chem.*, **44–46**, 519 (1961).

112. J. Minoux, *Rev. Gén. Caoutch.*, **39**, 779 (1962); *Makromol. Chem.*, **61**, 22 (1963).

113. D. N. Crammond et al., *Eur. Polym. J.*, **2**, 107 (1966).

114. D. Margerison et al., *Trans. Faraday Soc.*, **64**, 1872 (1968).

115. J. M. Alvarino et al., *Eur. Polym. J.*, **8**, 53 (1972).

116. F. M. Bower and H. W. McCormick, *J. Polym. Sci.*, **A1**, 1749 (1963).

117. B. W. Brooks, *Chem. Commun.*, **1967**, 68.

118. G. G. Eberhardt and W. A. Butte, *J. Org. Chem.*, **29**, 2928 (1964).

119. A. W. Langer, Jr., *Trans. N.Y. Acad. Sci.*, **27**, 741 (1965).

120. A. W. Langer, Jr., *Polym. Prepr.*, **7**, No. 1, 132 (1966).

121. J. N. Hay et al., *J. Chem. Soc., Faraday I*, **68**, 1227 (1972).

122. S. Kume et al., *Makromol. Chem.*, **84**, 137, 147 (1965); **98**, 109 (1966).

123. S. Kume, *Makromol. Chem.*, **98**, 120 (1966).

124. B. J. McElroy and J. H. Merkley, U.S. Pat. 3,678,121 (July 18, 1972).

125. R. C. Morrison and C. W. Kamienski, U.S. Pat. 3,725,368 (April 3, 1973).

126. C. W. Kamienski and J. F. Eastham, U.S. Pat. 3,742,077 (June 26, 1973).

127. C. W. Kamienski and J. H. Merkley, U.S. Pat. 3,751,501 (August 7, 1973).

128. A. Eisenberg and A. Rembaum, *J. Polym. Sci.*, **B2**, 157 (1964).

129. R. H. Michel and W. P. Baker, *J. Polym. Sci.*, **B2**, 163 (1964).

130. A. Gatzke, *J. Polym. Sci.*, *A-1*, **7**, 2281 (1969).

131. J. Geerts et al., *J. Polym. Sci.*, *A-1*, **7**, 2805 (1969).

132. J. Geerts et al., *J. Polym. Sci. A-1*, **7**, 2859 (1969).

133. A. A. Korotkov, *Angew. Chem.*, **70**, 85 (1958).

134. G. V. Rakova and A. A. Korotkov, *Dokl. Akad. Nauk SSSR*, **119**, 982 (1958); *CA*, **53**, 4089i (1959).

135. A. A. Korotkov and G. V. Rakova, *Vysokomol. Soedin.*, **3**, 1482 (1961); *Polym. Sci., USSR*, **3**, 990 (1962).

136. M. Morton and F. R. Ells, *J. Polym. Sci.*, **61**, 25 (1962).

137. A. F. Johnson and D. J. Worsfold, *Makromol. Chem.*, **85**, 273 (1965).

138. D. J. Worsfold, *J. Polym. Sci.*, *A-1*, **5**, 2783 (1967).

139. J. P. Oliver et al., *J. Am. Chem. Soc.*, **88**, 4101 (1966).

140. R. S. Hanmer and H. E. Railsback, *Rubber Age*, 73 October 1964.

141. J. R. Haws, *Rubber Plast. Age*, **46**, 1144 (1965).

142. C. F. Woffard and H. L. Hseih, *J. Polym. Sci.*, *A-1*, **7**, 461 (1969).

143. H. L. Hsieh, *J. Polym. Sci.*, *A-1*, **8**, 533 (1970).

144. British Pat. 994,726 (June 10, 1965).

145. R. Milkovich, S. African Pat. 280,712 (May 15, 1963).

146. A. W. van Breen and M. Vlig, *Rubber Plast. Age*, **47**, 1070 (1966).

147. M. Morton et al., *J. Polym. Sci.*, *C*, No. 26, 99 (1969).

148. J. F. Beecher et al., *J. Polym. Sci.*, *C*, No. 26, 117 (1969).

149. M. Matsuo, *Jap. Plast.*, **2**, 6 (1968).

150. P. R. Lewis and C. Price, *Nature*, **223**, 494 (1969).

151. E. Campos-Lopez et al., *Macromolecules*, **6**, 415 (1973).

152. E. Pedemonte and G. C. Alfonso, *Macromolecules*, **8**, 85 (1975).

153. M. Morton, in S. Aggarwal, ed., *Block Polymers*, Plenum, New York, 1970, p. 1.

154. L. J. Fetters and M. Morton, *Macromolecules*, **2**, 453 (1969).

155. J. M. G. Cowie and P. M. Toporowski, *Eur. Polym. J.*, **4**, 621 (1968).

156. D. J. Worsfold and S. Bywater, *J. Polym. Sci.*, **26**, 299 (1957).

157. D. P. Wyman and I. H. Song, Makromol. Chem., **115**, 64 (1968).

158. R. P. Zelinski and C. W. Strobel, U.S. Pat. 3,108,994 (1963).

159. R. P. Zelinski and H. L. Hsieh, U.S. Pat. 3,078,254 (1963).

160. G. E. Coates et al., *Organometallic Compounds*, Vol. 1, Methuen, London, 1967.

161. A. V. Tobolsky and C. E. Rogers, *J. Polym. Sci.*, **40**, 73 (1959).

162. L. J. Fetters, unpublished data.

163. T. L. Smith and R. A. Dickie, *J. Polym. Sci.*, *C*, No. 26, 163 (1969).

164. G. Karoly, in S. Aggarwal, ed., *Block Polymers*, Plenum, New York, 1970, p. 153.

165. D. J. Worsfold et al., *C.R. Acad. Sci. (Paris)*, **1968,** 267, 374.

166. L. J. Fetters et al., unpublished observations.

167. J. E. Figueruelo and D. J. Worsfold, *Eur. Polym. J.*, **4**, 439 (1968).

168. J. E. Figueruelo and A. Bello, *J. Macromol. Sci.-Chem.*, **A3**, 311 (1969).

169. M. Morton et al., *Macromolecules*, **4**, 11 (1971).

170. M. Morton et al., *4th Synthetic Rubber Symposium*, Rubber and Technical Press, Ltd., London, 1969, p. 70.

171. L. J. Fetters et al., *J. Appl. Polym. Sci.*, **16**, 2079 (1972).

172. R. E. Cunningham and M. R. Trieber, *J. Appl. Polym. Sci.*, **12**, 23 (1968).

173. M. Morton et al., *J. Polym. Sci.*, **57**, 471 (1962).

174. T. A. Orofino and F. Wenger, *J. Phys. Chem.*, **67**, 566 (1963).

175. J. A. Gervasi and A. B. Gosnell, *J. Polym. Sci.*, *A-1*, **4**, 1391 (1966).

176. T. Altares, Jr. et al., *J. Polym. Sci.*, **A3**, 4131 (1965).

177. J. C. Meunier and R. Van Leemput, *Macromol. Chem.*, **142**, 1 (1971).

178. J. E. Herz and C. Strazielle, *C.R. Acad. Sci. (Paris)*, **C272**, 747 (1971).

179. S. P. S. Yen, *Macromol. Chem.*, **81**, 152 (1965).

180. T. Masuda et al., *Polym. Prepr.* **12**, 346 (1971).

181. J. E. L. Roovers and S. Bywater, *Macromolecules*, **5**, 385 (1972).

182. D. J. Worsfold et al., *Can. J. Chem.*, **47**, 3379 (1969).

183. A. Kohles et al., *Eur. Polym. J.*, **8**, 627 (1972).

184. D. Decker and P. Rempp, *C.R. Acad. Sci. (Paris)*, **261**, 1977 (1965).

185. R. P. Zelinski and C. F. Woffard, *J. Polym. Sci.*, **A3**, 93 (1965).

186. H. H. Meyer and W. Ring, *Kaut. Gummi Kunstst.* **10**, 526 (1971).

187. L. J. Fetters, unpublished results.

188. L. J. Fetters and M. Morton, *Macromolecules*, **7**, 552 (1974).

189. L.-K. Bi, Ph.D. Thesis, University of Akron, Ohio, 1974; L.-K. Bi and L. J. Fetters, *Macromolecules*, **8**, 90 (1975).

190. C. Price et al., *Polymer*, **13**, 333 (1972).

191. M. Morton et al., *Macromolecules*, **6**, 186 (1973).

192. A. G. Kitchen and F. L. Szaila, U.S. Pat. 3,639,517 (February 1, 1972); L. M. Fodor et al., in R. D. Deanin, ed., *New Industrial Polymers*, ACS Symposium Series 4, American Chemical Society, Washington, D.C., 1974, p. 37.

193. L. J. Fetters and L.-K. Bi, unpublished results.

194. A. V. Tobolsky and C. E. Rogers, *J. Polym. Sci.*, **40**, 73 (1959).

195. F. C. Foster and J. L. Binder, *Adv. Chem. Ser.*, No. 17, 7 (1957).

196. H. Morita and A. V. Tobolsky, *J. Am. Chem. Soc.*, **79**, 5853 (1957).

197. R. S. Stearns and L. E. Forman, *J. Polym. Sci.*, **41**, 381 (1959).

198. C. A. Uraneck, *J. Polym. Sci.*, A-1, **9**, 2273 (1971).

199. T. A. Antkowiak et al., *J. Polym. Sci.*, A-1, **10**, 1319 (1972).

200. S. Bywater, *Adv. Polym. Sci.*, **4**, 66 (1965).

201. L. E. Forman, in J. P. Kennedy and E. Tornqvist, eds., *Polymer Chemistry of Synthetic Elastomers*, Wiley-Interscience, New York, Part II, p. 491.

202. A. V. Tobolsky and C. E. Rogers, *J. Polym. Sci.*, **38**, 205 (1959).

203. F. Schué et al., *Macromolecules*, **3**, 509 (1970).

204. D. J. Worsfold and S. Bywater, *Can. J. Chem.*, **42**, 2884 (1964).

205. M. Morton and J. Rupert, unpublished results.

206. S. Brownstein et al., *Macromoleculees*, **6**, 715 (1973).

207. E. R. Santee et al., *J. Polym. Sci., Polym. Lett. Ed.*, **11**, 449 (1973).

208. E. R. Santee, Jr. et al., *Rubber Chem. Technol.*, **46**, 1156 (1973).

209. V. D. Mochel, *J. Polym. Sci.*, A-1, **10**, 1009 (1972).

210. E. R. Santee, Jr. et al., *Polym. Lett.*, **11**, 453 (1973).

211. F. Schué et al., *J. Polym. Sci.*, **B7**, 821 (1969).

212. F. Schué et al., *Macromolecules*, **3**, 509 (1970).

213. M. Morton et al., *J. Polym. Sci.*, **B9**, 61 (1971).

214. M. Morton et al., *Macromolecules*, **6**, 181 (1973).

215. M. Morton et al., *J. Polym. Sci.*, **B10**, 561 (1972).

216. M. Morton and L. A. Falvo, *Macromolecules*, **6**, 190 (1973).

217. J. Sledz et al., *Makromol. Chem.*, **176**, 459 (1975).

218. S. L. Aggarwal et al., in *Block and Graft Copolymers*, Syracuse University Press, Syracuse, New York, 1973, Ch. 9, p. 157.

219. S. L. Hsu et al., *J. Chem. Phys.*, **50**, 1482 (1969).

220. P. LaChance and D. J. Worsfold, *J. Polym. Sci., Polym. Chem. Ed.*, **11**, 2295 (1973).

G. Dall'Asta

Centro Sperimentale, Snia-Viscosa
Milano, Italy

Chapter 6 Ring Opening Polymerizations

1 INTRODUCTION

One of the most recent developments in the field of polymer syntheses is the ring opening polymerization of cycloolefins, which may be schematized as follows:

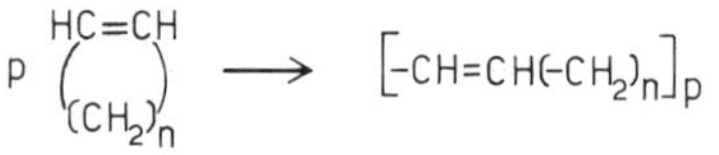

$$p \begin{pmatrix} HC=CH \\ (CH_2)_n \end{pmatrix} \longrightarrow \left[-CH=CH(-CH_2)_n \right]_p \tag{1}$$

Many different types of cycloolefins have been polymerized by ring opening. They may be classified in four different groups:

1. Unsubstituted monocyclic monoolefins. Cyclopentene is the most representative example of this group.
2. Unsubstituted monocyclic di- or multiolefins, such as 1,5-cyclooctadiene and 1,5,9-cyclododecatriene.
3. Substituted monocyclic mono-, di-, or multiolefins, the substituents generally being hydrocarbon groups such as methyl or phenyl. 3-Methylcyclobutene, 3-methylcyclooctene, and 1-methyl-1,5-cyclooctadiene are characteristic examples.
4. Substituted or unsubstituted bi- or multicyclic mono-, di-, or multiolefins, i.e., containing condensed rings, at least one of which contains a double bond. The best known examples of such monomers are norbornene, norbornadiene, ethylidene norbornene, and dicyclopentadiene.

Elastomers are generally characterized by a substantially linear and flexible polymer chain. All monomers of the fourth group yield ring opening polymers retaining in the repeating unit all rings of the monomer except one, as may be illustrated for norbornene:

$$\text{P} \quad \text{(structure)} \quad \rightarrow \quad \text{(structure)}-HC=CH-\text{(structure)}-HC=CH- \qquad [2]$$

Since such ring structures impart stiffness to the polymer chain, no useful elastomers are obtained by ring opening polymerization of monomers of group 4. Our considerations will therefore be restricted to the monocyclic olefins (groups 1–3).

One of the most interesting features of ring opening polymerization is its versatility, which allows the preparation of a large variety of polymer structures with elastomeric properties. Such variations may be achieved, e.g., by variation of the (a) type of double bonds (cis and/or trans); (b) distance between subsequent double bonds, related to the ring size of the monomer; (c) type, position, and number of substituents; (d) monomer units by copolymerization of different cycloolefins; (e) polymerization of cyclodiolefins having only one substituent (e.g., 1-methyl-1,5-cyclooctadiene), which results in alternation of substituted and unsubstituted repeat units; or (f) by a special type of grafting, to be described shortly.

Many of the possible elastomeric structures are of academic interest only since only a few cycloolefins will be available in the near future at a price that will make them attractive as monomers for commercial production of elastomers. One such structure is cyclopentene, whose polymerization will cover to a large extent the topics treated in this chapter.

From an economic standpoint, the most attractive starting material for cyclopentene is the C_5 stream from the steam cracking of virgin naphtha, which contains considerable amounts of both cyclopentene and cyclopentadiene. Selective solvents for the recovery of butadiene from C_4 streams by extractive distillation, e.g., N-methyl-2-pyrrolidone (1), dimethylformamide, furfural, aniline, acetonitrile, or formylmorpholine, may also be used to recover cyclopentene and cyclopentadiene from the C_5 stream. An alternative is the dimerization of cyclopentadiene to dicyclopentadiene, either alone or combined with extractive distillation (2). The selective hydrogenation of the monomeric cyclopentadiene to yield cyclopentene is easily achieved over supported hydrogenation catalysts.

Other economically attractive cycloolefins are cyclooctene, cyclododecene, 1,5-cyclooctadiene, and 1,5,9-cyclododecatriene. These are available from the cyclo di- or trimerization of butadiene. Selective hydrogenation leads to the corresponding monoolefins, the ring opening polymers of which have interesting elastomeric properties. The ring opening polymers of 1,5-cyclooctadiene and 1,5,9-cyclododecatriene, on the contrary, are substantially similar to the 1,4-polybutadienes and can hardly compete with them from an economic or quality standpoint.

2 RING OPENING CATALYSTS

2.1 General Considerations

The ring opening polymerization of cycloolefins is closely related to the so-called disproportionation of open chain olefins. A number of catalysts, belonging to two main classes, are common to both reactions: (a) supported transition metal oxide or carbonyl catalysts, and (b) transition metal halide combined with organometallic compounds (Ziegler-Natta systems) or with Lewis acids. The first class has been widely used to promote the disproportionation of open chain olefins, but exhibits comparatively little activity in ring opening polymerization. The second class is generally very active in both reactions, but has been applied predominantly to ring opening polymerization.

Ring opening polymerization of cycloolefins and olefin disproportionation of open chain olefins are not only similar with respect to the catalysts used, but were developed simultaneously and independently before their common features were detected.

The first catalyzed ring opening polymerization of a low strain cycloolefin was disclosed in 1957 in one example of a DuPont patent (3). Alumina supported and hydrogen reduced MoO_3, activated with $LiAlH_4$, was shown to induce (at 100°C) ring opening polymerization of cyclopentene to polypentenamer "showing trans and considerable cis unsaturation." Another example in the same patent reported that this catalyst also induced the polymerization of cyclohexene to "unsaturated low molecular weight polymer." However, further investigation of many ring opening catalysts has up to now not confirmed the ability of cyclohexene to undergo ring opening polymerization.

The catalytic activity of this catalyst is extremely poor and only trace amounts of polypentenamer are obtained. This is probably why this early discovery did not stimulate further investigations and why the elastomeric properties of polypentenamers remained unknown for a number of years.

Research work in the years between 1957 and 1962 was primarily devoted to ring opening polymerization of highly strained cycloolefins, such as norbornene (4) and cyclobutene (5, 6). The studies with norbornene showed the possibility of inducing ring opening polymerization with typical Ziegler-Natta catalysts containing titanium as transition metal.

However, such polymers do not show elastomeric properties. On the other hand, the ring opening polymerization of cyclobutene yielded polymers having a 1,4-polybutadiene (polybutenamer) structure, and hence elastomeric properties. They were obtained, e.g., with a $TiCl_4$–$AlEt_3$ catalyst, along with saturated polymers derived from addition polymerization of the cyclobutene double bond. Hence the general observation was anticipated that the ring opening polymerizations of highly strained cycloolefins, unlike those of low strained cycloolefins, frequently are not selective reactions.

Catalysts of practical importance, based on molybdenum or tungsten as transition metals, were discovered in the early 1960s. In the 1961–1964 period, Banks disclosed molybdenum hexacarbonyl, tungsten hexacarbonyl, or molybdenum oxide, supported on alumina, to be catalysts for the disproportionation of open chain olefins (7–10). In 1963, Natta, Dall'Asta, and Mazzanti discovered the high catalytic activity of Ziegler-Natta catalysts, prepared from tungsten or molybdenum halides and organometallic compounds, in the selective ring opening polymerization of cyclopentene (11, 12).

Further evolution of the two related reactions has shown that a series of transition metals are common components of the two classes. In addition to tungsten and molybdenum catalysts, niobium, tantalum, rhenium, titanium, zirconium, ruthenium, osmium, and iridium also proved efficient in both reactions (Table 1).

Table 1 Transition Metals as Components of Catalysts for the Ring Opening Polymerization of Cycloolefins and for Olefin Disproportionation Showing Relative Positions in the Periodic Table

Metathesis Reaction	Transition Metals as Catalyst Components[a]							
Ring opening polymerization of low strain cycloolefins		(Ti)	(V)					(Ni)
		(Zr)	Nb	Mo		Ru		
			Ta	W	Re		Ir	
Ring opening polymerization of high strain cycloolefins		Ti	(V)	(Cr)				(Ni)
		Zr		Mo		Ru	(Rh)	(Pd)
				W		Os	Ir	
Olefin disproportionation		(Ti)	(V)	Cr				Ni
		(Zi)	Nb	Mo		Ru	Rh	Pd
	La		Ta	W	Re	Os	Ir	

[a] Metals in parentheses exhibit poor activity in limited cases.

The problem of stereospecificity of the catalysts in ring opening polymerization concerns mostly the configuration of the double bonds in the polymers, which is not necessarily the same as in the starting cycloolefin monomer. This configuration is generally determined by the polymerization catalyst, which may be stereospecific (giving predominantly cis or trans double bonds in the polymer) or not (giving both cis and trans double bonds in the polymer).

Iso- or syndiotacticity in the ring opening polymers, on the contrary, is not a problem of catalyst stereospecificity. Ring opening polymerization, unlike addition polymerization of open chain olefins, does not create tertiary carbon atoms if they do not already exist in the cycloolefin monomer. In such a case, a stereoselective catalyst could select one of the two antipodes of a substituted cycloolefin (e.g., 3-methylcyclobutene) over the other antipode. However, the position of such a substituent generally appears too distant from the reaction site (i.e., the double bond) to allow the catalyst to exert its hypothetical stereoselectivity.

2.2 Tungsten and Molybdenum Based Ziegler-Natta Type Catalysts

The tungsten and molybdenum based Ziegler-Natta type systems introduced in 1963 by Montecatini, S.p.A. (11, 12) not only constituted the first economically feasible way of preparation of polypentenamers, but also allowed preparation of different cis/trans ratios of the pentenamer double bonds ranging from more than 95% cis to more than 85% trans. Along with different catalyst systems, the authors elucidated the chemical and physical properties of polypentenamers, and recognized the excellent elastomeric properties of high *trans*-polypentenamer as pure gum and as sulfur vulcanizates (12).

According to this disclosure, active ring opening catalysts for cyclopentene are prepared under nitrogen from tungsten or molybdenum salts and organometallic compounds of metals of groups IA, II, and III of the periodic table, especially those of aluminum, beryllium, and lithium. High valency W or Mo halides, such as WCl_6, WCl_5, $MoCl_5$, or MoF_5Cl, are preferred, but compounds containing organic ligands, such as $MoCl_2(OC_6H_5)_3$ or $MoO_2(CH_3-CO-CH-CO-CH_3)_2$, are also used.

Preferred organometallic compounds are trialkylaluminum, dialkylaluminum halides, alkylaluminum dihalides, and dialkylaluminum hydrides.

The catalyst components are generally used at transition metal/organometallic compound molar ratios of between 1:1 and 1:5.

The polymerizations are preferably carried out in the $-30°C$ range in bulk or in a hydrocarbon diluent such as toluene or n-heptane.

Catalyst systems such as $WCl_6/AlEt_3$, $WCl_6/AlEt_2Cl$, $MoF_5Cl/AlEt_3$, or $MoCl_2(OC_6H_5)_3/AlEt_2Cl$ are stereospecific, inducing formation of essentially *trans*-polypentenamer. The catalyst systems $MoCl_3/AlEt_3$ or $MoCl_5/Al(n\text{-hexyl})_3$ are even more stereospecific and promote formation of all *cis*-polypentenamer. Systems such as $MoO_2(CH_3COCHCO-CH_3)_2/AlEt_2Cl$, $MoCl_5/AlH(i\text{-Bu})_2$, $WCl_6/BeEt_2$, or WCl_6/Li n-Bu are only poorly or not at all stereospecific, and yield polymers with large amounts of both *cis*- and *trans*-pentenamer units.

These investigations showed three important features of such ring opening catalysts: (*a*) ring opening catalysts of the Ziegler-Natta type may be stereospecific; (*b*) the type of stereoregularity induced in the polymers is governed in a complex and sophisticated way by at least four different catalytic factors: the transition metal, the type and number of ligands bound to it, the metal of the cocatalyst, and the ligands attached to it; and (*c*) ring opening polymerization of cyclopentene is achieved with high yields with soluble catalytic systems also, such as $MoO_2(CH_3-COCHCOCH_3)_2/AlEt_2Cl$, thus suggesting that catalysis proceeds in the homogeneous phase.

In 1964 the same authors used the two-component tungsten and molybdenum based Ziegler-Natta catalysts for the ring opening polymerization of higher membered cycloolefins (13, 14). Operating in the -30 to $+30°C$ temperature range, they succeeded in polymerizing cycloheptene, cyclooctene, and cyclododecene to high molecular weight polyalkenamers. Cyclohexene could not be polymerized. In addition to the previously mentioned tungsten salts, some new important compounds were disclosed, such as tungsten oxytetrachloride, tungsten hexafluoride, tungsten hexaphenoxide, and molybdenum trichloride.

The ring opening polymerization of higher membered cycloolefins differed from the analogous polymerization of cyclopentene. High polymer yields are obtained only in the presence of tungsten catalysts, whereas the molybdenum based systems exhibit only a poor activity. This anticipated a more general observation, i.e., that the only catalysts able to convert higher membered cycloolefins to high molecular weight polyalkenamers are tungsten based Ziegler-Natta systems. Moreover, unlike cyclopentene, the higher membered cycloolefins are stereospecifically converted to high *trans*- (and not to high *cis*-) polyalkenamers by the action of catalysts such as $MoCl_5/Al(n\text{-hexyl})_3$. A fifth factor, the cycloolefin ring size, must be added to the above-mentioned four factors regulating the type of stereoregularity.

The same two-component tungsten based Ziegler-Natta systems was successfully employed by Calderon in the ring opening polymerization of cyclic di- or trienes (15). Systems such as $WCl_6/Al(i$-Bu$)_3$, $WCl_6/AlH(i$-Bu$)_2$, $WCl_6/AlEtCl_2$, or $WCl_6/Li\ n$-Bu in n-pentane were used to promote the polymerization of 1,5-cyclooctadiene and 1,5,9-cyclododecatriene to polybutenamer. Such polymers have substantially the structure of 1,4 polymers of butadiene, but differ from these in the absence of any type of side groups.

In 1967 Calderon et al. (16) showed increasing catalytic activities at ambient temperature, in the order $AlEtCl_2 > AlEt_2Cl > AlEt_3$, for organoaluminum compounds in combination with WCl_6. In addition to the known polymers of cyclooctene, cyclododecene, 1,5-cyclooctadiene, and 1,5,9-cyclododecatriene, the authors also prepared ring opening polymers of substituted cycloolefins such as 3-methyl- or 3-phenylcyclooctene.

The behavior of some other Ziegler-Natta catalysts of the two-component WCl_6 type in the ring opening polymerization of cyclopentene has been more recently described by some authors of the Bayer group (17). Depending on the cocatalyst and on the particular cocatalyst/WCl_6 ratio, polypentenamers could be obtained in a wide range of cis/trans molar ratios. Trans contents of 85–90% are characteristic of the $Al(i$-Bu$)_3/WCl_6$ (1.4:1), $HSnEt_3/WCl_6$ (1.75:1), and $Li_3[WEt_6]/WCl_6$ (4.8:1) systems. Trans contents of 55–65% result from the $(HSiOMe_3)_4/WCl_6$ (1.6:1), and $(HSiOMe_3)_8/WCl_6$ (0.4:1) systems. Low trans contents of 25–30% are obtained in the presence of the $Na_3[W(C_6H_5)_6]/WCl_6$ (0.95:1) system.

The ring opening polymerization of cyclopentene in the presence of the previously (13) disclosed WF_6/organoaluminum compound catalysts was thoroughly reexamined in that paper (17). In the case of the $Al_2Et_3Cl_3/WF_6$ and the $AlEtCl_2/WF_6$ systems in toluene at $-30°C$, a continuous variation of the cis/trans ratio in the polypentenamer from 85:15 to 10:90 was possible, as shown in Fig. 1. At higher temperatures ($-10°C$) the trans content increases considerably. Thus, temperature variation during polymerization is a tool for the preparation of block copolymers built up of blocks of cis- and $trans$-pentenamer units.

A significant contribution to understanding whether ring opening polymerization of cycloolefins or the related disproportionation proceeds in the homogeneous or in the heterogeneous phase has been recently given by Zuech et al. (18, 19). Different α and internal olefins, at about $0°C$, could be disproportionated by the action of a series of soluble catalyst systems, prepared from nitrosyl complexes of molybdenum or tungsten and organoaluminum cocatalysts. $Al_2Me_3Cl_3$ and $AlEtCl_2$ proved more active as cocatalysts than $AlEt_3$ or $AlEt_2Cl$. Examples of

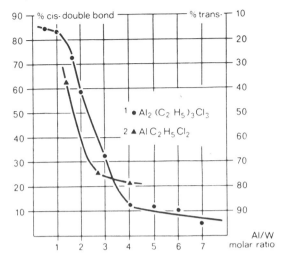

Figure 1 Polypentenamer microstructure as a function of the Al/W ratio in $WF_6/Al_2Et_3Cl_3$ (●) and $WF_6/AlEtCl_2$ (▲) catalysts (20% monomer solutions in toluene; temperature $-30°C$). Reprinted from Günther et al. (17), by courtesy of Hüthig and Wepf Verlag, Basel.

suitable transition metal compounds are $[(C_6H_5)_3P]_2(NO)_2MoCl_2$, $(C_5H_5N)_2(NO)_2MoCl_2$, and $[(C_6H_5)_3P]_2(NO)_2WCl_2$.

The two-component Ziegler-Natta type catalysts, especially those based on tungsten halides, though active in promoting ring opening polymerizations, were not completely satisfactory for industrial applications owing to their relative instability, their poor reproducibility, and their high sensitivity to even small differences in catalyst composition. One of the main causes of this behavior is the strong Friedel-Crafts character, especially of those catalysts containing halogenated organoaluminum compounds as cocatalysts. As a consequence, alkylation reactions have been observed, e.g., the formation of cycloalkylarenes from cycloolefins and aromatic solvents. Pampus et al. (17) also reported substitution chlorination of toluene solvent by WCl_6 according to

$$2\,WCl_6 + \langle\bigcirc\rangle\text{-}CH_3 \xrightarrow[30-50°C]{h\vartheta} 2\,WCl_5 + \overset{Cl}{\langle\bigcirc\rangle}\text{-}CH_3 + HCl \qquad [3]$$

as well as substitution chlorination of tetrachloroethylene:

$$2\,WCl_6 + Cl_2C{=}CCl_2 \xrightarrow[100°C]{h\vartheta} 2\,WCl_5 + Cl_3C{-}CCl_3 \qquad [4]$$

Such reactions are generally accompanied by color changes of the WCl_6 solutions.

A fundamental development, resulting in more stable catalysts and more reproducible reactions, was disclosed in 1965 by Dall'Asta and

Carella (20). Having found that monomer freshly distilled under nitrogen could not be converted to polypentenamer as easily as older monomer, which was presumably contaminated by traces of air, they investigated a series of hydroperoxides, peroxides, and peracids as third catalyst components (with $WCl_6/AlEt_2Cl$) and later other oxygen-containing compounds. Small amounts of alcohols (e.g., methanol, ethanol, n-butanol), phenols, molecular oxygen, hydrogen peroxide, or water are examples of compounds containing O–O or O–H bonds, which they also found to activate strongly the polymerization of cycloolefins such as cyclopentene (20), cycloheptene (20), cyclooctene (20), cyclodecene (21), and cyclododecene (20), as well as the copolymerization of such cycloolefins (22). The most suitable catalyst components proved to be the halides and oxyhalides of tungsten and molybdenum, and the organometallic compounds or hydrides of group II or III metals (e.g., Al, Be, Zn, Mg, Ca) with preference for trialkylaluminum, dialkylaluminum chloride, alkylaluminum dichloride, and diethylberyllium. Polymerizations were preferably carried out under nitrogen in hydrocarbon or in bulk at temperatures in the -50 to $+50°C$ range.

At present, such three-component tungsten based Ziegler-Natta catalysts are the most efficient and economically important for ring opening polymerization of low strain cycloolefins. In the analogous molybdenum catalysts, the activating ability of the oxygen-containing compound is considerably less pronounced. The three-component tungsten catalysts are more reproducible, more stable, give faster polymerization rate, increased conversion, and with lower catalyst consumption than the binary systems. On the other hand, stereospecificity is only slightly or not at all increased.

The order of addition of the catalyst components is important. The oxygen-containing activator should not be added to the organometallic compound, but joined to the tungsten salt (WCl_6, $WOCl_4$, etc.) and allowed to react for some time before adding the organometallic cocatalyst. In bulk polymerization, monomer, tungsten salt, and oxygenated activator are allowed to react together. A relatively slow reaction is observed, favored by temperature and characterized by heat evolution, dissolution of the tungsten halide, and formation of a typical dark red solution. These observations suggest that a soluble tungsten complex, containing an oxygenated compound and cycloolefin, is formed. The activator is used in a $1:0.1$ to $1:2$ molar ratio, preferably in a $1:0.5$ to $1:2$ molar ratio with respect to the tungsten salt. Upon adding the organometallic compound the mixture turns brown and polymerization starts immediately.

Analogous three-component tungsten based Ziegler-Natta catalysts were reexamined by Calderon and Judy (23) for the ring opening polymerization of cyclooctene, 1,5-cyclooctadiene, and 1,5,9-cyclododecatriene in hydrocarbon solution. Tungsten halides, such as the hexachloride or oxytetrachloride, were used in combination with different trialkylaluminum or alkylaluminum chlorides as cocatalyst, and activators such as straight chain or branched alcohols (including ethanol, allyl alcohol, glycerol, and benzyl alcohol), phenol, thiophenol, cumyl hydroperoxide, or water. The catalysts are preferably formed by first reacting tungsten hexachloride with an alcohol in a hydrocarbon and then adding the organoaluminum compound. However, the authors were not aware of the type of chemical reaction that takes place between tungsten halide and alcohol (26).

One particular three-component tungsten catalyst, prepared by reacting 1:1 molar ratios of WCl_6 and ethanol in benzene at room temperature and then adding 4 moles of ethylaluminum dichloride, is very active and has been widely used by Calderon et al. and later by other authors for the study of different ring opening polymerizations and olefin disproportionation reactions. In 1967 Calderon et al. (24, 25) investigated the disproportionation of internal olefins, and in 1968 they studied (26) the ring opening polymerization of cycloolefins, using in both cases the $WCl_6/C_2H_5OH/AlEtCl_2$ catalyst mentioned above, thereby contributing to the understanding of the relationships between these two types of reactions.

The activator capacity of peroxides, hydroperoxides, alcohols, and other oxygen-containing compounds in tungsten based Ziegler-Natta catalysts in the ring opening polymerization of cyclopentene was investigated in 1968 by a research team from Farbenfabriken Bayer. Nützel and co-workers (27) reported an efficient catalyst prepared as follows: Tungsten hexachloride is suspended in a cyclopentene-toluene mixture at ambient temperature until the original blue-green solution turns red-brown, thus indicating the formation of a soluble complex containing two cyclopentene molecules bound to one WCl_6. After addition of 0.3–0.5 moles of cyclopentene hydroperoxide, the mixture is cooled to $-10°C$ and polymerized by adding triisobutylaluminum. Under such conditions cyclopentene hydroperoxide increases polymer conversion to a considerably greater extent than other hydroperoxides and peroxides, without affecting the stereospecificity of the $WCl_6/Al(i\text{-}Bu)_3$ catalyst.

According to Witte and associates (29), halogenated alcohols and phenols are more efficient activators than simpler alcohols and phenols. Examples of such compounds are HOH_2CCH_2Cl, HOH_2CCH_2Br, ClH_2C-$CHOHCH_2Cl$, HOH_2CCCl_3, $HO(CH_2)_4Cl$, $o\text{-}ClC_6H_4OH$, $o\text{-}ClC_6H_{10}OH$, and $o\text{-}IC_6H_{10}OH$. The activation capability of peroxides is not limited to organic peroxides. Even Na_2O_2 or BaO_2 increases polypentenamer yield when used to promote a $WCl_6/Al(i\text{-}Bu)_3$ catalyst in toluene solution (28).

New information on the structure of three-component Ziegler-Natta catalysts based on tungsten halides was acquired through the investigations of the Bayer group (17). In the absence of light, WCl_6 solutions in toluene are reduced to WCl_5 by catalytic amounts of peroxides or *tert*-butyl hypochlorite. Moreover, the activating capability of peroxides is related to their decomposition temperature, and the optimum Al/W ratio depends on the peroxide structure.

We are still far from understanding the chemical significance of the action of peroxides in three-component tungsten catalysts. The reaction between alcohols and tungsten halides is much better known (17). Tungsten hexachloride in toluene at room temperature reacts with alcohols to form soluble alkoxytungsten halides:

$$WCl_6 \begin{cases} \xrightarrow{HOH_2C\text{-}CH_2Cl} Cl_5WOCH_2\text{-}CH_2Cl + HCl & [5a] \\ \xrightarrow{2\ HOH_2C\text{-}CH_2Cl} Cl_4W(OCH_2\text{-}CH_2Cl)_2 + 2\ HCl & [5b] \end{cases}$$

α-Halogen in the alcohols confers higher stability to such solutions. Dialkoxytungsten tetrahalides, for which the polymerization rate is maximum, are more stable than monoalkoxytungsten pentachlorides. The latter may decompose with time to form Cl_4WO and RCl. The characteristic red color of the WCl_6-alcohol solutions is most likely that of the alkoxytungsten halide. Hence at least two important functions may be attributed to alcohol: solubilization of the tungsten halide and stabilization of the tungsten component of the catalyst. The more specific role in the alkoxytungsten halide-monomer complex is not yet known.

Alternative pathways for the reaction between tungsten or molybdenum halides and alcohols or other oxygen-containing compounds are suggested by other observations. Clark and Wentworth (30) formulate the reaction between ammonium salts of tungsten halides and alcohols as chlorine/alkoxy exchange and simultaneous complexation of excess alcohol:

$$[(n\text{-}C_4H_9)_4N]_3W_2Cl_9 + 6\ ROH \rightarrow W_2Cl_4(OR)_2 \cdot (ROH)_4 + 2\ HCl + [(n\text{-}C_4H_9)_4N]^+Cl^- \qquad [6]$$

High valency tungsten and molybdenum halides are further known to act as strong scavengers for bonded oxygen. Thus, according to Kepert and Mandyczewsky (31), molybdenum pentachloride exchanges chlorine with the oxygen contained in arsine (or phosphine) oxides:

$$MoCl_5 + (C_6H_5)_3AsO \longrightarrow [(C_6H_5)_3AsCl]^+[MoOCl_4]^- \qquad [7]$$

or in cyclic ethers such as dioxane:

$$MoCl_5 + \text{[dioxane]} \longrightarrow MoOCl_3 + ClH_2C\text{-}CH_2\text{-}O\text{-}CH_2CH_2Cl \qquad [8]$$

According to Günther et al. (17), analogous alkoxytungsten halides may be prepared by reaction of tungsten hexachloride with an appropriate epoxide in toluene at room temperature, a reaction which inter alia avoids the formation of hydrochloric acid, as shown for epichlorohydrin:

$$WCl_6 + 2\ H_2C\text{-}CH\text{-}CH_2Cl \longrightarrow Cl_4W[OCH(CH_2Cl)_2]_2 \qquad [9]$$

In combination with organoaluminum compounds, these alkoxy-tungsten halides are very efficient polymerization catalysts for cyclopentene. According to Pampus et al. (32, 33), different epoxides may be used, e.g., ethylene oxide, propylene oxide, 1-butene oxide, epichlorohydrin, epibromohydrin, styrene oxide, or phenoxypropylene oxide. After reaction with tungsten halides (WCl_6, $WOCl_4$, WBr_5), they are combined with trialkylaluminum or alkylaluminum chlorides at a 1:0.6 to 1:1.5 W/Al molar ratio.

Other oxygen-containing compounds are efficient activators of the tungsten halide catalysts, e.g., tert-butyl hypochlorite (34) or nitrobenzenes such as 1,3-dinitro-2,5-dichlorobenzene (35). Chlorinated hydrocarbons, such as hexachlorocyclopentadiene, 2-chlorobutadiene, 1,2-dichloroethylene, or vinyl chloride (36), are less efficient activators. Bis(pyridine)molybdenum tetracarbonyl and methylaluminum sesquichloride activated by quaternary ammonium compounds have also been proposed (37) for ring opening polymerization of cyclopentene.

Table 2 lists most of the three-component tungsten catalysts proposed for the ring opening polymerization of low strain cycloolefins.

Even though the different classes of promotors (peroxides, hydroperoxides, alcohols, epoxides, hypochlorites, nitroarenes, oxygen, water) exhibit common features in acting as a catalyst component, the reactions they undergo with the other components are probably different. This may be deduced, for example, from the quite different maxima of polymer conversion as a function of the activator/WCl_6 molar ratio shown by the three different activators 1,3-dinitro-2,5-dichlorobenzene, cyclopentadiene hydroperoxide, and tert-butyl hypochlorite (Fig. 2).

The role of the organometallic cocatalyst is also obscure. In analogy with typical titanium or vanadium based Ziegler-Natta catalysts, one may argue that the halogen or other ligands of the tungsten compound are exchanged with alkyl groups of the cocatalyst, and that the relative instability of the W–C bond leads to reductive decomposition of tungsten

Table 2 Three-Component Catalyst Systems Proposed for the Ring Opening Polymerization of Low Strain Cycloolefins

Monomer	Catalytic System			Solvent	Ref.
	Transition Metal Compound	Activator	Cocatalyst		
Cyclopentene	WCl_6	O_2; air; H_2O_2; $(CH_3)_3COOH$; $C_6H_5C(CH_3)_2OOH$; 3-hydroperoxycyclopentene; Na_2O_2; BaO_2; $(C_6H_5COO)_2$; $((CH_3)_3CO)_2$; H_2O; C_2H_5OH; $n\text{-}C_4H_9OH$; ClH_2CCH_2OH; BrH_2CCH_2OH; $ClH_2CCHOHCH_2Cl$; Cl_3CCH_2OH; $ClC_6H_{10}OH(o, p)$; $ClC_6H_4OH(o, m)$; $Cl(CH_2)_4OH$; $(CH_3)_3COCl$; C_5Cl_6; $H_2C\!-\!CH_2$; $H_3C\!-\!HC\!-\!CH_2$; $ClH_2C\!-\!HC\!-\!CH_2$; $BrH_2C\!-\!HC\!-\!CH_2$; $H_5C_2HC\!-\!CH_2$; $C_6H_5\!-\!HC\!-\!CH_2$; CH_3COOH; (epoxides with O bridge)	$AlEt_3$; $Al(i\text{-}Bu)_3$; $AlEt_2Cl$; $AlEtCl_2$	Bulk, benzene, toluene	17, 20, 27, 28, 29, 32, 33, 34, 35, 36
	WBr_5	1,3-dinitro-2,5-dichlorobenzene $H_5C_2HC\!-\!CH_2$ (with O bridge)	$AlEtCl_2$	Toluene	33

WOCl₄	$(C_6H_5COO)_2$; $H_2C\!-\!CH_2$ (epoxide with O)	$AlEt_2Cl$; $Al(i\text{-}Bu)_2Cl$	Bulk, toluene	20, 33
WCl_2	$(C_6H_5COO)_2$	$AlCl_3$	Bulk	42
$WCl_2 \cdot AlCl_3$	$(C_6H_5COO)_2$	$AlEt_2Cl$; $AlEt_3Cl_3$; $Al_2Et_3Cl_3$; $AlEtCl_2$	Bulk	41
$NaW(C_6H_5)_6$	3-Hydroperoxycyclopentene	$SnCl_4$	Toluene	46
$MoCl_5$	$((CH_3)_3CO)_2$	$AlEt_3$	Bulk	20
$Mo(pyr)_2(CO)_1$	$(C_4H_9)_4NCl$	$Al_2Me_3Cl_3$	Chlorobenzene	37
$TaCl_5$	$((CH_3)_3CO)_2$; H_2O; C_2H_5OH; CH_3COCH_2-$COCH_3$; $ClH_2CCOOCH_3$; $(CH_3)_3COCl$; $H_2C\!-\!CH_2$ (epoxide with O); $H_3CHC\!-\!CH_2$ (epoxide with O); $ClH_2CHC\!-\!CH_2$; $CH_2\!=\!CHCl$: $CH_2\!=\!CCl_2$; $ClHC\!=\!CHCl$; $Cl_2C\!=\!CHCl$; $Cl_2C\!=\!CCl_2$; 3-chlorocyclopentene; C_6H_5Cl; C_6H_5Br; $C_6H_4Cl_2(o.p)$; $CuCl_2$; $2H_2O$; $FeCl_3$	$AlEtCl_2$; $Al_2Et_3Cl_3$; $Al(i\text{-}Bu)_3$	Chlorobenzene; toluene	17, 33, 34, 36, 54, 55
$ReCl_5$	$(CH_3)_3COCl$	$Al(i\text{-}Bu)_3$	Chlorobenzene	57

Table 2 (Continued)

Monomer	Catalytic System				
	Transition Metal Compound	Activator	Cocatalyst	Solvent	Ref.
Cycloheptene; cyclooctene; cyclodecene; cyclododecene; 1,5-cyclooctadiene; 1,5,9-cyclododecatriene	WCl_6	O_2; $C_6H_5C(CH_3)_2OOH$: $(C_6H_5C(CH_3)_2O)_2$; $(C_6H_5COO)_2$; H_2O; CH_3OH; C_2H_5OH; $i\text{-}C_3H_7OH$; $t\text{-}C_4H_9OH$: $CH_2=CH-CH_2OH$; $CH_2OH-CHOH-CH_2OH$; C_6H_5OH; $C_6H_5CH_2OH$; $C_6H_5C(CH_3)_2OH$; $t\text{-}C_9H_{19}SH$; C_6H_5SH	$AlEt_2Cl$; $AlEtCl_2$; $Al_2Et_3Cl_3$; $Al(i\text{-}Bu)_3$; $Al(i\text{-}Bu)_2Cl$; $BeEt_2$; $ZnEt_2$	Bulk, benzene, toluene, chlorobenzene, pentane, hexane, cyclohexane	20, 23
1-Methyl(ethyl-chloro)-1,5-cyclooctadiene; 1,2-dimethyl-1,5-cycloocta-diene	$WOCl_4$ $W(C_8H_{12})(CO)_4$ $TaCl_5$ WCl_6	$(C_6H_5C(CH_3)_2O)_2$ O_2, Br_2, I_2, $BrCN$ C_2H_5OH C_2H_5OH	$AlEt_2Cl$ $AlCl_3$, $AlBr_3$, $AlEtCl_2$ $AlEtCl_2$ $AlEtCl_2$	Bulk Benzene Chlorobenzene Benzene	21 50 55 99
Cycloolefin-cycloolefin copolymeriza-tion	WCl_6 $WOCl_4$	3-Hydroperoxycyclopentene; $(C_6H_5COO)_2$ $(C_6H_5COO)_2$; $(C_6H_5C(CH_3)_2O)_2$	$Al(i\text{-}Bu)_3$; $AlEt_2Cl$ $AlEt_2Cl$	Bulk, toluene Bulk	22, 27 22, 81
Cycloolefin polymers modified by olefins or con-jugated dienes	WCl_6 $WOCl_4$	$(C_6H_5COO)_2$; $CH_2=CHCl$; $CH_2=CCl_2$; $ClHC=CHCl$; $CH_2=CH-CCl=CH_2$ $(C_6H_5COO)_2$	$Al(i\text{-}Bu)_3$; $AlEt_2Cl$; $AlEtCl_2$ $AlEt_2Cl$	Bulk, toluene Bulk	36, 100, 101 101

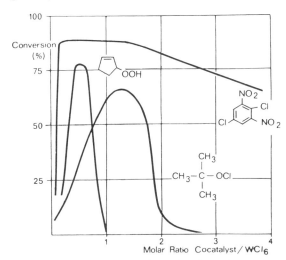

Figure 2 Activation of the WCl$_6$/Al(i-Bu)$_3$ catalyst in cyclopentene polymerization by 1,3-dinitro-2,5-dichlorobenzene, cyclopentene hydroperoxide, and t-butyl hypochlorite (1.4 mmoles W/100 g monomer; Al/W molar ratio 1.35; polymerization time 1 hour; temperature −10 to 0°C). Reprinted from Günther et al. (17), by courtesy of Hüthig and Wepf Verlag, Basel.

and formation of vacancy sites. When adding an organometallic compound to the tungsten halide-activator-monomer complex, the red color does immediately turn brown, indicating the formation of a new species, but no experimental data concerning the valency of tungsten in the catalytic species are known. Wang and Menapace (38), on the basis of the observation that the maximum activity of a WCl$_6$/Li n-Bu olefin disproportionation catalyst is reached at a 1 : 2 W/Li molar ratio, propose the existence of a W(IV) catalytic intermediate. Similar conclusions were drawn by Pampus and co-workers (39) based on spectroscopic examination of the WCl$_6$/SnEt$_4$ system. According to Hughes (40), who studied the kinetics of the olefin disproportionation induced by the (C$_5$H$_5$N)$_2$(NO)$_2$MoCl$_2$/Al$_2$Et$_3$Cl$_3$ system, the catalytic species is a zero valent molybdenum and the role of the organometallic cocatalyst is to create coordination vacancies.

As will be illustrated in the next section some tungsten systems that do not contain an organometallic cocatalyst are efficient catalysts for the ring opening polymerization of cyclopentene. In this case the reduction to a zero valent species appears quite unlikely. Partial reduction of the transition metal by the action of other reaction partners, such as monomer, solvent, or activator, is suggested by the findings (17) that catalytic amounts of $tert$-butyl hypochlorite promote the reduction of WCl$_6$ to

WCl_5, and that toluene may be chlorinated by WCl_6 with formation of WCl_5.

The ring opening polymerizations of low strain cycloolefins are highly specific reactions, especially with the three-component tungsten based Ziegler-Natta catalysts, whereas the analogous polymerizations of highly strained cycloolefins are far less specific. We will limit our considerations on this topic to cyclobutene, a highly strained cycloolefin that yields polymers having elastomeric properties. According to Natta et al. (41), catalytic systems prepared from tungsten or molybdenum salts and organoaluminum compounds induce the formation of 1,4-butadienic as well as cyclobutadienic units. The former arise from ring opening polymerization, whereas the latter are the result of addition polymerization of the double bonds. The catalytic systems prepared from WCl_6 and $AlEt_3$ (1:3) or from $MoO_2(CH_3-CO-CH-CO-CH_3)_2$ and $AlEt_3$ (1:5) are highly active, unlike the $MoCl_5/AlEt_3$ (1:3) and $MoCl_3/AlEt_3$ (1:2.5) systems, which are poorly active. All polybutadiene polymers contain nearly equimolecular amounts of cis-1,4 and trans-1,4 units, indicating lack of stereospecificity. The saturated polycyclobutene polymers are sometimes crystalline, suggesting the presence of at least two types of catalytic systems: one nonstereospecific which yields 1,4 polybutadienes, and the other stereospecific, yielding crystalline polycyclobutene. These findings imply a quite different behavior of such two-component tungsten and molybdenum catalysts toward high and low strain cycloolefins.

2.3 Metal Alkyl-Free Tungsten and Molybdenum Halide Catalysts

Development of metal alkyl-free tungsten and molybdenum catalyst systems began after noting that halogenated alkylaluminum compounds were frequently more effective cocatalysts than nonhalogenated metal alkyls.

Mixtures of organoaluminum halides and aluminum trialkyls were initially proposed as cocatalysts by Natta, Dall'Asta, and Mazzanti (42). Later, Dall'Asta and Carella (43) discovered the strong catalytic activity of systems prepared from tungsten halides (WCl_6, $WOCl_4$, WCl_2) and halides of group II and III metals (such as $AlCl_3$, $AlBr_3$, or $ZnCl_2$) in the ring opening polymerization of cycloolefins such as cyclopentene or cyclooctene used either with or without oxygen-containing activators. Particularly active catalysts are obtained from the very exothermic reaction of tungsten hexachloride and aluminum powder:

$$3WCl_6 + 4Al \rightarrow 3WCl_2 + 4AlCl_3 \qquad [10]$$

The reaction mixture may also contain smaller amounts of metallic tungsten, tungsten hexachloride, and different intermediate halides.

Tungsten halides as well as aluminum halides are strong Lewis acids. Obviously, their mixture also has Friedel-Crafts character. Consequently, the polyalkenamers obtained with such catalysts do not have a very regular structure. Apart from the low stereoregularity, alkylations, transannular polymerizations (in the case of medium sized cycloolefins), and cyclization reactions may considerably alter the polymer structure. The latter reaction is analogous to the known cyclizations of polybutadiene or natural rubber, induced by Lewis acids. In the case of polyalkenamers it may be recognized by the disappearance of a considerable part of the double bonds. Such polyalkenamers exhibit reduced elastomeric properties and for this reason the metal alkyl-free catalysts do not look very promising for commercial production.

Ring opening polymerization of cyclopentene can be induced by tungsten halides without any cocatalyst or promoter (43). Thus, low conversions to substantially atactic polypentenamer are obtained by prolonged contact of cyclopentene (at room temperature) with WCl_6, WBr_5, or $WOCl_4$. This is somewhat comparable to the finding that propylene may be polymerized (with low conversion) to isotactic polypropylene in the presence of titanium trichloride which is metal alkyl free, but contains $AlCl_3$. The difference lies in the fact that in the case of the tungsten halides any participation of the aluminum species is rigorously excluded. Very low polymer conversion and lack of stereospecificity in the absence of an aluminum species suggest that two of the main functions of aluminum (alkyl or halide) are to increase catalytic activity strongly and to confer stereospecificity to the catalyst.

Metal alkyl-free tungsten and molybdenum catalysts were also proposed by Marshall and Ridgewell for the ring opening polymerization of cyclopentene, cycloheptene, cyclooctene, cyclododecene, 1,5-cyclooctadiene, and 1,5,9-cyclododecatriene (44, 45). High polymer yields are obtained in the absence of solvent by using $WCl_6/AlBr_3$, $WCl_6/AlCl_3$-anisolate, or $MoCl_5/AlBr_3$ catalysts. Polyalkenamers are atactic or exhibit a small prevalence of cis over trans double bonds. Again, cyclohexene could not be polymerized. Metal alkyl-free catalysts such as those described above were also used by Judy (46) for ring opening polymerization of cyclooctene in benzene solution.

Catalysts prepared from π-allyl compounds of tungsten (or other transition metals) and Lewis acids, though containing an organometallic component, may be associated with the class of metal alkyl-free Friedel-Crafts catalysts since no σ–C–M bonds are present and the π-allyl–metal bond does not seem to be directly involved in the polymerization mechanism. Examples of such catalysts are those proposed by Nützel et al. (17, 47) for the polymerization of cyclopentene: $W_2(\pi\text{-allyl})_4/WCl_6$, which yields substantially atactic, and $Cr(\pi\text{-allyl})_3/WCl_6$, which yields predominantly cis-polypentenamer. Other similar catalysts were described by Kromer et al. (48) for the polymerization of cyclopentene, cyclooctene, and cyclododecene: $W(\pi\text{-crotyl})_4/AlBr_3$; $W(\pi\text{-crotyl})_4/TiCl_4$; $W(\pi\text{-crotyl})_4/MoCl_5$; $W(\pi\text{-allyl})_4/AlBr_3$; $Zr(\pi\text{-methallyl})_4/WCl_6$; $Cr(\pi\text{-allyl})_3/WF_6$. Acidic π-allyl complexes of tungsten were also used by Kromer et al. (49) for the ring opening polymerization of cyclobutene to yield cis-1,4- or trans-1,4-polybutadienes. Catalysts obtained from the reaction of tungsten halides (WCl_6, WBr_5, WCl_5OR) and lithium, calcium, or aluminum metal (17) should also belong to the Friedel-Crafts type. Three-component catalysts, prepared from 1,5-cyclooctadiene tetracarbonyl tungsten, aluminum trihalides, and O_2, Br_2, I_2, or BrCN as activators, are effective in ring opening polymerization of cyclooctene and cyclododecene (50).

2.4　Supported Tungsten and Molybdenum Catalysts

Supported tungsten and molybdenum catalysts are very important for the disproportionation of open chain olefins (e.g., 51) but have only limited importance for the related ring opening polymerization of cycloolefins. Depending on the transition metal compound and the particular carrier, they have optimum activity at temperatures from above 100° up to 500°C. The polymer yield and the molecular weights are generally low.

According to Alkema and Van Helden (52), Cyclooctene is oligomerized at 110°C to low molecular weight polyoctenamers in the presence of alumina supported MoO_3—CoO catalyst, previously activated in a nitrogen stream at 525°C. At higher temperatures, methylcycloheptene, ethylcyclohexane, and vinylcyclohexane as well as alkylation products such as cyclooctylcyclooctene are observed.

Catalysts obtained by supporting tungsten hexachloride on previously calcinated (at 500°C) α-alumina or zirconium oxide carrier were also proposed for the olefin disproportionation reaction and (without examples) for the polymerization of cycloolefins (53).

2.5　Niobium and Tantalum Based Ziegler-Natta Type Catalysts

Niobium and tantalum based Ziegler-Natta type catalysts were first prepared by Uraneck and Trepka (54) from the pentachlorides and used with trialkylaluminum at 1:1 to 1:2 molar ratios. Polymerization may be carried out either in bulk or in hydrocarbon or chlorinated solvents at -50 to $+10°C$.

Niobium catalysts are considerably less active than tantalum catalysts, whereas analogous vanadium systems are substantially inactive as ring opening catalysts. This corresponds to the trend observed in the group VI transition metal halides: chromium is inefficient, molybdenum is medium active, but tungsten is highly active. However, tantalum systems are less efficient than the analogous tungsten catalysts. This is in agreement with the observation (54) that, among the low strain cyclomonoolefins, only cyclopentene—but not cyclooctene—was polymerized via ring opening. Polypentenamers obtained in the presence of the tantalum catalysts are less stereoregular than those obtained by analogous tungsten systems. Their trans double bond content is generally about 65–75%. Some cyclization may be recognized by the observation that the actual total double bond content is about 92–95% of the theoretical.

Tantalum based two- and three-component Ziegler-Natta type catalysts were also proposed for the polymerization of cycloolefins by Günther and associates (55). Cyclopentene (or cyclopolyenes such as 1,5-cyclooctadiene or 1,5,9-cyclododecatriene) were polymerized to substantially *trans*-polyalkenamers at 0 to +20°C, in the presence of $TaCl_5/AlEtCl_2$ alone or with oxygen-containing activators such as ethanol, acetylacetone, *tert*-butyl hydroperoxide, propylene oxide, or epichlorohydrin. The highest polymer conversions were obtained with $TaCl_5/AlEtCl_2$/epichlorohydrin in chlorobenzene solvent (17). $TaBr_5$ is ineffective as a catalyst component.

According to the same authors (56) olefins containing labile chlorine (e.g., 3-chlorocyclopentene, α-chloroacetone, allyl chloride), transition metal chlorides (e.g., $CuCl_2 \cdot 2H_2O$ or $FeCl_3$), or *tert*-butyl hypochlorite (34) may act as activators of $TaCl_5/AlEtCl_2$ (or $Al_2Et_3Cl_3$) catalysts.

2.6 Rhenium Oxide and Halide Catalysts

According to Turner and co-workers (57) a catalyst, prepared by supporting ammonium perrhenate on alumina and activated by heating to 580°C in a dry air stream, oligomerizes cyclooctene and cyclododecene above 140°C to oligomers containing up to 10 monomeric units, but in agreement with the general trend of supported catalysts, was unable to yield high polymers from cycloolefins.

High polymers were obtained from cyclopentene in the presence of Ziegler-Natta type rhenium catalysts. According to Günther et al. (58), the catalysts prepared from rhenium pentachloride and triisobutyl-aluminum (or diethylaluminum chloride), at -30 to $-40°C$, yield essentially *cis*-polypentenamer with relatively low conversion.

Dall'Asta and Meneghini (59) found that rhenium hexachloride forms more active catalysts than the pentachloride but the organometallic compound strongly influences the stereospecificity. In analogy with the $ReCl_5$ case, cyclopentene is polymerized by the $ReCl_6/Al(i\text{-}Bu)_3$ catalyst to essentially *cis*-polypentenamer but $ReCl_6/AlEt_3$ induces formation of essentially *trans*-polypentenamer, and $ReCl_6/AlEt_2Cl$ (or $AlEtCl_2$) formation of atactic polypentenamer. The activity of the $ReCl_6/AlEt_3$ system in ring opening polymerization of cyclooctene is very low.

2.7 Titanium, Zirconium, and Vanadium Based Ziegler-Natta Catalysts

With titanium based Ziegler-Natta catalysts (5) the polymers obtained from cyclobutene exhibit different chemical and steric structure depending on the type of catalyst and on the polymerization conditions: $TiCl_4/AlEt_3$ (1:3) in *n*-heptane at $-50°C$ promotes formation of predominantly *cis*-1,4-polybutadiene (5); $TiCl_4/AlEt_3$ (1:2) in toluene at $-10°C$ yields 1,4-polybutadiene having a 1:2 cis/trans ratio (6); $TiCl_3 \cdot AlEt_3$ (1:3) in *n*-heptane at $+45°C$ induces formation of saturated cyclobutylene, *cis*-1,4-butenamer, and *trans*-1,4-butenamer units at a 2:1:2 ratio (6). Polymer conversions are generally quantitative.

According to Dall'Asta and his co-workers (11, 12), analogous titanium and zirconium catalysts ($TiCl_4/AlEt_3$; $TiBr_4/AlEt_3$; $ZrCl_4/AlEt_3$), used in the -30 to $+20°C$ range, exhibit very low activity in the polymerization of low strain cycloolefins such as cyclopentene, but high chemical and steric specificity, as indicated by the formation of essentially pure *trans*-polypentenamer.

The very different behavior of titanium catalysts with respect to high and low strain cycloolefins is even more anomalous if one considers that, as a general trend, first row transition metal catalysts (vanadium, chromium, nickel) promote addition polymerization of cycloolefins, but exhibit very little tendency for ring opening, whereas third row transition metal catalysts (tungsten, tantalum, rhenium, osmium, iridium) are typical ring opening systems.

Vanadium catalysts, especially those containing organoaluminum compounds as cocatalysts, polymerize and copolymerize cycloolefins such as cyclobutene (5, 6), cyclopentene (60), and higher cycloolefins (61) via addition polymerization. Ring opening activity in vanadium catalysts is poor but was observed in the polymerization of cyclobutene and 3-methylcyclobutene in the presence of halogen-free catalysts [$V(CH_3CO\text{-}CHCOCH_3)_3/AlEt_3$; $VO(OC_4H_9)_3/AlEt_3$] (6) or systems containing strongly dehalogenating metal alkyls [VCl_4 (or $VOCl_3$)/Li *n*-Bu] (62).

2.8 Group VIII Transition Metal Catalysts

Group VIII transition metal halides or coordination complexes, alone or in combination with organometallic compounds, were introduced in 1961 by Rinehart et al. (63) for the polymerization of butadiene and later for the polymerization of high strain cycloolefins such as norbornene (64, 65), cyclobutene (66), and their derivatives. However, they proved completely inefficient in the case of low strain cycloolefins such as cyclopentene.

Among the second and third row transition metals (osmium, iridium, ruthenium, rhodium, and palladium) only ruthenium was examined with cycloolefins yielding elastomeric ring opening polymers such as cyclobutene and its derivatives. Natta et al. (67) found that ruthenium trichloride hydrate in ethanol stereospecifically converts cyclobutene to pure *trans*-1,4-polybutadiene, and 3-methylcyclobutene to essentially *trans*-1,4-polypentadiene. The same catalyst, in water, yields substantially atactic 1,4-polybutadiene or 1,4-polypentadiene.

3 MECHANISM OF RING OPENING POLYMERIZATION

3.1 Ring Strain and Steric Hindrance—Retention of the Double Bond

There are two main pathways of cycloolefin polymerization: ring opening to unsaturated alkenamer units [11a], and addition to the double bond with formation of saturated cycloalkane ring units [11b]:

$$-HC=CH-(CH_2)_n- \qquad [11a]$$

$$-HC-CH- \atop (CH_2)_n \qquad [11b]$$

The different behavior of the various cycloolefins is governed mainly by three factors: (*a*) steric hindrance; (*b*) thermodynamic implications; and (*c*) type of catalyst. From the standpoint of steric hindrance, cycloolefins behave analogously to open chain olefins with internal double bonds. Steric hindrance of the monomer generally opposes homopolymerization of cycloolefins via addition polymerization. This is essentially true for all low strain cycloolefins, i.e., five- or higher membered rings with the exception of five-membered endocycles (norbornene and derivatives) and *trans*-cyclooctene. In the latter, high ring strain is due to the trans double bond and favors addition polymerization (68) promoted by radical initiators. However, addition polymerization of low strain cyclics, induced by typical addition polymerization catalysts, becomes possible by reducing steric hindrance, e.g., by copolymerization with ethylene (60, 61) or by polymerization of charge transfer complexes of cyclopentene with polar monomers such as acrylonitrile (69) or sulfur dioxide (70).

In high strain cycloolefins (three-, four-, and endocyclic five-membered rings) the steric hindrance is less pronounced and thermodynamic factors may contribute to overcome it. Actually, addition homopolymers of cyclopropene, cyclobutene, norbornene, and dicyclopentadiene are known.

There are two enthalpy contributions that favor addition polymerization of cycloolefins: disappearance of a ring double bond with formation of two single bonds in the repeat unit, and partial release of ring strain due to the transformation of a rigid cycloalkene into a less rigid cycloalkane system. In the case of ring opening polymerization (if no cis/trans isomerization is involved) the enthalpy term contribution to bond transformations is zero, since the number and type of bonds in the monomer are equal to those of the corresponding alkenamer unit of a polymer chain. However, ring strain is entirely released.

The tendency of small rings to addition polymerization is easily understood on considering the strong increase of ring strain and the decrease in steric hindrance with decreasing ring size. Thus, cyclopropene (71) homopolymerizes exclusively and 1-methylcyclopropene (72) copolymerizes exclusively via addition to the double bond. Even if largely determined by the particular catalyst used, the pathway taken in polymerization of the intermediate four- (6) and endocyclic five-membered (4, 73, 74) rings is in many instances unspecific. In this case both types of polymerization are actually favored from the thermodynamic point of view.

On the basis of these considerations, the high specificity of the ring opening polymerization of low strain cycloolefins may be understood. Actually, steric hindrance generally prevents addition polymerization of such cycloolefins. [However, cyclopentene was reported (75) to yield trace amounts of prevailingly saturated polymers of uncertain structure in the presence of Ziegler-Natta catalysts.] The extreme ease with which low strain cycloolefins are polymerized by ring opening is surprising if one merely considers the thermodynamics of the overall reaction. Before discussing this point in detail, its seems convenient to consider some other aspects.

One problem concerns whether or not the cycloolefinic double bonds are shifted during ring opening polymerization. Random shifting may be excluded on the basis of the observation, confirmed by ozonolysis as well as by x-ray studies on polyalkenamers (12), that the polymers always contain monomeric units with an unchanged number of methylene groups. This finding also demonstrates that ring opening polymerization, at least of the unsubstituted cycloolefins, results in head-to-tail succession of monomeric units. Analogous conclusions were made by Calderon et al. (16) for polybutenamers obtained from 1,5-cyclooctadiene.

The chemical structure of the ring opening polymers of substituted cycloolefins allows us to exclude a constant shift of the double bonds during polymerization. For example, 3-methylcyclobutene yields polymers with a 1,4-pentadiene (67), and 1-methylcyclobutene yields polymers having a 1,4-isoprene (76) derived structure, indicating that the methyl group and double bond retain their relative positions in the polymer:

$$\begin{array}{c} HC=CH \\ | \quad | \\ H_2C-C-CH_3 \\ | \\ H \end{array} \longrightarrow -HC=CH-\overset{\overset{\displaystyle CH_3}{|}}{C}H-CH_2- \qquad [12]$$

$$\begin{array}{c} \overset{\displaystyle CH_3}{\diagup} \\ HC=C \\ | \quad | \\ H_2C-CH_2 \end{array} \longrightarrow -HC=\overset{\overset{\displaystyle CH_3}{|}}{C}-CH_2-CH_2- \qquad [13]$$

These findings are also strong evidence against a cationic mechanism for ring opening polymerization, even in the case of metal alkyl-free catalysts. Friedel-Crafts agents (e.g., $AlCl_3$) in the absence of a transition metal such as tungsten do not induce ring opening polymerization, analogously to anionic agents (Lin-Bu, $AlEt_3$, $AlEtCl_2$) or radical initiators, which are inactive alone.

3.2 Sites of Ring Opening—Transalkylidenation Mechanism

A key to understanding the mechanism of ring opening polymerization is in knowing the site of ring cleavage. There are three bonds that should be considered: (a) the cycloolefinic double bond; (b) the single bond α to the double bond; (c) the single bond β to the double bond. In early papers only alternatives (b) and (c) were considered. Efforts were made to demonstrate the difference between the catalytic ring opening polymerization and the uncatalyzed thermal ring cleavage of cycloolefins, which had been demonstrated to proceed by opening of the single bond in the β position. Thus, cis-3,4-dimethylcyclobutene at 175°C isomerizes in a conrotatory electrocyclic reaction to cis,trans-1,4-dimethyl-1,3-butadiene (77):

$$[14]$$

On the basis of a series of experimental investigations on ring opening polymerization of cyclopentene and 3-methylcyclobutene, Natta et al. (12, 65) were able to rule out hypothesis (c) and, in analogy with the mechanism proposed by Truett et al. (4) for the ring opening polymerization of norbornene [2], suggested the α single bond [hypothesis (b)] as the probable site of ring cleavage. A similar mechanism involving opening of the α single bond was recently reproposed by Marshall and Ridgewell (45).

Calderon et al. (26) in 1968 considered the ring opening polymerization of cycloolefins as a special case of the disproportionation reaction of open chain olefins, and suggested the double bond itself as the site of ring cleavage [hypothesis (a)]. This hypothesis would not be in contradiction to results of Natta et al. (12, 67), which excluded hypothesis (c).

The disproportionation reaction of open chain olefins, which was first disclosed by Banks and which led to the Phillips "triolefin" process, allows the conversion of olefins to an equilibrium mixture comprising homologs of both shorter and longer carbon chain olefins. This reaction has been termed "olefin disproportionation" by Banks, "olefin dismutation," or "ethenolysis" when using ethylene and internal olefins as reaction partners (78) and "metathesis" or "transalkylidenation" (24).

This reaction was interpreted by Bradshaw and co-workers (78) as a simultaneous cleavage of two double bonds and rearrangement via a "quasi cyclobutane" intermediate, as shown for the disproportionation of 1-butene to ethylene and 3-hexene:

$$
\begin{array}{l}
C{=}C{-}C{-}C \\
C{=}C{-}C{-}C
\end{array}
\rightleftharpoons
\begin{array}{l}
C\cdots C{-}C{-}C \\
C\cdots C{-}C{-}C
\end{array}
\rightleftharpoons
\begin{array}{l}
C \\
\parallel \\
C
\end{array}
+
\begin{array}{l}
C{-}C{-}C \\
\parallel \\
C{-}C{-}C
\end{array}
\qquad [15]
$$

The analogous reaction between a cycloolefin and an open chain olefin [16], later described by Ray and Crain (79):

$$
\qquad\qquad\qquad\qquad [16]
$$

is strong experimental evidence for mechanism [15] and may suggest a relationship between olefin disproportionation and ring opening polymerization. All these reactions were carried out using alumina supported MoO_3–CoO catalysts.

Calderon et al. (24, 25) then showed that analogous disproportionation could also be obtained in the presence of a three-component tungsten based Ziegler-Natta catalyst of the type previously described by Dall'Asta and Carella (20) for the ring opening polymerization of cycloolefins. The disproportionation of 2-butene and 2-butene-d_8, promoted by a

$WCl_6/C_2H_5OH/AlEtCl_2$ catalyst, resulted in a $1:2:1$ equilibrium mixture of 2-butene, 2-butene-d_4, and 2-butene-d_8:

$$[17]$$

thus contributing new experimental evidence to the transalkylidenation mechanism in the case of open chain olefins.

The same group (26) postulated for the ring opening polymerization an analogous transalkylidenation mechanism, involving coordination of two cycloolefins to tungsten, followed by cleavage of the two double bonds and rearrangement via a quasi-cyclobutane intermediate [18]:

$$[18]$$

This mechanism implies: (a) cleavage of the cycloolefin at the double bond; and (b) chain growth via macrocycles of increasing size.

Experimental evidence for chain growth via macrocycles had been given in 1967 by Alkema and Van Helden (52), who oligomerized cyclooctene using an alumina supported MoO_3–CoO catalyst and isolated multiunsaturated monocyclic olefins from the reaction mixture. These were hydrogenated to 24-, 32-, 40-, 48-, 56-, 64-, and 72-membered cycloalkanes, which correspond to multiples of the starting olefin. In 1968 Wasserman et al. (80) and Scott et al. (26) isolated up to 120-membered, analogous oligomers from the low molecular weight fractions of a polyoctenamer, obtained in the presence of the $WCl_6/C_2H_5OH/AlEtCl_2$ catalyst.

Although the macrocyclic structure of polyalkenamers was demonstrated only for low molecular weight fractions, chain growth via macrocycles is now generally accepted for the ring opening polymerization of cycloolefins, at least in the case of the most typical tungsten and molybdenum catalysts. However, chain growth via macrocycles is not a demonstration of ring opening at the double bond. The existence of macrocycles is in fact consistent with ring opening at any one of the cycloolefinic (single or double) bonds, provided the succession of the monomeric units is head to tail.

A direct experimental demonstration of the cleavage of the double bond in ring opening polymerization has recently been given by Dall'Asta and Motroni (81, 82) based on the examination of random copolymers of [1-^{14}C]cyclopentene and cyclooctene, prepared using a $WOCl_4$/benzoyl peroxide/$AlEt_2Cl$ catalyst. The ozonization and successive reductive decomposition of such ozonides yielded 1,5-pentane diol and 1,8-octane diol. Since the radioactivity-gas chromatographic analysis revealed that the whole radioactivity was retained in 1,5-pentane diol, it must be concluded that the bond cleaved in the polymerization of such cycloolefins was the double bond. If the bond cleaved had been one of the single bonds, about 50% of the radioactivity should be found in the C_5 and about 50% in the C_8 glycol, as illustrated in scheme [19].

1) RING CLEAVAGE AT THE DOUBLE BOND:

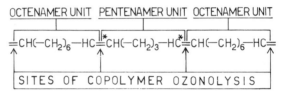

EXPECTED GLYCOLS:

$$HOH_2C(CH_2)_6CH_2OH + HOH_2\overset{*}{C}(CH_2)_3\overset{*}{CH_2}OH + HOH_2C(CH_2)_6CH_2OH \qquad [19]$$

2) RING CLEAVAGE AT THE ALPHA SINGLE BOND:

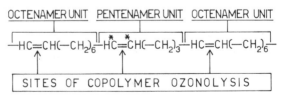

EXPECTED GLYCOLS:

$$HOH_2C(CH_2)_6\overset{*}{CH_2}OH + HOH_2\overset{*}{C}(CH_2)_3CH_2OH + HOH_2C(CH_2)_6CH_2OH$$

The random structure of such copolymers, which is the premise for validity of scheme [19], was shown by partial ozonization and successive reductive decomposition of the ozonides, which yielded large amounts of monounsaturated C_{13} glycol. In another paper (83) the same authors extended this approach to the highly strained cycloolefins and to molybdenum, titanium, and ruthenium catalysts. The results showed the general validity of the metathesis scheme in transition metal catalyzed ring opening polymerization.

3.3 The Transition State

The four-center quasi-cyclobutane intermediate was originally proposed by Bradshaw and co-workers (78) who deliberately avoided the term "mechanism of reaction," which would imply a deeper knowledge of the reaction than they in fact possessed. However, successive authors (e.g., 25, 38) have substantially adopted the quasi-cyclobutane intermediate.

Mol and associates (84), on the basis of disproportionation experiments with ^{14}C-labeled propylene showing that [2-^{14}C] and [3-^{14}C] were retained by 2-butene, and [1-^{14}C] by ethylene, proposed (85) an alternative mechanism. The catalyst was thought to abstract two hydrogens from each of two coordinated propylene molecules, thus forming a π bonded cyclobutadiene intermediate, which is then broken with recombination of hydrogen. The inconsistency of such a mechanism was demonstrated by Crain (86), who showed that 2,3-dimethyl-2-butene, an olefin without hydrogen attached to the carbon atoms of the double bond, was cleaved by ethylene on a MoO_3–Al_2O_3 catalyst to yield exclusively isobutene.

One of the drawbacks of the quasi-cyclobutane intermediate is the unreactivity of cyclobutane rings in the presence of metathesis catalysts, along with the fact that cyclobutane rings are essentially absent in the reaction products, although their formation would be conceivable according to Mango and Schachtschneider (87). Furthermore, cyclobutane rings would imply severe steric restrictions to the transition metal coordination complex, especially for ring opening polymerizations where the four substituents are crowded macromolecules.

An alternative mechanism of the metathesis transition state put forth in 1969 by Natta and Dall'Asta (88) considers separation of the four carbon atoms of two coordinated double bonds into a tetracarbene complex:

[20]

Such a transition state would avoid the steric hindrance and nonbonding interactions of the quasi-cyclobutane scheme.

In 1971, Lewandos and Pettit (89), on the basis of a molecular orbital treatment, reconsidered a carbene intermediate for ethylene metathesis. According to their proposal, ethylene reacts with the transition metal to form a bisethylene complex that reorganizes to a multicentered system in which the bonding is most conveniently described as resulting from the interaction of a basic set of metal atomic orbitals and four methylenic units (scheme [21]). Retraction of the latter along the x axis would lead to the starting materials, and separation along the y-axis to disproportionated products:

[21]

The transformation of the two π bonded olefins into four metal bonded carbenes implies donation of four additional electrons to the metal. In order to avoid excess electron donation beyond the inert gas configuration, the tetracarbene intermediate probably requires replacement of some other ligand. In accordance with this scheme Lewandos and Pettit (90) also showed that the reaction of 4-nonene with toluene tricarbonyl tungsten (0) results in metathetic scrambling of the olefin moieties only in conditions favoring replacement of a CO ligand, whereas in conditions in which this requirement is not satisfied essentially double bond isomerization occurs.

The carbene transition state also allows one to interpret the observations made by Smidt et al. (91) in the $PdCl_2$ catalyzed reaction of oxygen with ethylene to form acetaldehyde (Wacker olefin oxidation process). In this process, as in the analogous oxidation of higher olefins there are obtained, along with the expected aldehyde, smaller amounts of aldehydes or ketones containing one or two carbons less than the starting olefins. Thus, the side products of 1-butene, 2-butene, and isobutene oxidation are propionaldehyde, acetaldehyde, and acetone, respectively. We may therefore conclude that olefins undergo side reactions involving cleavage of the double bond and that the following nucleophilic attack by the hydroxy ion yields the carbonyl compounds corresponding to one olefin moiety. An analogous interpretation may be invoked to explain the formation of methyl chloride as a side product of ethylene oxidation. In this case a bicarbene species stemming from one coordinated ethylene may be visualized as the transition state. From this point of view, metatheses and oxidative ethylene scission appear as special cases of a general reaction, in which one, two, or even three complexed double bonds are cleaved with formation of different carbene intermediates.

Experimental support of the carbene hypothesis has recently been given by Cardin et al. (92) using electron-rich olefins, in which the double bond cleavage is facilitated. Induced by rhodium metathesis catalysts such olefins underwent metathetical disproportionation, and a complex containing a rhodium-carbene bond was isolated from the reaction products.

3.4 Thermodynamic and Kinetic Control of the Reaction

As already pointed out, ring opening polymerizations differ essentially from conventional addition polymerizations with regard to the bonds involved. In the former the overall type and number of bonds of the monomers are the same as in the resulting polymer; in the latter for each disappearing double bond of the monomer two new single bonds are formed in the polymer. This determines different thermodynamic situations. Addition polymerizations are generally antientropic, but enthalphy favored. Ring opening polymerizations are generally favored by enthalpy and in some cases also by entropy. In these latter the enthalpy term mainly derives from the release of ring strain: about 10 kcal/mole for highly strained rings, 3–5 kcal/mole for higher membered cycloolefins, and zero for cyclohexene.

Entropy involved in ring opening polymerizations appears to be negative for small rings up to cyclohexane, close to zero for cycloheptene, but positive for higher membered cycles. This is the result of the superposition of three different sources of entropy change. The negative translational entropy, characteristic of any polymerization, is very high for small cycles, but becomes less negative for larger ones. Conversely, torsional and vibrational entropies, which are always positive in ring opening polymerizations, decrease to a much smaller extent when going from small to large rings. Thus, the negative translational entropy prevails over the other two entropies in small rings up to cyclohexene, but the positive torsional and vibrational entropies prevail over the negative translational ones in larger rings. Consequently, e.g., the ring opening polymerization of cyclopentene is antientropic but enthalpy determined and that of cyclooctene is enthalpy and entropy favored, whereas that of cyclohexene would be weakly antientropic and thermoneutral. For this reason, cyclohexene is the only one of the unsubstituted cycloolefins unable to undergo ring opening polymerizations. On the contrary, when subjecting polyhexenamer chains to metathesis catalysts one observes the reverse reaction to yield cyclohexene along with small amounts of linear hexenamer oligomers (19, 37, 93).

In acyclic olefins the attainment of a statistical metathesis equilibrium is essentially entropy determined (25) since the reaction is thermoneutral, i.e., the reaction enthalpy is zero (if one neglects cis/trans isomerization of the double bonds).

The ring opening polymerization of weakly strained cycloolefins is an equilibrium reaction comprising monomer-oligomer, oligomer-polymer, and ring-chain equilibria. The negative enthalpy term, as a result of release of ring strain, shifts these equilibria away from the statistical composition. This equilibrium composition is temperature dependent. Ofstead and Calderon (94) investigated the temperature dependence of the monomer-polymer equilibrium for trans-polypentenamer in the 0–30°C range. Limiting conversions varying according to the temperature from 50 to 80% polypentenamer were obtained after short times by operating in solution with the $WCl_6/EtOH/AlEtCl_2$ catalyst. In accordance with the equilibrium nature of these polymerizations, gas chromatography of the low molecular weight fractions revealed the presence of substantial amounts of all $(C_5H_8)_n$ terms from monomer to nonvolatile oligomer. On the basis of the equilibrium monomer concentration at various temperatures the following thermodynamic quantities were calculated: $\Delta H_p = -4.4$ kcal/mole, $\Delta S_p^{\circ} = -14.9$ cal/mole $\cdot$ deg, and ceiling temperature about 150°C. The reversibility of the monomer-polymer equilibrium was shown by temperature change during polymerization, which results in shift of the conversion curve to that corresponding to the new temperature, as well as by depolymerization of purified polypentenamer induced by the same metathesis catalyst.

Studies reported by Pampus et al. (39) on the thermodynamics of ring opening polymerization of cyclopentene induced by $WCl_6/SnEt_4$ are in partial contrast with those of the foregoing authors. This contrast may in part be attributed to the fact that the latter catalyst avoids side reactions of the Friedel-Crafts type, which occur easily with the former catalyst. Temperature dependence of the monomer-polypentenamer equilibrium and its reversibility with temperature have been confirmed, but n-mers having n values ranging from 2 to 50 were absent or present only in trace amounts. One may therefore deduce that some of the reported depolymerization results are catalyst specific side reactions.

Thermodynamically controlled cis/trans equilibria of the double bond in olefins resulting from metatheses have been reported by several authors. From studies of the steric course of the olefin metathesis of pure *trans-* and *cis*-2-pentene, Calderon (25, 95) concluded that the formation of trans and cis isomers is provided for by the transalkylidenation step itself, and is not the outgrowth of an independent isomerization reaction. This view was not generally accepted. These results are actually in contrast with those for numerous stereospecific ring opening polymerizations. Thus, the cis/trans ratio of ring opening polymerizations appears to be kinetically rather than thermodynamically controlled.

It is not yet clear whether this statement has to be limited to the stereocontrol of such polymerizations or may, at least in part, be extended to monomer–polymer equilibrium and to molecular weight control. Pampus and co-workers (96) reported that the initial steps of cyclopentene polymerization induced by WCl_6/epichlorohydrin/$AlEt_2Cl$ catalyst yielded very high molecular weight polypentenamer rich in cis double bonds, but that with increasing conversion both the molecular weight and the cis content rapidly decreased.

Examination of the early stages of 2-pentene metathesis, induced by $Pyr_2(NO)_2MoCl_2/AlEtCl_2$ catalyst, led Hughes (97) to the conclusion that the steric structure of the initially formed olefins is dependent on the configuration of the starting 2-pentene and hence that olefin metathesis is kinetically controlled. *cis*-2-Pentene preferentially disproportionates to *cis*-2-butene and *cis*-3-hexene, but *trans*-2-pentene preferentially disproportionates to *trans*-2-butene and *trans*-3-hexene. However, after a short time isomerization of the double bonds becomes predominant. Disproportionation and isomerization rates decrease simultaneously, thus indicating that the two processes are intimately connected and probably occur on the same catalyst.

Different results were obtained by Dall'Asta and Stigliani (98) on *cis*- and *trans*-2-pentene by using a four-component $WCl_6/(ClH_2C)_2CHOH/Al_2Et_3Cl_3/(C_6H_5)_3P$ catalyst. *cis*-2-Pentene is metathesized essentially to *trans*-2-butene, *trans*-2-pentene, and *trans*-3-hexene. On the contrary, *trans*-2-pentene is essentially unreactive even after 1 day at 25°C. The authors concluded that, at least with this particular catalyst in which the isomerization capacity for double bonds is essentially suppressed, trans double bonds are virtually unable to undergo metathesis, whereas the cis double bonds are stereospecifically metathesized to the corresponding trans olefins.

On the basis of the results on 1,5-hexadiene metathesis induced by the same catalyst, Dall'Asta and co-workers (93) concluded that in the presence of such catalysts metathesis is an intimately stereospecific, kinetically controlled process, at least as far as double bond configuration is concerned.

3.5 Influence of Substituents on the Course of Polymerization

A further problem of considerable importance for the understanding of the mechanism of ring opening polymerizations arises with substituted cyclo-olefins. 3-Methylcyclobutene (67), 3-methylcyclopentene (17), 3-methyl-cyclooctene, and 3-phenylcyclooctene (16), containing the substituent at a carbon atom far from the double bond, are easily polymerized in the presence of conventional tungsten based ring opening catalysts. An exception is 3-isopropylcyclopentene, which could not be polymerized (17).

Cycloolefins having the substituent on a carbon atom of the double bond, such as 1-methylcyclopentene (17), are not polymerized, except for the highly strained 1-methylcyclobutene, which is polymerized by tungsten catalysts (76). Polymers with trisubstituted double bonds, such as *cis*-1,4-polyisoprene, unlike *cis*-1,4-polybutadiene and polypentenamers, are completely unaffected with regard to both cis/trans ratio and molecular weight, when treated with the $WCl_4[OCH(CH_2Cl)_2]_2/AlEt_2Cl$ catalyst (96).

Furthermore, as reported by Pampus et al. (96), the interaction of acyclic olefins with growing polypentenamer chains, which results in mixed transalkylidenations (as explained below), strongly decreases with increasing degree of substitution of the double bond of the olefin in the order 1-butene > *cis*-2-butene > *trans*-2-butene > isobutene > 2-methyl-2-butene. The last, characterized by a trisubstituted double bond, has only very poor activity.

All these remarks lead to the conclusion that the occurrence of the transalkylidenation reaction is largely dependent on steric hindrance around the double bond. Trisubstitution substantially prevents trans-alkylidenation. Such an interpretation, however, conflicts with the observation of Crain (86), who showed that even a tetrasubstituted double bond, like that of 2,3-dimethyl-2-butene, is disproportionated with ethylene in the presence of a Mo catalyst. One must assume that steric restrictions are less important in the case of small olefins than they are for polymers.

An interesting study on the relative reactivity of two double bonds in the same monomer was reported by Ofstead (99). 1-Methyl-1,5-cyclooctadiene (MC), 1,2-dimethyl-1,5-cyclooctadiene (DMC), 1-ethyl-1,5-cyclooctadiene (EC), and 1-chloro-1,5-cyclooctadiene (CC) were ring opening polymerized, by the $WCl_6/C_2H_5OH/AlEtCl_2$ catalyst, to substantially alternating copolymers of butadiene and substituted butadiene:

$$P \underset{}{\overset{R}{\bigcirc}} \longrightarrow \left[-CH_2CH=CH-CH_2CH_2\overset{R}{C}=CH-CH_2- \right]_P \qquad [22]$$

Retention of configuration at the substituted double bond in MC and CC polymers suggests lack of participation of that bond in the transalkylidenation reaction. This indicates that ring opening polymerization primarily occurs at the unsubstituted double bond. In the case of DMC, the absence of isoprene units, which would be due to interaction between bisubstituted and unsubstituted double bonds, also confirms the substantial unreactivity of the former.

Mass spectroscopic examination of the oligomers of MC (but not of the DMC, EC, and CC) polymerization, however, revealed the presence of small amounts of "sesquioligomers," increasing with polymer conversion; this indicates some reactivity of the trisubstituted bond also.

All ring opening polymers of 3-substituted cycloolefins reported so far, even if essentially stereoregular around the double bond, are amorphous and presumably atactic. This may be interpreted to mean that no stereoselectivity is exerted by the catalyst with respect to one of the two monomer antipodes, or that a hypothetical stereoselectivity is destroyed by racemization during the polymerization step.

3.6 Reactivity of the Polyalkenamer Double Bonds

Since any carbon–carbon double bond, and not only those of the cycloolefin monomers, is a possible reaction site for new transalkylidenations, it follows that already formed macrocycles may undergo intra- and intermolecular ring opening reactions. The intramolecular reaction, according to [23], leads to the lowering of average molecular weight:

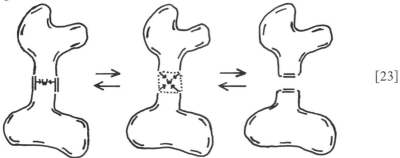

[23]

The intermolecular transalkylidenation between macrocycles, which is the reverse reaction, obviously leads to an increase of the average molecular weight. As a result of these two reactions, molecular weight distribution is not only determined by the kinetics of chain growth but undergoes redistribution as long as the catalyst is active. Such intra- and intermolecular reactions become more important with an increasing degree of polymerization, i.e., with the increase of the concentration of macrocycles. An experimental demonstration of such intramolecular rearrangements was given by Scott et al. (26). The mass spectrum of the low molecular weight fractions of a ring opening polymer of 1,5-cyclooctadiene consisted not only of a series of "whole" multiples of the monomer ($[C_8H_{12}]_n$) but also of a second series of sesquioligomers ($[C_8H_{12}]_n-C_4H_6$).

Not all macromolecules of a polyalkenamer prepared by ring opening polymerization are macrocycles. Under conditions of practical synthesis, many of the macromolecules, analogously to cyclopentene in reaction [16], undergo cleavage by the action of open chain olefins, present as impurities or arising from partial decomposition of Ziegler-Natta catalysts. Such a cleavage leads to linear macromolecules with unsaturated chain ends, as illustrated in scheme [24] for the case of ethylene:

An analogous effect is produced by conjugated dienes. The action of a second olefin or diene molecule on the linear polyalkenamer should result in degradation of molecular weight. Conjugated dienes are the most efficient degradation agents in ring opening polymerization of cycloolefins. This effect of small quantities of conjugated dienes was discovered in 1965 by Dall'Asta (100), who employed it for regulation of molecular weight in cycloolefin polymerization. Larger amounts of conjugated diene lead to oily, highly unsaturated hydrocarbons (101). In 1969 two Bayer patents and a Huls patent disclosed the use of α-olefins (102, 103) and of unconjugated dienes (104) as molecular weight regulators in cyclopentene polymerization. According to Pampus et al. (105) the molecular weight of polypentenamer may be regulated even by post-treatment of the polymer with α-olefins or conjugated dienes in the presence of tungsten catalysts.

These reactions on polyalkenamers are not limited to macrocyclic structures. Open chain macromolecules such as polyalkenamers, 1,4-polybutadienes, and even 1,2-polybutadienes undergo the same reactions in the presence of tungsten catalysts. The degradation of polyalkenamers or 1,4-polybutadienes to oligomeric products may be achieved either with an excess of tungsten catalyst, as described by Calderon (106), or by the combined action of catalyst and large amounts of an acyclic olefin. Thus, 2-butene in the presence of the $WCl_6/C_2H_5OH/AlEtCl_2$ catalyst nearly totally converts cis-1,4-polybutadiene to 2,6-octadiene and 2,6,10-dodecatriene (107). Conversely, reaction of cyclooctene with cis-2-butene yields a homologous series of compounds consisting of dodecadiene, eicosatriene, octacosatetraene, and hexatriacontapentaene (107). The analogous reaction between cycloolefins (cyclopentene, cyclooctene, 1,5-cyclooctadiene, 1,5,9-dodecatriene) and acyclic olefins (propylene, 2-butene, 1-pentene, 2-pentene) in the presence of $WOCl_4/AlEt_2Cl$ and $WOCl_4/Sn(C_4H_9)_4$ was recently reinvestigated by Hérisson and Chauvin (108).

Styrene-butadiene copolymers are converted by 2-butene in the presence of $WCl_6/C_2H_5OH/AlEtCl_2$ (109) to 5-phenyl-2,8-decadiene and 4-phenylcyclohexene, the latter according to:

$$[25]$$

This reaction provides a direct experimental demonstration that failure of cyclohexene to polymerize via ring opening does not mean failure to participate, but rather monomer-polyhexenamer equilibrium in favor of the monomer, as shown by the fully realized reverse reaction from polymer to cyclohexene structures.

Polymerization of cyclopentene in the presence of 1,2-polybutadiene (96), according to the principles of the transalkylidenation reaction, yields graft copolymers of a new type:

$$[26]$$

3.7 The Synthesis of Catenanes and Nectinodanes

An interesting aspect of ring opening polymerization is the possibility offered by this route to prepare interlocked rings called catenanes and nectinodanes. Recently, Wolovsky (110) and Ben-Efraim et al. (111) simultaneously described a statistical one-step approach to such structures based on intramolecular transalkylidenation of macrocycles. If this reaction on cycloolefins beyond a certain minimum size occurs under conditions in which at least a part of such rings is twisted to a degree of 360° [27] or 540° [28], catenanes or nectinodanes, respectively, may result:

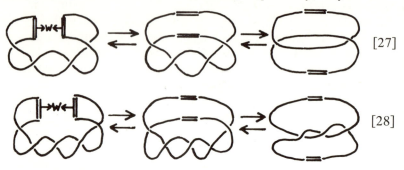

[27]

[28]

The structures contain "topological" (mechanical) bonds which may be cleaved only by cleavage of a chemical bond.

Cyclododecene, or an even larger ring olefin, is the starting material. It is treated with a ring opening catalyst, e.g., $WCl_6/C_2H_5OH/AlEtCl_2$, under conditions favoring the formation of oligomeric macrocycles. Obviously, the presence of acyclic olefins must be avoided. The presence of catenanes in the low molecular weight fractions of the reaction mixture was revealed by mass spectrometric examination. The "distillate" (5×10^{-6} torr), after gradual appearance and disappearance of products of increasing mass, upon further distillation at higher temperatures reveals reappearance of lower multiples of the starting olefin. This observation was interpreted as a result of fragmentation due to thermal cleavage of interlocked rings. It is interesting to note that the two rings composing catenanes are approximately equal in ring size, $C_{72}+C_{72}$, $C_{84}+C_{72}$, $C_{84}+C_{84}$ being favored. This suggests, according to Ben-Efraim et al. (111), a nonrandom conformation of the macrocycles at the moment of intramolecular transalkylidenation. The same authors (111) also observed, by fragmentation of the polymeric fractions of this polydodecenamer, mass peaks indicating the presence of catenane structures. Such structures may considerably influence some properties of polyalkenamers.

4 RING OPENING POLYMERIZATION PROCESSES

Two ring opening polymerization processes are described in some detail in the literature: the polymerization of cyclopentene to essentially *trans*-polypentenamer in the presence of three-component tungsten based

Ziegler-Natta catalysts, and the polymerization of cyclopentene to all *cis*-polypentenamer in the presence of a two-component molybdenum based Ziegler-Natta catalyst. Cis/trans ratios reported in the literature are often confusing because of the use of differing infrared absorption coefficients (see Section 5).

4.1 Polymerization of Cyclopentene to Essentially *trans*-Polypentenamer

The most convenient methods for preparation of *trans*-polypentenamer use some of the three-component tungsten catalysts described earlier (Section 2.2). The tungsten compound is generally a high valency halide, especially tungsten hexachloride. The oxygen-containing activators are hydroperoxide or preferably chlorinated alcohols or epoxides. They are combined with the tungsten hexachloride prior to the addition of monomer and organometallic cocatalyst.

The most efficient cocatalysts are organoaluminum compounds. Trialkyls or dialkyl chlorides are preferred over monoalkyl dichlorides or alkylaluminum sesquichlorides, as they avoid side reactions better, especially of the Friedel-Crafts type. As a matter of fact, the more highly chlorinated the aluminum compound, the easier the cyclization reactions occur, with partial disappearance of the double bonds and deterioration of the elastomeric properties.

A detailed description of solution polymerization of cyclopentene was given by the Bayer research group (17). Tungsten hexachloride is allowed to react in toluene with 2-chloroethanol or epichlorohydrin according to schemes [5a] and [9], respectively. Solutions of $WCl_5(OCH_2CH_2Cl)$ or $WCl_4[OCH(CH_2Cl)_2]_2$ (0.3 M) are used, the latter being more stable. Polypentenamer conversion with time in the presence of these two tungsten compounds and diethylaluminum chloride cocatalyst in toluene at 0°C is shown in Fig. 3. Ultimate polymer conversion and polymerization rate strongly increase with increasing activator content.

The comparison of two different organoaluminum cocatalysts in combination with $WCl_4[OCH(CH_2Cl)_2]_2$ shows higher polymerization rate and higher polymer conversion in the case of the chlorinated cocatalyst (Fig. 4).

Molecular weights are regulated by adding small amounts of 1-butene to the monomer. In the case of the $WCl_4[OCH(CH_2Cl)_2]_2/Al(i-Bu)_3$ catalyst, the molecular weights, as determined by solution viscosity a few minutes after polymerization begins, become constant with polymer conversion (Fig. 5). A substantially constant solution viscosity at different polymer conversions is also observed with this catalyst in the absence of

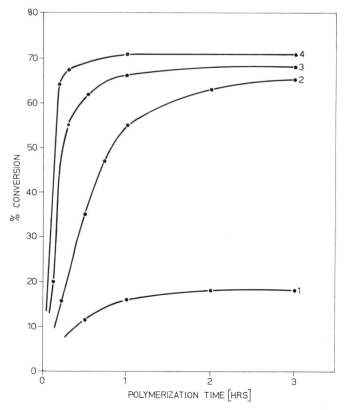

Figure 3 Rate of polymerization of cyclopentene to *trans*-polypentenamer in the presence of different tungsten compounds (solvent, toluene; temperature 0°C). 1, $WCl_6/Al(i\text{-}Bu)_3$; 1.0 mmole W/100 g monomer; 2, $WCl_5(OCH_2CH_2Cl)/AlEt_2Cl$; 0.4 mmole W/100 g monomer; 3, $WCl_{4.75}(OCH_2CH_2Cl)_{1.25}/AlEt_2Cl$; 0.4 mmole W/100 g monomer; 4, $WCl_4[OCH(CH_2Cl)_2]_2/AlEt_2Cl$; 0.4 mmole W/100 g monomer. Reprinted from Günther et al. (17), by courtesy of Hüthig and Wepf Verlag, Basel.

molecular weight regulators. However, if triisobutylaluminum is substituted by diethylaluminum chloride, a strong decrease of solution viscosity as a function of polymer conversion is surprisingly observed (Fig. 6). Such a decrease is particularly remarkable at high polymerization rates. The degradation of the already formed polymer is caused by the still active catalyst as shown by the fact that if, at the end of polymerization, new monomer is added, polymerization recommences with increase of the overall molecular weight of the polymer.

1-Butene acts as a molecular weight regulator in the case of both aluminum cocatalysts. A linear relationship between solution viscosity and 1-butene concentration is observed in both cases (Fig. 7). Chlorinated alkenes as molecular weight regulators yield polymers containing a mean value of one chlorine atom per macromolecule, thus confirming the mechanism of molecular weight regulation.

Molecular weights of polypentenamers may also be influenced by the Al/W ratio of the catalytic system as well as by the W/monomer ratio. In the case of the $WCl_4[OCH(CH_2Cl)_2]_2/Al(i\text{-}Bu)_3$ catalyst the influence of these factors on molecular weight is small, but is more pronounced in the corresponding system containing $AlEt_2Cl$ as cocatalyst.

The regulation of molecular weight of polypentenamer may also be achieved without olefins or diolefins simply by addition of small amounts of water (112). For the $WCl_5(OCH_2CH_2Cl)/AlEt_2Cl$ catalyst, the dependence of molecular weight on added water is shown in Fig. 8. In this

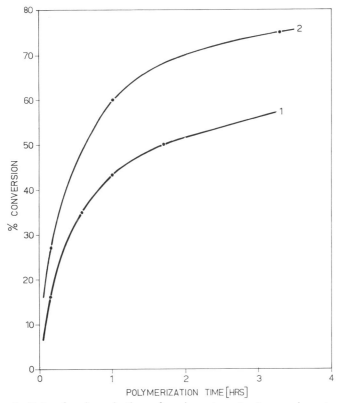

Figure 4 Rate of polymerization of cyclopentene to *trans*-polypentenamer as a function of the alkylaluminum compound. 1, $WCl_4[OCH(CH_2Cl)_2]_2/Al(i\text{-}Bu)_3$; 0.5 mmole W/100 g monomer; 2, $WCl_4[OCH(CH_2Cl)_2]_2/AlEt_2Cl$; 0.4 mmole W/100 g monomer. Reprinted from Günther et al. (17), by courtesy of Hüthig and Wepf Verlag, Basel.

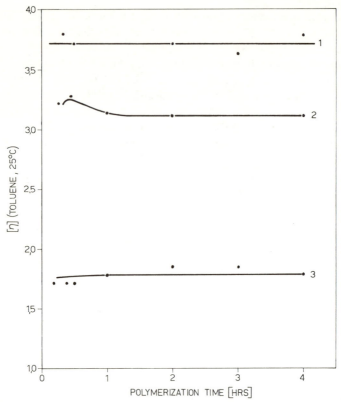

Figure 5 Solution viscosity of *trans*-polypentenamer versus polymerization time at different levels of molecular weight regulator. Catalyst $WCl_4[OCH(CH_2Cl)_2]_2/Al(i\text{-}Bu)_3$. 1, 100 ppm 1-butene referred to monomer; 2, 500 ppm 1-butene referred to monomer; 3, 800 ppm 1-butene referred to monomer. Reprinted from Günther et al. (17), by courtesy of Hüthig and Wepf Verlag, Basel.

particular case, water addition also modifies molecular weight distribution and results in better polymer processability.

Another extensively studied catalyst system for the preparation of *trans*-polypentenamer is that obtained from tungsten hexachloride, cyclopentene hydroperoxide, and triisobutylaluminum (17). The optimum Al/W molar ratios for this catalyst system are in the 1.25–1.75 range. Efficient regulation of the molecular weight is achieved by adding small quantities of conjugated dienes or α-olefins such as 1-butene or 1-pentene. Like the alkoxytungsten halide hydroperoxide catalyst, the molecular weight of polypentenamer as a function of polymer conversion soon reaches a maximum and then steadily decreases.

4.2 Polymerization of Cyclopentene to All *cis*-Polypentenamer

The bulk polymerization of cyclopentene to all *cis*-polypentenamer in the presence of the two-component system $MoCl_5/AlEt_3$ has recently been described in detail (82). It is the most stereospecific ring opening polymerization now known. Unlike tungsten halide based systems a catalyst activator is not necessary in order to obtain fair polymer conversions and reproducible results.

As shown in Table 3, stereospecificity of the $MoCl_5/AlEt_3$ catalyst is essentially unaffected by temperature as long as one operates below $-10°C$; at higher temperatures, stereoregularity of polypentenamer is considerably lower. Polymer conversion shows a pronounced maximum in the

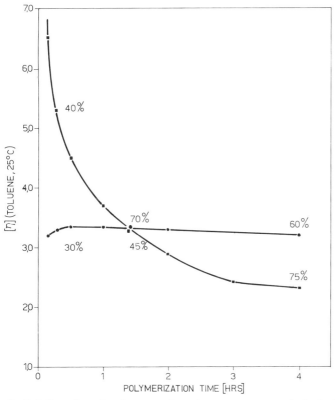

Figure 6 Solution viscosity of *trans*-polypentenamer versus polymerization time for two different alkylaluminum catalysts (the percent values indicate the degree of conversion). ■, $WCl_4[OCH(CH_2Cl)_2]_2/AlEt_2Cl$; ●, $WCl_4[OCH(CH_2Cl)_2]_2/Al(i\text{-Bu})_3$. Reprinted from Günther et al. (17), by courtesy of Hüthig and Wepf Verlag, Basel.

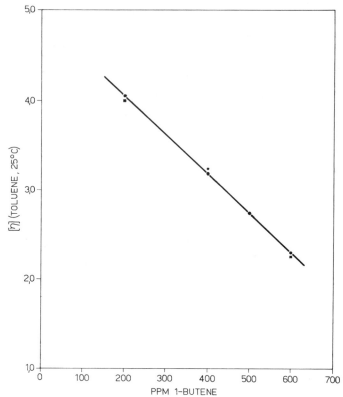

Figure 7 Molecular weight regulation by 1-butene in the polymerization of cyclopentene to *trans*-polypentenamer. ■, WCl₄[OCH(CH₂Cl)₂]₂/Al(*i*-Bu)₃ (1:2.2 molar ratio); ●, WCl₄[OCH(CH₂Cl)₂]₂/AlEt₂Cl (1:2.0 molar ratio). Reprinted from Günther et al. (17), by courtesy of Hüthig and Wepf Verlag, Basel.

−40°C range. Molecular weight, as indicated by solution viscosity, is not remarkably influenced by temperature, but is strongly affected by catalyst concentration (Table 4). Optimum polymer conversion, along with high stereospecificity, is observed at a 500:1 monomer/Mo ratio. At higher catalyst concentrations, the polymer conversion is considerably reduced. This fact was attributed by the authors to the formation of low molecular weight fractions, which are not precipitated from a benzene solution by methanol.

The influence of the Mo/Al ratio was investigated over a wide range (Table 5). This parameter does not considerably affect stereospecificity of the catalyst and molecular weight of polypentenamer, but strongly influences polymer conversion, which is maximum in the 1:1 to 1:3 range of Mo/Al molar ratios.

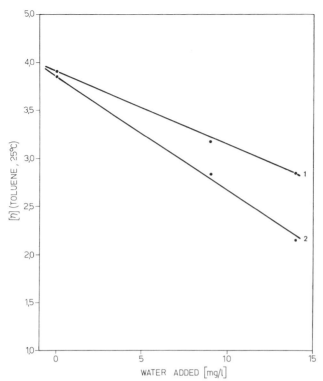

Figure 8 Molecular weight regulation by water in the polymerization of cyclopentene to *trans*-polypentenamer induced by $WCl_6(OCH_2CH_2Cl)/AlEt_2Cl$ catalyst (0.5 mmole W/100 g monomer). 1, W/Al molar ratio 1:2.8; 2, W/Al molar ratio 1:2.6. Reprinted from Günther et al. (17), by courtesy of Hüthig and Wepf Verlag, Basel.

Table 3   Polymerization of Cyclopentene to *cis*-Polypentenamer as a Function of Temperature[a,b]

Temperature (°C)	Polymer Yield (%)	Trans Double Bonds (%)	$[\eta]$ (Toluene, 30°C) (dl/g)
−80	3.3	0.6	—
−55	14.6	1.9	2.2
−40	40.8	0.7	1.7
−30	27.0	1.2	1.6
−10	13.3	2.3	2.0
+10	2.0	8.1	2.1
+30	0.3	14.6	2.7

[a] Reprinted from Dall'Asta and Motroni (82), by courtesy of Hüthig and Wepf Verlag, Basel.
[b] Bulk polymerization; time 4 hours; $MoCl_5/AlEt_3$ molar ratio 1:2.5; monomer/$MoCl_5$ molar ratio 500:1.

Table 4   Polymerization of Cyclopentene to *cis*-Polypentenamer as a Function of Catalyst Concentration[a,b]

Monomer/MoCl$_5$ Molar Ratio	Polymer Yield (%)	Trans Double Bonds (%)	$[\eta]$ (Toluene, 30°C) (dl/g)
125:1	15.3	0.6	0.8
250:1	29.8	0.5	1.0
500:1	40.8	0.7	1.7
1000:1	25.1	0.7	3.9
2000:1	12.3	0.6	4.6
4000:1	0.1	4.2	—

[a] Reprinted from Dall'Asta and Motroni (82), by courtesy of Hüthig and Wepf Verlag, Basel.
[b] Bulk polymerization; time 4 hours; temperature −40°C; MoCl$_5$/AlEt$_3$ molar ratio 1:2.5.

Aging the catalyst results in a considerable decrease of polymer conversion, but does not influence the stereoregularity (Table 6).

Figure 9 illustrates the polymer conversion versus time dependence for bulk polymerization of cyclopentene at −30 and −40°C. Polymerization rate is higher at −40°C and after 8 hours attains 50% conversion. The

Table 5   Polymerization of Cyclopentene to *cis*-Polypentenamer as a Function of the Mo/Al Ratio[a,b]

MoCl$_5$/AlEt$_3$ Molar Ratio	Polymer Yield (%)	Trans Double Bonds (%)	$[\eta]$ (Toluene, 30°C) (dl/g)
1:0.25	<1	0.9	—
1:0.5	10.7	2.0	2.3
1:1	41.4	0.6	1.8
1:1.5	49.2	0.9	1.6
1:2	45.3	0.6	1.8
1:2.5	47.2	0.7	1.4
1:3.5	7.4	0.6	1.1
1:5	3.0	1.1	1.0
1:10	2.3	1.0	1.9

[a] Reprinted from Dall'Asta and Motroni (82), by courtesy of Hüthig and Wepf Verlag, Basel.
[b] Bulk polymerization; time 4 hours; temperature −40°C; monomer/MoCl$_5$ molar ratio 500:1.

Table 6 Influence of Catalyst Aging on the Polymerization of Cyclopentene to *cis*-Polypentenamer[a,b]

Aging Time (minutes)	Polymer Yield (%)	Trans Double Bonds (%)	$[\eta]$ (Toluene, 30°C) (dl/g)
<1	36.5	1.1	1.6
40	25.4	1.1	1.5
115	16.0	1.3	1.9
300	12.5	1.1	2.2

[a] Reprinted from Dall'Asta and Motroni (82), by courtesy of Hüthig and Wepf Verlag, Basel.
[b] Monomer/toluene volume ratio 4 : 1; catalyst prepared in toluene; polymerization time 3 hours; temperature −30°C; monomer/MoCl$_5$ molar ratio 500 : 1; MoCl$_5$/AlEt$_3$ molar ratio 1 : 2.5.

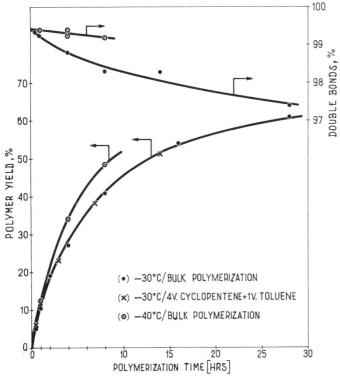

Figure 9 Dependence of polymer conversion and polymer microstructure on time in the cyclopentene polymerization to *cis*-polypentenamer in the presence of the MoCl$_5$/AlEt$_3$ catalyst (molar ratios: cyclopentene/MoCl$_5$ = 500 : 1; MoCl$_5$/AlEt$_3$ = 1 : 2.5). ●, −30°C, bulk polymerization; ×, −30°C, 4 volumes cyclopentene + 1 volume toluene; ⊙, −40°C, bulk polymerization. Reprinted from Dall'Asta and Motroni (82), by courtesy of Hüthig and Wepf Verlag, Basel.

presence of small amounts (20%) of aromatic solvent does not affect polymerization rate much. A small decrease of cis double bond content appears at $-30°C$, but essentially not at $-40°C$.

5 STRUCTURAL AND PHYSICAL PROPERTIES OF STEREOREGULAR POLYALKENAMERS

5.1 Structural Features (Ozonolysis, NMR, and Infrared Measurements)

The general behavior and chemical reactivity of polyalkenamers are in several respects similar to those of 1,4-polybutadienes. A detailed description of the reactions leading to derivatives of polyalkenamers is outside the scope of this chapter. Our considerations will be limited to the following structural features of polyalkenamers: (a) demonstration that the macromolecules are made up of alkenamer units; (b) head-to-tail arrangement of the monomeric units; (c) presence of saturated cyclic monomeric units; (d) strictly unbranched structure; and (e) determination of cis and trans double bonds.

The polyalkenamer structure of the polymers obtained from cycloolefins in the presence of the catalysts described in Section 2 was demonstrated early. Two methods were used: ozonolysis combined with NMR and/or infrared spectroscopy and x-ray studies. The latter method is described in the next section.

Ozonolysis was applied to different polyalkenamers such as *trans*-polypentenamer, *trans*-polyoctenamer, and pentenamer-octenamer copolymers. *Trans* polypentenamer, prepared in the presence of two-component tungsten catalysts, was ozonized and the ozonides oxidatively decomposed with hydrogen peroxide in CCl_4 (12). The dicarboxylic acids formed were glutaric (C_5) and smaller amounts of succinic (C_4) acid, the latter likely a product of secondary oxidation of the former. Higher membered dicarboxylic acids were absent. This demonstrates not only the presence of pentenamer units in the polymer, but also the head-to-tail succession of such units. In symmetric cycloolefins such succession is implicit in the transalkylidenation operating in ring opening polymerization (described in Section 3). Any other type of ring opening, involving cleavage of single bonds, could in principle also yield head-to-head and tail-to-tail successions, which are to be excluded on the basis of these results.

Oxidative decomposition of ozonides is generally accompanied by side reactions, which allow conversion of only part of the alkenamer units into dicarboxylic acids. In order to avoid these side reactions, reductive decomposition of ozonides with $NaBH_4$, as first described by Hubert (113) for cycloolefin monomers, was adopted in successive investigations (81, 82). For *trans*-polypentenamer, *trans*-polyoctenamer, and pentenamer–octenamer copolymers, prepared in the presence of three-component tungsten catalysts, this method yielded glycols having the same number of carbon atoms as the cycloolefins used for the polymer preparation. Moreover, reductive decomposition of partially ozonized polyalkenamers, in addition to these glycols, yielded mono-unsaturated glycols derived from two neighboring alkenamer units: $HOH_2C(CH_2)_3CH\!=\!CH(CH_2)_3CH_2OH$ (from polypentenamer), $HOH_2C(CH_2)_6CH\!=\!CH(CH_2)_6CH_2OH$ (from polyoctenamer), $HOH_2C(CH_2)_3CH\!=\!CH(CH_2)_6CH_2OH$ (in addition to the two above; from pentenamer–octenamer copolymers). These dimeric glycols are further proof of the polyalkenamer as well as of the head-to-tail structure of the polymers.

Once ascertained that the transalkylidenation mechanism is operative in ring opening polymerizations, the type of succession ceases to be a problem in the case of symmetric cycloolefins. However, for unsymmetric rings such as monosubstituted cycloolefins it still remains a problem. Monomers such as 3-methylcyclooctene may in principle undergo head-to-tail [29a] or head-to-head [29b] arrangements:

$$\longrightarrow \ =CH(CH_2)_5\overset{\overset{\displaystyle CH_3}{|}}{CH}\!-\!CH\!=\!CH(CH_2)_5\overset{\overset{\displaystyle CH_3}{|}}{CH}\!-\!CH\!= \qquad [29a]$$

$$\longrightarrow \ =CH(CH_2)_5\overset{\overset{\displaystyle CH_3}{|}}{CH}\!-\!CH\!=\!CH\!-\!\overset{\overset{\displaystyle CH_3}{|}}{CH}(CH_2)_5CH\!= \qquad [29b]$$

Dall'Asta et al. (114) recently found an essentially regular structure in ring opening polymers of 3-methylcyclooctene and their derivatives, suggesting that head-to-tail successions predominate. This is probably due to steric reasons, which favor a coordination of two monomers to the catalytic center having the substituents at opposite sides of a hypothetical line joining the two double bonds and the catalytic center and analogous to the arrangement in equation [29a].

The approximate number of alkenamer units in polyalkenamers was evaluated in early research work (12, 14, 16) by infrared spectroscopy. This research indicated a substantial alkenamer structure and about one double bond for each monomeric unit but owing to uncertainty in infrared absorption coefficients, it was only a rough estimate.

NMR spectroscopy revealed in a more quantitative manner to what extent alkenamer units are the building stones of the ring opening polymers. Calderon et al. (16, 115) showed for a series of polyalkenamers (polyoctenamer, polydodecenamer) the presence of three different ^{1}H NMR signals at 5.38, 1.95, and 1.30 ppm (from TMS) attributed to vinylene, allyl, and other methylene protons, respectively, in a ratio in agreement with a polyalkenamer structure. In the case of polybutenamers (from 1,5-cyclooctadiene or 1,5,9-cyclododecatriene) there were only two signals, attributed to vinylene and allyl protons, respectively. The absence of other methylene protons is further proof of the absence of double bond migration and of the head-to-tail structure of the polymers.

The ratios between the different types of protons indicate the essential polyalkenamer structure of the polymers and the substantial absence of saturated units, which could originate from addition or transannular polymerization of cycloolefin monomers or, more likely, from cyclization reactions. In polyalkenamers obtained from catalysts containing trialkyl-aluminum or dialkylaluminum chloride and an oxygenated activator, saturated units are present in amounts of 0–3% of the original double bonds. However, in polymers prepared in the presence of more acidic catalysts (containing alkylaluminum dichloride or aluminum trichloride as cocatalyst) such saturated units are more numerous and may even exceed 20% of the original double bonds.

Infrared and NMR spectroscopy also allowed proof of the absence of any type of side groups (alkyl groups or vinyls, except those due to possible chain ends). Analogous results were obtained from the gas-liquid chromatographic examination of low molecular weight fractions. This means that polyalkenamers originated by ring opening polymerization of cycloolefins possess a strictly unbranched structure. This fact is reflected in the rheological properties and processability of these polymers.

With regard to the quantitative determination of cis and trans double bonds in polyalkenamers by infrared analysis many contrasting results are reported in the literature. This is mainly because absorption coefficients for bands characteristic of cis and trans double bonds of 1,4-polybutadienes have frequently been applied to polyalkenamers without verifying their validity for such polymers. Moreover, it has been shown (14) that the absorption coefficients of the band at 1405 cm^{-1}, characteristic of the cis double bond, and of that at 965 cm^{-1}, characteristic of trans double bonds, vary considerably within homologous series of polymers.

A method for the quantitative determination of cis and trans double bonds and of the total amount of double bonds in a homologous series of polyalkenamers, recently proposed by Tosi and co-workers (116), uses different absorption coefficients for each polyalkenamer term for both cis and trans double bonds. Cis double bonds are determined on the basis of the infrared band at 1405 cm^{-1} using CCl$_4$ solutions of the polymers. Trans double bonds are determined on the basis of the infrared band at 965 cm^{-1} using CS$_2$ solutions. The total amount of double bonds is measured by NMR analysis.

Infrared spectroscopy was also used by Ciampelli et al. (117) for the comparison of a series of amorphous and crystalline *trans*-polyalkenamers. Several sharp bands appearing in the infrared spectra of crystalline polymers, but disappearing in the melted polymers, arise from crystallinity. Some of these bands are due to chain regularity and are found in all crystalline *trans*-polyalkenamers, independently of the crystalline modification in which the polymers crystallize. Some other bands are characteristic of a given crystalline modification. Thus, the band in the 730 cm^{-1} region, which is caused by rocking vibrations of the methylene groups, is split in the odd polyalkenamers, crystallizing in an orthorhombic lattice and in the monoclinic, but not in the triclinic modification of even *trans*-polyalkenamers. The similarity of chain packing of the orthorhombic and monoclinic modifications of these polymers and the comparably small chain-chain distance in the crystal lattice suggest that, analogously to polyethylene but unlike most of the known crystalline polymers, the latter type of crystallinity bands is due to crystalline lattice effects, i.e., to chain-chain interactions.

The infrared spectra of crystalline and amorphous *trans*-polypentenamers are given in Fig. 10.

5.2 X-Ray Examination

The results concerning the alkenamer structure of the monomeric units of polyalkenamers and their head-to-tail arrangement, obtained by ozonolysis, NMR, and infrared spectroscopy, were confirmed by x-ray spectroscopy (12, 14). Detailed studies of the crystal structure of the whole series of known *trans*-polyalkenamers were reported by Natta and co-workers (118–125). No studies concerning polymers of the cis series are known so far. This is because of the very low melting temperature of some members of the series, their slow crystallization kinetics, and, in some cases, the difficulty in obtaining highly cis-tactic polyalkenamers.

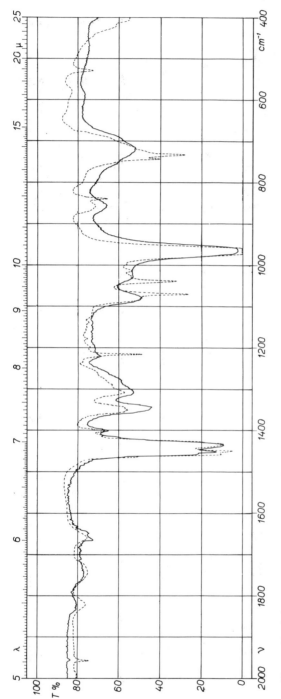

Figure 10 Infrared spectrum of *trans*-polypentenamer in the 2000–400 cm⁻¹ region (——, amorphous; –––, crystalline modification I). Reprinted from Ciampelli et al. (117), by courtesy of Pergamon Press Ltd., Oxford.

The presence of an odd or an even number of methylene groups in the recurring unit, i.e., between two subsequent double bonds of the polymer chains, introduces different symmetries into the macromolecule, which result in very different chain conformations. For this reason stereoregular polyalkenamers of the odd and the even series are treated separately.

Two members of the odd series were examined in detail: *trans*-polypentenamer and *trans*-polyheptenamer. On the basis of general conformational considerations, the existence of two different chain conformations, to which two different crystalline modifications should correspond, could be predicted for these polymers. They both contain two monomeric units in the crystallographic repeating unit. The first corresponds to an $S(2/1)m$ chain symmetry and is characterized by a twofold screw axis parallel to the chain axis, by a symmetry center on each double bond, and by a mirror plane perpendicular to the screw axis. This form has been called modification I (14). The other conformation (modification II), corresponding to a $t\ i\ c$ symmetry, is characterized by a glide plane parallel to the c axis, by a symmetry center on each double bond, and by a twofold axis normal to the glide plane.

Modification II has never been obtained with certainty, but some evidence in the infrared spectra of a particularly well-annealed sample of *trans*-polyheptenamer makes its existence quite probable. *trans*-Polypentenamer and *trans*-polyheptenamer normally crystallize in modification I. This chain conformation is shown in Fig. 11 for *trans*-polypentenamer. Such chains are packed in orthorhombic unit cells (118–120), which are similar to that of the orthorhombic modification of polyethylene (125), as illustrated in Table 7.

Only one chain conformation was foreseen for the even members of *trans*-polyalkenamers (14). It is characterized by $t\ i$ symmetry and contains a symmetry center as the only symmetry element. Only one monomeric unit is contained in the crystallographic repeating unit. Detailed examinations in the even series were reported for *trans*-polyoctenamer, *trans*-polydecenamer, and *trans*-polydodecenamer. Figure 12 shows the chain conformation of *trans*-polyoctenamer. Such polymers

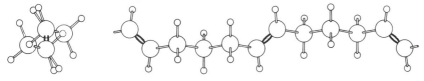

Figure 11 Chain conformation of *trans*-polypentenamer in the crystalline state (modification I) end and side views. Reprinted from Natta and Bassi (119), by courtesy of John Wiley & Sons, Inc., New York.

Table 7 Comparison Between the Unit Cell Parameters of Odd *trans*-Polyalkenamers and of Orthorhombic Polyethylene[a]

	trans-Polypentenamer	*trans*-Polyheptenamer	Orthorhombic Polyethylene
a (Å)	7.28 ± 0.1	7.40 ± 0.1	7.40 ± 0.1
b (Å)	4.97 ± 0.05	5.00 ± 0.05	4.93 ± 0.05
c (Å)	11.90 ± 0.1	17.10 ± 0.15	2.53 ± 0.02
	(chain axis)	(chain axis)	(chain axis)
Z	$4(C_5H_8)$	$4(C_7H_{12})$	$4(CH_2)$
dx (g/cm^3)	1.05	1.01	1.01
Space groups	D_{2h}^{16} or C_{2v}^{9}	D_{2h}^{16} or C_{2v}^{9}	D_{2h}^{16} or C_{2v}^{9}

[a] Reprinted from Natta et al. (125), by courtesy of Pergamon Press Ltd., Oxford.

show polymorphism. Two main modifications were reported, differing in the space packing of the polymer chain.

The first, called modification III (14), is characterized by triclinic unit cells (123, 124). The space packing of the chains is considerably different from that of the orthorhombic unit cells described above; this difference is reflected in the different solid state behaviors (as is illustrated in Section 5.4). Table 8 reports a comparison of the three known members of this series (125).

The second, called modification IV (14), is characterized by monoclinic unit cells (121, 122). The space packing of the chains in this case is quite similar to that of the orthorhombic odd polyalkenamer lattices and hence to polyethylene. This similarity is reflected in some solid state properties of the polymers (vide infra). A comparison (125) of the monoclinic unit cells of *trans*-polyoctenamer, *trans*-polydecenamer, and *trans*-poly-dodecenamer is made in Table 9.

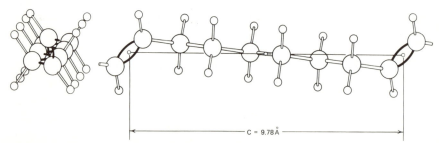

C = 9.78 Å

Figure 12 Chain conformation of *trans*-polyoctenamer in the crystalline state, end and side views. Reprinted from Bassi and Fagherazzi (123), by courtesy of Pergamon Press Ltd., Oxford.

Table 8   Comparison Between the Unit Cell Parameters of Monoclinic Even trans-Polyalkenamers[a]

	trans- Polyoctenamer	*trans* Polydecenamer	*trans-* Polydodecenamer
$a \sin \beta$ (Å)	7.40 ± 0.05	7.40 ± 0.05	7.40 ± 0.05
b (Å)	5.00 ± 0.05	5.00 ± 0.05	5.00 ± 0.05
a (Å)	7.43 ± 0.05	7.42 ± 0.05	7.43 ± 0.10
c (Å)	9.90 ± 0.10	12.40 ± 0.10	14.85 ± 0.15
	(chain axis)	(chain axis)	(chain axis)
β	$95°10' \pm 15'$	$94°10' \pm 15'$	$93°30' \pm 15'$
Z	$2(C_8H_{14})$	$2(C_{10}H_{18})$	$2(C_{12}H_{22})$
dx (g/cm³)	1.00	1.00	1.00
Space group	$P2_1/a$	$P2_1/a$	$P2_1/a$

[a] Reprinted from Natta et al. (125), by courtesy of Pergamon Press Ltd., Oxford.

Modification IV normally prevails in the upper members of the even series and modification III in the lower ones. The conversion of one into the other is generally achieved by treatment with particular solvents. Evidence for further crystalline modifications of the even series has been found (126).

Stereoregular *trans*-polyalkenamers are generally characterized by high

Table 9   Comparison Between the Unit Cell Parameters of Triclinic Even trans-Polyalkenamers[a]

	trans- Polyoctenamer	*trans-* Polydecenamer	*trans-* Polydodecenamer
$a \sin \beta$ (Å)	4.20 ± 0.05	4.26 ± 0.05	4.26 ± 0.05
$b \sin \alpha$ (Å)	4.88 ± 0.05	4.91 ± 0.05	4.91 ± 0.05
a (Å)	4.34 ± 0.05	4.40 ± 0.05	4.40 ± 0.05
b (Å)	5.41 ± 0.05	5.39 ± 0.05	5.39 ± 0.05
c (Å)	9.78 ± 0.1	12.30 ± 0.1	14.78 ± 0.1
α	$64°25' \pm 25'$	$65°30' \pm 25'$	$65°30' \pm 25'$
β	$104°50' \pm 25'$	$104°30' \pm 25'$	$104°30' \pm 25'$
γ	$118°35' \pm 25'$	$118°25' \pm 25'$	$118°25' \pm 25'$
Z	$1(C_8H_{14})$	$1(C_{10}H_{18})$	$1(C_{12}H_{22})$
dx (g/cm³)	1.00	0.98	0.99
Space group	$P\bar{1}$	$P\bar{1}$	$P\bar{1}$

[a] Reprinted from Natta et al. (125), by courtesy of Pergamon Press Ltd., Oxford.

three-dimensional order of the crystal lattice, as indicated by the appearance of numerous reflections in the equatorial and layer lines of well-oriented fiber spectra. Copolyalkenamers, prepared by random copolymerization of different odd- and even-membered cycloolefins (22); are also crystalline. Actually, alkenamer units of different length may isomorphically replace themselves in the crystalline lattice (127). However, such isomorphous systems, if compared to the corresponding crystalline homopolymers, exhibit lower melting points, slower crystallization rate, and only bidimensional order in x-ray fiber spectra, as indicated by the presence of sharp equatorial reflections but diffused intensities in the layer lines (127).

5.3 Solution Behavior and Molecular Weights

Stereoregular cis- and trans-polyalkenamers are generally soluble in a number of organic solvents. However, the solubility decreases with increasing degree of crystallinity and size of the monomeric unit, i.e., the length of the methylene sequence. Cis- and trans-tactic polypentenamers are dissolved, at room temperature, by aliphatic, cycloaliphatic, aromatic, and chlorinated hydrocarbons, but are insoluble in low boiling alcohols and ketones (12, 128). Acetone dissolves low molecular weight fractions.

Crystalline trans-polyheptenamer, trans-polyoctenamer, and trans-polydecenamer are generally soluble in cycloaliphatic, aromatic and chlorinated hydrocarbons, but poorly soluble or insoluble in n-heptane, dioxane, tetrahydrofuran, diethyl ether, and anisole, in addition to low boiling alcohols and ketones (14). A 1:1 n-hexane/2-propanol mixture extracts low molecular weight fractions (26). Highly crystalline trans-polydodecenamer is generally insoluble in these solvents, but is dissolved by cycloaliphatic, aromatic, and chlorinated solvents above 50–60°C.

Molecular weight distribution in trans-polypentenamers, prepared by solution polymerization of cyclopentene in the presence of three-component tungsten based Ziegler-Natta systems, is sufficiently broad to favor good processability. According to Pampus et al. (96), molecular weight distribution of such polymers, as expressed by changing solution viscosity, increases to a certain extent with increasing polymer conversion (Fig. 13). This was attributed by the authors to the fact that very high molecular weight polymers, formed in the initial states of polymerization, are partially cleaved to lower molecular weights during progress of the. polymerization.

In the case of cis-polypentenamer, molecular weight fractionation was achieved (128) in a thermal gradient elution column using a proper benzene/isopropanol mixture. By comparing the $[\eta]/\bar{M}_n$ ratio of cis- and

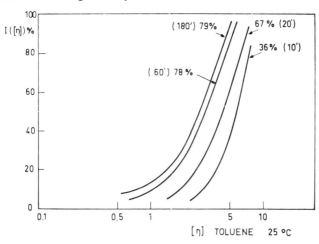

Figure 13 Molecular weight distribution of *trans*-polypentenamer obtained in solution polymerization. (The values on the curves indicate the percent conversion and the relative polymerization time.) Reprinted from Pampus et al. (96), by courtesy of SETCO, Paris.

trans-polypentenamer fractions of approximately the same molecular weight ($\bar{M}_n$, osmometric; $[\eta]$ in toluene at 30°C), Dall'Asta and co-workers (129) observed considerably higher $[\eta]/\bar{M}_n$ ratios for *trans*- than for *cis*-polypentenamer. This behavior may be attributed to a somethat higher stiffness of the *trans*-polypentenamer as compared to the *cis*-polypentenamer chain, caused by different polymer-solvent interaction or by reduced conformational changes around the C–C single bond α to the trans double bond.

The higher ratios between Mooney viscosity and solution viscosity of *trans*-polypentenamer, compared to those of *cis*-1,4-polybutadiene (17), may also be caused by this supposed higher stiffness.

5.4 Solid State Properties

Second-order transition temperatures of *cis*- and *trans*-polypentenamer were measured by different techniques. For *cis*-polypentenamer, T_g values of −105°C (shear modulus), −114°C (DTA), and −135°C (flexural modulus, forced vibrations) were reported (128). The differences may be ascribed, at least in part, to the different methods used and to the fact that in one case (DTA) the sample was crystallized and in the other it was not. For *trans*-polypentenamer the T_g values of −97°C (82) and −90°C (17), both measured by DTA, were reported.

For both stereoregular polypentenamers the second-order transition temperatures are far below what is required for practical rubbers. The T_g value of *cis*-polypentenamer, exceptionally low even compared to other rubbers (Chapter 9), suggests good low temperature performance for this polymer.

Second-order transition temperatures were not reported for higher membered polyalkenamers. However, the NMR line width as a function of temperature showed, for a series of *trans*-polyalkenamers (polypentenamer, polyheptenamer, polyoctenamer. polydecenamer, polydodecenamer), characteristic transitions, which are intermediate between those of polypentenamer and linear polyethylene (130). Since the T_g of the latter is presumed to be below −100°C, this suggests that the T_g values of *trans*-polyalkenamers higher than polypentenamer are in the −100°C ·region.

A first-order transition temperature (T_m) of *cis*-polypentenamer (99% cis double bonds) was found (128) at −41°C, and for *trans*-polypentenamer (85% trans double bonds) at +23°C (dilatometric) (12) or +18°C (DTA) (82).

The two polymers differ considerably not only in melting temperature but also in crystallization rate. *trans*-Polypentenamer crystallizes readily when cooled below 0°C or even at room temperature if stretched, whereas *cis*-polypentenamer crystallizes only after prolonged annealing at −75°C (128).

Cis unsaturated polymers generally exhibit considerably lower T_m values than the corresponding trans isomers as with other known stereoelastomers such as 1,4-polybutadiene, 1,4-polyisoprene, isotactic 1,4-polypentadienes, and 1,4-poly-2-chlorobutadienes. In the case of the T_g values this difference is much smaller. The T_g/T_m ratios (in degrees Kelvin) of *cis*- as well as *trans*-polypentenamer follow the general rule of Boyer; i.e., they range from 0.6 to 0.7 (Table 10).

The comparison of T_g and T_m values of the two stereoregular polyalkenamers with those of the corresponding stereoregular 1,4-polybutadienes shows (Table 10) considerably lower values for the former polymers. Considering the analogous structure of the two classes of polymers, these differences may be ascribed to the different chain symmetry and crystal structure, caused by the supplementary methylene group per monomeric unit or to the complete absence of side vinyl groups in the polypentenamers.

Very low glass transition temperature and melting temperature in the range of room temperature are both features exceptionally favorable to the elastomeric properties of *trans*-polypentenamer. The crystallization rate at different temperatures is also an important aspect. Unless

Table 10   Comparison of the T_g, T_m, and T_g/T_m Values of Polypentenamers and 1,4-Polybutadienes (Differential Thermal Analysis Values)

	Polypentenamers			1,4-Polybutadienes		
	T_g (°C)	T_m (°C)	T_g/T_m (°K)	T_g (°C)	T_m (°C)	T_g/T_m (°K)
Cis	−114	−41	0.69	−95	+2	0.63
Trans	−97[a]	+18	0.61	−80	+76[b]	0.55

[a] Polypentenamer containing ≥85% trans double bonds.
[b] Isophasic transition from modification I to II.

stretched, *trans*-polypentenamer does not crystallize at room temperature within reasonable times. At 0°C, on the contrary, pure *trans*-polypentenamer crystallizes readily in the unstretched state. This property improves building tack of rubber mixes and causes self-reinforcement, which results in high gum green strength and consequently in good processability, as discussed in Section 6. On the other hand, a too strong crystallization trend at temperatures above 30°C is not desired for an elastomer, because this would impair elastomeric properties. However, the rate of crystallization of *trans*-polypentenamer, which is much higher than that of any other general purpose elastomer in the −10 to +10°C range, may be regulated by varying the trans content, as shown by Haas and Theisen (131). Table 11 reports half-times of crystallization for some general purpose rubbers and for *trans*-polypentenamers having different trans contents.

Table 11   Half-Times of Crystallization of Some Elastomers[a]

Elastomer	$t_{1/2}$ at 0°C (hours)
cis-1,4-Polybutadiene (>95% cis)	Several weeks
cis-1,4-Polyisoprene (>98% cis)	≫100
Natural rubber (masticated)	50
trans-Polypentenamer (93% trans)	0.3
(90% trans)	0.8
(89% trans)	13
(87% trans)	45

[a] Reprinted from Haas and Theisen (131), by courtesy of Verlag für Radio-Foto-Kinotechnik GMBH, Berlin.

Table 12   Parameters of Crystallization Kinetics of
***trans*-Polypentenamer[a,b]**

Crystallization Temperature (°C)	$t_{1/2}$ (minutes)	n^c	K^c
0	20	2.22	8.1×10^{-4}
2	36	2.36	1.5×10^{-4}
4	71	2.63	8.9×10^{-6}
6	125	2.48	4.0×10^{-6}
8	300	2.52	3.8×10^{-7}
10	710	2.52	4.4×10^{-8}
12	1740	2.48	7.1×10^{-9}
14	4270	2.27	4.0×10^{-9}

[a] Reprinted from Capizzi and Gianotti (132), by courtesy of Hüthig and Wepf Verlag, Basel.
[b] Polypentenamer having 85% trans double bonds.
[c] According to the equation $(V_\infty - V_t)/(V_\infty - V_0) = \exp(-Kt^n)$.

Kinetics of crystallization at different temperatures was studied by Capizzi and Gianotti (132) dilatometrically on a high *trans*-polypentenamer that contained, according to an improved infrared analysis (116), 85% trans and 12% cis double bonds and had an $\bar{M}_n$ of 96,000. Half-times of crystallization at different temperatures in the 0 to +14°C range, as well as K and n values for the Avrami (133) equation as modified by Mandelkern, are reported in Table 12.

Capizzi and Gianotti (132) also studied the dependence of melting temperature (T_m) on crystallization temperature (T_x) to determine the thermodynamic melting temperature according to the method proposed by Hoffman and Weeks (134). The thermodynamic melting temperature, obtained by the intercept of the T_m and T_x straight lines of the T_m/T_x diagram, is 34°C.

By using the cryoscopic method proposed by Flory (135) with three different diluents, Gianotti (132) also determined the fusion enthalpy and entropy values of the same *trans*-polypentenamer sample. A mean value of $\Delta H_u = 2870$ cal/mole resulted, which allowed the calculation of a degree of crystallinity of about 25%. Fusion entropy ΔS_u of 9.87 or 9.36 eu, corresponding to 2.47 or 2.34 eu per C–C single bond, was found depending on whether the experimental or the thermodynamic melting temperature was used.

The melting temperatures of higher membered *trans*-polyalkenamers increase on increasing the length of the methylene sequence of the monomeric units, as shown by Natta et al. (14). The authors, plotting the crystallographically determined T_m values of a series of polyalkenamers containing from 85 to 94% trans double bonds (Table 13), versus $1/N$ (N, number of methylene units in the methylene sequence of the monomeric unit), showed an approximately linear dependence for the series going from *trans*-polypentenamer through the higher membered *trans*-polyalkenamers, to polyethylene. However, this relationship is valid only for polymer modifications in which the chains are packed similarly to those in orthorhombic polyethylene, i.e., in odd *trans*-polyalkenamers (orthorhombic unit cells) and in even *trans*-polyalkenamers crystallizing in modification IV (monoclinic unit cells). Even *trans*-polyalkenamers crystallizing in triclinic unit cells (modification III) have quite different chain packing and exhibit higher T_m values.

Thermodynamic quantities of *trans*-polyoctenamer result from a study by Calderon and Morris (115), who investigated the dependence of T_m on the trans content. Surprisingly, polyoctenamers having trans contents down to 40% were crystallizable. From the extrapolation of the T_m/trans content plot and from the Flory plot for which the cis unsaturations are

Table 13 Crystallographic Melting Temperatures of a Homologous Series of *trans*-Polyalkenamers

trans-Polyalkenamer[a]	n, in —CH═CH—(CH$_2$—)$_n$	Crystalline Modification	Melting Temperatures (°C)
Polyethylene	∞	I: orthorhombic	130
Polydodecenamer	10	III: triclinic	?
		IV: monoclinic	80
Polyoctenamer	6	III: triclinic	67
		IV: monoclinic	62
Polyheptenamer	5	I: orthorhombic	51
Polyhexenamer[b]	4	III: triclinic	61
		IV: monoclinic	41
Polypentenamer	3	I: orthorhombic	23[c]
Polybutenamer[d]	2	I: monoclinic	76[e]
		II: pseudohexagonal	145

[a] The polyalkenamers having $n = 3$–10 show a trans double bond content of 85–94%.

[b] alternating 1,4-butadiene-ethylene copolymer.

[c] dilatomeric value.

[d] 1,4-Polybutadiene.

[e] Isophasic transition from modification I to modification II.

considered to be randomly distributed and noncrystallizable, a crystalline melting temperature of 73°C, a heat of fusion of 3520 cal/mole, and an entropy of fusion of 10.2 cal/mole·°K were deduced for a 100% *trans*-polyoctenamer. Diluent method estimates made earlier by Flory (135) were, respectively, 60°C, 4800 cal/mole, and 14.4 cal/mole·°K.

An analogous approach for the determination of thermodynamic solid state properties of a series of *trans*-polyalkenamers was used by Gianotti and Capizzi (136). The melting point depression method of Flory, in the presence of α-chloronaphthalene as diluent, applied to polyoctenamers, polydecenamers, and polydodecenamers containing 43–80, 55–83, and 72–89% of trans double bonds, respectively, allowed estimates of T_m values for 100% *trans*-polyalkenamers. The figures, along with those of enthalpy and entropy of fusion, are compared in Table 14 with those of polyethylene and *trans*-1,4-polybutadiene.

The crystallization rates of polyoctenamers having different trans contents were given by Calderon (15). Trans contents of 78, 74, and 41% correspond to half-times of crystallization, at 17.9°C below the melting point, of 16, 32, and 3000 minutes, respectively.

A morphological characterization of dilute solution grown single crystals of polydecenamer and polydodecenamer was carried out by Keller and Martuscelli (137). In spite of the macrocyclic structure of part of the macromolecules, chain folded lamellar single crystals were found. The fold length could be varied as a function of crystallization temperature and trans content. Incorporation of at least some of the cis double bonds in the crystal interior, with formation of lattice defects, could be deduced.

Keller and Martuscelli (138) also examined the behavior of single crystals of *trans*-polydecenamer and *trans*-polydodecenamer under oxidative attack by ozone and, to a lesser extent, by nitric acid, by following the changes in the concentration of double bonds and the distribution of

Table 14 Extrapolated Thermodynamic Parameters of Fusion of Some *trans*-Polyalkenamers[a]

Polymer	Crystalline Modification	T_m (°K)	ΔH_u (cal/mole)	ΔS_u (cal/mole · deg)	ΔS/bond (cal/mole · deg)
Polyethylene	Orthorhombic	411	1920	4.68	2.34
Polydodecenamer	Monoclinic	357	9840	27.60	2.50
Polydecenamer	Monoclinic	353	7850	22.20	2.47
Polyoctenamer	Triclinic	350	5680	16.20	2.32
1,4-Polybutadiene (I)	Monoclinic	370[b]	3300	8.90	2.90

[a] Reprinted from Gianotti and Capizzi (136), by courtesy of Pergamon Press Ltd., Oxford.
[b] Extrapolated melting temperature of crystalline modification I.

molecular length. Double bonds were attacked deep within the crystals without a clear distinction between fold surface and crystal interior. Nevertheless, the attack does not follow a random pattern, but leads to a crystalline dimeric residue, highly resistant in spite of the double bond. This result is in certain aspects similar to that reported by Dall'Asta and Motroni (81, 82) for the solution ozonization of octenamer–pentenamer copolymers, where considerable quantities of unsaturated dimeric species appeared.

6 POLYALKENAMER ELASTOMERS

6.1 Preliminary Considerations

Even though 1,4-polybutadienes may be considered as polyalkenamers, only polymers originating by ring opening polymerization of cycloolefins are described here. All unsubstituted polyalkenamers, independent of the length of the methylene sequence and of the type of double bonds, are potential elastomers. Actually, the single bonds in the position α to the double bond, the presence of the methylene sequence, and the complete absence of side groups are elements that highly favor chain flexibility. Moreover, the low glass transition temperatures do not impair the elastomeric properties.

However, two main factors limit the number of polyalkenamers having good elastomeric properties: crystallization temperature and rate of crystallization. As pointed out in the preceding section, an excellent compromise between these two factors is met in *trans*-polypentenamer. High *cis*-polypentenamer, on the contrary, owing to its exceedingly low crystallization temperature and very low rate of crystallization, has poor gum properties in the temperature range of conventional rubber applications, but shows interesting elastomeric properties at low temperatures.

trans-Polyalkenamers higher than polypentenamer combine a high rate of crystallization with melting temperatures considerably higher than room temperature. For this reason they exhibit properties intermediate between those of a plastic and those of an elastomer. However, by substituting a certain amount of trans with cis double bonds, melting temperature and rate of crystallization are lowered. A compromise between these two factors can lead to good elastomeric properties in the room temperature range, provided that the length of the methylene sequences and the cis/trans ratio are well balanced. Use of copolymers of different cycloolefins also results in reduced melting temperatures and rate of crystallization. However, the use of higher polyalkenamers as elastomers is only at an early state of evolution; thus our considerations on elastomers will primarily concern stereoregular *trans*- and *cis*-polypentenamers. The preparation and properties of the elastomeric polyalkenamers were recently reviewed by Dall'Asta (139).

6.2   *trans*-Polypentenamer Elastomers

The elastomeric properties of *trans*-polypentenamer and its use in sulfur vulcanized rubber were first described and claimed by Natta, Dall'Asta, and Mazzanti (11, 12). In more recent times extensive studies on *trans*-polypentenamer elastomers have been reported by some authors at Farbenfabriken Bayer (17, 131, 140).

Stereoregular *trans*-polypentenamer elastomers have many outstanding properties, the most important of which are (*a*) high green strength of the pure gum and even in highly oil extended mixes; (*b*) excellent dispersion of vulcanization agents and fillers; (*c*) excellent processability; (*d*) high extendibility with oil and black; (*e*) exceptionally high cross-linking yield; (*f*) high mechanical properties of the vulcanizates; (*g*) comparatively high aging resistance; and (*h*) high abrasion resistance.

Disadvantages as a general purpose rubber and for tire applications are fewer: (*a*) hardening at temperatures below $-20°C$; (*b*) low road holding characteristics; and (*c*) low skid resistance.

trans-Polypentenamer and natural rubber are the only general purpose elastomers for which the highly desired compromise between low crystallinity in the unstretched state and crystallization under stretch is realized. As a result, *trans*-polypentenamer exhibits high green strength and high tack, presumably caused by self-reinforcement due to ready crystallization under mechanical deformation. As shown by these authors (17, 131), green strength of unvulcanized, black filled tire treads of *trans*-polypentenamer is as high as that of natural rubber (Fig. 14). The high reinforcement is also retained in high black loaded vulcanizates. That self-reinforcement is not only due to the presence of black is shown (12) by the curve of unreinforced sulfur vulcanizate (Fig. 15). Even under conditions at which natural rubber loses its reinforcement characteristics, i.e., in the oil extended state, *trans*-polypentenamer still maintains high green strength and tack.

The tack of carcass stocks of *trans*-polypentenamer (TP) (loaded with 30 phr of GPF black) at room temperature even exceeds the high levels characteristic of natural rubber (Table 15), as shown by Haas and Theisen (131). Different methods were proposed by these authors to avoid exceedingly high rates of crystallization in the vulcanized rubber, which

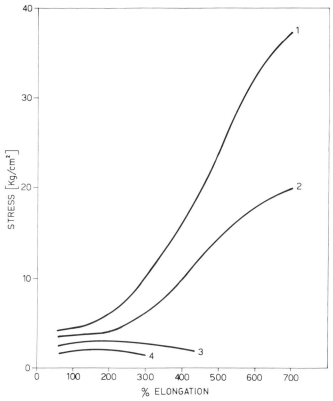

Figure 14 Green strength of uncured 50 phr HAF black loaded tire tread stocks (at 23°C): 1, *trans*-polypentenamer (TP); 2, natural rubber (NR); 3, SBR 1500; 4, *cis*-1,4 polybutadiene (Ti-BR 1220). Reprinted from Haas and Theisen (131), by courtesy of Verlag für Radio-Foto-Kinotechnik GMBH, Berlin.

would prejudice the elastomeric properties, especially at temperatures below −20°C:

1. Adjustment of the trans content during synthesis to slightly below the value usually obtained.
2. Oil extension, especially with crystallization hindering oils.
3. Copolymerization with small amounts of a different cycloolefin.
4. Induction of controlled isomerization of trans to cis double bonds during sulfur cure. This method appears the most elegant one, since it allows retention of the maximum trans content, and hence high green strength and tack, in the unvulcanized elastomer, and lowers it only during vulcanization, i.e., after processing. Such controlled isomerization

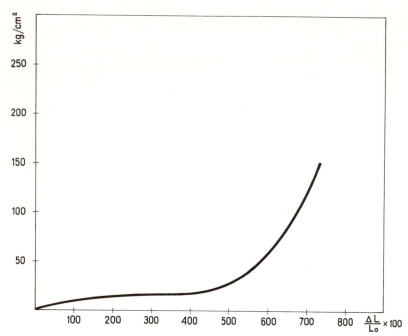

Figure 15 Stress-strain diagram for unreinforced sulfur vulcanized *trans*-polypentenamer (ML_{1+4} = 104, at 100°C).

Table 15 Tack of Carcass Stocks of Different Rubbers[a]

Rubber[b]	Separation Strength[c] (g)
NR	980
IR	850
EN-EPDM	200
BR 1220	350
SBR 1500	600
TP	>1000

[a] Reprinted from Haas and Theisen (131), by courtesy of Verlag für Radio-Foto-Kinotechnik GMBH, Berlin.
[b] Compounded with 30 phr GPF black.
[c] Wallace tackmeter test; storage 2 hours at 23°C, 50% relative air humidity; sheet thickness 2 mm; contact under 540 g pressure; time of contact 10 seconds (Bayer method).

is achieved (17) in the sulfur cure in the presence of zinc stearate, but does not take place if the sulfur cure is carried out with zinc oxide plus stearic acid as activators.

trans-Polypentenamer shows excellent processability both on the open mill and in the Banbury. The presence of low molecular weight fractions favors the ready formation of a smooth bank. Higher molecular weight fractions prevent cold flow and increase the shear gradient during mixing. This allows rapid incorporation and homogeneous distribution of ingredients and fillers. As shown by the Mooney/temperature plot of Fig. 16, taken from Günther et al. (17), *trans*-polypentenamer at normal processing temperatures (80–140°C) retains almost the same high Mooney viscosity as at conventional extruding temperatures (60–80°C). This confers on the extrusion mixes the high dimensional stability required for the buildup of tire carcasses and ensures the steep shear gradient that favors optimum dispersion of high levels of ingredients and fillers. The exceptionally high extendibility of *trans*-polypentenamer rubbers is favored by the linearity of the polymer chains (absence of side groups) and by a broad molecular weight distribution.

The extrusion behavior of *trans*-polypentenamer, as indicated by the Garvey Die extrusion indices, is better than that of *cis*-1,4-polybutadiene rubber and close to that of SBR.

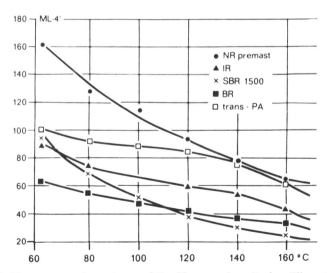

Figure 16 Temperature dependence of the Mooney viscosity for different elastomers. □, *trans*-Polypentenamer; ▲, *cis*-1,4-polyisoprene; ●, natural rubber (masticated); ×, SBR 1500; ■, *cis*-1,4-polybutadiene. Reprinted from Günther et al. (17), by courtesy of Hüthig and Wepf Verlag, Basel.

Table 16 Behavior of *trans*-Polypentenamer and Other General Purpose Rubbers in Sulfur Vulcanized Tire Treads[a]

Recipe	NR	IR	NR	SBR 1500	BR	SBR/BR 1:1	TP
Polymer	100	100	100	100	100	100	100
ISAF black	45	45	75	75.	75	75	75
Aromatic oil	5	4.5	35	35	35	35	45
N-Cyclohexyl-2-benzothiazyl-sulfenamide (Vulcacit CZ)	0.5	0.5	0.6	1.4	1.3	1.3	0.6
Sulfur	2.5	2.5	2.5	1.8	1.8	1.8	1.8
Stearic acid	3	3	3	3	3	3	3
Zinc oxide	3	3	3	3	3	3	3
p-Phenylenediamine	1	1	1	1	1	1	1
Phenyl-α-naphthylamine	1	1	1	1	1	1	1
Properties of Uncured Stock							
ML_4(100°C)	83	92	58	65	86	—	102
Green strength (kg/cm²)	21	—	9	5	3	3	20
Elongation at break (%)	610	—	900	510	400	470	700

Properties of Vulcanizate:
  Cure 30 minutes at 145°C

Tensile strength (kg/cm²)	245	235	165	170	145	165	200
M_{300} (kg/cm²)	105	95	95	95	80	90	115
Rebound at 75°C (%)	53	51	32	32	48	40	47
Compression set (70 hours, 100°C)%	56	51	66	55	49	52	44
Heat buildup (°C)[b]	21	—	Destroyed	39	58	38	30
Abrasion (mm³)	131	160	177	155	70	110	105
Skid test	52	—	55	55	41	49	48
Tear strength (Pohle) (kg/4 mm)	50	—	25	26	30	23	25

[a] Reprinted from Günther et al. (17) and Haas and Theisen (131), by courtesy of Hüthig and Wepf Verlag, Basel, and Verlag für Radio-Foto-Kinotechnik GMBH, Berlin, respectively.
[b] Goodrich Flexometer (stroke 0.0875 in.; 10 kg; 24 rpm; 100°C).

Table 17 Influence of Plasticizers on Low Temperature Crystallization of High *trans*-Polypentenamer and Other Elastomers[a]

Recipe	1	2	3	4	5	6
Natural rubber (RSS1)	100	—	—	—	—	—
cis-1,4-Polybutadiene (Buna CB 11)	—	100	—	—	—	—
High *trans*-polypentenamer (TP)	—	—	100	100	—	—
Oil extended TP:						
50 parts oil–100 parts TP	—	—	—	—	150	—
100 parts oil–100 parts TP	—	—	—	—	—	200
Stearic acid	3	3	3	3	3	4
Zinc oxide	3	3	3	3	3	4
ISAF black	50	50	50	75	85	120
Aromatic plasticizer	5	10	10	42	3	—
p-Phenylenediamine	1	1	1	1	1	1.3
Phenyl-α-naphthylamine	1	1	1	1	1	1.3
Sulfenamide (Vulcacite CZ)	0.5	1.1	0.45	0.5	0.7	1.2
Sulfur	2.3	1.6	1.6	1.6	2.2	2.4

Properties of Uncured Stock

Tack (Wallace) at 20°C, $p = 0.55$ kg		>1000	350	>1000	>1000	>1000	600
Green strength (kg/cm²)		16.5	3.9	14.0	9.5	13	7

Vulcanizate Properties

Tensile strength (kg/cm²)		220	110	175	135	145	145
Elongation at break (%)		535	505	505	550	460	440
Abrasion (mm³) (Emery No. 60, DIN 53 516)		89	20	23	21	32	44
Shore A hardness at −5°C at time (hours):	0	64	64	66	62	62	65
	0.5	68	67	69	67	66	70
	1	68	67	75	67	68	70
	4	68	67	86	68	72	70
	24	68	67	88	69	72	71
$\Delta(H_{24} - H_0)$		4	3	22	7	10	6
Shore A hardness at −20°C at time (hours):	0	64	64	66	62	62	65
	0.3	73	69	96	75	68	75
	1	74	70	97	80	72	78
	2	74	70	97	86	76	78
	5	74	70	97	86	79	78
$\Delta(H_5 - H_0)$		10	6	31	24	17	13

[a] Reprinted from Haas et al. (140).

trans-Polypentenamer is readily vulcanized with all conventional vulcanization agents used for highly unsaturated elastomers, especially those based on sulfur and accelerators or sulfur donors and accelerators (11, 12, 131, 140). High cure rate is obtained in the sulfur cure at conventional temperatures with comparably low sulfur and accelerator levels. Even at high temperatures at prolonged times little or no reversion is observed.

A set of mechanical and dynamic properties of sulfur vulcanized *trans*-polypentenamer tire treads (17, 131) is compared in Table 16 with those of conventional rubbers. Unlike natural rubber, *trans*-polypentenamer attains optimum characteristics when highly black and oil loaded. Tensile properties, moduli, and hardness are excellent even at these high extension levels.

The ability to vary vulcanizate characteristics of *trans*-polypentenamer tire tread stocks by varying vulcanization parameters has been described in detail by Haas and Theisen (131). They showed how, by adopting high vulcanization temperatures (170°C), short vulcanization times, low zinc oxide, stearic acid, and sulfenamide accelerator, and medium (2.0 phr) sulfur levels, excellent mechanical properties can be achieved along with good abrasion resistance.

The reduction of low temperature crystallization, a drawback in *trans*-polypentenamer vulcanizates, was examined by Haas and co-workers (140). Table 17 shows how, by employing high levels of aromatic or naphthenic oils in addition to black, the crystallization trend of *trans*-polypentenamer vulcanizates, as expressed by the increase in hardness at −5 or −20°C during storage, may be reduced to levels comparable to those of natural rubber, without impairing the elastomeric properties.

According to unpublished results from our laboratory, aging resistance of *trans*-polypentenamer vulcanizates at 100°C is better than that of *cis*-1,4-polybutadiene and close to that of SBR vulcanizates. Analogously, *trans*-polypentenamer exhibits considerable stability to ultraviolet irradiation, which results merely in cis-trans isomerization of the double bonds but, unlike 1,4-polybutadienes, in no appreciable chain scission (141). This has been attributed to the fact that the rupture of a CH_2–CH_2 bond yields two allylic radicals in polybutadienes, but only one in polypentenamer. This confers higher str ..gth to this bond, which is the most vulnerable in polybutadienes.

6.3 *cis*-Polypentenamer Elastomers

cis-Polypentenamer elastomers were investigated by Dall'Asta and Scaglione (128). The absence of crystallization, even under high elongation, in the temperature range of conventional applications, disfavors the

Table 18   Physical Properties of All *cis*-Polypentenamer Vulcanizates at Different Temperatures[a]

Property	Temperature[b] (°C)				
	+23	−20	−50	−70	−90
Tensile strength (kg/cm²)	168	148	225	284	392
Elongation at break (%)	500	420	490	490	495
100% modulus (kg/cm²)	24	25	32	37	61
200% modulus (kg/cm²)	48	53	64	79	126
300% modulus (kg/cm²)	88	93	113	142	213
Tear strength (kg/cm)	51	50	53	82	133

[a] Reprinted from Dall'Asta and Scaglione (128), by courtesy of the Division of Rubber Chemistry, Inc., American Chemical Society, Lancaster, Pennsylvania.
[b] Recipe: polymer, 100; PBNA, 1.5; stearic acid, 2; zinc oxide, 5; ISAF black, 50; N-cyclohexyl-2-benzothiazylsulfenamide, 0.8; sulfur, 1.0; cure 40 minutes at 150°C.

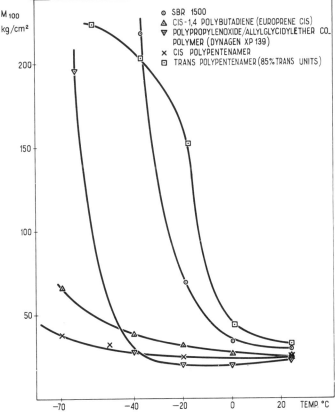

Figure 17   Modulus (100%) versus temperature curves of different elastomers. Reprinted from Dall'Asta and Scaglione (128), by courtesy of the Division of Rubber Chemistry, Inc., American Chemical Society, Lancaster, Pennsylvania.

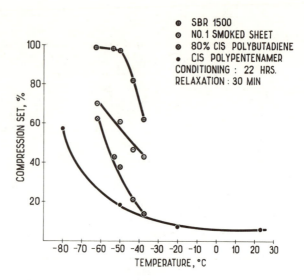

Figure 18 Low temperature compression set of different elastomers. Reprinted from Dall'Asta and Motroni (82), by courtesy of Hüthig and Wepf Verlag, Basel.

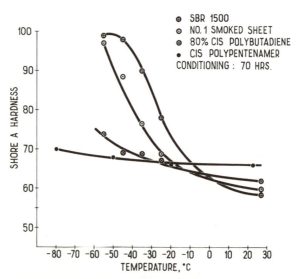

Figure 19 Shore A hardness versus the temperature for different elastomers. Reprinted from Dall'Asta and Motroni (82), by courtesy of Hüthig and Wepf Verlag, Basel.

358

mechanical characteristics of *cis*-polypentenamer. Thus, pure gum as well as black and oil extended vulcanizates exhibit markedly lower characteristics compared to those of *trans*-polypentenamer. However, optimum properties and high rate of cure without considerable inversion are achieved for *cis*-polypentenamer at even lower sulfur and accelerator levels than for *trans*-polypentenamer.

The very low glass and melting temperatures, unlike the slow crystallization rates, favor good low temperature characteristics of *cis*-polypentenamer elastomers. As shown in Table 18, carbon black reinforced sulfur vulcanizates reveal a dramatic increase of tensile strength and moduli at temperatures below −50°C, while elongation remains constant.

Reinforced *cis*-polypentenamer vulcanizates exhibit much better low temperature properties than typical low temperature rubbers such as polypropylene oxide/allyl glycidyl ether copolymer or even *cis*-1,4-polybutadiene, as illustrated by the 100% modulus/temperature plot (Fig. 17) and by the comparison of the temperatures of retraction (128).

Compression set is excellent in *cis*-polypentenamer rubber at low temperature (Fig. 18). The percent deformation increases to a high extent only at −60 to −80°C, unlike other rubbers which undergo analogous behavior at considerably higher temperatures. Figure 19 shows the only slight deviation of hardness from the value at room temperature through the whole range down to −80°C.

REFERENCES

1. See, e.g., *Chem. Eng.*, September 23, 1968, p. 172.
2. See, e.g., *Jap. Chem. Q.* IV-IV, 1968, p. 36.
3. H. Souza Eleuterio, U.S. Pat. 3,074,918 (to Du Pont, Priority June 20, 1957).
4. W. L. Truett et al., *J. Am. Chem. Soc.*, **82,** 2337 (1960).
5. G. Dall'Asta et al., *Makromol. Chem.*, **56,** 224 (1962).
6. G. Natta et al., *Makromol. Chem.* **69,** 163 (1963).
7. R. L. Banks, Belgian Pat. 620,440 (to Phillips Petroleum Co., Priority July 31, 1961).
8. R. L. Banks, Belgian Pat. 642,916 (to Phillips Petroleum Co., Priority January 23, 1964).
9. R. L. Banks, British Pat. 993,710 (to Phillips Petroleum Co., Priority May 21, 1963).
10. R. L. Banks and G. C. Bailey, *Ind. Eng. Chem. Prod. Res. Dev.*, **3,** 179 (1964).
11. G. Natta et al., French Pat. 1,394,380 (to Montecatini S.p.A., Priority April 10, 1963).
12. G. Natta et al., *Angew. Chem.*, **76,** 765 (1964); *Angew. Chem., Int. Ed.*, **3,** 723 (1964); presented by G. Dall'Asta at the Gordon Research Conference, Santa

Barbara, California, January 1964, and at the Makromolekular Colloquium, Freiburg, March 1964; *Chem. Eng. News,* April 6, 1964, p. 42.

13. G. Natta et al., Italian Pat. 733,857 (to Montecatini S.p.A., Priority January 17, 1964).

14. G. Natta et al., *Makromol. Chem.,* **91,** 87 (1966).

15. N. Calderon, Belgian Pat. 679,624 (to The Goodyear Tire & Rubber Co., Priorities April 16, 1965 and August 3, 1965).

16. N. Calderon et al., *J. Polym. Sci.,* A-1, **5,** 2209 (1967).

17. G. Günther et al., *Angew. Makromol. Chem.,* **14,** 87 (1970); presented by G. Pampus at the G.D.C.H. Meeting, Bad Nauheim, April 1970, and by F. Haas at the 97th ACS Meeting of the Division of Rubber Chemistry, Washington D.C., May 1970; see also *Rubber Chem. Technol.,* **43,** 1116 (1970).

18. E. A. Zuech, *J. Chem. Soc., Commun.,* **1968,** 1182.

19. E. A. Zuech et al., *J. Am. Chem. Soc.,* **92,** 528 (1970).

20. G. Dall'Asta and G. Carella, British Pat. 1,062,367 (to Montecatini S.p.A., Priority February 11, 1965).

21. G. Dall'Asta and R. Manetti, *Eur. Polym. J.,* **4,** 145 (1968).

22. G. Dall'Asta et al., French Pat. 1,545,643 (to Montecatini Edison S.p.A., Priority July 15, 1966).

23. N. Calderon and W. A. Judy, British Pat. 1,124,456 (to The Goodyear Rubber & Tire Co., Priorities March 28, April 6, and April 14, 1966).

24. N. Calderon et al., *Tetrahedron Lett.,* **1967,** 3327.

25. N. Calderon et al., *J. Am. Chem. Soc.,* **90,** 4133 (1968).

26. K. W. Scott et al., 155th ACS Meeting, San Francisco, April 1968, Division of Industrial and Engineering Chemistry, Preprint 54; *Advances in Chemistry Series,* No. 91, American Chemical Society, Washington, D.C., 1969, p. 399.

27. K. Nützel et al., French Pat. 2,002,570 (to Farbenfabriken Bayer, Priority February 24. 1968).

28. K. Nützel and F. Haas, French Pat. 2,008,180 (to Farbenfabriken Bayer, Priority May 9, 1968).

29. J. Witte et al., French Pat. 2,014,139 (to Farbenfabriken Bayer, Priority May 24, 1968).

30. P. W. Clark and R. A. D. Wentworth, *Inorg. Chem.,* **8,** 1223 (1969).

31. D. L. Kepert and R. Mandyczewsky, *J. Chem. Soc.* (A) **1968,** 530.

32. G. Pampus and J. Witte, French Pat. 2,005,701 (to Farbenfabriken Bayer, Priority April 6, 1968).

33. N. Schön et al., French Pat. 2,012,675 (to Farbenfabriken Bayer, Priority July 10, 1968).

34. W. Oberkirch et al., Belgian Pat. 740,769 (to Farbenfabriken Bayer, Priority October 25, 1968).

35. N. Nützel et al., French Pat. 2,039,196 (to Farbenfabriken Bayer, Priority April 15, 1969).

36. W. Oberkirch et al., Belgian Pat. 742,391 (to Farbenfabriken Bayer, Priority November 29, 1968).

37. H. W. Ruhle, French Pat. 2,074,042 (to Esso Research & Engineering Co., Priority December 19, 1969).

38. J. Wang and H. R. Menapace, *J. Org. Chem.*, **33**, 3794 (1968).

39. G. Pampus et al., 164th National ACS Meeting, New York City, August 1972, Preprints, Division of Polymer Chemistry, p. 880.

40. W. B. Hughes, *J. Am. Chem. Soc.*, **92**, 532 (1970).

41. G. Natta et al., *Makromol. Chem.*, **69**, 163 (1963).

42. G. Natta et al., Italian Pat. 805,709 (to Montecatini S.p.A., Priority October 20, 1965).

43. G. Dall'Asta and G. Carella, Italian Pat. 784,307 (to Montecatini Edison S.p.A., Priority August 2, 1966).

44. P. R. Marshall and B. J. Ridgewell, French Pat. 2,002,522 (to The International Synthetic Rubber Co., Priority February 23, 1968).

45. P. R. Marshall and B. J. Ridgewell, *Eur. Polym. J.*, **5**, 29 (1969).

46. W. A. Judy, French Pats. 2,016,360 and 2,016,364 (to The Goodyear Tire & Rubber Co., Priorities August 26, 1968).

47. N. Nützel et al., French Pat. 2,003,451 (to Farbenfabriken Bayer, Priority March 7, 1968).

48. V. A. Kromer et al., French Pat. 2,017,515 (Priority September 6, 1968).

49. V. A. Kromer et al., International I.U.P.A.C. Symposium on Macromolecules, Budapest, 1969, Abstract 4/31.

50. E. A. Ofstead, French Pat. 2,066,187 (to The Goodyear Tire & Rubber Co., Priority October 27, 1969).

51. G. C. Bailey, in H. Heinemann, ed., *Catalysis Reviews*, Vol. 3, Dekker, New York, 1971, p. 37.

52. H. J. Alkema and R. Van Helden, British Pat. 1,118,517 (to Shell Int. Res. Maatschappij, Priority May 26, 1967).

53. Shell Int. Maatschappij, Neth. Pat. Appl. 69,08733 (Priority June 10, 1968).

54. C. A. Uraneck and W. J. Trepka, Belgian Pat. 705,673 (to Phillips Petroleum Co., Priority October 27, 1966).

55. G. Günther et al., French Pat. 2,003,536 (to Farbenfabriken Bayer, Priority March 8, 1968).

56. W. Oberkirch et al., French Pat. 2,011,428 (to Farbenfabriken Bayer, Priority June 21, 1968).

57. L. Turner et al., British Pat. 1,105,565 (to The British Petroleum Co., Priority May 20, 1965).

58. G. Günther et al., French Pat. 2,025,142 (to Farbenfabriken Bayer, Priority December 3, 1968).

59. G. Dall'Asta and P. Meneghini, Italian Pat. 859,384 (to Montecatini Edison S.p.A., Priority March 28, 1969).

60. G. Natta et al., *Makromol. Chem.*, **54**, 95 (1962).

61. G. Dall'Asta and G. Mazzanti, *Makromol. Chem.*, **61**, 178 (1963).

62. G. Dall'Asta, *J. Polym. Sci.*, *A-1*, **6**, 2397 (1968).

63. R. E. Rinehart et al., *J. Am. Chem. Soc.*, **83**, 4864 (1961); **84**, 4145 (1962).

64. F. W. Michelotti and W. P. Keaveney, *J. Polym. Sci.*, *A*, **3**, 895 (1965).

65. L. Porri et al., *Chim. Ind.* (Milan), **46**, 428 (1964).

66. G. Natta et al., *J. Polym. Sci.*, *B*, **2**, 349 (1964).

67. G. Natta et al., *Makromol. Chem.*, **81**, 253 (1965).
68. J. K. Hecht, *J. Polym. Sci.*, *B*, **6**, 333 (1968).
69. J. K. Hecht, *J. Polym. Sci.*, *B*, **7**, 31 (1969); *A-1*, **8**, 2181 (1970).
70. Y. Yamashita et al., 159th National ACS Meeting, Houston, February 1970, Division of Industrial and Engineering Chemistry, Preprint 73.
71. K. B. Wiberg and W. J. Bartley, *J. Am. Chem. Soc.*, **82**, 6375 (1960).
72. S. Iwasaki et al., *J. Polym. Sci.*, *A-1*, **6**, 2441 (1968).
73. F. W. Michelotti and W. P. Keavaney, *J. Polym. Sci.*, *A*, **3**, 895 (1965).
74. G. Dall'Asta et al., *Makromol. Chem.*, **130**, 153 (1969).
75. J. Boor et al., *Makromol. Chem.*, **90**, 26 (1966).
76. G. Dall'Asta and R. Manetti, *Atti Accad. Naz. Lincei, Rend. Cl. Sci. Fis. Mat. Natur.*, **41** (8), 351 (1966).
77. R. E. K. Winter, *Tetrahedron Lett.*, **1965**, 1207.
78. C. P. C. Bradshaw et al., *J. Catal.*, **7**, 269 (1967).
79. G. C. Ray and D. L. Crain, French Pat. 1,511,381 (to Phillips Petroleum Co., Priority February 23, 1966).
80. E. Wasserman et al., *J. Am. Chem. Soc.*, **90**, 3286 (1968).
81. G. Dall'Asta and G. Motroni, *Eur. Polym. J.*, **7**, 707 (1971).
82. G. Dall'Asta and G. Motroni, *Angew. Makromol. Chem.*, **16/17**, 51 (1971), presented by G. Dall'Asta at the Meeting of GDCH-Fachgruppe Makromolekulare Chemie, Bad Nauheim, April 4, 1970.
83. G. Dall'Asta et al., *J. Polym. Sci.*, *A-1*, **10**, 1601 (1972).
84. J. C. Mol et al., *J. Chem. Soc., Commun.* **1968**, 633.
85. J. C. Mol. et al., *J. Catal.*, **11**, 87 (1968).
86. D. L. Crain, *J. Catal.*, **13**, 110 (1969).
87. F. D. Mango and J. H. Schachtschneider, *J. Am. Chem. Soc.*, **89**, 2484 (1967).
88. G. Natta and G. Dall'Asta, "Elastomers from Cyclic Olefins," in J. P. Kennedy and E. Törnqvist, eds., *Polymer Chemistry of Synthetic Elastomers*, Wiley-Interscience, New York, 1969, pp. 703–726.
89. G. S. Lewandos and R. Pettit, *Tetrahedron Lett.*, **1971**, 787.
90. G. S. Lewandos and R. Pettit, *J. Am. Chem. Soc.*, **93**, 7087 (1971).
91. J. Smidt et al., *Angew. Chem.*, **74**, 93 (1962); R. Jira and W. Freiesleben, private communication.
92. D. J. Cardin et al., *J. Chem. Soc., Commun.*, **1972**, 927.
93. G. Dall'Asta et al., 164th National ACS Meeting, New York City, August 1972; Preprints, Division of Polymer Chemistry, pp. 910–913; *Chim. Ind. (Milan)*, **55**, 1428 (1973).
94. E. A. Ofstead and N. Calderon, *Makromol. Chem.*, **154**, 21 (1972).
95. N. Calderon, *J. Makromol. Sci. C*, **7**, 105 (1972).
96. G. Pampus et al., *Rev. Gén. Caoutch.*, **47**, 1343 (1970), presented by J. Witte at Conference Int. du Caoutchouc, Paris, June 1970.
97. W. B. Hughes, *J. Chem. Soc., Commun.*, **1969**, 431.
98. G. Dall'Asta, XXIVth International Congress of Pure and Applied Chemistry, Hamburg, September 1973; *Pure Appl. Chem. Addit. Publ.*, **1**, 133 (1974).

99. E. A. Ofstead, 4th International Synthetic Rubber Symposium, London, September 1969.

100. G. Dall'Asta, Belgian Pat. 691,503 (to Montecatini Edison S.p.A., Priority December 21, 1965).

101. G. Dall'Asta and R. Manetti, Italian Pat. 789,585 (to Montecatini Edison S.p.A., Priority December 21, 1965).

102. K. Nützel et al., French Pat. 2,039,195 (to Farbenfabriken Bayer, Priority April 15, 1969).

103. R. Streck and H. Weber, D.O.S. 1,945,358 (to Chem. Werke Hüls, Priority September 8, 1969).

104. G. Pampus et al., French Pat. 2,070,802 (to Farbenfabriken Bayer, Priority December 10, 1969).

105. G. Pampus et al., French Pat. 2,040,131 (to Farbenfabriken Bayer, Priority march 27, 1969).

106. N. Calderon, French Pat. 1,576,325 (to The Goodyear Tire & Rubber Co., Priority September 1, 1967).

107. Shell Int. Res. Maatschappij, Neth. Pat. Appl. 69,08,732 (Priorities June 10 and 17, 1968).

108. J. L. Hérisson and Y. Chauvin, *Makromol. Chem.*, **141**, 161 (1971).

109. L. Michailov and H. J. Harwood, 160th National ACS Meeting, Chicago, September 1970, Division of Org. Coatings and Plastics Chemistry, Abstract 26.

110. R. Wolovsky, *J. Am. Chem. Soc.*, **92**, 2132 (1970).

111. D. A. Ben-Efraim et al., *J. Am. Chem. Soc.*, **92**, 2133 (1970).

112. J. Witte et al., French Pat. 2,069,443 (to Farbenfabriken Bayer, November 11, 1969).

113. A. J. Hubert, *J. Chem. Soc.*, 4088 1963.

114. G. Dall'Asta et al., Italian Pat. 889,242 (to Montecatini Edison S.p.A., January 13, 1970).

115. N. Calderon and M. C. Morris, *J. Polym. Sci.*, *A-2*, **5**, 1283 (1967).

116. C. Tosi et al., *J. Polym. Sci.*, *A-2*, **11**, 529 (1973).

117. F. Ciampelli et al., *Eur. Polym. J.*, **4**, 497 (1968).

118. G. Natta and I. W. Bassi, *Atti Accad. Naz. Lincei, Rend. Cl. Sci. Fis. Mat. Nature*, **38** (8), 315 (1965).

119. G. Natta and I. W. Bassi, International Symposium on Macromolecular Chemistry, Prague, 1965, Preprint 496; *J. Polym. Sci.*, *C*, **16**, 2551 (1967).

120. G. Natta and I. W. Bassi, *Eur. Polym. J.*, **3**, 33 (1967).

121. G. Natta and I. W. Bassi, *Eur. Polym. J.*, **3**, 43 (1967).

122. G. Natta et al., *Eur. Polym. J.*, **3**, 339 (1967).

123. I. W. Bassi and G. Fagherazzi, *Eur. Polym. J.*, **4**, 123 (1968).

124. G. Fagherazzi and I. W. Bassi, *Eur. Polym. J.*, **4**, 151 (1968).

125. G. Natta et al., *Eur. Polym. J.*, **5**, 239 (1969).

126. I. W. Bassi, private communication.

127. G. Motroni et al., *Eur. Polym. J.*, **9**, 257 (1973).

128. G. Dall'Asta and P. Scaglione, *Rubber Chem. Technol.*, **42**, 1235 (1969).

129.   G. Dall'Asta et al., *Makromol. Chem.*, **154,** 279 (1972).

130.   A. Chierico et al., *Eur. Polym. J.*, **5,** 115 (1969).

131.   F. Haas and D. Theisen, *Kaut. Gummi Kunstst.* **23,** 502 (1970).

132.   A. Capizzi and G. Gianotti, *Makromol. Chem.*, **157,** 123 (1972).

133.   H. Avrami, *J. Chem. Phys.*, **7,** 1103 (1939); **8,** 212 (1940).

134.   J. D. Hoffman and J. J. Weeks, *J. Res. Nat. Bur. Stand.*, **66,** 13 (1962).

135.   P. J. Flory, *J. Chem. Phys.*, **17,** 223 (1949); **15,** 684 (1947).

136.   G. Gianotti and A. Capizzi, *Eur. Polym. J.*, **6,** 743 (1970).

137.   A. Keller and E. Martuscelli, *Makromol. Chem.*, **141,** 189 (1970).

138.   A. Keller and E. Martuscelli, *Makromol. Chem.*, **151,** 169 (1972).

139.   G. Dall'Asta, *Rubber Chem. Technol.*, **47,** 511 (1974).

140.   F. Haas et al., French Pat. 2,006,105 (to Farbenfabriken Bayer, April 11, 1968).

141.   M. A. Golub, International Conference on Transformation of Polymers, Bratislava, June 1971.

Giovanni Crespi, Alberto Valvassori, and Umberto Flisi

Montedison, S.p.A.
Milano, Italy

Chapter 7   Olefin Copolymers

1   INTRODUCTION

Random, amorphous, saturated ethylene–propylene copolymers and unsaturated ethylene–propylene–diene terpolymers constitute a very interesting class of synthetic elastomers. Among the many reasons responsible for such an interest, the most important ones are the low cost and easy availability of the starting monomers and the excellent resistance of the polymers to chemical agents and to aging owing to their saturated or predominantly saturated nature.

Extensive research has been carried out in this field by research teams the world over working to establish the best conditions of synthesis and vulcanization, the relationships between structure and properties, and different fields of application. Such research work has been responsible for the rapid development and success of these elastomers, which sometimes may advantageously replace other already established rubbers.

2   SYNTHESIS OF SATURATED ETHYLENE–PROPYLENE COPOLYMERS

2.1   General Remarks

Some difficulties arise in ethylene-propylene copolymerization because the reactivities of the two monomers are quite different, and catalysts that are suitable for ethylene polymerization are not generally so for propylene polymerization.

Thanks to the research accomplished at the Polytechnic Institute of Milan, under the direction of Professor G. Natta and with the cooperation of a Montecatini-Edison research team, a number of suitable anionic coordinated catalyst systems were singled out (1–4). Among them, the best are those obtained by reaction of organometallic compounds of aluminum [e.g., $Al(i\text{-}C_4H_9)_3$, $Al(C_2H_5)_2Cl$, $Al_2(C_2H_5)_3Cl_3$, $Al(C_2H_5)Cl_2$]

with hydrocarbon soluble vanadium compounds (e.g., $VOCl_3$, VCl_4, vanadium triacetylacetonate, vanadyldiacetylacetonate, and alkyl esters of vanadic or chlorovanadic acid). These catalysts are essentially homogeneous and yield copolymers with optimum elastomeric properties, and without crystallinity or homopolymers. It is also possible to use hydrocarbon insoluble vanadium compounds (e.g., VCl_3) (3), but the copolymers obtained are scarcely homogeneous (5); in fact, both hydrocarbon soluble and insoluble fractions with a broad distribution of composition and molecular weights are formed. Furthermore, as revealed by x-ray analysis, some of the insoluble fractions have crystallinity of polyethylenic and polypropylenic types. Such heterogeneity may be attributed to the existence in these catalysts of two or more types of active sites with different reactivity ratios (6–11). As a consequence of this heterogeneity, the elastomeric properties of such copolymers are poor. In the synthesis of copolymers, at least one of the catalyst components should be halogenated (12, 13), as is true for propylene homopolymerization. However, α-olefin copolymers can be obtained in satisfactory yields by using a halogen-free catalyst component only when the catalyst preparation and/or the polymerization reaction are carried out in the presence of a halogen-containing hydrocarbon (14).

The catalyst may be prepared in the absence of olefins and subsequently introduced into the polymerization reactor, or it may be formed in situ by separately introducing the catalyst components into the reactor containing the olefins. Catalyst activity is initially high and decreases more or less markedly with time; hence industrial scale production, which requires high polymer yields per unit of catalyst, prefers catalysts prepared in situ.

Polymerization temperature and pressure may be chosen within a wide range. However, temperature generally ranges between -20 and $+40°C$, and pressure between 1 and 20 atm. The synthesis may be carried out either in solution or in suspension. In the former case, the reaction medium consists of an aliphatic, aromatic, or halogenated hydrocarbon which dissolves the copolymer formed; in the latter case, the reaction medium does not dissolve the copolymer (15). In either case, the polymerization may follow either a discontinuous (1, 5, 16) or a continuous process (17, 18).

The reactivities of ethylene and propylene are quite different; hence the monomer composition in the liquid phase differs from that of the copolymer formed. Means for keeping the composition in the liquid phase constant with time must be adopted in order to obtain a copolymer with a constant composition. For example, it is strictly necessary to maintain adequate stirring and accurate temperature control.

2.2 Factors Influencing Composition

If the concentration of the two olefins in the reaction mixture and the values of the reactivity ratios are known, the composition of an ethylene-propylene copolymer can be calculated by the well-known copolymerization equation:

$$\frac{m_1}{m_2} = \frac{M_1}{M_2} \cdot \frac{r_1 M_1 + M_2}{r_2 M_2 + M_1} \qquad [1]$$

where the subscripts 1 and 2 refer to ethylene and propylene, respectively; M_1 and M_2 are the molar fractions of ethylene and propylene in the reaction mixture; and m_1 and m_2 are the molar fractions of ethylene and propylene in the copolymer. The reactivity ratios of ethylene (r_1) and propylene (r_2) are connected to the rate constants of the four possible types of addition of the two monomers to the growing chains (according to the scheme reported in Table 1) by the relationships

$$r_1 = \frac{k_{11}}{k_{12}}, \qquad r_2 = \frac{k_{22}}{k_{21}}$$

Table 2 reports the reactivity ratios for the ethylene-propylene copolymerization in the presence of different catalyst systems. These reactivity ratios are virtually constant within a certain range of temperature (3, 13, 19). As shown in Table 2, ethylene is far more reactive than propylene, as may be expected for an anionic copolymerization mechanism. The product of the reactivity ratios is close to unity, as generally occurs in ionic copolymerizations.

As discussed later in more detail, some authors (20) have proposed that, at least in some catalyst systems, the propylene unit may be inserted between the metal and the last monomer unit which had entered the chain in two ways:

$$M-R-P+C_3H_6 \longrightarrow M-CH_2-\underset{\underset{\textstyle CH_3}{|}}{CH}-R-P$$

$$M-\underset{\underset{\textstyle CH_3}{|}}{CH}-CH_2-R-P$$

where M is the metal atom bound to the polymer chain P, and R is the last monomer unit in the chain, i.e.,

$$-CH_2-CH_2- \qquad \text{or} \qquad \underset{\underset{\textstyle CH_3}{|}}{CH_2-CH}- \qquad \text{or} \qquad -\underset{\underset{\textstyle CH_3}{|}}{CH}-CH_2-$$

Table 1   Addition of Monomers to Growing Polymer Chains

Type of Addition	Rate Constant
$\rightsquigarrow m_1 + M_1 \rightarrow \rightsquigarrow m_1 m_1$	k_{11}
$\rightsquigarrow m_1 + M_2 \rightarrow \rightsquigarrow m_1 m_2$	k_{12}
$\rightsquigarrow m_2 + M_1 \rightarrow \rightsquigarrow m_2 m_1$	k_{21}
$\rightsquigarrow m_2 + M_2 \rightarrow \rightsquigarrow m_2 m_2$	k_{22}

In such cases, the ethylene–propylene copolymerization, owing to the two possible orientations of the propylene unit, may be considered as a terpolymerization.

The composition of an ethylene–propylene copolymer is determined primarily by the concentration of the two olefins in the liquid phase and by the type of catalyst. All other conditions being the same, the copolymerization time (21), catalyst aging, and catalyst concentration do not exert any influence on copolymer composition (1–4, 13).

Table 2   Monomer Reactivity Ratios for Ethylene-Propylene Copolymerization in the Presence of Different Catalyst Systems

Catalyst System	$r_{C_2H_4}$	$r_{C_3H_6}$	Ref.
$VOCl_3 + Al(C_6H_{13})_3$	17.95	0.065	1
$VCl_4 + Al(C_6H_{13})_3$	7.08	0.088	2
$VCl_3 + Al(C_6H_{13})_3$	5.61	0.145	3
$VAc_3{}^a + Al(C_2H_5)_2Cl$	15.0	0.04	13
$TiCl_4 + Al(C_6H_{13})_3$	33.36	0.032	4
$TiCl_3 + Al(C_6H_{13})_3$	15.72	0.110	4
$TiCl_2 + Al(C_6H_{13})_3$	15.72	0.010	4
$VO(OR)Cl_2 + Al(C_2H_5)Cl_2$	17.5	0.05	19
$VO(OR)_3, VO(OR)_2Cl, VO(OR)Cl_2 + AlR_2Cl, AlRCl_2{}^b$	17–28	c	5
$VOCl_3, VCl_4 + AlR_3, AlR_2Cl, AlRCl_2{}^b$	17–28	c	5
$VO(O\text{-}n\text{-}C_4H_9)_3 + Al(i\text{-}C_4H_9)_2Cl$	22.00	0.046	20
$VO(OC_2H_5)_2Cl + Al(i\text{-}C_4H_9)_2Cl$	18.90	0.056	20
$VO(OC_2H_5)Cl_2 + Al(i\text{-}C_4H_9)_2Cl$	16.80	0.055	20
$VO(OC_2H_5)_3 + Al(i\text{-}C_4H_9)_2Cl$	15.00	0.070	20
$VCl_4 + HCl_2Al \cdot O(C_2H_5)_2$	10.7	0.022	18
$VCl_4 + Al(C_2H_5)_2Cl$	13.7	0.02	21

[a] $VAc_3 = $ vanadium triacetylacetonate.

[b] Any combination of an aluminum and vanadium compound.

[c] The product $r_{C_2H_4}$ $r_{C_3H_6}$ is equal to 1.

The molar ratio between alkylaluminum compound and transition metal compound does not generally influence the copolymer composition (1, 3, 4, 13, 22). However, with the catalyst system $VOCl_3 + Al(i\text{-}C_4H_9)_3$, Ichikawa and Kogure (23) noticed that the Al/V ratio does not influence composition in aliphatic solvents but if tetrachloroethylene or benzene is used, the copolymer composition varies, to a limited extent, within a certain range of the Al/V ratio. Composition also depends on the Al/V ratio when polymerizing in cyclohexane with the catalyst system $VCl_4 + Al(i\text{-}C_4H_9)_3$ (24).

With catalyst systems prepared by reaction of VCl_4 or $TiCl_4$ with reducing compounds of the $HX_2Al \cdot B$, H_2AlY, $H_3Al \cdot B$ series (where $X = Cl$, $Y = -N(CH_3)_2$, and $B = Lewis$ base) (25), the copolymer composition is generally constant within a narrow range of the Al/V and Al/Ti ratios in the region approaching highest catalyst activity. At higher ratios, the ethylene content increases sharply. It follows that, at higher Al/V and Al/Ti ratios, the active sites that derive from further reduction of $TiCl_4$ or VCl_4 differ from the active centers obtained at lower ratios.

Some authors have observed that, for a given transition metal compound, the type of organometallic compound exerts little influence on the relative reactivities of ethylene and propylene. Using VCl_4, all other conditions being the same, the copolymer composition does not vary if $Al(i\text{-}C_4H_9)_3$, $Zn(C_6H_5)_2$, $Zn(n\text{-}C_4H_9)_2$, and CH_3TiCl_3 are used as organometallic compounds (26). If $Al(i\text{-}C_4H_9)_3$ is used as a reducing compound and the transition metal compound is varied, the relative reactivity of propylene increases in the order $HfCl_4 < ZrCl_4 < TiCl_4 < VOCl_3 < VCl_4$. Such an increase seems to be connected to the increase of electronegativity of the transition metal.

The variation of the relative reactivities of ethylene and propylene on varying the type of the transition metal, and the constancy of such reactivities on varying the type of reducing compound suggest that propagation occurs on the transition metal without direct participation of the reducing compound (24). However, such a hypothesis cannot be generally valid, according to the data reported by Zambelli et al. (27), who carried out copolymerization runs at $-78°C$ in the absence of solvent, using VCl_4 with different organometallic compounds $[Al(C_2H_5)_2Cl, Al(C_2H_5)_2Cl\text{-}anisole, Al(C_2H_5)_3, and Ga(C_2H_5)_3]$.

According to Junghanns and co-workers (28), the relative reactivity of propylene increases considerably in the presence of some vanadium compounds (e.g., VCl_4 and $VOCl_3$) if $Al(C_2H_5)Cl_2$ or $Al_2(C_2H_5)_3Cl_3$ is used as reducing compound instead of $Al(C_2H_5)_3$ or $Al(C_2H_5)_2Cl$. These results contrast with those observed by Ichikawa and Kogure (23), who used $VOCl_3$ with different $Al(C_2H_5)_2Cl$ and $Al(C_2H_5)Cl_2$ mixtures.

2.3 Molecular Weight Control

Intrinsic viscosity (for the correlation between intrinsic viscosity and molecular weight, see Section 6) is a very important property for many applications of ethylene–propylene copolymers. Since these elastomers do not undergo any considerable degradation during the usual rubber industry processing, molecular weight must be controlled ahead of time in the synthesis phase. The main operating parameters that influence the intrinsic viscosity may be summarized as follows.

2.3.1 *Temperature*

With a number of catalyst systems, all other conditions being the same, the intrinsic viscosity of the copolymers increases on decreasing the polymerization temperature (22, 28). With the catalyst system $VCl_4 + HCl_2Al \cdot O(C_2H_5)_2$ in heptane, a substantially constant intrinsic viscosity was observed at temperatures between $+10$ and $-20°C$ (19). Analogous results were obtained with the same catalyst system in copolymerization runs carried out in the absence of solvent in the range $+20$ to $-18°C$ (29).

2.3.2 *Ethylene Propylene Ratio in the Reacting Phase*

On increasing the ethylene/propylene ratio, the intrinsic viscosity increases (see, e.g., refs. 1, 5, 12, and 22).

2.3.3 *Copolymerization Time*

Some increase of the intrinsic viscosity was observed in the presence of a few catalyst systems, e.g., $VCl_4 + Al(C_2H_5)_2Cl$ (22), vanadium triacetyl-acetonate $+ Al(C_2H_5)_2Cl$ (4), $VOCl_3 + Al(C_2H_5)Cl_2$ or $Al_2(C_2H_5)_3Cl_3$ (30, 28), and $VOCl_2(OR) + AlRCl_2$ (30). In other cases, e.g., with $VCl_4 + Al(C_6H_{13})_3$ aged at 60°C for 30 minutes and used at $+25°C$ (31), and with the catalyst system $VCl_4 + HCl_2Al \cdot (OC_2H_5)_2$ in the absence of solvent (19), the polymerization time did not seem to exert any appreciable influence on the intrinsic viscosity.

2.3.4 *Catalyst Concentration*

All other conditions being the same, the intrinsic viscosity of the copolymers decreases on increasing the catalyst concentration (see, e.g., refs. 13 and 22). Such a decrease may be partly explained by the hypothesis of a transfer reaction between the growing chains and the organometallic compound. On the contrary, with the catalyst system $VCl_4 + HCl_2Al \cdot (OC_2H_5)_2$ in benzene, no influence of catalyst concentration was observed (19).

2.3.5 Type of Aluminum Alkyl

In the presence of some vanadium compounds (e.g., VCl_4 and $VOCl_3$) at room temperature, the intrinsic viscosity decreases in the order (28) $Al(C_2H_5)_3 > AlEtCl_2 > Al_2Et_3Cl_3 > AlEt_2Cl$. Analogous results were obtained using mixtures of $AlEt_2Cl$ and $AlEtCl_2$ with $VOCl_3$ at an Al/V ratio of 20 (23).

2.3.6 Aluminum/Vanadium Molar Ratio

In the presence of some catalyst systems, e.g., vanadium triacetylacetonate + $Al(C_2H_5)_2Cl$ (13) or $VCl_4 + Al(C_6H_{13})_3$ (32) aged at 60°C for 30 minutes and subsequently used at 25°C, at constant concentration of the vanadium compound, the intrinsic viscosity decreases on increasing the Al/V ratio. A slight decrease of viscosity on increasing the Al/V ratio was also observed with the catalyst system $VCl_4 + Al(C_2H_5)_2Cl$ (22). Such a decrease may be attributed to a transfer process between the growing chains and the aluminum compound.

If the copolymerization is carried out in heptane with the catalyst system $VOCl_3 + Al(C_2H_5)_2Cl$ (23), the intrinsic viscosity decreases when the Al/V ratio varies from 1 to 3 and remains virtually constant between 3 and 10; on the contrary, in hexane and in benzene, the highest intrinsic viscosities are found at an Al/V ratio of 5 and 15, respectively.

A maximum in intrinsic viscosity values was also observed by Kelly et al. (32) with the catalyst system $VOCl_3 + Al(C_2H_5)Cl_2$ at an Al/V ratio of 15.

With the catalyst system $VCl_4 + HCl_2Al \cdot (OC_2H_5)_2$, in the absence of solvent, a maximum intrinsic viscosity was observed at an Al/V ratio of 3 (33). Using the same catalyst in benzene, the intrinsic viscosity increased with increasing Al/V ratio (19).

2.3.7 Aging of the Catalyst

In the presence of some catalyst systems, e.g., vanadium triacetylacetonate + $Al(C_2H_5)_2Cl$ prepared at −20°C (4) and $VOCl_3 + AlEtCl_2$ or $Al_2(C_2H_5)_3Cl_3$ prepared at +30°C (28), the intrinsic viscosity increases with aging time. However, with the system $VCl_4 + HCl_2Al \cdot (OC_2H_5)_2$ (19), aging time exerts a minor influence.

2.3.8 *Molecular Weight Regulators*

A copolymer with the desired intrinsic viscosity may be obtained more easily by adding, in the reaction phase, a compound that, by interacting with the growing chains in competition with the monomers by transfer or termination, reduces the copolymer molecular weight. Several substances were proposed, e.g., organic nitrates and nitrites (34), nitrous compounds (35), quaternary ammonium salts, and alkyl or aryl ammonium halides (36) and elementary sulfur (37).

The best molecular weight regulators, especially for industrial scale productions, are hydrogen (38) and zinc alkyls (39). These are quite effective regulators and are preferred because they do not appreciably decrease the catalytic activity. In other words, hydrogen and zinc alkyls are chain transfer agents that stop the growing chains but initiate new chains whereas, e.g., alkyl halides stop chain growth but do not initiate any new chains.

As in propylene homopolymerization, the molecular weight regulation done by zinc alkyls may be a simple exchange of alkyl groups between the alkyls of dialkyl zinc and the polymeric chains growing on the active sites:

$$[Cat]P + ZnR_2 \rightarrow [Cat]R + ZnRP$$

For hydrogen, the molecular weight modification may be a hydrogenolysis of the complex catalyst-polymer chain:

$$[Cat]P + H_2 \rightarrow [Cat]H + PH$$

2.4 Ethylene–Propylene Copolymerization Results in the Presence of Different Catalyst Systems

Junghanns et al. (28) studied the influence exerted by parameters connected with catalyst preparation on the copolymerization rate. Runs were carried out with the catalyst systems $VOCl_3 + Al_2(C_2H_5)_3Cl_3$ and $VOCl_3 + Al(C_2H_5)Cl_2$. If the catalyst is prepared in the absence of monomers and kept under nitrogen atmosphere for some time before adding it to the polymerization reactor containing the olefins, its activity decreases on increasing the aging time. Such a loss of activity is even more marked if the catalyst is prepared in the presence of monomers and kept under nitrogen before use. The decrease of activity is barely influenced by temperature, at least in the range between 20 and 40°C, and depends on the Al/V ratio. The rate of deactivation of the catalyst prepared in the absence of monomers is lower as the Al/V ratio increases from 5 to 40. On increasing the aging time of the catalyst, the valence of vanadium gradually decreases. The polymerization rate increases on increasing the C_2/C_3 ratio, the temperature, and the concentration of the olefins.

With the same catalyst systems, the increase of the Al/V ratio causes an increase in catalyst efficiency (32). By operating with an Al/V ratio equal to 20 and using mixtures of $Al(C_2H_5)_2Cl$ and $Al(C_2H_5)Cl_2$ in different ratios, a maximum catalyst efficiency occurs at a molar ratio of 40:60.

With the catalyst system vanadium triacetylacetonate + $Al(C_2H_5)_2Cl$ prepared and used at 25°C (13), the catalyst efficiency depends on the ratio Al/V and shows a maximum for an Al/V ratio of 3.5.

The decrease of catalytic activity with time is very rapid at 25°C, and is slower when the catalyst is prepared and used at lower temperatures. At −20°C, the increase of catalyst activity is remarkable when the Al/V ratio increases from 3 to 4 but negligible as the ratio increases from 4 to 20. At −20°C the yields of copolymer, obtained in a specific time and with catalysts aged under the same conditions, are directly proportional to the catalyst concentration; i.e., the copolymerization rate is first order with respect to the catalyst concentration. Yield and total olefin concentration are directly proportional.

All other conditions being the same, the polymerization rate increases considerably on increasing the ethylene/propylene ratio (in solution), as in the case of other catalyst systems (see, e.g., refs. 5, 23, and 28).

An increase in the copolymer yield on decreasing the temperature of copolymerization was also observed with the catalyst system $VCl_4 + Al(C_2H_5)_2Cl$ prepared in the presence of olefins (22). With this catalyst at −10°C, the amount of copolymer obtained in a given time increases rapidly on increasing the Al/V ratio. The polymerization yield, expressed as weight of polymer produced per unit weight of total catalyst used, reaches a maximum at Al/V molar ratios of 3–4.

On the contrary, yield with respect to vanadium content alone increases very rapidly up to ratios of 4–5; at higher ratios the variation is less marked. This is in agreement with the hypothesis that the number of active sites is proportional to the concentration of vanadium salt, and that is why increasing the Al/V ratio beyond a certain limit decreases yield at constant total catalyst.

The amount of polymer formed increases very rapidly in the first minutes of polymerization; it tends to a virtually constant value with time as a consequence of a strong deactivation of the catalyst. The catalyst deactivation occurs according to second-order kinetics (5, 28, 40, 41).

Runs carried out in solution with the catalyst system $VCl_4 + HCl_2Al \cdot (OC_2H_5)_2$ gave evidence of a maximum catalyst activity at an Al/V ratio of 2.4 (19). At ratios either above or below 2.4, the reduction of VCl_4 with respect to the optimum valence is either exceeded or insufficient, with a consequent decrease of catalyst activity. On varying the polymerization temperature between +30 and −30°C, this catalyst system maintains maximum activity for an Al/V ratio of about 2.5. With other catalyst systems (see, e.g., ref. 13), an interdependence exists between temperature and Al/V ratio; at lower temperatures, a higher ratio is required to obtain optimal activity, the interaction among the catalyst constituents being attenuated at lower temperatures. This may be connected with the strong reducing action of $HCl_2Al \cdot (OC_2H_5)_2$ both at room temperature and at −70°C. On decreasing the polymerization temperature, the copolymer yields increase; the polymerization rate (at constant concentration of VCl_4 and of monomers) increases with decreasing temperature in the range between 62 and 25°C. Below room temperature, the polymerization rate decreases on decreasing temperature.

As shown by these data, the increase of the rate constants of the copolymerization process with temperature is counterbalanced by the strong decay of the catalytic activity; this is the reason why, for temperatures higher than 25°C, the net result is a decrease in rate with increasing temperature. For runs carried out below room temperature, the decrease of rate constants for the copolymerization process on decreasing temperature prevails over the higher activity of the catalyst systems at low temperatures.

A series of systematic copolymerization runs was carried out by Natta et al. (31) with the catalyst system $VCl_4 + Al(C_6H_{13})_3$. The results may be summarized as follows:

1. The catalyst activity decreases with time. However, if the catalyst is aged for 30 minutes at 60°C and subsequently used at a lower temperature (0–40°C), the catalytic activity is constant with time (Fig. 1). The catalyst stabilized in this way shows a maximum activity at an Al/V ratio of about 2.5.

2. The copolymerization rate is directly proportional to the catalyst concentration and, at a constant ratio between the two monomers, to the total monomer concentration in the liquid phase.

3. For a given total monomer concentration, the copolymerization rate increases very rapidly with an increase of ethylene concentration (Fig. 2).

4. The activation energies of the four elementary processes of copolymerization are essentially the same (6600 cal/mole) and equal to the activation energies in the homopolymerization of ethylene and of propylene with the same catalyst system.

In some copolymerizations carried out in the presence of catalysts acting through an anionic coordinated mechanism [e.g., copolymerization of

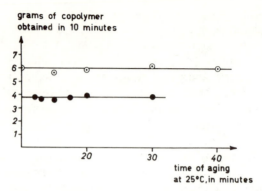

Figure 1 Independence of the activity of catalysts prepared from Al(C₆H₁₃)₃ and VCl₄, aged at 60°C for 30 minutes, on the aging time at 25°C. Experimental conditions: polymerization temperature 25°C; absolute pressure 1 atm; solvent, *n*-heptane, 410 cc; moles C_3H_6/moles C_2H_4 in the gaseous feed = 2.15. ●, V (grams of vanadium present in the catalyst system) = 0.0299; polymerization time 10 minutes. ○, $V =$ 0.0499; polymerization time 6 minutes. From Natta et al. (31), by courtesy of John Wiley & Sons, Inc., New York.

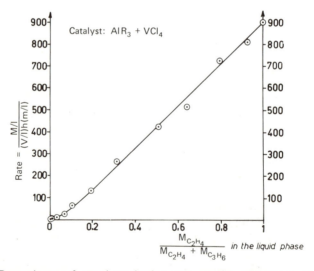

Figure 2 Dependence of copolymerization rate on the percentage of dissolved ethylene (moles) referred to the total moles of both monomers. Experimental conditions: catalyst prepared and aged for 30 minutes at 60°C; moles Al(C₆H₁₃)₃/moles VCl₄ = 2.5; polymerization temperature 25°C; absolute pressure 1 atm; M is the sum of ethylene and propylene moles polymerized in h hours; m is the sum of moles of dissolved olefins; V is grams of vanadium present in the catalyst system; l is the volume of the reacting liquid phase in liters. From Natta et al. (31), by courtesy of John Wiley & Sons, Inc., New York.

376

propylene with butene (42), of some vinyl aromatic monomers (43), and of ethylene with propylene in the presence of catalysts prepared from solid Ti or V halides (43)] the ratio between the polymerization rates of monomers 1 and 2 is virtually equal to r_1 (or to the reciprocal of r_2). This does not occur for ethylene and propylene polymerization in the presence of the stabilized catalyst system $Al(C_6H_{13})_3 + VCl_4$. In this case the ratio between the polymerization rates of ethylene and of propylene is much higher than the reactivity ratio of ethylene. Neither this fact nor the considerable increase in the copolymerization rate observed on increasing the ethylene/propylene ratio in the reacting phase can be explained merely on the basis of the values of the propagation constants. It becomes necessary to assume that the total number of growing chains changes when the ratio of ethylene to propylene is varied. This variation in the number of growing chains can be explained if active sites are present which are capable of initiating ethylene, but not propylene, polymerization (44). This phenomenon was also confirmed by direct measurement of the number of active centers (capable of originating growing chains) present in the homopolymerization of ethylene and propylene in the presence of the catalyst systems $VCl_3 + AlR_3$ and $TiCl_4 + AlR_3$. A further proof is given by the capability of preparing ethylene–propylene copolymers with the aid of catalysts that cannot initiate[1] propylene homopolymerization but are active in the homopolymerization of ethylene.

Assuming a variation of the number of growing chains, the ratio between the homopolymerization rates of ethylene and of propylene is of the same order of magnitude as the reactivity ratio of ethylene (43) if all polymerization rates are referred to the same concentration of growing chains.

A catalyst system similar to the industrial catalysts used to obtain "high yield" polyethylene has recently been found. It consists of a titanium halide supported on a magnesium halide either entirely or partially complexed with a Lewis base. Such a catalyst exhibits extremely high efficiency and, though heterogeneous, it yields products free from crystallinity and with properties comparable to those obtained in the presence of homogeneous catalysts (45).

2.5 Reactivators

On combining V compounds with alkylaluminum compounds there is a progressive reduction of the vanadium compound with time with consequent decrease of the catalytic activity, until an inactive bivalent state is reached.

Carrick et al. (46) reported that VCl_4 is completely reduced to the bivalent state, when mixed with $Al(i\text{-}C_4H_9)_3$ for 15 minutes at Al/V ratios higher than 1, but its reduction is only partial for ratios ranging between 0.4 and 1.

Vanadium is completely reduced to the bivalent state on mixing $VOCl_3$ with $Al(i\text{-}C_4H_9)_3$ for 15 minutes at Al/V ratios higher than 2. With $VO(O\text{-}n\text{-}C_3H_7)_3$ mixed with $Al(i\text{-}C_4H_9)_3$ at an Al/V ratio of 7, the reduction is slower; after 1 hour, the percentage of bivalent V is negligible and slowly increases with time, reaching 72% after about 72 hours. Gumboldt et al. (48) and others (12, 30, 47) found a rapid initial reduction of $VOCl_3$ with different aluminum alkyl compounds to compounds containing vanadium in a valence state intermediate between 2 and 3. The rapid initial reduction is followed by a slower reduction to the inactive bivalent state.

By adding suitable proportions of hexachlorocyclopentadiene to the two catalyst constituents, the inactive vanadium compound may be reoxidized with a consequent increase in catalyst efficiency. The reactivating action of hexachlorocyclopentadiene was examined in other vanadium compounds (vanadium dichloride, vanadium trichloride, vanadium tetrachloride, vanadium triacetylacetonate, dicyclopentadienylvanadium dichloride, and dibenzoylvanadium) combined with different organoaluminum compounds. Bivalent vanadium compounds are inactive, but become active on addition of hexachlorocyclopentadiene. Penta-, tetra-, and trivalent vanadium compounds become inert when the vanadium is reduced to the bivalent state, but are reactivated on the addition of hexachlorocyclopentadiene. Zerovalent vanadium compounds which are inactive in polymerization are not activated by hexachlorocyclopentadiene. An analogous effect was also observed with several other compounds, such as benzotrichloride (49), methyl trichloroacetate, thionyl chloride, hexachlorocyclopentanone (50), dichloroarylphosphine (51), N-nitrous diphenylamine, β-methylanthraquinone (35), triacetylacetonates of Co, Fe, Mn, and Cr (34), nitrobenzene, nitropropane (52), PCl_3, PBr_3, PI_3 (53), ethylperchlorocrotonate (54), and ethyl dichloromalonate (55).

3 SYNTHESIS OF UNSATURATED ETHYLENE–PROPYLENE–DIENE TERPOLYMERS

3.1 General Remarks

Binary, saturated ethylene–propylene copolymers can be vulcanized only with particular recipes based on organic peroxides. In order to obtain unsaturated elastomers that could be vulcanized with the traditional sulfur recipes extensive work was done on terpolymerization of ethylene and propylene with a suitable diene or triene. The use of two or more dienes in the polymerization with ethylene and propylene was also described (56, 57). A review of EPDM chemistry through 1973 has recently been published by Cesca (58).

3.2 Choice of the Third Monomer

The amount of unsaturation in terpolymers is usually low (1–2 mole %), being required only to provide cure sites for sulfur vulcanization. The third monomer should be easily prepared and of low cost. Additional criteria are based on reactivity both in polymerization and in vulcanization.

In regard to polymerization behavior, the third monomer should (*a*) contain at least two olefin units, only one of which should be polymerizable while the other is suitable for vulcanization; (*b*) contain the residual olefin unit or units outside the main chain in order to avoid any chain cleavage and reduction of the excellent chemical stability of the saturated chain; (*c*) enter isolated so as to be uniformly distributed along the main chain, in order to provide vulcanizates with good properties; (*d*) exhibit a high copolymerization rate and conversion and not interfere with the polymerization of ethylene and propylene; and (*e*) not be too high in molecular weight in order to have a favorable unsaturation/weight ratio and to simplify finishing operations after polymerization. The first condition excludes diolefins having the same reactivity in both double bonds, unless the introduction of branching or cross-linking is desired. The second condition excludes acyclic conjugated dienes such as butadiene, the 1,4 enchainment of which would provide poor ozone resistance. Conjugated dienes also give troubles such as catalyst deactivation and homopolymer formation. Vinyl type double bonds are readily incorporated into the growing chains, whereas the vinylidenic or internal ones are inert in polymerization.

The most important requirements for vulcanization are (*a*) high cure rate and yield with sulfur and accelerators, and (*b*) capability of producing elastomeric networks with satisfactory mechanical and dynamic properties, high abrasion, and aging resistance.

A large number of diolefins, both linear and cyclic, have been examined as third monomers for EPDM rubbers (59); some are reported in Fig. 3. Despite the number that can be used, only three are employed commercially (ethylidene norbornene, dicyclopentadiene, and 1,4-hexadiene).

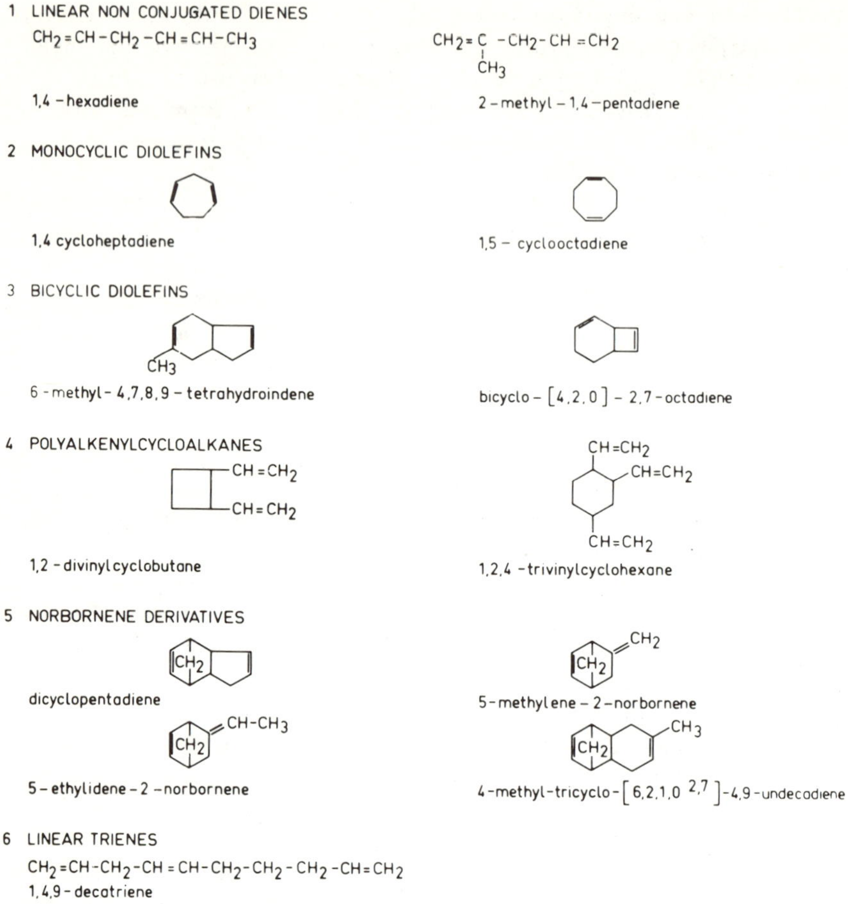

1 LINEAR NON CONJUGATED DIENES

$CH_2 = CH-CH_2-CH=CH-CH_3$

1,4 –hexadiene

$CH_2 = C\ -CH_2-CH=CH_2$
     $|$
     CH_3

2 – methyl – 1,4 –pentadiene

2 MONOCYCLIC DIOLEFINS

1,4 cycloheptadiene

1,5 – cyclooctadiene

3 BICYCLIC DIOLEFINS

CH_3

6 – methyl – 4,7,8,9 – tetrahydroindene

bicyclo – $[4,2,0]$ – 2,7 –octadiene

4 POLYALKENYLCYCLOALKANES

$CH=CH_2$

$CH=CH_2$

1,2 – divinylcyclobutane

$CH=CH_2$
$CH=CH_2$
$CH=CH_2$

1,2,4 – trivinylcyclohexane

5 NORBORNENE DERIVATIVES

CH_2

dicyclopentadiene

CH_2

CH_2 $CH-CH_3$

5 – ethylidene – 2 –norbornene

CH_2

CH_2

5 – methylene – 2 –norbornene

CH_2 CH_3

4 – methyl-tricyclo – $[6,2,1,0\ ^{2,7}]$ – 4,9 –undecadiene

6 LINEAR TRIENES

$CH_2 = CH-CH_2-CH = CH-CH_2-CH_2-CH_2-CH=CH_2$
1,4,9 – decatriene

Figure 3 Unsaturated monomers particularly studied for EPDMs. From G. Crespi, A. Valvassori, V. Zamboni, and U. Flisi, *Chim. Ind. (Milan)*, **55**, 130 (1973), By courtesy of SpA Editrice di Chimica, Milan.

The most extensively used termonomer is ethylidene norbornene (ENB). It is prepared by Diels-Alder adduction of butadiene and cyclopentadiene, giving vinylnorbornene which is then catalytically isomerized to ethylidene norbornene (60). ENB is characterized by a high reactivity in copolymerization, where the addition occurs at the strained double bond of the bicycloheptenic ring (Fig. 4) (61). The high cure rate of ENB is its most striking characteristic and derives from the particular reactivity of the hydrogen in the position α to the double bond (62).

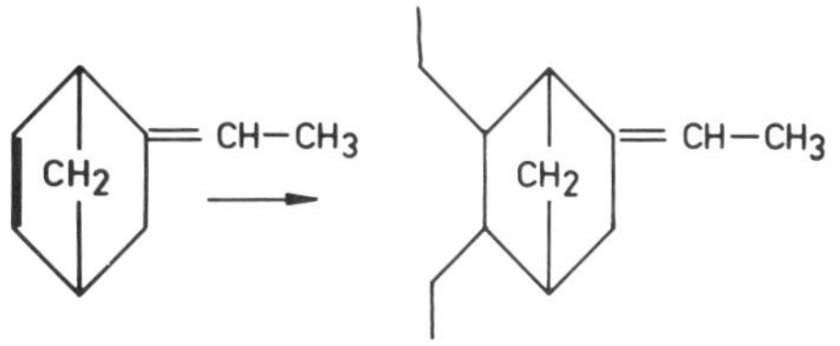

Figure 4 Incorporation of ethylidene norbornene into the EP copolymer. From G. Crespi, A. Valvassori, V. Zamboni, and U. Flisi, *Chim. Ind.* (*Milan*), **55**, 130 (1973), by courtesy of SpA Editrice di Chimica, Milan.

Second in commercial importance is 1,4-hexadiene (1,4-HD), which is the least reactive of the three and is prepared by codimerization of ethylene and butadiene (63). It enters the chain primarily through the terminal vinylic double bond, but some inter-intra-molecular polymerization can occur with formation of saturated cyclic units and consequent loss of unsaturation (64). This reduces the vulcanization yield, which combined with the lower cure rate makes 1,4-HD inferior to ENB.

The reactivity ratio of hexadiene with propylene is rather low (65), which allows its uniform distribution along the polymer chain and consequently the production of very good vulcanizates.

Dicyclopentadiene (DCPD) is certainly the cheapest of the three and shows a high copolymerization rate, but presents two drawbacks. These are its slow vulcanization and the small difference in reactivity in its two double bonds. This latter characteristic leads to the formation of long chain branching or even of gels. Branching, besides reducing the curing efficiency, markedly influences the rheological behavior of the polymer and consequently its processability, being associated with the elastic memory (66). DCPD may also be beneficial in some instances, as in EPDM blends with diene rubbers, because it imparts a high ozone resistance to the terpolymer. This is why DCPD–EPDMs are still mainly used in applications where ozone resistance is of fundamental importance.

The vulcanization behavior of EPDMs is illustrated in Fig. 5, where terpolymers with the same molar composition but containing different third monomers are compared. It appears evident that ENB is superior both in vulcanization rate and yield.

3.3 Polymerization

In general, the operating conditions influencing the synthesis of binary copolymers exert similar action on the unsaturated terpolymers. The synthesis of unsaturated terpolymers may be carried out in solution (17)

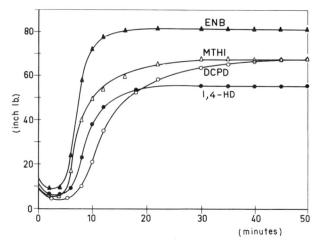

Figure 5 Oscillating disk rheometer curves of compounds of EPDMs containing different third monomers. From G. Crespi, A. Valvassori, V. Zamboni, and U. Flisi, *Chim. Ind. (Milan)*, **55**, 130 (1973), by courtesy of SpA Editrice di Chimica, Milan.

and in suspension (18). In the former case the polymer is soluble. The solution viscosity depends mainly on the average molecular weight of the elastomers (related to the Mooney viscosity of the rubber), temperature, and the type of solvent.

Figure 6 shows the variations of viscosity in cyclohexane at 38°C as a function of concentration for two terpolymers having a Mooney viscosity of about 100 but with different propylene contents. The viscosity increases so rapidly with concentration, as to make control of the reaction practically impossible with concentrations higher than 7–8 wt %.

In the suspension process, the viscosity of the reaction slurry is not appreciably increased by high solids content and it is possible to have high productivity for reactor volume unit without problems of stirring, heat, and mass transfer.

In suspension, cross-linking (originating from the fact that both diene double bonds are involved in the polymerization) is particularly high; however, this drawback may be avoided by the use of suitable termonomers. The best catalysts are still those obtained by the reaction of hydrocarbon soluble vanadium compounds with alkylaluminum compounds. The polymer intrinsic viscosity is influenced by the same factors as described earlier. In the ethylene–propylene–dicyclopentadiene terpolymerization with the catalyst system $VOCl_3 + (C_2H_5)_2Cl$ or $Al_2(C_2H_5)_3Cl_3$, the intrinsic viscosity of terpolymers obtained in aromatic solvents is lower than that of terpolymers obtained in aliphatic solvents

(16). Analogous results were reported by Cunningham (67) who operated with the catalyst system vanadium triacetylacetonate or vanadyl diacetylacetonate plus $Al_2(C_2H_5)_3Cl_3$.

The polymerization rate is lower with terpolymers than with the binary ethylene–propylene copolymers. All other conditions being the same, rates depend on the type and concentration of diene in the reaction mixture. On increasing the concentration of diene, the unsaturation content increases and both polymerization rate and intrinsic viscosity of the terpolymer are simultaneously reduced (68).

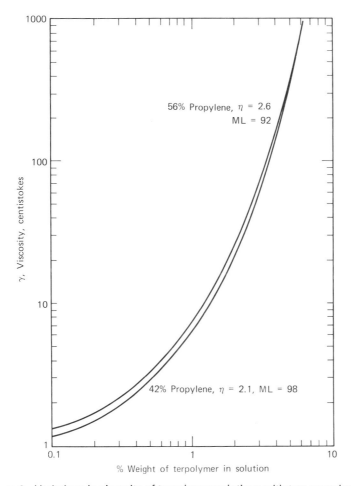

Figure 6 Variations in viscosity of terpolymer solutions with two propylene contents in cyclohexane at 38°C. From Crespi and Di Drusco (18), by courtesy of Gulf Publishing Co. Houston, Texas.

Terpolymerization runs carried out in n-pentane at $-60°C$ with the catalyst system $VOCl_3 + Al(i\text{-}C_4H_9)_3$, using dicyclopentadiene, both endo and exo, as termonomer showed that the endo form decreases the polymer yield more than the exo (69).

With some types of dienes, if the concentrations of ethylene and of propylene are kept constant and the third monomer's concentration is increased, the ethylene content in the terpolymer increases. This phenomenon was observed e.g., for linear nonconjugated diolefins (69), 1,5-cyclooctadiene (70), dicyclopentadiene (61, 71), and isopropylidene-tetrahydroindene (68, 72). However, different catalytic combinations gave different results with different dienes (73).

The evaluation of the reactivity ratios of these dienes referred to propylene, confirmed the high reactivity of the norbornene double bonds (16, 74, 75).

4 CHARACTERISTIC QUANTITIES OF THE COPOLYMERS

4.1 Definitions and General Properties

For an adequate understanding and for the correlation of physical properties with molecular structure of ethylene–propylene copolymers, it is necessary to know on the one hand the overall composition, i.e., the abundance of the two types of monomer units, and on the other hand the length distribution of the sequences (successions of identical monomer units) forming the chains.

The nature of these two quantities is quite different. Composition may be considered as an "exact" quantity in the sense that it may be measured by suitable experimental techniques. The result of such measurement shows, within the limits of accuracy, the actual situation existing in the copolymer.

On the contrary, sequence distribution may be inferred from experimental data only on the basis of statistical considerations. The whole sequence distribution may be deduced if (a) the amount of isolated units of one monomer is experimentally evaluated, and (b) we assume a well-specified copolymerization mechanism, which is described by one parameter only, i.e., the product of reactivity ratios, $r_1 r_2$. This is tantamount to assuming that the addition of monomer units to growing chains is uniquely influenced by the nature of the unit added last.

4.2 Characteristic Parameters of Composition

The mathematical parameters expressing the composition of a copolymer are the fractions of monomer 1, molar m_1 or weight w_1; the fractions of

Table 3 Conversion of the Representative Quantities for Copolymer Composition

	m_1	w_1	f_m	f_w
m_1		$\dfrac{M_2 w_1}{M_1+(M_2-M_1)w_1}$	$\dfrac{f_m}{1+f_m}$	$\dfrac{M_2 f_w}{M_1+M_2 f_w}$
w_1	$\dfrac{M_1 m_1}{M_2+(M_1-M_2)m_1}$		$\dfrac{M_1 f_m}{M_2+M_1 f_m}$	$\dfrac{f_w}{1+f_w}$
f_m	$\dfrac{m_1}{1-m_1}$	$\dfrac{M_2 w_1}{M_1(1-w_1)}$		$\dfrac{M_2}{M_1}f_w$
f_w	$\dfrac{M_1 m_1}{M_2(1-m_1)}$	$\dfrac{w_1}{1-w_1}$	$\dfrac{M_1}{M_2}f_m$	

monomer 2, $m_2 = 1 - m_1$ and $w_2 = 1 - w_1$, respectively, and the ratio (molar f_m or weight f_w) between monomer 1 and monomer 2. The conversion among these four quantities is reported in Table 3 (M_1 and M_2 are the molecular weights of the two monomers).

4.3 Characteristic Parameters of Distribution

The quantity that, under the restrictive hypotheses shown in Section 4.1, uniquely defines the whole distribution of the sequences is the product of the reactivity ratios $r_1 r_2$.

Although the earlier literature on ethylene–propylene copolymerization contains a large body of data on reactivity ratios, more sophisticated recent approaches question the significance of these parameters for actual copolymerizations. One of the first experimental facts to suggest a need for more involved treatment of the copolymerization kinetics is the wide variation of compositional homogeneity seen on fractionation of copolymers prepared with catalytic systems with different $r_1 r_2$ (as evaluated by the usual Lewis-Mayo or Fineman-Ross techniques). It is clear that the heterogeneity of the catalyst and the resulting presence of more than one active polymerization site play an important role. The problem has been tackled independently by Tosi et al. (76, 77) and Cozewith and Ver Strate (65). According to the latter authors, "a correlation of polymerization kinetics with polymer sequence distribution or physical properties in the general case where multiple species exist must await proper fractionation of the polymer produced by the individual species, evaluation of their individual $r_1 r_2$, and the prediction of the properties of the blend; otherwise, the reactivity ratios are best used as parameters to correlate the polymerization data."

A further insight into the mechanism of ethylene–propylene copolymerization may be found in recent work by Zambelli et al. (20). Starting from the experimental observation that ethylene–propylene copolymers prepared with syndiotactic specific catalysts contain both head-to-tail and tail-to-head oriented propylene units, these authors regard the copolymerization as an actual terpolymerization (and the propylene polymerization as a binary copolymerization), and consider the consequences of this fact on the copolymerization kinetics. The most important results emerging from this study are that at least in the presence of syndiotactic specific catalysts: (*a*) the kinetics of copolymerization is deceptively simple, owing to the high relative reactivity of ethylene, and does not give information on the copolymer structure; (*b*) literature data concerned with the statistics of the distribution of. the units are valid as far as they rely on spectroscopic information rather than on kinetic measurements; and (*c*) the $r_1 r_2$ values determined by spectroscopic analysis are real distribution indices, but they are not strictly related to the Lewis-Mayo equation.

4.4 Informational Entropy as a Measure of Copolymer Randomness

For $r_1 r_2 = 1$ the arrangement of the two types of units along the chain is completely random; i.e., the copolymer shows the highest disorder in the distribution of the sequence lengths. As soon as $r_1 r_2$ departs from unity, a tendency is observed toward the formation of longer sequences of identical monomer units if $r_1 r_2 > 1$, and of shorter sequences if $r_1 r_2 < 1$. But which is the amount of "order" for a given $r_1 r_2$? Or, in other words, how much must $r_1 r_2$ be shifted from unity to make detectable a given variation of the copolymer randomness?

These questions may be answered in terms of a quantity recently introduced, the so-called informational entropy per monomer unit (78–80). It is defined as the ratio between the natural logarithm of the number of possible ways of arrangement of the sequences present in a copolymer and the number of monomer units contained in it.

Informational entropy may be calculated in two different ways: on the basis of combinatorial analysis (78), by a direct evaluation of the number of distinct arrangements obtained by mutual permutation of the sequences of the two monomers (two arrangements that differ only in the exchange of the same monomer must be considered indistinguishable, i.e., they count as one only), or else, on the basis of the methods of information theory (79) (hence the name informational entropy), by evaluating the amount of information gained in knowing whether a given position of the chain is occupied by a unit 1 or by a unit 2. (An alternative name might be "informational multiplicity," to avoid confusion with thermodynamic entropy.)

The expression for informational entropy per monomeric unit versus fractional composition and the product of the reactivity ratios is

$$s = \frac{1}{f_m + 1} \left(\frac{K}{2 r_1 r_2 + K} \ln r_1 r_2 + f_m \ln \frac{K+2}{K} + \ln \frac{2 r_1 r_2 + K}{2 r_1 r_2} \right) \qquad [2]$$

where

$$K = f_m - 1 + [(f_m - 1)^2 + 4 r_1 r_2 f_m]^{1/2}$$

For a given value of $r_1 r_2$, informational entropy behaves as shown in Fig. 7. For low values of $r_1 r_2$, the curve of informational entropy versus

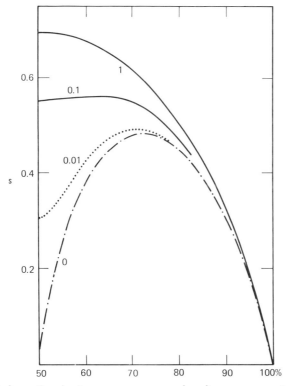

Figure 7 Informational entropy, s per monomeric unit versus percentage of copolymer composition for different values of $r_1 r_2$. The curve portion for compositions lower than 50% is the mirror image of those shown. From C. Tosi, *NMR Basic Principles and Progress*, Vol. IV, 1971, p. 129, by courtesy of Springer-Verlag, Berlin.

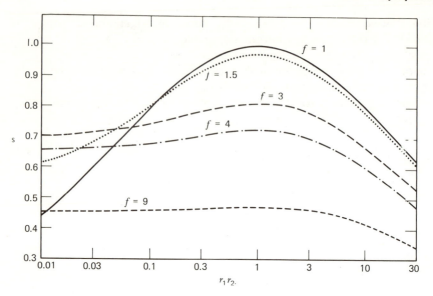

Figure 8 Representation of the informational entropy per monomeric unit versus the product of reactivity ratios for different values of f_m (i.e., for different comonomer ratios). From C. Tosi, *NMR Principles and Progress*, Vol. IV, 1971, p. 129, by courtesy

composition takes on a qualitatively different form, and at the limit of $r_1 r_2 = 0$, it goes to zero both for the 50% copolymer and for the two homopolymers, and has a maximum ($s \sim 0.48$) at a composition of about 72.4%. This behavior may be easily interpreted: At 100% there is only one sequence formed by one type of unit, so that no exchanges are possible and $s = 0$; at 50%, since $r_1 = r_2 = 0$, there is the alternating copolymer ... 121212..., consisting only of isolated units of both monomers, and no exchange is possible ($s = 0$); for intermediate compositions, only one of the reactivity ratios is equal to 0, say r_1, so that the copolymer consists of isolated units of monomer 1 alternating with sequences of any length of monomer 2: ... 12122212212222121221 .... It is therefore obvious that informational entropy differs from zero and is higher the farther the composition is from the two extreme values $f_m = 0$ and $f_m = 1$. At constant composition, informational entropy is the highest for $r_1 r_2 = 1$ and decreases slowly for $r_1 r_2 \to \infty$ and even more slowly for $r_1 r_2 \to 0$ (Fig. 8).

Equation [2] allows the calculation of the degree of randomness of the copolymers and therefore a quantitative comparison of copolymers prepared in the presence of different catalyst systems. Since the degree of randomness varies slightly in copolymers having products of reactivity ratios of the same order of magnitude, it follows that the statistical properties of these copolymers are very similar.

4.5 Graphical Representations of the Copolymers

Before describing the experimental methods for the determination of composition and sequence distribution in copolymers, we briefly review how these quantities can be represented in graphical form. Representations have been proposed based on systems of triangular coordinates (81) and Cartesian orthogonal coordinates (82).

In the first type, the fractions f_{11} of bonds 1–1, f_{22} of bonds 2–2, and $(f_{12} + f_{21})$ of bonds 1–2 and 2–1 are plotted on the sides of an equilateral triangle (Fig. 9). Copolymers with the same composition are then represented by vertical straight lines and copolymers with the same product of reactivity ratios by curves of "constant distribution" as shown in Fig. 9.

In the second type of representation, the ratios $y = f_{22}/(f_{22} + f_{21})$ and $x = f_{11}/(f_{11} + f_{12})$ are plotted on the axes of a system of orthogonal coordinates (Fig. 10). The curves of constant composition are the straight lines $y = f_m x + (1 - f_m)$, passing through point $A = (1, 1)$ and forming with AY the angle $\alpha = \tan^{-1} f_m$. The curves of constant distribution are

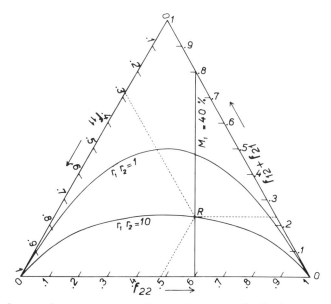

Figure 9 Curves of constant composition and constant distribution in the Igarashi diagram. From Tosi (82), by courtesy of Hüthig Wepf Verlag, Basel.

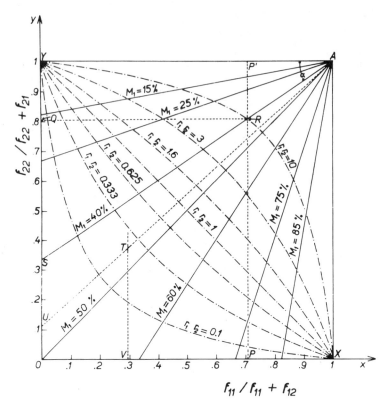

$$f_{11} \,/\, f_{11} + f_{12}$$

Figure 10 Representation of composition and sequence distribution in the copolymers according to Tosi. From Tosi (82), by courtesy of Hüthig Wepf Verlag, Basel.

equilateral hyperbolas passing through points $X = (1, 0)$ and $Y = (0, 1)$ and having equation $y = [r_1 r_2(1-x)]/[x + r_1 r_2(1-x)]$. For $r_1 r_2 = 1$, the representative curve is the diagonal XY.

5   CHARACTERISTIC QUANTITIES OF THE TERPOLYMERS

The mathematical characterization of the terpolymers is much more complex than that of the copolymers. There are eight dependent variables, two of which define the composition (either molar or weight fractions, or ratios, of the monomers) and six the reactivity ratios of the three binary copolymerizations, appearing in the terpolymerization equations as such and not in the form of products.

The   most   general   terpolymerization   equations   for   relative concentrations were proposed in 1944 by Alfrey and Goldfinger (83):

$$m_1 : m_2 : m_3 = M_1 A \left(\frac{M_1}{r_{31}r_{21}} + \frac{M_2}{r_{21}r_{32}} + \frac{M_3}{r_{31}r_{23}} \right)$$

$$: M_2 B \left(\frac{M_1}{r_{12}r_{31}} + \frac{M_2}{r_{12}r_{32}} + \frac{M_3}{r_{32}r_{13}} \right) \qquad [3]$$

$$: M_3 C \left(\frac{M_1}{r_{13}r_{21}} + \frac{M_2}{r_{23}r_{12}} + \frac{M_3}{r_{13}r_{23}} \right)$$

where, for the sake of brevity:

$$A = M_1 + \frac{M_2}{r_{12}} + \frac{M_3}{r_{13}}, \qquad B = \frac{M_1}{r_{21}} + M_2 + \frac{M_3}{r_{23}},$$

$$C = \frac{M_1}{r_{31}} + \frac{M_2}{r_{32}} + M_3$$

M and m are the molar concentrations in the feed and in the terpolymer, and the r_{ij} are the reactivity ratios.

If among the r_{ij} the relationship

$$r_{12}r_{23}r_{31} = r_{21}r_{32}r_{13}$$

holds, equations [3] may be replaced by the simpler equations:

$$m_1 : m_2 : m_3 = M_1 A : M_2 B \frac{r_{21}}{r_{13}} : M_3 C \frac{r_{31}}{r_{13}} \qquad [4]$$

proposed in 1962 by Valvassori and Sartori (84, 85) and later, independently, by Ham (86).

A particularly meaningful case of validity of equations [4] occurs in ionic copolymerizations (42, 87), in which the product of reactivity ratios is often equal to unity; in this case, the usual value of the products $r_{12}r_{23}r_{31}$ and $r_{21}r_{32}r_{13}$ is also unity. This implies that, if the reactivity ratios of the binary copolymerizations 1–2 and 1–3 are known, the reactivity ratios of copolymerization 2–3 can be calculated:

$$r_{23} = \frac{r_{13}}{r_{12}}, \qquad r_{32} = \frac{r_{31}}{r_{21}} \qquad [5]$$

Equations [5] were used for example, for the evaluation of the reactivity ratios of the anionic coordinated copolymerization of propylene with butene-1 (42), knowing the reactivity ratios of both these monomers with ethylene. If the three binary copolymerizations are ionic, the terpolymerization equations [4] are simplified further:

$$m_1 : m_2 : m_3 = M_1 : \frac{M_2}{r_{12}} : \frac{M_3}{r_{13}}$$

The deduction of formulas relating to the sequence distribution in terpolymers is formally analogous to that in copolymers. The problem was recently dealt with by Tosi (88, 89), who gave mathematical expression to the main characteristic quantities and calculated the various fractions and length of sequences and informational entropy.

Contrary to what occurs for copolymers, all these formulas for terpolymers involve not only the reactivity ratios but also the composition of the terpolymer and of the feed. Thus, whereas the copolymerization equation allows the easy transformation of relationships expressed versus the composition of feed into relationships expressed versus the composition of the copolymer, and vice versa, this transformation is practically impossible in the case of terpolymers, owing to the complexity of equations [3] and [4] connecting the two groups of variables.

6 MOLECULAR WEIGHT, $[\eta]$–M RELATIONSHIPS, AND MOLECULAR WEIGHT DISTRIBUTION

The copolymerization mechanism and physical properties of copolymers have been the topic of a number of reports. Their solution properties have rarely been investigated, although the study of dilute solutions of polymeric materials proves particularly useful for the determination of both molecular weight and molecular weight distribution. These parameters are directly connected to mechanical properties and processability, and their importance in the characterization of all polymeric materials is therefore undisputed. For copolymers, other important parameters are average composition, homogeneity, and microstructure. The absolute methods for molecular weight determination include osmometry, light scattering and ultracentrifugation.

Viscosimetry is a relative method but molecular weights may be determined simply and quickly by intrinsic viscosity measurements on dilute solutions. This method only requires previous knowledge of the K' and a constants of the Mark-Houwink relationship

$$[\eta] = K' \cdot M^a \tag{6}$$

which may be determined experimentally by comparison of molecular weight M measured by some absolute method, and intrinsic viscosity $[\eta]$ on polymer fractions of negligible molecular weight dispersity. The proper use of equation [6] for copolymers always assumes good compositional homogeneity of the starting sample; in this way, the solubility of the fractions depends only on the molecular size.

This situation, of course, occurs strictly for homopolymers; copolymers usually show the superposition of two types of distributions deriving from the dispersity of both molecular weight and composition; thus narrow fractions can hardly be prepared by the usual methods. Sometimes, such difficulties can be overcome, e.g., for copolymers having an azeotropic composition or for statistical copolymers synthesized under very particular conditions (very low conversions, etc.).

The few data reported in the literature mostly concern copolymers of styrene with different comonomers, e.g., methyl methacrylate (90, 91), acrylonitrile (92), butadiene (93), and dimethylitaconate (94).

In the case of ethylene–propylene copolymers, the problem was solved by approximate theoretical methods (95) based on the Flory-Fox treatment for polymers (96) applied to copolymers by Stockmayer et al. (97). The treatment allows obtaining, to a first approximation, the constants for relationship [6], which are essential for the determination of the molecular weight $\bar{M}_v$ from measurements of intrinsic viscosity in tetralin at 135°C.

Table 4 shows the K' constants calculated for a composition series, assuming that the value of a is constant and equal to 0.74. Values intermediate between those reported may be easily obtained by interpolation.

It follows that the determination of the molecular weight distribution of a copolymer is subject to the same limitations as those just mentioned for

Table 4 K' **Constants of Relationship [6] for Ethylene–Propylene Copolymers Containing Different Mole Percentages of Ethylene (% E) ($a = 0.74$)**

% E	$K' \times 10^4$	% E	$K' \times 10^4$
5	2.020	55	3.020
10	2.115	60	3.140
15	2.205	65	3.260
20	2.295	70	3.385
25	2.390	75	3.515
30	2.485	80	3.645
35	2.585	85	3.790
40	2.690	90	3.940
45	2.795	95	4.240
50	2.910	—	—

the relationships between intrinsic viscosity and molecular weight. Interpretation of the fractionation schemes must pay special regard to the compositional inhomogeneity of the sample.

All the usual methods of fractionation used for homopolymers can also be used for copolymers, but the methods preferred are those that allow the separation of fractions, on which it is possible to determine not only the molecular weight but also the composition of every fraction.

A description of both types of variance may sometimes be obtained by simultaneously using various techniques, e.g., gel permeation chromatography, osmometry, light diffusion, and ultraviolet spectrophotometry (98).

In the particular case of EPM copolymers and EPDM terpolymers, some molecular weight distributions may be deduced, for example, from the papers by Natta and Crespi (99), Phillips and Carrick (9), and Baranwal and Jacobs (100). The results of the fractionation of an EPM copolymer having a propylene content of 10% and the relative curve of molecular weight distribution are reported in Table 5 and Fig. 11. Table 6 shows the good homogeneity of composition of an EPDM terpolymer

Table 5 Fractionation of an Ethylene–Propylene Copolymer, 10.0 mole % Propylene, Made with a VCl_4–$(C_6H_5)_4Sn$–$AlBr_3$ Catalyst

Fraction No.	Weight (%)	Cumulative Weight (%)	Propylene (mole %)	$[\eta]$ (dl/g)	M
1	2.5	1.2	—	~0.1	1,800
2	11.2	8.1	10.9	0.26	6,400
3	9.8	18.6	10.9	0.42	12,000
4	13.8	30.4	11.0	0.57	19,000
5	14.3	44.4	10.1	0.75	27,000
6	6.5	54.8	10.4	0.86	32,000
7	5.2	61.0	9.4	0.94	37,000
8	5.0	65.8	10.2	1.04	43,000
9	5.4	71.0	9.4	1.14	48,000
10	4.1	75.7	10.5	1.18	50,000
11	4.4	80.0	9.4	1.27	55,000
12	6.9	85.6	10.2	1.48	68,000
13	9.6	94.0	9.4	1.85	92,000
14	1.3	99.3	—	1.72	83,000
				0.885^a	
Unfractionated copolymer			10.0	0.884	(M_v)
				0.870	34,000

^{a} Weight average intrinsic viscosity.

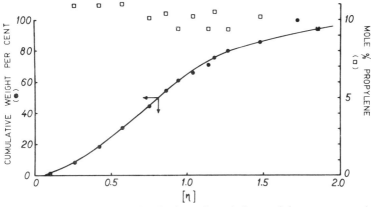

Figure 11 Molecular weight distribution: Cumulative weight percent and mole percent, propylene in an ethylene-propylene copolymer. From Phillips and Carrick (9), by courtesy of the American Chemical Society.

both with regard to unsaturation and the content of the two main comonomers.

The thermodynamic properties of unperturbed copolymer chains have recently been evaluated (101) according to a theory based on a pseudo-stereochemical equilibrium among the different monomer units. The calculated unperturbed dimensions agree well with the experimental data

Table 6 Composition of Various Fractions of a Terpolymer Having a Total Content of 1.7 Double Bonds per 100 Monomeric Units and 63 mole % of Ethylene

Fraction No.	% of Total Terpolymer	C_2 (mole %)	Double Bond Content per 100 Monomeric Units
1	13.8	64.0	1.68
2	21.6	64.0	1.68
3	28.0	64.0	1.68
4	33.0	64.0	1.68
5	42.7	64.0	1.65
6	51.7	64.0	1.65
7	61.5	64.0	1.65
8	68.0	63.0	1.72
9	73.5	63.0	1.72
10	81.0	63.0	1.72
11	86.0	61.5	1.67
12	91.5	59.0	1.78
13	98.5	60.5	1.70

obtained by measurements in θ solvents, thus proving the correctness of the assumptions made for our polymers, i.e., the constancy of $r_1 r_2$ value and the statistical distribution of the two monomers in the chains.

7 ANALYTICAL PROBLEMS

7.1 Methods for the Determination of Composition

The composition of ethylene–propylene copolymers and the related terpolymers may be determined by various techniques, a detailed description of which is beyond the scope of this chapter. For an exhaustive discussion of the applications of infrared spectroscopy not only in the analytical field but also in all aspects of ethylene–propylene copolymers, see Tosi and Ciampelli (102).

7.1.1 *Radiochemical Analysis* (33)

This method is based on the fact that, when copolymerizing propylene with ^{14}C-labeled ethylene, the weight percentage of the ethylene units in the copolymer is given by the ratio between the specific activities of the copolymer and of polyethylene, obtained by homopolymerization of radioactive ethylene. This is very accurate (reproducibility within 1–2% of the amount of ethylene present) and very useful for the preparation of standard copolymer samples to be used for calibration with other methods. It is scarcely or never used for routine analyses.

The composition determined by radiochemical analysis is *source based*; i.e., it refers to the source which yields the macromolecule (the feed mixture of $CH_2\!=\!CH_2$ and $CH_3\!-\!CH\!=\!CH_2$), but tells nothing about the repeating units that occur in the copolymer and how they are linked within the chains.

7.1.2 *Infrared Analysis* (9, 103–114)

This method is most frequently used owing to its practicability, although the data it supplies are less accurate than those given by the radiochemical method.

The composition determined by infrared analysis is *structure based*; i.e., it results from measurements done on the units actually present in the copolymer. It coincides with the source based composition only if the copolymer is exclusively formed by *ideal constitutional units* (102), i.e.,

$$-CH_2-CH_2- \quad \text{and} \quad -CH_2-\underset{\underset{CH_3}{|}}{CH}-$$

units, respectively.

Infrared analysis may be carried out on films, on molten samples (if the bands are sensitive to the presence of crystallinity or paracrystallinity), or in solution.

As a general rule, methods using infrared bands originating from C–H stretching vibrations are less sensitive to the nature of the catalytic system used to prepare the copolymers than methods based on bending vibrations (114). The latter are more influenced by the conformation of the polymeric chain.

Greater care must be taken when applying to terpolymers the analytical methods set up for copolymers since the third monomer may have some bands interfering, often to an indefinable extent, with those used for the analysis. In this case, the examination of binary copolymers of ethylene and/or propylene with the third monomer may be of considerable help.

Infrared spectroscopy is the most widely used method for determining the third monomer in terpolymers. The wavelengths of the bands used for the analysis of the most common termonomers are listed here. Bands at about 6 μ derive from C=C stretching vibrations, whereas the bands with higher wavelength derive from the out-of-plane deformation of the hydrogen atoms on the double bonds:

	Wavelength (μ)
Methyltetrahydroindene	12.56
Ethylidene norbornene	5.93
Dicyclopentadiene	6.22
1,5,9-Decatriene	10.98
1,5-Cyclooctadiene	15.05
Isobutenylnorbornene	11.28 (or 6.07)
Methylpentadiene	11.28
Methylene norbornene	11.48
trans-1,4-Hexadiene	10.35
Butenyl norbornene	6.03

7.1.3 *Nuclear Magnetic Resonance*

^{1}H NMR has rarely been used for the analysis of ethylene–propylene copolymers, not only owing to the slight difference among the chemical

shifts of the various types of protons but also to the possible superposition of the peaks of different protons.

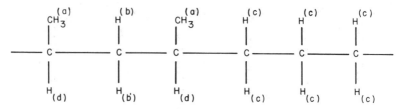

(a) = 0.856 ppm from TMS
(b) = 0.859 ppm from TMS
(b') = 1.26 ppm from TMS
(c) = 1.22 ppm from TMS
(d) = 1.55 ppm from TMS

Comparison of the spectra of isotactic polypropylene and those of the ethylene–propylene copolymers, if the chemical shifts and the values of the coupling constants $J_{bb'} = 13$ Hz and $J_{bd} = 6.5$ Hz are considered, shows that the complete separation of the methyl protons (a) from the methylenic ones (b) is impossible when some propylene units belong to isotactic sequences.

The analysis of the ethylene molar fraction in terpolymers with *trans*-1,4-hexadiene (which is the same as in the copolymers) was described by Dudek and Bueche (115). It is based on the ratio between the area under the curve for the methyl protons and that under the curve for the methylene and methine protons (which cannot be completely resolved, one from the other). The percentage of ethylene is determined by comparison with a theoretical curve obtained by assuming that the terpolymer contains only ethylene and propylene.

An analytical method for block copolymers was described by Porter (116). Spectra were recorded at $200 \pm 10°C$ on solutions at concentrations below 10% in diphenyl ether. At higher ethylene contents, the accuracy of the method is about 10%.

A time averaging method was proposed (117) for analysis of the third (diene) monomer. The method gives a sufficiently characteristic signal to allow its identification and moreover, since the third monomer initially contains two double bonds differing in structure and reactivity, the bond used in copolymerization may be distinguished from the bond remaining for subsequent use in vulcanization. The chemical shifts of olefinic protons of a number of termonomers are as follows:

1,5-Cyclooctadiene 4.55τ
Dicyclopentadiene 4.55τ

1,4-Hexadiene	4.7τ
Methylene norbornene	5.25τ; 5.5τ
Ethylidene norbornene	4.8τ; 4.9τ

In contrast to ^{1}H NMR, ^{13}C NMR provides subtle structural information on ethylene–propylene copolymers, and will probably solve many ambiguities that still exist in sequence studies. For a detailed description of both the experimental technique and the assignment of the spectrum, the reader is referred to an excellent work by Carman and Wilkes (118), who found the ^{13}C chemical shifts of ethylene–propylene copolymers to be very sensitive to monomer sequence distribution. Methylene resonances were interpreted in terms of methylene sequence lengths, and tertiary carbon resonances were interpreted in terms of propylene centered triad sequences. They also derived a formula, based on copolymerization theory, for calculating the reactivity ratios product directly from the copolymer composition and the ratio of contiguous to isolated propylene sequences.

Examination by ^{13}C NMR of the microstructure of ethylene–propylene copolymers, prepared in the presence of syndiospecific vanadium based catalyst systems, also provided evidence on the mechanism of steric and arrangement control (119).

7.1.4 Mass Spectroscopy of the Pyrolysis Products

A method for the analysis of ethylene–propylene copolymers based on the pyrolysis of the copolymer and the subsequent analysis by mass spectrograph of the products obtained was proposed by Bua and Manaresi in 1959 (120).

7.1.5 Halogenation with Iodine Monochloride

By this method it is possible to determine the third monomer in terpolymers, on the basis of the main reaction:

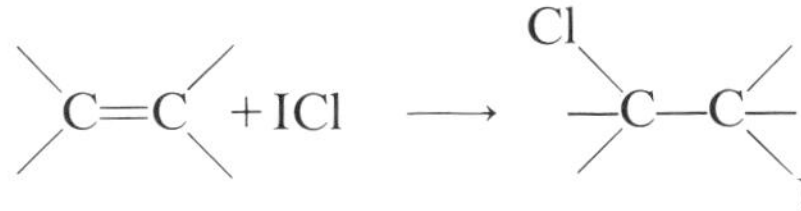

This method was applied to polymers by Lee et al. in 1948 (121); for application to ethylene and propylene terpolymers, see Tunnicliffe et al. (122). The method is quite delicate, and may easily lead to high results owing to the possible occurrence of secondary substitution and elimination reactions.

7.1.6 *Pyrolysis and Gas Chromatography*

This method consists of the static or dynamic pyrolysis of the polymer and subsequent separation by gas chromatography of the fragments obtained. At the exit of the gas chromatograph the various fragments may be identified by various techniques, such as mass spectroscopy, infrared analysis, and NMR.

Pyrolysis may take place by heating in an oven or on an electrical resistance, by heating induced by electromagnetic field (123) (method of the Curie point), or by laser beam (124). The three methods give different results: secondary reactions are especially reduced in the second and third ones, the time of heating being only a few milliseconds.

Fragmentation occurs (125) by initiation and β scission or unzipping followed by intermolecular and intramolecular hydrogen transfer. The pyrolysis products may be hydrogenated to simplify the chromatogram interpretation.

Among the most significant work concerning the C_2–C_3 copolymer is that of Van Schooten and Evenhuis (126) who set up a method for the determination of the C_3 content in the C_2–C_3 copolymer, from the ratio of the peaks n-$C_7/2MC_7 + 4MC_7$ (M is methyl, n is normal, C_y is cyclo, and C_i is a straight chain of i carbon atoms). This method is independent of the catalyst system used; the intensities of the ratios 2–$4MC_7/n$-C_4, 2–$5MC_6/n$-C_4, $4MC_9/n$-C_4 versus the percentage of C_3 indicate the amount of head-to-tail, head-to-head, and tail-to-tail sequences, respectively, in the copolymer.

In the case of terpolymers, Van Schooten (126) reports some examples in which sharp peaks characteristic of the termonomer are obtained by pyrolysis:–MC_yC_5 from methylnorbornadiene; –C_yC_5 from DCPD; –MC_yC_5 from methylcyclopentadiene dimer. In the case of the C_2–C_3–dicyclopentadiene terpolymer, he reports an example of calibration, by plotting the ratio of C_yC_5/n-C_4 versus the ozone number.

7.2 Methods for the Determination of Sequence Distribution

The analytical methods for sequence distribution are based on the evaluation of the number of units of one or both monomers present in sequences of a given length. Such an analysis may be carried out directly, e.g., on the basis of both position and intensity of a spectral band characteristic of a particular group of monomer units, or indirectly, e.g., by cracking of the copolymer and subsequent gas chromatographic analysis of the products obtained from cracking. The infrared spectrum

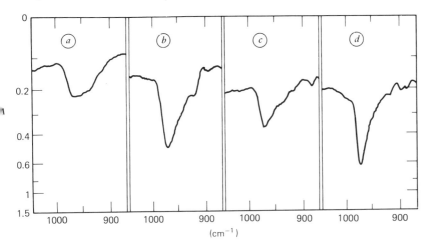

Figure 12 Infrared spectra of various ethylene-propylene copolymers with films of constant thickness: (a) f_m (propylene/ethylene) = 0.43, $r_1 r_2 = 0.27$; (b) $f_m = 1.5$, $r_1 r_2 = 0.27$; (c) $f_m = 0.43$, $r_1 r_2 = 1.15$; (d) $f_m = 1.5$, $r_1 r_2 = 1.15$. From Tosi et al. (130) by courtesy of Pergamon Press Ltd., Oxford.

gives considerable help in the study of the sequence distribution. Actually, two spectral regions, that between 10 and 11 μ, characteristic of the rocking vibrations of the methyl group, and that between 13 and 14 μ, characteristic of the rocking vibrations of the methylenic group, appear completely different (Fig. 12) in copolymers having the same composition, but prepared in the presence of different catalysts. The possibility of obtaining information on sequence distribution from the infrared spectrum was first pointed out by Natta et al. (127); they observed a qualitative agreement between the shape of the band between 13 and 14 μ and the fraction of the ethylene sequences with different length. Shortly thereafter, Bucci and Simonazzi (128) examined this work from a qualitative point of view. On the basis of the absorptivities of suitable model substances, they proposed a method for quantitative determination of the ethylene sequences of different length present in the copolymer. The results obtained were in qualitative agreement with the distribution curves drawn by the statistical laws of copolymerization.

The rocking vibrations of the methyl groups were employed by Valvassori et al. (129, 130). On the basis of the absorptivities of the bands at 10.30 and at 10.68 μ deduced from two model substances, i.e., atactic polypropylene in which all propylene units are in a sequence, and hydrogenated natural rubber in which all propylene units are isolated between ethylene units, these authors defined a "distribution index" ϕ, which corresponds to the ratio between the number of propylene–ethylene and propylene–propylene bonds. Such a correspondence allows the deduction of an equation for the product of reactivity ratios:

$$r_1 r_2 = \frac{1}{f_m} \left(\frac{1}{\phi^2} - \frac{f_m - 1}{\phi} \right)$$

where f_m (molar ratio between propylene and ethylene) can be obtained from measurements on the same infrared spectrum.

Sequence distribution of monomeric units in ethylene–propylene copolymers was studied by computer analysis of the infrared spectra by Drushel et al. (131). These authors singled out a position in the complex band between 10 and 11 μ which they attributed to sequences of two propylene units. However, their results are questionable since it seems that the ratios found among the sequences of different length cannot be deduced from any statistical model of copolymerization.

The distribution of monomer units in ethylene–propylene copolymers was also studied by Kissin et al. (132), who considered the two different steric configurations possible for propylene. Theoretical data were correlated with the data obtained from the infrared spectrum on the basis of the ratio between the absorbances of the bands at 973 and 1380 cm^{-1}.

A study of the blockiness of ethylene–propylene copolymers by pyrolysis was reported by Michajlov et al. (133). According to these authors, the relative amounts of linear and branched hydrocarbons obtained from the chromatogram of the pyrolysis products are, for the same composition, an index of the copolymer blockiness. This work gives evidence of the sharp differences existing between amorphous and crystalline copolymers, as well as among the amorphous copolymers differing in composition by five propylene units.

A method for determination of the distribution of the monomers, based on pyrolysis and mass spectrometry of the volatile products from it (n-butane and n-pentane), was also developed by Nencini et al. (134).

8 PHYSICAL PROPERTIES

Ethylene–propylene copolymers and terpolymers are materials of exceptional interest and versatility owing to the number of variables that can alter their characteristics; hence they are particularly suitable for the study of the relationships between structure and properties (135). In fact, unlike many other synthetic elastomers, their structural parameters, i.e., molecular weight, molecular weight distribution, monomer composition, and monomer sequence distribution, may all be varied during manufacture. This offers advantages, both technical and scientific since it is possible to choose the best combinations of the structural variables to obtain the performance characteristics most convenient for a given use, and to study the influence of different structural parameters on the physical and technological properties.

8.1 Characteristics of the Raw Materials

The raw terpolymers do not differ considerably from the bipolymers, the third monomer content usually being so low that it does not influence the molecular structure considerably. Thus copolymers here will mean both the binary and ternary ones unless otherwise specified.

8.1.1 Rheological Properties

Rheological properties are of major interest because they are connected with processability (136) since most operations performed before vulcanization involve flow and shear treatments. The general aspects of processability of ethylene–propylene elastomers will not be discussed here (but see Chapter 11). We will focus attention on the relationships between rheological behavior and molecular structure, and in particular on the structural parameters, e.g., composition and microstructure, which are primary in distinguishing these polymers from other elastomers.

The scientific studies on this topic are fairly recent (137–140); their development was promoted by the necessity of tackling processability problems of stereo rubbers, which are acquiring an ever larger commercial importance. These synthetic elastomers generally have a relatively narrow molecular weight distribution, different microstructure, and are more linear than either natural rubber or emulsion SBR; hence users accustomed to traditional rubbers met with new handling problems. EP copolymers are perhaps unique in that they are barely degraded on milling (100), thanks to their outstanding chemical stability. When ethylene–propylene copolymers are linear and amorphous, they show a rheological behavior typical of linear hydrocarbon rubbers; hence their processability does not differ markedly from that of other stereo rubbers.

The rheological parameters best characterizing elastomer processability (141) are the bulk viscosity, in particular the Mooney viscosity ML_4 (142), and the maximum relaxation time τ_m, which is associated with elastic memory (140, 66). The dependence of these parameters on molecular weight is the same for EP elastomers as for other unbranched linear hydrocarbon elastomers (143). The Mooney viscosity increases with about the first power, and τ_m with the 3.6–3.8 power of the molecular weight

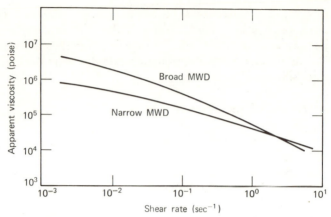

Figure 13 Apparent viscosity as a function of shear rate for EPDMs of different molecular weight distribution. From Kresge et al. (144), by courtesy of Rubber & Technical Press Ltd., Tenterden, Kent.

(66, 143, 144). Increasing the width of the molecular weight distribution increases the relaxation time τ_m (143) and the deviation from Newtonian behavior (145) of the viscous flow; this last effect is shown in Fig. 13.

Effect of Composition and Microstructure. The effect of composition and of sequence distribution is related to the possibility of the existence of supermolecular structures caused by ethylene associations.

Tokita and Scott (66) studied the dependence of τ_m on composition, for both bipolymers and terpolymers. They found that at room temperature there is no dependence for ethylene contents below 60 wt %, whereas at ethylene/propylene ratios above 60:40, τ_m increases sharply (Fig. 14). This sudden increase of the relaxation time, found in both bipolymers and terpolymers, is connected with the ethylene association and with a true crystallinity for higher ethylene contents (143). A similar trend also occurs for the dependence of viscosity on composition (144), at temperatures below 100°C.

These effects are rather normal if we consider that the ethylene associations cause a strong increase in the frictional interchain forces on which the rheological behavior primarily depends. Obviously, the occurrence of such phenomena of pseudocrystallinity depends on temperature; their influence decreases with increasing temperature, at the same ethylene content. This behavior is not observed in the high propylene copolymers, since the Ziegler-Natta catalysts, which are typical of the copolymerization, do not produce stereoregular polypropylene sequences.

The influence of composition on viscoelastic behavior is also evident in the stress-strain curves of crude polymers. The stress-strain behavior of uncured materials can be regarded as an aspect of processability and is important for several applications. Figure 15 shows some stress-strain curves of uncured ethylene–propylene elastomers with different molecular structures (146); ethylene-rich polymers generally show a behavior like that of curve 1, characterized by a high tensile strength, due to stress induced crystallization. Such curves are influenced not only by composition but also by other factors, such as molecular weight distribution, presence of fillers and/or oils, and molecular microstructure.

The influence of the monomeric sequence distribution, expressed by the product of the reactivity ratios $r_1 r_2$, on tensile properties is given in

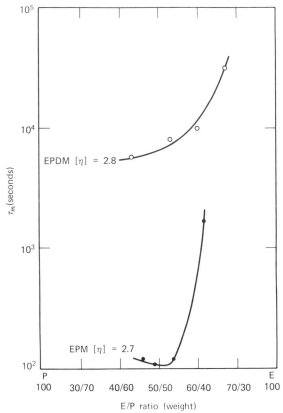

Figure 14 Effect of E/P ratio on the relaxation time τ_m for EPM and EPDM–DCPD. From N. Tokita and R. Scott, *Rubber Chem. Technol.* **42,** 944 (1969), by courtesy of the Division of Rubber Chemistry, American Chemical Society, Akron, Ohio.

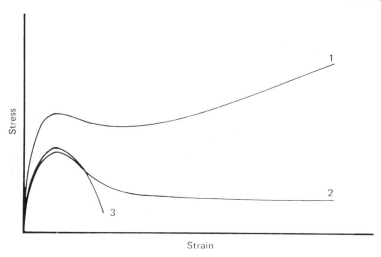

Figure 15 Stress-strain curves of uncured EP rubbers of different compositions. 1, Ethylene-rich samples; 2, propylene-rich samples; 3, samples with 40–55 wt % of propylene. From Ballini et al. (146).

Fig. 16, where the stress-strain curves of samples with approximately the same composition are shown. The sample with the highest value of $r_1 r_2$ shows a crystallization upturn due to the presence of ethylene sequences long enough to produce association phenomena.

Hence when studying the dependence of rheological properties on composition, the influence of microstructure and of temperature must always be taken into account, since the presence of ethylene sequences of varying lengths exerts great influence on these properties.

Model polymers of different microstructural homogeneity have been synthesized and their properties are being widely investigated in order to obtain clear relationships with their structure (135). The preparative scheme is shown in Fig. 17. The 1:1 model, consisting of regularly alternating ethylene and propylene units, was obtained by hydrogenation of polyisoprene (147). The 2:1 model, where two ethylene units regularly alternate with a propylene one, was obtained by hydrogenating an alternating copolymer of butadiene and propylene (148). The 3:1 E/P model was obtained by hydrogenation of polymethyloctenamer (149). The 5:1 E/P model was obtained by hydrogenation of polymethyldodecenamer (150).

Effect of the Third Monomer. The third monomer, present in small amounts in terpolymers, does not remarkably influence rheological properties, unless the diene participates in the copolymerization and

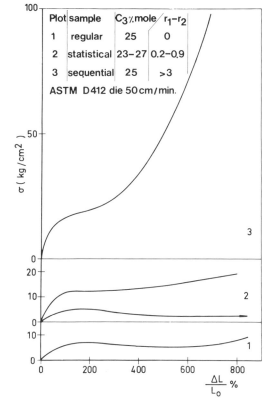

Figure 16 Stress-strain curves of uncured medium ethylene copolymers. From G. Crespi, A. Valvassori, V. Zamboni, and U. Flisi, *Chim. Ind.* (*Milan*), **55**, 130 (1973), by courtesy of SpA Editrice di Chimica, Milan.

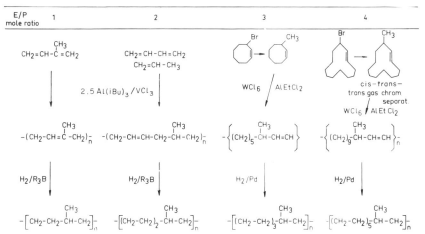

Figure 17 Scheme of various methods for synthesis of ethylene-propylene model copolymers.

influences the molecular weight distribution or gives rise to branching or even to cross-linking. Long chain branching is formed in terpolymers when using third monomers having only a small difference in reactivity between its two double bonds (66). This is the case, e.g., with dicyclopentadiene which easily gives branched terpolymers with values of τ_m much higher than those of the corresponding copolymers (66), the overall composition and other characteristics being the same.

Branching leads to higher zero shear viscosity and to larger deviations from Newtonian flow (140), at the same molecular weight. Cross-linking is a negative factor for processability since the presence of gel in elastomers makes any operation difficult and mainly hinders the homogeneous incorporation of curing ingredients.

Entanglements. Since entanglements act as labile cross-links, they exert considerable influence on viscoelastic properties. Their best known effect is to cause the Newtonian viscosity to vary with (high) molecular weight following a 3.4 power law (151). Below a critical molecular weight corresponding to the spacing between entanglements, viscosity varies approximately linearly with M.

Entanglements influence relaxation, creep, and vulcanizate properties (152). The spacing between two consecutive entanglements in ethylene–propylene copolymers, determined by measurements of relaxation (153) and creep (115) and by mechanical-dynamic properties (154), is surprisingly short. The number N_e of chain bonds between junction points was found to range between 30 and 70, depending on the method of analysis. In this case composition probably exerts some influence, since in ethylene-rich polymers the measured value may represent a spacing between very small crystallites rather than between entanglements. However, the shortest distance of 31 bonds was found in a copolymer with an ethylene/propylene ratio of 16:84 (154).

A much higher value of N_e was found by calculations based on the difference between the cross-link density measured by physical methods and that predicted theoretically on the basis of the amount of cross-linking agent used (155). The value of 182, found here for N_e, approaches the usual value for other elastomers. Entanglements and/or chain associations must be taken into particular account when the cross-link density is being investigated (156), since their effects may give misleading results, especially when composition is varied.

8.1.2 *Thermodynamic Properties*

For many rubber applications, the most important properties refer to transitions. For example, if the glass transition temperature T_g is low, a rubber may be used down to low temperatures.

T_g influences rheological properties according to the well-known equation by Williams, Landel, and Ferry (WLF):

$$\ln \frac{\eta_0}{\eta} = \frac{A(T - T_g)}{B + (T - T_g)} \qquad [7]$$

where η_0 is the viscosity at T_g, η is the value at temperature T, A and B are two constants that, for a number of noncrystalline polymers, equal 40 and 52, respectively.

T_g also influences mechanical properties since polymers with high T_g are stiffer at room temperature. However, neither the molecular weight nor the molecular weight distribution exerts a considerable effect, since T_g changes only at very low molecular weights (157). Composition also does not exert much influence, at least between 20 and 50 mole % of propylene (149, 158, 159), which is the area of practical interest. For these compositions, T_g varies between -55 and $-60°C$. The other transition that affects mechanical properties is crystallization, which is of particular interest in connection with association phenomena.

In rubbers, "crystallinity" indicates crystallinity that is present in the elastomer in the undeformed state. It is a drawback for many applications, since it makes rubber stiff, hard, and barely processable. Crystallization which occurs only under stress (stress induced crystallinity), on the other hand, is a typical phenomenon for, e.g., natural rubber; i.e., crystallinity at room temperature appears only at high stretching ratios and disappears when the sample is released. This makes natural rubber softly elastic at low deformations and very strong at high deformations. The phenomenon in ethylene–propylene elastomers is rather complex because both situations are almost always present (160, 161) and depend on both composition and microstructure. As a general trend, random copolymers with an ethylene content between 80 and 60 mole % are practically amorphous under x-ray examination (162). When stretched, they give an x-ray diffraction pattern revealing crystallinity.

The type of order attained by the macromolecules in stretched samples may be described as a disordered structure connected with that of orthorhombic polyethylene. The unit cell of polyethylene is enlarged in this case along the "a" axis by the entering propylene units, which also causes a crystallinity decrease. The degree of crystallinity at room temperature may vary from about 75% to zero over a composition range from pure polyethylene to a 60 mole % ethylene copolymer (163). Degree of crystallinity and melting temperature also depend strongly on monomer sequence distribution. In fact, while microstructurally heterogeneous copolymers contain some crystallinity at room temperature with an ethylene content around 70 mole %, the 3 : 1 regular model (75% ethylene) (Fig. 17) is amorphous under the same conditions (135).

Unlike true crystallization under stress, that of ethylene–propylene copolymers is not completely reversible. This may be related to the large sequence distribution, which probably produces different types of crystallites, where the unit cell may be more or less enlarged depending on the propylene content. Such heterogeneity of the crystallites is demonstrated by the broad melting range for the copolymers. Values of melting temperature, T_m, calculated from stress-temperature curves, have, in fact, revealed, as shown in Fig. 18, a variation of 50°C, passing from a very heterogeneous sample to a regular one.

Stress induced crystallization also depends on composition and markedly on sequential heterogeneity, as has been found by thermoelastic measurements (161) on vulcanized EPDMs with different composition.

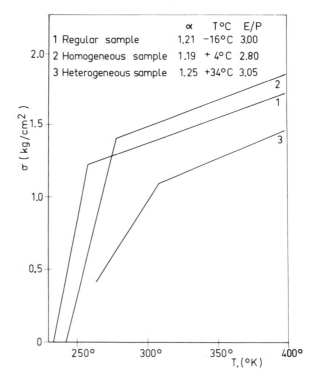

Figure 18 Stress-temperature curves of samples having similar composition but different homogeneity. α is the stretching ratio, T the crystallization temperature and E/P the ethylene/propylene ratio. From G. Crespi, A, Valvassori, V. Zamboni, and U. Flisi, *Chim. Ind.* (*Milan*), **55**, 130 (1973), by courtesy of SpA Editrice di Chimica, Milan.

Unlike natural rubber, where stress induced crystallization depends strongly on stretching ratio, it occurs at very low elongation in EPDMs showing this phenomenon as if crystallization nuclei were present in the undeformed state.

8.1.3 Chemical Stability

The crude ethylene–propylene copolymers are predominantly saturated and therefore, compared to other elastomers, show a higher stability toward oxidation, heat, and aging. After addition of small amounts of stabilizers, they may be stored for long times, and can stand considerable changes of temperature. Stability depends on the purity of the polymers; in fact, catalytic residues markedly reduce resistance to aging. Cold flow, another important characteristic for storage presents no particular problem for these elastomers.

8.2 Vulcanization Behavior

Rather than a description of the vulcanization of ethylene–propylene elastomers, this will be a discussion of the structural parameters affecting their behavior in this process. Vulcanization behavior with details of the effects of curatives is described in Chapter 8.

It is of fundamental importance to distinguish bipolymers from terpolymers, since the greatest difference between the two groups of materials consists in just the method of cross-linking. Owing to their saturated nature, bipolymers cannot be cured by the conventional systems based on accelerated sulfur. It was this drawback of bipolymers that led to the preparation of unsaturated terpolymers, which have had the largest commercial success. On the other hand, peroxide cross-linking of ethylene–propylene elastomers offers some advantages (146) over sulfur cures in some applications, e.g., high temperature cure, the production of vulcanizates with good resistance to both high temperature and compression set, and production of sulfur-free vulcanizates for electrical uses.

8.2.1 Cross-linking of Copolymers

The copolymers are vulcanized with organic peroxides, generally with coagents (164), among which sulfur is the most widely used (165) because it gives excellent characteristics to the vulcanizates (166–168). The cross-linking efficiency of peroxide, both in the absence and in the presence of coagents, depends on the polymer composition. It equals 70% in copolymers having 25 mole % propylene but falls below 10% when the propylene content is 90 mole % (166). The low cross-linking efficiency of propylene-rich copolymers is due to polymer chain scission, particularly in regions where sequences of two or more propylene units are present (169). Chain rupture, in fact, probably occurs by β scission on the tertiary radical bound to other tertiary carbons in β position on both sides (166).

Since sulfur is the most widely used coagent, there are several studies of its effect in increasing the cross-linking efficiency of peroxide (164–172). As revealed by these investigations, sulfur exerts its largest influence on efficiency when it is present at the level of one atom per molecule of peroxide (165); at this ratio it actually decreases the amount of peroxide required to obtain a given cross-link density (172). The effect of the amount of sulfur on the tensile properties is given in Fig. 19. It was also found that sulfur enters into the network forming mono- or polysulfidic cross-links (173, 174), which supply the vulcanizates with good tensile properties (167, 168) and stability to oxidation (171).

Sulfur takes part in cross-linking by a mechanism that consists mainly of reactions that stabilize the radicals produced by attack at tertiary carbons, thus preventing scission (170, 175). This may be deduced not only from the higher efficiency and the presence of sulfidic cross-links,

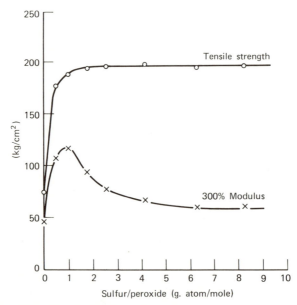

Figure 19 Effect of sulfur/peroxide ratio on tensile properties of EPM vulcanizates. From Natta et al. (33), by courtesy of the Division of Rubber Chemistry, American Chemical Society, Akron, Ohio.

but also from the fact that its synergistic action is observed mainly in propylene-rich polymers, i.e., where scission is the highest. An excellent description of the curing mechanisms of EPM elastomers with organic peroxides in the absence or presence of adjuvants has been given by Baldwin and Ver Strate in their outstanding review (175) on EP elastomers.

Other methods, different from those employing peroxides, have been used to study the cross-linking of bipolymers; some of these methods are based on a partial chemical transformation of the polymer, as may be obtained by chlorination, chlorosulfonation, and grafting with unsaturated acids. A detailed description of these methods is given in an earlier review on polyolefin elastomers (33).

8.2.2 Cross-linking of Terpolymers

Terpolymers may also be cross-linked with peroxides, in which case the cross-linking efficiency is far higher than in the corresponding bipolymers (169, 176). In general, such a higher efficiency is attributed to the fact that unsaturation may give chain reactions, even though this conclusion has been questioned by some authors, who carried out vulcanizations on hydrogenated terpolymers (177). These authors attribute the higher efficiency with the terpolymers to the presence of branching. The capability of unsaturation to give chain reactions in peroxide cure is supported by the fact that the most efficient termonomers are negatively influenced by the presence of sulfur (178). In fact, in this case sulfur stabilizes the peroxide induced radicals and reduces efficiency, preventing chain reactions.

As already mentioned, vulcanization with sulfur plus accelerators is much more interesting and important than that with peroxide, from the technological point of view. EPDM elastomers offer an ideal system for the study of the vulcanization process since the active centers for cross-linking may have different reactivities and may be introduced in the polymer in different ways and in variable amounts. Furthermore, the low concentration of these centers, separated by relatively long inert segments, allows the obtainment of less complex cross-linked structures than in highly unsaturated rubbers. Baldwin et al. (169) gave an example of the amount of useful information that may be obtained from the study of these polymers for investigating the vulcanization process.

Referring to the reaction mechanism it was found (179) that one cross-link is formed for each two double bonds initially present, as happens in highly unsaturated rubbers. The presence of unsaturation was found to be a necessary although not sufficient requisite, since the presence of allylic hydrogen is also a prerequisite for effective vulcanization (169).

The kinetics of EPDM vulcanization is very important from both the scientific and practical points of view. Its practical importance derives from the necessity of obtaining terpolymers curing fast enough to have cure compatibility with traditional rubbers such as SBR or natural rubber.

The curing rate can be improved by either choosing a suitable termonomer or increasing the unsaturation level (180, 181). Much work has been done and numerous papers have been published on the problem. In one of the earliest papers, Crespi and Arcozzi (182) assumed that the reaction had a second-order dependence on unsaturation.

The kinetic interpretation of the curing behavior of EPDMs, as of other rubbers, is complicated by the simultaneous occurrence of various reactions; (a) cross-linking that predominates in the initial phase, in which the polysulfidic cross-links are mainly formed; (b) bond cleavage occurring at longer times and leading to reversion; and (c) conversion of the polysulfidic cross-links to the monosulfidic ones.

An attempt to study this complex phenomenon taking into account all its steps was carried out by Fujimoto and Nakade (183), who compared EPDMs containing the most common termonomers and concluded that ENB is the fastest one. A comparison of the curing curves of terpolymers containing a variety of termonomers is shown in Fig. 5, where ENB clearly prevails over the others. ENB termonomer not only has a high cure rate but also a high cross-linking efficiency, and gives networks very stable to vulcanization reversion. Even the presence of a very fast curing termonomer such as ENB is not sufficient to raise the cure rate to the level of the traditional rubbers without increasing its content to three or four times its original concentration to produce the so-called fast curing EPDMs (181, 184–186).

8.2.3 Covulcanization with High Unsaturation Rubbers

The use of ethylene–propylene elastomers in blends with highly unsaturated rubbers, such as natural rubber or SBR, for the main purpose of increasing their ozone resistance (187), is having ever increasing success and represents an area of major commercial interest.

Some difficulties are found in covulcanizing EPDMs with unsaturated rubbers since the unsaturation contents are so different. The problems become self-evident on examining the curves of Fig. 20 (188) where the physical properties of the vulcanizates are plotted versus the blend composition. There is a drop of physical mechanical properties of the blends in comparison with those of the vulcanizates obtained from pure polymers. This drop is dramatic at high EPDM contents, whereas it is not

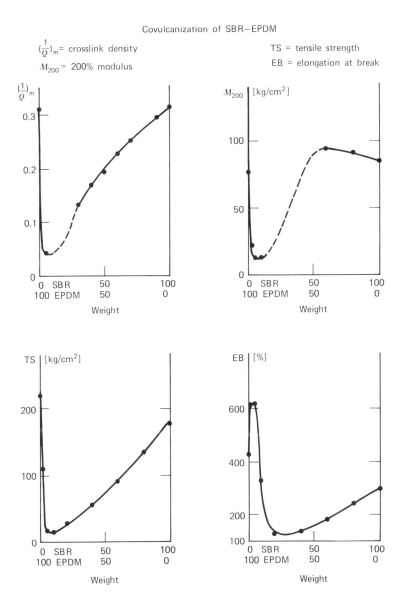

Covulcanization of SBR—EPDM

$(\frac{1}{Q})_m$ = crosslink density

M_{200} = 200% modulus

TS = tensile strength

EB = elongation at break

Figure 20 Covulcanization of SBR–EPDM. From Kerrutt et al. (188), by courtesy of Verlag für Radio-Foto Kino-Technik, Berlin.

very strong in blends containing up to 20–30 parts of terpolymer. At these concentrations, the terpolymer is not easily extractable from the vulcanizate with solvents.

This does not mean that it is covulcanized with the unsaturated rubber because the saturated copolymer, which certainly does not covulcanize, is also insolubilized by these vulcanization procedures. In fact only a small percentage of EPM is extractable from sulfur cured blends containing up to 20% of copolymer (188). Neither solvent extraction nor tensile modulus gives sure evidence of covulcanization; other characteristics, such as tensile strength and dynamic properties, are more indicative.

The difficulty of covulcanizing EPDMs with diene rubbers such as NR, BR, SBR, or CR, derives from at least three main factors: (a) the lack of miscibility with many rubbers, for thermodynamic reasons (189) and because of the low polarity (190) of EPDMs; (b) the low cure rate of EPDMs, due to their low unsaturation; and (c) the tendency of curatives to migrate from the EPDM phase, due to their better solubility in other rubbers (190–192). All these factors preventing EPDMs from covulcanizing with other elastomers have been widely investigated.

Thus the cure rate was improved by choosing a proper termonomer (183) and by increasing the amount of unsaturation (180, 181, 184, 185). These efforts have led to the "fast curing EPDMs." Important studies were also devoted to the problem of curative solubility and migration, and new accelerators were proposed (193) as a possible solution. The question of covulcanization was also tackled with other approaches, such as using halogenated EPDMs (194) or grafting accelerators onto the EPDM chain (195). All these approaches and others (196) are helping to solve the very important problem of compatibility with diene rubbers, which is a requisite for the use of EPDM as a general purpose elastomer.

8.3 Properties of the Vulcanizates

Like the general purpose rubbers (e.g., SBR), EP elastomers show good mechanical-dynamic properties, and are quite superior in important properties, such as resistance to ozone, chemical agents and heat.

No great difference exists between the mechanical-dynamic properties of co- and terpolymers, apart from the fact that peroxide vulcanization binds the chains with C–C bonds, and in sulfur vulcanization polysulfidic cross-links are obtained. Sulfur cross-links are also obtained in the copolymers, when the curing recipe involves peroxide and sulfur. The two classes of materials differ markedly in their resistance to aging and to ozone.

Table 7 shows a comparison between the main resistance properties of the EP elastomers and those of other commercial elastomers.

Table 7 Properties of Vulcanizates of EP Rubber Compared to Those of Other Elastomers

Resistance[a]	EP Rubber	Natural Rubber	SBR	Chloroprene Rubber	Nitrile Rubber	Chloro-sulfonated Polyethylene	Butyl Rubber	Silicone Rubber
Aging	5	3	3	5	3	5	5	5
Sunlight	5	2	2	5	2	5	5	5
Ozone	5	0	2	4	1	5	4	5
Heat	4	1	2	3	3	4	3	5
Flame	0	0	0	5	0	5	0	2
Acids	5	3	3	4	3	5	4	3
Alkalis	5	4	4	4	3	5	5	5
Aliphatic solvents	1	1	1	4	5	4	2	1
Aromatic solvents	0	0	0	1	3	1	2	0
Chlorinated solvents	0	0	0	0	2	1	1	0
Brake liquids	5	5	5	3	1	3	5	3
Vegetable oils	4	4	4	5	5	5	4	5

[a] 0 = very poor; 1 = poor; 2 = fair; 3 = good; 4 = very good; 5 = outstanding.

8.3.1 *Physical and Technological Properties*

The mechanical properties are markedly influenced by the ethylene content and obviously also by the regularity of distribution of the monomeric sequences.

High ethylene content polymers are stiffer at room temperature and therefore present higher elastic moduli and hardness at the same cross-link density (152). Furthermore, the possibility of crystallization under stretching allows the attainment of higher tensile strengths. That is why ethylene-rich terpolymers may be filled with larger amounts of oil and carbon black (146) for the same vulcanizate properties with consequent large economical advantage (197). However, too high an ethylene content and above all a nonhomogeneous sequence distribution, which results in blockiness, worsens the dynamic properties and tension set.

Hence, the ethylene-rich polymers show better overall mechanical-dynamic properties, provided the monomer distribution is as regular as possible and the ethylene content is not so high as to result in appreciable crystallinity in the undeformed state. Abrasion resistance may be taken as an example of a technological property that depends on polymer structure. It is higher in terpolymers than in copolymers, owing to the cure system (146). It depends considerably on composition and on homogeneity, being higher in ethylene-rich polymers (199), especially in the structurally homogeneous ones.

Properties are also influenced by the degree of unsaturation, in the sense that the higher the unsaturation, the higher is the maximum cross-link density attainable. Terpolymers with a higher degree of unsaturation yield vulcanizates with a lower compression set. This property is well known to depend on cross-link type, being lower in the vulcanizates containing mainly monosulfidic bonds (198). With the same vulcanization recipe, polymers with a higher degree of unsaturation contain a higher fraction of monosulfidic cross-links than less unsaturated ones (184) at the same cure times.

The possibility of controlling both concentration and type of cross-links by adjusting the level and type of unsaturation, as well as the vulcanization recipe, is a peculiar property of EPDMs, distinguishing them from highly unsaturated rubbers.

8.3.2 *Aging and Ozone Resistance*

During thermooxidation of copolymer vulcanizates, balancing phenomena of degradation and of cross-linking occur and properties do not vary considerably. On the other hand, in terpolymers, cross-linking reactions prevail and mechanical properties vary much more strongly, as may be seen in Fig. 21.

Aging properties of the vulcanizates also depend considerably on the cure ingredients (171). In copolymers cross-linked with peroxide alone, the vulcanizates are very resistant to heat in inert ambient, but degrade rapidly in hot air. Instead, if in a sulfur plus peroxide cure, the sulfur is present in such ratios as to yield polysulfidic cross-links, the vulcanizates are more stable to thermooxidative aging, especially at long times. This different behavior is shown in Fig. 22, which plots relaxation curves of a copolymer cross-linked with peroxide alone and with peroxide plus increasing amounts of sulfur, respectively. The curve of the vulcanizate with 0.75 sulfur reveals the characteristic behavior of networks containing polysulfidic cross-links, which are thermally labile. At high temperatures, some of these cross-links break down and some become monosulfidic; bound sulfur resulting from these reactions would act as an antioxidant in the vulcanizate (171). The cleavage of the polysulfidic bonds corresponds to the initial phase of rapid stress decay in the relaxation curve; in the subsequent phase the constant stress reveals the stabilizing action of sulfur.

In copolymers, the ethylene/propylene ratio remarkably influences the aging resistance, with a higher ethylene content resulting in a higher stability. In terpolymers, the ethylene content does not exert much influence, at least within the range of commercial compositions, whereas the type of third monomer is of great importance, as may be seen in Fig. 21.

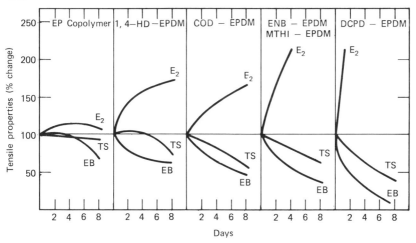

Figure 21 Oven aging at 150°C for EPM and EPDM vulcanizates. From Ballini et al. (146).

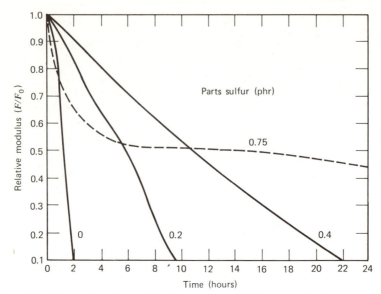

Figure 22   Stress relaxation and sulfur content of EPM gum vulcanizates. From Dunn (171), by courtesy of the Division of Rubber Chemistry, American Chemical Society, Akron, Ohio.

In terpolymer aging, cross-linking reactions generally prevail in the initial stage over degradation reactions, although the behavior depends markedly on the vulcanization system. In peroxide vulcanizates some termonomers such as ethylidene norbornene and 1,4-hexadiene show a balance between degradation and cross-linking; in others, especially dicyclopentadiene (DCPD), cross-linking clearly prevails. In the sulfur vulcanizates postcuring occurs for all EPDMs; however, here also, DCPD differs quantitatively (201). Similar peculiar behavior of DCPD and other analogous termonomers, such as cyclooctadiene (COD), is also observed in the ozone resistance of terpolymers.

It should be emphasized that all ethylene–propylene elastomers, both saturated and unsaturated show an exceptional stability to ozone; this is why they are mixed with the highly unsaturated rubbers to increase their stability. Economic and technical reasons support the use of blends rather than of EPDM alone for certain applications, such as tire sidewalls (187, 196). In these blends, the EPDM is about 20–30% of the total rubber. At the 30% level, protection is good with all terpolymers, whereas with 20% it is marginal, in the sense that the resistance of the blend depends on the type of terpolymer and on the preparation conditions of the blend.

When, in addition to ozone resistance, good mechanical dynamic properties are required, it must be borne in mind that an EPDM content of 30% depresses the dynamic properties, unless a covulcanization state is attained.

Thus, if a standard terpolymer must be used, a concentration of 20% is more likely to maintain good vulcanizate properties and the ozone resistance will be good only with particular termonomers, such as DCPD. In these terpolymers the protective action is due to branching (146), the presence of which was demonstrated by rheological measurements (66, 146).

The connection between good ozone resistance and branching is proved by the following test: A standard nonprotective terpolymer becomes very effective at 20% concentration when it is previously branched with a small amount of peroxide (146). Saturated copolymers were much less effective than terpolymers, a result attributed to a less uniform dispersion of EPM in the unsaturated rubber (202).

9 APPLICATIONS

The saturated copolymers find fewer applications than the terpolymers since they cannot be vulcanized with the traditional sulfur based recipes and aging properties are lower, in particular for antiozone uses. Hence their use is justified only when the peroxide cure is preferred, as in the manufacture of electric cables or in high temperature cure. For these reasons, the uses of the saturated copolymers are very limited and tend to decrease continuously, whereas those of the terpolymer are rapidly expanding. Therefore, the applications described here generally refer to EPDMs, except for the particular cases already mentioned. EPDMs may be used both as general purpose and specialty rubbers, in a wide range of applications (203, 204). Only the most important ones will be mentioned here, i.e., those in the automotive industry, electrical industry appliances, and a few others.

9.1 Automotive Industry (204, 205)

A distinction must be made between tire and nontire applications.

9.1.1 *Tires* (199, 206–209)

Great efforts are being devoted to the development of EPDM in the tire field, since its future as a general purpose rubber depends greatly on its adoption in the tire market. All EPDM tires are good in comparison with those made of traditional rubbers (206); however, processing problems hinder a wide penetration in this market. Larger possibilities of success exist in specialty tires, e.g., tractor and snow tires, where processing is less important.

EPDMs are already widely used in blends with unsaturated rubbers for white and black sidewalls (196, 210, 211), which require special protection against ozone and sunlight aging. Actually blends, which are currently more easily obtained with the fast curing terpolymers, seem to be a good entry for EPDMs in the tire market.

9.1.2 *Nontire Applications* (204, 205)

This is the largest outlet for EPDM. EPDMs ozone and heat resistance is very useful for automotive parts exposed to high under-the-hood temperatures, and its aging and weathering properties are excellent for window channeling, weather stripping, boots, and covers, radiator and air conditioning hoses, body seals, bumpers and insulators, gaskets, mountings, etc.

9.2 Appliances

Outlet hoses and seals of washing and dishwashing machines undergo the action of hot detergent solutions, and those of dryers the action of very hot air. Such uses make appliances a high volume field of application, second only to the automotive industry.

9.3 Electrical Industry (212, 213)

In addition to their outstanding stability and resistance, the EP rubbers exhibit good electrical properties, such as dielectric strength, insulation constant, electrical stability in water, and, above all, exceptional corona resistance. Some electrical properties of EP elastomers are shown in Fig. 23, in comparison with those of other polymers. Peroxide vulcanization is preferred for these applications; therefore, the lower cost copolymer may compete advantageously with the terpolymer.

9.4 Injection Molding

EPDM elastomers are particularly suitable for applications of this type. They may be filled with very large amounts of oil and black and are resistant to cure reversion (214). This latter property is very important, because in injection molding vulcanization occurs at a temperature higher than usual in order to increase fluidity of the compounds and the rate of cure.

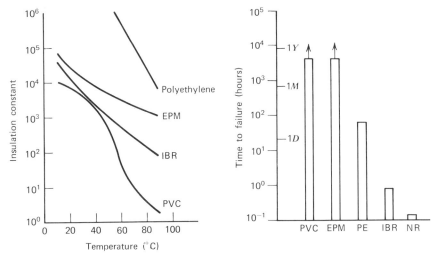

Figure 23 Electrical properties of EP rubbers and other polymers (insulation constant versus temperature and partial discharge resistance). From Pedretti (212), by courtesy of the Division of Rubber Chemistry, American Chemical Society, Akron, Ohio.

9.5 Other Applications

The good chemical resistance of EP elastomers (see Table 7) makes them competitive with special rubbers, such as polychloroprene and chlorosulfonated polyethylene. For example, it may be conveniently used for antiacid coatings of industrial floors and for conveyor belts, especially those conveying hot and corrosive materials.

Among many other applications are those for footwear, fabric coatings, bridge pads, rug underlay, extruded closed-cell sponge (186), mastics, sealing tapes, paint films (215), seals and gaskets, and industrial, steam, and garden hose.

The increasing need for chemical and ozone resistant rubbers for seals, expansion joints, and vibration insulators is expected to lead to many new engineering applications for EPM and EPDM.

10 ETHYLENE–BUTENE-1 COPOLYMERS

In addition to propylene, other α-olefins, and particularly butene-1, have been examined in copolymerization with ethylene. According to Natta et al. (216) the best catalysts for the synthesis of ethylene–butene-1 amorphous copolymers over a wide range of composition, are prepared from vanadium compounds and aluminum alkyl compounds.

The difference between the reactivities of ethylene and butene-1 is even more marked than between ethylene and propylene. The reactivity ratios for ethylene and butene-1, with the catalyst system $VCl_3 + Al(C_6H_{13})_3$, are 26.96 and 0.043, respectively; when a catalyst prepared from VCl_4 and $Al(C_6H_{13})_3$ is used, they are 29.60 and 0.019, respectively. In the presence of these catalyst systems, operations in heptane and all other conditions being the same, the composition of ethylene–butene-1 copolymers is independent of the polymerization time, the time elapsed between catalyst preparation and use, the catalyst concentration, and the Al/V ratio.

The copolymer composition depends on the catalyst age in runs carried out in benzene and with high concentrations of ethylene in the reacting phase, in the presence of soluble catalysts prepared by reaction of bis(cyclopentadienyl)titanium dichloride with $Al(C_2H_5)_2Cl$ (217) and $Al(i-C_4H_9)_2Cl$ (218).

Catalyst aging before the addition of the monomers also exerts a marked influence on the copolymerization rate, the copolymer yield, and molecular weight. These results are explained by the hypothesis that, in the in situ catalysts, the active species present differ from those in aged catalysts.

In copolymerization runs carried out in liquid butene in the presence of the catalyst system $VCl_4 + Al(C_6H_{13})_3$ at temperatures ranging between 0 and 60°C, Seidov et al. (219) found that, by increasing the concentration of ethylene in the liquid phase, the copolymerization rate increases. The copolymer yield and the molecular weight decrease on increasing the temperature of copolymerization. Investigations on copolymerization in the absence of solvent were extended to the catalyst system vanadium triacetylacetonate + $Al(i-C_4H_9)_2Cl$ (220). Copolymerization runs were carried out at temperatures ranging between 10 and 60°C under pressures between 5.5 and 11 atm and with Al/V ratios between 8 and 30. At a constant temperature of 30°C and at a constant Al/V ratio of 8, on increasing pressure (e.g., on increasing the concentration of ethylene) an increase is observed in the copolymer yield, the copolymerization rate, and the molecular weight of the copolymer obtained.

At a constant concentration of ethylene in the liquid phase and at a constant Al/V ratio of 8, on increasing temperature, the copolymer yield and the molecular weight decrease, the copolymerization rate increases, whereas the copolymer composition remains unaltered.

By operating at 30°C and at 6 atm, on increasing the Al/V ratio, yield with respect to the catalyst used decreases, as well as the molecular weight, and the copolymerization rate increases, whereas the copolymer composition remains unaltered.

In the presence of the catalyst system $VCl_4 + Al(i\text{-}C_4H_9)_3$ and operating in liquid butene at 25°C, Gromova et al. (221) prepared a series of elastomeric ethylene–butene-1 copolymers with butene-1 contents between 27 and 76 mole %. All copolymers obtained, except those with a butene content of about 60 mole %, are crystalline at room temperature. A correlation between copolymer composition and glass transition temperature has been reported.

Completely amorphous ethylene–butene copolymers with butene contents ranging between 10 and 47 mole % were obtained with the catalyst system $VCl_4 + Al(C_2H_5)_2Cl + anisole$ in runs carried out at $-78°C$ in liquid butene (27).

The reactivity ratios for ethylene and butene-1 are 35.5 and 0.006, respectively. The product of the reactivity ratios is considerably lower than unity; hence, with the catalyst systems mentioned above, the monomer units of the copolymer show a strong tendency to alternance. This was also observed with the catalyst system $VCl_4 + Al(C_2H_5)_2Cl$. On the other hand, with the systems $VCl_4 + Al(C_2H_5)_3$ and $VCl_4 + Ga(C_2H_5)_3$, the product of the reactivity ratios in the ethylene–butene-1 copolymerization is near unity; consequently, the monomeric units are distributed at random. Analogous results were obtained in the ethylene–propylene and propylene–butene copolymerizations.

Hence a tendency to alternance should be found when passing from nonstereospecific to syndiospecific catalysts. In this respect, isospecific catalysts behave as nonstereospecific catalysts.

REFERENCES

1. G. Natta et al., *Chim. Ind. (Milan)*, **39,** 733 (1957).
2. G. Mazzanti et al., *Chim. Ind. (Milan)*, **39,** 825 (1957).
3. G. Natta et al., *Chim. Ind. (Milan)*, **40,** 717 (1958).
4. G. Natta et al., *Chim. Ind. (Milan)*. **40,** 896 (1958).
5. C. H. Lucach and H. H. Spurlin, in G. Ham, ed., *Copolymerization*, Wiley-Interscience, New York, 1964, p. 115.
6. W. L. Carrick et al., *J. Am. Chem. Soc.*, **82,** 3883 (1960).
7. S. Murahashi et al., *Bull. Chem. Soc. Japan*, **33,** 1170 (1960).
8. C. G. Overberger and F. Ang, *J. Am. Chem. Soc.*, **82,** 929 (1960).
9. G. Phillips and W. Carrick, *J. Am. Chem. Soc.*, **84,** 920 (1962).
10. G. Phillips and W. Carrick, *J. Polym. Sci.*, **59,** 401 (1962).
11. H. Wesslau, *Makromol. Chem.*, **26,** 102 (1958).
12. G. Bier, *Angew. Chem.*, **73,** 186 (1961).

13. G. Natta et al., *J. Polym. Sci.*, **51**, 411 (1960).

14. U.S. Pat. 3,340,240 (to Montecatini, 1967).

15. Italian Pat. 581,418 (to Montecatini, August 27, 1958).

16. R. German et al., *Kaut. Gummi*, **19**, 67 (1966).

17. P. Brennan, *Chem. Eng.*, **72**, (14), 94 (1965).

18. G. Crespi and G. Di Drusco, *Hydrocarbon Processing*, **48**, 103 (1969).

19. W. Marconi et al., *Chim. Ind. (Milan)*, **46**, 1131 (1964).

20. P. Locatelli et al., *Makromol. Chem.*, **176**, 1121 (1973).

21. A. P. Firsov et al., *Visokomol. Soedin.*, **8**, 1860 (1966).

22. E. Giachetti et al., *Chim. Ind. (Milan)*, **48**, 1037 (1966).

23. M. Ichikawa and A. Kogure, *J. Chem. Soc. Japan (Ind. Chem. Sect.)*, **68**, 535 (1965); *Int. Chem. Eng.*, **5**, 724 (1965).

24. F. Y. Karol and W. L. Carrick, *J. Am. Chem. Soc.*, **83**, 2654 (1961).

25. W. Marconi et al., *Chim. Ind. (Milan)*, **46**, 1287 (1964).

26. W. L. Carrick et al., *J. Am. Chem. Soc.* **82**, 1502 (1960).

27. A. Zambelli et al., *Makromol. Chem.*, **115**, 73 (1968).

28. E. Junghanns et al., *Makromol. Chem.*, **58**, 18 (1962).

29. W. Marconi and S. Cesca, *Chim. Ind. (Milan)*, **47**, 30 (1965).

30. G. Bier et al., *Makromol. Chem.*, **58**, 43 (1958).

31. G. Natta et al., *J. Polym. Sci.*, **51**, 429 (1961).

32. R. J. Kelly et al., *Ind. Eng. Chem. Prod. Res. Dev.*, **1**, 210 (1962).

33. G. Natta et al., *Rubber Chem. Technol.*, **36**, 1583 (1963).

34. British Pat. 1,132,517 (to Uniroyal, November 6, 1968).

35. British Pat. 1,128,508 (to Uniroyal, September 29, 1968).

36. Belgian Pat. 683,828 (to U.S. Rubber, September 15, 1966).

37. French Pat. 1,555,802 (to Goodyear, December 23, 1968).

38. G. Natta et al., *Chim. Ind. (Milan)*, **41**, 519 (1959).

39. G. Natta et al., *Chim. Ind. (Milan)*, **43**, 871 (1961).

40. A. Schindler, *Makromol. Chem.*, **70**, 94 (1963).

41. W. E. Smith and R. G. Zelmer, *J. Polym. Sci.*, **A1**, 2587 (1963).

42. G. Mazzanti et al., *Chim. Ind. (Milan)*, **42**, 468 (1960).

43. D. Sianesi et al., *Chim. Ind. (Milan)*, **41**, 1176 (1959).

44. A. Valvassori et al., *Makromol. Chem.*, **60**, 46 (1963).

45. Italian Pat. 956,396 (to Montecatini Edison, October 10, 1973).

46. W. L. Carrick et al., *J. Am. Chem. Soc.*, **82**, 5319 (1960).

47. Yu. Obloi et al., *Vysokomol. Soedin.*, **5**, 939 (1965).

48. A. Gumboldt et al., *Makromol. Chem.*, **101**, 229 (1967).

49. U.S. Pat. 3,301,834 (to Hercules, 1963).

50. French Pat. 1,517,185 (to Farbwerke Hoechst, February 5, 1968).

51. French Pat. 1,553,808 (to Soc. Natl. Petrol. D'Aquitaine, December 9, 1968).

52. British Pat. 1,132,023 (to Uniroyal, October 30, 1968).

53. British Pat. 1,081,696 (to Uniroyal, August 31, 1967).

54. French Pat. 1,569,833 (to Farbwerke Hoechst, April 5, 1968).

55. Italian Pat. 967,145 (to Montecatini Edison, February 28, 1974).

56. U.S. Pat. 3,444,146 (to Montecatini Edison, May 13, 1969); French Pat. 1,564,113 (to DuPont, March 10, 1969).

57. H. Blümel, *Kaut. Gummi*, **21**, 547 (1968).

58. S. Cesca, *J. Polym. Sci., Macromol. Rev.*, **10**, 1 (1975).

59. A. Valvassori and N. Cameli, in S. A. Miller, ed., *Ethylene and Its Industrial Derivatives*, Benn, London, 1969, p. 425.

60. U.S. Pats. 3,347, 944 (to Union Carbide, October 17, 1967) and 3,535,395 (to Goodrich, October 20, 1970).

61. E. K. Gladding et al., *Ind. Eng. Chem., Prod. Res. Dev.*, **1**, 65 (1962).

62. F. P. Baldwin et al., *Rubber Chem. Technol.*, **43**, 522 (1970).

63. R. G. Miller et al., *J. Am. Chem. Soc.*, **89**, 3756 (1967).

64. R. Hank, *Kaut. Gummi*, **18**, 295 (1965).

65. C. Cozewith and G. Ver Strate, *Macromolecules*, **4**, 482 (1971).

66. N. Tokita and R. Scott, *Rubber Chem. Technol.*, **42**, 944 (1969).

67. R. E. Cunningham, *J. Polym. Sci., A-1*, **4**, 1203 (1966).

68. S. Cesca et al., *Chim. Ind. (Milan)*, **50**, 171 (1968).

69. N. D. Zavorokhin, *Vestn. Akad. Nauk Kaz. SSR*, **23**, 76 (1967); *CA*, **67**, 22638z (1967).

70. G. Natta et al., *Rubber Plast. Age*, **46**, 683 (1965).

71. G. Sartori et al., *Rubber Chem. Technol.*, **38**, 620 (1965).

72. S. Cesca et al., *Chim. Ind. (Milan)*, **50**, 1203 (1968).

73. D. L. Christman and G. I. Keim, *Macromolecules*, **1**, 358 (1968).

74. M. Samuels and K. W. Wirth, *Rubber Age*, **99**, 73 (1967).

75. S. Cesca et al., *J. Polym. Sci.*, **A1**, 1575 (1971).

76. C. Tosi, *Chim. Ind. (Milan)*, **53**, 458 (1971).

77. T. Simonazzi et al., *Quad. Ric. Sci.*, **84**, 206 (1973).

78. C. Tosi et al., *J. Polym. Sci.*, **C22**, 1085 (1969).

79. P. Corradini and C. Tosi, *Eur. Polym. J.*, **4**, 227 (1968).

80. C. Tosi, *Makromol. Chem.*, **138**, 299 (1970); **150**, 199 (1971); **176**, 453 (1975); *NMR—Basic Principles and Progress*, **4**, 129 (1971); *Quad. Ric. Sci.*, **84**, 216 (1973).

81. S. Igarashi, *Polym. Lett.*, **1**, 359 (1963).

82. C. Tosi, *Makromol. Chem.*, **108**, 307 (1967).

83. T. Alfrey and G. Goldfinger, *J. Chem. Phys.*, **12**, 322 (1944).

84. A. Valvassori and G. Sartori, *Rend. Ist. Lomb. Sci. Lett.*, **A96**, 107 (1962).

85. A. Valvassori and G. Sartori, *Polym. Lett.*, **5**, 345 (1967).

86. G. E. Ham, *J. Polym. Sci.* **A2**, 2735, 4169, 4181 (1964).

87. K. F. O'Driscoll, *J. Polym. Sci.*, **A2**, 4201 (1964).

88. C. Tosi, *Eur. Polym. J.*, **6**, 161 (1970).

89. C. Tosi, *Eur. Polym. J.*, **8**, 91 (1972).

90. A. Dondos and H. Benoit, *Makromol. Chem.*, **118**, 165 (1968).

91. L. A. Urricki and R. Simha, *Macromolecules*, **1**, 505 (1968).

92. Y. Shimura, *J. Polym. Sci., A-2*, **4**, 423 (1966).

93.   T. Homma and J. Fujita, *J. Appl. Polym. Sci.*, **9**, 1701 (1965).

94.   J. Velickovic et al., *Makromol. Chem.*, **129**, 203 (1969).

95.   G. Moraglio, *Chim. Ind. (Milan)*, **41**, 984 (1959).

96.   P. J. Flory and T. G. Fox, *J. Am. Chem. Soc.*, **73**, 1904 (1951).

97.   W. H. Stockmayer et al., *J. Polym. Sci.*, **16**, 517 (1955).

98.   E. G. Owens and I. G. Cobler, Fourth International Seminar G.P.C., Miami Beach, Florida, May 22, 1967.

99.   G. Natta and G. Crespi, *J. Polym. Sci.*, **61**, 83 (1962).

100.   K. Baranwal and H. L. Jacobs, *J. Appl. Polym. Sci.*, **13**, 797 (1969).

101.   S. Bruckner et al., *Eur. Polym. J.*, **10**, 347 (1974).

102.   C. Tosi and F. Ciampelli, *Adv. Polym. Sci.*, **12**, 87 (1973).

103.   T. Gössl, *Makromol. Chem.*, **42**, 1 (1960).

104.   G. Bucci and T. Simonazzi, *Chim. Ind. (Milan)*, **44**, 262 (1962).

105.   F. Ciampelli et al., *Chim. Ind. (Milan)*, **44**, 489 (1962).

106.   W. E. Smith et al., *J. Polym. Sci.*, **61**, 39 (1962).

107.   H. V. Drushel and F. A. Iddings, *Anal. Chem.*, **35**, 28 (1963).

108.   J. N. Lomonte, *Polym. Lett.*, **1**, 645 (1963).

109.   J. E. Brown et al., *Anal. Chem.*, **35**, 2172 (1963).

110.   P. J. Corish and M. E. Tunnicliffe, *J. Polym. Sci.*, **C7**, 187 (1964).

111.   G. Garrasi, *Chim. Ind. (Milan)*, **47**, 52 (1965).

112.   R. M. Bly et al., *Anal. Chem.*, **38**, 217 (1966).

113.   T. Takeuchi et al., *Anal. Chem.*, **41**, 184 (1969).

114.   C. Tosi et al., *Makromol. Chem.*, **120**, 225 (1968).

115.   T. J. Dudek and F. Bueche, *J. Polym. Sci.*, **A2**, 811 (1964).

116.   R. S. Porter, *J. Polym. Sci.*, *A-1*, **4**, 189 (1966).

117.   P. R. Sewell and D. W. Skidmore, *J. Polym. Sci.*, *A-1*, **6**, 2425 (1968).

118.   C. J. Carman and C. E. Wilkes, *Rubber Chem. Technol.*, **44**, 781 (1971).

119.   A. Zambelli, *NMR—Basic Principles and Progress*, **4**, 101 (1971); A. Zambelli et al., *Macromolecules*, **4**, 475 (1971); F. A. Bovey, M. C. Sacchi, and A. Zambelli, *ibid.*, **7**, 752 (1974).

120.   E. Bua and P. Manaresi, *Anal. Chem.*, **31**, 2022 (1959).

121.   T. S. Lee et al., *J. Polym. Sci.*, **3**, 66 (1948).

122.   M. E. Tunnicliffe et al., *Eur. Polym. J.*, **1**, 259 (1965).

123.   W. Simon and W. Kriemler, *J. Gas Chromatogr.*, **5**, 53 (1967).

124.   O. F. Folmer and L. V. Azarraga, *J. Chromatogr. Sci.*, **1969**, 665.

125.   J. Van Schooten and J. K. Evenhuis, *Polymer*, **6**, 343 (1965).

126.   J. Van Schooten and J. K. Evenhuis, *Polymer*, **6**, 561 (1965).

127.   G. Natta et al., *Chim. Ind. (Milan)*, **42**, 125 (1960).

128.   G. Bucci and T. Simonazzi, *J. Polym. Sci.*, **C7**, 203 (1964).

129.   F. Ciampelli and A. Valvassori, *J. Polym. Sci.*, **C16**, 377 (1967).

130.   C. Tosi et al., *Eur. Polym. J.*, **4**, 107 (1968).

131.   H. V. Drushel et al., *Anal. Chem.*, **40**, 370 (1968).

132. Yu. V. Kissin et al., *Vysokomol. Soedin.*, **A9**, 1374 (1967); translated in *Polym. Sci., USSR*, **9A**, 1541 (1967).

133. L. Michajlov et al., *Polymer*, **8**, 325 (1968).

134. G. Nencini et al., *Polym. Lett.*, **3**, 483 (1965).

135. G. Crespi et al., *Chim. Ind. (Milan)*, **55**, 130 (1973).

136. M. Mooney, "The Rheology of Raw Elastomers," in F. Eirich, ed., *Rheology*, Vol. 2, Academic Press, New York, 1958.

137. J. L. White and N. Tokita, *J. Appl. Polym. Sci.*, **9**, 1929 (1965).

138. N. Tokita and J. L. White, *J. Appl. Polym. Sci.*, **10**, 1011 (1966).

139. J. L. White and N. Tokita, *J. Appl. Polym. Sci.*, **11**, 321 (1967).

140. J. L. White, *Rubber Chem. Technol.*, **42**, 257 (1969).

141. E. G. Kontos, *Rubber Chem. Technol.* **43**, 1082 (1970).

142. R. Ninomiya and G. Yasuda, *Rubber Chem. Technol.*, **42**, 714 (1969).

143. G. Ballini et al., Proceedings of the First Meeting of the Italian Society of Rheology, Siena, May 13–15, 1971, Vol. 2, p. 273.

144. E. N. Kresge et al., 4th International Synthetic Rubber Symposium, Church House, Westminister, London, October 1969, p. 58.

145. C. K. Shih, *Trans. Soc. Rheol.*, **14**, 83 (1970).

146. G. Ballini et al., Paper presented at the International Rubber Conference, Moscow, November 1969.

147. F. L. Ramp et al., *J. Org. Chem.*, **27**, 4368 (1962).

148. J. Furukawa, *Angew. Makromol. Chem.*, **23**, 189 (1972).

149. G. Gianotti et al., *Makromol. Chem.*, **149**, 117 (1971).

150. G. Dall'Asta, *Makromol. Chem.*, **154**, 1 (1972).

151. R. S. Porter and J. F. Johnson, *Chem. Rev.*, **66** (1), 1 (1966).

152. U. Flisi et al., Paper presented at the International Rubber Conference, Moscow, November 1969.

153. A. V. Tobolsky and M. Takahashi, *J. Appl. Polym. Sci.*, **7**, 134 (1963).

154. J. R. Richards et al., *J. Polym. Sci.*, **B2**, 197 (1964).

155. F. P. Baldwin et al., *Rubber Chem. Technol.*, **42**, 1167 (1969).

156. V. Zamboni et al., *Rubber Chem. Technol.*, **44**, 1109 (1971).

157. M. C. Shen and A. Eisenberg, *Rubber Chem. Technol.*, **43**, 95 (1970).

158. R. F. Boyer, *Plast. Polym.*, **41**, 15 (1973).

159. J. J. Maurer, *Rubber Chem. Technol.*, **42**, 110 (1969).

160. U. Flisi et al., *Kaut. Gummi Kunstst.*, **22**, 154 (1969).

161. U. Flisi et al., *Rubber Chem. Technol.*, **44**, 1093 (1971).

162. I. W. Bassi et al., *Eur. Polym. J.*, **6**, 709 (1970).

163. G. Ver Strate and Z. W. Wilchinsky, *J. Polym. Sci.*, *A-2*, **9**, 127 (1971).

164. L. P. Lenas, *Ind. Eng. Chem.*, *Prod. Res. Dev.* **2**, 202 (1963).

165. E. Di Giulio and G. Ballini, *Kaut. Gummi*, **15** (1), WT6 (1962).

166. A. E. Robinson et al., *Ind. Eng. Chem.*, *Prod. Res. Dev.*, **1**, 78 (1962).

167. L. D. Loan, *J. Polym. Sci.*, *A2*, 3053 (1964).

168. L. D. Loan, *Polym. Lett.*, **2**, 59 (1964).

169. F. P. Baldwin et al., *Rubber Chem. Technol.*, **43**, 522 (1970).

170. G. Natta et al., *Rubber Plast. Age*, **42**, 53 (1961).

171. J. R. Dunn, *Rubber Chem. Technol.*, **41**, 304 (1968).

172. U. Flisi and G. Crespi, *J. Appl. Polym. Sci.*, **12**, 1947 (1968).

173. G. G. Wanless et al., *Rubber Chem. Technol.*, **35**, 118 (1962).

174. C. G. Moore and B. R. Trego, *J. Appl. Polym. Sci.*, **8**, 1957 (1964).

175. F. P. Baldwin and G. Ver Strate, *Rubber Chem. Technol.*, **45**, 709 (1972).

176. L. D. Loan, *Rubber Chem. Technol.*, **40**, 149 (1967).

177. J. Cornell et al., *Rubber World*, **152** (1), 66 (1965).

178. K. Hummel and J. Desilles, *Kaut. Gummi*, **15**, WT 492 (1962).

179. H. K. Frensdorff, *Rubber Chem. Technol.*, **41**, 316 (1968).

180. M. E. Samuels and K. H. Wirth, *Rubber Age*, **99**, 73 (1967).

181. G. F. Figini, *Rev. Gén. Caoutch. Plast.*, **47**, 41 (1970).

182. G. Crespi and A. Arcozzi, *Chim. Ind. (Milan)*, **46**, 151 (1964).

183. K. Fujimoto and S. Nakade, *J. Appl. Polym. Sci.*, **13**, 1509 (1969).

184. F. D. Shaw, Jr., ACS Rubber Division, Cleveland, Ohio, April 1968.

185. K. H. Wirth, *Rubber World*, **158** (2), 61 (1968).

186. D. R. Filburn and L. Spenadel, *Rubber Age*, **102** (11), 37 (1970).

187. Z. T. Ossefort and E. W. Bergstrom, *Rubber Age*, **101** (9), 47 (1969).

188. G. Kerrutt et al., *Kaut. Gummi Kunstst.*, **22**, 413 (1969).

189. J. E. Cattan et al., *Rubber Chem. Technol.*, **44**, 814 (1971).

190. J. B. Gardiner, *Rubber Chem. Technol.* **43**, 370 (1970).

191. J. B. Gardiner, *Rubber Chem. Technol.*, **41**, 1312 (1968).

192. J. B. Gardiner, *Rubber Chem. Technol.*, **42**, 1058 (1969).

193. R. P. Mastromatteo et al., *Rubber Chem. Technol.*, **44**, 1065 (1971).

194. R. T. Morrissey, *Rubber Chem. Technol.*, **44**, 1025 (1971).

195. K. C. Baranwal and P. N. Son, *Rubber Chem. Technol.*, **47**, 88 (1974).

196. A. H. Speranzini and S. J. Drost, *Rubber Chem. Technol.*, **43**, 482 (1970).

197. R. W. Hallman et al., Paper presented at the 99th Meeting of the ACS Rubber Division, Miami Beach, Florida, April 1971.

198. F. P. Baldwin, *Rubber Chem. Technol.*, **43**, 1040 (1970).

199. K. Satake et al., *J. Inst. Rubber Ind.*, **5**, 233 (1971).

200. E. Di Giulio and A. Guglielmino, *Proc. Inst. Rubber Ind.*, **12**, 190 (1965).

201. K. C. Baranwal and G. A. Lindsay, *Rubber Chem. Technol.*, **45**, 1334 (1972).

202. J. C. Ambelang et al., *Rubber Chem. Technol.*, **42**, 1186 (1969).

203. R. F. McCabe, *Rubber Plast. Age*, **45**, 1492 (1964).

204. N. Keck, *Rubber Age*, **105** (9), 43 (1973).

205. L. M. Glanville, *Chem. Ind.*, **1974**, 255.

206. R. G. Arnold et al., Paper presented at the ACS Rubber Division, Cleveland, Ohio, April 1968.

207. K. Satake et al., *J. Inst. Rubber Ind.*, **4**, 21 (1970).

208. K. Satake et al., *J. Inst. Rubber Ind.*, **4**, 71 (1970).

209. K. Satake et al., *J. Inst. Rubber Ind.*, **5**, 104 (1971).

210. H. J. Leibn et al., Paper presented at the 99th meeting of the ACS Rubber Division, Miami Beach, Florida, April 1971.

211. H. J. Herzlich et al., Paper presented at the 99th Meeting of the ACS Rubber Division, Miami, Florida, 1971.

212. G. Pedretti, *Rubber Plast. Age,* **48,** 1091 (1967).

213. E. O. Forster and L. Spenadel, *Rubber Age,* **54,** 39 (1973).

214. W. G. De Pierri, Jr., and J. R. Hopper, *Rubber Chem. Technol.,* **42,** 1321 (1969); W. G. De Pierri, *ibid.,* **42,** 1336 (1969).

215. R. D. Singer et al., *Ind. Eng. Chem., Prod. Res. Dev.,* **10,** 287 (1971).

216. G. Natta et al., *Chim. Ind. (Milan),* **41,** 764 (1959).

217. R. E. Wiman and I. D. Rubin, *Makromol. Chem.,* **94,** 160 (1966).

218. I. D. Rubin, *J. Polym. Sci., A-1,* **5,** 1119 (1967).

219. N. M. Seidov et al., *Dokl. Akad. Nauk SSSR,* **164,** 826 (1965).

220. N. M. Seidov et al., *Azerb. Khim. Zh.,* No. 3, 46 (1966).

221. V. N. Gromova et al., *Vysokomol. Soedin.,* **A9,** 1123 (1967).

Derek A. Smith
Queen Mary College
University of London
London, England

Chapter 8   Network Formation

1   INTRODUCTION

This chapter deals primarily with what Kraus (1) has called "the two most important processes in rubber technology": vulcanization and reinforcement. Both are concerned not only with the conversion of essentially

433

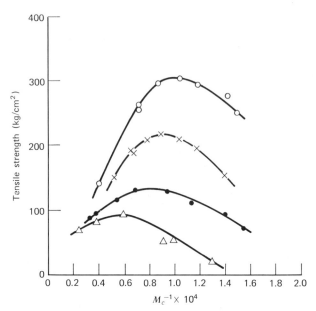

Figure 1 Influence of cross-link type and concentration on the tensile strength of a gum natural rubber vulcanizate. Vulcanizing system: $\bigcirc$, accelerated sulfur; $\times$, TMTD/Sulfurless; ●, peroxide; $\triangle$, high energy radiation. M_c is molecular weight between cross-links. From Mullins (2).

linear, flowable polymers of very high molecular weight into reticulated structures exhibiting high elasticity but also with the technically important properties of these network structures: adequate tensile and tear strength, extensibility, hardness, wearing qualities (abrasion resistance), rebound resilience, dynamic response over a specified range of frequency/temperature, fatigue life, and so on. It is now widely recognized that different types of network structure are associated with considerable differences in vulcanizate properties (2) (Fig. 1).

A considerable body of knowledge exists in connection with the vulcanization and reinforcement of natural rubber (NR) and, to a lesser extent, emulsion styrene–butadiene rubber (SBR). This is partly on account of NR having a longer history than the synthetic rubbers but also because of the existence of research organizations devoted to its study, especially the MRPRA, England (3). NR exhibits a high degree of stereoregularity, containing at least 97% *cis*-1,4-isoprenic residues (and probably 100%) linked in head-to-tail configuration. This technological knowledge, which is profusely documented and has been extensively reviewed (4), is therefore of great help in providing a background to which we can relate the technological behavior of the synthetic stereoregular elastomers. However, in both vulcanization and reinforcement there are areas of conflicting interpretation and even of conflicting experimental observation; it is therefore desirable to preface this chapter with a brief summary of the author's beliefs.

1.1 Vulcanization, with Special Reference to Natural Rubber

Conversion of linear polymer to network results in several fundamental changes in the physical properties of the polymer. In particular, (a) the polymer becomes insoluble in its normal solvents; instead it swells, often considerably but to an "equilibrium volume"; and (b) the load-bearing characteristics improve since the network cross-links (net-links) largely prevent viscous flow of the molecular chains.

However, most important to rubber technology is the development of reversible high elasticity (5), a property peculiar to very long, flexible molecular chains which have been lightly cross-linked to prevent their slippage under load. Many physical properties of rubber, but most of all the tensile properties, depend intimately on the concentrations of network junctions or the related "network chain density" (ν) present in the rubber after vulcanization.

As a general rule, satisfactory technical materials are only obtained when the average molecular weight between junction points (M_c) is large; highly cross-linked polymers of low initial molecular weight are often brittle.

1.1.1 Network Chain Density and Vulcanization Efficiency

The physical behavior of networks has been adequately described using thermodynamic and statistical mechanical arguments (5). These have necessarily been idealized, and the departures from predicted or model behavior shown by real vulcanizates are attributed (6) either to invalidity of certain simplifying assumptions made in the derivation of the high elasticity equations or to the presence of defects in the real network. Defects take the form of either permanent or slipping entanglements (7), which have some of the characteristics of net-links, and chain ends or loops formed by intramolecular linkage which are believed to dilute the network in rather the same way that could be achieved by vulcanizing in the presence of an unreactive hydrocarbon process oil (8).

The elastic theory is reasonably satisfactory when applied to two-dimensional extension (or the equivalent compression) of rubber networks, but somewhat less adequate, at least in its simple Gaussian form, in describing the force-extension curve for unidirectional extension (9), the traditional mode of rubber testing. Phenomenological descriptions of the tensile strain region have been put forward (10), and, in principle, it is possible to measure the degree of vulcanization of a rubber network, ν_{phys}, by making measurements of its dry or swollen equilibrium extension modulus. Such data are to be found in the literature (11).

Accurate measurements of ν_{phys} are important in the assessment of vulcanization efficiency. Because of the imperfections of structure such as chains ends and entanglements and also because the vulcanization reactions involving a given chain cannot distinguish between reactive centers in other chains (intermolecular reaction, cross-linking) and those in their own chain (intramolecular reaction, formation of cyclic structures, i.e., closed loops), there is no simple theoretical relationship between the extent of vulcanization reaction which has taken place within the assembly of rubber chains and the resultant "physically effective network chain density" ν_{phys}. Where such a relationship exists, it has been established experimentally as in Moore and Watson's work (12) with natural rubber.

Using the same polymer, one may compare vulcanizing systems by measuring the extent of chemical reaction necessary to achieve given values of ν_{phys} and so obtain a calibration of observed swelling or modulus data in terms of *chemical contribution to vulcanizate density* $(\rho M_c^{-1})_{\text{vulc}}$. Use of chemical probes as analytical tools has enabled C. G. Moore and his co-workers (13) to establish for the natural rubber-sulfur system: (*a*) the number of sulfur atoms combined with the network for each physically effective net-link produced; and (*b*) the approximate distribution of sulfur between mono-, di-, and polysulfidic cross-links.

While simple mixtures of rubber and sulfur may yield networks with sulfur/cross-link ratios as high as 40–55, use of appropriate accelerators and activators together with lower curing temperatures can reduce this ratio to about 2–3. Moore described the latter systems as "efficient," which indeed they are in terms of sulfur utilization per cross-link formed.

1.1.2 *Technological Assessment of Curing Systems*

A curing system that is chemically efficient is not necessarily technologically effective. The peroxide-natural rubber and peroxide-butadiene rubber systems illustrate this well, being of high efficiency in terms of molecules reacted per net-link formed (14), but objectionable for other reasons.

Among the important criteria establishing the effectiveness of a curing system are (*a*) the ready availability of the nontoxic ingredients in a form allowing easy handling and ready dispersion in a polymer; (b) controllability with respect to activity of the system in various ranges of temperature from ambient (to avoid precure) through processing temperatures (control of scorch) to achievement of rapid reaction at curing temperatures (from about 140 up to about 210°C); and (c) absence of deleterious effects such as modification of the polymer chain which might reduce its flexibility or resistance to aging; formation of toxic, discoloring, staining, blowing, or odoriferous products; or interaction with other ingredients of the compound (e.g., antioxidants), rendering them ineffective.

Additionally, an effective cure system yields products of adequate strength, wear and tear resistance, appropriate high or low modulus, etc., and all these most economically.

We have proposed the characterization of a technical curing system in terms of overall kinetics (15) (Fig. 2) of cross-linking and scission reactions and including study of temperature and concentration effects. This has been extended by Coran (16) and co-workers to include a detailed description of the induction or delay period shown by many industrially important systems. Even for a single system, the measurement of these

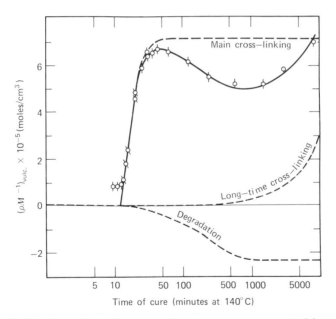

Figure 2   Kinetic analysis of an overall curing curve, represented by experimental points, in terms of an induction period followed by three additive reactions. (Data for NR 100.0; zinc oxide 5.0; stearic acid 1.0; MBTS 1.0; sulfur 2.0.) Full line is computed from the kinetic treatment (15).

important performance parameters by conventional methods is long and tedious and there is clearly a case here for the use of automatic methods. The method has also been applied to determine kinetic parameters for the sulfur-free thiuram vulcanization of NR and *cis*-1,4-polybutadiene (17).

1.1.3 *Technological Indices of Network Chain Density*

While the more fundamental swelling (18), equilibrium modulus (19), and reticulometer (20) measurements may often be made for reference purposes, simpler, empirical tests have commonly been used in the rubber industry to assess degree of cure, and though known to be imprecise, they yield useful comparative data.

Among the usual tests are the Mooney scorch test (21); also standard press cures for a limited range of times followed by measurement of 300% modulus, tensile strength, and elongation at break. A development of the greatest importance has been the introduction of equipment for the continuous measurement of some function of the shear modulus of a rubber compound throughout the curing period (22). The results are displayed in the form of a real cure curve which is suitable for the direct evaluation of several of the important kinetic parameters. However, despite their excellent reproducibility, existing instruments suffer from two principal disadvantages: (*a*) a requirement for relatively large sample sizes; and (*b*) lack of any simple proportionality between the recorded cure index and some more fundamental index of net-link concentration.

1.2 Sulfur Curing

It is hardly surprising that sulfur vulcanizing systems for natural rubber have held their own for more than a century in the face of strenuous competition from the numerous sulfurless systems which have been developed in recent times, since the properly formulated system sulfur/accelerator(s)/activator(s) satisfies the majority of technical criteria outlined earlier and moreover does so more cheaply than any currently available alternative system.

Alone, sulfur is unsatisfactory for technical curing on account of its low overall vulcanization rate, low optimum modulus, and because, in addition to the desirable reticulation effect, extensive main chain modification of the rubber occurs which leads to a deterioration of several properties. At least one property is improved, however, the resistance to crystallization stiffening, particularly in the temperature region of maximum crystallization rate at around −25°C. Used alone in NR, 6–8 phr sulfur is usually required with cure conditions of 2–4 hours at 160°C. The uncured stocks and vulcanizates usually show an unacceptably heavy sulfur bloom. Blooming occurs when the solubility limit of the sulfur in rubber is exceeded, for example on cooling a rubber stock saturated with sulfur at milling temperatures (say 80°C). The excess sulfur thrown out of solution migrates to the surface of the stock to form a crystalline layer which reduces the tack and interferes with plying-up and similar operations. Sulfur is only satisfactory as a curative if properly dispersed, and in some synthetic rubbers this has proved difficult. Use of special mixing methods can often result in better batch to batch reproducibility and higher average values for the vulcanizate tensile properties (22) (see Section 2).

The principal present-day vulcanization accelerators are amine derivatives modified to reduce toxicity, improve factory handling, and widen the range of curing characteristics. With the sole exception of the rarely used xanthates, the principal vulcanization accelerators are solid, fairly low melting point derivatives of either primary aromatic or secondary aliphatic amines. They normally require the presence of activators, usually a metal oxide (zinc oxide) and a fatty acid (stearic acid); in the absence of a metal oxide, cures may be quite rapid but of small extent. This latter property has been turned to good use in the deliberate omission of zinc oxide from aldehyde-amine accelerated compounds to give low modulus vulcanizates (23).

Considerable differences in behavior exist between the basic accelerators (guanidines, aldehyde-amines, etc.) and the acidic accelerators, particularly the very important thiazole group (24). The latter were developed in the early 1930s, and the past 40 years have seen but one major addition to the range of available materials, the introduction of amine-thiazole adducts, the sulfenamides. The sulfenamides give longer induction or delay periods combined with a shorter time to (higher) optimum modulus although specific results will be influenced by the nature and concentrations of the other ingredients of the sulfur vulcanizing system (15). Recent developments have been concerned less with development of new materials than with further investigation of the sulfur/accelerator/activator ratios which are now known to exert a profound effect on both rate and extent of network formation.

1.2.1 Thiazole and Sulfenamide Acceleration

2-Mercaptobenzothiazole (MBT), benzothiazyl disulfide (MBTS), and N-cyclohexyl-2-benzothiazylsulfenamide (CBS) comprise about 70% of the total accelerator usage in the United Kingdom rubber industry, and thus these three systems have been studied exhaustively both from the practical and fundamental points of view.

In natural rubber, conventional ratios of the main components of the vulcanizing system intended for normal temperature curing (in the range 140–160°C) are (25):

NR	100.0
Zinc oxide	3–6
Stearic acid	1–2
Sulfur	2–4
Accelerator	1.5–0.5

while SBR requires less sulfur but a higher accelerator concentration. In NR, progressive reduction of sulfur to about 1 phr and increase of accelerator to 2 phr yields improvements in processing safety and oven aging behavior without much change in modulus or hardness, but, unless modifications are made to the nature or concentration of the activators, use of sulfur/accelerator ratios much below 0.5 often leads to vulcanizates with reduced resistance to cut growth and lower tear strength.

Vulcanizates are virtually free from staining and discoloration problems and show a good plateau which assists the even curing of thick sections and allows the selection of a compromise technical cure for good all-around physical properties. All thiazole accelerators show powerful antioxidant activity and perform well in the presence of fine particle size carbon blacks. The high present-day processing temperatures for black stocks prevent the use of MBT (scorchy) and discourage use of MBTS since delay periods for curing systems containing this accelerator are generally much shorter than those for sulfenamide systems. The principal advantages of sulfenamide systems, long delay periods and high modulus, are obtained only at greater material cost than that of the parent thiazole and, in the case of CBS, with the risk of premature decomposition of the accelerator at high mixing temperatures. At even greater cost, it is possible to obtain special sulfenamide accelerators with greater delayed action, and better storage and processing stability than CBS. The accelerating effect and scorch behavior of these materials are determined by two principal factors: pK strength of the free base and the amount of steric hindrance at the amino nitrogen atom (26).

1.2.2 Comparison of MBT with MBTS Acceleration

Using conventional systems extended to include very low sulfur/accelerator ratios and in the temperature range 110–140°C, Russell and co-workers (15) were able to show that (a) at the same weight concentrations, the induction period for MBTS is always greater than that for MBT (i.e., MBTS is a delayed action thiazole); (b) following the induction period, the rate of reticulation using MBTS is always lower than that for MBT (i.e., the MBTS system is slower curing); (c) for the sulfur/accelerator ratios studied, the MBTS system always gives a higher optimum value of v (a tighter cure) than does MBT; and (d) the activation energies for reticulation for systems containing the two accelerators (their coefficients of vulcanization) are of similar magnitude, but there are small differences in reversion characteristics. These studies were extended to higher temperatures by Redding and Smith (17).

Until recently, no detailed characterization of the delay period in thiazole and sulfenamide acceleration had been reported. However, the study of Coran (16) and co-workers has enabled the single characteristic parameter t_0, the induction period, to be analyzed in terms of constituent chemical processes. This study has been extended to a consideration of the effects of compounding variables on the associated rate coefficients, the conclusions being summarized (after Coran) in Table 1. One of the most interesting features of the results is to establish that, in general, increase in accelerator concentration improves processing safety.

1.2.3 Other Acceleration

In natural rubber, the guanidines can be used with high sulfur levels (ca. 3 phr) and normal activators to give rather slow curing, plateau containing curves, but it is necessary to avoid compounding with adsorptive fillers and to protect the vulcanizates by addition of suitable antioxidants. Under these conditions, the optimum modulus can be increased by increasing the accelerator concentration from a minimum of about 0.5 phr up to 2.0 or even higher. In general, present-day sulfur vulcanizing systems for NR make use of DPG or the safer processing phthalate salt as a secondary accelerator, but rarely as a primary accelerator.

The activation energy for reticulation in sulfur–DPG systems is probably higher than for sulfur-thiazole systems: Using NR, Robinson and Pinfold (27) obtained a value of 27 kcal/mole and Scheele et al. (28) 30 kcal/mole for two specific DPG accelerated formulations. Accepted values for MBT– or MBTS–sulfur systems lie between 20 and 23 kcal/mole.

Although simple amines may occasionally still find use in compound formulations, there are advantages in first condensing these with aldehydes to produce less volatile, less toxic oligomers. A typical member of the group, butyraldehyde–aniline (BA), is an orange liquid with a characteristic smell which is retained by the vulcanizate. BA–sulfur systems stain and discolor, cure slowly with a well-defined plateau, and are

Table 1   Concentration Effects in Delayed Action Systems[a] (16)

An increase in the concentration of	Corresponds to	Having the Technological Consequences:
Sulfur	Small increases in k_1 and k_2	Shorter delay period with somewhat faster cross-link formation.
Fatty acid	Increase in k_1 and k_4/k_3 but a decrease in k_2	Little change in scorch but a slower rate of cross-link formation (though more abrupt in onset)
Accelerator	Decrease in k_2 and k_4/k_3 but an increase in k_2	Little change in scorch but a faster rate of cross-link formation (though less abrupt in onset)

[a] Based on the postulated reaction sequence:

$$A \xrightarrow{k_1} B \xrightarrow{k_2} B^* \xrightarrow{k_3} Vu$$

(accelerator and monomeric polysulfides)   (polymeric polysulfides)   (X-link precursor)   (X-link)

$$A + B^* \xrightarrow{k_4} \beta B \quad \text{(quench reaction responsible for delay)}$$

very economical for curing thick section, heavily loaded compounds since the system is not readily activated by trace impurities, nor is it readily adsorbed and rendered inactive by the filler surface. In the presence of normal activators, fairly high modulus compounds are produced, but omitting zinc oxide and increasing sulfur level gives low modulus vulcanizates with high breaking elongation. The less discoloring aldehyde–amine accelerators may be used in this way to prepare cut thread for elasticated garments.

Dithiocarbamates and thiuram sulfides are valuable as nondiscoloring secondary accelerators of low toxicity; they are used particularly in the formulation of rapid curing mixes for less reactive synthetic rubbers such as the saturated hydrocarbon types.

Thiuram disulfides, the oxidation products of dithiocarbamates, in common with their sulfurated derivatives, the thiuram tetrasulfides, act as sulfur donors and are capable of vulcanization without added elemental sulfur. The disulfides are particularly suitable for use with low concentrations of sulfur (0.5 phr or less) and normal activators in the formulation of heat resistant rubbers. While it is possible to formulate using the less active tetraethylthiuram disulfide (TETD) as sole accelerator, more particularly for thin section brightly colored products such as bathing caps, tetramethylthiuram disulfide (TMTD) and tetramethylthiuram monosulfide (TMTM) are more usually employed as boosters for thiazoles and sulfenamides. MBTS–TMTM–sulfur systems show fair delayed action and are economically competitive with sulfenamides, although they give lower modulus vulcanizates.

1.2.4 Activation and Retardation

It was first shown clearly by Barton and Hart (29) that both the rate and extent of reticulation achieved using a thiazole system are largely determined not only by the sulfur/accelerator ratio but also by the concentration of fatty acid activator. Moore and co-workers (30) showed that high fatty acid concentrations result in slower cures but of greater ultimate extent, the mono- and disulfidic net-links being characterized by higher thermal stability than those formed in inefficient systems of lower fatty acid content. However, the ability of the more complex polysulfidic type net-links to break and reform (31) in different configurations in order to accommodate applied stress is a valuable feature of inefficient systems and may be responsible for their continued technological popularity.

Occasionally, it may be desirable to reduce the rate of reticulation, particularly where high processing temperatures introduce an abnormally high scorch risk. Ideally, addition of a retarder would reduce the cure rate at processing temperatures but not at cure temperatures, and it is even possible to imagine a material that would retard at processing temperatures yet increase the rate of network formation (i.e., behave as a special kind of activator) during higher temperature curing. Claims for behavior of either kind were dispelled by the careful studies of Juve and Shearer (32), well summarized in Craig's review (33). Salicylic acid, phthalic anhydride, diphenylnitrosamine, and benzoic acid were tested in MBT, MBTS/BA, and TMTM accelerated sulfur systems; all retarded uniformly over the range of temperatures investigated (115–150°C).

1.3 Vulcanization Not Involving Elemental Sulfur

Nonsulfur systems may be divided into two groups, according to whether or not an unsaturated center is required in the elastomer.

1.3.1 *Requiring an Unsaturated Centre*

Besides the various sulfur donors such as the thiuram di- and tetrasulfides referred to earlier, natural rubber can be reticulated by means of other chemicals: *p*-quinonedioxime and certain of its derivatives (in the presence of a higher oxide of lead), sulfur monochloride, and certain dimethylol phenolic resins. However, if sulfur curing is not used, a principal advantage of working with an unsaturated polymer disappears and it is often preferable to use a chemically saturated elastomer which is inherently more resistant to oxidative deterioration. An interesting alternative is provided in the case of ethylene–propylene rubbers which are available either in the form of fully saturated copolymers or as sulfur vulcanizable terpolymers (EPDM) which contain unsaturated side chains. Their vulcanization is discussed in Chapter 7 and in Section 6 of this chapter.

1.3.2 *Not Requiring an Unsaturated Center*

It is often neither feasible nor desirable to convert a saturated to an unsaturated hydrocarbon chain simply for cure purposes since curatives are available which are capable of reacting satisfactorily with saturated chains. Chief among these are the organic peroxides.

The earliest studies of peroxide curing were concerned with an unsaturated polymer, natural rubber. Ostromislensky (34) first pointed out the disadvantages of curing with dibenzoyl peroxide in comparison with sulfur: the proportion of peroxide required for a tight cure was high, benzoic acid, a decomposition product, bloomed to the vulcanizate surface, and the resultant physical properties and aging were generally poorer. A reappraisal of peroxide cures by Braden and Fletcher (35) led to the conclusion that dicumyl peroxide was more satisfactory as a practical curative, particularly since it functioned well in the presence of reinforcing blacks.

The efficiency of peroxides measured as cross-links formed per molecule reacted is high (36), being of the order 1.0 for dicumyl peroxide/NR in the range 110–150°C, although Bristow (37) has presented data showing a reduction in efficiency as the cure temperature is lowered from 160 to 100°C. He attributes this reduction to the different temperature dependence of competitive reactions between cumyloxy and methyl radicals derived from the peroxide with nonrubber model compounds on the one hand and with rubber hydrocarbon on the other, rather than to any change in the ratio of cross-linking to chain scission. In *cis*-BR and SBR, very high efficiencies are reported; these are attributed (38) to a chain reaction possibly involving the main chain double bonds (Section 3.2).

Although of high chemical efficiency dicumyl peroxide suffers from several practical disadvantages. Vulcanizates often have sticky surfaces and retain the smell of reaction products (cumyl alcohol, acetophenone); effective accelerators and retarders have not been found to impart the necessary flexibility in choice of time and temperature for the wide range of industrial cure requirements; since the most powerful antioxidants are radical traps, they reduce both the rate and extent of reticulation; and there exist general problems in connection with handling and dispersion. In attempts to overcome some of these difficulties, alternative aliphatic peroxides have been developed, including 2,5-dimethyl-2,5-*tert*-butyl-peroxyhexane, use of which leads to more volatile and less unpleasant smelling products (39). However, general use of peroxide cures for unsaturated rubbers is restricted to transparent vulcanizate applications; rubbers that must contain no sulfur compounds whatsoever; and networks where thermally stable carbon–carbon cross-links show some definite advantage over the traditional $C-S_x-C$ sulfur vulcanizate structure.

Reaction of peroxides with saturated hydrocarbon structures is less rapid but nevertheless technologically more useful, since it enables networks to be formed from a wide range of polymers which are difficult to reticulate satisfactorily in other ways. Thus polyethylene, polyvinyl chloride, ethylene–propylene and ethylene–vinyl acetate copolymers, and fluorocarbon, polyurethane, and silicone rubbers are among the common rubbers which respond to this type of cure. In all cases, the function of the peroxide is to abstract hydrogen from a methylene (or methyl) group; the residual radical structure then determines the ratio of cross-linking to scission. For polyethylene, cross-linking efficiencies are high. Increasing methyl substitution favors scission so that polyisobutene (or low unsaturation butyl rubber) degrades in preference to forming a network. EPM occupies an intermediate position with contributions from both reticulation and scission.

In addition to the diene rubbers, butyl, EPM, and atactic polypropylene can be cured more successfully by peroxides with sulfur added. While it was formerly presumed that the sulfur suppressed scission, so favoring cross-linking, Loan (40) has shown that, for EPM, addition of sulfur leads to only a small increase in network chain density but a very large increase in strength, which he attributes to the formation of labile, sulfur-containing cross-links rather than to any large change in reticulation efficiency (see Section 6.1).

1.3.3 Ionic Cross-Linking

The principles of ionic cross-linking have been understood for many years, more particularly in connection with the metal oxide curing of carboxyl modified rubbers. The product behaves as a network at normal temperatures but the polymer may be softened by heat when the cross-links weaken sufficiently to permit processing. Different opinions have been advanced to explain the very high strength of these vulcanizates: Mullins (41) proposes that since ionic cross-links are readily broken and reformed, this lability permits accommodation of stresses by spreading the load equally over a larger number of network chains. Halpin and Bueche (42) believe that the high strength is a natural result of the concentration distribution of this particular type of cross-link in the rubber.

1.4 Reinforcement, Particularly of NR and SBR

The classical picture of the rubber network used in the preceding section envisages an assembly of randomly linked chains joined together at a small number of junction points sufficient to form a three-dimensional network but insufficient to restrain seriously the individual network chains from undergoing thermally induced configurational changes of a random nature. It is assumed that under an applied stress at constant temperature the network will undergo strain such that the chains take up new average configurations in the oriented network which may be represented, at least in model terms, using thermodynamics and statistical mechanics appropriate for equilibrium conditions. It is now being recognized that the constituent network chains of bulk rubber vulcanizates which are unswollen do not respond to applied stress as rapidly as had once been supposed so that equilibrium strained configurations are unlikely ever to be attained in many rubber product applications.

Better understanding of polymer rheology has established the relative importance of both viscous and elastic contributions to network behavior, and many time dependent phenomena are now being explained in terms of breakage and reformation of chain entanglements (43) and chain slippage which are not concepts of equilibrium theories. Studies of networks, particularly their moduli and fracture behavior sometimes described as ultimate properties, over a wide range of temperature and strain rate, lead to the following somewhat simplified conclusions:

1.   If the elastomer is amorphous and at a temperature well above T_g, the ultimate strength is low and the extension at break is only about 10% of that corresponding to full extension of the network chains (44).

2.   Reduction of temperature or increase of extension rate may lead to manyfold increases in both strength and extensibility, these increases

resulting from local viscosity increases which increase the rate of dissipation of energy during extension.

3. Failure processes are initiated at flaws in the network of macroscopic dimensions, but if a crack does start to grow, it does so more slowly under conditions in which the rubber is hysteretic, that is, where there exists a mechanism for dissipation of energy.

4. Natural rubber, and other elastomers of suitably regular structure, will orient under strain (stress crystallize), the crystalline domains forming more readily in regions of high stress such as are to be found close to flaws of sufficient size to initiate crack growth.

5. Rubbers of less regular structure, such as SBR, do not have available to them a crystalline hysteresis mechanism for energy dissipation, with the result that the gum networks are weak.

6. Addition of fine particle size fillers of a suitable chemical nature, particularly selected carbon blacks, modifies the stress relaxation mechanism (45) and increases the energy dissipation at the tip of a crack, thus reducing its growth rate (46).

7. Other domains of harder material in a soft, more liquidlike rubber matrix may likewise increase strength such as are present, for instance, in the block copolymers discussed in Section 4.

Physical properties are considered in more depth and from other viewpoints in Chapters 10–13.

1.4.1 Polymer-Filler Interaction

The interaction of elastomers with reinforcing fillers has been comprehensively reviewed by Kraus (1), and Smith has commented on the technological implications of recent work (47).

Fillers may be particulate in form, resinous, or fibrous. Particulate fillers added primarily to improve certain physical properties of the polymeric material are often called reinforcing fillers. Resinous fillers are organic macromolecules related to rubbers and plastics but usually of substantially lower molecular weight. Some types are covulcanizable .with the elastomer and become chemically bonded to the elastic network. Fibrous fillers, added as relatively short fibers dispersed on conventional mixing machinery, appear to become distributed through the rubber in the form of a more rigid network which reduces flow leading to a high apparent hardness and, under some conditions, to excellent resistance to abrasive wear. The fibers may be organic, particularly cellulosic (lignin, cotton flock, etc.), and special properties can be imparted through the introduction of "whiskers".

1.4.2 *Nature of Filled Compounds*

The polymer-filler blend is often described as a complex, heterophase mixture. Its properties are determined not only by interaction between polymer and filler surface but also by aggregation-disaggregation of filler units within the compound, that is, the formation and disruption of filler structure. Such structure may result from particle–particle surface interactions, chemical reactions of resins, or entanglements of fibers.

Criteria for producing a properly reinforced filled elastomer compound include: (*a*) preparing the filler in a suitably disaggregated form; and (*b*) taking special steps appropriate to the filler and the rubber to disperse the one in the other uniformly (including wetting of the filler surface). Mixing with fillers requires doing work on the rubber, this work being greater for the more finely divided additives. Processing advantages which accrue include lower swell and shrinkage than would be expected merely from dilution of the polymer, smoother surface finish, and much better flow and/or maintenance of dimensions in nearly every shaping operation. As the rubber content is reduced, behavior improves until a point is reached when the viscosity of the rubber-filler mixture is too high for flow to occur easily. This corresponds to the maximum volume loading for this particular filler-polymer combination. For important particulate fillers such as the fine particle size carbon blacks, this difficulty occurs at disappointingly low loading values, but the workable loading range may be considerably extended by the simultaneous addition of processing aids (softeners) such as aromatic or naphthenic hydrocarbon oils.

Filled vulcanizate properties may differ considerably from those of the parent gum vulcanizate. Indentation hardness and modulus may increase, the latter several-fold. While it seems obvious that the wear resistance of (soft) polymers will be improved by the addition of (relatively hard) fillers, this naive principle cannot explain abrasion loss in quantitative terms. In particular, frequency-temperature conditions have to be specified closely in order to understand why a gum vulcanizate that wears out quickly in the form of a tire tread, may yet last for years as the lining for a gravel delivery hose, an application for which conventional passenger tread rubber is less than adequate.

The progressive replacement of elastic rubber by nonelastic filler leads to an expected reduction of resilience and increase in hysteresis (energy dissipated during a deformation cycle, leading to heat formation). Heat buildup in tire sections is related to this property, although to some extent it also measures the ability of the compound to dissipate heat (e.g., to the atmosphere). This dissipation is facilitated by the increase in thermal conductivity (diffusivity) resulting from replacement of insulating rubber by thermally conductive carbon black (or zinc oxide).

1.5 Particulate Reinforcement

For spherical particles that interact only in pairs (i.e., which lack structure) and which are completely wetted by the rubber, Smallwood and Guth's development of the Einstein relationship predicts an increase of modulus with filler concentration of the form

$$E = E_0(1 + 2.5\phi + 14.1\phi^2)$$

where E is the Young's modulus of the filled rubber, E_0 that of the rubber matrix, i.e., the rubber network between the filler particles, and ϕ is the volume fraction of filler. Strictly speaking E and E_0 can only replace the correct viscosities, η and η_0, for small strains. Since at normal filler loadings the ratio of stresses with and without filler is nearly constant, it is possible at constant extension ratio to use the approximation

$$\sigma \simeq \sigma_0(1 + 2.5\phi + 14.1\phi^2)$$

where σ and σ_0 refer to measured stress values (rubber moduli). The term in parentheses may be regarded as a *strain amplification factor* which describes the effect on modulus due to the addition of an inactive (i.e., noninteracting) particulate filler. Experimental verification of the equation has been made using nonreinforcing fillers such as glass beads, medium thermal black, and calcium carbonate.

In practice, reinforcing fillers produce an increase in modulus in excess of that predicted by the equation just given (48). The reasons for this might be associated with bonding between particles and the rubber network. Kraus (49) has proposed quantitative relationships for the two extreme cases, nonadhesion and perfect adhesion; for the latter case,

$$\frac{(v_r)_0}{v_r} = 1 - m\phi(1 - \phi)$$

where

$$m = 3c[1 - (v_r)_0^{1/3}] + (v_r)_0 - 1$$

and v_r is the volume fraction of rubber in the filled vulcanizate (network) swollen to equilibrium in a suitable liquid; $(v_r)_0$ is the v_r for the same vulcanizate containing no filler; ϕ is the volume fraction of filler in the filled vulcanizate; and c is a constant. The equation predicts a linear dependence of $(v_r)_0/v_r$ on $\phi(1 - \phi)$, which is observed in some cases, particularly for the important tire tread filler, high abrasion furnace black, as shown in Fig. 3. From the negative slope m, a value of c may be computed which is characteristic of the filler but independent of the polymer, the swelling liquid, or the degree of cure. The value $c = 1.17$ was taken to characterize the reinforcing potential of HAF black.

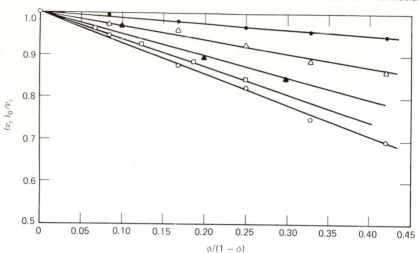

Figure 3 Dependence of vulcanizate swelling on the volume loading of carbon black (HAF). O, SBR/cyclohexane [$(V_r)_0 = 0.189$]; ●, SBR/isooctane [$(V_r)_0 = 0.548$]; △, SBR/n-heptane [$(V_r)_0 = 0.371$]; □, NR/n-heptane [$(V_r)_0 = 0.277$]; ▲, polybutadiene/n-heptane [$(V_r)_0 = 0.330$]. From Kraus (1).

These observations are consistent with a picture of the unfilled rubber matrix with swelling behavior indistinguishable from that of a gum vulcanizate in which float particles of black that are so firmly bonded to the neighboring rubber chains that swelling at the particle surface is completely restricted. Such a picture is quite consistent with strong adsorption forces known to operate between polymer and particulate filler. Nevertheless, it appears to represent a serious oversimplification of the real state of affairs; for instance, neither Kraus' equation nor an alternative (nonequivalent) treatment by Lorenz and Parks (50) can explain M. Porter's observations, quoted by Saville and Watson (51), that the presence of HAF black increases the cross-linked sulfide yield in accelerated sulfur/rubber model (2-methylpent-2-ene) experiments. For efficient systems with high accelerator/sulfur ratios, where cross-link yields are already very high, Porter finds that the black exerts little or no influence on the yield of chemical cross-links [X]. He proposes the simple equation

$$[X]_{apparent} = [X]_{actual}(1 + c\phi)$$

where $[X]_{apparent}$ is the value of chemical cross-link concentration obtained using the Mullins, Moore, and Watson (52) relationship. c is an alternative parameter characteristic of the filler; for HAF black, $c = 3.63$.

Porter's equation is claimed to be more consistent with that of Lorenz and Parks than with that of Kraus.

While it is thus possible to describe favorable systems by use of semiempirical single-parameter equations, one cannot help feeling that the actual polymer-particle interaction picture must be vastly more complicated. For instance, the numerous publications cited previously together with those of Payne, Mullins, and co-workers have established the hysteretic contribution to particulate reinforcement. Wake (53) has reviewed surface chemical evidence based on heats and entropies of adsorption; he believes that whereas inorganic, nonreinforcing fillers adsorb unsaturated hydrocarbons at fixed sites, the adsorption sites on the carbon black surface are mobile, thus allowing the possibility of local redistribution of stress by slippage of diene rubber molecules without desorption, molecular rupture, or void formation. Silica would be expected to occupy an intermediate position with partially mobile adsorption sites; this seems to be consistent with its rather less generally satisfactory reinforcement behavior.

While there is general agreement between the physical (hysteretic) and surface chemical viewpoints, a number of significant chemical observations are not yet fully included in present treatments, in particular the observation by Watson (54) of the radical nature of carbon- or silica-elastomer reactions during cold milling leading to cross-linked gel. Such reactions would be anticipated from the established presence of hydroxyl, carbonyl, and quinone groups at the surface of the black particles which can interact with polymer radical chain ends (55). The importance of these groups has been emphasized technologically by the use of promoters of filler reinforcement (56), many of which increase the bound rubber content of the mix and lead to easier processing and/or improved vulcanizate properties. Nevertheless, data have been accumulated which suggest that deliberate deactivation of these supposed radical-reactive sites has little or no effect on the reinforcing behavior of the black (57), a conclusion that has been contested (58).

Existing evidence certainly favors a physical rather than chemical origin of reinforcement. It has been observed that 4% of phenyl-β-naphthylamine (PBN) imparts the same high modulus to rubber vulcanizates as some 25% of a reinforcing black and that subsequent solvent extraction of the PBN reduces the modulus to the gum value (59). Payne's classical experiment, which "missed out" the rubber (60), showed that high modulus can still be obtained in the low extension region using blends of reinforcing black with a saturated hydrocarbon oil, thus indicating the importance of interaction between filler particles, a *structure* effect. From this it would seem that neither rubber nor carbon black is needed to demonstrate reinforcement, unless we adopt Brennan and Jermyn's view (61) that high extension modulus is influenced primarily by bound rubber, while low extension modulus only is determined by viscoelastic effects and particulate structure.

1.5.1 Measurement of Reinforcing Ability

Integral with the problem of explaining reinforcement is that of measuring it with some satisfactorily reproducible test or tests. Ultimate failure tests are limited in value by excessive dependence on test piece imperfections and by lack of a satisfactory theoretical background, now remedied to some extent by the work of Halpin and Bueche (45). Estimation of tensile modulus is likewise fraught with difficulty on account of the Mullins effect (a reduction in modulus in successive stress–strain cycles which indicates progressive changes in the material under test). This was formerly associated with the breakage under stress of rubber-filler bonds, but studies of gum vulcanizates have revealed that the effect occurs in the absence of filler to an almost proportionate extent (62). It thus appears more likely to be short network chains or labile cross-links which are breaking, although such explanations do not account satisfactorily for the partial recovery of modulus after a relaxation period.

The general situation has been discussed by Harwood et al. (63) who suggest several contributory mechanisms which might effect stress softening: (a) rearrangement of the rubber network associated with slippage of chain entanglements and nonaffine displacement of network junctions in the rubber matrix, which may occur in gum vulcanizates and in the gum phase of vulcanizates containing fillers; (b) changes in the state of aggregation and structure of the dispersed reinforcing filler materials; (c) slippage of rubber chains anchored to the filler surface, or breakage of (possibly covalent) rubber-filler chemical bonds; (d) breakage of weak cross-links, for instance hydrogen bond or ionic types; and (e) displacement of large particles through the rubber matrix-hydrodynamic effects.

Since the very lability of structures that constitute the polymer matrix may be the chief clue to reinforcement behavior (64), much interest has been aroused in the quantitative contributions of mechanisms (a)–(e) in technologically important systems. An example of what can be learned from model systems designed to isolate individual factors is Morton and Healy's recent study (65) of SBR reinforcement by glassy polystyrene or styrene–butadiene copolymers blended with the rubber in emulsion. Fillers of particle size between 35 mμ (similar in diameter to, say, FEF black) and 650 mμ were used and could be bonded or nonbonded to the elastomer at will (as shown by solubility studies). The authors concluded that increase in tensile strength of the elastomer by addition of filler resulted from a viscoelastic mechanism, the temperature-strain rate data obeying the Williams, Landel, and Ferry (WLF) equation. The finer particle size filler showed the greater strengthening effect, but the increase in tensile strength with filler content was reduced by the introduction of chemical bonds between filler and elastomer. The presence of the (model) filler did not influence the cross-link density of the elastomer. These model conclusions are consistent with those of Oberth (66) except insofar as the latter author regards filler-polymer adhesion as a necessary prerequisite for, if not a primary factor in determining, reinforcement.

1.5.2 Reinforcement by Carbon Black

Until about 1960, passenger treads were based on (oil extended) SBR, high abrasion furnace (HAF) black, and hydrocarbon process oil. Truck and bus tire treads have incorporated substantial quantities of natural rubber and/or, more recently, cis-polybutadiene which are less hysteretic than SBR, so that the thicker section tires run proportionately cooler. The choice of HAF rather than, say, super abrasion furnace (SAF) black was based on considerations of the balance between tread wear and other important tire properties, for instance hardness and resilience, processing, and cost. Substantial changes have taken place in tread compounding which are resulting in higher performance at lower material cost. Polybutadienes of high average molecular weight are capable of tolerating very high loadings of process oils and the more reinforcing blacks (ISAF, SAF), and the resulting treads show high wear values. Newer high structure oil furnace blacks, now covered by ASTM classifications, have proved most useful in these applications since their use in cis-polybutadiene/oil extended SBR blends enables a substantial increase in the oil loading without deterioration of processing or properties (67); the high structure seems to be associated with better dispersibility. Similar results (68) obtained in EPDM would favor this polymer for tread applications if other difficulties, both legal and technical, could be overcome. Some studies have been made of polymer influence on polymer-black interaction; for instance, solution SBR gives higher viscosity and carbon gel than the series 1500 emulsion polymers (69).

1.5.3 Nonblack Particulate Reinforcement

The principal characteristics of nonblack reinforcement were listed by Moakes and Pyne (70), and many more recent reviews are available, particularly that of Bachmann et al. (71) and the short chapter by Sellers and Toonder (72). While it seems fairly clear that the characteristics of nonblack reinforcement are broadly similar to those outlined previously, i.e., modification of viscous properties and formation of a chemically bound gel phase (73), detailed studies of rubber-filler interaction reveal certain differences. For instance, Thomas and Moore's study (73) of the effect of fine particle size additives on the viscosity of a model polydimethylsiloxane oil has suggested a mechanism involving solvation followed by aggregation to form microgel. Their data indicate similar behavior for silicas and for carbon blacks of similar particle size, except that black is rather more effective in increasing the viscosity. Southwart and Hunt (74) using a poly(methyl vinyl)siloxane rubber with variously treated pyrogenic silicas have established some unusual features not previously observed in carbon black/hydrocarbon rubber studies. They propose a mechanism involving binding of the silica particles by rearrangement between silanol groups on the surface of the filler particles and the siloxane groups of a relatively small proportion of the polymer chains.

It seems reasonable that the key difference in reinforcing behavior between silica and black lies in the different chemisorptive interaction of groups present at the filler surface with the different elastomer structures.

Precipitated aluminum and calcium silicates are generally of larger particle size than the finer silicas; they are easier to process and rather less reinforcing as judged by conventional tests such as static modulus, tear, and abrasion resistance. All these fillers, but most of all the fine silicas, may gain in reinforcement when subject to hot mixing (with or without promoters) or other forms of heat treatment of the compound (76), the effects being most apparent in improvement of the dynamic properties of the vulcanizates.

Silicas and calcium silicates retard vulcanization to a serious extent although aluminum silicate is rather less objectionable in this respect. This retardation was formerly attributed to adsorption of curative ingredients, either sulfur or accelerators (or both), by the filler, although modern views (77) favor a deactivation mechanism involving interaction of the filler with the zinc-accelerator-sulfur complexes which are postulated vulcanization intermediates. The traditional additives—triethanolamine and diethyleneglycol—added in loadings of some 10% on the filler content have now largely been displaced by various waxy polyethylene glycols (Carbowax); use of these enables silica and silicate compounds to be cured with fairly conventional sulfur and accelerator levels (e.g., sulfur 3, MBTS 1, DOTG 1.5 with NR 100, precipitated silica 60) to yield vulcanizates with exceptionally high tear strengths (silica) and with other mechanical properties roughly comparable to those of SRF (silicate) or FEF (silica) black vulcanizates.

Comprehensive data on the performance of these fillers (and precipitated calcium carbonate) in various elastomers have been tabulated by Westlinning and Fleischhauer (78). They record that the greatest use of silicas and silicates is in sole and heel compounds, including black products that are merely pigmented with the black to avoid scuff-marking. Other uses include silicas with carbon blacks in tires to control crack initiation and growth and to improve tear resistance (of polybutadiene) and cutting and chipping of SBR off-the-road tires; silicas in translucent rubbers intended to simulate traditionally accepted high quality crepe; and silicates in rubbers for the food and pharmaceutical industries. Other nonblack fillers are less effective reinforcing agents.

1.5.4 *Secondary Network Reinforcement*

Resins added to rubbers are generally of fairly high molecular weight; they may be capable of cross-linking to form an independent network, possibly joined to the rubber network at a few points, or they may be thermoplastic and exert their influence simply by entanglement with the rubber chains. Examples of the first group are the oil soluble phenolic resins added to NR, SBR, or NBR; examples of the second group are the high styrene/butadiene resins (virtually incurable at 85% styrene level) added to NR or SBR, or vinyl chloride polymers mixed with NBR. There is considerable interest at the present time in the development of reinforcing resins for use with ethylene–propylene co- and terpolymer rubbers; a recent claim (79) is for a resinous copolymer of styrene and 2-ethylhexyl acrylate which is claimed to increase tensile and tear strength, elongation at break, modulus, and hardness.

Cellulose fibers are among the older established fillers in the polymer industry; flock is used extensively in hard rubber products such as solid tires for hand trucks and has been reviewed by Pacitti (80). Ground waste leather (81), granulated cork (82), and inorganic fibers such as asbestos and glass reinforced rubber (83) are finding many uses.

1.6 Summary—Network/Reinforcement Phenomena

We may now summarize the principal factors which we believe influence the measured modulus and other network dependent properties of a filled rubber vulcanizate:

1. Microchain structure: relaxation behavior, T_g, ease of formation of crystalline or glassy domains, chemical reactivity.

2. Macrochain structure: molecular weight and molecular weight distribution, which affect the concentration and distribution of network imperfections such as chain ends and entanglements.

3. Extent and nature of cross-linking/scission reactions: the concentration of cross-links which, in the absence of concomitant chain scission determines the number of network chains and hence the equilibrium modulus (temperature dependent), the distribution of cross-links if not random, and the lability of cross-links that may give way under stress and/or at elevated temperatures. These factors will be determined in turn by the composition of the vulcanizing system, reactivity of the polymer, and time and temperature of cure.

4. The concentration and nature of added filler: the volumetric replacement of soft extensible rubber by a hybrid material, normal strain amplification, polymer-filler bonding interactions, and filler particle/aggregate shape, size, and surface reactivity.

5. Time (rate) and temperature of measurement which, besides earlier citations, in the case of wholly viscoelastic processes will be related by a shift factor in accordance with the WLF transform.

6. Degree of dispersion, both of vulcanizing system and filler particles or aggregates, which ideally should be sufficient to yield a homogeneous material at the molecular level but which in practice can probably not be relied on much below the submicron range (see Section 8). In fact, Section 2 of this chapter quotes a good example of the difficulties that may arise from variation in the degree of dispersion which sometimes makes it difficult to compare the work of investigators using different mixing techniques.

The precise influence of each of these variables on vulcanizate properties and the subtleties of their interactions are not yet fully elucidated but the framework they represent is now sufficiently established to be of value in discussion of the special behavior of individual stereoregular elastomers.

2 ISOPRENE RUBBERS: VULCANIZATION*

Early work with synthetic polyisoprene rubbers showed the general similarity of their vulcanization behavior to that of natural rubber. The chief differences are (a) the absence of equivalent degrees of cross-linking (modulus values) in gum stocks at equal cures; (b) the slightly poorer tensile stress-strain properties and substantially lower tear strength and

* See also chapter 4.

abrasion resistance of the synthetic rubbers even with optimization of the compounding formulas; and (c) the greater difficulty of dispersing curatives and fillers in the new rubbers.

The lower modulus was attributed to the absence of the nonrubber constituents of natural rubber and could be partially overcome by inclusion in the synthetic compound formulations of equivalent materials, for instance 1.5 phr of 5% triethanolamine/95% soya lecithin. Black vulcanizates showed acceptable modulus values although these were still substantially lower than those for equivalent NR compounds (84). There are also differences in the behavior of polyisoprenes made with different catalyst systems, as will be described below.

Early commercial (alkyllithium type) polyisoprene rubbers appeared to be softer than natural rubber during the normal mill mixing cycle. This meant that the rubber suffered less reduction of molecular weight during mastication and was believed to be a prime cause of bad dispersion of compounding ingredients and at least a contributory cause of the poorer vulcanizate properties when compared with NR. It was later realized that the lower temperatures of mixing the softer rubber also led to further difficulties in dispersion of, particularly, high melting point accelerators. The latter materials (as well as sulfur and zinc oxide) were usually masterbatched (preferably in the tougher NR) and the dispersion was often further improved by hot milling and by resting the mixed stock for 24 hours before cure to allow time for dispersion by diffusion. Systematic studies of these differences lagged behind their practical solution in the technical service laboratory. Nevertheless the results focused attention on the differences in dispersibility of ingredients (particularly curatives) and structural purity of the polymer (1,4-*cis* content).

Himes and Sullivan (85) studied the first of these problems and showed that when liquid or low melting point accelerators were used, efficient mixing methods were employed, or particulate fillers were present to aid dispersion by their grinding action, differences between natural and synthetic polyisoprene rubbers in respect to their curing behavior were minimal and of little technological consequence. Under conditions where adequate dispersion was not achieved, network flaws developed in overcured regions of the curative-rich synthetic rubber, resulting in significant reductions in tensile and tear strength. It was also shown that the nonrubber constituents (e.g., proteins) absent from the synthetic rubber could be satisfactorily replaced by addition of aldehyde-amine types of accelerators. A combination of N-cyclohexyl-2-benzothiazylsulfenamide (CBS, 0.3 phr), Trimene Base (0.4 phr), with sulfur (1.5 phr) was found to be particularly satisfactory for use with IR. The two accelerators are readily dispersible (CBS, mp. 94°C; Trimene Base, liquid). A fuller study of the effect of nonrubber constituents on factory processing has been made by Gregg and Macey (86).

Optimum strength values were obtained at relatively low sulfur levels (ca. 1.5 phr) whereas NR required up to 3 phr sulfur for optimum strength in similar systems. The fall in strength of synthetic cis-1,4-polyisoprene (IR) networks vulcanized with more than 1.5 phr sulfur was attributed to nonrandom distribution of cross-links since swelling measurements showed equal cross-link densities in networks of IR and NR produced using identical recipes, times, and temperatures of cure. In fact this was shown to result from poor sulfur dispersion which may be overcome, at least up to levels of 3 phr, by use of special high temperature masterbatching techniques or by incorporation of fillers.

The question of structural purity of IR in relation to network properties was tackled systematically by Bruzzone and associates (87). They first extended the curative study of Himes and Sullivan to separate compounding and related dispersion differences from genuine network differences related to the structures of different polyisoprene chains.

Table 2 Raw Elastomer Properties (87)

	Polymer		
	A Natural Rubber RSS No. 1	B Polyisoprene Al–Ti[a]	C Polyisoprene Li Alkyls[b]
1,4-Cis (%)	97	96	92.5
1,4-Trans (%)	1	0.5	1.5
1,2 (%)	—	—	—
3,4 (%)	2	3.5	6
Mooney viscosity ML_{1+4} (100°C)			
(a) Initial	96.5	95	64.5
(b) After 10 passes[c]	81.5	57	75
Gel content (%)			
(a) Initial	14	23	1
(b) After 10 passes	Nil	Nil	Nil
Intrinsic viscosity in toluene at 30°C			
After 10 passes (dl/g)	4.9	3.4	4.3

[a] Polyisoprene obtained with a catalytic system based on aluminum and titanium compounds.

[b] Polyisoprene obtained with a catalytic system based on lithium alkyls.

[c] At 60°C, mill nip 4×10^{-3} in.

Table 2   (Continued)
Gum Stock Vulcanization Recipes (phr)

	Polymer		
	A	B	C
Polymer	100	100	100
"Pepton" 65[a]	—	—	0.25
Lecithin—TEA[b]	—	—	1.50
Stearic acid	3.00	3.00	3.00
Antioxidant 2246	1.00	1.00	1.00
Zinc oxide	5.00	5.00	5.00
Sulfur	[d]	[d]	[d]
CBS[c]	[d]	[d]	[d]
Vulcanization temperature: 144.5°C			

[a] Zinc-2-benzamidothiophenate.
[b] Lecithin 95% + triethanolamine 5%.
[c] N-cyclohexyl-2-benzothiazylsulfenamide
[d] As shown in Fig. 4.

Black Stock Vulcanization Recipe (phr)

Polymer	100
Stearic acid	3
Zinc oxide	5
HAF	50
Aromatic oil	5
BLE	1
Sulfur	0.5–3.5
CBS	0.5–1.5
Curing temperature (°C)	144.5
Curing time (minutes)	10–60

The polyisoprene rubbers used together with gum stock vulcanization recipes are listed in Table 2. The results of curing at 144.5°C are recorded as contour graphs for tensile strength, modulus, and rebound resilience as functions of sulfur concentration, CBS concentration, and cure time. Figure 4 shows the tensile strength results, the rapid fall in tensile strength at higher sulfur levels for the 92% low cis content polymer "C" confirming the earlier results. It would seem that optimum tensile strength, modulus, and rebound figures for gum vulcanizates of this lithium alkyl type of polymer are all significantly lower than for NR "A" or for the Al–Ti (Ziegler-Natta) type IR, polymer "B." Differences between Himes and Sullivan's results and those for the substantially identical polymer "C" may have arisen from more thorough mixing and milling techniques of the former.

However, other results suggest that the differences shown between polymers of different cis content are real, i.e., not merely a reflection of inhomogeneous cure resulting from poor dispersion. In particular, results obtained for black vulcanizates (Fig. 5), where the filler would be expected to aid dispersion, still show an increase of strength at optimum cure with cis content; similar results have been obtained for laboratory tear strength and abrasion resistance where the 96% cis polymer "B" was inferior to NR.

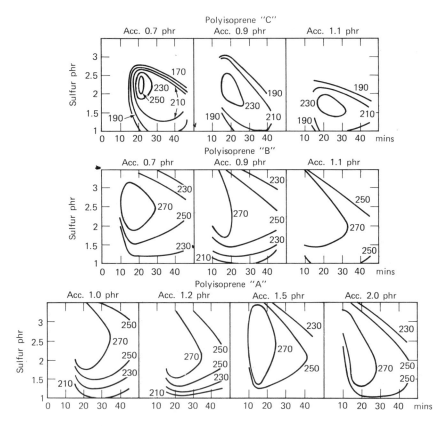

Figure 4 Gum tensile strength contour lines as a function of sulfur and curing time. From Bruzzone et al. (87).

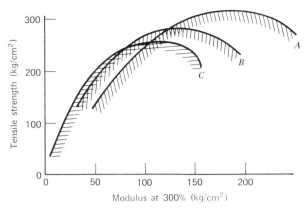

Figure 5 Black stock envelope curves, tensile strength, and modulus at 300% elongation. From Bruzzone et al. (87).

It seems probable that little isomerization occurs during the normal curing of IR and that the structural purity of the starting polymer does have some residual effect on ultimate network properties even in the presence of reinforcing filler. We can therefore explain the higher modulus and tensile strength of 96% cis polymers at optimum cure in terms of their easier crystallization under stress. At low degrees of cure, 92% cis rubber shows significantly better tensile and tear behavior than the higher cis rubbers, including NR, and in view of the small dependence of T_g on cis content, it seems unlikely that this can be explained in terms of differences in chain mobility. The Al/Ti polymer "B" showed much less heat buildup on flexing, i.e., was less hysteretic than NR.

Bruzzone et al. attribute the effects at low cure to the greater perfection and absence of flaws in the IR network which is related to the narrow molecular weight distribution of lithium based IR. At higher cross-link densities the distribution becomes less important and properties are more related to ability to crystallize under stress. Other important factors might be the distribution of the stereoregular sequences in each type of chain. For example, it has been determined that 15–20% of head-to-head structures are found in free radical IR compared to 1–3% in lithium or Ziegler IR, still a significant structural impurity. The chain reversal is postulated to occur at points of 3,4 addition (88).

Finally, mention should be made of vulcanization of *trans*-1,4-polyisoprene, a balata substitute recommended for use in golfball covers (89). Low curing temperatures are required both to prevent degradation of the elastic thread winding and particularly to retain the desirable "click" sound of the vulcanizate cover. The crystalline melting point of the polymer is at 64–66°C; if curing is carried out at a higher temperature, the cross-links hinder crystallization and some of the click is lost. Surprisingly, it was found possible to produce adequate vulcanizates from both balata and synthetic *trans*-IR by immersing molded covers containing pigment, ultraaccelerators, and zinc oxide activator (but no curative) in powdered sulfur at 44°C for several days. This is the only curing process known to the author which relies on the diffusion of a solid curative into (partially) crystalline polymer. There are, of course, other, faster curing procedures for golfball covers.

A useful state-of-the-art summary of work in progress on isoprene rubber was published in 1973 (90).

3 BUTADIENE RUBBERS: VULCANIZATION

3.1 Sulfur Vulcanization

3.1.1 *Cis-Trans Isomerization*

Polybutadienes differ from polyisoprenes in that they appear to undergo significant cis-trans isomerization when heated with sulfur (91). Shipman and Golub (92), using infrared techniques, studied the spectral shifts when thin films of *cis*-polyisoprene and *cis*-polybutadiene (also deuterated polymers) were heated with 15 phr sulfur at 140°C for 1–20 hours under nitrogen. In the case of polyisoprene, development of strong absorption bands at 10.4 and 14.0 μ observed by earlier workers in studies of NR vulcanization could not be assigned wholly to cis-trans isomerization but appeared to arise principally from the presence of cyclic sulfides of the type envisaged in the polar chain type vulcanization mechanism of Bateman et al. (93), together with conjugated double bond sequences arising from shifts by free radical processes as suggested by Linnig and Stewart (94).

In contrast, the 10.4 and 14.0 μ bands in the spectrum of sulfur-reacted *cis*-polybutadiene are definitely associated with cis-trans isomerization, further evidence being adduced from other changes in spectra of deuterated polymers. The mechanism of the isomerization is not yet fully established. One suggestion is that the first step is the formation of polymeric thiyl radicals:

$$\sim\!\!CH_2\!-\!CH\!=\!CH\!-\!\overset{\cdot}{C}H\!\sim + \cdot S_x \longrightarrow \sim\!\!CH_2\!-\!CH\!=\!CH\!-\!CH\!\sim$$
$$\underset{\overset{|}{\underset{\cdot}{S}_x}}{}$$

which could then induce isomerization in a way similar to that proposed for monomeric thiyl radicals (91). However, these would have to be formed and act effectively in very small concentrations, otherwise a significant double bond shift would be observed in the polybutadiene spectrum.

An alternative suggestion by analogy with sulfur-diene reaction mechanisms (93) involves addition to a double bond of a persulfenium ion, TS_a^+, where T is an alkyl or alkenyl group, to form the transition complex

$$\sim\!CH_2\!-\!CH\!-\!\!\!-\!\!\!-\!CH\!-\!CH_2\!\sim$$
$$TS_a^+$$

If the TS_α^+ entity is readily detached before reaction to cyclic sulfide can take place, reformation of the double bond can take place in either a cis or trans configuration, of which the trans form is the more stable and therefore more probable. The absence of a similar mechanism for polyisoprene can be explained by the much firmer attachment of the TS_a^+ to the methyl dominated double bond. Kinetic data favor the polar rather than the radical mechanism.

Essentially similar results have been reported by Bishop (95) who studied the isomerization of high cis-polybutadiene by reaction with sulfur, dibenzthiazyl disulfide, or dicumyl peroxide. Using elemental sulfur at an initial concentration of 1.69 moles/liter, the cis content fell from ~94 to ~70% in about 140 minutes at 160°C; both high cis and high trans polymers were found to isomerize. Starting with high cis polymer, kinetics were (pseudo) first order over a wide range of temperature and sulfur concentration; the rate coefficient was accurately proportional to $[S_x]_{initial}$ and the apparent activation energy for the process was 30.1 kcal/mole.

Peroxide and sulfur initiated isomerizations were proposed to occur by somewhat different radical mechanisms, both involving interaction with main chain double bonds. It was suggested that sulfur forms a π complex which dissociates to a freely rotating radical whereas the peroxide mechanism is presumed to involve direct radical addition at the double bond as shown in the diagram on page 464.

Polybutadiene containing initially about 95% cis units may be transformed after *extensive* reaction with sulfur into a polymer with about 35% cis units. Although accelerated sulfur cures do not result in such extensive isomerization (96) the differences in initial cis content of the various commercially available polymers (ranging from about 40 to about 97%), while significant in respect to processability, may be less important in sulfur vulcanizates, particularly those requiring lengthy curing cycles, provided that the mechanism of isomerization is operative in the presence of accelerators and activators. Where this is the case, we would not expect 96% cis-polybutadiene to yield a readily crystallizable self-reinforcing vulcanizate like high cis-polyisoprene unless cured rapidly at low temperatures.

3.1.2 Influence of Chain Structure on Vulcanizate Properties

Blümel (96), quoting work by Madge (97), believes that the degree of isomerization in accelerated vulcanizates is quite low (not greater than about 7%) and so a matching of the configurational contents of the different polybutadienes under practical vulcanization conditions does not take place. He presents data for 14 polybutadienes of different structure produced in Ziegler–Natta, alkyllithium, alkali metal initiated bulk and emulsion systems, in comparison with cold SBR and NR (Table 3). The small changes in cis content on vulcanization at 134°C and somewhat larger change at 180°C are shown for typical polymers in Table 4 which also lists some black vulcanizate properties. The number of commercially available polybutadienes continues to multiply. Additional useful characterization data have been published by Curchod (98).

The following generalizations may be made by way of comparison among the vulcanizates (95):

1. High cis content rubbers have higher tensile strength, tear resistance, and elongation at break but there is no difference in modulus. The overall results are not as good as for cold SBR or NR.

2. Low cis content rubbers have higher hardness and rebound resilience, particularly for the rubbers with narrow molecular weight distribution.

3. The resilience of solution polybutadiene vulcanizates is significantly better than that of NR or SBR at 22°C but of the same order as SBR and slightly inferior to NR at 75°C.

4. Laboratory abrasion resistance deteriorates with decreasing cis content, but all solution polybutadienes were considerably better than NR or SBR.

5. Crack growth is relatively rapid in all polybutadiene vulcanizates (see Section 5).

To summarize the effects of structure, it seems that the cis content of the raw polymer is reflected to some extent in the properties of the vulcanizate but that other effects such as average molecular weight, molecular weight distribution, branching, and degree of cure always accompany changes in cis content and may be of greater importance in determining these properties. In practice such differences as exist among the different polybutadienes disappear when they are used in blends.

3.1.3 Network Properties and Cross-link Structure

An extensive study of the influence of structure including cross-link density on the properties of nine polybutadiene vulcanizates made from polymers of varying (95–5%) cis and varying (0.3–93%) trans contents has been published by Short, Kraus, and co-workers (99, 100). Gum and black stocks were cured to fixed values of network chain density (as determined by swelling), progressive increase in sulfur content being

Table 3 Polymer Analysis (96)

Rubber No.	Preparation, Grouped by Process	Hydrocarbon Composition of polymer (%)	Total Ash[a] (%)	Non-rubber Components (%)	Configuration or Structure[b] (%) Cis-1,4	Trans-1,4	1,2 (3,4)	Gel Content[c] (%)	RSV[d]	Molecular Weight[e] ×10⁻³	Inhomogeneity[f]	Mol. Wt. Dist.[g]	Configuration (%)	Configurational distribution[h]
1	Solution polymerization with Ziegler–Natta catalysts	100C₄H₆	0.15	<2	98	1	1	1	2.7	370	0.6			
2		100C₄H₆	0.10	<2	97	2	1	1	2.7	380	0.7			
3		100C₄H₆	0.20	<2	94	1	5	1	3.0	390	0.3			
4		100C₄H₆	0.17	<2	94	2	4	2	3.0		0.4			
5		100C₄H₆	0.20	<2	93	3	4	4	3.0		0.3			
6		100C₄H₆	0.40	<2	89	6	5	2	2.4		0.7			
7		100C₄H₆	0.10	<2	85	14	1	5	2.8	420	1.5			
8		100C₄H₆	0.29	>10	98	1	1	3	2.3	360	0.7			
9		100C₄H₆	0.60	>10	97	2	1	3	1.8	350	0.8			
10	Solution polymerization with alkali metal compounds	100C₄H₆	<0.1	<2	41	49	10	1	2.6	280	0.2			
11		100C₄H₆	<0.1	<2	42	49	9	1	2.9	350	0.3			
12	Bulk polymerization with alkali metals[i]	100C₄H₆	<0.1	<2	41	49	10	1	2.9		0.3			
13		100C₄H₆	1–2	<4	20	40	40	45	2.1		2.6			
14	Emulsion polymerization with redox catalyst[j]	100C₄H₆	0.12	5–10	10	70	20	2	2.7	450	1.6			
15	Cold rubber[j]	77C₄H₆ 23C₈H₈	0.3	5–10	15	69	16	1	2.2	500	0.9			
16	Natural rubber[j]	100C₅H₈	0.5	4–7	96	0	(4)	5	5.2	2500				

[a] Ash at 550°C according to ASTM D 1416–58a T Section 11–15.

[b] Data refer to polymer-hydrocarbon (and in the case of No. 15 to the C_4H_6 portion; for No. 16 the data apply to the proportion of 3,4 (not 1,2) structure.

[c] Insoluble in toluene, determined through centrifuging of a solution containing 0.2 g/100 cm³ of polymer (drying for 2 hours at 90°C in vacuo).

[d] Reduced solution viscosity, $RSV = [\eta] = (1/c) \cdot [(\eta - \eta_0)/\eta_0]$, measured after gel separation in an Ostwald viscometer in toluene, at 27°C; concentration as in preceding footnote.

[e] Measured by light scattering.

[f] Inhomogeneity $U = (M_w - M_n)/M_n$; M_w = weight average molecular weight; M_n = number average molecular weight.

[g] Differential molecular weight distribution $dI(M)/dM$, determined by solution fractionation with n-hexane/isopropanol mixtures.

[h] Configurational distribution, determined through infrared analysis according to footnote b, of the fractions obtained according to footnote g. The curves were obtained by plotting the infrared measurement points for the individual molecular weight fractions, beginning with the low molecular weights, going from bottom to top.

[i] Except for the values for ash content and nonrubber components, the data are valid only for the sol portion separated from the gel.

[j] The data for molecular weight, distribution, and inhomogeneity hold true only for the polymer-hydrocarbon portion.

Table 4 Cis-Trans Isomerization During Vulcanization

Rubber No.	Initial Product: Amount of			Configuration or Structure of the Product after Vulcanization at 134°C[a] (Polymer = 100; ZnO = 5; Vulcanization Agents as Indicated)					
				S = 1; CBS[b] = 1.5			S = 2; CBS[b] = 0.75		
	Cis-1,4 (%)	Trans-1,4 (%)	1,2 (%)	Cis-1,4 (%)	Trans-1,4 (%)	1,2 (%)	Cis-1,4 (%)	Trans-1,4 (%)	1,2 (%)
1	98	1	1	(92)	(6)	(2)	97 (92)	1 (7)	2 (1)
4	96	1	3	(91)	(5)	(4)	96 (89)	1 (8)	3 (3)
10	43	47	10	(37)	(52)	(11)	43 (41)	48 (50)	9 (9)

[a] Numbers in parentheses are for 180°C.
[b] CBS is N-cyclohexyl-2-benzothiazylsulfenamide.

468

Static Vulcanizate Properties

Rubber No.	Tensile Strength (kg/cm²)	Elongation at break (%)	Modulus at 300% Elongation, (kg/cm²)	Permanent Set (%)	Shore Hardness	% Rebound at		Tear Resistance (kg)
						22°C	75°C	
1	169	470	90	8	60	54	57	13
2	153	435	87	8	57	50	54	11
3	167	520	79	8	63	54	56	13
4	163	555	70	9	60	53	56	12
5	165	490	85	10	61	54	57	13
6	153	485	80	7	59	52	56	15
7	148	500	55	7	55	47	53	16
8	165	505	69	9	58	52	56	14
9	155	435	81	10	56	48	51	11
10	145	455	78	10	65	54	55	10
11	140	455	71	8	69	56	58	12
12	128	405	77	7	62	57	59	9
13	117	365	92	9	58	38	47	14
14	167	355	129	8	65	45	54	10
15	227	505	110	10	61	40	55	17
16	260	530	105	20	60	44	60	27

Table 4 (Continued)

Dynamic Properties of Vulcanizates

Rubber No.	Abrasion[a]	Crack Growth[b] Number of Flexes for Tear Propagation from 2 mm to			Terminal Deformation Load (kg)	Deformation Load (%)	Flexometer Behavior[c] Running Time[d] (minutes)	Equilibrium Temperature (°C)
		10 mm	15 mm	20 mm				
1	18	3,000	6,000	9,000	24.6	12	14	130
2	21	3,700	6,500	9,500	31.0	6	9	143
3	20	1,500	3,200	5,800	27.5	8	14	143
4	22	1,500	3,000	7,000	26.0	11	19	124
5	21	2,000	4,000	8,000	25.0	6	14	127
6	22	2,900	8,200	18,000	30.4	13	22	125
7	24	2,000	4,500	7,000	19.9	25	9	147
8	20	2,300	5,000	10,000	26.8	5	16	138
9	20	2,500	6,000	15,000	23.1	7	11	147
10	28	1,400	3,500	11,500	34.3	10	30	118
11	27	3,000	6,000	13,000	33.5	7	30	132
12	28	2,000	3,000	6,000	36.0	10	30	129
13	110	2,000	3,500	5,300	23.8	8	14	166[e]

14	47	1,200	3,300	8,000	23.9	20	26	131
15	90	13,000	30,000	53,000	22.4	32	24	138
16	105		115,000		9.7	46	16	115

[a] Abrasion test according to AP method, load 3 kg.
[b] Test with De Mattia machine, according to DIN 53522.
[c] Test with St. Joe Flexometer according to ASTM D 623–52 T, deformation amplitude 6 mm.
[d] Measured at a deformation amplitude of 8 mm.
[e] Test specimen already destroyed at measurement.

Recipe of the Compound

Rubber	100
Stearic acid	2
ZnO RS	3
HAF black	47.50
Aromatic plasticizing oil	8
Phenyl-2-naphthylamine	1
Phenylenediamine derivative	1
Wax	1
Sulfur	2
N-cyclohexyl-2-benzothiazylsulfenamide	1

necessary with increasing trans content of the polymer. In addition, a study was made of black vulcanizates covering a range of cross-link density.

Vulcanizates of all 1,4-polybutadienes exceeding 15% cis content are completely rubbery at ordinary temperatures. BRs of 93% trans content yield vulcanizates that are tough, leathery, and crystalline at 80°F but which become rubbery at moderately elevated temperatures, tread type stocks showing high modulus, tensile strength, and hardness but only fair hysteresis behavior. The tendency to crystallize readily at low temperatures disappears between 87 and 82% cis; below 82% cis the polymers remain rubbery down to the brittle point (−85°C) and all have similar properties with the exception of high vinyl structures which are deficient in tensile strength and low temperature flexibility (Figs. 6a–d).

It is evident that crystallization is necessary for development of high tensile strength in gum vulcanizates and that, surprisingly, this does not appear to detract from the excellent resilience. To achieve this, vulcanization recipes are usually of the low sulfur (1.5 phr) type. As with polyisoprene (Section 2), higher sulfur levels seem to lead to poorer vulcanization properties. Because of the inherently low modulus of black stocks and their low sulfur requirement, together with the tendency for polybutadiene rubber to continue to cross-link during overcure (101) rather than to revert (degrade), tire compounds may be given extended cures which further improve resilience and reduce heat buildup at the cost of some sacrifice in tensile strength and elongation at break.

Smith and Willis (102) provide similar cure data for a low cis-BR and show that this polymer has a slower cure rate and longer induction period than NR. Low sulfur and high accelerator levels are recommended, the ratio being critical in determining some network properties. Much of the data in their paper refer to blends of BR with NR or SBR.

The influence of cross-link structure has received little attention because of the difficulties in making meaningful determinations. Recently, Gregg (103) has identified and estimated the cross-link terminal structures of monosulfide cross-links using a methyl halide cleavage technique applied to an unusual polybutadiene vulcanizate containing 2-benzothiazyl-N-morpholinyl disulfide as curing agent and other special additives. Treatment with phenyllithium confirmed that the network contained some 40% monosulfidic and 60% di- and polysulfidic cross-links. Identification of terminal structures involved kinetic analogies with model compound work (104).

Polybutadiene differs from NR during the conversion of the initially formed polysulfidic cross-links. In NR these are reduced to mono and disulfidic cross-links and similar amounts of ZnS are formed. In BR no

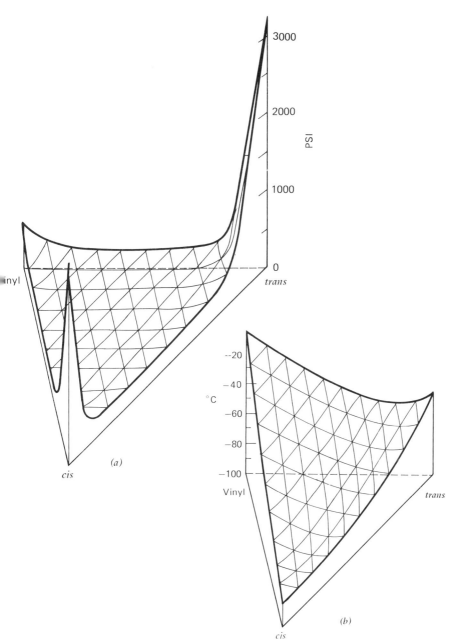

Figure 6 (*a*) Tensile strength of polybutadiene gum vulcanizates at 25°C (100). (*b*) Glass transition temperatures as determined by a modified Gehman torsional stiffness test (100). (*c*) Tensile strength of polybutadiene tread type vulcanizates at 25°C (100). (*d*) 300 Percent modulus of polybutadiene tread stocks at 25°C. The extreme trans corner of the diagram is not defined because samples do not reach 300% elongation before breaking (100).

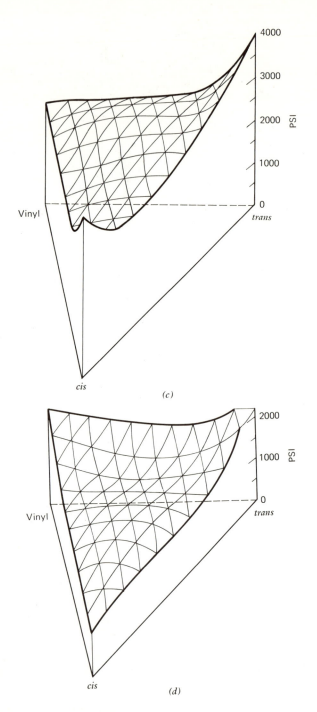

Figure 6 (Continued)

significant amount of monosulfidic cross-links forms and very little of the reacted sulfur is present as ZnS. It was proposed that the accelerator combines with the BR to prevent desulfuration of the dialkenyl polysulfides to monosulfides (as in NR). Instead vicinal cross-linking proceeds. Nitrogen analysis of BR vulcanizates (nitrogen from the accelerator) supports this contention (105).

3.2 Nonsulfur Vulcanization

A number of workers (106–108) have studied the peroxide vulcanization of polybutadiene rubbers and shown that the yield of cross-links is much higher than the approximate equivalence (one molecule of peroxide reacted ≡ one chemical cross-link) established for the di-*tert*-butyl peroxide/NR system by Moore and Watson. Scheele and co-workers (109) suggested that the high yield resulted from subsequent "polymerization" through the vinyl side groups, and van der Hoff (108) added to this the possibility of participation by main chain double bonds from 1,4 units which would be expected to be more reactive than those in polyisoprene rubber. Certainly the differences between BR and NR cannot be explained as due to a different initiation reaction with the peroxide which appears to decompose according to first-order kinetics at the same rate in NR and BR (Fig. 7). Hümmel and Kaiser (110) present data for cures at

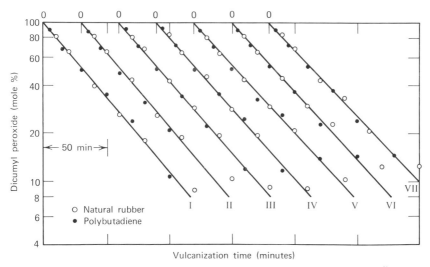

Figure 7 Peroxide decrease in polybutadiene and natural rubber according to a first-order time law (at 145°C). Millimoles peroxide/100 g mix: I, 35; II, 30; III, 25; IV, 20; V, 15; VI, 10; VII, 5. Zero point offset for each peroxide level. From Hümmel and Kaiser (110).

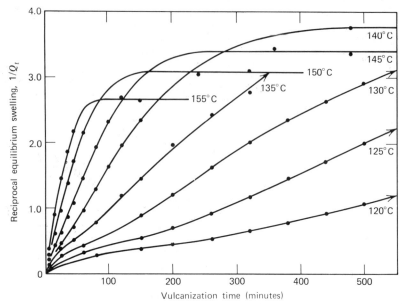

Figure 8   Cross-linking of polybutadiene; dependence of reciprocal equilibrium swelling on vulcanization time (10 mmoles cumylperoxide/100 g mix) (110).

temperatures from 120 up to 155°C, from which it is clear that the ultimate level of cross-linking, representing complete reaction of the peroxide (dicumyl), falls with increasing cure temperature presumably due to the increasing contribution of chain scission reactions (Fig. 8).

Further work by Loan (111) using 97% *cis*-BR and dicumyl peroxide leads to efficiency values (cross-links formed per molecule of peroxide decomposed) of 10.5 for BR and 12.5 for emulsion SBR (compared to ~15 for a 95% *cis*-BR in Hümmel and Kaiser's work). Surprisingly Loan reported that in the SBR cross-linking study only 78% of the peroxide could be accounted for as the sum of the cumyl alcohol resulting from hydrogen abstraction together with acetophenone plus methane resulting from decomposition of the peroxy radical followed by hydrogen abstraction by the methyl radical. The molar ratio ketone/alcohol was about 0.07. Since the dicumyl peroxide was almost fully decomposed under the conditions of cure (10 minutes at 153°C), the fate of the remaining 22% peroxy radicals is a mystery which merits further investigation not only in SBR but also in BR.

Loan explains the observed first-order dependence on peroxide concentration according to the following reaction scheme:

$$ROOR \xrightarrow{k_d} 2RO\cdot$$

$$RO\cdot + \text{\textasciitilde}CH_2-CH=CH-CH_2\text{\textasciitilde} \xrightarrow{k_1} \text{\textasciitilde}CH_2-\underset{\underset{(R_1\cdot)}{|}}{\overset{\overset{RO}{|}}{C}}H-\dot{C}H-CH_2\text{\textasciitilde}$$

$$R_1\cdot + \text{\textasciitilde}CH_2-CH=CH-CH_2\text{\textasciitilde} \xrightarrow{k_2} \text{\textasciitilde}CH_2-\overset{\overset{RO}{|}}{C}H-CH-CH_2\text{\textasciitilde}$$
$$\text{\textasciitilde}CH_2-\underset{\underset{(R_2\cdot)}{|}}{C}H-\dot{C}H-CH_2\text{\textasciitilde}$$

$$\xrightarrow{k_3} \text{\textasciitilde}CH_2-\overset{\overset{RO}{|}}{C}H-CH_2-CH_2\text{\textasciitilde}$$
$$\text{\textasciitilde}\dot{C}H-CH=CH-CH_2\text{\textasciitilde}$$
$$(R_3\cdot)$$

$$RO\cdot + \text{\textasciitilde}CH_2-CH=CH-CH_2\text{\textasciitilde} \xrightarrow{k_4} ROH + R_3\cdot$$

$$2R_3\cdot \xrightarrow{k_t} \begin{matrix} \text{\textasciitilde}CH-CH=CH-CH_2\text{\textasciitilde} \\ | \\ \text{\textasciitilde}CH-CH=CH-CH_2\text{\textasciitilde} \end{matrix}$$

He regards as unlikely the participation of pendant unsaturated groups unless they are of unexpectedly high reactivity because of their very low concentration. Since there is a small amount of decomposition of peroxy radicals:

$$\phi C(CH_3)_2O\cdot \rightarrow \phi COCH_3 + CH_3\cdot$$

$RO\cdot$ must be replaced by $CH_3\cdot$ in a portion of the reactions.

Making the usual stationary state assumption for radical concentrations leads to the rate expression

$$k_2[R_1\cdot][B] = k_1 k_2 k_d[P][k_3(k_1 + k_4)]^{-1}$$

indicating the observed first-order dependence on peroxide concentration.

A number of other nonsulfur curing systems have been investigated but not developed industrially. Scheele et al. (109) draw special attention to the use of dibenzimidazoyl disulfide (DBIS) in conjunction with a diisocyanate, such as naphthalene-1,5-diisocyanate, in the vulcanization of both high *cis*- and high *trans*-BRs.

A mechanism involving reaction with vinyl side groups is tentatively proposed:

$$HN{=}RS{-}SR{=}NH \longrightarrow 2HN{=}RS\cdot$$
(DBIS)

$$\sim CH_2{-}\underset{\underset{CH{=}CH_2}{|}}{CH}{-}CH_2\sim \xrightarrow{+HN{=}RS^\circ} \sim CH_2{-}\underset{\underset{\cdot CH{-}CH_2{-}SR{=}NH}{|}}{CH}{-}CH_2\sim$$

$$2 \sim CH_2{-}\underset{\underset{\cdot CH{-}CH_2{-}SR{=}NH}{|}}{CH}{-}CH_2\sim \xrightarrow{R(NCO)_2} \sim CH_2{-}\underset{\underset{\cdot CH{-}CH_2{-}SR{=}N}{|}}{CH}{-}CH_2\sim$$

$$\begin{array}{c} | \\ CO \\ | \\ NH \\ | \\ R \\ | \\ NH \\ | \\ CO \\ | \\ \cdot CH{-}CH_2{-}SR{=}N \\ | \\ \sim CH_2{-}CH{-}CH_2\sim \end{array}$$

This gives no indication of the fate of the branched polymeric radicals. The cure rates achieved are too low for practical applications.

In summary, butadiene rubbers are rather slow curing to form networks, having some properties similar to those of NR, and some to those of SBR (see also Chapter 4). Overcure is preferred, since undercure leads to poor resilience and other undesirable properties. The tear strength of gum vulcanizates is low and even in reinforced compounds it is still often advisable to use BR in blends with NR or SBR.

4 SBS AND SIS ELASTOMER NETWORKS

Butadiene–styrene copolymers may be prepared using organolithium initiators as described in Chapter 4 in several forms (112): (*a*) "uniformly random," in which the random distribution of monomer residues is maintained throughout the chains; (*b*) "random," in which the distribution varies along a chain without giving rise to a detectable block of styrene residues; (*c*) "block," in which at least one block of S units can be detected; and (d) "ideal block," in which all the chains are constructed from compositionally pure blocks of butadiene and styrene residues.

While there appears to be little to distinguish the vulcanization/reinforcement behavior of types (a) and (b) from that of emulsion copolymers with similar styrene contents (113), the block copolymers may exhibit the phenomenon of self-reinforcement when each chain contains a styrene block at each end and is rubbery when the styrene content is not too high (say about 25 mole %, as in emulsion SBR). Such structures consist of an amorphous continuous phase, poly-butadiene, with polystyrene domains as described in Chapter 5. If the S blocks are terminal to the B chain, the structure will consist of S domains linked by B chains whose ends are embedded in the large multifunctional rigid S structures which are analogous to and, it appears, quite similar in size to particles of a conventional reinforcing filler such as carbon black (114).

As one would anticipate, these block copolymers exhibit two maxima in their plot of loss tangent versus temperature corresponding to the T_g values of the B (or I) and S blocks (see Fig. 24 in Chapter 11). Below $(T_g)_{B/I}$, the copolymer is glassy; between $(T_g)_{B/I}$ and $(T_g)_S$, the B or I chains are rubbery but end-linked by the rigid S domains to form a rubbery "vulcanizate"; above $(T_g)_S$, the copolymer is thermoplastic and may be thermoformed. Chapter 11 has further details.

Whereas the S blocks are unreactive, the B or I blocks may be chemically cross-linked. Childers and Kraus (112) point out that above $(T_g)_S$ such a cross-linked SBS copolymer will behave as though it is a polybutadiene with abnormally long unreactive chain ends. A similar result is obtained below $(T_g)_S$ for the swelling of cured SBS block polymers. Using a swelling agent that is a good solvent for the S ends, namely toluene, plots of swelling volume versus network chain density fall on a single curve regardless of styrene content. Substitution of a poor solvent for polystyrene, n-heptane, immediately shows the restraining effect on swelling of the unsolubilized S domains. With such solvents, plots such as those described above depend on the styrene block content. Under conditions where the S domains are hard they reinforce, but the reinforcing effect diminishes with increasing chemical cross-linking.

Probably a substantial contribution to the strength of these copolymers comes from entanglement of the B or I chains leading to an increase in effective network chain density. Holden et al. (114) have suggested that when equilibrium modulus or swelling measurements are used to establish the elastically effective network chain density, the value obtained should be identified with the *entanglement* length rather than with the average B or I block length between the S domains. They argue that the firm anchoring of the rubbery chain ends in the rigid domains largely prevents disentanglement under stress and allows the residual trapped entanglements to function as tetrafunctional cross-links even in the absence of chemical cross-linking. In an interesting attempt to quantify these "crosslinking" and "reinforcement" effects, these authors combined the Mooney equation with that of Guth-Smallwood to obtain

$$f = (\rho RTM_c^{-1} + 2C_2\lambda^{-1})(\lambda - \lambda^{-2})(1 + 2.5\phi_S + 14.1\phi_S^2)$$

where the symbols have their usual meanings except that ϕ_S is now the volume fraction of polystyrene domains. Experimental values of reduced force were plotted against reciprocal extension ratio to give a straight line from which were calculated $M_c = 2 \times 10^4$ and $2C_2 = 2.2 \times 10^6$ dynes/cm^2, which seem plausible in comparison with the values for NR vulcanizates. However, many assumptions are implicit in the treatment: applicability of the Mooney equation to the amorphous phase, and (even if it is obeyed) that $2C_1 = \rho RTM_c^{-1}$; and applicability of the Guth-Smallwood equation, which implies spherical, rigid particles (i.e., domains) wetted by but not bonded to the amorphous elastic network.

The question of adhesion has been tackled by Bishop and Davison (115) using the Kraus treatment (Section 1.5) applied to their microswelling data. Values of the Kraus parameter c are compared (Table 5) with his previous values for various carbon blacks, suggesting that the reinforcing power of the S blocks is similar to that of high abrasion furnace black. Again assumptions were made: (a) that $(v_r)_0$ is constant for a series of polymers of different B/S ratios and block lengths; (b) that swelling in the elastic phase of the network is the same as in the unfilled network (i.e., B or I homopolymer, so that literature values of χ can be substituted in the Flory-Rehner equation); (c) swelling is restricted close to S domains; and (d) the domain-elastic bonds resist swelling stresses.

Table 5 Values of c (Kraus Equation) for SIS and SBS Domains Compared to Various Reinforcing Carbon Blacks (115)

	m^2/cm^3	c
SIS domains	100–200	1.11
SBS domains	200–250	1.18
FT black	25	0.92
SRF black	50	1.15
HAF black	135	1.17
SAF black	250	1.27
EPC black	205	1.26
Acetylene black	105	1.35

Some doubt has been cast on the spherical shape and rigidity of polystyrene domains. Smith and Dickie (116) studied the stress-strain properties of Kraton 1101 (SBS copolymer) and a similar polymer, Thermolastic 226, which contains plasticizer and inorganic pigments, over a wide range of strain rate and temperature. They observed a Mullins effect in that successive stress-strain curves were not superimposable since the material suffered an apparent softening due, it was suggested, to modification or disruption of the S domain structure. A similar softening might result from the pulling out of chain ends from S domains as postulated by Bishop and Davison to explain their observation of incremental swelling (swelling creep). Likewise, the swelling results could be due to slow penetration of the S domains by a relatively poor solvent leading to disruption.

These authors rejected the applicability of the Guth-Smallwood equation on the grounds of its prediction, for an increase in styrene content from 22 to 29 vol %, of only a 30% increase in modulus, whereas G', the storage shear modulus at 1 Hz, increases more than fivefold over this change in composition (117). Smith and Dickie proposed that this remarkable increase results from a change in domain morphology with styrene content including a progressive increase in the contour lengths of the domains and the extent of network structure. For rodlike particles, the modified equation

$$E = E_0(1 + 0.67 f \phi_S + 1.62 f^2 \phi_S^2)$$

should be used where the length/diameter ratio f may be large and may also change with deformation due to plastic flow.

Bishop and co-workers believe the high strength in these block copolymers reflects the labile nature of slipping entanglements which allow redistribution of stress uniformly over a large proportion of network chains. Smith and Dickie believe that the controlling step in the overall rupture process involves viscoelastic disruption of the S domains in the vicinity of a crack during its slow growth stage. The slowly growing crack needs to disrupt several domains before it is large enough and has sufficient elastic stored energy for rapid growth. The picture is therefore of a finely divided, dispersed, particulate phase with sufficient ductility to deform and rupture in regions of high stress concentration so that energy may be dissipated at the tip of a slowly growing crack. According to this explanation, particle-polymer adhesion is of no consequence which appears to be supported by the experiments of Morton et al. (65), who showed that conventional vulcanizates containing either 25 vol % of 35 mμ polystyrene spheres or 35 vol % of 48.5 mμ spheres have tensile strengths essentially the same as those of Kraton 1101 over the temperature range −25 to +40°C. Essentially similar conclusions were reached by Morton and co-workers (118), while Brunwin et al. (119) draw attention to S domain interactions.

A systematic study of flow properties as a function of block structure showed that, at constant $\bar{M}_w$, branching decreases viscosity, but it is the length of the terminal blocks rather than the average molecular weight which governs viscoelastic behavior (120). Continued study of the physical properties of ABA polymers may well shed light on many of the mysteries of rubber reinforcement.

5 DIENE RUBBERS: REINFORCEMENT

Aside from the possible existence of structural blocks discussed in Section 4, there seems in principle little to distinguish the particulate reinforcement of stereo diene elastomers from that of their random counterparts, Nevertheless, the average molecular weights and molecular weight distributions of commercially available solution elastomers differ markedly from those of their emulsion or natural equivalents, and these variations lead to differences in reinforcement behavior which are of technological or economic importance. There are also significant differences in polymer breakdown during mastication and mixing procedures.

5.1 Polymer-Filler Interaction

It is generally recognized that polyisoprenes differ from polybutadienes in respect to their response to mastication. Synthetic or natural *cis*-polyisoprenes undergo mechanical degradation of the chains at low temperatures (say below 100°C), but at higher temperatures oxidative chain scission reactions become more and more important. In contrast, polybutadienes are resistant to shear breakdown at low temperatures, but may be oxidized at high temperatures (say above 140–150°C) with resultant scission and concomitant cross-linking (gelation). This difference in response to shear was attributed by Kraus and Rollman (121) to the absence from solution BR of a really high molecular weight fraction as is found in the broad distribution polymers such as emulsion BR or SBR, or in natural rubber. They carried out mastication experiments on *cis*-polybutadienes with standard and bimodal distributions and showed that the latter polymer was somewhat susceptible to low temperature shear degradation (Table 6).

Table 6 Response of Standard and Bimodal Distribution Polymers to Mastication in Air (121)

Polymer	Stock Temperature Range (°C)[a]	Mastication Time (minutes)	Torque at 100 rpm (m-kg) Initial	Final	Inherent Viscosity Initial	Final	Gel Initial	Final
Standard	25–105	10	2.02	2.04	2.51	2.51	0	0
	140–171	9	2.00	1.24	2.51	1.85	0	0
Bimodal	25–115	10	2.85	1.90	3.95	3.84	0	0
	25[b]	25	—	—	3.95	3.19	0	0
	140–173	4	1.75	2.13	3.95	3.30	0	0
	140–183	20	1.71	1.14	3.95	1.00	0	20

[a] Jacket (initial) temperature to maximum or dump temperature.
[b] Masticated on 2 in. roll mill, 20 mil gap, 18 rpm, friction ratio 1.2 : 1.

This difference in behavior could influence reinforcement characteristics since it leads to differences in polymer interaction with blacks. When reinforcing carbon black is mixed into a rubber some of the rubber is bound to the surfaces of the carbon particles, and this in turn influences the molecular weight distribution of the residual (free) rubber phase with which it may interact through entanglement coupling. The amount of bound rubber is related to its physical adsorbability but also to the number of polymeric radicals resulting from mechanical shear which are available to react with suitable groupings on the particle surfaces (122).

Brennan and Jermyn (123) have suggested the use of a modified Guth-Smallwood equation to describe the behavior of black reinforced cis-polybutadienes in which the volume fraction of filler, ϕ, is replaced by the volume fraction of (filler + bound rubber), Φ. This equation showed better agreement with experimental results than the unmodified equation. However, Kraus (1) has pointed out that an equation of this type does not allow for observed increases of viscosity (and modulus) with decreasing particle size, with filler structure (particle–particle interactions), or with bound rubber where this modifies the composition of the free rubber phase. These effects may be related; for example, given a constant volume fraction of filler, smaller particles are sited closer together and so may interact more strongly. Although Brennan and co-workers (124) have correlated maximum Brabender torque with bound cis-polybutadiene rubber content for a series of blacks differing in particle size and structure, this would appear to apply only to such polymers and mixing conditions as avoid shear induced degradation. Kraus's data for a

Table 7 Effect of Black on Mooney Viscosity of Broad Distribution Polybutadiene (1)

	Inherent Viscosity[a]	Gel or "Bound Rubber"[a] (%)	Swelling Index[a]	MS—$1\frac{1}{2}$ (212°F)
Original rubber	3.60	0	—	38
As compounded[b]	2.49[c]	9.2	75	50
After 2 minute cold remill	2.01[c]	17.6	78	26

[a] In toluene, 25°C.
[b] Mixed at 240°F, no remill: HAF black 50; highly aromatic oil 5, rosin acid 5.
[c] Soluble rubber.

broad distribution BR mixed with HAF black show no correlation between carbon gel and limiting viscosity number *of the soluble portion* of the rubber (Table 7).

Later work by Kraus and Gruver (125) has quantified the preferential adsorption of the high molecular weight fractions of broad distribution BR or SBR on black surfaces. For narrow distribution polymers, the adsorption was proportional to $M^{1/2}$. These data refer to physical adsorption, not to chemical bonding, and extensive chain scission will lead to an increase of bound rubber by grafting to the carbon black surface.

5.2 Properties of Reinforced Vulcanizates

5.2.1 Polybutadiene

It is obviously of interest to discuss the practical reinforcement of stereoregular diene rubbers as exemplified by their performance in tire treads. Published data and, one suspects, a large proportion of the confidential reports in the archives of the tire manufacturers, are concerned with blends of BR with SBR or NR because of the loss in skid properties of high BR content treads. In addition, early investigators did not always recognize the unusual sensitivity of the wear index for *cis*-polybutadiene/black vulcanizates to minor changes in the severity of road test conditions, although the generally better performance relative to regular SBR, OE-SBR, and NR treads under high severity conditions had been recognized for some time. The careful technological studies of

Davison and co-workers (126) quantified this superiority (Fig. 9) but also showed an unexpectedly large effect of pavement surface texture on wear rating under otherwise similar severity conditions (e.g., speeds, loads, braking, acceleration, cornering).

Evstratov et al. (127) have made similar observations; they attribute the better tread wear performance of *cis*-polybutadiene rubber (SKD) under high severity conditions to the lack of dominance of pattern abrasion which leads to the more rapid failure of other diene rubbers. They suggest that BR wears principally by a fatigue abrasion mechanism.

It is important to choose the curing system for tread stocks, particularly sulfur and accelerator levels, with care. Studies by Harris and co-workers (128) using Taktene 1200 (95% cis; $ML_4/100°C$: 30) and 1220 (95% cis; $ML_4/100°C$: 45) have shown that quite small changes in concentration can make an appreciable difference to abrasion index (Fig. 10). It will be interesting to see such data extended to different black levels, with careful measurement and control of cross-link density. The results showed substantially different contour lines for BR/OE-SBR blends. *cis*-Polyisoprene does not show this extreme sensitivity to sulfur level.

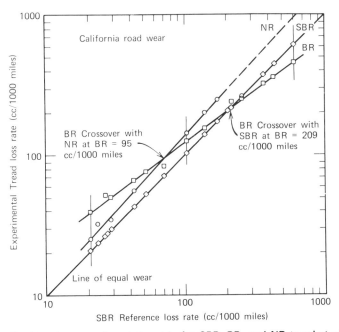

Figure 9 Abrasion-severity relationship for SBR, BR, and NR tread stocks, 50 HAF, Northern California road tests. From Davison et al. (126).

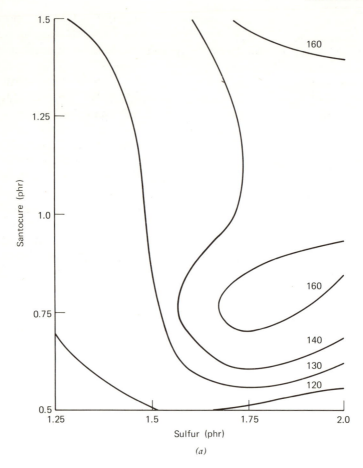

Figure 10   Sulfur accelerator study of *cis*-BR (Polysar Taktene 1200). Contours are (*a*) abrasion index (% SBR control), low severity level; (*b*) same, high severity level; (*c*) heat buildup (°F); (*d*) rebound (%). Santocure is *N*-cyclohexy benzothiazylsulfenamide (CBS). From Harris et al. (128).

Considerable success has been achieved using oil extension of high molecular weight BR. Very high levels of oil and black may be incorporated in formulating economical high performance tread stocks. Sarbach et al. (129) conducted laboratory and road wear evaluations of heavy duty tire treads based on blends of natural rubber and OE-BR. At the level 30 BR/70 NR, substitution of OE-BR for the entire BR content improved the wear rating, resistance to groove cracking, chipping, and chunk-out.

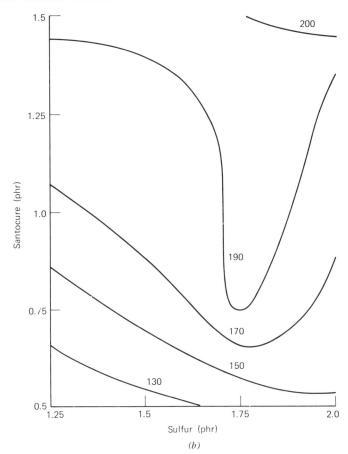

(b)

Figure 10   (Continued)

5.2.2   *Polyisoprene*

Synthetic *cis*-polyisoprene is of interest as a substitute for NR in heavy duty tire treads. Krol's study has indicated that with proper attention to formulation, similar properties may be obtained from both the synthetic and natural polymers (130).

5.2.3   *Solution SBR*

Solution polymerized (random) SBR is also of considerable interest because the lower 1,2 content (below 10%) and the greater linearity are reflected in a heat buildup that is lower than that for emulsion SBR.

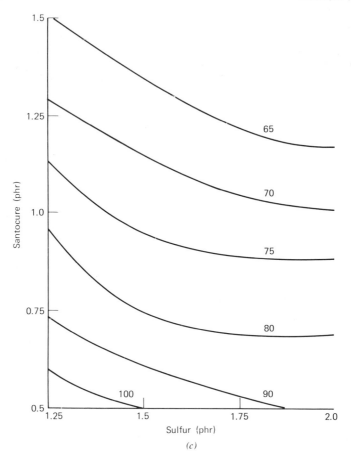

Figure 10   (Continued)

Properties of the solution polymers are, overall, marginally better than for their emulsion counterparts. Processing of solution SBR is more difficult than for emulsion SBR. Processing is said to be easier than that of BR and the wear is similar, thus partially overcoming the necessity for using blends.

6   ETHYLENE–PROPYLENE RUBBERS: VULCANIZATION

Suitable vulcanization methods (131) depend on the structure of the elastomer:

1.   For ethylene–propylene copolymer rubbers (EPM) it is necessary

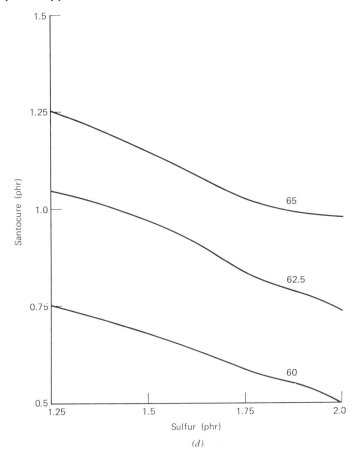

(d)

Figure 10   (Continued)

to use highly reactive vulcanizing agents which are capable of reacting with a paraffin type chain.

2.   Alternatively, it is possible to introduce more reactive functional groups into the polymer molecule either by chemical modification of EPM or by introducing a third monomer into the chain (EPDM).

For practical purposes, the only real industrial interests are in (*a*) peroxide vulcanized copolymer and (*b*) sulfur vulcanized terpolymer where the third monomer is such as to introduce side chain saturation. A comprehensive review has been published (132). The reader should also refer to the appropriate sections of Chapter 7.

6.1   Vulcanization of EPM

Peroxide curing of EPM appears to proceed through the following type of reaction:

$$ROOR \rightarrow 2RO\cdot$$

$$RO\cdot + P\text{—}H \rightarrow ROH + P\cdot$$

$$2P\cdot \rightarrow P\text{—}P$$

For some peroxides, *secondary decomposition* of the radicals formed by homolytic scission of the peroxide bond may occur. Usually this is the rate determining step for the curing process. Such is the case for dicumyl peroxide:

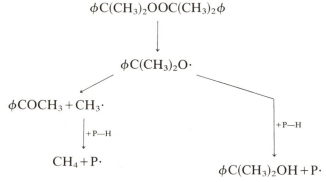

The methyl radicals are extremely reactive as hydrogen abstractors so that the cross-link yield is not influenced.

The reaction is complicated by the alternative action of the radicals in causing chain scission. The efficiency of cross-linking (yield of cross-links per molecule of peroxide dissociated) is related to the propylene content (133) (Chapter 7, Section 8.2.1). Although peroxides have been widely regarded as quantitative cross-linking agents in natural rubber, that is, the reaction path can be followed by measurement of changes in network chain density and analysis of peroxide decomposition products equivalent to the amount added, this does not appear to be the case in ethylene–propylene copolymers. Although first-order rates were obtained for peroxide cure in polyethylene, Harpell and Walrod (134) found that the cure reaction in both EPM and EPDM is more complex and leaves significant amounts of unreacted peroxide in the vulcanizate. They compared the reactivity of several peroxides and found that quinoline type antioxidants interfered least with the cure.

Studies with ^{14}C-tagged dicumyl peroxide have shown that solvent extraction of EPM vulcanizates does not completely remove radioactive species which (saving exchange with ^{12}C in the polymer structure) must be assumed to remain attached to the polymer chain as modifications. Infrared absorption bands in the 14–14.5 μ region suggest the presence of aromatic groups in the vulcanizate (135).

Addition of sulfur to the cross-linking agent results in improved network properties, particularly tensile strength. This was explained by Natta et al. (131) as due to stabilization of the polymeric radicals through which scission was suppressed and cross-linking enhanced by formation of polysulfidic cross-links additional to the carbon–carbon cross-links resulting from mutual termination of the unstabilized radicals. The optimum modulus (indicative of maximum cross-link density) is obtained at a 1:1 ratio of gram-atoms of sulfur to moles of peroxide (see Fig. 19 in Chapter 7). The maximum increase in cross-link density in sulfur-containing systems (about 30%) is insufficient to explain the higher strength, there being evidence (133) that sulfur-containing vulcanizates are stronger than peroxide vulcanizates even at the same value of the index of cross-link density, v_r. [Zamboni et al. (136) have expressed reservations about the use of the swelling method in measuring cross-link density.] Since sulfur is combined with the network and is presumed to be present in cross-links, it has been postulated (40) that the lability of the polysulfidic cross-links must account for the higher strength. If this is correct, the number of such bonds responsible for such a large change in properties must be quite small since they appear to be difficult to detect by chemical analysis. Nevertheless, there is a substantial difference in stress relaxation behavior between peroxide and peroxide/sulfur cures as shown by Loan's results (133) which support the labile cross-links hypothesis. Imoto and co-workers (137) have done further work in an attempt to clarify the position.

Other additives that improve the cross-link yield in peroxide systems include polyfunctional monomers. They appear to act by adding to the tertiary radical, thereby stabilizing it with respect to scission (138) but still allowing it to abstract hydrogen from another polymer molecule. Lenas (139) has evaluated many peroxide cross-linking coagents, and found ethylene dimethacrylate, divinylbenzene, diallyl itaconate, and triallyl cyanurate the most efficient monomer additives, although the allyl monomers were less effective in the presence of large quantities of fillers. A liquid polybutadiene resin with a high vinyl content was also efficient but inferior to the best monomers at the highest filler loadings, possibly because the latter are also able to increase the network density by polymerization. All these materials are better coagents than sulfur (which is ineffective in mineral filled stocks) and avoid the product odor associated with its use. Their employment usually enables a reduction of peroxide level to be made while maintaining a given degree of cure.

Peroxide cross-linking has technical disadvantages of time and temperature limitations, and also, in the case of aryl peroxide curatives, of undesirable stickiness and residual odor in the vulcanized product. Strenuous efforts have been made to develop alternative curing systems. For instance, certain chlorinated compounds have been found effective in both EPM and EPDM (140); these include trichloromethane sulfonyl chloride, trichloromelamine, quinone-N-chlorimide, octachloropropane, octachlorocyclopentene, and other perhalogenated compounds, often in the presence of metal oxides, sulfur, or sulfur donors (131, 141, 142).

Curing systems for butyl and other saturated rubbers generally rely on small quantities of main chain unsaturation or other reactive groupings, and many attempts have been made to introduce suitably reactive vulcanization sites into the saturated EP chain. These include (a) *chlorosulfonation*, with salt formation by reaction with polyvalent metal oxides to form cross-links (cf. Hypalon); (b) *grafting* of, for example, maleic anhydride in the presence of a radical initiator which modifies the cross-linking action of the initiator and adds to the chain active carboxyl groups which are reactive with polyvalent metal oxides, the ionic cross-links having desirable lability; and (c) *chlorination/dehydrochlorination*, which introduces reactive unsaturation into the polymer main chain and so may impair resistance to oxidation and ozone attack (143). Russian work has shown that brominated EP rubber can be sulfur vulcanized by nucleophilic interaction with the combined bromine to give dehydrobromination, which accompanies cross-linking. Alternatively, zinc oxide cures are possible, giving C–O–C cross-links (144).

In general, better results are obtained by synthesis of terpolymers having vulcanization reactive unsaturation other than in the main chain; these are discussed in the following section. Radiation cures of EPM are feasible, best results being obtained in the presence of about 0.5 phr of sulfur (145).

6.2 Vulcanization of EPDM with Sulfur

6.2.1 *Technological Considerations*

Broadly speaking, EPDM polymers have been available commercially with propylene contents between 30 and 53 mole % (commonly 40%), the termonomer being tetrahydroindene (THI), cyclooctadiene (COD), dicyclopentadiene (DCP), 1,4-hexadiene (HD), or 5-methylene-2-norbornene (MN) (commonly sufficient to give 1.5–3.1% unsaturation pendant to the main chain) and with limiting viscosity numbers between 1.9 and 3.4 (raw Mooney viscosity, $ML_4/100°C$: 55–105), the higher values permitting oil extension of the rubber. The properties of vulcanizates made from such a wide range of polymers will themselves vary considerably, it being observed for instance that EPDM containing dicyclopentadiene is slower curing (146, 147) but tends to age better than HD or MN terpolymers. Fujimoto et al. (148) showed that reactivity of the unsaturation with dicumyl peroxide is in the order MN > DCP > HD. With a conventional accelerated sulfur cure system it is in the order MN > HD > DCP.

The amount of unsaturation is also very important. Natta et al. (131) showed that with a constant cure time and recipe the modulus rose and elongation fell as the unsaturation (cyclooctadiene) in an EPDM increased. Baldwin et al. (149) also studied the effects of different types of unsaturation on cure and decided few generalizations could be made. Such differences must be borne in mind in interpreting the general comments on the vulcanization of EPDM which follow. Recently, fast curing trienes (EPTM) have been introduced into EP polymers by Cesca et al. (150), making for still additional varieties of cure behavior.

Since the amount of termonomer is low, the physical properties of networks are similar to those of EPMs of similar propylene content and molecular weight; since unsaturation is absent from the main chain, the network is virtually saturated and so resists oxidation. The main technological difference between EPM and EPDM is in the cure itself, in the ability to use accelerated sulfur systems. A spate of papers appeared during 1962–1963 to which reference should be made for details of basic curing systems (151–156) for the various types of elastomers. The following general points emerged:

1.   The best general purpose EPDM curing system is based on mercaptobenzothiazole (0.5), tetramethyl thiuram monosulfide (1.5), sulfur (1.5), with metal oxide/fatty acid activators, using cure temperatures in the range 150–170°C (cf. systems for butyl rubber). Faster cures without substantial impairment of vulcanizate properties may be obtained by adding to this system a dithiophosphate activator. Increasing sulfur levels up to a limit of about 2.5 phr lead to higher modulus, hardness, resilience, and set with lower elongation at break and poorer aging.

2.   Sulfur donor systems are not advantageous, giving poorer properties at high curing temperatures and poorer retention of properties on aging.

3.   Physical property levels, particularly tensile strength, fall with increasing cure temperature.

4.   Certain special purpose butyl rubber curing systems, particularly the halogen activated resin cure and the dibenzo-p-quinonedioxime/red lead system, are less effective (for some polymers, ineffective) in EPDM

and appear to offer no technological advantages over accelerated sulfur recipes.

5. Heat treatments of filled compounds in the presence of p-quinone-dioxime lead to increased network chain densities. Mooney viscosity, modulus, hardness, resilience, and electrical resistivity are increased; scorch time, elongation, tear strength, and compression set are reduced (157).

6.2.2 Mechanism of Cure and Network Structure

Mechanistic studies of the sulfur cure have not been numerous. Crespi and Arcozzi (158) made the assumption of a cross-linking reaction, second order in unsaturation, competing with a first-order scission reaction. However, they used an excess of sulfur, so the problem has been reexamined by German and co-workers (159) using a wider range of sulfur levels. They studied the effect on cross-linking (and reversion) kinetics of (a) polymer unsaturation, in relation to sulfur level; (b) molecular weight distribution; (c) ethylene–propylene ratio; (d) nature of termonomer; (e) curing temperature, in the range 130–180°C; and (f) sulfur/accelerator ratio. It was concluded that the apparent first-order rate coefficient for and efficiency of cross-linking remained fairly constant for different unsaturation levels provided that the levels of sulfur and other curing ingredients were increased in proportion to the percentage of unsaturation (Table 8). As would be expected, the removal of low molecular weight fractions increases the efficiency of vulcanization (Table 9), but since at least three types (single and double peaked) of molecular weight distribution may be obtained (159) in different polymerization systems, the effect is not a simple one. In black stocks, at least, the degree of cross-linking decreases with increasing propylene content although the rate coefficient for cross-linking is unaffected.

Table 8   Vulcanization Parameters for Constant Sulfur/Unsaturation Ratio (159)

Unsaturation (wt %)	Sulfur[a] (phr)	$10^4 M_c^{-1}$	$10^2 k_1$	Efficiency
3	1.5	1.19	2.68	0.56
4	2.0	1.90	2.49	0.66
6	3.0	2.43	2.56	0.58
11	5.5	4.55	2.53	0.58

[a] Other vulcanization ingredients changed proportionally with sulfur.

Table 9   Improvement of Vulcanization Efficiency by Fractionation (159)

Polymers	$10^4 M_c^{-1}$	$10^2 k_1$	α	Vulcanizate Soluble (%)
A (whole polymers)	1.02	2.7	0.16	20
A_1 (low molecular fraction removed)	1.44	2.5	0.22	9.4
A_2 (middle fraction)	1.40	2.5	0.22	2.9
A_3 (high molecular fraction)	2.30	3.5	0.35	4.0

For three terpolymers containing, respectively, DCP, COD, and HD, activation energies for cure lay in the range 22–27 kcalmole, values similar to those found for sulfur curing of unsaturated rubbers. Surprisingly, these authors report lack of reversion in DCP terpolymers up to a cure temperature of 160°C, although COD and HD terpolymers reverted significantly at 150°C and above; this DCP result appears to be in conflict with other literature data. Lyons et al. (160) have shown by stress relaxation techniques that stable sulfur cross-links, predominantly mono- and disulfidic, can be generated in 5-MD terpolymer on prolonged cure, and that these are extremely resistant to oxidative scission. The HD terpolymer, Nordel 1040, of relatively low viscosity molecular weight, was found to have a relatively high fraction of polymer chains not attached to the network (soluble after cure), suggesting either nonrandom distribution of termonomer residues within the chains and/or substantial scission during cure (159). Low sulfur/accelerator ratios gave faster cures and the extent of cross-linking increased with sulfur content as found in earlier technological studies. In addition to increasing the degree of cross-linking, use of high molecular weight polymer of low propylene content assists the development of high modulus vulcanizates.

A comprehensive study of HD terpolymer cure has been published by Frensdorff (161) involving examination of cross-link type and concentration introduced by a number of dithiocarbamate (and similar) accelerated sulfur curing systems. Using a Monsanto rheometer, pseudo-first-order cross-linking was observed except that the first few minutes of each vulcanization following an induction period did not conform. Results in this region might have been influenced by lack of achievement of platen temperature, a well-known disadvantage of the standard cavity/rotor system (162, 163). On the basis of rheometer measurements and equilibrium swelling determinations before and after cross-link cleavage reactions, it was concluded that curing is a two-stage process: an initial fast reaction, followed by a slower pseudo-first-order stage. A similar suggestion was put forward by Wolfe (164). Rate coefficients were little influenced by the concentration of accelerator, sulfur, or zinc oxide but were nearly proportional to polymer unsaturation, although Wolfe, working at lower accelerator concentrations, found complex kinetic dependence on the concentrations of the curing ingredients which was explained in part as due to the limited solubility of zinc dimethyldithiocarbamate at the cure temperatures (150, 160, 170°C).

In a further study of tetramethyl thiuram di- and monosulfides as accelerators for HD terpolymer, Wolfe found kinetically more complex behavior which appears to require elucidation through probe studies.

Baldwin et al. (165) have attempted to determine from swollen compression modulus measurements a value for the entanglement network chain density using Langley's equation (166). Certain of their assumptions seem to require further scrutiny but it is noteworthy that values for entanglement spacings obtained by their equilibrium method are of the same order as those obtained for other rubbers (cis-BR, NR, SBR) using dynamic rheological techniques.

6.3 Vulcanization by Other Systems

Although the efficiency of EPDM peroxide cures depends on the nature and amount of unsaturation (Section 6.2.1), peroxide/sulfur (or other coagent) systems similar to those used for EPM appear to be of value when a particularly stable cross-linked structure is desired. Formulations using ethylene dimethacrylate are given in an early paper by Howarth and associates (167).

It has also been shown that addition of polymethylene polyphenyl isocyanate (2–3 phr) to mineral filled HD terpolymer (and other unsaturated elastomer) compounds may result in increased strength, higher modulus, and reduced elongation at break (168).

7 REINFORCEMENT OF ETHYLENE–PROPYLENE COPOLYMERS AND TERPOLYMERS

A considerable number of articles on compounding such as that by Schoenbeck (169), have established the general similarity in behavior of these rubbers to other amorphous, noncrystallizing rubbers such as emulsion SBR. Table 10 gives typical data for vulcanizates based on an early HD terpolymer but most properties of present-day commercial elastomers would not differ greatly in similar compounds. One observes the well-known necessity for a fine particle size carbon black or precipitated silica to impart high tensile strength, although quite reasonable laboratory tear strength values seem to be achieved with some mineral fillers. The compression set of many mineral filled compounds is unacceptably high, as is the Mooney viscosity, which may be allied with a very short scorch time.

The parallel with SBR is a close one which has been evaluated more thoroughly by Dudek and Bueche (170). Using the same type of elastomer as Schoenbeck, tensile strength measurements were carried out on gum and black reinforced networks over a range of strain rates and temperatures and compared with similar data for SBR 1502 vulcanizates. In each case, the maximum of the tensile strength versus cross-link density plot decreased in magnitude with strain rate but occurred at the same value of cross-link density irrespective of strain rate. Activation energy for both rubbers in creep-rupture experiments was the same. SBR and EPDM containing 30 phr HAF black showed similar temperature dependence of tensile strength up to about 120°C, above which oxidative scission led to more rapid deterioration of SBR (Fig. 11). The dependence of strength on strain rate in this temperature range was surprisingly small (Table 11).

Detailed comparisons of filler effects over a range of loadings are to be found in manufacturers' compounding literature. For instance, Hansen and Church (171) have reported property versus loading curves for MT, GPF, APF, HAF–LS, and HAF–HS blacks in SBR 1500, NR, EPDM, butyl, NBR, and neoprene. Subject to processability at the high Mooney viscosities involved, the black tolerance of EPDM without development of excessive stiffness in the vulcanizate permits loadings even of HAF black up to 120–160 phr compared with 70–80 phr in the other hydrocarbon rubbers. This high loadability is also noticeable in figures for breaking strain where although the highest elongation at break (EB) values (over 700%) are not achieved, so slow is the fall of EB with filler loading that values of 200–300% are retained at loadings of 160 phr HAF (corresponding to 80–90 hardness). Tear strength is higher than that of SBR or butyl rubber but below the optimum range for NR; the same holds true for rebound resilience, the rate of fall with filler loading being close to that for NR. Laboratory abrasion resistance is substantially better than for NR, vulcanizates reinforced with 50–70 phr HAF–HS giving results similar to those for equivalent SBR vulcanizates. Compression set at 70°C is considerably lower than for NR or butyl, but cut growth of EPDM is the worst for the group of rubbers and but little improved by use of low structure blacks.

Useful as these general indications are, it must be remembered that substantial differences in technological behavior exist among different grades of EPM and EPDM, as shown in Table 12 using data taken from a supplier's report (172).

Table 10 (169)

Part A Comparison of Carbon Black Fillers at Equal Hardness

	A	B	C	D	E	F
HD–EPDM	100	100	100	100	100	100
Zinc oxide	5	5	5	5	5	5
Stearic acid	1	1	1	1	1	1
MT carbon black	130	—	—	—	—	—
SRF carbon black	—	85	—	—	—	—
FEF carbon black	—	—	60	—	—	—
HAF carbon black	—	—	—	50	—	—
SAF carbon black	—	—	—	—	50	—
EPC carbon black	—	—	—	—	—	55
Circosol 2XH	20	20	20	20	20	20
2-Mercaptobenzo-thiazole	0.5	0.5	0.5	0.5	0.5	0.5
Tetramethyl thiuram monosulfide	1.5	1.5	1.5	1.5	1.5	1.5
Sulfur	1.5	1.5	1.5	1.5	1.5	1.5
Mooney viscosity (ML_4, 212°F) 4 minute reading	76	80	84	84	85	84
Mooney scorch (MS@250°F) Minimum reading	33	35	37	37	37	36
Minutes to 10 point rise	45+	45+	45+	38	34	37
Cure: 20'/320°F						
Modulus@100% (psi)	250	300	280	240	200	175
Modulus@300% (psi)	510	1225	1350	1380	1090	740
Tensile strength (psi)	775	1625	2000	2800	3350	3150
Ultimate elongation (%)	520	420	430	460	530	640
Hardness (Durometer A)	66	67	65	62	63	62
Instron tear (pli)	18	22	21	20	30	35
Graves tear (pli)	105	165	180	160	170	195
Yerzley resilience (%)	72	72	74	74	67	67
Compression set, Method "B" (%) 22 hours@158°F	13	13	13	12	15	22
22 hours@212°F	58	56	54	51	56	64

Table 10   (Continued)

Part B   Comparison of Mineral Fillers at Equal Hardness

	G	H	I	J	K	L	M
HD–EPDM	100	100	100	100	100	100	100
Zinc oxide	5	5	5	5	5	5	5
Stearic acid	1	1	1	1	1	1	1
Atomite whiting	200	—	—	—	—	—	—
Soft clay	—	130	—	—	—	—	—
Hard clay	—	—	120	—	—	—	—
Mistron vapor	—	—	—	110	—	—	—
Silene EF	—	—	—	—	65	—	—
Zeolex 23	—	—	—	—	—	75	—
Hi-Sil 233	—	—	—	—	—	—	50
Circosol 2XH	20	20	20	20	20	20	20
Diethylene glycol	—	—	—	—	—	—	3
2-Mercapto- benzothiazole	0.5	0.5	0.5	0.5	0.5	0.5	0.5
Tetramethyl thiuram monosulfide	1.5	1.5	1.5	1.5	1.5	1.5	1.5
Sulfur	1.5	1.5	1.5	1.5	1.5	1.5	1.5
Mooney viscosity (ML_4, 212°F) 4 minute reading	68	84	96	54	125	123	113
Mooney scorch (MS @ 250°F) Minimum reading	29	35	43	21	59	58	51
Minutes to 10 point rise	45+	12	8	26	14	12	13
Cure: 20'/320°F Modulus @ 100% (psi)	100	230	230	290	230	250	150
Modulus @ 300% (psi)	150	290	330	400	325	520	310
Tensile strength (psi)	475	1200	1800	1700	1600	1600	2925
Elongation at break (%)	580	790	820	720	730	740	790
Hardness (Duro- meter A	60	60	61	62	68	66	62
Instron tear (pli)	7	40	89	47	36	52	91
Graves tear (pli)	49	131	169	150	160	160	144
Yerzley resilience	60	69	64	67	67	65	59
Compression set, Method "B" (%) 22 hours @ 158°F	45	42	47	37	28	28	30

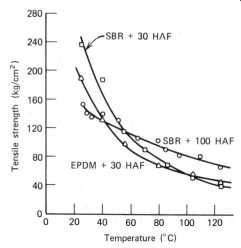

Figure 11 Temperature variation of tensile strength measured at a strain rate of
0.8 min^{-1} for EPDM and SBR containing the indicated amounts of HAF black. From
Dudek and Bueche (170).

The ultimate test of a black reinforced rubber is in a tire tread but tire
manufacturers have been loath to publish their test results. However, one
recent paper, referring to Japanese road tests, quotes a wear rating of 90
for a EPDM/ISAF passenger tread against a SBR/BR/ISAF control
(173). It is suggested by the authors that an oil extended EPDM rubber
would have shown a higher wear rating. It is widely believed that the
hesitation of tire manufacturers to use EPDM in tires is related to
problems in building the tire rather than the quality of the tire once built.

**Table 11   Tensile Strength at Three Strain Rates for Reinforced SBR and
EPDM at 25 and 86°C   (170)**

	Strain Rate	Tensile Strength (kg/cm^2)	
Elastomer	(min^{-1})	25°C	86°C
SBR+30 HAF[a]	22.0	316	112
	5.0	295	89
	0.48	230	65
EPDM+30 HAF[b]	22.0	260	95
	5.0	256	86
	0.48	220	69

[a] Cured 30 minutes at 290°F.
[b] Cured 60 minutes at 290°F.

Table 12 Effect of Grade of Elastomer on Vulcanizate Properties   (172)

Black Elastomer	300% Modulus (psi)[a]			Tensile Strength (psi)[a]			Resilience[b] (%)			Tear Strength[c] (lb/in.)			Abrasion Resistance[d]		
	1	2	3	1	2	3	1	2	3	1	2	3	1	2	3
SAF	890	1500	2450	3170	3460	4020	57.9	63.1	64.7	200	240	270	137	154	140
ISAF	1010	1650	2180	3270	3410	3860	59.0	64.7	67.2	210	210	280	138	142	136
HAF	880	1540	2320	2620	3050	3170	62.9	69.6	71.0	180	210	260	100	100	100
SCRF	590	1150	1730	3010	3710	3830	58.9	66.4	68.2	210	210	260	108	119	116
CRF	580	1100	1850	2780	3280	3320	61.1	68.0	68.5	190	190	280	95	100	108
EPC	580	970	1670	2680	3140	3250	59.9	66.9	67.5	180	220	300	87	95	96
FEF	1050	1540	2270	2220	2350	2910	64.7	70.6	73.8	190	210	260	75	67	77

Compound Formulations

1:	EPM	100.0		2:	HD-EPDM	100.0		3:	DCP-EPDM	100.0
	Black	50.0			Black	50.0			Recipe: Same as 2	
	Zinc oxide	5.0			Zinc oxide	5.0				
	Calcium stearate	1.0			Stearic acid	1.0				
	Foxam 40 oil	10.0			Foxam 40	10.0				
	Sulfur	0.32			TMTM	1.5				
	Dicumyl peroxide	2.7			MBT	0.5				
					Sulfur	1.5				

Test methods

[a] ASTM D412–61T at 75°F.

[b] Goodyear-Healey pendulum and ASTM D1054–55.

[c] ASTM D624–5 (Die C).

[d] Akron angle abrader, relative to HAF = 100 for each polymer.

8 BLENDS

The preceding sections of this chapter have attempted to characterize the network formation and reinforcement behavior of individual stereoregular elastomers in isolation and to compare this behavior with that of elastomers such as NR and emulsion SBR. In practice, large quantities of stereoregular elastomers are used in blends.

Walters and Keyte (174) first showed clearly by phase contrast and electron microscopy that the individual elastomers constituting a blend maintain their identities in zones of between 0.5 and 70 μ size, depending on the blending technique (Table 13). Heterogeneity is observed even within blends of chemically similar elastomers (e.g., NR/IR, Table 14) but the zone size is reduced by blending polymers of similar Mooney viscosity; for example, a high molecular weight elastomer may be oil extended to approximately the Mooney viscosity level of the softer rubber. Nevertheless, no practical technique exists for reducing the zone size to near molecular dimensions, the results of solution mixing being no better than those obtained using conventional rubber machinery. Similar observations were made on the polybutadiene/polystyrene system (175).

One must therefore accept that there will be present in blends two (or more) interlocking networks between which compounding ingredients, whether soluble or insoluble, will in general be unevenly distributed both within each phase and between phases. Whereas the physical properties of gum vulcanizates do not appear to be greatly affected by such heterogeneity, the properties of filled stocks may fare less well.

Corish (176) has stated requirements that must be fulfilled to allow estimation of blend properties by interpolation between values for the constituent elastomers:

1. Rubbers should have similar values of solubility parameter and also yield blends whose vulcanizates exhibit a single glass transition temperature.

Table 13 Polybutadiene Blended with NR (174)

Molecular Weight	% Cis	Zone Size (μ)
1,200,000	94	3–70
1,000,000	94	2–40
750,000	93	0.5–2
500,000	92	1–4
280,000	92	1–13

Table 14 Heterogeneity in Polymer Blends (174)

Polymers Blended (50/50)	Mooney Plasticities 4 minutes at 100°C	Zone Size (μ)
High Mooney SBR/NR	90/53	6
NR/SBR	53/50	2
Neoprene/SBR	53/52	0.5
Buna 85/NR	45/52	0.5
High *cis*-BR/NR	50/53	0.5
trans-BR/NR (milled hot)	—	2
SBR-40/SBR-10	75/68	1
NR/polyisoprene	53/50	2
NR/butyl	53/45	1
SBR/BR	50/49	2
NR/factice	—	4
NR/polyethylene (milled hot)	—	2

2. Rubbers should be of similar viscosity.

3. Mixing methods should be capable of giving good disaggregation of compounding ingredients and satisfactory distribution between the constituent rubbers.

4. Rubbers should be of similar curative requirement and cure rate and have curative systems capable of giving interlocked networks.

5. Suitable choice of plasticizer, which may offset slight deficiencies in (1) and (2) should be made.

Callan and co-workers (177) showed that blends of butyl rubber with EPDM contained an inhomogeneous distribution of added black which preferred the terpolymer zones; similarly, with blends of *cis*-BR and IR at black loadings up to 20 phr, most of the carbon resided in the polybutadiene (178).

Hess et al. (179) have extended these studies to NR/BR blends containing up to 40 phr of ISAF black, of relevance to passenger tread compositions. Use of ordinary mixing procedures in which the black was added to the preblended polymers resulted in about 50% in the BR phase at 10 phr black and up to 80% in the BR phase at 40 phr black. In 50:50 NR/BR blends, by use of separate masterbatches each containing 20 phr ISAF, the final distribution of black was 41:59. Using an NR black masterbatch only, this changed to 60:40 and a "promoted" NR black masterbatch raised the black distribution to 82:18 in favor of NR. These results were little affected by the presence of up to 30 phr process oil.

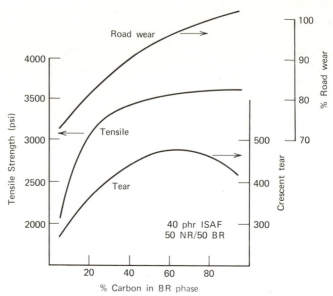

Figure 12 Effect of carbon black distribution on vulcanizate properties. From Hess et al. (179).

It was found possible by further variations of technique to mix $1:1$ NR:BR/40 phr ISAF compounds with between 7 and 95% of the black located in the BR zones. The physical properties of such blends including tread wear are shown in Fig. 12 which indicates that ordinary mixing, which puts most of the carbon into the polybutadiene component, gives the best results.

There is little doubt that a viscoelastic explanation for this and other interesting properties of blends will be forthcoming in due course; until then we can only comment that Nature, assisting the traditional procedures of the rubber factory, seems to be doing quite a good job for us!

REFERENCES

1. G. Kraus, *Rubber Chem. Technol.*, **38**, 1070 (1965).
2. See, e.g., L. Mullins, in *Proceedings of the NRPRA Silver Jubilee Conference, Cambridge, England, 1964*, Maclaren, London, 1965, Ch. 7, p. 99.
3. Natural Rubber Producers' Research Assoc., Welwyn Garden City, England, now the Malaysian Rubber Producers Research Assoc. (MRPRA).
4. Introductory accounts are given by M. Gordon, *High Polymers: Structure and Physical Properties*, 2nd ed., Iliffe, London, 1963, Ch. 5; and D. A. Smith, *Addition Polymers: Formation and Characterization*, Butterworth, London, 1968, Ch. X.

5. For fuller discussions, see L. R. G. Treloar, *Physics of Rubber Elasticity*, 2nd *ed.*, Oxford University Press, London and New York, 1958; P. Meares, *Polymers: Structure and Bulk Properties*, D. Van Nostrand Co. Ltd, London, Toronto, New York, Princeton (N.J.), 1965

6. See, e.g., J. Scanlon, *J. Polym. Sci.*, **43**, 501 (1960).

7. R. S. Porter and J. F. Johnson, *Chem. Rev.*, **66** (*i*), 1 (1966).

8. B. Meissner, *J. Polym. Sci.*, **C16**, 781 (1967); B. Meissner, I. Klier, and S. Kucharik, *ibid.*, **C16**, 793 (1967).

9. L. R. G. Treloar, *Trans. Faraday Soc.*, **40**, 59 (1944).

10. See ref. 5 and L. Bateman, ed., *Chemistry and Physics of Rubberlike Substances*, Maclaren, London, 1963, Ch. VII.

11. L. Mullins, *J. Polym. Sci.*, **19**, 225 (1956); *J. Appl. Polym. Sci.*, **2**, 1 (1959).

12. C. G. Moore and W. F. Watson, *J. Polym. Sci.*, **19**, 237 (1956); also G. M. Bristow et al., *J. Polym. Sci.*, **A3**, 3893 (1965).

13. See, e.g., a review by C. G. Moore in ref. 2, p. 167.

14. L. O. Amberg and W. D. Willis, *Proceedings International Rubber Conference*, *Washington, D.C.*, 1959, p. 565; G. Kraus, *J. Appl. Polym. Sci.*, **7**, 1257 (1963).

15. R. M. Russell et al., *Proceedings Fourth Rubber Technology Conference, London, 1962*, Institution of the Rubber Industry, London, 1963, p. 150.

16. A. Y. Coran, *Rubber Chem. Technol.*, **37**, 668, 673, 679, 689 (1964); **38**, 1 (1965); *Rubber Dev.*, **18**, No. 1, 4 (1965).

17. V. Ducháček, *J. Appl. Polym. Sci.*, **15**, 2079 (1971); *Rubber Chem. Technol.*, **45**, 945 (1972); R. B. Redding and D. A. Smith, *J. Inst. Rubber Ind.*, **4**, 198 (1970); *Rubber Chem. Technol.*, **44**, 1316 (1971).

18. B. Ellis and G. N. Welding, *Techniques in Polymer Science, Monograph No. 17*, Soc. Chem. Ind., London, 1963, p. 46.

19. M. C. Morris, *J. Appl. Polym. Sci.*, **8**, 545 (1964).

20. D. A. Smith, *J. Polym. Sci.*, **C16**, 525 (1957); R. B. Redding and D. A. Smith, I.U.P.A.C. Macromolecular Conference, Toronto, 1968, Preprint A10.11 (Pt. I); *J. Polym. Sci.*, *C*, **30**, 491 (1970).

21. See, e.g., J. R. Scott, *Physical Testing of Rubbers*, Maclaren, London, 1965, p. 48.

22. *Continuous Measurement of the Cure Rate of Rubber*, A.S.T.M. Special Tech. Publ. No. 383, Philadelphia, Pennsylvania 1965.

23. See, e.g., R. D. Heap, *Trans. Inst. Rubber Ind.*, **41**, 127 (1965).

24. M. Jones, *Trans. Inst. Rubber Ind.*, **11**, 38 (1935).

25. *Monsanto Rubber Chemicals Handbook*, August 1965, p. 27.

26. J. O. Harris, Lecture given in Part II of a course on Advanced Rubber Technology, Ryerson Polytechnical Institute, Toronto, Canada, 1965.

27. R. W. Robinson and R. N. F. Pinfold, *Trans. Inst. Rubber Ind.*, **39**, 26 (1963).

28. W. Scheele et al., *Kaut. Gummi Kunstst.*, **10**, WT185 (1957); **12**, WT205 (1959) (with English version).

29. B. C. Barton and E. J. Hart, *Ind. Eng. Chem.*, **42**, 671 (1950); **44**, 2444 (1952).

30. C. G. Moore, ref. 13, p. 182.

31. L. Mullins, ref. 2, p. 224. For contrary evidence, see J. Lal and K. W. Scott, *J. Polym. Sci.*, **C9**, 113. (1965).

32. A. E. Juve and R. Shearer, unpublished work.

33. D. Craig, *Rubber Chem. Technol.*, **30**, 1291 (1957).

34. I. Ostromislensky, *J. Russ. Phys. Chem. Soc.*, **47**, 1885 (1915).

35. M. Braden and W. P. Fletcher, *Trans Inst. Rubber Ind.*, **31**, 155 (1955); J. Scanlan and D. K. Thomas, *J. Polym. Sci.*, **A1**, 1015 (1963).

36. L. D. Loan and J. Scanlan, *Rubber. Plast. Age*, **44**, 1315 (1963); D. K. Thomas, *J. Appl. Polym. Sci.*, **6**, 613 (1962).

37. G. M. Bristow, *J. Appl. Polym. Sci.*, **9**, 3255 (1965).

38. G. Kraus, *J. Appl. Polym. Sci.*, **7**, 1257 (1963); D. Reichenbach, *Kaut. Gummi Kunstst.*, **18** (*i*), 9 (1965).

39. R. T. Vanderbilt Co., Varox.

40. L. D. Loan, *R.A.P.R.A. Bull.*, **17**, 197 (1963); *J. Polym. Sci.*, **A2**, 3053 (1964).

41. L. Mullins, ref. 2, p. 224, based on work by A. G. Thomas.

42. J. C. Halpin and F. Bueche, *J. Polym. Sci.*, **A3**, 3935 (1965).

43. R. S. Porter et al., *Rubber Chem. Technol.*, **41**, 1 (1968).

44. T. L. Smith, *J. Polym. Sci.*, **A1**, 3597 (1963); *J. Appl. Phys.*, **35**, 3142 (1964); T. L. Smith and J. E. Frederick, *J. Appl. Phys.*, **36**, 2996 (1965).

45. J. C. Halpin and F. Bueche, *J. Appl. Phys.*, **35**, 3142 (1964); J. C. Halpin, *Rubber Chem. Technol.*, **38**, 1007 (1965).

46. E. H. Andrews and A. Walsh, *Proc. Phys. Soc.*, **42**, 72 (1958); E. H. Andrews, *J. Appl. Phys.*, **32**, 542 (1961); *Rubber Chem. Technol.*, **36**, 325 (1963).

47. D. A. Smith, *Rubber J.*, **150** (vii), 38; (viii), 39; (ix), 54 (1968).

48. L. Mullins and N. R. Tobin, *Rubber Chem. Technol.*, **39**, 799 (1966).

49. G. Kraus, *J. Appl. Polym. Sci.*, **7**, 861 (1963).

50. O. Lorenz and C. R. Parks, *J. Polym. Sci.*, **50**, 299 (1961).

51. B. Saville and A. A. Watson, *Rubber Chem. Technol.*, **40**, 111 (1967).

52. L. Mullins, *J. Polym. Sci.*, **19**, 225 (1956); C. G. Moore and W. F. Watson, *J. Polym. Sci.*, **19**, 237 (1956).

53. W. C. Wake, *Kaut. Gummi Kunstst.*, **18**, S.654 (1965) (in English).

54. W. F. Watson, in G. Kraus, ed., *Reinforcement of Elastomers*, Wiley-Interscience, New York, 1965, Ch. 8.

55. V. A. Garten and G. K. Sutherland, *Proceedings Third Rubber Technology Conference*, Institution of the Rubber Industry, London, 1964, p. 536.

56. J. O. Harris and R. W. Wise, in ref. 54, Ch. 9.

57. E. Papirer, Thesis, Strasbourg, 1962, quoted in *Carbon Black Abstr.*, **14**, 799 (1965).

58. W. M. Hess et al., *Kaut. Gummi Kunstst.*, **20**, 135 (1967) (in German).

59. R. Stiehler, private communication, 1962; A. R. Payne, *J. Appl. Polym. Sci.*, **11**, 383 (1967).

60. A. R. Payne, *J. Colloid Sci.*, **19**, 744 (1964).

61. J. C. Brennan and T. E. Jermyn, *J. Appl. Polym. Sci.*, **A9**, 2749 (1965).

62. J. A. C. Harwood et al., *J. Polym. Sci.*, **B3**, 119 (1965).

63. J. A. C. Harwood et al., *J. Inst. Rubber Ind.*, **1**, 17 (1967).

64. E. M. Dannenberg, *Trans. Inst. Rubber Ind.*, **42**, 26 (1966); B. B. Boonstra, *J. Appl. Polym. Sci.*, **11**, 389 (1967).

65. M. Morton and J. C. Healy, ACS Division of Rubber Chemistry, Akron, Ohio, September 1967, Paper 26; M. Morton, J. C. Healy, and R. L. Denecour, *Proceedings Fifth International Rubber Conference, Brighton, 1967*, Institution of the Rubber Industry, London, 1968, p. 175.

66. A. E. Oberth, *Rubber Chem. Technol.*, **40**, 1337 (1967).

67. R. R. Juengel and T. D. Bolt, *Tech. Service Lab. Bull. LD-7*, Cabot Corp., Cambridge, Massachusetts, 1967; B. Kastein, Jr., *Rubber Age*, **96**, 724 (1965).

68. C. J. Kochmit, ACS Division of Rubber Chemistry, Akron, Ohio, May 1966, Paper 54.

69. F. W. Barlow, *Rubber J.*, **147** (ix), 30 (1965).

70. R. C. W. Moakes and J. R. Pyne, *Proc. Inst. Rubber Ind.*, **1**, 151 (1954).

71. J. H. Bachmann et al., *Rubber Chem. Technol.*, **32**, 1286 (1959).

72. J. W. Sellars and F. E. Toonder, in ref. 54, Ch. 13.

73. D. K. Thomas and B. B. Moore, *Report No. 66061*, Royal Aircraft Establishment, Farnborough, England, 1966.

74. D. W. Southwart and T. Hunt, *J. Inst. Rubber Ind.*, **2**, 77, 79, 140 (1968); **3**, 249 (1969); **4**, 74, 77 (1970).

75. D. Pocknell et al., *Proceedings of the Fourth International Synthetic Rubber Conference, London, 1969*, Rubber Technical Press, London, p. 5; D. Baker et al., *Polymer*, **9**, 437 (1968).

76. A. M. Gessler et al., *Rubber Age*, **78**, 73 (1955).

77. D. B. Russell, ACS Division of Rubber Chemistry, Akron, Ohio, May 1966, Paper 34; J. R. Creasey, D. B. Russell, and M. P. Wagner, *Rubber Chem. Technol.*, **41** (5), 1300 (1968).

78. H. Westlinning and H. Fleischhauer, in ref. 54, Ch. 14.

79. Borg Warner Corp., British Pat. 1,019,352 (1961).

80. J. Pacitti, *RAPRA Review*, March 1962.

81. M. Dayne et al., French Pat. 1,424,003 (1964).

82. *Tech. Inf. Sheet 34*, Nat. Rubber Prod. Res. Assoc., Welwyn G. C., England, 1964.

83. For example, J. E. Hauck, *Mater. Design Eng.*, **61** (v), 104 (1965); Goodyear Tire & Rubber Co., *Rubber Age*, **96**, 753 (1965), and other references.

84. F. W. Hannsgen, *Rubber Plast. Age*, **42**, 166 (1961).

85. G. R. Himes and R. D. Sullivan, *Rubber Age*, **95**, 65 (1964).

86. E. C. Gregg, Jr., and J. H. Macey, *Rubber Chem. Technol.*, **46**, 47 (1973).

87. M. Bruzzone et al., *Rubber Plast. Age*, **46**, 278 (1965).

88. M. J. Hackathorn and M. J. Brock, *Rubber Chem. Technol.*, **45**, 1295 (1972).

89. F. B. Swinney, *Polysar Tech. Rept. No. 64: 7A*, Polymer Corp., Sarnia, Canada, 1964.

90. International Symposium on Isoprene Rubber, Moscow, USSR, November 20–24, 1972; *Abstr.*, *Rubber Chem. Technol.*, **46**, 555 (1973).

91. M. A. Golub, *J. Polym. Sci.*, **25**, 373 (1957).

92. J. J. Shipman and M. A. Golub, *J. Polym. Sci.*, **58**, 1963 (1962).

93. L. R. Bateman et al., *J. Chem. Soc.*, **1958**, 2846; *J. Appl. Polym. Sci.*, **1**, 257 (1959).

94. F. J. Linnig and J. E. Stewart, *J. Res. Nat. Bur. Stand.*, **60**, 9 (1958).

95. W. A. Bishop, *J. Polym. Sci.*, **55**, 827 (1961).

96. H. Blümel, *Kaut. Gummi Kunstst.*, **16**, 571 (1963) (in German); *Rubber Chem. Technol.*, **37**, 408 (1964) (English translation).

97. E. W. Madge, *Chem. Ind. (London)*, **42**, 1806 (1962).

98. J. Curchod, *Rubber Chem. Technol.*, **43**, 1367 (1970).

99. J. N. Short et al., *Rubber Plast. Age*, **38**, 880 (1957).

100. J. N. Short et al., *Rubber Chem. Technol.*, **32**, 614 (1959).

101. A. V. Tobolsky and A. Mercurio, *J. Appl. Polym. Sci.*, **2**, 186 (1952).

102. W. A. Smith and J. M. Willis, *Rubber Age*, **87**, 815 (1960).

103. E. C. Gregg, Jr., *J. Polym. Sci.*, **C24**, 303 (1968).

104. E. C. Gregg, Jr., and S. E. Katrenick, *Rubber Chem. Technol.*, **43**, 549 (1970).

105. J. D. Skinner, *Rubber Chem. Technol.*, **45**, 182 (1972).

106. K. Hümmel et al., *Kaut. Gummi Kunstst.*, **14**, WT171 (1961) (with English version).

107. W. Scheele, *Kaut. Gummi Kunstst.*, **15**, WT482 (1962) (with English version).

108. B. M. E. van der Hoff, *Ind. Eng. Chem., Prod. Res. Dev.*, **2**, 273 (1963).

109. W. Scheele et al., *Kaut. Gummi Kunstst.*, WT57 (1962) (with English version).

110. K. Hümmel and G. Kaiser, *Kaut. Gummi Kunstst.*, **16**, 426 (1963) (in German); *Rubber Chem. Technol.*, **38**, 581 (1965) (English translation).

111. L. D. Loan, *J. Appl. Polym. Sci.*, **7**, 2259 (1963).

112. C. W. Childers and G. Kraus, *Rubber Chem. Technol.*, **40**, 1183 (1967).

113. H. L. Hsieh, *Rubber Plast. Age*, **46**, 394 (1965).

114. G. Holden et al., *J. Polym. Sci.*, **C26**, 37 (1969).

115. E. T. Bishop and S. Davison, *J. Polym. Sci.*, **C26**, 59 (1969).

116. T. L. Smith and R. A. Dickie, *J. Polym. Sci.*, **C26**, 163 (1969).

117. H. Hendus et al., *Kolloid-Z.*, **216–217**, 110 (1967).

118. M. Morton et al., *J. Polym. Sci.*, **C26**, 99 (1969).

119. D. M. Brunwin et al., *J. Polym. Sci.*, **C26**, 135 (1969).

120. G. Kraus et al., *J. Polym. Sci.*, A-2,⁴9, 1839 (1971); *Rubber Chem. Technol.*, **45**, 1005 (1972).

121. G. Kraus and K. W. Rollman, *J. Appl. Polym. Sci.*, **8**, 2585 (1964); *Rubber Chem. Technol.*, **38**, 493 (1965).

122. W. F. Watson, *Ind. Eng. Chem.*, **47**, 1281 (1955); and ref. 54.

123. J. J. Brennan and T. E. Jermyn, *J. Appl. Polym. Sci.*, **9**, 2749 (1965).

124. J. J. Brennan et al., *J. Appl. Polym. Sci.*, **8**, 2687 (1964).

125. G. Kraus and J. J. Gruver, *Rubber Chem. Technol.*, **41**, 1256 (1968).

126. S. Davison et al., *Rubber World*, **151** (v), 81 (1965); *Rubber Chem. Technol.*, **38**, 457 (1965).

127. V. F. Evstratov et al., *Kauch. Rezina*, **25** (ii), 9 (1966) (in Russian); *Rubber Chem. Technol.*, **40**, 685 (1967) (*English translation*).

128. I. W. E. Harris et al., *Polysar Taktene Prod. Bull.*, Polymer Corp., Sarnia, Canada, undated, p. 45.

129. D. V. Sarbach et al., *Rubber Chem. Technol.*, **40**, 766 (1967).

130. L. H. Krol, *Rubber Age*, **45**, 1341 (1964).

131. G. Natta et al., *Rubber Chem. Technol.*, **36**, 1583 (1963).

132. F. P. Baldwin and G. Verstrate, *Rubber Chem. Technol.*, **45**, 709 (1972).

133. L. D. Loan, *Rubber Chem. Technol.*, **40**, 149 (1967).

134. G. A. Harpell and D. H. Walrod, *Rubber Chem. Technol.*, **46**, 1007 (1973).
135. G. Ballini et al., *Rubber Chem. Technol.*, **39**, 521 (1966).
136. V. Zamboni et al., *Rubber Chem. Technol.*, **44**, 1109 (1971).
137. M. Imoto et al., *Rubber Chem. Technol.*, **43**, 424, 431 (1970).
138. A. E. Robinson et al., *Ind. Eng. Chem., Res. Dev.*, **1**, 78 (1962).
139. L. P. Lenas, *Ind. Eng. Chem., Prod. Res. Dev.*, **2**, 202 (1963).
140. P. E. Wei and J. Rehner, Jr., *Rubber Chem. Technol.*, **35**, 133, 1091 (1962).
141. P. E. Wei and J. Rehner, Jr., U.S. Pat. 3,136,734 (June 1964, appl. March 1962).
142. J. V. Fusco, *Rubber World*, **148** (ii), 48 (1963).
143. H. S. Makowski et al., *Ind. Eng. Chem., Prod. Res. Dev.*, **3**, 282 (1964).
144. A. A. Dontsov et al., *Vysokomol. Soedin.*, **A11**, 2630 (1969); *Rubber Chem. Technol.*, **44**, 721 (1971).
145. J. Lal and J. E. McGrath, *Rubber Chem. Technol.*, **36**, 248 (1963).
146. G. Natta et al., *Rubber Chem. Technol.*, **36**, 988 (1963).
147. G. Sartori et al., *Rubber Chem. Technol.*, **38**, 620 (1965).
148. K. Fujimoto and K. Wataya, *Rubber Chem. Technol.*, **43**, 860 (1970); K. Fujimoto and S. Nakade, *ibid.*, **43**, 411 (1970).
149. F. P. Baldwin et al., *Rubber Chem. Technol.*, **43**, 522 (1970).
150. S. Cesca et al., *Rubber Chem. Technol.*, **48**, 747 (1975).
151. E. K. Gladding et al., *Ind. Eng. Chem., Res. Dev.*, **1**, 65 (1962).
152. I. D. Roche, *Rubber Age*, **93**, 921 (1963).
153. J. M. Mitchell et al., *Rubber World*, **148** (viii), 52 (1963).
154. H. H. Waddell et al., *Rubber Age*, **94**, 427 (1963).
155. H. Blümel et al., *Kaut. Gummi Kunstst.*, **16**, 369 (1963).
156. H. E. Haxo et al., *Rubber Age*, **94**, 255 (1963).
157. J. W. Youren, private communication, 1970.
158. G. Crespi and A. Arcozzi, *Rubber Chem. Technol.*, **38**, 590 (1965).
159. R. German et al., *Kaut. Gummi Kunstst.*, **19**, 67 (1966) (in German); *Rubber Chem. Technol.*, **40**, 569 (1967) (English translation).
160. P. F. Lyons et al., *Rubber Chem. Technol.*, **40**, 1634 (1967).
161. H. K. Frensdorff, *Rubber Chem. Technol.*, **41**, 316 (1968).
162. R. B. Redding and D. A. Smith, *J. Polym. Sci., C*, **30**, 491 (1970).
163. M. J. Downing and J. E. Stuckey, *Rubber J.*, **151** (i), 48 (1969).
164. J. R. Wolfe, *Rubber Chem. Technol.*, **40**, 760 (1967); *J. Appl. Polym. Sci.*, **12**, 1167, 1183 (1968).
165. F. P. Baldwin et al., *Rubber Chem. Technol.*, **42**, 1167 (1969).
166. N. R. Langley, *Macromolecules*, **1**, 348 (1968).
167. J. J. Howarth et al., *Rubber World*, **148** (viii), 69 (1963).
168. J. R. Wolfe and J. R. Albin, *Rubber Chem. Technol.*, **37**, 927 (1964).
169. M. A. Schoenbeck, *Rubber Age*, **92**, 75 (1962).
170. T. J. Dudek and F. Bueche, *J. Appl. Polym. Sci.*, **8**, 555 (1964).
171. E. B. Hansen and F. C. Church, *Tech. Rept. RG-127*, Cabot Corp., Cambridge, Massachusetts, undated.

172.   *Tech. Rept. RG-111*, Cabot Corp., Cambridge, Massachusetts, undated.

173.   K. Satake et al., *J. Inst. Rubber Ind.*, **4,** 21 (1970).

174.   M. H. Walters and D. N. Keyte, *Trans Inst. Rubber Ind.*, **38,** 40 (1962); *Rubber Chem. Technol.*, **38,** 62 (1965) (corrected version of the same paper).

175.   A. J. Curtius et al., *Rubber Chem. Technol.*, **45,** 1582 (1972).

176.   P. J. Corish, *Rubber Chem. Technol.*, **40,** 324 (1967).

177.   J. E. Callan et al., *Rubber World*, **151,** 60 (1965).

178.   W. M. Hess et al., Carbon-Elastomer Adhesion, ACS Division of Rubber Chemistry, Miami Beach, Florida, May 1965.

179.   W. M. Hess et al., *Rubber Chem. Technol.*, **40,** 371 (1967).

J. R. Dunn

Polysar, Limited
Sarnia, Ontario

Chapter 9 Aging and Degradation

1 INTRODUCTION

Aging may be defined as an unwanted deterioration in the physical properties of a polymer due to the action of heat, radiation, oxygen, ozone, or mechanical work, either separately or in combination. Degradation and cross-linking of elastomers during storage or service both cause aging. Degradation, i.e., reduction in molecular weight of the polymer main chain or reduction in cross-link density from scission of the main chain or the cross-links, results in reduction of modulus and tensile strength. Cross-linking impairs raw polymer processing, raises the modulus and lowers the elongation of vulcanizates, and causes permanent set of a deformed elastomer in its deformed position.

The aging behavior of an elastomer is determined primarily by the amount of unsaturation and branching in the main polymer chain, and this depends, in turn, on the nature of the monomers of which it is composed. Consequently, much of the subject matter of this chapter is of a general nature; for a more detailed discussion, the reader is referred to recent books devoted to the topic (1–5). In the sections that follow, the causes of aging are considered individually with specific reference to their effect on stereoregular elastomers. It is now recognized that different modes of aging involve different chain reaction mechanisms (4, 5). Methods of combating the various types of aging are also examined.

2 THERMAL AGING

Purely thermal aging of unvulcanized elastomers in the complete absence of light and oxygen may be regarded as commercially unimportant since storage temperatures are well below the decomposition temperature of the polymer backbone. Golub and Gargiulo (6) found on heating cis-1,4- and trans-1,4-polyisoprene and polybutadiene under vacuum that cis-trans isomerization and cyclization occurred in all four polymers. Polybutadiene shows cis-trans isomerization at about 200°C, well below the region of cyclization and decomposition (350°C). For polyisoprene these reactions occur only at pyrolytic conditions (>300°C) and are accompanied by isoprene and dipentene formation.

At lower temperatures (55–150°C) thermal oxidation of 1,4-polyisoprene results in a marked loss of $C(CH_3)=CH—$ double bonds and the formation of cyclic structures via $RO_2\cdot$ radicals, and is accompanied by the appearance of epoxy, peroxide, hydroperoxide, and alcohol groups distributed along the oxidized chain (7).

The thermal stability of vulcanizates in the absence of light and oxygen is important in considering the behavior of large rubber articles at service temperatures. Under these circumstances the stability of the cross-link is important as well as the stability of the polymer backbone. Extensive studies by Moore et al. (8) using low molecular weight model compounds have indicated a "maturing" of cross-links on heating. Studies involving "molecular probes" which either cleave polysulfidic cross-links or remove sulfur from them (8–10) confirm that on prolonged heating sulfur present in polysulfidic cross-links produces further cross-linking of the polymer. This, in turn, results in increase in hardness and permanent set. Murakami and Tamura (11) reported that natural rubber degradation in air occurred by distinguishable thermal and oxidative random main chain scission, cross-link scission, and exchange reactions.

The solution to problems of thermal aging may lie in the choice of vulcanization system. The so-called efficient vulcanization systems that have been extensively studied for natural rubber (12) are stable to thermal aging. Unfortunately, the unaged tensile properties of these vulcanizates are inferior to those of conventional sulfur vulcanizates.

3 OXIDATIVE AGING

3.1 Mechanism of Oxidation

Oxygen in conjunction with thermal, radiant, or mechanical energy is the principal cause of elastomer aging. The autooxidation of hydrocarbons has been extensively studied and reviewed (1, 13–17). It is well established that this is a free radical chain process involving the cleavage of C–H bonds rather than addition of oxygen to the double bond of an olefin (18). The active hydrogen in an olefin is the one in the α position with respect to the double bond since the radical formed on its removal is stabilized by resonance.

In its simplest form the accepted scheme for uninhibited hydrocarbon oxidation is

$$
\begin{array}{llll}
\text{Initiation:} & \text{production of } R\cdot \text{ or } RO_2\cdot \text{ radicals} & \text{(a)} \\
\text{Propagation:} & R\cdot + O_2 \rightarrow ROO\cdot & \text{(b)} \\
& ROO\cdot + RH \rightarrow ROOH + R\cdot & \text{(c)} \\
\text{Termination:} & 2R\cdot \longrightarrow & & \text{(d)} \\
& R\cdot + ROO\cdot \longrightarrow & \text{Nonradical} & \text{(e)} \\
& ROO\cdot + ROO\cdot \rightarrow & \text{products} & \text{(f)}
\end{array}
$$

If the rate constants for reactions (a)–(f) are r_i, k_2, k_3, k_4, k_5, and k_6, respectively, the overall rate of oxidation for long kinetic chains is

$$
\frac{dO_2}{dt} = r_i k_3 k_6^{-1/2}(RH) \frac{k_2 k_6^{-1/2}(O_2)}{k_3 k_4^{-1/2}(RH) + k_2 k_6^{-1/2}(O_2)} \qquad [1]
$$

At oxygen pressures in excess of 100 torr, the rate of oxidation is generally independent of oxygen pressure, and stationary state kinetics leads to a rate equation:

$$\frac{dO_2}{dt} = r_i^{1/2} k_3 k_6^{-1/2} (RH) \qquad [2]$$

Thus the rate of oxidation can be seen to be controlled by the stability of the active C–H bond. Bolland (19), studying the benzoyl peroxide catalyzed oxidation of substituted propenes,

$$CH_3—CH{=}CH_2$$
$$\text{(a)} \quad \text{(b)} \quad \text{(c)}$$

at 45° evaluated the effect on k_3 of substituents in positions (a), (b), and (c). One important deduction was that replacement of one or two hydrogen atoms at positions (a) or (c) increases reactivity by $(3.3)^n$ where n is the total number of substituents. Thus, for example, the overall rate of oxidation of polyisoprene would be expected to be greater than that of polybutadiene.

Norling and co-workers (20) studied the relative vulnerability of several polymers to the hydroperoxide forming reaction (c). They examined the rate of oxygen uptake at 1 atm oxygen and 60–100°C of thin lightly carbon–carbon cross-linked vulcanizates. With the external initiator, the initiation step becomes

$$\text{initiator} \xrightarrow{k_d} 2R'\cdot \qquad \text{(a')}$$

$$R'\cdot + RH \xrightarrow{k_1} R\cdot + R'H \qquad \text{(a'')}$$

and

$$\frac{dO_2}{dt} = \frac{k_3}{k_6^{1/2}} [2e_i k_d(I)]^{1/2} (RH) + 2e_i k_d(I) \qquad [3]$$

where e_i is the efficiency of initiation, i.e., the fraction of radicals from benzoyl peroxide which attack the hydrocarbon, and x is the number of molecules of oxygen absorbed in initiation and termination steps. The expression $k_3 e_i^{1/2}(RH)/k_6^{1/2}$, denoted as R_0, indicates the relative values of k_3 as long as the initiator efficiency and the rate of termination do not change in going from one rubber to another. Some values of R_0 evaluated by plotting $(dO_2/dt)_0/2k_d I_0$ as a function of $(2k_d I_0)^{-1/2}$ are shown in Table 1.

These data indicate, as would be expected, similar ease of hydroperoxide formation in synthetic *cis*-polyisoprene and in natural rubber. They also indicate that the ease of hydroperoxide formation is similar for a wide variety of polymers. The authors rationalize this by pointing to other hydrogen abstraction reactions which are relatively insensitive to the

Table 1 Relative Values of k_3 for Various Polymers (20)

Polymer	Source	Relative R_0 (at 80°C)
Natural rubber	Pale crepe (Uniroyal)	7.0
cis-1,4-Polyisoprene	Isoprene rubber (Shell)	7.0
Styrene–butadiene	Stereon (Firestone)	5.4
Poly(propylene oxide)	Dynagen XP-139 (General Tire)	5.9
EPM	1. Experimental (Goodyear)	
	2. Enjay EPR-404	2.1
EPDM	Enjay EPT-3509	2.6
Amorphous polypropylene	Experimental (Hercules)	4.3

effect of substituents. Since the polymers were used without special purification, the assumption that chain termination follows reaction (f) may not be justified. It is evident from Table 1 that saturated polymers such as EPM and polypropylene are subject to the same oxidation reaction as are polydienes. The tertiary hydrogen in propylene based polymers is labile, although to a lesser extent than the α-methylenic hydrogen of polydienes. Tobolsky et al. note that EPM is relatively more prone to hydrogen abstraction than polypropylene. The EPM contains only 35 mole % propylene and the R_0(propylene unit concentration) is 1.3 times as great in EPM as in polypropylene. This difference is attributed to steric hindrance in polypropylene. An increased oxidizability of EPDM is noted which is attributed to the additional influence of the double bond.

3.2 Mechanism of Oxidative Degradation and Cross-Linking

The simple hydrocarbon oxidation mechanism represented by reactions (a)–(f) does not, per se, explain the degradation (cross-link or chain scission) of polymers during aging. There is abundant evidence that both cross-linking and scission occur (21–23) and Table 2, taken from Hofmann's review (21), demonstrates that the predominant effect depends on the type of polymer under study.

If α-methylenic reactivity is not too great, some addition of $RO_2\cdot$ to an olefinic double bond may be possible. Thus, two more reactions may be added to the scheme (24):

$$ROO\cdot + CR_2' {=} CR_2'' \rightarrow RO_2{-}CR_2'{-}\dot{C}R_2'' \qquad (c')$$

$$RO_2{-}CR_2'{-}\dot{C}R_2'' + O_2 \rightarrow RO_2{-}CR_2'{-}CR_2''{-}\dot{O}_2 \qquad (b')$$

Table 2 Benzene Solubility and Intrinsic Viscosity of Elastomers After 30 Hours in Air at 125°C (21)

	% Solubility		Intrinsic Viscosity	
Elastomer	Before	After	Before	After
NR	78	98	5.62	0.47
IR	78	98	1.06	0.28
BR	85	39		
SBR	79	44		

Bolland and Hughes (25) discuss the case of 1,5-dienes such as 2,6-dimethylocta-2,6-diene, squalene, and polyisoprenes where hydroperoxide and diperoxide formation proceed consecutively in the same molecule through intermolecular addition of a peroxy radical to a double bond:

$$- CH_2 - \overset{\overset{\displaystyle CH_3}{|}}{C} = CH - CH_2 - CH_2 - \overset{\overset{\displaystyle CH_3}{|}}{C} = CH - CH_2 - \quad \xrightarrow{\;RO_2\;}$$

$$- CH_2 - \overset{\overset{\displaystyle CH_3}{|}}{C} = CH - CH_2 - CH - \overset{\overset{\displaystyle CH_3}{|}}{C} = CH - CH_2 - \; + \; RO_2H \quad \xrightarrow{\;O_2\;}$$

$$- CH_2 - \overset{\overset{\displaystyle CH_3}{|}}{C} = CH - CH_2 - \underset{\underset{\displaystyle O - O}{|}}{CH} - \overset{\overset{\displaystyle CH_3}{|}}{C} = CH - CH_2 - \quad \xrightarrow{\quad}$$

$$- CH_2 - \overset{\overset{\displaystyle CH_3}{|}}{C} = CH - CH_2 - \underset{\underset{\displaystyle O}{|}}{CH} - \underset{\underset{\displaystyle O}{|}}{\overset{\overset{\displaystyle CH_3}{|}}{C}} - CH - CH_2 - \quad \xrightarrow{\;O_2\;}$$

$$- CH_2 - \overset{\overset{\displaystyle CH_3}{|}}{C} = CH - CH_2 - \underset{\underset{\displaystyle O}{|}}{CH} - \underset{\underset{\displaystyle O}{|}}{\overset{\overset{\displaystyle CH_3}{|}}{C}} \overset{O - O}{\diagup} CH - CH_2 -$$

which now, taking the role of $RO_2\cdot$, attacks another chain. Mayo has pointed out that β-peroxyalkoxy radicals are unstable (26). However, breakdown of the peroxide-hydroperoxide proposed by Bolland and

Hughes would not immediately lead to chain scission. Thus Mayo proposed the following scheme for polyisoprene:

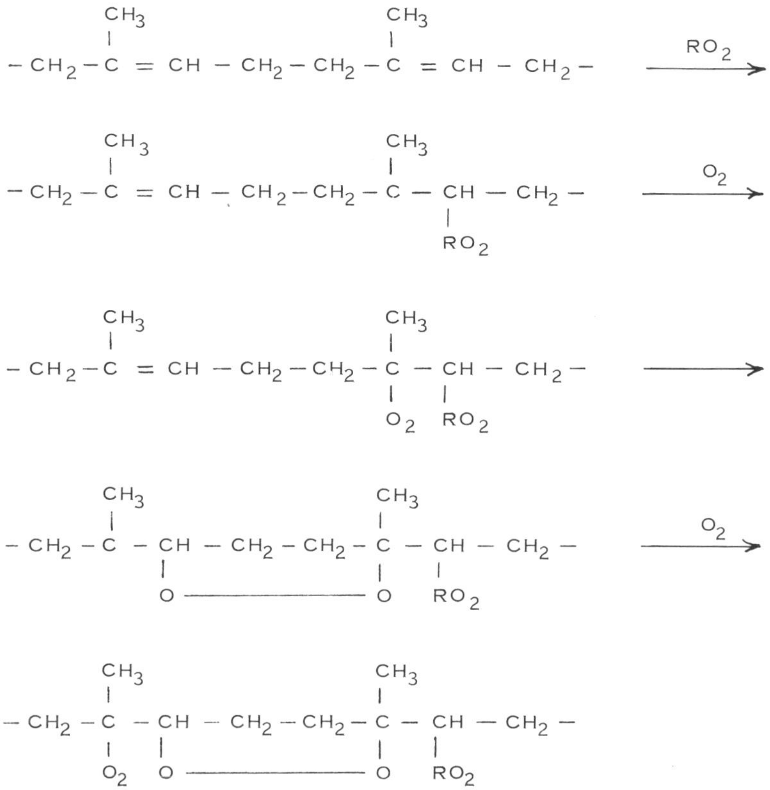

$$
\underset{\text{CH}_2}{-}\text{CH}_2-\overset{\overset{\displaystyle CH_3}{|}}{C} = CH - CH_2 - CH_2 - \overset{\overset{\displaystyle CH_3}{|}}{C} = CH - CH_2 - \qquad \xrightarrow{\;RO_2\;}
$$

$$
-CH_2 - \overset{\overset{\displaystyle CH_3}{|}}{C} = CH - CH_2 - CH_2 - \underset{\underset{\displaystyle RO_2}{|}}{C} - CH - CH_2 - \qquad \xrightarrow{\;O_2\;}
$$

$$
-CH_2 - \overset{\overset{\displaystyle CH_3}{|}}{C} = CH - CH_2 - CH_2 - \underset{\underset{\displaystyle O_2 \quad RO_2}{|\quad|}}{C} - CH - CH_2 - \qquad \xrightarrow{\qquad}
$$

$$
-CH_2 - \overset{\overset{\displaystyle CH_3}{|}}{C} - CH - CH_2 - CH_2 - \underset{\underset{\displaystyle O \rule{1cm}{0.4pt} O \quad RO_2}{|\qquad\qquad| \quad |}}{C} - CH - CH_2 - \qquad \xrightarrow{\;O_2\;}
$$

$$
-CH_2 - \overset{\overset{\displaystyle CH_3}{|}}{C} - CH - CH_2 - CH_2 - \underset{\underset{\displaystyle O_2 \quad O \rule{1cm}{0.4pt} O \quad RO_2}{|\quad|\qquad\qquad|\quad|}}{C} - CH - CH_2 -
$$

Mayo went on to suggest that the peroxide radical itself would be fairly stable below 100°C and that it must be converted into an alkoxy radical.

This conversion might arise via hydrogen abstraction and subsequent decomposition of the rubber hydroperoxide into alkoxy radical and OH, i.e., during an initiation step. It might also arise by addition to a double bond:

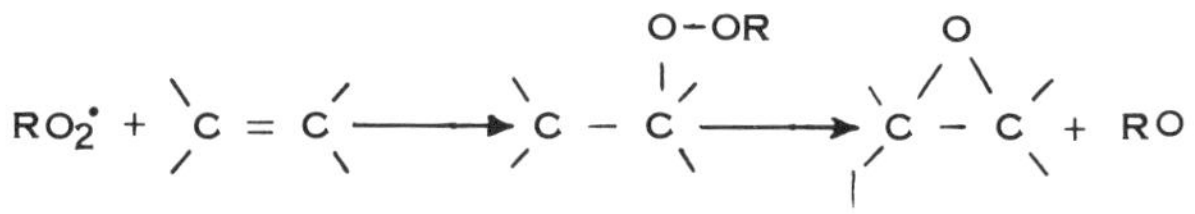

$$
RO_2^{\bullet} + \underset{/}{\overset{\backslash}{C}} = \underset{\backslash}{\overset{/}{C}} \longrightarrow \overset{\overset{\displaystyle O-OR}{|}}{\underset{\backslash}{\overset{/}{C}}} - \underset{\backslash}{C} \longrightarrow \underset{|}{\overset{\backslash \; / \; \overset{\displaystyle O}{\backslash}}{C}} - \underset{\backslash}{\overset{/}{C}} + RO
$$

Finally, it could arise by interaction of two peroxide radicals to give two alkoxy radicals during a termination step.

The reaction responsible for oxidative scission of polyisoprenes would be

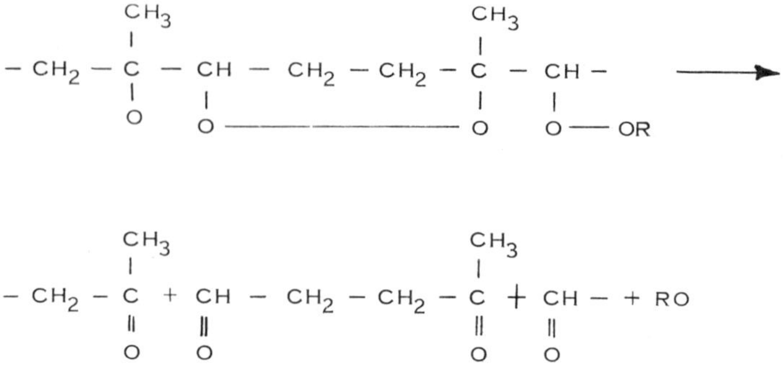

$$
\begin{array}{ccccccccc}
& & CH_3 & & & & & CH_3 & \\
& & | & & & & & | & \\
-CH_2 & - & C & - CH & - CH_2 & - CH_2 & - & C & - CH - \\
& & | & | & & & & | & | \\
& & O & O & \text{------------} & & & O & O - OR
\end{array}
\longrightarrow
$$

$$
\begin{array}{ccccccccc}
& & CH_3 & & & & & CH_3 & \\
& & | & & & & & | & \\
-CH_2 & - & C & + CH & - CH_2 & - CH_2 & - & C & + CH - + RO \\
& & \| & \| & & & & \| & \| \\
& & O & O & & & & O & O
\end{array}
$$

The primary anticipated low boiling product is levulinaldehyde and the other products are a cleaved rubber chain and an alkoxy radical which could continue the chain.

Mayo's mechanism raises the question of why an $RO_2\cdot$ radical attacks a double bond in the first place. It explains the formation of levulinaldehyde as a major product in the extensive studies of natural rubber oxidation made by Bevilacqua (27–29) but does not directly account for the formation of formic acid. Bevilacqua himself proposed a mechanism that involved a normal hydrogen abstraction of an α-methylenic carbon atom followed by an isomerization (30, 31):

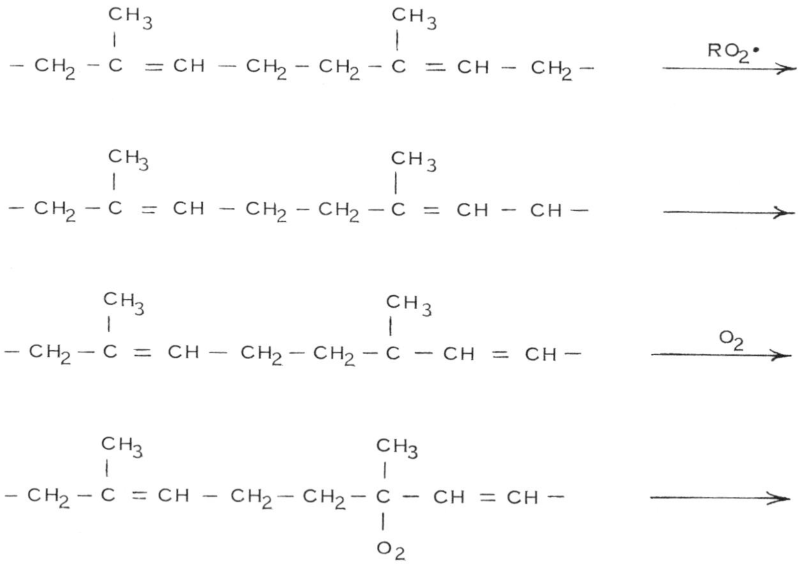

$$
\begin{array}{ccccccccc}
& CH_3 & & & & CH_3 & & \\
& | & & & & | & & \\
-CH_2 - C & = CH & - CH_2 & - CH_2 & - C & = CH & - CH_2 - & \xrightarrow{RO_2\cdot} \\
\end{array}
$$

$$
\begin{array}{ccccccccc}
& CH_3 & & & & CH_3 & & \\
& | & & & & | & & \\
-CH_2 - C & = CH & - CH_2 & - CH_2 & - C & = CH & - CH - & \longrightarrow \\
\end{array}
$$

$$
\begin{array}{ccccccccc}
& CH_3 & & & & CH_3 & & \\
& | & & & & | & & \\
-CH_2 - C & = CH & - CH_2 & - CH_2 & - C & - CH & = CH - & \xrightarrow{O_2} \\
\end{array}
$$

$$
\begin{array}{ccccccccc}
& CH_3 & & & & CH_3 & & \\
& | & & & & | & & \\
-CH_2 - C & = CH & - CH_2 & - CH_2 & - C & - CH & = CH - & \longrightarrow \\
& & & & & | & & \\
& & & & & O_2 & & \\
\end{array}
$$

$$
\begin{array}{ccc}
& \overset{\displaystyle CH_3}{\underset{|}{}} & \overset{\displaystyle CH_3}{\underset{|}{}} \\
-CH_2-C-CH-CH_2-CH_2-C-CH=CH- & & \xrightarrow{\ O_2\ } \\
& |\underline{\quad\quad O\quad\quad\quad\quad\quad O\quad} &
\end{array}
$$

$$
\begin{array}{ccc}
& \overset{\displaystyle CH_3}{\underset{|}{}} & \overset{\displaystyle CH_3}{\underset{|}{}} \\
-CH_2-C-CH-CH_2-CH_2-C-CH=CH- & & \\
& \underset{|}{O_2}\ \ \underline{O\quad\quad\quad\quad\quad O} &
\end{array}
$$

The scission products from such an intermediate would be levulinal-dehyde, formaldehyde, and a severed rubber chain. The oxidation of formaldehyde would yield formic acid while oxidation of some levulinal-dehyde would account for acetic acid and CO_2. Thus, Bevilacqua's mechanism accounts for his major products but, as Mayo points out, involves an unexpected isomerization and does not necessitate formation of levulinaldehyde as a primary product. Tobolsky and Mercurio (32) considered yet another route to scission in polyisoprenes:

$$
\begin{array}{c}
\overset{CH_3}{\underset{|}{}}\qquad\qquad\overset{CH_3}{\underset{|}{}}\\
-CH_2-C=CH-CH_2-CH_2-C=CH- \qquad \xrightarrow{\ RO_2\ }
\end{array}
$$

$$
\begin{array}{c}
\overset{CH_3}{\underset{|}{}}\qquad\qquad\overset{CH_3}{\underset{|}{}}\\
-CH_2-C=CH-CH-CH_2-C=CH- \qquad \xrightarrow{\ O_2\ }
\end{array}
$$

$$
\begin{array}{c}
\overset{CH_3}{\underset{|}{}}\qquad\qquad\overset{CH_3}{\underset{|}{}}\\
-CH_2-C=CH-CH-CH_2-C=CH- \qquad \longrightarrow \\
\underset{|}{}O_2
\end{array}
$$

$$
\begin{array}{c}
\overset{CH_3}{\underset{|}{}}\qquad\qquad\overset{CH_3}{\underset{|}{}}\\
-CH_2-C=CH-CH-CH_2-C-CH- \qquad \xrightarrow{\ O_2\ } \\
\underline{\quad O\quad\quad\quad\quad O\quad}
\end{array}
$$

$$
\begin{array}{c}
\overset{CH_3}{\underset{|}{}}\qquad\qquad\overset{CH_3}{\underset{|}{}}\\
-CH_2-C=CH-CH-CH_2-C-CH- \\
\underline{\quad O\quad\quad\quad\quad O_2\quad}
\end{array}
$$

They proposed that RO· radicals would result from a termination reaction between two RO$_2$· radicals and scission would then occur thus:

$$
\begin{array}{ccccccc}
& \text{CH}_3 & & & & \text{CH}_3 & \\
& | & & & & | & \\
-\text{CH}_2-\text{C} & = \text{CH}-\text{CH}-\text{CH}_2-\text{C}-\text{CH}-\text{CH}_2- & \longrightarrow \\
& | & & & & | & \\
& \text{O} &\rule{2cm}{0.4pt}& & & \text{O} &
\end{array}
$$

$$
\begin{array}{cccc}
\text{CH}_3 & & \text{H} & \\
| & & | & \\
-\text{CH}_2-\text{C} & = \text{CH}-\text{C} & = \text{O} + \text{CH}_3\,\text{CCH}_2 + \text{HC}-\text{CH}_2- \\
& & \parallel \qquad\quad \parallel & \\
& & \text{O} \qquad\quad \text{O} &
\end{array}
$$

The acetonyl radical could then give rise to observed products such as acetone, pyruvaldehyde, and 2,5-hexanedione.

All polydienes have numerous structural defects such as 1,2 or 3,4 linkages, cyclic structures, head-to-head or tail-to-tail linkages, and conjugated polyenes which exist at the ends or along the length of the chains and whose proportion increases during vulcanization. Although their concentration may be small, some of them and especially the conjugated groups can play a significant role in the initiation stages of oxidation. The formation of 4-methyl-4-vinyl butyrolactone and 4-hydroxy-2-butanone as well as levulinaldehyde, methacrolein, and methyl vinyl ketone from the oxidation of polyisoprene was suggested by Morand to arise from rapid decomposition of these conjugated polyenes (33).

Since one single mechanism cannot account for the many products of polyisoprene oxidation it is entirely possible that more than one of the suggested mechanisms are operative. Furthermore, scission may occur during more than one of the reaction steps. Bell (34) showed that if the probabilities of scission during propagation, termination, and initiation steps were W_p, W_t, and W_i, respectively, then for a kinetic chain length n:

$$\frac{1}{\epsilon} = W_p + \frac{W_t}{n} + \frac{W_i}{n} \qquad [4]$$

where ϵ is the number of moles of oxygen absorbed per scission event. Scission during initiation was considered unlikely since ϵ is not affected by the mode of initiation and there is no scission on heating oxidized natural rubber in the absence of oxygen.

Thus

$$\frac{1}{\epsilon} = W_p + \frac{W_t}{n} \qquad [5]$$

If scission occurs in both propagation and termination steps, then ϵ will decrease as the kinetic chain length n is decreased. Bell and Tiller (35) found that in carefully purified vulcanizates ϵ at 100°C was 22 and was little affected by temperature. The value of ϵ decreased rapidly as the concentration of a radical trapping antioxidant was raised. As will be discussed later the antioxidant introduces a new termination step:

$$RO_2 \cdot + AH \rightarrow ROOH + A \cdot \qquad (g)$$

which reduces the kinetic chain length. This effect of antioxidant level on ϵ is to be expected if W_t is appreciably greater than W_p. Thus scission during a termination step is regarded as highly probable. Earlier workers had found lower and temperature dependent values for ϵ (32, 36) and Bell and Tiller conclude that some antioxidant must have been present in their samples. If so, any scission mechanism involving combination of $RO_2 \cdot$ radicals would have been out of the question in the earlier studies.

The discussion of polyisoprene degradation thus far has not considered the effect of cross-links. A number of pieces of evidence indicate that in the presence of a carbon-carbon cross-link, such as may be introduced by cross-linking with dicumyl peroxide or by radiation curing, the oxidative scission of the main chain proceeds by the mechanism just described and is unaffected by the cross-link. This is consistent with the cross-link being a chemically stable entity and the chemistry of the main chain being unchanged by the introduction of such cross-links (8). Degradation of antioxidant-free peroxide vulcanized natural rubber has been studied by continuous stress relaxation in oxygen at elevated temperatures (37, 38). The rate of decrease in stress was autocatalytic and could be accounted for by the simple mechanism outlined in reactions (a)–(f). Intermittent relaxation measurements showed that very little cross-linking occurred on aging. Dunn et al. and Mullins showed that the molecular weight between cross-links was not changed as a peroxide vulcanizate was aged, i.e., there was no cross-link scission (39, 40).

Horikx (41) showed by investigation of the solubility and swelling of oxidized, vulcanized rubber that the hydrocarbon chain must be broken during oxidation. Bevilacqua (28, 42) subsequently found the same scission efficiency and volatile reaction products in the oxidation of unfilled NR vulcanized with either peroxide or efficiently ultilized sulfur and in a peroxide vulcanizate containing carbon black. This again suggested that main chain degradation by a mechanism such as that discussed earlier was a central feature of vulcanized polyisoprenes. However, it did not eliminate the possibility of some cross-link scission.

Colclough and co-workers recently suggested that oxidative scission of monosulfidic cross-links in natural rubber could occur in the following way (43):

$$
\begin{array}{cc}
-\,\mathrm{C} \;=\; \mathrm{CH} - \mathrm{CH} - \mathrm{CH_2} - & -\,\mathrm{C} \;=\; \mathrm{CH} - \mathrm{CH} - \mathrm{CH_2} - \\
\;\;|\qquad\qquad\qquad| & \;\;|\qquad\qquad\qquad| \\
\;\;\mathrm{CH_3}\qquad\qquad\;\; | & \;\;\mathrm{CH_3}\qquad\qquad\;\; | \\
\qquad\qquad\;\;\mathrm{S} & \qquad\qquad\;\;\mathrm{SO} \\
\qquad\qquad\;\; | & \qquad\qquad\;\; | \\
-\,\mathrm{C} \;=\; \mathrm{CH} - \mathrm{CH} - \mathrm{CH_2} - & -\,\mathrm{C} \;=\; \mathrm{CH} - \mathrm{CH} - \mathrm{CH_2} - \\
\;\;| & \;\;| \\
\;\;\mathrm{CH_3} & \;\;\mathrm{CH_3}
\end{array}
$$

(center arrow: $\xrightarrow[\text{OR } RO_2\cdot]{\text{ROOH}}$)

$$
\begin{array}{cc}
-\,\mathrm{C} \;=\; \mathrm{CH} - \mathrm{CH} - \mathrm{CH} - & -\,\mathrm{C} \;=\; \mathrm{CH} - \mathrm{CH} = \mathrm{CH} - \\
\;\;|\qquad\qquad\; \cdots|\cdots\cdots|\cdots & \;\;| \\
\;\;\mathrm{CH_3}\qquad\;\; \mathrm{S}=\mathrm{O}\;\;\; \mathrm{H} & \;\;\mathrm{CH_3} \\
\qquad\qquad\qquad\;\; | & \\
-\,\mathrm{C} \;=\; \mathrm{CH} - \mathrm{CH} - \mathrm{CH_2} - & -\,\mathrm{C} \;=\; \mathrm{CH} - \mathrm{CH} - \mathrm{CH_2} - \\
\;\;| & \;\;|\qquad\qquad\qquad| \\
\;\;\mathrm{CH_3} & \;\;\mathrm{CH_3}\qquad\qquad\;\; \mathrm{S} - \mathrm{OH}
\end{array}
$$

(arrow labeled HEAT)

CONDENSATION

$$
2 \left(\begin{array}{c} -\,\mathrm{C} \;=\; \mathrm{CH} - \mathrm{CH} - \mathrm{CH_2} - \\ \;\;|\qquad\qquad\qquad| \\ \;\;\mathrm{CH_3}\qquad\qquad \mathrm{S} - \mathrm{OH} \end{array} \right) \xrightarrow{\text{CONDENSATION}}
\begin{array}{c}
-\,\mathrm{C} \;=\; \mathrm{CH} - \mathrm{CH} - \mathrm{CH_2} - \\
\;\;|\qquad\qquad\qquad| \\
\;\;\mathrm{CH_3}\qquad\qquad\;\; \mathrm{SO} \\
\qquad\qquad\qquad\;\; | \\
\qquad\qquad\qquad\;\; \mathrm{S} \\
\qquad\qquad\qquad\;\; | \\
-\,\mathrm{C} \;=\; \mathrm{CH} - \mathrm{CH} - \mathrm{CH_2} - \\
\;\;| \\
\;\;\mathrm{CH_3}
\end{array}
$$

Thus two monosulfidic cross-links are replaced by one cross-link containing two sulfur atoms, and conjugated structures appear in the main polyisoprene chain. Treatment of a TMTD sulfurless vulcanizate of NR with *t*-butyl hydroperoxide, which does not degrade peroxide vulcanizates, was found to reduce the cross-link density and introduce disulfide cross-links. Diene and triene structures in the main chain were identified by UV absorption. According to Cuneen et al. (43, 44) two cross-links are broken for every main chain scission event. It is important to note that the cross-link scission is a consequence of attack by radicals or hydroperoxides involved in the main chain scission. Consequently, if the primary main chain oxidation is prevented, cross-link scission should also cease.

Cuneen's observation of a cross-linking reaction during the scission of monosulfidic cross-links is difficult to reconcile with the close similarity between continuous and intermittent stress relaxation curves for purified TMTD/sulfurless vulcanizates (37, 38).

The work concerning degradation mechanisms described thus far has been conducted using natural rubber but should apply to synthetic polyisoprenes also. Specific studies on other stereoregular elastomers such as polybutadiene, EPM, EPDM, and styrene–butadiene copolymers are rare. The same general type of oxidation mechanism described in equations (a)–(f) will still apply, but account must be taken of the formation of cross-links during aging. Dunn and Scanlan (37–39) discussed the oxidative stress relaxation of peroxide vulcanized *cis*-1,4-polybutadiene extracted to remove antioxidant. The rate of decrease in stress was autocatalytic and was consistent with degradation during one of the steps in the hydrocarbon oxidation cycle. Jaroszynska (45) has made similar observations and both of these publications emphasize a marked cross-linking reaction during aging indicated by the difference in stress relaxation between samples stretched intermittently and those under continuous stress (see Fig. 1). In the case of a carbon-carbon cross-linked network purified by extraction, the cross-linking cannot be attributed to continued vulcanization or to maturing of cross-links. It is known that polybutadiene may be thermally cross-linked at temperatures over 200°C, and a reaction of this type would account for the cross-linking of the polybutadiene vulcanizates in vacuo. However, there is an appreciably greater amount of cross-linking in oxygen (46). The cross-linking of polybutadiene probably occurs via an intermolecular version of reaction (c′):

$$-CH_2 - CH = CH - CH_2 - \; + \; -CH - CH = CH - CH_2 - \; \longrightarrow$$
$$\overset{|}{O_2}$$

$$-CH_2 - CH - CH - CH -$$
$$\overset{|}{O}$$
$$\overset{|}{O}$$
$$\overset{|}{-CH} - CH = CH - CH_2 -$$

which could lead to the "polymerization" of several polybutadiene molecules. Van der Hoff (47) and Loan (48) have shown that during the peroxide vulcanization of polybutadiene many cross-links may be produced following the removal of one α-methylenic hydrogen because the resulting polybutadienyl radical is not stabilized to the same extent as the corresponding radical from polyisoprene and readily adds to double bonds.

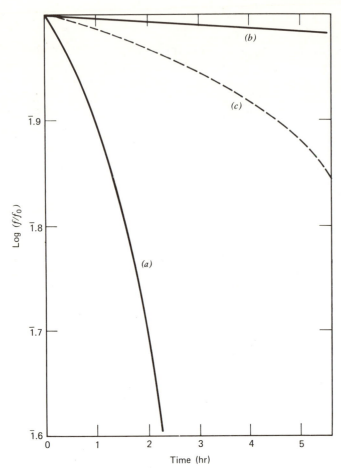

Figure 1 Stress relaxation of an extracted peroxide vulcanizate of polybutadiene in air at 80°C. (*a*) Continuous relaxation; (*b*) continuous relaxation in the presence of 0.2% of zinc dibutyldithiocarbamate or 2,2′-methylenebis(4-methyl-6-*tert*-butylphenol); (*c*) intermittent relaxation (14).

Bevilacqua (22) has discussed the oxidation of butadiene polymers. He infers from the work of Hargrave and Morris (49) that intermolecular addition of peroxy radicals is at least one quarter as frequent as hydrogen transfer. In addition, since the oxidation of diene polymers is pressure dependent up to near atmospheric pressure the concentration of alkenyl radicals is not negligible relative to peroxy radicals. Consequently R· radicals as well as RO_2· radicals may add to double bonds during polybutadiene oxidation and the situation would then be precisely that decribed for peroxide vulcanization.

Little evidence has been found concerning the nature of the scission step during polybutadiene degradation. Bevilacqua notes that if the step were analogous to that in polyisoprene, succinaldehyde should be produced in place of levulinaldehyde. He and other workers were able to detect only one carbon fragments, e.g., formic acid, during polybutadiene oxidation together with acetone and acetaldehyde. Russian workers attributed formic acid formation to oxidation of pendant vinyl groups, which would not produce scission (50):

$$RO_2\text{·} + CH_2 = \overset{\displaystyle |}{CH}- \longrightarrow RO_2 - CH_2 - \overset{\displaystyle |}{CH} - \xrightarrow{\;\; O_2 \;\;}$$

$$RO_2 - CH_2 - \overset{\displaystyle |}{\underset{\displaystyle |}{\underset{\displaystyle O_2}{CH}}}- \longrightarrow RO + HC\overset{\displaystyle ||}{\underset{\displaystyle O}{-}}OH + \overset{\displaystyle |}{CH}\overset{\displaystyle ||}{\underset{\displaystyle O}{-}}$$

However, Bevilacqua found the same yield of formic acid regardless of the concentration of vinyl groups and noted that it was well established as a product of main chain scission in polyisoprene.

Details of the mechanism for EPM and EPDM are also scant. Dunn (51) noted that the oxidation of antioxidant-free EPM at 150°C resulted in the formation of an oil which indicates that degradation is predominant in the raw polymer. Nezu (52) has drawn a similar conclusion regarding the oxidative scission of EPDM (Royalene 200) and notes that the propylene units are more subject to oxidation than the ethylene units. Laud and Stuckey (53) by stress relaxation studies showed that there was negligible cross-linking during the aging of peroxide cured EPDM.

Tobolsky et al. (54) studied the benzoyl peroxide catalyzed autooxidation of EPDM and polypropylene. Their conclusions are dependent on the assumption that the reaction proceeds according to reactions (a)–(f) and that chain termination is by combination of $RO_2\text{·}$ radicals. As stated earlier this will only be so as long as the polymer is free of antioxidant. Tobolsky compared the rate of oxygen absorption with the rate of scission as determined by measurements of continuous stress relaxation, with benzoyl peroxide present as initiator in both cases. He concluded that the rate of oxygen absorption is controlled by the rate of the propagation reaction (c), whereas the rate of scission is controlled by the rate of initiation or termination. He suggests the termination reaction

$$2\left(-CH_2 - \overset{\displaystyle \overset{CH_3}{|}}{\underset{\displaystyle \underset{O_2}{|}}{C}} - CH_2-\right) \longrightarrow 2\left(-CH_2 - \overset{\displaystyle \overset{CH_3}{|}}{\underset{\displaystyle \underset{O}{|}}{C}} - CH_2-\right) + O_2$$

$$-CH_2 - \overset{\displaystyle \overset{CH_3}{|}}{\underset{\displaystyle \underset{O}{|}}{C}} - CH_2 - \longrightarrow -CH_2 - \overset{\displaystyle \overset{CH_3}{|}}{\underset{\displaystyle \underset{O}{||}}{C}} + CH_2 -$$

Since this does not represent a termination step per se Tobolsky envisaged a reaction between primary peroxy radicals derived from ·CH$_2$~~ which Russell (55) showed to be more likely than propagation with this type of peroxy radical:

$$2\left(-CH_2 OO\cdot\right) \longrightarrow -CH_2 OH + -CH = O + O_2$$

The mechanism outlined thus far predicts that no low molecular weight fragments would be formed. Tobolsky's data indicate that some volatiles are produced and he proposed an oxidative unzipping reaction (54):

$$
\begin{array}{c}
CH_3 \\
| \\
\cdot OO - CH_2 - C - CH_2 - \\
| \\
H
\end{array}
\longrightarrow
\begin{array}{c}
CH_3 \\
| \\
\cdot O - CH_2 - C - CH_2 - + \tfrac{1}{2}O_2 \\
| \\
H
\end{array}
$$

$$
\begin{array}{c}
CH_3 \\
| \\
\cdot O - CH_2 - C - CH_2 - \\
| \\
H
\end{array}
\longrightarrow
HCHO + \begin{array}{c}
CH_3 \\
| \\
\cdot C - \\
| \\
H
\end{array}
$$

SIMILARLY:

$$
\begin{array}{c}
CH_3 \\
| \\
\cdot OO - C - CH_2 - \\
| \\
H
\end{array}
\longrightarrow
\begin{array}{c}
CH_3 \\
| \\
\cdot O - C - CH_2 - + \tfrac{1}{2}O_2 \\
| \\
H
\end{array}
$$

$$
\begin{array}{c}
CH_3 \\
| \\
\cdot O - C - CH_2 - \\
| \\
H
\end{array}
\longrightarrow
\begin{array}{c}
CH_3 - CH + \cdot CH_2 - \\
\| \\
O
\end{array}
$$

Miller et al. (56) found formaldehyde and acetaldehyde to be primary oxidation products of isotactic polypropylene.

It must be pointed out that EPM is appreciably more stable to thermooxidation than the polydienes. Vinogradov and Ivanova (57) found no change in viscosity of EPM up to 260°C whereas polybutadienes cross-linked rapidly at 120–150°C. In cis-polyisoprene, degradation occurred at 80–140°C but cross-linking predominated at 160°C.

Singlet oxygen (see also Section 3.4) is an extremely reactive form of molecular oxygen which can be produced from unactivated oxygen by light induced processes or from decomposition of ozone complexes or peroxides (58). It is very reactive with many functional groups because of its excess energy (about 22 kcal/mole above ground state, triplet oxygen). In spite of an extremely short half-life (about 0.1 second in air) it is suspected of being an intermediate in many oxidation processes because of its ability to react rapidly at room temperature with allylmethylene groups to form hydroperoxides. One proposed reaction is

Although singlet oxygen may be less important than ozone because of its comparatively low concentrations it may also be a key intermediate in autooxidation processes because it may be formed from peroxy radicals and contribute to the continuance of the oxidation chain reactions (58).

Much remains to be done to clarify the mechanism of autooxidation of various stereoregular polymers. The details no doubt vary with the microstructure of the polymer. However, it is clear that removal of chain carrying $RO_2\cdot$ radicals and harmless decomposition of polymer hydroperoxide ROOH, which is a potential chain initiator, provide the key to preventing oxidation and thereby halting both degradation and cross-linking.

Let us now consider agents which promote oxidation, and then the means of countering them.

3.3 Heavy Metal Catalysis of Oxidative Degradation

Since stereoregular polymers are likely to contain catalyst residues, the effects of metals such as Co, Ni, Ti, Li, and Al on their oxidative degradation are of the utmost importance. The effect of metals on the oxidation of olefins has been reviewed by Barnard et al. (14) and others (1, 59, 60). The dominant reaction is that proposed by Robertson and Waters (61):

$$ROOH + M^{n+} \rightarrow RO\cdot + OH^- + M^{(n+1)+} \qquad (h)$$

$$ROOH + M^{(n+1)+} \rightarrow ROO\cdot + H^+ + M^{n+} \qquad (i)$$

which is equivalent to a bimolecular decomposition of hydroperoxide

$$2ROOH \rightarrow RO\cdot + ROO\cdot + H_2O \qquad (j)$$

For cobalt ion the first of these steps is rapid and the second is slow. The ease of hydroperoxide breakdown should be controlled by the redox potential of the ion. For metals that cannot exist in two valency states a complex between catalyst and hydroperoxide in which the hydroperoxide is polarized toward homolytic fission is proposed.

The oxidation mechanism in the presence of metals having two valency states can be written

$$ROOH + M^{n+} \rightarrow RO\cdot + M^{(n+1)+} + OH^- \qquad (h)$$

$$ROOH + M^{(n+1)+} \rightarrow RO_2\cdot + M^{n+} + H^+ \qquad (i)$$

$$RO_2\cdot + RH \rightarrow RO_2H + R\cdot \qquad (c)$$

$$RO\cdot + RH \rightarrow ROH + R\cdot \qquad (k)$$

$$R\cdot + O_2 \rightarrow RO_2\cdot \qquad (b)$$

$$2RO_2\cdot \rightarrow \text{inactive products} \qquad (f)$$

If k_c' and k_c are the rate constants for reactions (h) and (i), respectively, and k_2, k_3, k_3', and k_6 the rate constants for reactions (b), (c), (k), and (f), then under stationary conditions for M^{n+} and $M^{(n+1)+}$ and when (RH) is large:

$$\frac{dO_2}{dt} = k_3(RH)[2k_c'k_c''(ROOH)(M)[k_c' + k_c'']^{-1}k_6^{-1}]^{1/2} \qquad [6]$$

In a nonpolar medium the hydroperoxide is likely to be associated with the metal and (ROOH)(M) can be replaced by a single term involving the concentration of the complex.

Uri (59) had suggested that the metal might react directly with oxygen:

$$M^{n+} + O_2 \rightarrow M^{(n+1)+} + O_2^-$$

or might complex with oxygen and subsequently yield initiator radicals. Barnard et al. (14) discounted these suggestions since reaction with oxygen is very fast at high pressure and since catalysis occurs only in those olefins that yield hydroperoxides on autooxidation and not in those that yield polymeric diperoxides (62).

The mechanism proposed for the metal catalysis of autooxidation would suggest that the relative activity of a series of metals should be independent of the substrate and depend only on the redox potential of the metal. This is not so even for simple hydrocarbons (63) and presumably the effect of metals is more complicated than has been suggested here. One possibility is that the catalytic entity is a complex between metal and substrate for which the redox potential is substrate dependent (64) and is a measure of the stability of the complex.

There are indications (65, 66) that transition metal ions, particularly in their lower valence states, may function as antioxidants as well as prooxidants. Ionic copper has been used as a heat stabilizer for nylon and cobaltous ions can inhibit the oxidation of tetralin. Labile metal chelate-oxygen complexes or reactions such as these are suggested:

$$ROO \cdot + Co^{2+} \rightarrow ROO^- + Co^{3+}$$

and

$$\text{\textasciitilde}CH_2\dot{C}H\text{\textasciitilde} + Cu^{2+} \rightarrow \text{\textasciitilde}CH{=}CH\text{\textasciitilde} + H^+ + Cu^+$$

Lee et al. (67) have studied the effect of adding metal stearates to a number of stereoregular polymers including cis-1,4-polybutadiene, solution SBR, EPDM, and propylene oxide rubber. The polymers were used without removal of antioxidant, but were substantially free of heavy metals except iron. The iron may have a greater synergistic effect with some metals than with others which may change their relative order of reactivity, and consequently this work does not provide as fundamental a foundation as desirable for the relative effect of different metals on aging. The parameters studied were the induction time for the metal catalyzed oxidation and rates of oxygen uptake. Rankings of the efficiency of metals in promoting oxidation are given in Table 3. These results serve to indicate the complexity of the situation. In the six examples here Co^{3+} always shows a strong catalytic effect but Cu^{2+}, Fe^{2+}, and Mn^{2+} do not always do so. Ions such as Pb^{2+} and Zn^{2+} are relatively damaging in a few cases.

This study (67) did not refer to the catalytic effect of Ti which is of major interest for stereoregular polymers. Kuz'minskii et al. (68) have added titanium stearate, titanium hydroxide, and titanium dioxide to polyisoprene and measured its activity. At levels of 0.0045%, beyond which the activity of titanium stearate becomes constant, the catalytic activity ranked

$$Cu > Fe > Ti$$

In later work the same group (69) reported that the presence of $Ti(OH)_4$ in lithium polyisoprene stabilized with PBNA or BHT increased the rate of oxidation by 80–150% at low antioxidant concentrations and about 30–65% at higher concentrations (>0.5 phr). The $Ti(OH)_4$ did not appear to form complexes with the antioxidants or to participate in redox reactions with them. It was proposed that the titanium took part by catalyzing peroxide decomposition [reaction (i)].

Rosik (70) found Ti^{2+} had little effect on the induction period of oxidation for cis-1,4-polybutadiene (JSR-BR01) prepared with a Ni

Table 3
A   Ranking of Effect of Metal in Reducing Induction Period (67)

cis-Polyisoprene	(Shell)	Mn > Co > Cu > Fe (Na, Sn, Ni inactive)
cis-Polybutadiene	(Phillips—Cis-4)	Fe ≥ Co > Ce ≥ Cu > Mn ≫ Sn > Na ≥ Pb (Zn, Ni inactive)
Polybutadiene	(Firestone—Diene	Co, Mn > Cu > Fe > Pb ≥ Ce ≥ Zn > Ni, Na (Sn inhibits)
Solution SBR	(Phillips—Solprene)	Mn > Co, Fe > Cu > Sn > Na (others inhibit)
EPDM	(Dupont—Nordel)	Co > Fe > Cu ≫ Mn > Ce ≫ Zn > Ni > Sn > Pb
Polypropylene oxide	(Dow)	Co > Cu > Ce > Sn > Ni > Pb > Na > Zn > Fe

B   Ranking of Effect of Metal on Oxidation Rate

cis-Polyisoprene	(Shell)	Co > Mn > Cu ≥ Zn > Na > Fe, Ni
cis-Polybutadiene	(Phillips—Cis-4)	Fe ≫ Ce > Pb > Sn, Co > Na ≥ Mn > Ni > Zn
Polybutadiene	(Firestone—Diene	Co > Fe > Cu > Mn > Pb > Zn > Ni > Na ≫ Sn (Sn inhibits)
Solution SBR	(Phillips—Solprene)	Co ≫ Ce, Cu > Ni, Mn > Zn > Na > Fe > Sn
EPDM	(Dupont—Nordel)	Co > Fe > Cu ≫ Na, Ni ≥ Mn ≥ Ce > Pb > Sn > Zn (Pb, Sn, and Zn inhibit)
Polypropylene oxide	(Dow)	Mn ≥ Co ≫ Zn > Ni > Pb, Sn > Ce ≥ Fe > Na > Cu (Ce, Fe, Na, and Cu inhibit)

catalyst. He ranked metals in the following order as to their effect in reducing induction period:

$$Co^{2+} > Mn^{3+}, V^{3+} > Ce^{3+} > Cu^{2+} > Ni^{2+} > Al^{3+}$$

Presumably the order of effectiveness depends on impurities present in the polymer being studied. The complicated synergism of the catalytic activity of metals was studied by May and Bsharah in the oxidation of antioxidant-free SBR (71) by measuring the temperature of the DTA peak during oxidation. When a metal was used alone it was at a concentration of 0.1 wt %, but when two were present 0.05% of each was added. Thus any positive or negative change may be regarded as true synergism providing one can neglect the differing atomic weights. The effect of adding a second metal is shown in Table 4. The effects are small but illustrate the confusion that reigns when trace metals are combined. Cobalt always acts as a positive synergist, but copper, iron, and manganese sometimes show a negative effect. Thus they may not always be oxidation promoters. On the other hand, zinc, which is frequently used in rubber compounding, can increase the catalytic activity of cobalt.

The effect of various metal ions on the oxidation of purified *cis*-polyisoprene in solution was studied by Mayo et al. (72). In the absence of AIBN or *t*-butyl hydroperoxide as initiator a long induction period is observed. The lower valence form of the metal often showed a net retarding effect under these conditions, bearing out the theory that metal ions act through decomposition of hydroperoxides rather than promoting direct reaction with oxygen. In the presence or absence of oxygen the addition of metal salts caused decomposition of peroxides and cleavage of chains. Iron does not destroy peroxides as rapidly as cobalt nor does it cause as rapid a chain initiation, yet is causes more chain cleavage. Manganese is more effective than iron in decomposing peroxide but is less effective in promoting scission and starting oxidation chains. Copper slightly accelerates oxidation rate, whereas calcium, zinc, and lead salts have little effect. Nevertheless these metals cause the lowest oxygen uptake per scission event. If the chain length were less in the presence of these metals, this phenomenon would be rationalized.

Table 4 Relative Synergistic Effects of Pairs of Metals as Compared with Metals Used Singly[a]

Original Metal	Ca	Ce	Co	Cu	Fe	Pb	Mn	Vo	Zn
				Added Metal[b]					
Ca		+	+	+	+	0	+	+	0
Ce	0		+	−	−	0	−	0	0
Co	0	−		−	0	0	−	+	+
Cu	0	−	+		−	+	−	+	0
Fe	−	−	+	−		−	−	−	−
Pb	−	+	+	+	−		+	+	+
Mn	0	−	+	−	0	0		0	0
Vo	−	0	+	+	0	−	−		0
Zn	−	+	+	+	+	+	+	−	

[b] +, Promotes; 0, no effect; −, retards.

3.4   Oxidation Initiated by Ultraviolet Radiation

The effect of ultraviolet light on vulcanized rubber is generally less important technologically than its effect on plastics which may be used out of doors as thin, often translucent sheets. Frequently rubber is compounded with high levels of carbon black, which is a most effective UV absorber. In light colored compounds the effect is confined to the surface. However, the crazed and chalky surface resulting from UV degradation is undesirable.

In the UV catalyzed degradation of simple olefins the usual sequence of reactions (a)–(f) occurs but with a unimolecular decomposition of a hydroperoxide as an initiation reaction (13):

$$ROOH \xrightarrow{h\nu} RO\cdot + \cdot OH \tag{1}$$

This initiation step can, in itself, be quite complex, with net rates of oxidation dependent on rubber structure, and radiation intensity and wavelength. Reaction (1) initiates a radical chain autooxidation and generally requires light in the UV range.

Photosensitized oxidations may also initiate such chains (73). These take place in the presence of some phtotosensitive material (dye, impurity, conjugated chain segment) which is excited by lower energy photons in the visible or near UV range. Singlet molecular oxygen excited from its triplet ground state by energy transfer from the excited triplet state (of the photosensitive chromophore) is believed to be the reactive intermediate. The hydroperoxides thus formed generate a conventional degradation chain reaction. Under ordinary circumstances singlet oxygen initiation is less likely to occur than, say, ozone initiated attack (58, 74). In the presence of sensitizing materials or in transparent articles it may assume importance.

Morand (75) found, in addition to the main UV effect, a series of maxima in the oxidation rates at several (near UV) wavelengths of the incident light. These occurred at about the same wavelength for IR, BR, and SBR both in the crude and cured states and were suggested as being due to chromophores in the polymer chains. Morand proposed a mechanism with energy absorption by isolated double bonds via a direct $S_0 \to T$ transition between the ground state and the first triplet state of π electrons even though such a transition is formally forbidden.

Dunn and co-workers (76) showed that the sequence of reactions (a)–(f), (1) applied in the oxidative stress relaxation of highly purified natural rubber under the influence of 365 mμ radiation. As in the case of thermal oxidation (38) the rate of stress decay was autocatalytic and could be accounted for by propagation reaction (b) or (c) of appreciable length but with termination by reaction (g), because of the presence of very small quantities of residual antioxidant. The relationships between stress f at time t and the initial stress f_0 for aging at constant elongation are

$$\frac{f}{f_0} = (1 - a + ae^{kt})^{-1} \quad \text{(photochemical)} \tag{7}$$

$$\frac{f}{f_0 - f} = \frac{a}{t} - b \quad \text{(thermal)} \tag{8}$$

These may be derived assuming unimolecular and bimolecular decomposition of peroxides, respectively.

During irradiation with UV light in the absence of oxygen cis-trans isomerization occurs in both *cis*- and *trans*-polyisoprene and 1,4 double bonds are lost while vinyl and vinylidene double bonds are formed (77, 78). An analogous observation was also made for polybutadiene (79). Rupture of C–C bonds between isoprene units followed by the formation of biradicals was found to be more important than the rupture of α-methylenic C–H bonds. This observation will only be of relevance to oxidative degradation if radicals are formed primarily by any of these routes rather than by degradation of hydroperoxide.

The photochemical degradation of rubber was recently reviewed by Morand (79), who proposed the following overall scheme for the primary processes in polyisoprene photodegradation (see overleaf).

Purified hydrocarbons do not absorb strongly in the near ultraviolet. However, —C=O groups that build up as oxidation proceeds will absorb more readily. Photochemical oxidative degradation of a polybutadiene rubber (Intene 45) resulted in the initial formation of α, β unsaturated carbonyls which were then further degraded (80).

3.5 Inhibition of Thermooxidative Degradation

Stabilization against thermal oxidation has been reviewed recently in considerable detail by Shelton (16, 81), Ingold (82), and Denisov (83). It was pointed out at the conclusion of Section 3.2 that the oxidative degradation and cross-linking of polymers can be inhibited by removing the active chain carrying $RO_2\cdot$ radicals or by harmlessly decomposing the hydroperoxide ROOH [reaction (g)]. Antioxidants must be added to the stereoregular polymers to protect the raw polymer during both finishing and storage. They must also be present in the vulcanizate to prevent aging in service. An extensive listing of stabilizers for raw polymers and antioxidants suitable for use in vulcanizates, together with their properties

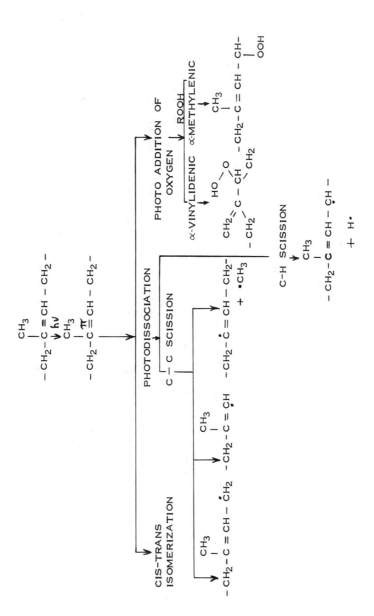

has been given in a review by Ambelang et al. (84). Phenols, amines, and phosphites are the most common types of antioxidant but a complex variety of additives has been suggested in the patent literature [for examples, see Scott (1)].

3.5.1 Phenolic Antioxidants

The simplest examples of radical scavenger are provided by the phenols. EPR studies (85) have confirmed that phenols and amines react with $RO_2\cdot$ radicals formed during the oxidation of NR, SBR, and polybutadiene and do not react with $R\cdot$ radicals. Bickel and Kooyman (86) showed that for hydroquinone each mole of inhibitor reacted with two initiator radicals.

This is an example of how phenolic antioxidants terminate chain reactions by hydrogen abstraction [reaction (g)]. The phenoxyl radical formed can either react with another alkylperoxy radical or dimerize. The dimer can then terminate two more radicals.

A linear relationship exists between the oxidation-reduction potentials of a series of phenols and the logarithm of their relative antioxidant efficiencies (97). However, while necessary, a low oxidation potential is not a sufficient condition for antioxidant activity. Using ESR, Westfahl et al. (87) found that the stability of the formed phenoxyl radical is important. By plotting the relative radical stability as a function of temperature they could rank 12 commercial antioxidants in good agreement with their activity in protecting EPDM. However, the ranking was unsatisfactory for polyisoprene. Further work (88) showed that subsequent secondary radicals resulting in formation of complex quinone methides influenced the antioxidant activity.

Shelton and Vincent (89), studying the oxidation of purified polyisoprene, demonstrated that replacing hydrogen in 2,6-di-t-butylphenol by deuterium resulted in an increase in rate. A similar deuterium isotope effect was found using N-phenyl-2-naphthylamine; such an effect of deuterated amine on the oxidation of SBR had previously been noted by Shelton and McDonel (90).

Howard and Ingold (91) showed that replacing hydroxyl hydrogen in a monophenol by deuterium increases the rate of oxidation of hydrocarbons in solution by a factor of 6–10. From this they concluded that hydrogen abstraction is the rate determining step. They also postulated that the resulting phenoxyl radical **2** subsequently reacts with a second peroxy radical to give peroxycyclohexadienones **3** which were identified as reaction products:

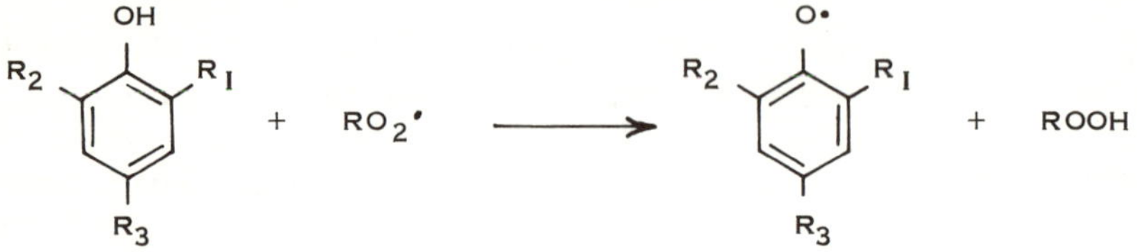

Hammond et al. (92) and other investigators (93, 94) also obtained products of this type, and it is well established that the hindered monophenols deactivate two RO_2· radicals for each molecule. The mechanism by which they do this has been the subject of much debate. Hammond et al. (92), finding the oxidation rate to be first order in peroxy radical and half-order in inhibitor, concluded that formation of a complex between phenol and peroxy radical was rate determining:

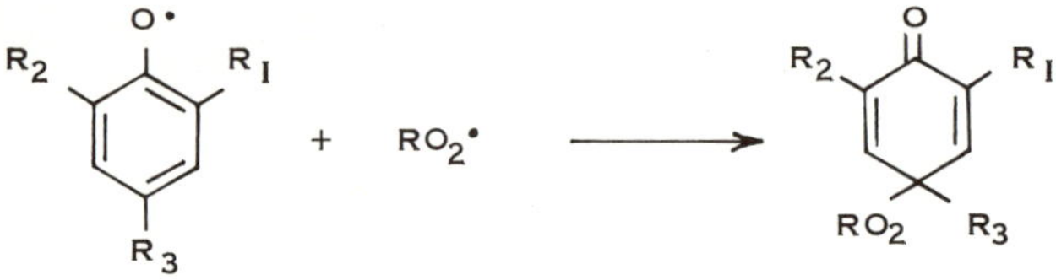

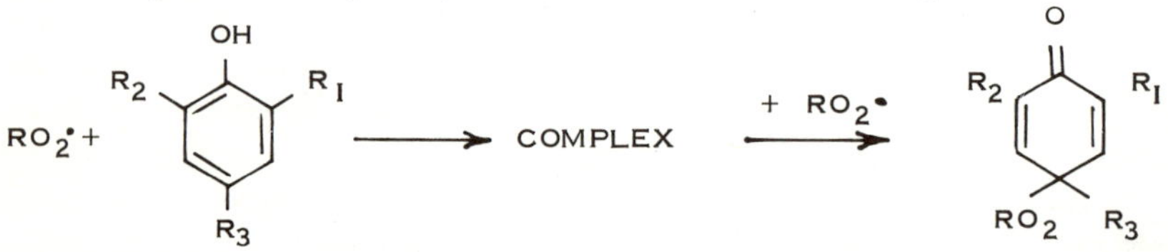

Tsurugi et al. (95) obtained an intermediate in the oxidation of NR or IR vulcanizates inhibited by substituted *p*-phenylenediamines. This could be extracted by a hydrochloric acid-ethanol mixture to yield extracts exhibiting the absorption spectra of Wurster cations of the corresponding amines. They concluded that the intermediate was a charge transfer complex between the amine and a rubber peroxy radical.

Coppinger (96) has reconciled the well-substantiated deuterium isotope effect with Hammond's kinetics by postulating that the complex is, in fact, a charge transfer complex:

$$\left[RO_2:^- \quad \begin{array}{c} OH \\ \text{(+)} \end{array} \right]$$

He attributed the isotope effect to differing contributions by hyperconjugation of hydrogen and deuterium to the stabilization of the charge transfer complex.

The behavior of p substituents in the phenol has been cited as evidence that lability of the hydroxyl hydrogen is important in the inhibition by phenols. Thus Bolland and tenHave (97) connected high antioxidant activity with low redox potential and Howard and Ingold (91) were able to relate it to the Hammett function. Desirable p substituents are alkyl and alkyloxy. Phenols commonly used as antioxidants are usually hindered with CH_3 or t-butyl groups in the ortho positions. In a recent review, Mahoney (98) pointed out that with an unhindered phenol chain transfer reactions of the type

$$RH + A \cdot \rightarrow AH + R \cdot \qquad (m)$$

can occur. Thus unhindered phenols may slow down oxidation but cannot prevent it.

This review also discusses the synergistic behavior of hindered and unhindered phenols, attributed to the regeneration of the nonhindered phenol (AH) by hydrogen abstraction from the hindered phenol (BH):

$$A \cdot + BH \rightleftharpoons AH + B \cdot \qquad (n)$$

The equilibrium in reaction (n) is far to the right since the O–H bond energy of BH is much lower than that of AH. Termination occurs by combination of the stable hindered phenol radicals.

Many of the phenol antioxidants used in stereoregular elastomers are bisphenols rather than the more volatile monophenols. The effectiveness of a number of bisphenols in enhancing the induction period during the air oxidation of white oil was examined by Low (99).

Grinberg et al. (100) compared the effectiveness of alkenebisphenols **4** and **5** with that of thiobisphenols **6** and **7** in promoting retention of physical properties of sulfur vulcanized NR.

4   **5**

6   **7**

They found compounds of type **4** to be more effective than those of type **5**. The preferred antioxidants of type **4** were 2,2-bis(4-hydroxyphenylbutylphenol) and styrenated 2,2-bis(4-hydroxyphenyl-propane). Contrary to Low, they found sulfur bridged compounds to be less effective than their alkene bridged analogs. This may reflect the ineffectiveness of primary sulfur based antioxidants in sulfur vulcanizates, which is discussed in Section 3.5.6. Once again ortho linked bisphenols of type **6** were more effective than para linked bisphenols of type **7**.

Kempermann (101) has recently discussed antioxidants of types **4** and **5** with particular emphasis on color development, which is more marked in bisphenols than in monophenols. He found generally that color development was most severe in the highly effective antioxidants, but this was not a hard and fast rule. He examined the relative effect on physical properties and color of NR vulcanizates of a number of bisphenols. Again antioxidants of type **4** were found to be superior to those of type **5**. 2,2-Methylenebis(4-methyl-6-*tert*-butylphenol) was the most efficient of type **4** but gave rise to strong pink coloration during aging. 2,2-Isobutylidene-bis(4,6-dimethylphenol), which generated much less color, was almost as effective.

O'Shea (102) has found that β,β'-bis(2-hydroxy-3-*tert*-butyl-5-methyl-benzylthio)diethyl ether is superior to conventional mono- and bisphenols for both color and antioxidant effectiveness in aging of both polyisoprene and polybutadiene. 1,2-Bis(3,5-di-*tert*-butyl-4-hydroxybenzylthio)ethane is also potentially useful.

Patel et al. (103) have compared the effectiveness of more complex phenols with that of a typical bisphenol and monophenol in stabilizing unvulcanized EPDM, BR, and emulsion SBR. Compound **8** was five times more effective in stabilizing EPDM at 150°C than either 2,6-di-*tert*-butyl-4-methylphenol or 2,2-methylenebis(4-methyl-6-*tert*-butylphenol).

$$\left(HO - \underset{C(CH_3)_3}{\overset{C(CH_3)_3}{\bigcirc}} - CH_2 - CH_2 - \overset{\overset{O}{\parallel}}{C} - O - CH_2 \right)_4 C$$

8

On the other hand, compound **8** was no more effective in preventing gel formation in polybutadiene at 100°C than the mono- or bisphenols. However, compound **9** was very effective.

$$HO - \underset{C(CH_3)_3}{\overset{C(CH_3)_3}{\bigcirc}} - NH - \underset{S \cdot C_8 H_{17}}{\overset{S \cdot C_8 H_{17}}{\bigcirc}}$$

9

The reasons why one polyphenol is preferable to another in antioxidant efficiency and color development are obviously still not well understood. There is need for a fundamental investigation in the interest of developing nondiscoloring antioxidants of greater potency.

There are a number of reasons why chain breaking antioxidants such as phenols and amines do not prevent oxidative degradation completely. The possibility of a chain transfer reaction (m) has already been pointed out. In addition, the end product of a phenolic antioxidant (compound **3**) is itself a peroxide which can decompose into initiating free radicals at 140°C (104). A direct reaction between antioxidant and oxygen to yield radicals [reactions (o) and (p)] has been recognized for several years, although its importance does not appear to be widely appreciated:

$$AH + O_2 \rightarrow A\cdot + HO_2\cdot \qquad \text{(o)}$$

$$A\cdot + O_2 \rightarrow AO_2\cdot \qquad \text{(p)}$$

Shelton and Cox postulated this mechanism to explain a marked dependence of rate of oxidation of NR vulcanizates on oxygen pressure (105, 106) and the existence of an optimum antioxidant concentration beyond which the oxidation rate increases with increasing antioxidant concentration (107).

Initiation by direct oxidation of antioxidant is also in keeping with the observed decrease in deuterium isotope effect with increasing temperature in the inhibited oxidation of purified polyisoprene (16, 81, 89). At increasing antioxidant level the rate of oxidation becomes greater in the presence of hydrogenated rather than deuterated antioxidants (see Table 5).

Table 5   Relative Initial Rates of Polyisoprene Oxidation (R_D/R_H) in the Presence of Hydrogenated and Deuterated 2,6-Di-*tert*-butyl-4-methylphenol (89)

Temperature (°C)	Inhibitor Concentration (moles/g)	R_D/R_H
60	4.41×10^{-5}	1.76
	13.2×10^{-5}	1.16
75	8.82×10^{-5}	1.56
	13.2×10^{-5}	0.92
90	4.4×10^{-5}	1.27
	13.2×10^{-5}	0.79

The inhibited oxidation of purified polyisoprene was found to proceed in three stages, each with a characteristic rate (9, 81, 89), as is shown in Fig. 2. The extent of oxygen uptake during the initial low rate stage was independent of temperature and inhibitor concentration, and it was concluded that a certain concentration of hydroperoxide needed to be built up before it contributed significantly to initiation. It was proposed

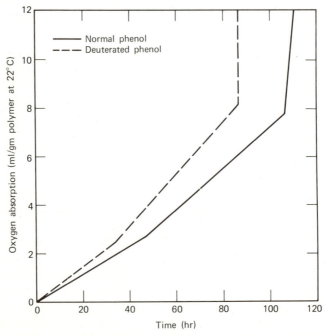

Figure 2   Oxidation of purified polyisoprene in the presence of normal and deuterated 2,6-di-*tert*-butylphenol (89).

that during the initial stage, initiation by direct attack of oxygen on the antioxidant was predominant.

Rosik (108) has also discussed initiation by direct attack of oxygen on antioxidant and suggests the additional reaction

$$HO_2\cdot + RH \rightarrow R\cdot + H_2O_2 \qquad (q)$$

In addition to the direct involvement of chain breaking antioxidants in initiation reactions their primary action is to convert $RO_2\cdot$ into $ROOH$. Consequently, hydroperoxide is regenerated and potentially able to initiate new chains.

3.5.2 Phosphite Antioxidants

A potentially more thorough method of combating oxidative degradation is to remove the $ROOH$. The most widely used of these peroxide decomposing antioxidants in synthetic elastomers is tris(nonylphenyl phosphite). The main use of trisalkylphenyl phosphites is in protection of raw polymers since they are not effective in cured stocks.

The chemistry of the action of phosphite antioxidants is believed to be

$$(RO)_3P + ROOH \rightarrow (RO)_3PO + ROH$$

Despite their popularity the mechanism of action of these stabilizers has not been studied in detail. It is reported that such studies are underway (104).

Synergism wherein a combination of inhibitors produces a greater effect than an equal total amount of either inhibitor used alone can be achieved by combining a radical scavenger (e.g., a phenol) with a peroxide decomposer (e.g., trialkyphenyl phosphite).

3.5.3 Amine Antioxidants

The commonly used amine antioxidants are hydrogen donors and terminate kinetic chains by hydrogen transfer to form $ROOH$ from $RO_2\cdot$ radicals. The newly formed antioxidant radical can terminate a second chain. As with the phenolics a kinetic deuterium isotope effect is observed when the labile hydrogen is replaced by deuterium. The rate of chain termination is slowed and the observed rate of oxidation is increased. Shelton observed this effect on SBR and cis-1,4-polyisoprene (16). At higher temperatures and higher antioxidant levels a reversal in the direction of the isotope effect was observed which was attributed to increased initiation by direct attack of O_2 on the antioxidant.

There is some evidence that amines act both as chain breakers and peroxide decomposers. Thus, Shelton and Cox (107) studied the oxidation of vulcanized rubber and found it necessary to invoke both chain transfer and peroxide decomposition in their reaction mechanism. Capp and Hawkins (109) showed that amines would react with tertiary hydroperoxides, with reactivity decreasing in the order tertiary > secondary > primary, and with aromatic amines being less reactive than aliphatic. The ability of amines to act as a radical trap has been discussed by Furukawa et al. (110). Furukawa showed that there was an excellent correlation between the oxidation potential of an amine and the relative tensile strength (TS)/(TS)$_0$ of a gum natural rubber vulcanizate containing the amine, where (TS)$_0$ and (TS) are the tensile strengths before and after 48 hours oven aging at 100°C. The optimum potential was about 0.4 V, which is close to the oxidation potential of most of the N,N'-diaryl-p-phenylenediamines. The ratio of the viscosity of an SBR solution after 5 hours oxidation at 100°C to its original viscosity was also studied as a function of the oxidation potential of the amine present (see Table 6).

Again an optimum is apparent around an oxidation potential of 0.4 although the results show a good deal of scatter. Amines are accepted as being the most efficient of antioxidants but their use is generally limited to elastomers that are to be black loaded since their reaction products are highly colored.

Table 6   Relationship Between Oxidation Potential and Viscosity Reduction on SBR Aging in Oxygen and Ozone (110)

Amine	Oxidation Potential (V)	Oxygen (η/η_0)	Ozone (η/η_0)
None	—	0.656	—
p-Phenylenediamine	0.18	0.890	—
N-Phenyl-N'-cyclohexyl-p-phenylene-diamine	0.24	0.896	0.920
2,2,4-Trimethyl-1,2-dihydroquinoline	0.47	0.912	—
6-Ethoxy-2,2,4-trimethyl-1,2-dihydro-quinoline	0.33	0.913	0.916
Diphenylamine	0.62	0.915	0.768
N-Phenyl-β-naphthylamine	0.54	0.931	—
N,N'-Diphenyl-p-phenylenediamine	0.35	0.934	0.914
N,N'-Di-β-naphthyl-p-phenylenediamine	0.40	0.988	—
N-Phenyl-p-phenylenediamine	0.22	—	0.886

Lorenz and Parks (111) found that diaryl-p-phenylenediamines were more efficient in preventing oxidation of NR vulcanizates than either alkylaryl- or dialkyl-p-phenylenediamines. They also concluded that the diaryl-p-phenylenediamines participated directly in a chain termination reaction whereas their dialkyl or arylalkyl analogs were readily oxidized and the oxidation products were the active inhibitors.

Tsurugi et al. (95) attributed the superiority of secondary amines over primary or tertiary amines to the ability of the charge transfer complex resulting from them to react with a second rubber peroxy radical.

3.5.4 Sulfur Bearing Antioxidants

Shelton has reviewed the role of several classes of organic sulfur compounds as preventive antioxidants (112). The activity of sulfides and disulfides appears to involve their reaction with hydroperoxides to form sulfoxides and thiolsulfinates which are the actual major contributors to the observed antioxidant effect.

The dithiocarbamates represent one class of additives that are very efficient antioxidants. They are of limited use in unvulcanized rubber since they are vulcanization accelerators. They have been shown to be potent inhibitors of the stress relaxation of peroxide vulcanized natural rubber (37, 38, 111). Zinc dimethyldithiocarbamate is responsible for the excellent aging of vulcanizates cross-linked with tetramethyl thiuram disulfide (113, 114), where it occurs as one of the products of vulcanization. It was shown that dithiocarbamates decompose cumene hydroperoxide in solution (111). Holdsworth and co-workers (115) examined the decomposition of cumene hydroperoxide in the presence of a number of dithiocarbamates and dithiophosphates, and concluded that the active peroxide decomposer is formed by an initial reaction between the metal complex and alkyl hydroperoxide, with quantitative precipitation of metal sulfate.

Copious evolution of SO_2 occurred when cumene hydroperoxide was reacted with zinc dimethyldithiocarbamate; it was attributed to the reaction:

12 decomposes into isothiocyanate and alcohol.

SO$_2$ was proposed as the active peroxide decomposer by the following *catalytic* reaction:

$$\underset{\underset{CH_3}{|}}{\overset{\overset{CH_3}{|}}{C_6H_5-C-OOH}} + SO_2 \rightarrow \underset{\underset{CH_3}{|}}{\overset{\overset{CH_3}{|}}{C_6H_5-C-O^+}} SO_3^- H \rightarrow C_6H_5OH + (CH_3)_2CO + SO_2$$

Scott also proposed that SO$_2$ is the active peroxide decomposer formed during inhibition by thiodipropionates (104, 116).

Marshall (117) showed that at least 20 moles of hydroperoxide were eventually decomposed by 1 mole of dilaurylthiodipropionate. He demonstrated that a number of sulfur-bearing compounds were peroxide decomposers but 4,4-thiobis(3-methyl-6-*tert*-butylphenol) was not one of these.

Brooks (118) has accounted for the formation of metal sulfates as follows:

$$(R_2N-\overset{\overset{S}{\parallel}}{C}-S)_2Zn \xrightarrow{ROOH} (R_2N-\overset{\overset{S}{\parallel}}{C}-\underset{\underset{O}{\parallel}}{S}-Zn-S-\overset{\overset{O}{\parallel}}{\underset{}{}}\overset{S}{\parallel}C-NR_2) \xrightarrow{ROOH}$$

$$\textbf{13}$$

$$R_2N-\overset{\overset{S}{\parallel}}{\underset{\underset{O}{\parallel}}{C}}-S-Zn-\overset{\overset{O}{\parallel}}{\underset{\underset{O}{\parallel}}{S}}-\overset{S}{\parallel}C-NR_2 \longrightarrow ZnSO_4 + R_2N-\overset{\overset{S}{\parallel}}{C}-S-\overset{\overset{S}{\parallel}}{C}-NR_2$$

$$\textbf{14} \qquad\qquad\qquad\qquad\qquad\qquad\qquad\qquad \textbf{15}$$

He proposed that the thiuram monosulfide **15** would react with ROOH to form carbamyl thiocarbamyl disulfide from which dithiocarbamate could be regenerated in the presence of zinc oxide and water.

The radiation cross-linking of rubber is inhibited by dithiocarbamates as well as by phenols (119). This suggests that these additives can act as both peroxide decomposers and radical traps and this internal synergism might account for their potency as antioxidants.

Colclough and Cuneen (120) examined the influence of dithiocarbamates, dithiophosphates, and xanthates on the α,α'-azobisisobutyronitrile catalyzed oxidation of squalene in which hydroperoxide formation is very small. They concluded that these additives were effective by chain termination as well as by peroxide decomposition. Since no active hydrogen was apparent in these additives they proposed chain termination by an electron transfer process:

$$(R_2-N-\overset{\overset{S}{\parallel}}{C}-S)_2Zn + RO_2\bullet \rightarrow (RO_2)^- (R_2N-\overset{\overset{S}{\parallel}}{C}-S)^+ -Zn-S-\overset{\overset{S}{\parallel}}{C}-NR_2$$

Burn (121) found that such a reaction would not account for his observed kinetics in the zinc dialkyldithiophosphate inhibited autooxidation of cumene and postulated hydrogen abstraction from the alkyl group as the radical trapping reaction. He pointed out that phenol produced during hydroperoxide decomposition would itself have an inhibiting effect and lead to autoinhibition. He expressed the view that no readily acceptable mechanism had been advanced for peroxide decomposition by dithiophosphates and pointed out that the nature of the metal is important and must be accounted for.

The use of 1,3-imidazoline-4-thione derivatives as stabilizers was described by Oberster et al. (122) in an attempt to find a stabilizer for 1,3-polydienes which is both effective and nondiscoloring. This is said to be the first completely aliphatic compound that is an effective antioxidant for synthetic rubber. The compound studied in detail was 7,14-diazadispiro(5.1.5.2)pentadecane-15-thione (DDPT). The mechanism of action of this compound was not discussed but it may well be an "internal synergist" since it contains both C=S and NH groups:

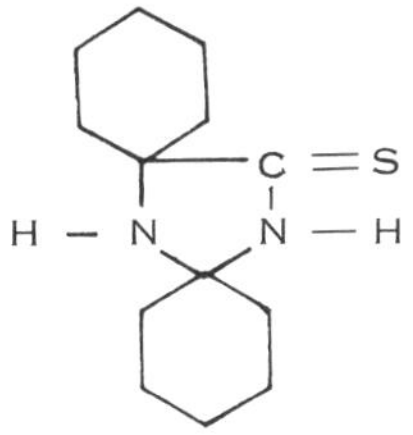

DDPT

16

Data for the stabilization of solution polymerized styrene-butadiene copolymer by DDPT, tris(nonylphenyl phosphite), and 2,6-di-*tert*-butyl-*p*-cresol (DBPC) are compared in Table 7.

Table 7   Stabilization of Solution SBR with 0.5 phr Additive[a] (122)

	Mooney Viscosity		Gel		Color	
	Unaged	Aged	Unaged	Aged	Unaged	Aged
DDPT	63	66.5	0	0	White	White
Tris(nonylphenol phosphite)	68.5	52.0	0	17.7	White	Yellow
DBPC	73	59.4	0	6.7	White	White

[a] Aged 2 days at 100°C.

3.5.5 Antioxidant Volatility and Bound Antioxidants

The volatility of antioxidants presents a serious problem particularly in the stabilization of styrene–butadiene–styrene block copolymers or in those butadiene-containing polymers that are to be used in plastic polyblends and subjected to high temperatures during processing. This problem was studied by Spacht and co-workers and Robinson and Dunn (123) who examined the effect of a number of antioxidants on the Mooney viscosity and gel content of purified SBS block copolymer, *cis*-polybutadiene, and SBR on milling at 160–210°C. Conventional antioxidants such as 2,6-di-*tert*-butyl-*p*-cresol, tris(nonylphenyl phosphite), and dilaurylthiodipropionate were found to be ineffective above 160°C. Bisphenols and *p*-phenylenediamines were more effective at higher temperatures, but nothing was very effective at 210°C. It was noted that radical trapping antioxidants were more effective in preventing cross-linking (gel formation) than the peroxide decomposing types. Thus, it can be postulated that cross-linking arises from attack of $RO_2 \cdot$ (or perhaps $R \cdot$) radicals on polybutadiene double bonds. Such radicals are trapped by hindered phenols.

A concept that overcomes some of the shortcomings of elastomer antioxidants has recently been explored by Cain and others at the MRPRA (124, 125). They noted that antioxidants tend to be leached out of tires by water in prolonged service and pointed out that the effect of leaching and the effect of antioxidant volatility at elevated temperatures could be overcome by binding the antioxidant moiety to the elastomer backbone.

In the patented process which they describe (126), the binding occurs through reaction of a nitroso group with a polymer double bond by the following reaction:

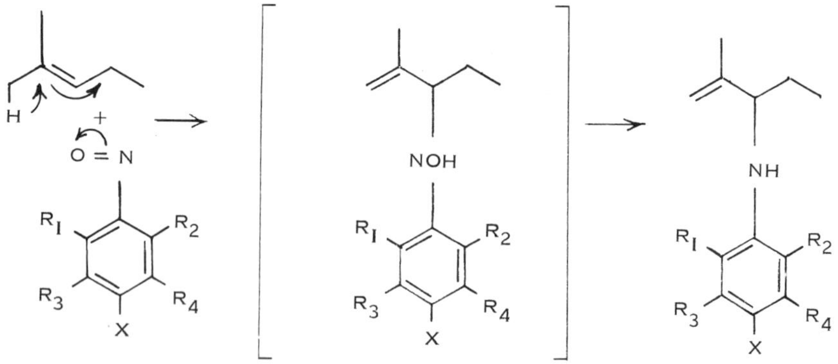

17   18

where R_1 to R_4 are hydrogen or hydrocarbon substituents and X is NH_2, NHR, NR_2, OH, or OR. The mechanism leading to the preferential formation of a substituted p-phenylenediamine or aminophenol (18) instead of a nitrone has been discussed recently by Knight and Saville (127). Reaction is said to occur at temperatures as low as 50°C, but preferably above 100°C. Typical nitroso compounds which may be bound to natural or synthetic rubbers during vulcanization are

N,N-dimethyl-p-nitrosoaniline	(DMNA)
N,N-diethyl-p-nitrosoaniline	(DENA)
p-nitrosodiphenylamine	(NDPA)
p-nitrosophenol	(NP)
2-methyl-4-nitrosophenol	(2-MNP)
3-methyl-4-nitrosophenol	(3-MNP)

NDPA is the preferred additive.

The fact that the antioxidant is chemically bound is illustrated by the effect of methanol/acetone/chloroform azeotrope extraction on the time (T_1) to absorb 1% w/w of oxygen in vulcanizates of various elastomers (124). The relevant examples are given in Table 8. It was also shown that the oxygen uptake after water extraction was about half as rapid in the presence of bound antioxidant as in the presence of conventional antioxidant. As with conventional antioxidants, the bound amines are appreciably more effective than bound phenols. Thus in a CBS vulcanized NR, one part of NDPA extended the time to 1% absorption of oxygen at 100°C to 48 hours before extraction and 59 hours after. In the presence of one part of p-nitrosophenol, the time to 1% absorption was 30 hours before extraction and 25 hours after. 2-Methyl- and 3-methyl-4-nitrosophenol behaved similarly to nitrosophenol itself.

Table 8   Efficiency of Bound Antioxidants (124)

		Time (T_1) To Absorb 1% Oxygen	
Polymer	Antioxidant	Unextracted	Extracted
NR	Bound NDPA	60	53
	Conventional IPPD	47	4
cis-Polyisoprene	Bound NDPA	59	54
	Conventional IPPD	69	10
Polybutadiene	Bound NDPA	25	31
	Conventional IPPD	25	11

Certain disadvantages to the bound antioxidant system are still to be overcome:

1. The products are colored and a water white bound antioxidant remains desirable.

2. Although the mobility of the antioxidant has been reduced, some can still be extracted and staining has not been completely eliminated.

3. The nitroso additives can affect the scorch stability of compounds, and the use of retarders with them has been recommended.

4. The nitroso additives act as peptizers if mixing is too prolonged.

Further activity in the field of bound antioxidants can be expected. The binding of PBNA to epoxidized polybutadiene to produce an effective antioxidant has been described (128).

Cain et al. (129) have suggested methods of synthesizing novel derivatives of *p*-phenylenediamine which involve the reaction of *p*-nitrosoanilines with low molecular weight olefins having a hydrogen atom on the α-methylenic carbon. Presumably high molecular weight phenolic antioxidants could be prepared in the same way using *p*-nitrosophenols. In fact, high molecular weight antioxidants of superior efficiency may be all that is required. Both Lorenz and Parks (111) and Tsurugi et al. (95) indicated that antioxidant moieties can become bound to rubber during aging.

Attempts to copolymerize monomers containing antioxidant moieties have so far been unsuccessful in solution polymerization systems.

3.5.6 Interaction Between Antioxidants and Vulcanizate Networks

A complicating factor in the protection of elastomer vulcanizates by antioxidants is the interaction between the vulcanizate network and the additives. It was shown by Dunn et al. that dithiocarbamates strongly inhibit the oxidative stress relaxation of peroxide vulcanized NR (37, 38) and of peroxide vulcanized polybutadiene (37). Nevertheless, dithiocarbamates have no protective action when introduced by swelling into conventional CBS accelerated NR vulcanizates or CBS accelerated polybutadiene vulcanizates of high sulfur (2.5%) content. Furthermore, α-naphthol is less effective in conventional CBS cures than in peroxide cures. This difference in behavior was attributed to an interchange reaction between dithiocarbamate and polysulfide. There is little doubt that the problem lay with the polysulfidic cross-link, since in efficient vulcanization systems which contain only mono- and disulfide cross-links (37), and in systems from which polysulfidic sulfur had been removed by reaction with triphenylphosphine (130), dithiocarbamates added by a swelling technique are excellent antioxidants.

Parks and Lorenz (131) examined oxygen uptake of diphenylguanidine accelerated sulfur (highly polysufidic), efficient mercaptobenzothiazole-sulfur, and peroxide vulcanizates in the presence of phenyl-β-naphthylamine antioxidant and found oxygen uptake to be greatest in the polysulfidic network. They attributed this to initiation of oxidation by perthiyl radicals formed on polysulfide scission and to activation of hydrogen atoms by cyclic sulfides. Lal (132) has made similar observations and shown that removal of polysulfides by treatment with triphenylphosphine results in reduced rates of oxidation.

The situation has been clarified by Bell and Cuneen (133) who studied the oxidation of protected and unprotected vulcanizates of highly purified NR and unpurified RSS1 (No. 1 Ribbed Smoked Sheet). They confirm that a polysulfide network (unaccelerated sulfur or CBS accelerated sulfur) in the presence of N-isopropyl-N'-phenyl-p-phenylenediamine and that made from RSS1 absorb oxygen more rapidly than TMTD/sulfurless or an efficiently vulcanized system and much more rapidly than a peroxide vulcanized system. On the other hand, in the absence of antioxidant, a purified NR containing an extracted TMTD/sulfurless network oxidizes rapidly and autocatalytically and a peroxide network oxidizes more slowly, but also autocatalytically. This difference is attributed to catalysis by conjugated dienes and trienes in the TMTD system. The polysulfidic (CBS) network initially oxidizes rapidly, but oxidation slows down after absorption of 0.5% oxygen, undoubtedly due to antioxidant effects of polysulfide oxidation products such as were observed using model compounds in squalene (14). A fundamental difference appears when antioxidant (natural or added) is present together with polysulfide.

Bell and Cuneen studied this further by adding a model trisulfide (**19**):

$$\begin{array}{c} \mathrm{C} \\ \diagdown \\ \qquad \mathrm{C} = \mathrm{C} - \mathrm{C} - \\ \diagup \\ \mathrm{C} \qquad\qquad\quad | \\ \qquad\qquad\qquad \mathrm{S}_3 \\ \mathrm{C} \qquad\qquad\quad | \\ \diagdown \\ \qquad \mathrm{C} = \mathrm{C} - \mathrm{C} - \\ \diagup \\ \mathrm{C} \end{array}$$

19

to RSS1 and purified rubber vulcanizates. In a peroxide cure of purified rubber, the trisulfide acted as an antioxidant—as it does in squalene. In RSS1, it accelerated autooxidation, and in both it markedly accelerated oxidation in the presence of conventional antioxidant. It thus appears that polysulfides (and to a lesser extent monosulfides) are antagonistic to conventional antioxidants. Model studies using saturated trisulfides suggest that these are not antagonistic to antioxidants. Thus the ideal situation is a vulcanizate with fully saturated polysulfide cross-links.

Bell and Cuneen do not suggest the mechanism by which antioxidants and polysulfides interact. (The reaction of antioxidants and polysulfides outside of rubber has not been shown.) Elucidation of this, as well as a fuller investigation of this antagonistic effect in other rubbers, would be of value to the development of improved aging. Meanwhile, if additives are to be effective, it is imperative to strive for vulcanizates that are free of unsaturated polysulfides.

Lorenz et al. (134) made the significant observation that in peroxide vulcanizates of polybutadiene to which diphenyl-p-phenylenediamine (DPPD) had been added, as much as 17.5% of the corresponding diimine was present at one stage of the oxidation, whereas <3% of the diimine was found in sulfur containing vulcanizates. They also noted that this diimine reacted with cis-polyisoprene to regenerate the diamine thus:

$$H_5C_6-N =\!\!\!\left\langle\!\!\bigcirc\!\!\right\rangle\!\!= N-C_6H_5 \; + \; -CH_2-\underset{\underset{CH_3}{\mid}}{CH}= CH-CH_2-\cdots \longrightarrow$$

$$H_5C_6-NH -\!\!\!\left\langle\!\!\bigcirc\!\!\right\rangle\!\!- NH-C_6H_5 \; + \; -CH= \underset{\underset{CH_3}{\mid}}{C}-CH= CH-$$

Lorenz et al. believed that the diimine could not be formed in a termination reaction, since it rarely appears in sulfur vulcanizates. However, another explanation could be that the diimine is formed in all cases, but that it reacts with polysulfides and is therefore not available to regenerate the diamine.

Lorenz et al. (134) also compared the oxidizability of TMTD/sulfurless cured cis-polybutadiene and cis-polyisoprene which contain similar amounts of bound sulfur. In the presence of DPPD, they found an optimum concentration of antioxidant which produced a minimum initial rate of oxidation. This minimum initial rate was one-fourth as much in polybutadiene as in polyisoprene (Fig. 3). It is interesting that the optimum level for DPPD was under 0.1 phr in polybutadiene and 0.3–0.35 phr in polysioprene. No explanation of these observations was offered. Much greater levels are normally added in practice to make sure that an effective quantity is present. Lorenz and co-workers suggest that the increased oxygen absorption at the higher levels is due to oxidation of the DPPD and not of the rubbers.

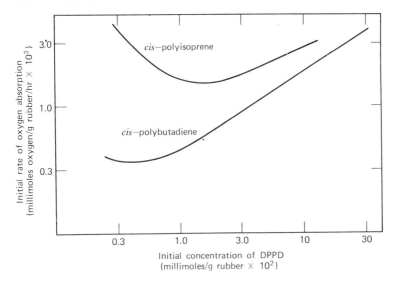

Figure 3   Effect of *N,N'*-diphenyl-*p*-phenylenediamine concentration on initial rate of oxygen uptake at 100°C by TMTD vulcanizates of *cis*-polyisoprene and *cis*-polybutadiene (134).

A comparison of aging of polyisoprene and polybutadiene was also made by Japanese workers (135). They concur that, in sulfur-accelerator vulcanizates, the continuous stress relaxation of NR vulcanizates is much more rapid than in those based on polybutadiene. However, they suggest that scission and oxidation of peroxide cured polybutadiene are more rapid than those of peroxide cured polyisoprene.

Borzenkova et al. (136) found that 2-mercaptobenzothiazole (MBT) in the presence of phenolic antioxidants was a weak inhibitor of the oxidation of SKI-3, *cis*-1,4-polyisoprene. Since the effect was dependent on the ratio of MBT to the phenolic it was conjectured that, at certain mole ratios, hydrogen bonds formed which inhibited mobility of the hydroxylic hydrogens and reduced their effectiveness.

3.6   Inhibition of Metal Catalyzed Oxidative Degradation

Lorenz et al. (134) allude to the importance of removing metallic catalyst residues or converting them to an inactive form. Chalk and Smith (63), in an extensive study of catalyzed cyclohexene autooxidation, showed that catalysis by cobalt or iron is inhibited by chelating agents with six or more functional groups sterically capable of filling the coordination shells, whereas it can be activated by chelating agents with less than six functional groups. Consequently, the chelating agent must be chosen with care to avoid doing more harm than good. Although it is preferable to remove metallic ions from the polymer altogether, for example by coagulation of the cement in alcohol or water containing a soluble complexing agent such as aminoacetic acid (137), it may be necessary to add complexing agents to the final compound. Amines are believed to act as metal deactivators as well as radical trapping antioxidants. Kuz'minskii et al. (138, 139) studied complex formation between copper or iron atoms and amines. The number of atoms of metal bound per molecule of amine was found to depend both on the metal and the amine.

Lee and co-workers (140) made a specific study of metal deactivators in polybutadiene. They added 0.1% of either iron or copper stearate to lithium catalyzed polybutadiene in the presence of oxamide, AGERITE HP (a mixture of phenyl-β-naphthylamine and diphenyl-p-phenylene-diamine), tetramethyl thiuram disulfide, and zinc dimethyldithiocar-bamate (ZDC). They found that ZDC was the best metal deactivator for both iron and copper; the only other additive tested which was useful was tetramethyl thiuram disulfide. The copper sequestering effect of dithio-carbamates is in keeping with the observation of Dunn and Scanlan (113) that copper dithiocarbamate is an outstanding antioxidant for peroxide vulcanized NR. Unfortunately, copper dithiocarbamate is deep brown and other dithiocarbamates are strongly colored also. The dithio-phosphates have been found to be similarly effective, and these cause less discoloration.

Lee et al. (67, 140) noted that metal catalysis can also be promoted by addition of stearic acid to rubber already containing stearates. This was attributed to formation of stearates from metals already present in the rubber and subsequent synergism between stearates of two different metals.

3.7   Inhibition of Ultraviolet Catalyzed Oxidative Degradation

The prevention of polymer degradation by UV radiation has been the subject of many reviews (e.g., 141–143), but the emphasis is frequently on the UV stabilization of polyolefins, where the problem is more acute. Cicchetti (144) has covered the subject very thoroughly and cites the following methods for achieving light stabilization: (*a*) absorption of most of the UV by additives (UV absorbers) and subsequent harmless dissipa-tion of the energy; (*b*) quenching of the excited state of the polymer molecule by transfer of electronic energy to a convenient acceptor; (*c*) inhibition of the propagation of autooxidation by phenols and amines; and (*d*) decomposition of peroxides (e.g., by dithiocarbamates), since it is peroxide decomposition which starts new chains.

Conventional antioxidants and most dithiocarbamates are poor UV stabilizers on their own, but have a synergistic effect in the presence of UV absorbers (145). Amine antioxidants have an initial catalytic effect on UV degradation by 365 mμ light (146). The onset of inhibition coincides with the formation of intense color, at which stage the amine reaction products presumably act as UV absorbers. Thus it appears that if the electronic energy is not removed in some harmless way, oxidation is too rapid for conventional antioxidants to handle.

An ideal UV absorber should absorb efficiently in the UV, but not at all in the visible, so that it introduces no discoloration. It should dissipate all the absorbed energy as heat. The most effective compounds known which satisfy these conditions are

2′-HYDROXYBENZOPHENONES

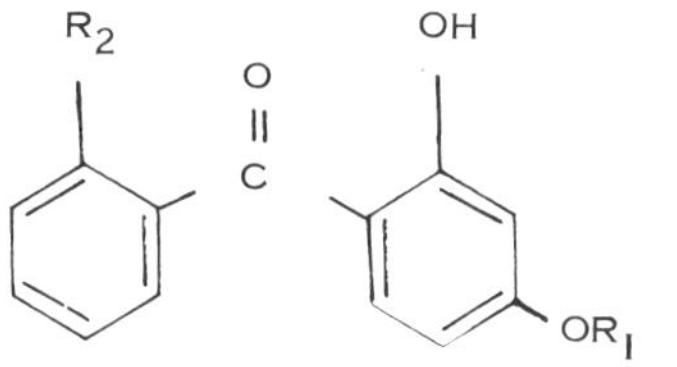

20

and

2′-HYDROXYBENZOTRIAZOLES

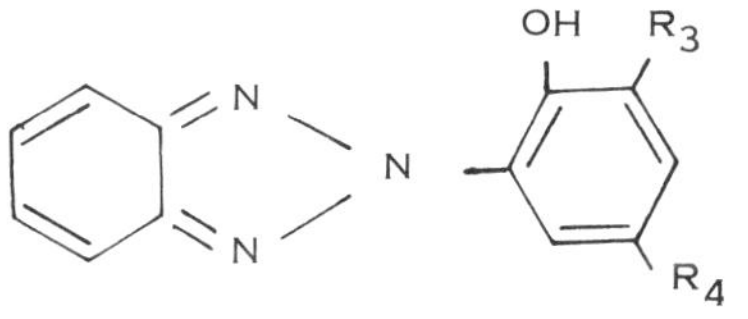

21

Both classes of additives have been effective in protecting gum vulcanizates of NR against UV light, especially in the presence of hindered phenols (145). The 2-hydroxybenzotriazoles were the most effective, but since they are expensive and are required in relatively large quantities they are not widely used in rubber.

The key to the activity of these UV absorbers is their ability to form an internal hydrogen bond (144) which permits energy dissipation by the following sequence of reactions (see overleaf).

The possibility of UV stabilization by quenching was discussed by Heskins and Guillet (147) who examined the UV degradation of ethylene-carbon monoxide polymers in the presence of 1,3-cyclooctadiene (COD). They found that COD exclusively quenched excited triplet states, which were involved in only 45% of the reaction. It has been suggested that UV protection by certain metal (particularly nickel) chelates is due to quenching of excited singlet and triplet states (144).

One of the simplest chelates is nickel dibutyldithiocarbamate which is effective against UV degradation of natural rubber (113, 145), as is nickel isopropylxanthate (148). Other isopropylxanthates, particularly that of chromium, also protect, but none of these systems is as effective in rubber as a hindered phenol-UV absorber combination and none is improved by the addition of a UV absorber (145). No study was made of a possible synergistic effect of hindered phenol with quencher, although such an effect is not unlikely.

4 OZONE DEGRADATION

4.1 Mechanism of Ozone Cracking

In 1926, Ira Williams (149) noted that "cracks appeared on all sides" of a stretched rubber sample exposed to ozone and that light was not necessary to cause this cracking. He also observed that "unstrained rubber is not affected by ozone." In 1931, Van Rossem and Talen (150) provided proof that ozone alone could give rise to cracking on outdoor exposure by exposing samples only at night and finding that cracking occurred in the absence of light. In fact, they found less cracking on direct exposure to light.

As in the case of oxidative aging, the amount of chemical unsaturation of the polymer itself primarily determines whether or not it is prone to ozone cracking. The fact that a polymer is stereoregular has little to do with its ozone resistance; most of the observations described here have been made on natural rubber, but are believed to be generally applicable.

Ossefort (151) has pointed out that chemical saturation is a necessary, but not sufficient, condition for inherent ozone resistance "since polysulfide rubbers and certain polyurethanes exhibit ozone cracking." Thus polyisoprene, polybutadiene, and SBR are subject to ozone degradation because of their unsaturated backbone. EPDM, on the other hand, has no backbone unsaturation and is not attacked by ozone (152).

Williams' statement that "unstrained rubber is not affected by ozone" is now known to be false. Erickson et al. (153) determined ozone absorption by SBR vulcanizates in the unstressed state and showed that it was initially rapid, but soon declined to a low value. Tucker (154) made a similar observation when studying the ozone uptake of open cell sponge of natural rubber or polychloroprene. He also found that no change in the infrared pattern could be seen when rubber films were laid down on salt crystals and exposed to an atmosphere containing 150 ppm ozone for 16 hours.

Electron microscope studies by Andrews and Braden (155, 156) showed that an oily layer of degraded polymer 10^{-5} cm thick is formed in unstrained, unprotected samples. This layer is sufficient to prevent attack by ozone on the underlying rubber so long as it is not strained. Thus, the surface of a sample exposed for 20 minutes is merely roughened whereas that of a sample exposed to ozone for only 1 minute under 50% strain shows incipient ozone cracks.

The protective effect of this thin surface film was demonstrated by Erickson et al. (153) in their study of the ozone absorption of SBR (see Fig. 4). If an unstrained sample already exposed to ozone and rested in the relaxed state was reexposed to ozone, the uptake rate was negligible. If the sample was stretched to break up the surface film, subsequently relaxed, and then reexposed to ozone, rapid uptake of ozone occurred until the protective film had reformed.

Salomon and van Bloois (157) have summarized the effect of strain on ozone cracking and identified the following four stages:

1. At very low stress there is no cracking.
2. Above a critical stress (158) cracking becomes very sensitive to

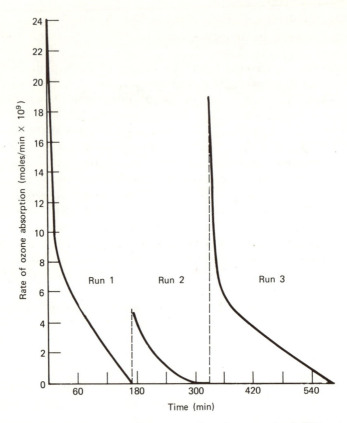

Figure 4   Rate of ozone absorption versus time for unstrained SBR vulcanizates (samples stretched between runs) (153).

increasing strain, and in this region the rate of crack growth increases exponentially.

3.   Above a certain critical strain the increase in crack growth rate ceases.

4.   At very high strain, cracking occurs almost explosively fast.

Newton (159), in a review of ozone degradation, noted that cracks appear in vulcanizates held at extensions of the order of 10%. At these low extensions there are few cracks, but they grow large and deep and represent   the most damaging effect of ozone. At higher extensions the number of cracks is larger but due to mutual interference and stress relaxation the cracks do not go deep into the rubber and so are less damaging.

The concept of a *critical stress* was developed by Braden and Gent (160). They found that below the critical stress a small cut in a rubber strip did not grow visibly during ozone exposure whereas above this critical stress the rate of growth was constant over a wide range of stress.

Braden and Gent (161) found that the critical stress is dependent on the length of the razor cut but not on the sample thickness, the ozone concentration, or the temperature. They showed that critical stress increased in proportion to the square root of Young's modulus in NR vulcanizates with varying degrees of cross-linking. The details in their various discussions of this work differ slightly, but the conclusion is clear—the critical property governing onset of cracking is the stored elastic energy in the test piece.

The factors affecting the rate of growth of a single cut once the critical stress had been exceeded were also examined by Braden and Gent (160). For rubbers prone to ozone attack, such as NR and SBR, growth rate was linearly dependent on ozone concentration and proceeded at a constant rate across the test piece. Growth rate was substantially independent of applied stresses, until stress approached levels sufficient to induce tearing in the absence of ozone. The same rate of growth was observed whether or not the crack was initiated by a razor cut, but without the initial cut it was necessary to employ a larger tensile load in order for the cut to form and grow. Neither black nor white fillers had an appreciable effect on the rate of crack growth. Some influence of impurities in the rubber and of products of vulcanization was detected by Braden and Gent. Cut growth rate in a dicumyl peroxide vulcanizate was found to be 0.08 mm/min, whereas that for a sulfenamide accelerated sulfur cure was 0.22 mm/min. Growth rate in the peroxide cure became comparable to that of the sulfur vulcanizate after hot acetone extraction. The effect was traced to the presence of oleic acid in the NR which remained in the peroxide vulcanizate. Vulcanization with tetramethyl thiuram disulfide also resulted in a decrease to one-third the original rate of cut growth. This was attributed to the presence of zinc dimethyldithiocarbamate as one of the reaction products resulting during TMTD vulcanization. The magnitude of the effect of materials such as oleic acid and zinc dimethyldithiocarbamate on the rate of ozone cracking is such that they are of limited practical use as additives. The observed decrease in crack growth rate with increasing degree of cross-linking (160) is also not sufficient for it to be used as a means of preventing ozone cracking.

While extracted vulcanizates of NR or SBR were found to exhibit similar rates of ozone cracking at room temperature, butadiene-acrylonitrile copolymers and butyl rubber were found to crack at only one-tenth the rate. Braden and Gent attributed this to differences in segmental

Table 9   Rates of Ozone Crack Growth at Different Temperatures[a] (160)

Base Polymer	Cut Growth (mm/min) at		
	2°C	20°C	50°C
NR	0.15	0.22	0.19
SBR	0.13	0.37	0.34
NBR	0.004	0.04	0.23
Butyl	—	0.02	0.16

[a] Ozone concentration 1.15 mg/liter.

mobility, since they found that vulcanizates of all four polymers approached similar rates of crack growth at elevated temperatures, as is shown in Table 9. They also showed that a crystalline polymer (e.g., gutta-percha) showed no cracking and that plasticization enhanced the rate of ozone cracking.

This work has been explored further by Gent and Hirakawa (162) who agree that the growth rate is quantitatively related to segmental mobility in both SBR and butyl rubber. Diffusion of ozone into the butyl rubber before reaction is the rate controlling step whereas in SBR there is less dependence of rate on ozone concentration and rates are somewhat higher at equivalent mobilities. This difference will likely apply in other cases where polymers with low and high levels of main chain unsaturation are compared.

As might be expected, cracking under the more practical circumstances of exposure of a stretched surface to ozone is more complex than the growth of a single crack. This has been discussed by Lake (163) who points out that the critical energy for crack growth can be attained at observable imperfections of about 10^{-3} cm in length, such as occur in well molded surfaces, at a 5% strain. As the strain is increased, the critical energy is attained at smaller and smaller flaws and, consequently, the number of cracks increases. At the same time the dimensions of the individual cracks decrease due to elastic interference between adjacent cracks. The rate of crack growth perpendicular to the surface was found to increase with ozone concentration, as in the case of single crack growth, but to be a maximum at the critical strain and to decrease as strain was increased. In addition, Cheetham and Gurney (164) demonstrated that the rate of attack increased as the velocity of the ozone carrying airstream increased.

Lake (163, 165) was able to explain these observations by a boundary layer theory. The layer of gas adjacent to the rubber surface will be depleted of ozone because of its interaction at sites at which the critical energy has been attained. With increasing strain more and more of such sites will exist and the available concentration of ozone at the surface will decrease. The process becomes controlled by the rate of diffusion of ozone to the surface. If the velocity of the ozone carrying air is increased, the availability of ozone at the surface is increased.

4.2 Prevention of Ozone Cracking

As a result of their investigations Braden and Gent (166) concluded that ozone degradation of elastomers could be minimized either by reducing the rate of crack growth or by increasing the critical condition. They suggested as means of achieving a reduced rate: (a) increasing the chemical resistance of the polymer to ozone attack; (b) increasing the internal viscosity of the polymer; (c) increasing the degree of cross-linking; and (d) including certain additives which retard crack growth in the mix formulation.

As means of increasing the critical condition they suggested; (a) employing an improved surface finish; (b) use of stiffer vulcanizates in constant stress applications and softer vulcanizates under constant strain; and (c) increasing the characteristic energy required for crack formation.

Braden and Gent (166, 167) showed that an increase in the critical energy could be achieved by the use of N,N'-dialkyl-p-phenylene-diamines but not by the diaryl- or the arylalkyl-p-phenylenediamines, nor by other types of antiozonants. N,N'-Dialkyl-p-phenylenediamines also reduce the rate of crack growth. Consequently, two mechanisms of protection probably apply in this specific case, a fact that does not seem to have been clearly appreciated in later investigations.

The increase in critical energy has been cited by Andrews (168) as the means by which EPDM acts as an antiozonant in unsaturated rubbers. The growth of a crack was believed to be interrupted at the EPDM interface and could only be reinitiated on the far side of the inert particle at high stresses. On the microscopic scale this results in a marked increase in critical stress. In practice this phenomenon is applied by the use of small quantities of EPDM as a nonstaining antiozonant in blends with NR, SBR, or polychloroprene (169–171). SBRs containing both EPDM and antiozonants are said to be especially resistant to attack (172).

4.2.1 Methods of Reducing Rate of Ozone Crack Growth

Crack growth can be reduced either by physical or chemical means. Physical protection is achieved by coating the surface of vulcanized rubber by a film of wax or oily material impervious to ozone. This can be achieved by painting a film on the surface but, more commonly, blends of paraffin and microcrystalline wax are added to the rubber compound. Best and Moakes (173) showed that the protection arises from a film which forms on the rubber surface when the solubility of wax in the rubber is exceeded. Ferris et al. (174), following a detailed investigation of the properties of fractionated waxes and an attempt to tailor antiozonant waxes, concluded that identification of such materials was more an art than a science. Protection by waxes is limited to static applications since the rate of diffusion is too low to enable a fresh wax film to form before ozone attack begins.

Among chemical agents which reduce rate of ozone crack growth the arylenediamines remain the most effective (176). These suffer from being highly colored and staining, but there is continued interest in the mechanism of their action because there is still no equally effective nonstaining material. The use of certain substituted dithioureas as nonstaining antiozonants has been reported but these are far less effective than the staining types. Oleic acid has been shown to reduce the rate of growth of a single crack in NR, and other carboxylic acids were found to have some effect also (175). In the same paper it was shown that zinc dithiocarbamates, particularly those of low molecular weight, reduced the rate of crack growth, as did 6-ethoxy-2,2,4-trimethyl-1,2-quinoline. These materials and the arylenediamines all appeared to reduce the rate of crack growth to about the same extent—fivefold at best. Braden and Gent discounted direct reaction with ozone as the mode of action since such diverse materials would not likely react with ozone at comparable rates. They noted that many of the materials used were antioxidants and hazarded that they might prevent oxidation reactions subsequent to molecular scission by ozone.

Andrews and Braden (156) studied the surface of natural rubber samples containing oleic acid using an electron microscope. They concluded that, whereas the ozonized surface of unprotected rubber is practically fluid, in the presence of oleic acid it is of low molecular weight but sufficiently coherent to support appreciable strain. This layer is still subject to ozone attack but at a slower rate than the rubber itself.

A detailed study of a highly effective antiozonant, N-phenyl-N'-isopropyl-p-phenylenediamine (PIPP), which, according to Braden and Gent (167) acts by reducing the rate of cracking rather than by increasing critical energy, has been made by Razumovskii and Batashova (177). They discounted the theory that an ozone-antiozonant reaction product was the true protective agent since addition of this to a polybutadiene film had no influence on molecular weight changes in ozone. They did not determine whether the concentration of the antiozonant-ozone reaction product used was representative of that present in the surface layers of a vulcanizate, and thus this conclusion is open to question. Working in CCl_4 solution they made the surprising observation that the relative reaction rate between PIPP and ozone is 50–100 times larger in the temperature range -78 to $-20°C$ than that of hex-1-ene and ozone; they drew a similar conclusion for the comparative rates of attack on PIPP and polybutadiene in solution. In later work Razumovskii (178) found a relationship between the rate constant of reaction with ozone and the ability to prevent crack growth in vulcanizates for a variety of antiozonants. This very potent "scavenging" effect of PIPP and other antiozonants would provide a simple explanation of their action in reducing the rate of cut growth but is not consistent with the fact that the rate of diffusion of antiozonant in rubber is too low by a factor of 10^6 to replenish that removed by reaction with ozone (179).

A mechanism for antiozonant action put forward by Loan and associates (180) is in keeping with a number of known facts. Unfortunately, many of the facts were established using antiozonants which affect only the rate of cracking, whereas Loan et al. discuss their mechanism in terms of reduction of critical stress which is not affected by most antiozonants (167). Their mechanism can be used effectively to explain the influence of antiozonants on the rate of crack growth, although it leaves some questions unanswered.

Loan et al. base their theory on the mechanism for attack of ozone on simple olefins propounded by Criegee (181, 182) namely

$$\begin{array}{c} \diagdown \\ C \diagup \end{array} = \begin{array}{c} \diagup \\ C \diagdown \end{array} + O^3 \rightarrow \begin{array}{c} \diagdown \quad \diagup \\ C - C \\ \diagdown \; \diagup \\ O^3 \end{array} \longrightarrow$$

22

23   **24**

$$\downarrow$$

$$- \left[C - O - O \right]_n -$$

25

The initially formed molozonide **22** decomposes well below room temperature to give the zwitterion **23** together with a carbonyl compound. These may subsequently react to give such entities as polymeric peroxide **24** and normal ozonide **25**.

The action of secondary amines as antiozonants is envisaged as a reaction with the zwitterion. Thus,

$$
R_2NH + R'-\underset{\underset{R''}{|}}{C}=\overset{+}{O}-O^- \longrightarrow R'-\underset{\underset{R''}{|}}{\overset{\overset{+}{R_2NH}}{\underset{|}{C}}}-O-O^- \longrightarrow R'-\underset{\underset{R''}{|}}{\overset{\overset{+}{R_2N}}{\underset{|}{C}}}-O-OH
$$

where R' is a polymer radical and R" is hydrogen or an alkyl radical. This reaction consummates scission of the polymer chain, which is in keeping with the reduction in molecular weight observed in the surface in the presence of polar solvents (183). It would prevent the formation of polyperoxide **24** or normal ozonide **25**. Loan et al. theorize that polyperoxide or normal ozonide may be subject to rapid degradation, and consequently crack growth is slowed down when they cannot be formed. Andrews and Braden (156) believe that antiozonants that reduce the rate of crack growth provide a coherent layer that can support considerable strain and which is attacked by ozone less rapidly than the original rubber. The ability to react with zwitterion would be in keeping with the use of 6-alkoxydihydroquinolines and p-alkoxyanilines as antiozonants.

Lorenz and Parks (184) found that up to 45% of an antiozonant was unextractable from rubber after exposure to ozone, in keeping with the prediction of the mechanism that the antiozonant will become bound to the polymer.

Furukawa et al. (110) pointed out that the antioxidant activity of an amine increases with oxidation potential (Table 6) and reaches a maximum at about 0.4 V, whereas antiozonant activity reaches a maximum at 0.25 V. The data in Table 6 show this point very clearly. Ratios of the viscosities of SBR solution before and after exposure (η/η_0) are shown for oxygen and for ozone treatment and indicate the change in protective power at different oxidation potentials. Those amines that are the most potent antiozonants have the greatest nucleophilicity, which fits well with the theory of Loan et al. (180).

4.2.2 Antiozonants Which Increases Critical Stress

As pointed out earlier, workers at MRPRA (167) found that dialkyl-p-phenylenediamines (e.g., N,N'-dioctyl-p-phenylenediamine, DOPPD) affected the critical conditions for crack growth. They subsequently examined the surface of ozonized samples containing DOPPD (156) and concluded that the behavior they found was typical of antiozonants which increase the critical energy. Since similar studies were not made using other p-phenylenediamines, it is not certain whether the behavior they observed is typical of antiozonants that increase critical energy or whether it is typical of all p-phenylenediamines.

Braden and Gent (167) found that the critical stress at a given DOPPD concentration was independent of time but highly dependent on ozone concentration. At a given ozone concentration, the critical stress increased markedly with DOPPD concentration. They concluded that the significant effect of DOPPD was due to a reaction with rubber ozonide to give a product that did not readily undergo scission or to relinking of severed polymer molecules. Direct reaction of ozone with DOPPD was seen as an undesirable side reaction accounting for the sensitivity of the effectiveness to increasing ozone concentration.

In their electron microscope study Andrews and Braden (156) detected a "brittle layer" formed on the surface of samples containing DOPPD and subjected to low concentrations of ozone. This layer was replaced by an oily layer at high ozone concentrations unless the sample was subjected to a pretreatment at low ozone level. They argued that the energy to rupture the coherent (brittle) layer would be larger than that to rupture the oily layer on a normally ozonized sample. Removal of DOPPD by direct reaction with ozone was cited as the reason brittle layer formation and associated protection were not found at high ozone concentration.

More recently, Lake (163) found that the highly colored layer formed on the surface of either ozonized NR or EPDM containing DOPPD could be removed mechanically or with acetone. He implied that this was the brittle layer found by Braden and Andrews (156) but identified it as the reaction product of DOPPD and ozone. It is hard to understand why the reaction product between DOPPD and ozone should increase critical stress while that between ozone and N-alkyl-N'-aryl-p-phenylenediamine should not. In addition, the ineffectiveness of the layer at high ozone concentration is difficult to explain although Lake suggested that the layer might be penetrated by ozone at sufficiently high concentration.

Andries et al. (185) examined ozone attack and antiozonant (DOPPD) protection on surfaces of raw and cured natural rubber using attenuated total reflectance infrared spectroscopy. When raw NR was ozonized a thick surface layer of ozonides and carbonyl compounds formed. When cured NR (without filler or DOPPD) was exposed esters and lactones also appeared to form, while the ozonide and carbonyl surface concentrations were considerably reduced. When the cured NR contained clay filler, ozone exposure brought an increase in the surface filler concentration—presumed to arise from ozone attack on the rubber resulting in its chain scission followed by relaxing of the rubber away from the filler to give it increased surface exposure. Similar behavior was observed in black loaded stocks.

With DOPPD present the surface film (up to 0.3 μ in thickness) appeared to be an ozonized liquid DOPPD of very complex structure but possibly containing nitroso, oxime, and other groups. There was no evidence of DOPPD on the surface of the unexposed controls or of ozonized NR on exposed stocks. The results were consistent with a scavenger mechanism or an ozonized protective layer mechanism.

It is apparent that the mechanism of antiozonant action is still far from clear. A nonstaining antiozonant that is effective in raising critical stress and reducing rate of crack growth is still awaited and there appears to be no reason why it should not be found.

5   MECHANOCHEMICAL DEGRADATION

This many faceted subject can be given only cursory treatment here but its involvement in fatigue failure in tires is discussed in Chapter 13, Section 4. Mechanical degradation of vulcanizates, which embraces flex cracking and some aspects of abrasion as well as fatigue failure, is generally harmful. The only possible exception is mechanical degradation as used in the reclaiming processes. On the other hand, mechanical breakdown of unvulcanized rubbers is frequently necessary and beneficial. Prior to the introduction of natural rubber with controlled Mooney viscosity it was the usual means of reducing NR to a workable state.

5.1   Mastication of Unvulcanized Rubbers

The processes involved in mechanicochemical degradation of raw rubber are well understood, thanks in large measure to studies carried out at MRPRA. This work has been thoroughly discussed by Bristow and Watson (186), and the whole field of mechanical degradation of polymers has recently been reviewed by Casale and co-workers (187). The *efficiency of mastication* i.e., the fractional increase in the number of molecules, is defined as $(1/M - 1/M_0)/(1/M_0)$ where M_0 and M are the initial and final number average molecular weight, respectively. This parameter yields a characteristic U-shaped curve as illustrated in Fig. 5, which is based on the work of Pike and Watson (188). These workers showed that the U-shaped curve can be resolved into two parts. The portion with the negative temperature coefficient was observed in the absence of oxygen if radical acceptors such as thiols were present. The efficiency of mastication of natural rubber at 60°C depended on the free radical acceptor used but at 120°C it was zero in all cases. The negative temperature coefficient and the absence of degradation above a certain temperature were attributed to a reduction in fluid viscosity to a point at which the energy input was no longer sufficient to rupture the chains. The presence of macroradicals at this stage is borne out by the mechanicochemical reaction of polymers and the initiation of polymerization by shear degraded polymers (189–191), and by chemical incorporation of free radical acceptors such as diphenylpicrylhydrazyl (192). The role of oxygen in the shear induced degradation is to act as a highly efficient radical trap.

The high temperature portion of Fig. 5 is attributed directly to oxidative scission. This process is undoubtedly accelerated by the shear. Betts

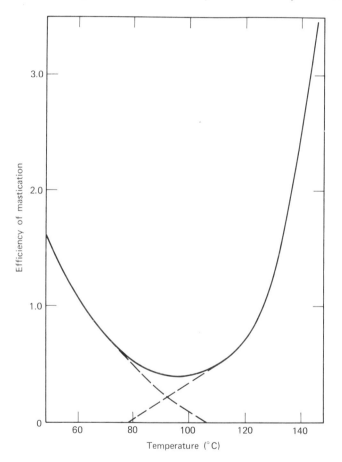

Figure 5 Efficiency of mastication of NR after 30 minutes Banbury mixing at various temperatures (as measured by the fractional increase in the number of molecules) (188).

et al. (193) have observed that the mastication of antioxidant free *cis*-polyisoprenes displays no negative temperature coefficient. Oxidative scission clearly predominates at all temperatures.

Bristow and Watson (186) believed that bulk strength was required in the rubber in addition to high viscosity. They noted that isomerized NR with 10% trans double bonds and *cis*-1,4-polybutadiene, which do not stress crystallize like NR, are not degraded on mastication. Kraus and Rollmann (194) confirmed that low temperature shear degradation occurs much less readily in typical *cis*-1,4-polybutadiene than in NR. When a polymer with a bimodal molecular weight distribution was masticated the high molecular weight portion was degraded as predicted by Bueche's theory (195). However, the degradation was slow, which suggests that some other factor such as stress crystallization might also be involved. Bristow (196) showed that low temperature degradation of synthetic *cis*-polyisoprene was also more rapid if a high molecular weight fraction was present.

All these observations indicate that, in contrast to high temperature breakdown, low temperature breakdown is not a random process. Bueche's theory predicts that cleavage will be concentrated near the center of a chain because it occurs through entanglements, and less than one chain in 10 will break as far as one-third of the way out from the center. Booth (197) examined the molecular weight distribution of degraded synthetic polyisoprene which initially had a narrow molecular weight distribution and concluded that "chain scission occurs at a point far removed from the ends but not necessarily near the center of the molecules." This observation does not differ materially from that made by Watson and his co-workers.

The fate of the macroradicals is worthy of some comment. Pike and Watson (188) found no change in NR subjected to mastication under nitrogen. They concluded that scission occurred between isoprene units and that the radicals formed were stabilized by resonance and subsequently recombined. Bristow (198) later showed that at molecular weights above 3×10^5 gelation occurred on anaerobic mastication. At a molecular weight of 10^6, 50% gel was present after a few minutes mastication. With *cis*-1,4-polybutadiene gelation is more likely to occur since the polybutadienyl radical is not resonance stabilized and can readily attack a double bond. Kraus and Rollmann (194) noted that *cis*-1,4-polybutadiene became branched on mastication at 140°C in either the presence or absence of oxygen. The lowering of molecular weight due to degradation and the introduction of branching both improved the raw polymer processability.

5.2   Mechanical Fatigue of Vulcanized Rubbers

The "fatigue" of vulcanized rubbers was concisely defined by Gent et al. (199) as "the gradual weakening of rubber specimens and eventual fracture brought about by repeated deformations much lower than the breaking strain," a definition that would apply equally well to fatigue of metals. Beatty (200) in reviewing the subject in 1964 took pains to emphasize the similarity between fatigue in metals and rubbers. He gave groove cracking in tires, tread and ply separations, and failures in motor mounts as typical examples of fatigue.

Beatty showed that for a black filled natural rubber vulcanizate tensile strength remained essentially constant after 10^7 cycles while modulus increased slightly. Ultimate elongation, on the other hand, decreased steadily from 525 to 350%. This phenomenon was observed in nitrogen as well as air and with all the rubbers tested, namely NR (black filled), SBR, CR, IIR, and NBR. Beatty postulated that "fatigue failure starts when the elongation or deformation to which the rubber is subjected exceeds *locally* the ultimate elongation or deformation of the rubber." He suspected that the changes he observed were due to nonoxidative cross-linking.

Scott (ref. 1, Ch. 10) in a useful review of the subject to about 1960 discussed the work of Slonimskii and his co-workers in some detail. Slonimskii et al. (201) demonstrated that mastication and fatigue were related. They showed that uncured polyisobutene, which is comparatively stable to oxidation, suffered reduced molecular weight when subjected to small repeated shear deformations for a prolonged period. Twenty to thirty minutes of micromilling resulted in a lowering of molecular weight comparable to that achieved by 5–15 days flexing. Benzoquinone, a radical scavenger, accelerated fatigue at low temperatures just as it accelerates degradation from mastication in the absence of oxygen.

James (202) showed that the fatigue resistance (number of flex cycles till break) at equal stored energy depended on the deformation frequency and the temperature. A WLF type of transform could be used to bring all the data for four SBR gum rubbers on to one master curve. It was concluded that the fatigue resistance is a viscoelastic phenomenon and so primarily determined by the T_g of the rubber. In stress crystallizing NR the fatigue life was almost independent of frequency and temperature; the behavior is not purely viscoelastic but much more dependent on chemical oxidation processes.

Dynamic ozone cracking and flex cracking were recognized as two distinct phenomena by Thornley (203). He pointed out that while strain was a common factor affecting both phenomena, the rate of development of flex cracks fell away very quickly below a certain strain whereas ozone cracking was apparent at quite low strains and, for the most part, did not increase with strain. Lake has reviewed the fracture mechanics (204) and the environmental factors (205) which affect fatigue life including the effect of swelling by liquids as well as attack by atmospheric oxygen and ozone. Lake and Lindley (206) differentiated clearly between two types of cut growth in rubber, each of which could occur under static or dynamic conditions: (*a*) mechanicooxidative cut growth due to mechanical rupture at the tip of a flaw; and (*b*) ozone cut growth due to primarily chemical scission.

However, there is a common factor between ozone cracking and flex cracking, namely the concept of tearing energy T, available for cut propagation (207–209). The tearing energy T is related to the growth of a cut of length c and the number of flexes n. The relation is described more completely in Chapter 13. The relationship between the rate dc/dn and T, referred to as the "cut growth characteristic," is complex, as can be seen by reference to Fig. 6, which is characteristic of conventionally sulfur vulcanized NR flexed in air (210). The curve can be divided into four portions:

1. Where $T \leqslant T_0$:

$$\frac{dc}{dn} = r \tag{9}$$

T_0 is the minimum tearing energy at which mechanicooxidative cut growth occurs and r is a constant. At lower tearing energies cut growth is due to ozone attack and occurs at a slow steady rate which depends on ozone concentration.

2. Where $T_0 \leqslant T \leqslant T_t$:

$$\frac{dc}{dn} = A(T - T_0) + r \tag{10}$$

T_t is a transition tearing energy in the region of 0.5 kg/cm, above which dependence of growth rate changes from linear to square law. A is a constant. If dc/dn is plotted against T for values below 0.5 kg/cm, T_0 can be determined from the intercept on the T axis (Fig. 6). The contribution of r, the cut growth due to ozone, rapidly becomes negligible once T_0 is exceeded.

3. Where $T_t \leqslant T < T_c$:

$$\frac{dc}{dn} = BT^2 \tag{11}$$

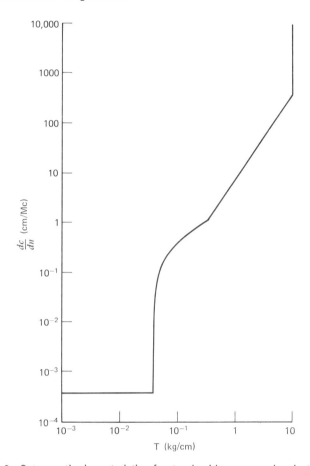

Figure 6 Cut growth characteristic of natural rubber gum vulcanizate (210).

T_c is the catastrophic tearing energy and B is a constant. This stage is not found with all vulcanizing systems.

    4.   Where $T = T_c$:

$$\frac{dc}{dn} = \infty \qquad [12]$$

Similar T versus dc/dn relationships were found for noncrystalline rubbers such as SBR. Again a minimum tearing energy T_0 was observed and for tearing energies from T_0 to about 0.3 kg/cm the rate varied linearly with T. In the third region the rate depended on the fourth power rather than the square of the tearing energy.

This work (209) demonstrated clearly that fatigue failure occurs by cut growth from flaws originally present in the rubber. In died-out dumbbells failure usually occurs from flaws in the cut edges, but it may occur from a flaw in a molded surface. Local inhomogeneities such as hard particles or small regions of abnormal cross-link density within the body of rubber may also initiate failure. This explains the observation by Akkerman (211) that cracks grow below the surface of rubber during flexing. The approximate size of the flaws can be deduced from the fatigue life of a vulcanizate at high tensile strains (206). The effective initial flaw size in NR is about 0.0025 cm while it is as much as twice this in synthetic rubbers. Lake and Lindley (212) attribute the difference between effective initial flaw size in NR and SBR to a larger amount of the rapid "smooth" cut growth which occurs before the tip of the tear roughens in SBR vulcanizates.

Lake and Lindley (212) note that although the same basic failure mechanism applies to both crystallizing and noncrystallizing rubbers there are important differences in their behavior:

1. Static cut growth and fatigue occur in noncrystallizing rubbers at moderate tearing energies. They occur in NR only at energies close to the catastrophic tearing energy.
2. In noncrystallizing rubbers cut growth and fatigue behavior are strongly temperature dependent, in NR they are not.
3. Cut growth rate in NR at high strains is proportional to the first or second power of the tearing energy, whereas for noncrystallizing rubbers it is proportional to the fourth power.

Andrews' (213) postulate of gross hysteresis induced by crystallization creating a "frozen" stress pattern at the cut tip explains these differences. It also explains why the flex life of crystallzing rubbers is much longer if the stress cycle does not pass through zero stress. As stress approaches zero the stress pattern "melts" and the cut grows when stress is applied once more. With noncrystallizing rubbers it is immaterial to cut growth rate in extension whether or not the stress cycle passes through zero. This was demonstrated by Beatty (200) (see Fig. 7) who plotted minimum strain against fatigue life for dumbbells of revolution of SBR and of NR. Under compression, the noncrystallizing rubber had the longer fatigue life.

Lindley (214) obtained good agreement between crack growth and fatigue data for noncrystallizing gum vulcanizates of SBR under nonrelaxing conditions. For SBR hysteresis is not markedly stress dependent and the effect of hysteresis is somewhat similar for all tearing energies. NR

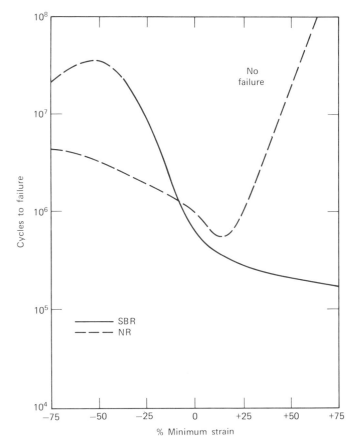

Figure 7 Effect of strain cycle on fatigue life of NR and SBR (200).

differs considerably since crystallization at high strain causes considerable hysteresis. As a result crack growth in NR depends in a very complex way on the maximum and minimum tearing energies during cyclic deformation.

T_0 may be termed the "mechanical cut growth limit" as it is the minimum tearing energy at which a flaw can be mechanically propagated. There is, likewise, a "mechanical fatigue limit" e_0, the strain below which flaws must grow by ozone cracking until T_0 is reached. e_0 can be quite readily determined since ozone induced growth is much slower than mechanicooxidative growth and there is an abrupt drop in fatigue life as e_0 is exceeded. Beatty (200) noted that EPDM had a longer time for crack initiation than other rubbers examined. This was attributed to the

Table 10   Effect of Atmosphere on Fatigue Life of Sulfur/MBTS Vulcanized Black Filled NR (200)

	Cycles to Failure
Nitrogen	No sign of failure at 8.5×10^6 cycles
Air (ozone free)	3.5×10^6
Air	2.2×10^6
Oxygen	2.5×10^6
Ozone (10 pphm)	1.6×10^6

ozone resistance of EPDM, although crack propagation in EPDM is very rapid.

The atmosphere is which flexing is carried out is of considerable importance. The role of ozone in cut growth was examined by Beatty (200) who obtained the data in Table 10 for fatigue life of sulfur/MBTS vulcanized black filled NR.

Beatty attributed the influence of ozone to three factors: (*a*) initiation of cracks at points of stress concentration; (*b*) lowering of ultimate elongation through chemical action (this is also the effect of oxygen on fatigue life); and (*c*) relief of major strains by formation of small cracks around defects. This may lead to an increase rather than a decrease in fatigue life.

Gent (215) had previously shown that fatigue life was substantially longer in vacuo than in air for both SBR and NR vulcanizates. Lake and Lindley (206) examined the cut growth of an antioxidant free NR vulcanizate in air and in vacuo compared with cut growth in air of a similar vulcanizate containing 1 phr PBNA (Fig. 8). At comparable tearing energy the rate of cut growth in air is three times that in vacuo. The cut growth of an oxidation and ozone resistant rubber such as butyl is the same in the presence or absence of air.

5.3   Chemical Prevention of Mechanical Fatigue of Vulcanizates

Since oxygen and ozone both promote cut growth and fatigue in unsaturated polymers it is to be expected that antioxidants and antiozonants would improve fatigue life. Lake and Lindley (206) showed that 1 phr PBNA reduced the rate of cut growth of an NR gum vulcanizate in air, although not to the level found in vacuo. The PBNA resulted in a 50% increase in T_0 for the vulcanizate in question. These authors noted that antioxidants such as N-isopropyl-N'-phenyl-p-phenylenediamine, N,N'-dioctyl-p-phenylenediamine, and PBNA which protect most effectively

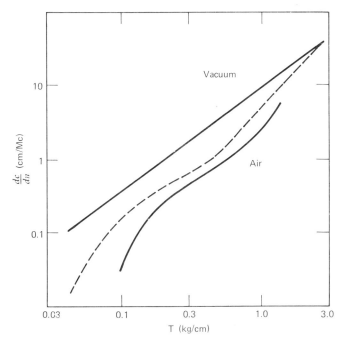

Figure 8 Cut growth characteristics of NR gum vulcanizate in air and in vacuo (broken line indicates protective action of 1 phr PBNA in air) (206).

against oxidative cut growth, also protect against dynamic ozone cut growth. Gent (215) and later Lloyd (216) showed that arylalkyl- and dialkyl-p-phenylenediamines were the most effective additives in prolonging fatigue life. It is significant that these are radical trapping antioxidants. Mercaptobenzimidazole and zinc dialkyldithiocarbamates do not prevent cut growth. On the other hand, a peroxide-decomposing antioxidant, tri-p-dodecylphenyl phosphite, has been claimed to be superior to diphenyl-p-phenylenediamine (217).

The chemical nature of the cross-link has been shown to have a profound influence on fatigue life. Lake and Lindley (206, 210) found that while T_0 was substantially the same for low modulus polysulfidic, monosulfidic, and peroxide vulcanizates of NR, T_0 was appreciably less for the polysulfidic than for the other vulcanizates in tightly cross-linked systems. This effect was attributed to the labile nature of polysulfidic cross-links which break readily and reform at equilibrium in the stressed state, thus relieving local stress concentrations. Similar observations had been made by Dogadkin et al. (218, 219) who also noted that the superiority in fatigue resistance of the polysulfidic cross-links was lost at elevated temperatures because of the poor oxidative stability of poly-sulfidic networks. The conclusion is borne out by the work of Cox and Parks (220) who found that antioxidant increased the fatigue life of a polysulfidic network fivefold while it merely doubled the fatigue life of a peroxide vulcanizate.

The challenge of obtaining good fatigue life and aging resistance in the same vulcanizate has been taken up by McCall (221). He compared the properties of NR cross-linked by a "semiefficient" vulcanization system comprising 0.7 part 2-morpholinothiobenzothiazole, 0.5 part dimor-pholinodisulfide, and 1.5 parts sulfur with those of "conventional" and "efficient" vulcanization systems (cf. ref. 12). The conventional vulcaniza-tion system contained 0.7 part 2-morpholinothiobenzothiazole and 2.5 parts sulfur while the "efficient" vulcanization system comprised 1.0 part each of N-cyclohexyl-α-benzothiazolesulfenamide, dimorpholinodis-ulfide, and tetramethyl thiuram disulfide and no sulfur. As shown in Table 11, the "conventional" and "semi-efficient" vulcanizates had almost identical fatigue lives before aging, whereas the fatigue life of the "effi-cient" vulcanizate was much smaller. After 2 days aging at 100°C the fatigue life of the conventional vulcanizate was comparable with that of the efficient vulcanizate, which was not appreciably affected by aging. The fatigue life of the semiefficient vulcanizate after aging was by far the best of the three systems. The semiefficient system also results in better retention of tensile strength on aging than the conventional system. The difference was attributed to "the more efficient usage of sulfur together with the avoidance of any monosulfide formation." Bristow and Tiller (222) showed that monosulfidic cross-links are not detrimental to fatigue life of an NR gum vulcanizate using a cure system based on zinc di-methyldithiocarbamate and a high level of sulfur. Blackman and McCall (223) found that the fatigue life of a comparable black filled vulcanizate, which contained 91% mono- and disulfide cross-links and only 9% poly-sulfidic cross-links, was very close to that of a conventional vulcanizate. Unfortunately, the aging of the dithiocarbamate accelerated vulcanizates

Table 11^a   Fatigue Life of Conventional, Semiefficient, and Efficient NR Vulcanizates Before and After 2 Days Aging at 100°C^b

	Conventional	Semiefficient	Efficient
Unaged	176	176	58
Aged	62	127	55

[a] From McCall (221, Fig. 3).
[b] Strain energy 10 kg/cm^2; 2 phr IPPD antidegradant in all stocks.

is poor, although the dithiocarbamates themselves are good antioxidants. Since the main chain of these vulcanizates is extensively modified these modifications were cited as possible sources of a stress relieving mechanism leading to good fatigue properties. A vulcanizate superior in both aging and fatigue has not yet been produced.

It is very obvious from this last section and throughout the chapter that the bulk of knowledge regarding elastomer degradation has been gained from studies of natural rubber. Although the assumption that synthetic elastomers made in solution will behave similarly is often true, there is room for fundamental study of the degradation of solution polymers in their own right. Only in this way will the relations between polymer structure and reactivity be clarified.

REFERENCES

1. G. Scott, *Atmospheric Oxidation and Antioxidants*, Elsevier, Amsterdam, 1965.
2. W. O. Lundberg, ed., *Autoxidation and Antioxidants*, Wiley-Interscience, New York, Vol. I, 1961; Vol. II, 1962.
3. M. B. Neiman, ed., *Aging and Stabilization of Polymers*, Consultants Bureau, New York, 1965.
4. L. Reich and S. S. Stivala, *Elements of Polymer Degradation*, McGraw-Hill, New York, 1971.
5. W. L. Hawkins, ed., *Polymer Stabilization*, Wiley-Interscience, New York, 1972.
6. M. A. Golub and R. J. Gargiulo, *Polym. Lett.*, **10**, 41 (1972).
7. M. A. Golub et al., *Rubber Chem. Technol.*, **48**, 953 (1975).
8. L. Bateman et al., in L. Bateman, ed., *The Chemistry and Physics of Rubberlike Substances*, Maclaren, London, 1963, Ch. 15.
9. B. Saville and A. A. Watson, *Rubber Chem. Technol.*, **40**, 100 (1967); C. G. Moore and M. Porter, *Rev. Gén. Caoutch.*, **39**, 1768 (1962); *Rubber Chem. Technol.*, **36**, 547 (1963).
10. D. S. Campbell and B. Saville, *Proceedings of the 5th Rubber Technology Conference*, London, 1967, p. 1.
11. K. Murakami and S. Tamura, *Polym. Lett.*, **11**, 317 (1973).
12. T. D. Skinner and A. A. Watson, *Rubber Age*, **99**, (11), 76 (12), 69 (1967).
13. J. L. Bolland, *Q. Rev. (London)*, **3**, 1 (1949); L. Bateman, *ibid.*, **8**, 147 (1954).
14. D. Barnard et al., in L. Bateman, ed., *The Chemistry and Physics of Rubberlike Substances*, Maclaren, London, 1963, Ch. 17.
15. P. M. Norling and A. V. Tobolsky, in R. T. Conley, ed., *Thermal Stability of Polymers*, Vol. I, Dekker, New York, 1970.
16. J. R. Shelton, *Rubber Chem. Technol.*, **45**, 359 (1972).
17. J. A. Howard, *Rubber Chem. Technol.*, **47**, 976 (1974).
18. J. L. Bolland and G. Gee, *Trans. Faraday Soc.*, **42**, 236, 244 (1946); R. Criegee et al., *Chem. Ber.*, **72B**, 1799 (1939).

19. J. L. Bolland, *Trans. Faraday Soc.*, **46**, 358 (1950).
20. P. M. Norling et al., *Rubber Chem. Technol.*, **38**, 1198 (1965).
21. W. Hofmann, *Rev. Gén. Caoutch. Plast.*, **45**, 73 (1968).
22. E. M. Bevilacqua, *J. Polym. Sci., C*, **24**, 285 (1968).
23. P. Delafield, ed., *J. Inst. Rubber Ind.*, **1**, 255 (1967).
24. J. L. Bolland and G. Gee, *Trans. Faraday Soc.*, **42**, 294 (1946).
25. J. L. Bolland and H. Hughes, *J. Chem. Soc.*, **1949**, 492.
26. F. R. Mayo, *Ind. Eng. Chem.*, **52**, 614 (1960).
27. E. M. Bevilacqua, et al., *J. Org. Chem.*, **25**, 1276 (1960).
28. E. M. Bevilacqua, *J. Am. Chem. Soc.*, **80**, 5364 (1958).
29. E. M. Bevilacqua, in W. O. Lundberg, ed., *Autoxidation and Antioxidants*, Wiley-Interscience, New York, 1962, Ch. 18.
30. E. M. Bevilacqua, *Rubber Age*, **80**, 271 (1956); *Science*, **126**, 386 (1957).
31. E. M. Bevilacqua and E. S. English, *J. Polym. Sci.*, **49**, 495 (1961).
32. A. V. Tobolsky and A. Mercurio, *J. Am. Chem. Soc.*, **81**, 5535 (1959).
33. J. L. Morand, *Rubber Chem. Technol.*, **47**, 1094 (1974).
34. C. L. M. Bell, *Trans. Inst. Rubber Ind.*, **41**, 212 (1965).
35. C. L. M. Bell and R. Tiller, *J. Appl. Polym. Sci.*, **11**, 1289 (1967).
36. E. M. Bevilacqua, *J. Chem. Soc.*, **77**, 5394, 5396 (1955); **79**, 2915 (1957).
37. J. R. Dunn and J. Scanlan, in L. Bateman, ed., *The Chemistry and Physics of Rubberlike Substances*, Maclaren, London, 1963, Ch. 18.
38. J. R. Dunn et al., *Trans. Faraday Soc.*, **55**, 667 (1959).
39. J. R. Dunn and J. Scanlan, *Trans. Faraday Soc.*, **57**, 160 (1961).
40. L. Mullins, *J. Appl. Polym. Sci.*, **2**, 257 (1959).
41. M. M. Horikx, *J. Polym. Sci.*, **19**, 445 (1956).
42. E. M. Bevilacqua, *J. Am. Chem. Soc.*, **81**, 5071 (1959).
43. T. Colclough et al., *J. Appl. Polym. Sci.*, **12**, 295 (1968).
44. J. I. Cuneen, *Rubber Rev.*, *Rubber Chem. Technol.*, **41**, 182 (1968).
45. D. B. Jaroszynska, Proceedings of the 5th Rubber Technology Conference, London, 1967, p. 59.
46. A. S. Kuzminskii et al., *Kauch. Rezina*, **26** (2), 15 (1967) (in Russian); *Sov. Rubber Technol.*, **26** (2), 15 (1967) (English translation).
47. B. M. E. van der Hoff, *Ind. Eng. Chem., Prod. Res. Dev.*, **2**, 273 (1963).
48. L. D. Loan, *J. Appl. Polym. Sci.*, **7**, 2259 (1963).
49. K. R. Hargrave and A. L. Morris, *Trans. Faraday Soc.*, **52**, 89 (1956).
50. A. S. Kuzminskii et al., *Rubber Chem. Technol.*, **29**, 573 (1956).
51. J. R. Dunn, *J. Macromol. Chem.* **1**, 739 (1966).
52. K. Nezu, *Nippon Gomu Kyokaishi*, **41**, 96 (1968).
53. L. I. Laud and J. E. Stuckey, *J. Inst. Rubber Ind.*, **3**, 29 (1969).
54. A. V. Tobolsky et al., *J. Am. Chem. Soc.*, **86**, 3925 (1964).
55. G. A. Russell, *J. Am. Chem. Soc.*, **79**, 3871 (1957).
56. V. B. Miller et al., *Polym. Sci. USSR*, **2**, 121 (1961).
57. G. V. Vinogradov and L. I. Ivanova, *Kauch. Rezina*, **27** (1), 15 (1968) (in Russian); *Sov. Rubber Technol.*, **27** (1), 16 (1968) (English translation).

58. D. J. Carlsson and D. M. Wiles, *Rubber Chem. Technol.*, **47,** 991 (1974).

59. N. Uri, in W. O. Lundberg, ed., *Autoxidation and Antioxidants*, Vol. I, Wiley-Interscience, New York, 1961.

60. N. M. Emanuel et al., *Angew. Chem., Int. Ed.*, **8,** 97 (1969).

61. A. Robertson and W. A. Waters, *Trans. Faraday Soc.*, **42,** 201 (1946).

62. W. Kern and H. Willersinn, *Makromol. Chem.*, **15,** 1 (1955).

63. A. J. Chalk and J. F. Smith, *Trans. Faraday Soc.*, **53,** 1235 (1957).

64. N. Uri, *Chem. Ind.*, **1967,** 2060; A. T. Betts and N. Uri, *Makromol. Chem.*, **95,** 22 (1966).

65. G. Scott, *Brit. Polym. J.*, **3,** 24 (1971).

66. C. Copping and N. Uri, *Discuss. Faraday Soc.*, **46,** 202 (1968).

67. L.-H. Lee et al., *J. Appl. Polym. Sci.*, **10,** 1699 (1966).

68. A. I. Andreeva et al., *Kauch. Rezina*, **27** (11), 9 (1968) (in Russian); *Sov. Rubber Technol.*, **27** (11), 11 (1968).

69. L. G. Angert et al., *Sov. Rubber Technol.*, **28** (4), 19 (1969).

70. L. Rosik, *Makromol. Chem.*, **16/17,** 209 (1971).

71. W. R. May and L. Bsharah, *Ind. Eng. Chem., Prod. Res. Dev.*, **9,** 73 (1970).

72. F. R. Mayo et al., *Rubber Chem. Technol.*, **41,** 271 (1968).

73. G. P. and J. J. Canva, *Rubber J.*, **September 1971,** 36.

74. M. L. Kaplan and P. G. Kelleher, *Rubber Chem. Technol.*, **45,** 423 (1972).

75. J. L. Morand, *Rubber Chem. Technol.*, **45,** 481 (1972).

76. J. R. Dunn et al., *Trans. Faraday Soc.*, **54,** 740 (1958).

77. M. A. Golub and C. L. Stephens, *J. Polym. Sci.*, *A-1*, **6,** 763 (1968).

78. M. A. Golub and C. L. Stephens, *J. Polym. Sci.*, *C*, **16,** (2), 765 (1967).

79. J. Morand, *Rev. Gén. Caoutch. Plast.*, **45,** 615 (1968) (in French).

80. S. W. Beavan and D. Phillips, *Eur. Polym. J.*, **10** (7), 593 (1974).

81. J. R. Shelton, in W. L. Hawkins, ed., *Polymer Stabilization*, Wiley-Interscience, New York, 1972, Ch. 2.

82. K. U. Ingold, "Inhibition of Autoxidation," in R. F. Gould ed., *Oxidation of Organic Compounds*, *Adv. Chem. Ser.* 75-1, American Chemical Society, Washington, D.C., 1968, p. 296.

83. E. T. Denisov, *Russ. Chem. Rev.*, **42,** 157 (1973).

84. J. C. Ambelang et al., *Rubber Rev.*, *Rubber Chem. Technol.*, **36,** 1497 (1963).

85. E. V. Reztsova and G. V. Chubarova, *Dokl. Akad. Nauk SSSR*, **170** (2), 383 (1968) (in Russian).

86. A. F. Bickel and E. C. Kooyman, *J. Chem. Soc.*, **32,** 64 (1955).

87. J. C. Westfahl et al., *Rubber Chem. Technol.*, **45,** 402 (1972).

88. J. C. Westfahl, *Rubber Chem. Technol.*, **46,** 1134 (1973).

89. J. R. Shelton and D. N. Vincent, *J. Am. Chem. Soc.*, **85,** 2433 (1963).

90. J. R. Shelton and E. T. McDonel, *J. Polym. Sci.*, **32,** 75 (1958).

91. J. A. Howard and K. U. Ingold, *Can. J. Chem.*, **40,** 1851 (1962); **41,** 1744, 2800 (1963); **42,** 2324 (1964).

92. G. S. Hammond et al., *J. Am. Chem. Soc.*, **77,** 3233 (1955).

93. T. W. Campbell and G. M. Coppinger, *J. Am. Chem. Soc.*, **74,** 1469 (1952).

94. A. F. Bickel and B. C. Kooyman, *J. Chem. Soc.*, **1953,** 3211.

95.   J. Tsurugi et al., *Rubber Chem. Technol.*, **44**, 857 (1971).

96.   G. M. Coppinger, *J. Am. Chem. Soc.*, **86**, 4385 (1964).

97.   J. L. Bolland and P. tenHave, *Discuss Faraday Soc.*, **2**, 252 (1947).

98.   L. R. Mahoney, *Angew. Chem., Int. Ed.*, **8**, 547 (1969).

99.   H. Low, *Ind. Eng. Chem., Prod. Res. Dev.*, **5**, 80 (1966).

100. A. A. Grinberg et al., *Kauch. Rezina*, **26** (3), 22 (1967) (in Russian); *Sov. Rubber Technol.*, **26** (3), 24 (1967) (English translation).

101. *T. Kempermann*, Paper given at ACS, Division of Rubber Chemistry, Cleveland, Ohio, October 1971; published as *Bayer Technical Notes for the Rubber Industry*, No. 45, 1972.

102. F. X. O'Shea, "Phenolic Polymer Stabilizers," in R. F. Gould, ed., *Stabilization of Polymers and Stabilizer Processes*, **85**, American Chemical Society, Washington, D.C., 1968, p. 126.

103. A. R. Patel et al., *J. Appl. Polym. Sci.*, **16**, 2751 (1972).

104. G. Scott, *Brit. Polym. J.*, **3**, 24 (1971).

105. J. R. Shelton and W. L. Cox, *Ind. Eng. Chem.*, **45**, 392 (1953).

106. J. R. Shelton and W. L. Cox, *Ind. Eng. Chem.*, **45**, 397 (1953).

107. J. R. Shelton and W. L. Cox, *Ind. Eng. Chem.*, **46**, 816 (1954).

108. L. Rosik, *Polym. Lett.* **5**, 1083 (1967).

109. C. W. Capp and E. G. E. Hawkins, *J. Chem. Soc.*, **1953**, 4106.

110. J. Furukawa et al., "Mechanism of Antioxidation and Antiozonization of Amines for Rubber," in R. F. Gould, ed., *Stabilization of Polymers and Stabilizers Processes*, Adv. Chem. Ser. 85, American Chemical Society, Washington, D.C., 1968, p. 110.

111. O. Lorenz and C. R. Parks, *Rubber Chem. Technol.*, **34**, 816 (1961).

112. J. R. Shelton, *Rubber Chem. Technol.*, **47**, 949 (1974)

113. J. R. Dunn and J. Scanlan, *Trans. Inst. Rubber Ind.*, **34**, 228 (1958).

114. W. P. Fletcher and S. G. Fogg, *Rubber J.*, **134**, 16 (1958); J. R. Dunn and J. Scanlan, *J. Appl. Polym. Sci.*, **1**, 84 (1959).

115. J. D. Holdsworth et al., *J. Chem. Soc.*, **1964**, 4692.

116. N. P. Neureiter and D. E. Brown, *Ind. Eng. Chem., Prod. Res. Dev.*, **1**, 236 (1962).

117. B. A. Marshall, "Hydroperoxide Decomposition by Some Sulphur compounds," in R. F. Gould, ed., *Stabilization of Polymers and Stabilizer Processes*, Adv. Chem. Ser. 85, American Chemical Society, Washington, D.C., 1968, p. 140.

118. L. A. Brooks, *Rubber Chem. Technol.*, **36**, 887 (1963).

119. J. R. Dunn, *Kaut. Gummi*, **14**, WT 114 (1961) (in German); *Rubber Chem. Technol.*, **34**, 910 (1961).

120. T. Colclough and J. I. Cuneen, *J. Chem. Soc.*, **1964**, 4790.

121. A. J. Burn, "Mechanism of Oxidation Inhibition by Zinc Dialkyl Dithiophosphates," in R. F. Gould, ed., *Oxidation of Organic compounds*, Adv. Chem. Ser. 75-1, American Chemical Society, Washington, D.C., 1968, p. 323.

122. A. E. Oberster et al., *Rubber Chem. Technol.*, **41**, 255 (1968).

123. K. J. Robinson and J. R. Dunn, Paper presented to Chemical Institute of Canada, Annual Meeting, Montreal, 1969; R. B. Spacht et al., *Rubber Chem. Technol.*, **37**, 210 (1964); **38**, 134 (1965).

124. M. E. Cain et al., *Rubber J.*, **150** (11), 10 (1968).

125. D. G. Lloyd and J. Payne, *Rubber News India*, **6** (9), 26 (1962).

126. M. E. Cain and B. Saville, British Pat. 1,185,896 (to Natural Rubber Producers Res. Assoc., 1970).

127. G. T. Knight and B. Saville, Cork Mechanisms Conference on Structure and Mechanism in Nitrogen Chemistry, Kinsale, Ireland, April 1971.

128. V. P. Kirpichev, A. I. Yakabchik, and G. N. Maglysh, *Vysokomol Soedin.*, **10A**, 2347 (1968) (in Russian); *Rubber Chem. Technol.*, **43**, 1225 (1970) (English translation).

129. M. E. Cain et al., *Chem. Ind.*, **1970**, 126.

130. J. R. Dunn et al., *J. Appl. Polym. Sci.*, **8**, 723 (1964).

131. C. R. Parks and O. Lorenz, *Ind. Eng. Chem., Prod. Res. Dev.*, **2**, 279 (1963).

132. J. Lal, *J. Polym. Sci., C*, **16**, 3391 (1968).

133. C. L. M. Bell and J. I. Cuneen, *J. Appl. Polym. Sci.*, **11**, 2201 (1967).

134. O. Lorenz et al., Proceedings of the 4th Rubber Technology Conference, London, 1962, p. 656.

135. K. Komuro and T. Saito, *Nippon Gomu Kyokaishi*, **38** (4), 246 (1965).

136. A. Ya. Borzenkova et al., *Sov. Rubber Technol.*, **29** (12), 17 (1972).

137. R. H. Mann and D. F. Hess, U.S. Pat. 3,234,173 (to Shell Oil, 1966).

138. A. S. Kuz'minskii et al., *Vysokomol. Soedin.*, **4**, 1682 (1962) (in Russian); *Polym. Sci., USSR*, **4**, 520 (1963) (English translation).

139. A. S. Kuz'minskii et al., *Kauch. Rezina*, **21** (4), 10 (1962) (in Russian); *Sov. Rubber Technol.*, **21** (4), 10 (1962) (English translation).

140. L.-H. Lee et al., *J. Appl. Polym. Sci.*, **10**, 1717 (1966).

141. A. M. Trozzolo, in W. L. Hawkins, ed., *Polymer Stabilization*, Wiley-Interscience, New York, 1972, Ch. 4.

142. J. E. Guillet et al., "Fundamental Processes in the Photodegradation of Polymers," in R. F. Gould, ed., *Stabilization of Polymers and Stabilizer Processes*, Adv. Chem. Ser. 85, American Chemical Society, Washington, D.C., 1968, p. 272.

143. Yu. A. Ershov et al., *Russ. Chem. Rev.*, **38** (2), 147 (1969).

144. O. Cicchetti, *Adv. Polym. Sci.*, **7**, 70 (1970).

145. J. R. Dunn, *J. Appl. Polym. Sci.*, **4**, 151 (1960).

146. J. R. Dunn, *Trans. Inst. Rubber Ind.*, **34**, 20 (1958).

147. M. Heskins and J. E. Guillet, *Macromolecules*, **1**, 97 (1968).

148. P. Karmitz, *Rev. Gén. Caoutch.*, **35**, 913 (1958).

149. I. Williams, *Ind. Eng. Chem.*, **18**, 367 (1926).

150. A. Van Rossem and H. W. Talen, *Kautschuk*; **7**, 79, 115 (1931) (in German); *Rubber Chem. Technol.*, **4**, 490 (1931) (English translation).

151. Z. T. Ossefort, "Ozone Resistance of Elastomeric Vulcanizates," in *Symposium on Effect of Ozone on Rubber*, ASTM Special Tech. Publ. 229, ASTM, Philadelphia, 1958, p. 39; *Rubber Chem. Technol.*, **32**, 1008 (1950).

152. J. E. Hauck, *Mater. Design Eng.*, **60**, 79 (1964).

153. E. R. Erickson et al., "A Study of the Reaction of Ozone with Polybutadiene Rubbers," in *Symposium on Effect of Ozone on Rubber*, ASTM Special Tech. Publ. 229, ASTM, Philadelphia, 1958, p. 11; *Rubber Chem. Technol.*, **32**, 1062 (1959).

154. H. Tucker, "The Reaction of Ozone with Rubber," in *Symposium on Effect of Ozone on Rubber*, ASTM Special Tech. Publ. 229, ASTM, Philadelphia, 1958, p. 30; *Rubber Chem. Technol.*, **32**, 269 (1959).

155. E. H. Andrews and M. Braden, *J. Polym. Sci.*, **55**, 787 (1961).

156. E. H. Andrews and M. Braden, *J. Appl. Polym. Sci.*, **7**, 1003 (1963).

157. G. Salmon and F. van Bloois, *Rubber Chem. Technol.*, **41**, 643 (1968).

158. M. Braden and A. N. Gent, *Proc. Inst. Rubber Ind.*, **8**, 88 (1961).

159. R. J. Newton, *J. Rubber Res.*, **14**, 27 (1945).

160. M. Braden and A. N. Gent, *J. Appl. Polym. Sci.*, **3**, 90 (1960).

161. M. Braden and A. N. Gent, *J. Appl. Polym. Sci.*, **3**, 100 (1960).

162. A. N. Gent and H. Hirakawa, *J. Polym. Sci.*, A-2, **5**, 157 (1967).

163. G. J. Lake, *Die Nederlandse Rubber Ind.*, **30** (24), 1 (1969); **31**, 172 (1970); *Rubber Chem. Technol.*, **43**, 1230 (1970).

164. I. C. Cheetham and W. R. Gurney, *Trans. Inst. Rubber Ind.*, **37**, 35 (1961).

165. G. J. Lake and A. G. Thomas, Proceedings of the 5th Rubber Technology Conference, London, 1967, p. 525.

166. M. Braden and A. N. Gent, *Kaut Gummi*, **14**, WT 157 (1961) (in German); *Rubber Chem. Technol.*, **35**, 200 (1962).

167. M. Braden and A. N. Gent, *J. Appl. Polym. Sci.*, **6**, 449 (1962).

168. E. H. Andrews, *J. Appl. Polym. Sci*, **10**, 47 (1966).

169. J. C. Ambelang et al., *Rubber Chem. Technol.*, **42**, 1186 (1969).

170. J. F. O'Mahoney, *Rubber Age*, **102** (3), 47 (1970).

171. L. Spenadel and R. L. Sutphin, *Rubber Age*, **102** (12), 55 (1970).

172. M. J. Nix, *Eur. Rubber J.*, **157** (6), 27 (1975).

173. L. L. Best and R. C. W. Moakes, *Trans Inst. Rubber Ind.*, **27**, 103 (1951).

174. S. W. Ferris et al., "Prevention of Ozone Attack with Wax," in *Symposium on Effect of Ozone on Rubber*, ASTM Special Tech. Publ. 229, ASTM, Philadelphia, 1958, p. 72.

175. G. T. Hodgkinson and C. E. Kendall, Proceedings of the 4th Rubber Technology Conference, London, 1962, p. 711.

176. W. L. Cox, "Chemical Antiozonants and Factors Affecting Their Utility," in *Symposium Effect of Ozone in Rubber*, ASTM Special Tech. Publ. 229, ASTM, Philadelphia, 1958, p. 52; *Rubber Chem. Technol.*, **32**, 364 (1959).

177. S. D. Razumovskii and L. S. Batashova, *Vysokomol. Soedin.*, A11, 588 (1969) (in Russian), *Polym. Sci., USSR*, **11**, 667 (1969) (English translation).

178. S. D. Razumovskii, private communication.

179. M. Braden, *J. Appl. Polym. Sci.*, **6**, 56 (1962).

180. L. D. Loan et al., *J. Inst. Rubber Ind.*, **2**, 73 (1968).

181. R. Criegee, *Rec. Chem. Prog. (Kresge-Hooker Sci. Lib.)*, **18**, 111 (1957).

182. R. Criegee, in J. O. Edwards, ed., *Peroxide Reaction Mechanisms*, Wiley-Interscience, New York, 1962, p. 29.

183. J. H. Gilbert, Proceedings of the 4th International Conference, London, 1962, p. 696.

184. O. Lorenz and C. R. Parks, *Rubber Chem. Technol.*, **36**, 194 (1963).

185. J. C. Andries et al., *Rubber Chem. Technol.*, **48**, 41 (1975).

186. G. M. Bristow and W. F. Watson, in L. Bateman, ed., *The Chemistry and Physics of Rubberlike Substances*, Maclaren, London, 1963, Ch. 14.

187. A. Casale et al., *Rubber Chem. Technol.*, **44**, 534 (1971).

188. M. Pike and W. F. Watson, *J. Polym. Sci.*, **9**, 229 (1952).

189. D. J. Angier and W. F. Watson, *Trans. Inst. Rubber Ind.*, **33**, 22 (1957).

190. R. J. Ceresa and W. F. Watson, *Trans. Inst. Rubber Ind.*, **35**, 19 (1959).

191. R. J. Ceresa, *Block and Graft Copolymers*, Butterworth, London, 1962.

192. G. Ayrey et al., *J. Polym. Sci.*, **19**, 1 (1956).

193. G. E. Betts et al., *Kauch. Rezina*, **28** (2), 11 (1969) (in Russian), *Sov. Rubber Technol.*, **28** (2), 14 (1969) (English translation).

194. G. Kraus and K. W. Rollmann, *J. Appl. Polym. Sci.*, **8**, 2582 (1964).

195. F. Bueche, *J. Appl. Polym. Sci.*, **4**, 101 (1960).

196. G. M. Bristow, *J. Polym. Sci.*, A, **1**, 2261 (1963).

197. C. Booth, *Polymer*, **4**, 471 (1963).

198. G. M. Bristow, *Trans. Inst. Rubber Ind.*, **38**, T 29, T 104 (1962).

199. A. N. Gent et al., *J. Appl. Polym. Sci.*, **8**, 455 (1964).

200. J. R. Beatty, *Rubber Rev.*, *Rubber Chem. Technol.*, **37**, 1341 (1964).

201. G. L. Slonimskii et al., *Starenie i Utomleriie*, *V.N.I.T.O. Rezinshchikov Conf.*, 1953, p. 100.

202. A. G. James, *Kaut. Gummi Kunstst.*, **26** (3), 87 (1973).

203. E. R. Thornley, Proceedings of the 4th Rubber Technology Conference, London, 1962, p. 682.

204. G. J. Lake, *Rubber Chem. Technol.*, **45**, 309 (1972).

205. G. J. Lake, *Rubber Age*, **104** (8), 30 (1972); **104** (10), 39 (1972).

206. G. J. Lake and P. B. Lindley, *Rubber J.*, **146** (9), 24 (10), 30 (1964).

207. R. S. Rivlin and A. G. Thomas, *J. Polym. Sci.*, **10**, 291 (1953).

208. A. G. Thomas, *J. Polym. Sci.*, **31**, 467 (1958).

209. A. N. Gent et al., *J. Appl. Polym. Sci.*, **8**, 455 (1964).

210. G. J. Lake and P. B. Lindley, *J. Appl. Polym. Sci.*, **9**, 1233 (1965).

211. F. H. D. Akkerman, *J. Appl. Polym. Sci.*, **7**, 1425 (1963).

212. G. J. Lake and P. B. Lindley, *J. Appl. Polym. Sci.*, **8**, 707 (1964).

213. E. H. Andrews, *J. Appl. Phys.*, **32**, 542 (1961).

214. P. P. Lindley, *Rubber Chem. Technol.*, **47**, 1253 (1974).

215. A. N. Gent, *J. Appl. Polym. Sci.*, **6**, 497 (1962).

216. D. G. Lloyd, *Fatigue Failure and Its Reduction in Natural Rubber*, Monsanto Ltd., Publication No. LA 24/1, 1966.

217. Z. N. Tarasova et al., *Kauch. Rezina*, **22** (10), 14 (1963) (in Russian); *Sov. Rubber Technol.*, **22** (10), 17 (1963) (English translation).

218. B. A. Dogadkin et al., Proceedings of the 4th Rubber Technology conference, London, 1962, p. 65.

219. Z. N. Tarasova and B. A. Dogadkin, "Chemical Nature of Crosslinks in Vulcanizates and Its Effect on the Strength and Elastic Properties and Fatigue Life of the

Vulcanized Rubber," in P. F. Badenov et al., eds., *Pneumaticheskie Shiny*, Izdatel'stvo Khimya, Moscow, 1969, NRPRA Translation No. 1676.

220. W. L. Cox and C. R. Parks, *Rubber Chem. Technol.*, **39**, 785 (1966).

221. E. B. McCall, *J. Rubber Res. Inst. Malaya*, **22**, 354 (1969).

222. G. M. Bristow and R. F. Tiller, *Kaut Gummi Kunstst.*, **23**, 55 (1970).

223. E. J. Blackman and E. B. McCall, *Rubber Chem. Technol.*, **43**, 651 (1970).

G. S. Trick

Research and Technology Branch
Department of Industry and Commerce
Government of Manitoba
Winnipeg, Canada

Chapter 10   Some Basic Physical Properties

1 INTRODUCTION

Any discussion on basic physical properties of stereoregular elastomers is hampered by the absence of materials that can really be regarded as standard; i.e., an extremely wide range of elastomers have been prepared using a variety of catalysts and experimental conditions. In most cases the principal concern has been the attainment of a practical rate of reaction in combination with the production of a material of such a molecular weight (as determined by bulk or dilute solution viscosity) that experience

583

indicates can be converted by conventional processing techniques to a finished product. Consequently, detailed investigations of basic physical properties, particularly as related to method of preparation, have rarely been made. In fact, the only polymer that falls within the scope of this book that can be regarded as reasonably well investigated is *cis*-1,4-polyisoprene. The synthetic polymer has not been thoroughly examined, but many of the measurements that have been made on natural rubber may also be applied to its synthetic analog, with, as shall be seen, a few important exceptions.

In somewhat similar fashion many measurements on emulsion SBR have been assumed to be applicable to the solution polymerized styrene–butadiene copolymers. As a result few experimental measurements have been reported on the commercially interesting solution SBRs. Although emulsion SBR data (1, 7) are often adequate, the effects, found in solution SBR, of narrow molecular weight distribution, differing vinyl contents, and styrene blockiness on basic physical properties have yet to be differentiated.

In general, it is the intention of this chapter to provide a short summary of the literature on a few of the basic properties that have been or can be measured on stereoregular polymers. As new polymers are prepared by different catalyst systems, they can then be subjected to similar measurements. This chapter is further restricted by not considering methods of determining microconfiguration such as infrared and NMR. Many of the results of such measurements and many of the basic properties of the bulk, raw polymers, such as molecular weight and its distribution, are considered in other chapters of this book. Emphasis is placed on properties that are considered to be of practical importance and discussion will not be presented on measurements that are aimed at advancing the theories of polymer behavior rather than providing specific information on a polymer of potential industrial importance.

2 GENERAL REVIEW ARTICLES

General surveys of the basic physical properties of elastomers, particularly as related to methods for the determination of molecular characteristics, have been made in a series of biannual review articles (2, 3). Other series of review articles (4–6), although primarily concerned with a broader range of properties and performance characteristics, also contain some information on basic physical properties. Other surveys contain similar data specifically related to butadiene polymers (7, 8), polymers of butadiene derivatives (9), isoprene polymers (10), and elastomeric ethylene copolymers (11). All of such general review articles provide a detailed survey of published work on measurements of polymer properties, both in the raw and cured states.

3 TRANSITION BEHAVIOR—GENERAL

High molecular weight materials have many common properties which arise from the generally similar behavior of long chains in the solid state. One of the most fundamental measurements on any polymer is a measurement of the temperature(s) at which solid state transitions occur since specific properties and the manner of usage depend to a large extent on the relation of these transition temperatures to the temperature at which the material is to be used. All polymeric materials will, at some temperature, undergo a glass transition (T_g) associated with a rather abrupt change from a plastic to a rubbery state (12, 13). In addition, many polymeric materials exhibit a first-order change at a temperature (T_m) resulting from the melting of polymer crystals to form an amorphous rubber. In general, a useful elastomer should have a T_g considerably below the temperature of application and have a structure such that crystallization, with its associated increase in hardness, does not take place on long-term standing of the finished product.

Detailed examination of the transition behavior of elastomers is of interest (a) to assess possible applications of the material and to predict where the material may be unsuitable, and (b) to provide information on microstructure which may be difficult to obtain by other techniques.

Although transitions in polymers may be examined experimentally by a wide variety of techniques, those most commonly used for detailed examination usually involve some direct measurement of specific volume, such as dilatometry, or of some parameter associated with specific heat, such as differential thermal analysis in its many forms. Different techniques may give somewhat different values for the transition temperatures since they may correspond to different rates of molecular motion (time scales). A number of widely mentioned values (1, 12–14) for T_g are given in Table 1; others are tabulated in Chapters 11 and 12. More detailed information is given below where some of the major types of stereoregular elastomers are discussed separately in terms of their transition behavior.

3.1 Transitions—Polyisoprene

Specific volume measurements on natural rubber have established a T_g of −72°C for this material. Since synthetic polyisoprenes may be expected to contain structures arising from somewhat different amounts of *trans*-1,4,

Table 1   Second-Order (Glass) Transition Temperatures of Some Polymers (1, 12–14, 19)

Polymer	T_g (°C)
trans-1,4-Polychloroprene	−45
Ethylene–propylene copolymer[a]	−37 to −60
SBR[a]	−50 to −65
cis-1,4-Polyisoprene	−70, −72
trans-1,4-Polyisoprene	−72, −60, −53
Butyl rubber	−70
Propylene oxide rubber	−75
trans-1,4-Polybutadiene	−80
cis-1,4-Polybutadiene	−95, −108
1,2-Polybutadiene	−12
trans-Polypentenamer	−97
Silicone rubber	−120
cis-Polypentenamer	−105, −135
Poly(2,3-dimethylbutadiene)	−11, −15
Polyethylene	−125
Polypropylene (atactic)	−20
Polypropylene (isotactic)	−10

[a] T_g depends on copolymer composition.

1,2, and 3,4 configurations, some variation of T_g is to be expected in the synthetic materials depending on their exact microstructures. Although other values have been reported it has been estimated (15) that the trans-1,4 polymer has a T_g similar to the cis-1,4 material. If V is the percentage of combined 1,2 and 3,4 content, then

$$T_g(°C) = -72 + 0.74\,V$$

(Adequate data are not available to establish separately the effects of 1,2 and 3,4 addition structures on the T_g.) Although other T_g values have been reported for the various configurations, this equation gives good estimates. The results exemplify the differences arising from different methods of measuring T_g. When the transition temperatures of purified balata and purified hevea were determined by shear modulus measurements with the torsion pendulum, results on both polymers indicated a T_g of −64°C. The difference between this value and the −72°C value determined dilatometrically for hevea is to be attributed to the faster time scale employed in the torsion pendulum measurements and is of the magnitude expected on the basis of the WLF equations (15).

The T_g is only the tip of the iceberg with regard to complexities associated with solid state transitions. Boyer has discussed many of these effects in detail (12, 13). Since such transitions are associated with the allowance of some hitherto restricted chain segmental motion, the T_g has been found, as anticipated, to change with degree of cross-linking in NR stocks (16) and in blends of polyisoprene with other compatible polymers (17). Blends of incompatible or semicompatible rubbers show two separate T_g values (17). Although cis-1,4-polyisoprene (or polybutadiene) did not show any transitions below its T_g, measurements of modulus and internal friction indicated one transition for trans-1,4-polyisoprene at 156°K (and two for trans-polybutadiene at 80–90 and 132°K). These were suggested to result from complex vibration and angular distortion to yield a crankshaft motion (18).

Much of the information on the first-order transition behavior of synthetic cis-1,4-polyisoprene must be inferred from measurements on natural rubber. Of all the elastomeric materials that show a first-order transition, none has been examined in nearly the detail or by the variety of techniques that have been applied to natural rubber. This situation arises from a number of factors, particularly the fact that natural rubber was found to undergo crystallization at a readily measurable rate at convenient temperatures and hence served as a model for the development of experimental and theoretical treatments of the crystallization of polymers in general. Much of the early quantitative work arose from investigations at the National Bureau of Standards; an interesting survey of such work has been presented (20). A detailed summary of a variety of experimental work on the crystallization of natural rubber has been made (21).

An example of the complexities encountered in measuring the first-order transition behavior of polymers in general, and of natural rubber in particular, is the observation that the semicrystalline polymer melts over a wide temperature range, with the temperature for disappearance of crystals depending markedly on the temperatures of crystal formation. The data of Bekkedahl and Wood (22) in Table 2 illustrate such an effect. Because of such complexities it is difficult to report a unique value for the

Table 2 Melting Measurements on Natural Rubber

Crystallization Temperature (°C)	Start of melting (°C)	Completion of Melting (°C)
−40	−34	−2.5
−30	−24.5	−1.5
−20	−15	+2
−10	−5	+7.7
0	+4.5	+14
10	+14	+23
15	+22	+32

melting point of natural rubber, although a calculated equilibrium melting point of 39.1°C has been reported (23).

Both NR and synthetic polyisoprene (Al–Ti), Ziegler) were shown to exhibit two associated melting points which depended on the previous crystallization temperature but not on gel or cross-linking. Both x-ray diffraction (24) and high resolution transmission electron microscopy (23) indicate that the two transitions have the same crystal structure and thus are morphological in origin. The lower melting point is attributed to thermodynamically less stable lamellar crystals with a higher interfacial free energy.

Extensive measurements have been made of the *rate* of crystallization of natural rubber at a fixed temperature, usually by following dilatometrically the decrease in volume as a function of time. For carefully purified natural rubber, the rate of development of crystallinity has been shown (25) to obey the equation:

$$C = C_0[1 - \exp(-kt^4)]$$

where C is the extent of crystallinity developed at time t, C_0 is the corresponding value after a long period of time, and k is a constant. The manner in which this equation fits typical experimental data is illustrated in Fig. 1. Because of the form of the equation, the half-life of crystallization ($t_{1/2}$) may be taken as a single parameter to represent the crystallization rate. Results at various temperatures (25) are summarized in Table 3 for purified natural rubber and plotted in Fig. 2 for both purified rubber and smoked sheet. The rate of crystallization reaches a maximum in the vicinity of -25°C, with the rate decreasing at higher and lower temperatures as either the melting temperature or the glass transition temperature is approached.

Despite the extensive experimental data available on the melting point

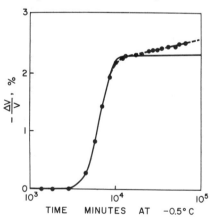

Figure 1 The progress of crystallization in purified natural rubber at -0.5°C, as the percent change in volume with time. The full curve is theoretical (25).

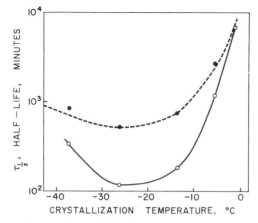

Figure 2 Dependence of the half-life of crystallization on the crystallization temperature for purified natural rubber (filled circles) and smoked sheet (open circles) (25).

and rate of crystallization of natural rubber, comparatively little information is available on the synthetically prepared *cis*-1,4-polyisoprenes. However, it has been found (26) that dilatometric rate of crystallization measurements at $-25°C$ provide a useful criterion for how closely the microstructure of a synthetic polymer approaches that of natural rubber (NR). In general (15) polyisoprenes prepared with a lithium type of catalyst show no evidence of crystallization even after 300 hours at $-25°C$, whereas those prepared with an AlR_3–$TiCl_4$ (Ziegler) type of catalyst have half-lives of around 25 hours at $-25°C$. Hence, the rate of crystallization of the synthetic polymers is markedly affected by the presence of small amounts of non-*cis*-1,4 configuration. This reduction in crystallization rate has been confirmed in isomerized NR; a similar effect is also produced by the addition of cross-links (27).

Making 50:50 blends of lithium polyisoprene and NR results in severely retarding the NR crystallization rates, although a similar mixture of NR with Alfin polyisoprene (high trans and 3,4) had no retarding effect. The lack of crystallizability of lithium polyisoprene was suggested to result from a change from head-to-tail *cis*-1,4 addition to tail-to-head addition whenever a 3,4 unit was formed during the polymerization reaction. This would result in poor packing and a lack of crystallizability of the segments in neighboring folded parallel chains (28). Mixtures of

Table 3 Rate of Crystallization of Purified Natural Rubber

Temperature (°C)	-37	-30	-25	-10	0
$t_{1/2}$ (minutes)	900	480	450	1000	8000

NR and lithium polyisoprene being compatible might be expected to cocrystallize, albeit at a slow rate. An incompatible Alfin polymer should have no effect.

As a further complication, fractional crystallization at $-78°C$ of a dilute (0.1%) solution of an Al–Ti catalyzed polyisoprene has allowed (26) the separation of two fractions which differ in their rates of crystallization in the dry state. Hence, the particular sample examined was not a truly random copolymer with all molecules of the same average composition of cis and non-cis units.

Using differential thermal analysis (29), a comparison has been made of the melting behavior of a number of polyisoprenes. As with natural rubber, it was found that the observed melting points depended on the temperature of crystallization and on the heating rate of the thermal analysis scan. Some of the experimental results obtained are given in Table 4.

A thorough investigation of the crystallization of synthetic polyisoprenes of high cis-1,4 content has arisen from observations on bulk flow behavior of such materials under selected conditions of temperature and shear rate. The first observations of this type were reported by Van der Vegt and Smit (30) who found for natural rubber an anomalous increase in viscosity with increasing shear rate at temperatures of 50°C or less. Similar measurements on a sample of synthetic polyisoprene did not show anomalous behavior. The difference was ascribed to differences in the shear induced crystallization of the two materials. A subsequent investigation (31) showed that the experimental conditions under which anomalous results were obtained could be directly related to the cis-1,4 content of the polyisoprene samples. A thorough investigation of such a phenomenon has been reported (32) using samples of polyisoprene containing 0–4% of 3,4 content and one sample with 6.2% trans-1,4. It was found that the experimental conditions under which crystallization took

Table 4 Melting Points of Polyisoprene Samples

Catalyst	% Cis-1,4	Upper Melting Transition (°C)[a]
Lithium	92	None
Ziegler	96	1.0
Ziegler	96	1.7
Ziegler	97	6.5
Natural rubber	98	13.5

[a] Samples crystallized 20 hours at −25°C, heated at 0.08°C/min.

place were a complex function of polymer microstructure, temperature, molecular weight, and geometry of the shear capillary. The ease of crystallization was markedly reduced as the *cis*-1,4 content was decreased, with the *trans*-1,4 and 3,4 units being equally effective in retarding crystallization. However, in some cases samples showed crystallization under shear even though they showed no tendency to crystallize under long-term storage at low temperatures. The experimental results confirm the general observation that of all the crystallizing polymeric materials that have been examined, the *cis*-1,4-polyisoprene structure shows the most sensitive response of its rate of crystallization to the structural purity of the chain. An important conclusion to be drawn from the work on shear induced crystallization is the realization that, although various samples of synthetic polyisoprene may show only small differences in *cis*-1,4 content, they may show marked differences in processing behavior if the shear conditions are such that induced crystallization may take place.

Stress-birefringence measurements (33) on NR, isomerized NR, and Ziegler *cis*-polyisoprene (Natsyn 200) have shown similar results. Based on the amounts of hysteresis found, the extent of crystallization on stretching was greatest for NR, less for Natsyn 200, and least for the anticrystallizing NR. It would be interesting to have similar measurements on lithium polyisoprene since it also crystallizes on stretching.

Among other measurements pointing out differences in the crystallization ability of natural rubber and *cis*-1,4-polyisoprene is a study (34) involving direct measurements of volume change of a cured sample resulting from crystallization during stretching. It was found that compared to the natural rubber sample it was necessary to stretch the synthetic material to a higher elongation to bring about measurable crystallization. On retraction the synthetic material retained crystallinity to a lower elongation than the natural polymer. A possible important variable that should be examined in measurements of this type is the effect of the state of cure upon the observed results.

3.2 Transitions—Polybutadiene

Because of the comparatively recent development of polymers with this structure there does not exist a long historical background of transition behavior measurements such as existed for the naturally occurring *cis*-1,4-polyisoprene. However, many of the experimental techniques and theories that were originally applied to natural rubber have also been applied to butadiene polymers.

With respect to glass transition behavior, the T_g is considerably influenced by the vinyl content but not by the relative amounts of cis and trans configuration. By differential thermal analysis it was found (35) that T_g (cis) = $-106°C$, T_g (trans) = $-107°C$, T_g (vinyl) = $-15°C$, so that the T_g could be expressed by

$$T_g(°C) = -106 + 91\ V$$

where V is vinyl content. A separate investigation (26) using a differential scanning calorimetry technique produced a slightly different relation:

$$T_g(°C) = -105 + 107\ V$$

For most applications, the difference between these equations is of no great importance.

There is some evidence from internal pressure and dilatometric measurements (36, 37) that cis-1,4-polybutadiene also undergoes another second-order transition at higher temperatures. The transition temperature was related to polymer structure and rose from 32 to 55°C as the cis-1,4 content increased from 33 to 94%, and was considered to account for the effect of temperature on the milling behavior of such materials.

Although cis-polybutadiene in the unstretched state is a rubbery amorphous material at room temperature, samples with sufficiently high cis-1,4 content harden rapidly from crystallization when cooled for a few minutes to around -20 to $-40°C$. After such treatment an x-ray crystal pattern develops and the melting temperature may be determined by dilatometry or thermal analysis. The exact value of the melting point shows considerable dependence on the catalyst system and experimental conditions employed in the polymer preparation. In general, the measured melting point shows a relation to the cis-1,4 content but in many cases a more complex situation has been found to exist. The complexities arise from the nonrandom placement of the non-cis units in the polymer chain. For example, samples prepared under certain conditions with an $AlR_2Cl/CoCl_2$ catalyst covered a wide range of cis-1,4 levels with no corresponding variations in the measured melting point (38). Such results indicate the presence of block copolymers, polymer blends, or a mixture of the two. (These are blocks or blends of different structural units arising from the same monomer.) On the other hand, samples of the same cis-1,4 content prepared with an AlR_3/TiI_4 catalyst had a considerably lower melting point, indicating a more random placement of copolymer units. To avoid the uncertainties and difficulties of preparing truly random copolymers with high cis-1,4 content, an indirect approach was made (39) consisting of isomerizing a starting material of high cis-1,4 content and hence introducing random trans-1,4 components. The isomerization can

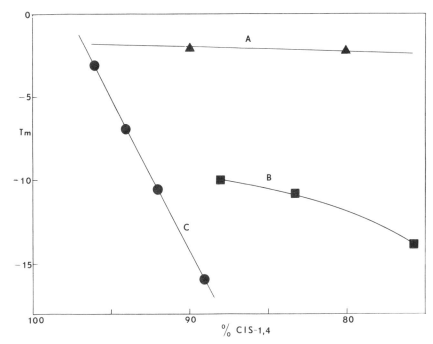

Figure 3 Melting point versus *cis*-1,4 content for polybutadienes made with various catalysts: *A*, AlR₂Cl/CoCl₂; *B*, AlR₃/TiI₄; *C*, from isomerization of high *cis*-polybutadiene (15).

be conveniently carried out photochemically with diphenyl disulfide as sensitizer (40). The measured melting points at a given *cis*-1,4 content are found (15) to be even lower than those obtained by a variety of methods of direct preparation. The complex relation between melting point and *cis*-1,4 content is illustrated in Fig. 3. Whether the nonrandom polymers are blocks or blends is difficult to establish, requiring either the selection of a solvent system that can selectively dissolve polymers with small structural differences or, alternatively, the use of a fractional crystallization technique employing low temperatures and very dilute solutions. In one case (41) a sample prepared with an AlR₃/TiCl₄ catalyst, when subjected to selective solvent extraction, allowed isolation of a fraction of increased *cis*-1,4 content, indicating some degree of blending of chains of different structural composition.

As in the case of all polymeric materials, the experimentally determined melting points depend not only on the precise structural composition but also on such factors as time and temperature of crystallization and rate of melting. The highest reported (42) T_m value is 12.5°C for a sample of over 99% *cis*-1,4 structure, which is as high as the value calculated with the Flory copolymer equation (43).

Differential thermal analysis measurements (44) on a sample of 96.4% *cis*-1,4 content are in general agreement with other observations and in addition show the presence of two distinct overlapping melting endotherms for samples that were cooled very rapidly and then warmed slowly. The significance of the two endotherms has not been established.

As with *cis*-1,4-polyisoprene, *cis*-1,4-polybutadiene is a material that lends itself readily to measurements of rates of crystallization by simple dilatometric techniques. As with the melting points, the data obtained with such measurements depend on the frequency and randomness of the non-*cis*-1,4 structures in the polymer chain. At −20°C the half-time of crystallization for a 99% *cis*-1,4-polybutadiene (with mp of +2.2°C) was 5 minutes. This polybutadiene made with a uranium catalyst was 40% crystalline (42). Some typical results (39) on samples where the *trans*-1,4 content was increased by isomerization are shown in Fig. 4 with half-life of crystallization as a function of temperature and *trans*-1,4 content. Similar data from a more extensive collection have been smoothed (39) and are summarized in Table 5.

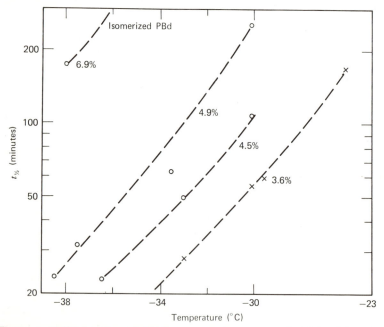

Figure 4   Half-time of crystallization versus temperature for 1,4-polybutadienes of different *cis*-1,4 content (based on results of ref. 39). Numbers on curves are percent *trans*-1,4 content.

Table 5   Crystallization Rates of cis-1,4-Polybutadienes

Trans Content[a] (%)	T_m (°C)	$t_{1/2}$ (minutes) at				
		−35°C	−32°C	−28°C	−26°C	−22°C
3.6	−8.0	18	32	87	160	1200
4.5	−10.5	30	60	170	—	—
4.9	−12.5	52	100	300	—	—
6.9	−16.0	350	650	—	—	—
21.7	—		Noncrystalline			

[a] All samples had approximately 5% vinyl content.

Table 5 illustrates a number of important points that are characteristic of crystallizing polymers in general: (a) the marked effect of random non-cis-1,4 structures on the measured melting point, (b) the marked effect of small changes in structure on the rate of crystallization at a fixed temperature, and (c) the marked effect of small changes in temperature on the rate of crystallization. Further analysis of the data (39) has shown that the decrease in rate of crystallization with increasing trans-1,4 content arises largely from the decreased melting point, and that comparison of the data at a fixed quantity of a thermodynamic function related to the degree of supercooling results in superposition of the rates for all samples except that containing 6.9% trans-1,4. Hence, only above some critical concentration of noncrystallizing units does the decrease in crystallization rate become greater than can be accounted for by the depression in melting point.

A secondary effect contributing to the form of the time dependence of the crystallization of cis-1,4-polybutadiene and to some extent also contributing to the half-life of crystallization has been shown (43) to arise when the polymer is subjected to a shearing action in a laboratory mill. The results observed are attributed to shear ordering of either the polymer itself or to impurities. In either case, the effects are small in comparison to those brought about by changes in microstructure.

Since the cis/trans ratio in 1,4-polybutadienes can be readily modified by photochemical isomerization (40) with a minimum of side reactions, it is possible, by starting with high cis and high trans samples, to prepare materials with any desired cis/trans content. The equilibrium composition obtained by such a technique is about 77% trans (45). Direct quantitative x-ray measurements reveal that crystallinity occurs at room temperature for materials of more than 75% trans-1,4 content. With less than this trans content no crystallinity at room temperature is present but it may be inferred from the literature than when the trans-1,4 content becomes less than about 20%, low temperature crystallization of the *cis*-1,4 structure will take place.

Because of their mixed structural composition, polybutadienes prepared with lithium type catalysts do not exhibit sufficiently long sequences of a given structural component to allow crystallization to take place. NMR and other techniques may allow estimates of the minimum average length of sequences needed to allow crystallization to take place (46).

3.3 Transitions—Ethylene-Propylene Rubbers

Normally, EPM materials are considered to be random copolymers with a sufficiently high propylene content to prevent crystallization. In practice, however, complex transition behavior may arise from complications that may or may not be related: (*a*) the exact overall composition of the copolymer, (*b*) the distribution of copolymer units in the chain, or (*c*) the presence of small amounts of crystallinity or short-range intermolecular order. The glass transition of EPM copolymers has been the subject of several reviews (47–49). It is concluded that if the ethylene content is less than 50 wt %, the T_g of the copolymers may be expressed by

$$T_g = 0.58 \frac{(-81 - T_g) W_2}{1 - W_2} - 20$$

where W_2 is the weight fraction of ethylene and temperatures are in degrees centigrade. Above 50 wt % ethylene, marked deviations from this relationship take place, with a T_g of $-58°C$ being observed for materials of quite different compositions. Data obtained are illustrated in Fig. 5. On the basis of available experimental data, it was concluded (47) that the complex T_g behavior arose from nonrandom copolymerization but that associated crystallization need not be taking place.

Baldwin and Ver Strate, in their extensive review of ethylene–propylene elastomers (48), discuss this behavior and the complications resulting from the onset of crystallization. The presence of diene termonomers in EPDM influences the T_g in ways depending on the individual diene but the effects are small because the diene concentration is low. As an example, ethylidene norbornene increased the measured T_g by about 0.8°C/wt % diene at levels below 10 wt %.

Historically, in the development of EPM rubbers, the monomer feed composition was simply adjusted so that sufficient propylene was incorporated to lead to a material that was amorphous at room temperature. However, the presence of small amounts of crystallinity can lead to desirable properties related to cold flow and processing behavior. It has been shown (50) that small amounts of crystallinity can lead to marked increases in the relaxation times from stress relaxation measurements and

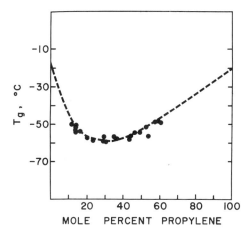

Figure 5 The glass transition temperature T_g versus composition for EPM fractions should be (47). The line is representative of other literature data.

that the relaxation times also change sharply above the melting point of the crystals. Other properties associated with small amounts of crystallinity in EPDM rubbers have been examined in detail (51). It was found that one polymer containing as little as 50 mole % ethylene exhibited a broad melting transition around $-20°C$. It was also found that the degree of crystallinity not only depended on ethylene content but was also strongly affected by the monomer reactivity ratios associated with various catalyst systems. The presence of crystallinity was evident in a marked response of bulk viscosity to temperature, low room temperature cold flow, high green strength, etc. Most commercial EPDMs exhibit some degree of crystallinity. Crystallization occurs with estimated minimum crystallizable sequence lengths of 10–20 main chain carbon atoms (52). It should be emphasized that polymers with small amounts of crystallinity can have definite advantages in terms of handling and processing, provided it is possible, during polymer preparation, to maintain careful control over the microstructure sequences that result in crystallinity. If such control cannot be maintained, different samples are likely to exhibit severe changes in processing behavior.

4 SOLUTION PROPERTIES

Most of the stereoregular elastomers have been the subject of some work on various solution properties, usually using such measurements as osmotic pressure, light scattering, and ultracentrifugation. In the past measurements were not made with an intent to characterize standard commercial materials but rather to provide information to test various theories of solution properties. There has been a resurgence of interest in such properties because of their utility in measuring molecular weight and its distribution by gel permeation chromatography (GPC). For example, GPC has been used (53) to study the differences between the Al/Ti and lithium polyisoprenes and to show that *unmasticated* natural rubber is bimodal but that its distribution is probably genetically controlled.

Information appropriate to this chapter will be discussed in terms of the particular polymers involved.

4.1 *cis*-1,4-Polyisoprene

One measurement of particular interest in characterizing polymers involves the relation between a dilute solution viscosity measurement and a viscosity related molecular weight average (M). Using fractions of natural rubber (54) the following relation was obtained:

$$[\eta] = 5.02 \times 10^{-4} M^{0.667} \quad \text{(toluene, 30°C)}$$

where $[\eta]$ is the intrinsic viscosity. Further measurements on carefully prepared samples of *cis*-1,4-polyisoprene made with butyllithium catalyst led to the following relations:

$$[\eta] = 2.00 \times 10^{-4} M^{0.728} \quad \text{(toluene, 30°C, ref. 55)}$$

$$[\eta] = 2.22 \times 10^{-4} M^{0.683} \quad \text{(isooctane, 30°C, ref. 55)}$$

The two equations quoted for toluene are considered to agree within experimental error. For measurements under theta conditions, i.e., conditions of solvent and temperature such that the second virial coefficient is zero, the following relation was obtained (56):

$$[\eta] = 1.2 \times 10^{-3} M^{1/2} \quad \text{(methyl n-propyl ketone, 14.5°C)}$$

For a sample of synthetic polymer of >97% *cis*-1,4 content, another equation (57) is in good agreement with those just given:

$$[\eta] = 1.9 \times 10^{-4} M^{0.745} \quad \text{(toluene, 30°C, ref. 57)}$$

Some of the complexities that arise in making solution measurements on polyisoprenes become evident from reported results (58) on light scattering and viscosity measurements on synthetic *cis*-1,4-polyisoprenes that led to the relation:

$$[\eta] = 5.9 \times 10^{-6} M^{1.01} \quad \text{(toluene, presumably 25–30°C)}$$

The last equation leads to quite different results from the preceding equations in toluene. The reason for these differences is not clear, but in general the literature on solution measurements on both natural rubber and synthetic polyisoprene points out the difficulties in making quantitative measurements of clearly defined significance. For example, both polymers contain (58) a material ("microgel") that markedly affects the light scattering properties of the solutions, with filtration only partially removing the component. In addition, solutions prepared in tetrahydrofuran yield a much lower polymer M_w value than those prepared in chloroform (58). On the other hand, a sample of natural rubber showed (8) microgel in light scattering measurements whereas the synthetic polymer behaved normally with $M_w = 2 \times 10^6$ and $M_n = 6.2 \times 10^5$. Such behavior has led to the suggestion that the synthetic polyisoprene "gels" are not networks but rather large aggregates containing a dense nucleus arising from a heterogeneous catalyst residue (59). It is probably because of experimental difficulties arising from macrogel, microgel, and branching that no thorough study has been reported on the molecular weight and molecular weight distribution of the various commercially available *cis*-1,4-polyisoprenes.

4.2 Polybutadiene

In sharp contrast to the *cis*-polyisoprene case, a very extensive study has been made (8) on the molecular weights and molecular weight distributions of a large number of commercial polybutadienes prepared by anionic, coordinated, and free radical catalyst systems. The work reported can be divided roughly into two parts: (*a*) the commercial samples, after purification, were dissolved in cyclohexane and measurements were made of limiting intrinsic viscosity, light scattering molecular weight, and osmotic pressure molecular weight, and (*b*) absolute alcohol was added to cyclohexane solutions of each purified polymer, and by a suitable choice of experimental conditions about 15 fractions of differing molecular weight were obtained. Each fraction was characterized in the same manner as was the whole polymer. In this way, integral and differential molecular weight distribution curves were established for each polymer. In association with the absolute molecular weight measurements, second virial coefficient data were generated. The experimental results reported are much too extensive to be quoted here and the original paper should be consulted for the actual results. As an extension of such work, it would be of interest to establish how well the various manufacturers control the molecular parameters of their product in their factories over extended periods.

To correlate intrinsic viscosity with molecular weight a number of experimentally determined relations are available. For polybutadiene prepared in hydrocarbon solvent with butyllithium, the following equations have been reported:

$$[\eta] = 1.45 \times 10^{-4} M^{0.76} \qquad \text{(benzene, 32°C, ref. 59)}$$
$$[\eta] = 6.11 \times 10^{-4} M^{0.659} \qquad \text{(toluene, 25°C, ref. 60)}$$
$$[\eta] = 1.56 \times 10^{-4} M^{0.78} \qquad \text{(toluene, 25°C, ref. 61)}$$
$$[\eta] = 2.17 \times 10^{-4} M^{0.75} \qquad \text{(toluene, 25°C, ref. 62)}$$

Despite the differences in the constants in the last three equations, they lead to reasonably similar values, but the first equation leads to significantly different results. For polymers with about 93–94% cis-1,4 configuration, the following equations have been established:

$$[\eta] = 1 \times 10^{-4} M^{0.77} \qquad \text{(benzene, 32°C, ref. 63)}$$
$$[\eta] = 1.3 \times 10^{-4} M^{0.772} \qquad \text{(toluene, 30°C, ref. 64)}$$
$$[\eta] = 3.39 \times 10^{-4} M^{0.688} \qquad \text{(toluene, 30°C, ref. 65)}$$

From further solution property measurements on similar polymers, several relations under theta conditions were also established (66):

$$[\eta] = 1.81 \times 10^{-3} M^{0.50} \qquad \text{(diethyl ketone, 10.3°C)}$$
$$[\eta] = 1.78 \times 10^{-3} M^{0.50} \qquad \text{(methyl isoamyl ketone, 12.6°C)}$$
$$[\eta] = 1.57 \times 10^{-3} M^{0.50} \qquad \text{(methyl n-propyl ketone, 59.7°C)}$$

For the high cis-1,4 content polybutadienes (97–98%) a complete investigation has been made (67) of solution properties, including a listing of solvents and nonsolvents for the polymer. A combination of osmotic pressure and viscosity measurements on fractions led to the following relations:

$$[\eta] = 3.05 \times 10^{-4} M^{0.725} \qquad \text{(toluene, 30°C)}$$
$$[\eta] = 3.37 \times 10^{-4} M^{0.715} \qquad \text{(benzene, 30°C)}$$

In addition, from determination of the osmotic second virial coefficient as a function of temperature for fractions in isobutyl acetate a theta temperature of 20.5°C was established (67). The corresponding viscosity-molecular weight relation is

$$[\eta] = 1.85 \times 10^{-3} M^{0.5} \quad \text{(isobutyl acetate, 20.5°C)}$$

By determining theta temperatures by either estimating the critical solution temperature for an infinite molecular weight polymer or by the effect of temperature on the osmotic second virial coefficient, the following relations were also obtained (65):

$$[\eta] = 1.90 \times 10^{-3} M^{0.50} \quad \text{(n-heptane, -1°C)}$$
$$[\eta] = 1.81 \times 10^{-3} M^{0.50} \quad \text{(n-propyl acetate, 35.5°C)}$$

Further relations are

$$[\eta] = 1.54 \times 10^{-3} M^{0.566} \quad \text{(toluene, 30°C, ref. 64)}$$
$$[\eta] = 3.24 \times 10^{-3} M^{0.57} \quad \text{(toluene, 25°C, ref. 68)}$$

It is clear from a plot of some of the reported viscosity-molecular weight relations that there is considerable disagreement among the reported results. In addition, some of the data are dependent on each other; for example, the results calculated for theta conditions (66) are subject to the validity of measurements in benzene (67). Some potential sources of error in generating such data include (*a*) failure to obtain sharp fractions if osmotic pressure measurements are made and (*b*) difficulties arising from branched structures which affect viscosity values markedly. Some of the problems encountered with branching and microgel in *cis*-1,4-polyisoprene also arise with *cis*-1,4-polybutadiene. For example, it has been shown (63) that the presence of branching markedly affects the second virial coefficient from light scattering while having no effect on the coefficient from osmotic pressure measurements. The presence of microgel also has considerable effect (59) on the light scattering molecular weight and second virial coefficient. There is some evidence (64, 69) that *cis*-1,4-polybutadienes prepared with cobalt based catalysts have considerable branching whereas polymers from a titanium based catalyst are linear.

With respect to molecular weight distribution measurements, a number of investigations have been made (69–73) using conventional precipitation, selective solubility, and gel permeation chromatography techniques on specific samples of polybutadienes. However, for results on commercial samples the measurements mentioned previously (8) should be consulted. In general, it is difficult to assess quantitatively the validity of either viscosity molecular weight data or molecular weight distribution data. However, as the quantitative importance of such parameters for determining the processing and performance characteristics of polymers becomes clearer, improvements in quantitative measurements will be warranted.

4.3 Ethylene–Propylene Rubbers

Several references to the intrinsic viscosity-molecular weight relationship of EPM and EPDM exist, but they are not always in agreement with one another, not always in the same solvent, not at the same temperature, nor do they cover the same composition range (see, e.g., the discussion in ref. 48). One relationship based on viscosity-molecular weight relations for polyethylene and polypropylene is apparently consistent with and satisfactory for use with the copolymers containing 50–60 wt % ethylene:

$$[\eta] = 3.15 \times 10^{-4} M^{0.74} \qquad \text{(tetralin, 135°C, ref. 74)}$$

5 POLYMER-SOLVENT INTERACTIONS

Although considerable insight into the interaction of a polymer with a given solvent can be obtained from measurements of various properties on dilute solutions (viscosity, osmotic pressure, light scattering, etc.), measurements made over a much broader concentration range provide information on solvent resistance of polymers and on the characteristics of vulcanizates. If precise details are desired on all the parameters affecting the solution behavior of a given polymer-solvent system over all concentrations, then a number of theories have been developed (75–78) to treat data on measurements of this type. However, a discussion of the relative merits of the various theories is beyond the scope of this chapter. Of more importance for present purposes is a discussion of measurements that will, at least to a good approximation, give information on the interaction of solvents with stereoregular polymers. The most generally applied technique for measuring such interaction involves combining equilibrium swelling measurements on a vulcanized rubber with some parameter characterizing the vulcanizate network (usually obtained from stress-strain measurements) by means of the following equation (79):

$$-\ln(1 - V_r) - V_r - \chi V_r^2 = \frac{2 V_0 C_1}{RT} V_r^{1/3} \qquad [1]$$

where V_r is the volume fraction of rubber at equilibrium swelling; V_0 is the molar volume of swelling agent; C_1 is a network parameter from elasticity measurements; and χ is the solvent-polymer interaction parameter. A modification of this equation (80) includes a term involving the functionality (f) of the cross-link points:

$$-\ln\left(1-V_r\right)-V_r-\chi V_r^2=\frac{2V_0C_1}{RT}\left(V_r^{1/3}-\frac{2V_r}{f}\right) \qquad [2]$$

Another useful experimental technique involves determining the activity (a_1) of the swelling agent from vapor pressure measurements as a function of the degree of swelling (V_r) of the raw polymer and applying the relation (75)

$$\ln a_1=\ln\left(1-V_r\right)+V_r+\chi V_r^2 \qquad [3]$$

Recently the application of gas-liquid chromatography to solvent-polymer interactions (81, 82) has resulted in the rapid generation of data on many solvents and into high polymer concentration ranges. Some theoretical questions still confront the method but its facility to present precise data is striking. So far its only pertinent use has been on raw NR (82).

Comments can now be made on the results of measurements of this type as applied to various stereoregular polymers. For the same experimental data, equations [1] and [2] will yield slightly different χ values. Experimentally it has been found that there is often a dependence of χ upon V_r. Compared to equation [1], equation [2] will lead to slightly higher χ values and show a greater dependence on V_r.

5.1 *cis*-1,4-Polyisoprene

As before, most of the actual experimental measurements have been made on natural rubber and the reasonable assumption is made that the same results apply to the synthetic material. Among all the measurements that have been made combining stress-strain and swelling measurements the most extensive appear to be those of Bristow (83) with the following results for natural rubber:

$$\chi = 0.411 \qquad \text{(decane, 25°C, equation [1])}$$

$$\chi = 0.40+0.20\,V_r \qquad \text{(decane, 25°C, equation [2])}$$

The manner in which the experimental results may be fitted by equation [1] is illustrated in Fig. 6. Much less extensive data give the following results:

$$\chi = 0.42 \qquad \text{(benzene, 25°C, equation [1])}$$

$$\chi = 0.41+0.20\,V_r \qquad \text{(benzene, 25°C equation [2])}$$

$$\chi = 0.425+0.20\,V_r \qquad \text{(heptane, 25°C, equation [1])}$$

$$\chi = 0.415+0.35\,V_r \qquad \text{(heptane, 25°C equation [2])}$$

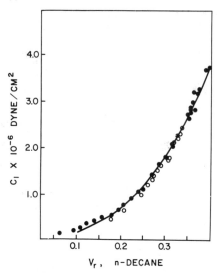

Figure 6 The relation between the elastic constant C_1 and equilibrium volume swelling V_r. The line is based on equation [1] with $\chi = 0.411$ (83). ●, Natural rubber; ○, cis-1,4-polyisoprene.

Similar measurements on synthetic cis-1,4-polyisoprene show minor differences from natural rubber but it is believed (83) that the differences arise from difficulties in treating the stress-strain data. Another set of experiments (84) does not give details of the range or number of experimental points used, but gives the following results:

$$\chi = 0.43 + 0.05 V_r \qquad \text{(toluene, 30°C, equation [2])}$$

$$\chi = 0.44 + 0.18 V_r \qquad \text{(benzene, 25°C, equation [2])}$$

$$\chi = 0.46 + 0.20 V_r \qquad \text{(heptane, 30°C, equation [2])}$$

which are in fair agreement with the more recent results quoted above.

Some of the complexities in evaluating χ values from equation [1] or [2], particularly in determining C_1, may be avoided by making vapor activity measurements over a range of concentrations and applying equation [3]. Considerable precision is required in the measurements but experimental methods for measuring vapor pressure and weight changes are readily available to attain the necessary precision. Measurements of this type (85) give, for V_r values of 0.5–0.8, a value of $\chi = 0.43$ for benzene at 25°C. A very thorough examination (86) of the natural rubber-benzene system, using high pressure osmometry and vapor activity measurements, yields, when the data are treated by theories which are the basis for equation [3], values approximately:

$$\chi = 0.39 + 0.04 V_r \qquad \text{(benzene, 25°C, equation [3])}$$

For general collections of experimental values for χ, sometimes at single

V_r values and sometimes covering a range of V_r, but including a wide variety of swelling agents, several references may be consulted (87–90).

5.2 Polybutadiene

Using stress-strain and swelling measurements on cured 94% cis-1,4-polybutadiene combined with equation [2], the following results were obtained (91):

$$\chi = 0.38 \qquad \text{(toluene, 25°C)}$$
$$\chi = 0.42 \qquad \text{(benzene, 25°C)}$$
$$\chi = 0.49 + 0.18\,V_r \qquad \text{(decane, 25°C)}$$

The experimental results for these and other swelling agent systems are illustrated in Fig. 7. Some similar measurements on 97% cis-1,4 polymer (92) gave a mean value of $\chi = 0.546$ in decane over a limited V_r range. The results are consistent with the equation above. Slightly different results were obtained (93) for 97–98% cis-1,4-polybutadiene in heptane

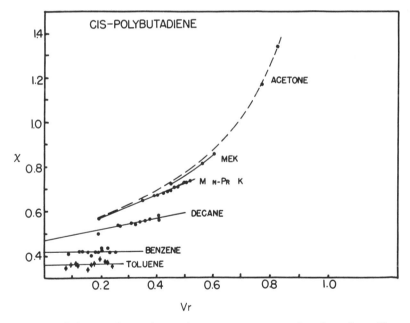

Figure 7 Solvent-polymer interaction parameters as a function of swelling ratio for cis-1,4-polybutadiene in various swelling agents (91).

at 20°C for sulfur vulcanizates:

$$\chi = 0.428 + 0.535\,V_r$$

and for resin/peroxide vulcanizates:

$$\chi = 0.428 + 0.51\,V_r$$

Results in benzene did not depend on the vulcanization method:

$$\chi = 0.33 + 0.638\,V_r$$

From measurements in heptane (94) of polybutadiene made from a butyllithium catalyst, the following equation results:

$$\chi = 0.37 + 0.52\,V_r \qquad \text{(heptane, equation [2])}$$

5.3 Ethylene–Propylene Rubbers

Standard measurements on stress-strain and swelling for an EPDM rubber containing 47 mole % propylene, lead to the results (95):

$$\chi = 0.44 \qquad \text{(heptane, 25°C, equation [2])}$$

$$\chi = 0.49 + 0.33\,V_r \qquad \text{(benzene, 25°C, equation [2])}$$

The same paper reports the χ values for a range of swelling agents for a single cured sample. Measurements on peroxide cured EPM polymers having from 30 to 60 mole % propylene showed (96) that within experimental error, all the results could be represented by

$$\chi = 0.485 + 0.256\,V_r \qquad \text{(benzene, 25°C, equation [2])}$$

in fair agreement with the previously cited results. Similar measurements on EPM of 47 mole % ethylene gave (97)

$$\chi = 0.493 + 0.26\,V_r \qquad \text{(benzene, equation [2])}$$

and for EPDM with 35 mole % propylene:

$$\chi = 0.513 + 0.18\,V_r \qquad \text{(benzene, equation [2])}$$

An examination has been made (98) on EPM samples of about 45 mole % propylene for five solvents over a range of temperatures. All calculations were based on the assumption of the correctness of the published results (96) for swelling in benzene. By such measurements χ values were obtained as a function of concentration and temperature. Direct measurements (99) on an EPDM containing 36 mole % propylene yielded

$$\chi = 0.354 + 0.083\,V_r \qquad \text{(heptane, 23°C, equation [1])}$$

Similar measurements on EPM of 45 mole % propylene (100) gave

$$\chi = 0.48 + 0.29 V_r \quad \text{(benzene, 25°C, equation [2])}$$

Some equilibrium vapor pressure-swelling measurements are also available (101) and among the results obtained are the following:

$\chi = 0.26$ (hexane, 23°C, equation [3])
$\chi = 0.30$ (carbon tetrachloride, 23°C, equation [3])
$\chi = 0.65$ (benzene, 23°C, equation [3])

Baldwin and Ver Strate (48) in reviewing the available data suggested a reasonable average relation for swelling in benzene at 25°C to be

$$\chi = 0.488 + 0.271 V_r$$

They regarded the data for swelling in heptane to be beyond reconciliation and experimental error. As a general observation, χ values quoted in the literature are sometimes in rather serious disagreement with each other. In some cases the disagreement may be traced to the method of treating the experimental data (e.g., whether equation [1] or [2] is employed) but in most cases the difficulties arise from experimental problems, particularly in obtaining and treating stress-strain data. Until some of these problems are solved, care must be taken to realize the limitations of applying some of the reported results and in the interpretation of their significance on the molecular level.

5.4 Cohesive Energy Densities

The preceding discussion has been directed toward specific polymer-solvent systems, but it is obviously desirable to be able to consider a more general case where it is possible, by combining two parameters each representing a given polymer and a given swelling agent, to predict the solvent-polymer behavior of the combined system. Such parameters can

Table 6

Polymer	δ (cal/ml)$^{1/2}$	Ref.
Natural rubber	8.10, 8.08	87
Natural rubber	8.03, 8.01	106
cis-1,4-Polybutadiene	8.5–8.6	107
cis-1,4-Polybutadiene	8.44–8.50	91
EPDM (47% propylene)	7.9	95
EPM (51 mole % propylene)	7.8–8.1	101

be derived from heat of mixing measurements. The parameter usually employed is the cohesive energy density (CED), or more commonly its square root (δ, solubility parameter). The manner in which such parameters may be employed to treat practical applications of solubility behavior has been the subject of a number of reviews (1, 102–105) including extensive collections of δ values for many polymers and swelling agents. Some reported δ values are listed in Table 6.

6 CURRENT PROBLEMS

From the preceding brief coverage it is apparent that a number of experimental techniques are available to examine what are considered to be significant physical properties of elastomers and that considerable effort has been expended on making such measurements on elastomers of technical importance. In many cases such determinations have been useful in advancing various fundamental theories for polymer behavior. However, a considerable gap still exists in efforts to transpose basic physical property measurements into quantities that may be useful in practical applications. For example, work on isothermal crystallization of elastomers has strengthened our understanding of the kinetics of polymer crystallization. However, although it has been well illustrated (32) that elastomer crystallization may occur under certain shear conditions, markedly affecting the flow behavior, it has not been possible to extrapolate the results of isothermal crystallization measurements into valid predictions of the role of crystallization in the flow of elastomers. By the same token, various molecular weight parameters of elastomers can be experimentally determined, but again it is difficult to convert such quantities into predictions of processing and performance behavior. An excellent description has been given (108) of the deformation and flow of elastomers as related to processing characteristics and will be the subject of further discussion in subsequent chapters of this book.

There is an appreciable gap in our data on fundamental properties of solution polymerized styrene–butadiene copolymers. This lack of information may, in part, result from the difficulties inherent in preparing random copolymers with solution catalysts. Such scattered data as exist relate to lithium catalyzed copolymers and the effect of microstructure on T_g (108–111) and to high vinyl butadiene content copolymers (112).

Among other specific problems in characterizing polymers the following are considered particularly important and yet lacking in a straightforward method of measurement: (a) branching in polymers that can be assessed in terms of the frequency of branch points and the size distribution of the branches; and (*b*) identification and characterization of microgel found in many elastomer systems. This material, not visible by normal observation, seriously affects filtration behavior and light scattering measurements on solutions and many bulk properties.

When these kinds of parameters are measurable, then, combined with current measurements and with information on shear induced crystallization, it should be possible to make predictions on processing behavior and properties of a finished product.

REFERENCES

1. J. Brandrup and E. H. Immergut, eds., *Polymer Handbook*, 2nd ed., Wiley-Interscience, New York, 1975.

2. C. W. Wadelin and G. S. Trick, *Anal. Chem.*, **39**, 239R (1967); **41**, 299R (1969); **43**, 334R (1971).

3. C. W. Wadelin and M. C. Morris, *Anal. Chem.*, **45**, 333R (1973); **47**, 327R (1975).

4. F. C. Weissert, *Ind. Eng. Chem.*, **61**, 53 (1969).

5. L. A. Wood, *Rubber Chem. Technol.*, **49**, 189, (1976).

6. A. R. Payne et al., *J. Institution of the Rubber Industry*, **1**, 17, 206 (1967).

7. W. M. Saltman, "Butadiene Polymers," in *Encyclopedia of Polymer Science and Technology*, Vol. 2, Wiley-Interscience, New York, 1965.

8. J. Curchod, *Rubber Chem. Technol.*, **43**, 1367 (1970).

9. W. M. Saltman and J. N. Henderson, "Butadiene Derivatives, Polymeric," in *Encyclopedia of Polymer Science and Technology*, Vol. 2, Wiley-Interscience, New York, 1965.

10. A. R. Bean et al., "Isoprene Polymers," in *Encyclopedia of Polymer Science and Technology*, Vol. 7, Wiley-Interscience, New York, 1967.

11. P. J. Canterino, "Ethylene Polymers," in *Encyclopedia of Polymer Science and Technology*, Vol. 6, Wiley-Interscience, New York, 1967.

12. R. F. Boyer, *Rubber Chem. Technol.*, **36**, 1303 (1963).

13. R. F. Boyer, *J. Macromol. Sci. Phys.*, **9** (2), 187 (1974); **8** (3), 503 (1973); *J. Polym. Sci., Symp.*, **50**, 189 (1975).

14. G. Dall'Asta, private communication.

15. K. W. Scott et al., *Rubber Plast. Age*, **42**, 175 (1961).

16. L. A. Wood et al., *Rubber Chem. Technol.*, **45**, 1388 (1972).

17. I. Yu. Zlatkevich and V. G. Nikolskii, *Rubber Chem. Technol.*, **46**, 1210 (1973).

18. R. J. Morgan et al., *Rubber Chem. Technol.*, **45**, 1033 (1972).

19. O. Kramer et al., *Rubber Chem. Technol.*, **48**, 164 (1975).

20. N. Bekkedahl, *Rubber Chem. Technol.*, **40** (3), xxxv (1967).

21. E. H. Andrews and A. N. Gent, in L. Bateman, ed., *The Chemistry and Physics of Rubberlike Substances*, Wiley, New York, 1963, Ch. 9.

22. N. Bekkedahl and L. A. Wood, *Ind. Eng. Chem.*, **33**, 381 (1941).

23. B. C. Edwards, *J. Polym. Sci.*, **13**, 1387 (1975).

24. H.-G. Kim and L. Mandelkern, *J. Polym. Sci.*, *A-2*, **10**, 1125 (1972).

25. A. N. Gent, *J. Polym. Sci.*, **18**, 321 (1955).

26. G. S. Trick, Goodyear Tire & Rubber Co., unpublished results.

27. E. H. Andrews et al., *Rubber Chem. Technol.*, **45**, 1315 (1972).

28. M. J. Brock and M. J. Hackathorn, *Rubber Chem. Technol.*, **45**, 1303 (1972).

29. W. Cooper and R. K. Smith, *J. Polym. Sci.*, *A*, **1**, 159 (1963).

30. A. K. Van der Vegt and P. P. A. Smit, *Adv. Polym. Sci.* (*S.C.I. London*), **26**, 313 (1967).

31. P. P. A. Smit and A. K. Van der Vegt, *Proceedings of the International Rubber Conference*, May 1967, p. 99.

32. V. L. Folt et al., *Rubber Chem. Technol.*, **44**, 1, 12, 29 (1971).

33. D. P. Mukherjee, *Rubber Chem. Technol.*, **47**, 1234 (1974).

34. Y. Tanaka and T. Hiroyuki, *Polym. J.* (*Japan*), **1**, 656 (1970).

35. W. S. Bahary et al., *Rubber Chem. Technol.*, **40**, 1529 (1967).

36. V. Bianchi and E. Bianchi, *Rubber Chem. Technol.*, **38**, 343 (1965).

37. E. Pedemonte and U. Bianchi, *Rubber Chem. Technol.*, **38**, 347 (1965).

38. C. Longiave et al., Belgian Pat. 573,680 (1959).

39. G. S. Trick, *J. Polym. Sci.*, **41**, 213 (1959).

40. M. Golub, *J. Polym. Sci.*, **25**, 373 (1957).

41. G. Natta et al., Belgian Pat. 559,676 (1958).

42. M. Bruzzone et al., *Rubber Chem. Technol.*, **47**, 1175 (1974).

43. J. C. Mitchell, *Polymer*, **8**, 369 (1967).

44. E. A. Collins and L. A. Chandler, *Rubber Chem. Technol.*, **39**, 193 (1966).

45. M. Berger and D. J. Buckley, *J. Polym. Sci.*, *A*, **1**, 2945 (1963).

46. V. D. Mochel, *J. Polym. Sci.*, *A-1*, **10** (4), 1009 (1972).

47. J. J. Maurer, *Rubber Chem. Technol.*, **38**, 979 (1965); J. J. Maurer, *ibid*; **42**, 110 (1969).

48. F. P. Baldwin and G. Ver Strate, *Rubber Chem. Technol.*, **45**, 709 (1972).

49. S. Cesca, *J. Polym. Sci.*, *Macromol. Rev.*, **10**, 1 (1975).

50. N. Tokita and R. Scott, *Rubber Chem. Technol.*, **42**, 944 (1969).

51. E. N. Kresge et al., 4th International Synthetic Rubber Symposium, London, 1969.

52. S. Davison and G. L. Taylor, *Rubber Chem. Technol.*, **46**, 30 (1973).

53. A. Subramaniam, *Rubber Chem. Technol.*, **45**, 346 (1972).

54. W. C. Carter et al., *J. Am. Chem. Soc.*, **68**, 1480 (1946).

55. W. H. Beattie and C. Booth, *J. Appl. Polym. Sci.*, **7**, 507 (1963).

56. P. J. Flory and H. L. Wagner, *J. Am. Chem. Soc.*, **74**, 195 (1952).

57. A. De Chirico, *Chim. Ind.* (*Milan*), **46**, 53 (1964).

58. P. W. Allen and G. M. Bristow, *J. Appl. Polym. Sci.*, **4**, 237 (1960).

59. W. Cooper et al., *J. Polym. Sci.*, **59**, 241 (1962).

60. G. Kraus, *J. Appl. Polym. Sci.*, **7**, 1257 (1963).

61. G. Kraus and G. A. Maczvgemba, *J. Polym. Sci.*, *A*, **2**, 277 (1964).

62. G. Kraus and J. T. Gruver, *J. Polym. Sci.*, *A*, **3**, 105 (1965).

63. W. Cooper et al., *J. Polym. Sci.*, **50**, 159 (1961).

64. B. M. E. Van der Hoff et al., *Rubber Plast. Age*, **36**, 821 (1965).

65. M. Takeda and R. Endo, *Rept. Prog. Polym. Phys.* (*Japan*), **6**, 37 (1963).

66. M. Abe and H. Fujita, *J. Phys. Chem.*, **69**, 3263 (1965).

67. F. Danusso et al., *J. Polym. Sci.*, **54**, 475 (1961).

68. H. Scott et al., *J. Polym. Sci.*, *A*, **2**, 3233 (1964).

69. I. J. Poddubnyi et al., *Makromol. Chem.*, **94**, 268 (1966).

70. G. Morrisey and T. W. Harvey, *Rubber Plast. Age*, **43**, 1110 (1962).

71. W. Ring and H. J. Cantow, *Makromol. Chem.*, **89**, 138 (1965).

72. J. M. Hulme and L. A. McLeod, *Polymer*, **3**, 153 (1962).

73. D. J. Harmon, *J. Polym. Sci.*, *C*, **8**, 243 (1965).

74. G. Moraglio, *Chim. Ind.* (*Milan*), **41**, 984 (1960).

75. P. J. Flory, *Principles of Polymer Chemistry*, Cornell University Press, Ithaca, New York, 1953.

76. S. H. Maron, *J. Polym. Sci.*, **38**, 329 (1959).

77. P. J. Flory, *J. Am. Chem. Soc.*, **87**, 1833 (1965).

78. D. Patterson, *Rubber Chem. Technol.*, **40**, 1 (1967).

79. P. J. Flory and J. Rehner, *J. Chem. Phys.*, **11**, 521 (1943).

80. P. J. Flory, *J. Chem. Phys.*, **18**, 108 (1950).

81. W. R. Summers et al., *Rubber Chem. Technol.*, **45**, 1655 (1972).

82. Y. B. Tewari and H. P. Schreiber, *Rubber Chem. Technol.*, **45**, 1665 (1972).

83. G. M. Bristow, *J. Appl. Polym. Sci.*, **9**, 1571 (1965).

84. G. Kraus, *Rubber World*, **135**, 67 (1956).

85. G. Gee et al., *Polymer*, **6**, 541 (1965).

86. A. E. Eichinger and P. J. Flory, *Trans. Faraday Soc.*, **64**, 2035 (1968).

87. G. M. Bristow and W. F. Watson, *Trans. Faraday Soc.*, **54**, 1731 (1958).

88. C. J. Sheehan and A. L. Bisio, *Rubber Chem. Technol.*, **39**, 149 (1966).

89. C. Booth et al., *Polymer*, **5**, 343 (1964).

90. C. Booth et al., *Polymer*, **5**, 353 (1964).

91. G. S. Trick, Paper presented at Gordon Research Conference on Elastomers, July 1963.

92. L. D. Loan, *J. Appl. Polym. Sci.*, **7**, 2259 (1963).

93. V. M. Gavrilov and A. G. Shvarts, *Rubber Chem. Technol.*, **47**, 303 (1974).

94. G. Kraus, *J. Appl. Polym. Sci.*, **7**, 1257 (1963).

95. T. J. Dudek and F. Bueche, *J. Polym. Sci.*, *A*, **2**, 811 (1964).

96. U. Flisi and G. Crespi, *J. Appl. Polym. Sci.*, **12**, 1947 (1968).

97. A. M. Hassan and L. N. Ray, *J. Appl. Polym. Sci.*, **15**, 145 (1971).

98. E. D. Holly, *J. Polym. Sci.*, *A*, **2**, 5267 (1964).

99. F. P. Baldwin et al., *Rubber Chem. Technol.*, **42**, 1167 (1969).

100. G. Natta et al., *Chim. Ind.* (*Milan*), **41**, 741 (1959).

101. H. K. Frensdorff, *J. Polym. Sci.*, *A*, **2**, 333 (1964).

102. H. Burrell, *Off. Digest*, **27**, 726 (1955).

103. J. L. Gardon, *J. Paint Technol.*, **38**, 43 (1966).

104. J. D. Crowley et al., *J. Paint Technol.*, **38**, 269 (1966).

105. C. M. Hansen, *J. Paint Technol.*, **38,** 104 (1967).
106. G. M. Bristow et al., *Trans. Faraday Soc.*, **54,** 1742 (1958).
107. S. K. Bhatnagar, *Makromol. Chem.* **122,** 82 (1969).
108. J. L. White, *Rubber Chem. Technol.*, **42,** 257 (1969).
109. G. Kraus et al., *J. Appl. Polym. Sci.*, **11** (8), 1581 (1967).
110. R. N. Kienle et al., *Rubber Chem. Technol.*, **44,** 996 (1971).
111. G. Kraus and C. J. Storey, *J. Polym. Sci.*, *A-2,* **10,** 657 (1972).
112. M. Hoffmann, *Makromol. Chem.*, **153,** 99 (1972).

GERARD KRAUS

Phillips Petroleum Company
Bartlesville, Oklahoma

Chapter 11   Viscoelastic Behavior of Solution Rubbers

1   INTRODUCTION

The viscoelastic behavior of the so-called stereo rubbers—polybutadiene, polyisoprene, and butadiene–styrene copolymers prepared by solution polymerization—is not, in general, dominated by their degree of structural regularity. Instead, the viscoelastic properties of these rubbers depend on the totality of structural characteristics produced by the polymerization reactions involved in their synthesis. Thus certain properties, such as the melt viscosity, depend almost entirely on chain architecture—molecular weight distribution and long chain branching—

rather than on structural regularity. This is easily appreciated by considering that one of the most immediate consequences of structural regularity is the ability of a polymer to crystallize. On the other hand, most of the technologically important properties of rubbers, while intimately connected with their viscoelasticity, are not fundamentally related to, or even favorably affected by, crystallinity. The influence of stereoregularity on the useful viscoelastic properties of elastomers is thus confined to its effects on segmental friction, free volume, and the position of the glass transition temperature. Unlike crystallization phenomena and the properties dependent on them, these variables do not vary dramatically with steric purity.

In this chapter, the elastoviscous behavior of unvulcanized polymer melts and its bearing on various aspects of processability are considered first. This is followed by a discussion of the viscoelastic behavior of vulcanizates and its broader significance with regard to such ultimate use properties as wear and traction of tires. Finally, a brief account is given of the viscoelastic behavior of block copolymers exhibiting properties of vulcanized rubbers without the benefit of chemical cross-linking.

2 ELASTOVISCOUS BEHAVIOR OF UNVULCANIZED POLYMERS

The viscoelastic response of a noncrystallizing, linear, high molecular weight polymer, when followed as a function of temperature (T), time (t), or frequency (ω) in dynamic experiments, passes through four more or less distinct zones of behavior: glassy, transition, rubbery, and terminal. These zones are traversed in the order listed as the temperature is increased (t or ω constant), as the time is increased (T constant), or as the frequency is decreased (T constant). The polymer "feels" hard and brittle in the glassy zone, leathery and tough in the transition zone, highly elastic in the rubbery zone, and becomes a viscous liquid in the terminal zone. However, even in the terminal zone, there is significant elastic response, which is commonly referred to as "melt elasticity." Polymers are processed almost entirely in the terminal zone, where viscous flow is the dominant response.

For two decades prior to the discovery of the stereo rubbers, the Mooney viscosity served virtually as the sole criterion for the plasticity and processability of elastomers. Even today, the Mooney value remains an indispensable quantity in all elastomer specifications. However, early observations on the new solution rubbers, particularly polybutadiene, showed clearly that the Mooney viscosity is an insufficient criterion for describing the flow behavior of these materials. The most spectacular manifestation of this is the phenomenon of cold flow. When blocks of 50 ML_4 Mooney viscosity SBR and early types of *cis*-polybutadiene were placed on a horizontal flat surface, the SBR was found to hold its shape indefinitely, while the polybutadiene spread into a puddle within a few days' time—clearly an intolerable situation from the standpoint of warehousing the material! Fundamental investigations of the melt flow of polybutadienes, prompted to a considerable extent by this phenomenon, have not only provided practical solutions to the cold flow problem but have contributed materially to our general knowledge of the melt flow of high polymers.

2.1 Rheology of Polymer Melts

2.1.1 Newtonian Viscosity

Polymer melts are fundamentally non-Newtonian viscous liquids with elastic properties. At low enough rates of shear, many polymers exhibit a well-defined Newtonian flow range. However, the existence of a Newtonian flow range is by no means assured; some polymers exhibit shear dependence down to the lowest shear rates attainable experimentally. The molecular weight dependence of the Newtonian viscosity of polymers has been the subject of much study and appears to be represented best by a semiempirical equation due to Fox and Allen (1):

$$\eta_0 = \left(\frac{N_a}{6}\right)\left(\frac{\overline{s_0^2}}{M}\right)(Z_c\rho)\left(\frac{Z_w}{Z_c}\right)^{3.4}\xi; \qquad Z_w > Z_c \qquad [1]$$

where η_0 is the limiting zero shear viscosity, N_a is Avogadro's number, ρ is the density, $\overline{s_0^2}/M$ is the mean square radius of gyration of the molecule per unit molecular weight (a near constant for a given polymer), Z_w is the weight average chain length expressed as the number of main chain atoms per molecule, Z_c is the critical value of Z for entanglement coupling, and ξ is the coefficient of segmental friction per chain atom. Equation [1] is confined to chain lengths in excess of Z_c. For shorter chains,

$$\eta_0 = \left(\frac{N_a}{6}\right)\left(\frac{\overline{s_0^2}}{M}\right)(Z_c\rho)\left(\frac{Z_w}{Z_c}\right)\xi; \qquad Z_w < Z_c \qquad [2]$$

The friction factor ξ at low Z is not independent of chain length. Consequently, the first power dependence on Z_w is not always observed, unless suitable corrections to constant ξ are made. This is of no particular importance in the present context, for we are interested primarily in high molecular weight polymers for which $Z_w > Z_c$ and ξ is substantially independent of chain length. For such polymers the Newtonian viscosity varies linearly with the 3.4th power of weight average molecular weight down to a critical molecular weight, below which a marked decrease in molecular weight dependence is observed. The two regimes are now generally accepted to represent flow with ($Z_w > Z_c$) and without ($Z_w < Z_c$) entanglement coupling.

None of the quantities in equation [1], with the exception of ξ, is strongly temperature dependent. The temperature dependence of η_0 is thus dominated by that of the friction factor ξ. Berry and Fox (2) have shown that this dependence can be expressed for a large number of polymers by an empirical equation similar to one originally proposed by Vogel (3):

$$\log \xi = \log \xi_0 + K\left(1 + \frac{T - T_g}{\Delta}\right)^{-1} \qquad [3]$$

where T_g is the glass transition temperature and K and Δ are empirical constants. The quantity ξ_0 is of the order of 10^{-11}–10^{-12} for many polymers. Many other empirical relationships have been used to describe the temperature dependence of polymer melt viscosity, including the well-known Williams, Landel, and Ferry (WLF) equation (4), which can be shown to be equivalent to the Vogel equation. A particularly simple rule has been proposed by Magill and Li (5). It states that the ratio of the viscosity at temperature T to the viscosity at a reference temperature $T_s = 1.24 T_g$ is a universal function of T/T_g, applicable to all polymers.

An interesting generalization regarding Z_c has been proposed by Fox and Allen (1, 2):

$$X_c \equiv \left(\frac{\overline{s_0^2}}{M}\right)(\rho Z_c) \cong 4 \times 10^{-15} \qquad [4]$$

The approximate constancy of this quantity allows some immediate predictions regarding the effects of long chain branching and dilution. Rearranging a linear chain to a branched configuration lowers its mean square radius of gyration to a fraction, g, of its former value, i.e.,

$$(\overline{s_0^2})_{br} = g(\overline{s_0^2})_{lin} \qquad [5]$$

In view of equations [4] and [5]

$$(Z_c)_{br} = \frac{(Z_c)_{lin}}{g} \qquad [6]$$

Reference to equation [1] shows that the effect of branching should be to reduce η_0 by a factor of $g^{3.4}$.

The effect of a diluent in relatively low concentration (such as in oil extended elastomers) can also be ascertained by use of equation [4]. Since

$\overline{s_0^2}/M$ is not seriously affected by the quality of the solvent in highly concentrated solutions (6),

$$\rho(Z_c)_{dry} = c(Z_c)_{soln} \qquad [7]$$

where c is the polymer concentration in grams per cubic centimeter. In addition, the diluent will alter the segmental friction factor ξ. Substituting into equation [1] one finds for the ratio of $\eta_0(\phi)$ at volume fraction of polymer ϕ $(= c/\rho)$ to the viscosity of undiluted polymer $\eta_0(1)$:

$$\frac{\eta_0(\phi)}{\eta_0(1)} \cong \phi^{3.4} \cdot \frac{\xi(\phi)}{\xi(1)} \qquad [8]$$

2.1.2 Non-Newtonian Viscosity

The shear rate dependence of the viscosity of high polymers is usually amenable to reduced variable treatments which are highly useful in generalizing experimental data. If we denote η as the viscosity at some shear rate $\dot{\gamma}$, then

$$\frac{\eta}{\eta_0} = f(\dot{\gamma}\eta_0 M^\alpha T^\beta) \qquad [9]$$

where η_0 is, as before, the limiting Newtonian viscosity ($\dot{\gamma} \to 0$), M is the molecular weight, T is the absolute temperature, and f is a characteristic function of the reduced variable $\dot{\gamma}\eta_0 M^\alpha T^\beta$. The exponents α and β are parameters, the values of which appear to differ from one polymer to another. There is evidence that α, $\beta = 0$ is at least a good approximation for many polymers (7), although theory (8) predicts $\alpha = 1$, $\beta = -1$. Intermediate values have also been used to fit experimental data (9). The theoretical reduced variable $\dot{\gamma}\eta_0 M/\rho RT$ is dimensionless and hence most pleasing aesthetically. Its success is generally limited to the regime of unentangled flow, $Z_w < Z_c$ (10).

It was once believed that the function f might be universal for all polymers (11). This does not appear to be true, even for monodisperse polymers (12). It is decidedly untrue for polymers exhibiting a distribution in molecular weight; in fact, f is a sensitive function of the molecular weight distribution (13). Various "universal curves" reported in the literature thus owe their success mainly to the fact that polymers of similar molecular weight distribution are compared.

Let us examine briefly some of the consequences of equation [9] and assume, for simplicity, that α, $\beta = 0$. As $\dot{\gamma} \to 0$, $\eta \to \eta_0$ and the viscosity becomes independent of shear rate. The departure from Newtonian behavior is a function of $\dot{\gamma}\eta_0$ only. In view of the strong dependence of η_0

on chain length and temperature, the onset of non-Newtonian flow will occur at lower shear rates as the molecular weight is increased or as the temperature is lowered. Further, the functional form of f is such as to hasten the onset of non-Newtonian flow as the molecular weight distribution is broadened.

2.1.3 Melt Elasticity

The rheological behavior of uncross-linked polymers is not described completely by their viscosity, but requires also a specification of the elastic properties of the melt. Melt elasticity is a factor of considerable importance in determining certain aspects of the processability of a polymer. Both viscous and elastic effects contribute to the non-Newtonian viscosity so that some information regarding melt elasticity can be inferred from the non-Newtonian flow curve. In general, the more pronounced the non-Newtonian behavior of a polymer, the greater is its melt elasticity.

Melt elasticity can be specified by any of the usual viscoelastic functions describing the phenomena of stress relaxation, creep, dynamic response to sinusoidally varying stresses or strains, or stress relaxation after cessation of steady flow (14). In the present discussion we confine our remarks mainly to creep under a steady stress and subsequent recovery from creep, to dynamic viscosity, and to normal stress effects. The shear creep compliance J in *linear* viscoelasticity (small strain) is defined as the ratio of strain to imposed (constant) stress and is composed of three terms:

$$J = J_g + J_d(t) + \frac{t}{\eta_0} \qquad [10]$$

Here J_g is the glassy compliance, $J_d(t)$ is the delayed elastic compliance, and t is time. The term t/η_0 is the contribution from viscous flow. J_g is negligibly small compared to the other terms for a polymer in the rubbery response region. The creep compliance is of obvious relevance to the phenomenon of cold flow where the polymer flows under its own weight. At relatively long times $J_d(t)$ reaches a steady value and all further deformation is by flow. This value, neglecting the small contribution of J_g, is termed the steady state compliance and is denoted simply by the symbol J_e. The deformation corresponding to J_e is fully recoverable and is a measure of the total melt elasticity. In many polymer melts the steady value of J_e is attained quite rapidly, the rate of attainment being a function of temperature and of the viscoelastic spectrum of the polymer. For monodisperse polymers of low molecular weight theory (14, 15) predicts that J_e should be proportional to molecular weight:

$$J_e = \frac{(2/5)M}{\rho RT} \qquad [11]$$

but for high molecular weight polymers J_e is more nearly independent of M (16, 17). However, it is very sensitive to molecular weight distribution. The theoretical prediction for polydisperse polymers (14) is

$$J_e = \frac{(2/5)M_{z+1}M_z}{M_w \rho RT} \qquad [12]$$

where M_w is the weight average molecular weight and M_z and M_{z+1} are the progressively higher moments of the molecular weight distribution. Equation [12] is difficult to compare with experiment because M_z and M_{z+1} are not easily obtained with high precision. However, the most remarkable prediction of equation [12], that J_e exhibits a maximum at some intermediate composition when two polymers differing in M are blended, has been amply confirmed (18, 19).

It should be pointed out that the theory underlying equations [11] and [12] is one for macromolecules in dilute solution, and its successful application to *undiluted* polymers is limited to relatively low molecular weights. The upper limit of applicability of equation [11] appears to be about three to five times Z_c (20).

When a sinusoidally varying shear strain is imposed on a polymer melt the resulting stress will have a component in phase with the *strain rate* (viscous) and a component out of phase with the *strain rate* (elastic). It is thus convenient to specify a complex dynamic viscosity

$$\eta^* = \eta' - i\eta'' \qquad [13]$$

where η' is the ratio of the component of stress in phase with the strain rate to the strain rate and η'' is the corresponding ratio for the out-of-phase component of the stress. The individual components of η^* are related to the dynamic storage (G') and loss moduli (G'') through the frequency ω:

$$\eta' = \frac{G''}{\omega}; \qquad \eta'' = \frac{G'}{\omega} \qquad [14]$$

The absolute magnitude of the dynamic viscosity, $|\eta^*|$, is of particular interest as it has been found to coincide to a first approximation with the non-Newtonian steady flow viscosity (21, 22), provided the shear rate $\dot{\gamma}$ is replaced by ω, i.e.,

$$|\eta^*(\omega)| = \eta(\dot{\gamma}) \qquad [15]$$

Examples of this will be shown below. This relationship establishes the already mentioned fact that the non-Newtonian flow curve reflects both the viscous and elastic response of the polymer. It was also pointed out that the function f in equation [9] is dependent on the molecular weight distribution, with broad distributions leading to increased shear dependence. In light of the analogy between dynamic viscosity and the non-Newtonian flow curve, this means that broad molecular weight distribution must also affect dynamic viscosity and hence melt elasticity at finite shear rates, in qualitative agreement with the prediction of equation [12] for the limiting elastic compliance.

In certain respects equation [15] offers serious conceptual difficulties. The complex dynamic viscosity is a description of linearly viscoelastic phenomena, whereas the non-Newtonian steady flow viscosity is ipso facto a nonlinear property. Linear behavior is confined to small strains (as in the usual dynamic experiment), but large elastic deformations usually accompany the non-Newtonian flow of polymers. The coincidence expressed by equation [15] is, therefore, truly astonishing. Taken as an experimental fact, however, it establishes a link between linear and nonlinear behavior and indicates that the phenomena of linear viscoelasticity have a direct bearing on nonlinear effects. This is important because almost all polymer processing involves large strains and nonlinear behavior. On the other hand, our understanding of viscoelastic effects in polymers is confined almost entirely to the simpler, linear behavior. Because of this link, we may have some confidence that conclusions from linear theory, e.g., the prediction of equation [12], will carry over, at least qualitatively, into practical processing behavior.

When a viscoelastic material is sheared, there develop, in addition to the shearing stress, normal stresses in the direction of flow (P_{11}), in the direction of the velocity gradient (P_{22}), and in the neutral direction (P_{33}). The principal normal stress difference ($P_{11} - P_{22}$) in the Newtonian flow region is proportional to the square of the shear rate (23–25):

$$P_{11} - P_{22} = K_0 \dot{\gamma}^2 \qquad [16]$$

The normal stress cofficient K_0 is directly related to a steady state compliance

$$J_{en} = \frac{K_0}{2\eta_0^2} \qquad [17]$$

The theory of second-order viscoelastic fluids identifies J_{en} with J_e of linear viscoelasticity (23, 24). Thus, normal stresses provide an alternative way for studying melt elasticity. In the non-Newtonian flow regime $P_{11} - P_{22}$ varies less rapidly with $\dot{\gamma}^2$ than indicated by equation [16].

An empirical rule for estimating J_e from the dynamic viscosity versus frequency curve has been formulated by Prest and co-workers (26). The rule also enables one to calculate J_e from non-Newtonian viscosity-shear rate data as long as equation [15] holds.

2.1.4 Flow in Cylindrical Tubes and Orifices

Because of its widespread use in viscometry of polymer melts and its important application in extrusion operations, we now review briefly the flow of a non-Newtonian fluid through a cylindrical tube of length L and radius R (27). If a pressure difference ΔP is established across the tube, the shear stress at the wall will be

$$\tau = \frac{\Delta P}{2[(L/R)+e]} \qquad [18]$$

where e is an "end correction" that is related to the recoverable shear strain, and hence to the melt elasticity. The end correction is readily evaluated from experiments at more than one L/R ratio, holding the rate of shear constant. If we define an average apparent shear rate D as

$$D = \frac{4Q}{\pi R^3} \qquad [19]$$

where Q is the volumetric flow rate, then it may be shown that the shear rate at the wall, $\dot{\gamma}$, is

$$\dot{\gamma} = \frac{(N+3)D}{4} \qquad [20]$$

where

$$N = \frac{d \log D}{d \log \tau} \qquad [21]$$

The viscosity is simply $\tau/\dot{\gamma}$. It is instructive to combine these relations into an explicit expression for the flow rate:

$$Q = \frac{\pi R^3 (\Delta P)}{2(N+3)\eta[(L/R)+e]} \qquad [22]$$

For a simple Newtonian liquid at moderate rates of shear, $e \cong 0$ (neglecting the kinetic energy and Couette corrections), $N = 1$ ($D = \dot{\gamma}$), $\eta = \eta_0$, and the expression reduces to Poiseuille's law.

In the flow of non-Newtonian polymer melts through very short tubes or dies, e is often of the same order of magnitude or even greater than L/R. Because of the partially elastic nature of the deformation, the extrudate recovers, giving rise to the phenomenon of die swell. The quantitative interrelationships between phenomena at the die entrance, melt elasticity and die swell, are complex, and are beyond the scope of this discussion.

In extrusion of a polymer melt there is a critical stress above which the extrudate becomes rough (28). This is not necessarily connected with an anomaly in the flow rate. The phenomenon is often called "melt fracture," but this term should probably be reserved to the instances in which a discontinuity in the flow rate occurs. This discontinuity is a manifestation of plug flow with slippage at the die wall (29). In the regime of plug flow the equation set forth here are no longer applicable.

2.2   Melt Rheology of Solution Rubbers

Natural rubber often contains a large gel fraction prior to mastication and is, undoubtedly, branched even after breakdown to complete solubility. Emulsion SBR, whose molecular weight distribution (MWD) is very broad, will actually form gel if the polymerization is carried to high conversion. In contrast, solution polybutadienes and butadiene–styrene copolymers are relatively free from long chain branching and exhibit much narrower molecular weight distributions, unless special techniques are employed to broaden the MWD and introduce branching. The characteristic flow behavior of these rubbers is thus dominated by branching and MWD.

2.2.1   Melt Flow of Narrow Distribution Linear Chain Polymers

When carried out under conditions favoring rapid initiation, the polymerization of butadiene with alkyllithium initiators leads to narrowly distributed $(1 < M_w/M_n < 1.5)$ polymers apparently free from long chain branching. These polymers contain approximately 52% trans, 40% cis, and 8% vinyl unsaturation and have a glass transition temperature of about $-99°C$. Viscosity measurements on these polymers have shown an extended Newtonian flow regime (30, 31) even at molecular weights of several hundred thousand. The Newtonian viscosity of these polymers (Fig. 1) follows the expected 3.4th power law in M_w down to a molecular weight of 5600, where the usual transition to a lower exponent occurs. The shear dependence of the steady flow viscosity is illustrated in Fig. 2. In these flow curves, the Newtonian regime is evidenced by linearity with unit slope. The curves of Fig. 2 have been reduced to a common reference temperature of 300°K. The reduction scheme follows from

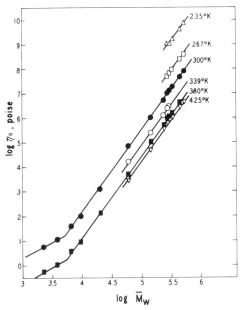

Figure 1 Dependence of Newtonian viscosity on molecular weight for linear polybutadiene prepared by butyllithium initiation (30). Reproduced by permission of the *Journal of Polymer Science.*

equation [9]. Since $\eta = \tau/\dot{\gamma}$, where τ is the shear stress,

$$\frac{\tau}{\dot{\gamma}\eta_0} = f(\dot{\gamma}\eta_0) \qquad [23]$$

taking $\alpha = \beta = 0$. Consequently, the reduced variable $\dot{\gamma}\eta_0$ is a function of the shear stress alone. Comparing measurements made at constant shear stress, but at different temperatures—T and a reference temperature (e.g., 300°K)—we have

$$\dot{\gamma}(T)\eta_0(T) = \dot{\gamma}(300)\eta_0(300) \qquad [24]$$

or

$$\dot{\gamma}(300) = \frac{\dot{\gamma}(T)}{a_T} \qquad [24a]$$

where $a_T \equiv \eta_0(300)/\eta_0(T)$.

An analysis of the temperature dependence of the viscosity in terms of equations [1] and [3] has been made by Berry and Fox (2), using the present data. The parameters are $\log \xi_0 = -10.90$, $K = 12.2$, and $\Delta = 50$.

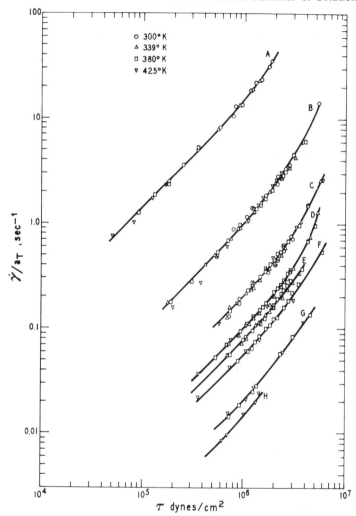

Figure 2 Flow curves for linear polybutadienes prepared by butyllithium initiation, reduced to 300°K (30). M_w ranges from 61,000 (A) to 524,000 (H). Reproduced by permission of the *Journal of Polymer Science*.

In the range covered by the data, the shear dependence follows an expression proposed by Ree and Eyring (32):

$$\frac{\eta}{\eta_0} = (\beta\dot{\gamma})^{-1} \sinh^{-1} (\beta\dot{\gamma}) \qquad [25]$$

where β is a relaxation time (not the β of equation [9]). This relaxation time turns out to be itself proportional to $M_w^{3.4}$. However, this fact is not predicted by the Ree-Eyring theory, so that adherence to equation [25] should not be taken as confirmation of the validity of the theory. Since β and η_0 have the same molecular weight dependence,

$$\eta_0 = kM_w^{3.4} \qquad [26]$$

$$\beta = k'M_w^{3.4} \qquad [27]$$

the parameter β in equation [25] can be expressed in terms of η_0:

$$\frac{\eta}{\eta_0} = \left(\frac{k'\eta_0\dot{\gamma}}{k}\right)^{-1} \sinh^{-1}\left(\frac{k'\eta_0\dot{\gamma}}{k}\right) \qquad [28]$$

This establishes a functional relationship for the reduced variable plot, equation [9], for the narrow distribution, linear polybutadienes. The ratio $k'/k = 5 \times 10^{-7}$, independent of temperature; i.e., k' evidently has the same temperature dependence as the Newtonian viscosity. The reduced variable master plot according to equation [28] is shown in Fig. 3, along with those for narrow distribution polybutadienes of 38% vinyl unsaturation (33) and butadiene–styrene copolymers (12) of 25% styrene content. For the latter, $k'/k = 6.5 \times 10^{-7}$ and 8×10^{-7}, respectively. For all these solution polymers, the master curves are displaced by over an order of magnitude toward higher values from the "universal" curve of Vinogradov and Malkin (7). This "universal" curve is followed by polymers of broader molecular weight distribution.

It is now easily seen why polymers of the present type exhibit the phenomenon of cold flow. To process such a polymer at shear rates available in conventional equipment, it is necessary to fix M_w in such a way as to provide adequate fluidity at practical processing temperatures. Because of the flatness of the flow curve in the intermediate range of shear rates, the viscosity does not increase much as shear rate is reduced. Consequently, the polymer flows even at the extremely small rates of

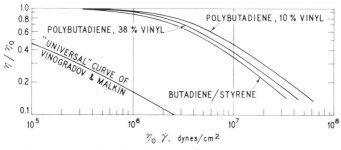

Figure 3 Reduced variable viscosity plots for narrow distribution polymers. (1) Polybutadiene, 10% vinyl; (2) polybutadiene, 38% vinyl; (3) butadiene-styrene random copolymer, 25% styrene.

shear encountered in flow under its own weight. Because of its greater shear sensitivity, a polymer of broader molecular weight distribution can still be processed at the same conditions of shear rate and temperature, even though it may have a much larger M_w. Because of the dependence of η_0 on $M_w^{3.4}$, cold flow is substantially reduced. Practically acceptable values of cold flow are generally achieved by broadening the molecular weight distribution and by the introduction of long chain branching.

2.2.2 Melt Flow of Narrow Distribution Branched Polymers

From what has been said in Section 2.1.1, one would expect that long chain branching (at constant molecular weight) would *decrease* the melt viscosity. Early experiments designed to overcome the cold flow of polybutadienes, however, showed consistently large increases in melt viscosity when small amounts of branching or cross-linking were introduced into the polymer molecule. In these experiments, the molecular weight distribution was, of course, also affected, but the broadening of the distribution did not appear to account entirely for the reduction in flow. To investigate this problem in a definitive manner, Zelinski and Wofford (34) prepared "star shaped" multichain polybutadienes by coupling active polymer lithium with methyltrichlorosilane and silicon tetrachloride. This kind of coupling broadens the molecular weight distribution in an absolute sense, but produces a narrowing of the relative distribution as expressed by the familiar M_w/M_n ratio. It is easily shown (35) that the final M_w/M_n of the coupled, branched polymer is

$$\frac{M_w}{M_n} = \frac{(M_w/M_n)_0 + p - 1}{p} \qquad [29]$$

where $(M_w/M_n)_0$ is the molecular weight ratio for the parent polymer and p is the number of chains coupled, i.e., 3 or 4. The value of g in equations [5] and [6] has been calculated theoretically for star shaped polymers (36):

$$g = \frac{3}{p} - \frac{2}{p^2} \qquad [30]$$

This equation yields $g = \frac{7}{9}$ for trichain and $g = \frac{5}{8}$ for tetrachain polymers.

The parent polymers used in the coupling experiments of Zelinski and Wofford (34) had, with the exception of the lowest molecular weight members of the series, M_w/M_n ratios of less than 1.5. The coupling reaction occurred in good yield and the coupled products had very nearly the expected M_w. The number average molecular weight of the branched rubbers was not measured, but from equation [29] the ratio M_w/M_n

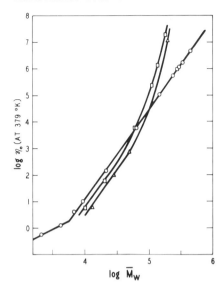

Figure 4 Dependence of Newtonian viscosity on molecular weight for branched polybutadiene (35). ○, Linear; □, trichain; △, tetrachain. Reproduced by permission of the *Journal of Polymer Science.*

should have been less than 1.3 for all polymers used in the rheological studies. Figure 4 shows the Newtonian viscosity obtained (35). At the lower molecular weights, the prediction of equations [1] and [6] is fulfilled:

p	$g^{3.4}$	$\eta_0(\mathrm{br})/\eta_0(\mathrm{lin})$
3	0.43	0.46
4	0.20	0.22

At higher molecular weights a dramatic rise in η_0 is evident for the branched polymers. A trichain polymer of $M_w = 303{,}000$ exhibited non-Newtonian flow over the entire range of experimentally realizable shear rates and no meaningful extrapolation to zero shear could be made. Creep data on this polymer suggested a limiting viscosity about two orders of magnitude higher than the last point in Fig. 4. Yet this same polymer exhibited a lower apparent viscosity than a linear polymer of the same M_w at high rates of shear. This behavior is illustrated in Fig. 5 with data on the polymers of Fig. 4, taken at a shear rate of 20 sec^{-1}. At this shear rate all the branched polymers fall below the viscosity curve of the linear rubbers. Thus the branched polymers not only exhibit high viscosity at low shear rates, but also show a much stronger tendency toward non-Newtonian behavior. Experiments on other branched polymers do

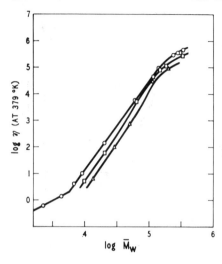

Figure 5 Viscosity at shear rate 20 sec^{-1} for polymers of Fig. 4 (35). O, Linear; □, trichain; △, tetrachain. Reproduced by permission of the *Journal of Polymer Science.*

not necessarily show the crossover in the Newtonian viscosity. A possible reason for this may be the extent of entanglement of the branches. In Fig. 4 the crossover occurs when the branch length reaches roughly three to four times the critical entanglement molecular weight M_c for the linear polymer. For polystyrene $M_c = 38,000$, so that no crossover might be expected short of 350,000 molecular weight (for trichain polymers) if it may be assumed that the same ratio of branch length to M_c applies as a criterion. Masuda et al. (37) found, for tetrastar polystyrene, $\eta_0 \sim M_w^{4.5}$, suggesting possible crossover at molecular weights higher than 1 million. Moreover, Suzuki and Kovacs (38) have reported crossover in multichain polystyrenes above 600,000 molecular weight. Similar results have been obtained with certain long chain branched polystyrenes of very low branching frequency (39). Thus, the behavior of branched polybutadienes is certainly not unique. Long and associates (40) demonstrated increased melt viscosities in branched polyvinyl acetates, and Chartoff and Maxwell (41) in polyethylene. The results on polybutadiene described here have been confirmed by several groups of investigators (31, 42, 43).

An interesting application of the increased shear sensitivity of star branched polybutadienes and butadiene–styrene copolymers has been described by Ghijsels and Mieras (44), who found that the logarithmic slope of the flow curve, $- d \log \eta / d \log t$ (at small t), was a linear function of the product of the average coupling functionality and the weight fraction of branched molecules. The authors found this observation useful in assessing differences in the degree of branching in a number of rubbers, although they are careful to point out the limitation of the method to polymers of relatively narrow molecular weight distribution.

2.2.3 *Randomly Branched Polybutadiene*

When branching is introduced at random rather than by coupling, such as when a small amount of divinylbenzene is copolymerized with butadiene, the molecular weight distribution changes along with the degree of branching attained. To determine whether an increase in viscosity is realized, similar to the effect observed in the multichain polybutadienes, it is necessary to separate somehow the effects of branching and MWD. Such an attempt has been made (33) and appears to confirm the behavior of the multichain branched polymers. A series of polybutadienes containing small amounts of divinylbenzene (DVB) were prepared under conditions giving a sharp MWD in the absence of DVB. The molecular weight distribution for such branched polymers is predictable by theory and was confirmed by fractionation experiments with a Baker-Williams column. The flow curves of the branched polymers were then compared with those of linear narrow MWD polymers of the same M_w and with calculated flow curves of linear polymers of the same MWD as the branched rubbers. The latter calculation was carried out as follows. According to equation [1] the mixture rule for Newtonian viscosity is

$$\eta_0^{1/3.4} = \sum w_i \eta_{0i}^{1/3.4} \qquad [31]$$

where w_i is the weight fraction of polymer of viscosity η_{0i}. This mixture rule was also found to apply to non-Newtonian viscosities when the comparison was made at *equal shear rate* (33). A test of this mixture rule is illustrated in Fig. 6. It is seen to be quite successful, although it is an oversimplification which fails for very broad distributions and at very high shear rates. This mixture rule was then used to sum equation [28] over the molecular weight distribution. The comparison of the three flow curves for the branched polymer, linear narrow MWD polymer, and linear equivalent MWD polymer is shown in Fig. 7. The number of cross-linked units per molecule (γ) in the branched polymer was 0.52 and $M_w = 145,000$. It is interesting that the flow curve for the branched polymer crosses the curves for both the linear polymers. The branched polymer is thus most viscous at very low shear rates, but least viscous at high shear rates, just as the high molecular weight star polymers of Figs. 4 and 5.

2.2.4 *Melt Flow of cis-Polybutadienes*

The polymers discussed so far are carefully synthesized laboratory samples, prepared to test the effects of narrowness of molecular weight distribution and branching on flow. They are not typical of commercial

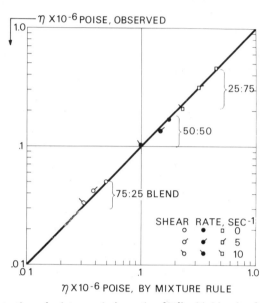

Figure 6 Illustration of mixture rule (equation [31]) with blends of polybutadienes of $M_w = 61,000$ and $278,000$ (33). The three points shown for each blend represent shear rates of 0, 5, and 10 sec^{-1}. Reproduced by permission of the Society of Chemical Industry.

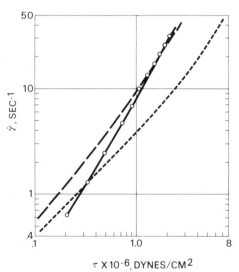

Figure 7 Flow curve for divinylbenzene branched polybutadiene ($\gamma = 0.52$) compared to flow curves of linear polymers of same M_w. Short dashes, narrow MWD; long dashes, MWD equivalent to branched polymer. Temperature 379°K (33). Reproduced by permission of the Society of Chemical Industry.

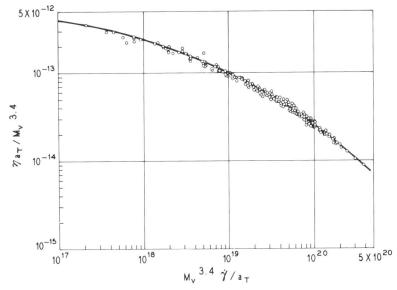

Figure 8 Reduced variable viscosity plot for *cis*-polybutadiene (iodine type). Data cover 10 polymers ranging from $M_v = 63,000$ to 404,000 and temperatures from 300 to 379°K (45). Reproduced by permission of the *Journal of Applied Polymer Science*.

alkyllithium polymerized rubbers, which have broader molecular weight distributions and which almost invariably are deliberately branched. We shall return to discuss commercial polymers at a later point. Instead, we now take up the behavior of *cis*-polybutadiene.

Although relatively narrow in MWD compared to such polymers as emusion SBR, *cis*-polybutadienes nevertheless exhibit considerable polydispersity, with M_w/M_n ranging typically from 2 to 4. Because of their breadth of MWD they do not exhibit the extended Newtonian flow regime of extreme narrow distribution polymers. However, certain types of *cis*-polybutadiene contain little long chain branching unless branches are introduced by design. Such polymers exhibit cold flow and some of the other characteristics of the linear alkyllithium polybutadienes, even if to a smaller extent. A reduced variable plot for this type of polymer is shown in Fig. 8. Since flow is non-Newtonian over the whole range of shear rates, it is not possible to use the reduced variable $\dot{\gamma}\eta_0$. If one assumes that a limiting Newtonian viscosity exists nevertheless, the reduction for chain length and temperature can be effected by employing $M_w^{3.4}/a_T$ in place of η_0. Thus

$$\frac{\eta a_T}{M_w^{3.4}} = f\left(\frac{\dot{\gamma} M_w^{3.4}}{a_T}\right) \qquad [32]$$

which illustrates the variables employed in Fig. 8. Actually, viscosity average molecular weights were used in Fig. 8, but this is unimportant since polymers of similar MWD are correlated so that the ratio M_v/M_w is constant. Of course, a fixed multiplier will not affect the success of the reduction scheme.

The effect of further broadening of the MWD and of the introduction of long chain branching on the steady flow viscosity is qualitatively as already described for the alkyllithiun polymers. For details the reader is referred to the original literature (45).

Vinogradov et al. have reported that the critical chain length for entanglement coupling, Z_c, increases with the degree of cis-enchainment (46).

2.2.5 Melt Elasticity of Polybutadienes

Dynamic viscosity data for several of the polymers described above have been obtained by J. K. Hughes, using a Weissenberg Rheogoniometer (47).

The behavior of three of these (Table 1) is described in some detail to illustrate the effects of molecular weight distribution and branching on the elastic response of these polymer melts. Figures 9–11 show plots of $|\eta^*|$ against frequency, superimposed on the steady flow viscosity versus shear rate curves of these polymers, together with the loss tangents

$$\tan \delta = \frac{\eta'}{\eta''} = \frac{G''}{G'} = \frac{J''}{J'} \qquad [33]$$

where the only quantities not already defined are the dynamic storage (J') and loss (J'') compliances. We note first of all the very good agreement between the dynamic and steady flow viscosities according to equation [15]. It is also clear that the dynamic data confirm the abnormally high limiting viscosity of the tetrachain polymer. The molecular weight of the tetrachain polymer B is very nearly that at the crossover in Fig. 4.

Table 1 Polybutadienes Investigated by Dynamic Methods

Polymer	Type	Cis	Trans	Vinyl	$M_w/1000$	M_w/M_n	Long Chain Branching
A	Alkyllithium	40	52	8	179	1.2	None
B	Alkyllithium	40	52	8	113	1.2	Tetrastar
C	High cis	95	3	2	342	3.4	Possibly some, random

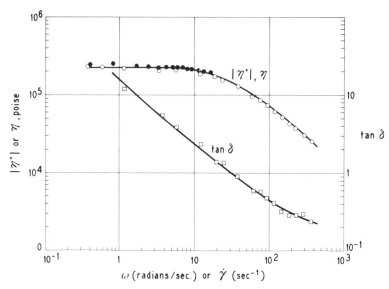

Figure 9 Dynamic and steady flow viscosities for linear butyllithium polybutadiene ($M_w = 179,000$) at 379°K. Data of Hughes (47).

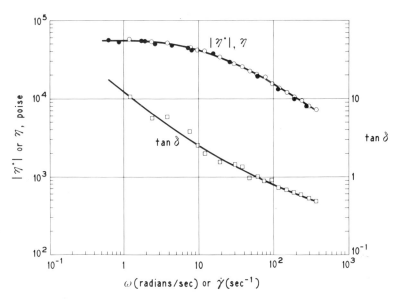

Figure 10 Dynamic and steady flow viscosities for tetrachain polybutadiene ($M_w = 113,000$) at 379°K. Data of Hughes (47).

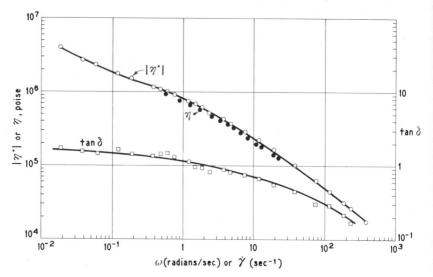

Figure 11   Dynamic and steady flow viscosities for *cis*-polybutadiene ($M_w = 342{,}000$) at 379°K. Data of Hughes (47).

Consequently, the limiting viscosities of A and B should be in proportion to the 3.4th power of their M_w. They are very close to this. In spite of the difference in molecular weight between A and B, there is little difference in tan δ. However, since η and $|\eta^*|$ of the branched polymer are five times larger (by a factor of $g^{-3.4}$) than predicted by equations [1] and [6], it follows that η'' must also be correspondingly larger. In other words, the melt elasticity increases along with the viscosity in the anomalous viscosity regime of branched polymers. The loss tangent of the *cis*-polybutadiene lies considerably lower over the entire frequency range. It is easily shown that the dynamic storage compliance is related to $|\eta^*|$ and the phase angle δ by the equation

$$J'(\omega) = \frac{\cos \delta}{\omega |\eta^*|} \qquad [34]$$

Since $J'(\omega)$ approaches J_e, the steady state compliance, in the limit of zero frequency (14), it is of interest to recast the results of Figs. 9–11 in terms of $J'(\omega)$. This has been done in Fig. 12. Although the data cannot be extrapolated to obtain J_e (except perhaps in the case of polymer A), it is quite clear that $J'(\omega)$ is much smaller for the linear, narrow distribution polymer. In particular, it is smaller than that of the tetrachain polymer in spite of its greater molecular weight.

Valentine, Ferry, and co-workers (48, 49) measured viscoelastic properties of a linear polybutadiene practically identical to A, along with three other samples of higher molecular weight. Their loss tangents are reproduced in Fig. 13, together with Hughes' data for polymer A, temperature shifted to 25°C. The agreement is quite good, considering that different polymer preparations are involved. Valentine et al. also found that the dynamic storage compliance $J'(\omega)$ did not change rapidly with molecular weight so that most of the change in tan δ is due to a decrease in the loss compliance, for at low frequencies $J''(\omega) \rightarrow 1/\omega\eta$ and η increases as $M_w^{3.4}$. From their data, Ferry and Valentine were able to calculate the steady state compliance J_e. They found values increasing regularly from 0.19×10^{-6} to 0.39×10^{-6} cm^2/dyne (at 25°C) in going from $M_w = 180,000$ to $M_w = 510,000$. The value for J_e indicated by Fig. 12, polymer A, would be somewhat larger. A larger value (1.0×10^{-6} cm^2/dyne) was also found by Plazek (50) on Valentine's 180,000 molecular weight sample. Unpublished normal stress measurements by the author on two relatively low molecular weight narrow distribution linear polybutadienes gave $J_e = 0.25 \times 10^{-6}$ cm^2/dyne at $M_w = 52,000$ and $J_e = 0.30 \times 10^{-6}$

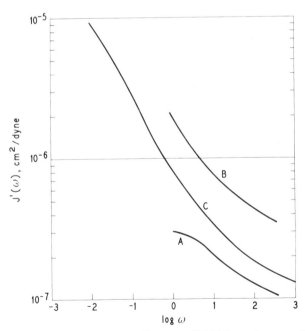

Figure 12 Dynamic storage compliance at 379°K for polymers of Figs. 9–11. A, linear BuLi polybutadiene; B, tetrachain BuLi polybutadiene; C, cis-polybutadiene.

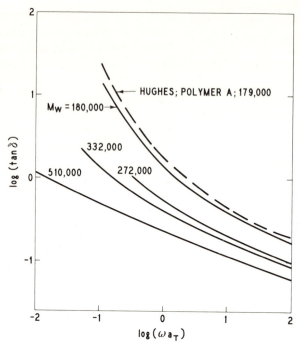

Figure 13   Loss tangents for linear, narrow distribution (BuLi) polybutadienes in the terminal zone. Solid lines are data of Valentine et al. (48, 49).

$cm^2/dyne$ at $M_w = 75,000$, both at 100°C. All these results lie well below the prediction of equation [11]. In spite of the general lack of quantitative agreement, it must be concluded that the melt elastic complicance of *linear, narrow molecular* weight distribution polybutadienes is indeed small, even at appreciable molecular weights.

When Valentine's sample A was branched lightly by cross-linking short of the gel point, using small amounts of sulfur and accelerator, the results of Table 2 were obtained (49, 50). The amount of cross-linking was clearly insufficient to produce a marked broadening of the MWD, but the effect on J_e is nevertheless spectacular and the observed values are now *greater* than those predicted by theory. The viscosities are also far greater than expected from the $M_w^{3.4}$ law. The results of Valentine et al. thus confirm in every respect results reported earlier by J. T. Gruver and the author (35). In Section 2.2.2 reference was made to a trichain polybutadiene of M_w 303,000 with enormously high viscosity at very low shear rates. Creep experiments on this sample gave very high recovery; in fact, the extremely high viscosity was calculated from the permanent set

Table 2 Steady State Compliance of Branched Polybutadienes (25°C)

Polymer	$M_w/1000$	$M_n/1000$	$J_e \times 10^6$ (cm²/dyne) Observed	$J_e \times 10^6$ (cm²/dyne) Calculated (eq. [12])	$\eta_0 \times 10^{-6}$ (poise)
A (linear)	180, 170	150	0.19	3.6	5.07[a]
A-I (branched)	160	110	>11.0	4.5	>16.7
A-II (branched)	230	180	>22.3	6.3	>416

[a] Value appears high. Other values are 2.6×10^6 (Plazek), 2.5×10^6 (Gruver and Kraus).

which constituted only a minor part of the total deformation in the creep test. Steady state was not reached in the duration of the test so that J_e could not be estimated. It obviously must have been considerable. Similar results were found with artificially branched cis-polybutadienes (45).

The steady state compliance of a narrow distribution tetrachain polybutadiene ($M_w = 119,000$) obtained from normal stress measurements at 100°C was found to be 1.4×10^{-6} cm²/dyne, substantially greater than for linear polymer of this molecular weight (51). More recently Masuda et al. (37) have shown that star and comb shaped polystyrenes show much higher steady state compliances than their linear counterparts.

In summary, it may be stated with some confidence that in spite of the experimental difficulties in measuring melt elasticities and the lack of quantitative agreement with theoretical predictions applicable to low molecular weight polymers, long chain branching in *high molecular weight* polymers increases elastic compliance. The increased elastic compliance obtained on broadening of the MWD (see curve C, Fig. 12) is, of course, well established (18, 19). The combined effects of branching and molecular weight distribution have been described by White and Tokita (52).

Vinogradov and associates (53, 54) have studied the dynamic viscoelastic behavior of linear, narrow distribution polybutadienes and polyisoprenes in the entanglement plateau region (see also Section 3.2). The length of the entanglement plateau, in units of reduced frequency, increased greatly with polymer molecular weight. Of particular interest is the authors' observation that the discontinuity of the flow curve, log $\dot{\gamma}$ versus log τ, observed in extrusion through cylindrical capillaries (Section 2.1) occurs at a shear stress that is approximately equal to 45% of the storage modulus in the entanglement plateau (54). The authors propose that at this critical stress the polymer becomes incapable of displaying fluidity and behaves as a cross-linked network. The "cross-links" are, of course, chain entanglements.

2.2.6 *Effect of Diluents*

The effect of diluents up to 50% in concentration on the melt flow of raw rubbers is of great interest in connection with their oil extension. According to equation [8] the Newtonian viscosity of an oil extended polymer will be approximately related to that of the dry rubber by the 3.4th power of the volume fraction of polymer and by the ratio of the friction factors. The segmental friction factor ξ is a function of free volume, and many plasticizers function primarily by adding free volume to the system, thereby reducing ξ. However, this is not always the case when extender oils are added to rubbers. Many of these oils are glass-forming liquids with glass transitions considerably greater than those of the rubbers to which they are added, particularly in the case of polybutadienes (see also Table 1 in Chapter 10):

	T_g (°C)
Butyllithium polybutadiene	−99
cis-Polybutadiene (95% cis)	−112
Highly aromatic extender oil	−44
Naphthenic extender oil	−64

If the familiar concept of the glass transition as an iso-free volume state (55) is extended to hydrocarbon oils, then it is clear that there will be instances, especially at relatively low temperature, where the "plasticizer" will decrease the free volume of the system and hence raise ξ. It should also increase T_g, a fact that has been confirmed (Table 3). Under these

Table 3   Effect of Extender Oils on T_g of Polybutadiene

Polymer Type	Oil	ϕ	T_g (°C)
Bu–Li	—	1.000	−99
	Naphthenic	0.837	−97
	Naphthenic	0.674	−92
	Naphthenic	0.535	−90
	Naphthenic	0.340	−83
	Highly aromatic	0.676	−92
95% cis	—	1.000	−112
	Naphthenic	0.774	−109
	Naphthenic	0.672	−106
	Naphthenic	0.505	−103
	Naphthenic	0.338	−96

conditions, the extender oil is in a sense an antiplasticizer. It will still reduce the viscosity, but only through the factor $\phi^{3.4}$ which, in effect, represents the loosening of the entanglement network (equation [8]). At temperatures far enough above T_g of both the polymer and diluent, their *relative* contributions of free volume to the system will become more nearly equal. The quantity $\xi(\phi)/\xi(1)$ may then approach unity and the ratio $\eta_0(\phi)/\eta_0(1)$ will be nearly equal to $\phi^{3.4}$. This evidently happens with linear polybutadienes and several different plasticizers at 100°C (56). Under these conditions the viscosity becomes independent of the nature of the plasticizer! The more usual case is, of course, for the plasticizers to exert specific effects on segmental friction, lowering it drastically whenever T_g (diluent) $\ll T_g$ (polymer), and the temperature of measurement is not too far above T_g (polymer).

The non-Newtonian flow of plasticized elastomers can be described by reduced variable treatments (56). Equation [9] applies with the inclusion of the polymer concentration or volume fraction in the reduced independent variable:

$$\frac{\eta}{\eta_0} = f(\eta_0 \dot{\gamma} \phi^{-2}) \qquad [35]$$

Temperature and molecular weight reduction are effected through the Newtonian viscosity η_0. The inclusion of ϕ^{-2} in the independent variable is essentially empirical, but it is highly sucessful in superimposing the data on the master curve for the dry rubber. When Newtonian viscosities are not realized in the experiments, an empirical shift factor α_ϕ is used in place of η_0. The shift factor α_ϕ is equivalent to $\eta_0(\phi)/\eta_0(1)$, and the ratio $\alpha_\phi/\phi^{3.4}$ is then roughly proportional to the ratio of friction factors. A theory by Graessley et al. (57) suggests the ϕ^{-2} dependence in equation [35] to be the result of a highly entangled system; for less entangled molecules, the dependence should approach ϕ^{-1}.

The effect of dilution on the low shear melt viscosity of high molecular weight branched polybutadienes is greater than is observed with linear polymers, i.e., η_0 (br)$/\eta_0$ (lin) decreases from a value larger than unity to less than unity (31, 58), usually at rather modest concentrations of diluent. In dilute solutions η_0 (br) is always less than η_0 (lin).

2.2.7 Melt Flow of Random Copolymers

The conclusions reached from investigations of the rheological behavior of polybutadienes have been found to apply equally to random copolymers of butadiene and styrene. This is not surprising when one considers

Table 4 Flow Governing Parameters for Butadiene–Styrene Copolymers and Homopolymers

Parameter	Butadiene–Styrene Copolymer[a]	Polybutadiene[b]	Polystyrene
$T_g(°K)$	223	178	373
$(\overline{s_0^2}/M) \times 10^{18}$	11.8	12.6	7.6
Z_c	370	330	600
K	13.5	12.2	13.0
Δ	40	50	60
$\log \xi_0$	−10.42	−10.90	−11.05

[a] 25% styrene, 75% butadiene (19% cis, 47% trans, 34% vinyl).
[b] 52% trans, 40% cis, 8% vinyl.

that the unit of flow in long chain polymers contains several monomer units. In a truly random copolymer, not only do all chain segments "see" the same environment but the segments themselves all have substantially the same composition. To this we add the rather obvious fact that polybutadienes are in reality also copolymers, composed of cis-1,4, trans-1,4, and 1,2 addition units.

Narrow distribution, linear butadiene–styrene random copolymers show the extended Newtonian flow region (Fig. 2) of other such polymers (12). The viscosity-temperature relation for butadiene–styrene copolymers is similar to that of polybutadiene and, for that matter, of polystyrene. Table 4 gives a comparison of the parameters contained in equations [1]–[3]. Because the values of the Vogel (2, 3) parameters are not very different, the principal difference is derived from the position of T_g: At fixed temperatures the copolymer, being closer to its T_g, will show a greater temperature coefficient of viscosity. The higher T_g of the copolymer has other consequences. In oil extended polymers, we no longer have the condition prevailing in 1,4-polybutadienes, that T_g (polymer) < T_g (diluent), which tends to diminish the plasticizing effect. The copolymer thus responds more readily to plasticization.

2.2.8 Melt Flow of Nonrandom Sequence Copolymers

In a binary block copolymer containing long block sequences, there will obviously be more than a single kind of flow unit. Because segmental motion of the two units will have different temperature dependences, simple shear rate-temperature reduction of viscosity data is no longer to be expected. This is illustrated in Figs. 14 and 15, comparing flow curves

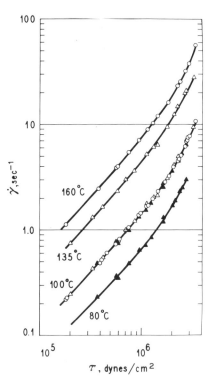

Figure 14 Flow curves of random butadiene-styrene solution copolymer (25% styrene). Data for 80, 135, and 160°C have also been superimposed on the 100°C curve. Note the excellent success of the temperature reduction (59). Reproduction by permission of the *Journal of Applied Polymer Science.*

for random and block polymers of 25% styrene content (59). Lack of superposability is even observed in certain random copolymers in which composition is variable along the polymer chain. Such polymers result from slight "underrandomization" in alkyllithium copolymerizations and show no block styrene by chemical analysis. However, their short-range blockiness may be revealed by their NMR spectra.

The interesting question of whether a block or a random polymer (at equal M_w) has the higher viscosity is answered in Fig. 16. In this series of alkyllithium polymers initiator level was constant and only the amount of randomizer was varied. Although molecular weights were not measured, it was known that the polymerization conditions used produced polymers of nearly equal molecular weight (slightly *lower*, the higher the block content). The data show that the viscosity decreases with increasing randomization. Furthermore, the observed trend cannot be due to the increased vinyl content that accompanies randomization, in this example 6.5% for polymer F-5, increasing to 45.8% for F-1. Although 1,2 enchainment shortens the chain, the effect is more than overcome by an increase in segmental friction. The glass transition of polybutadienes is

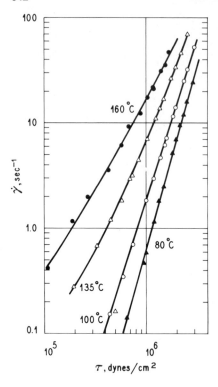

Figure 15 Flow curves for a two-sequence block copolymer of butadiene (75%) and styrene (25%). Note the lack of temperature superposability (59). Reproduced by permission of the *Journal of Applied Polymer Science.*

quite sensitive to vinyl content, increasing from $-99°C$ for 8% vinyl to $-7°C$ for essentially all vinyl polymer. At fixed temperature, the closer proximity to T_g strongly increases the viscosity as vinyl content is increased, for the Vogel parameters (equation [3]) are quite insensitive to vinyl content (61). Thus, for example, at equal M_w polybutadiene of 38% vinyl unsaturation has about 1.5 times the Newtonian viscosity of 10% vinyl polymer at 106°C (33).

Statistical thermodynamic considerations (62) lead one to expect the two-phase domain morphology of butadiene–styrene block copolymers (see Section 4) to persist in the melt, i.e., above T_g (styrene). The flow behavior of these melts is, indeed, consistent with such a structure. Below T_g(styrene) the association of styrene blocks clearly converts a simple two-sequence block polymer molecule into something resembling a star shaped multichain polymer. The melt flow of such rubbers then resembles that of branched polymers. The block polymers show no cold flow even at quite low molecular weights. They also show considerable shear sensitivity (see Fig. 15, 80°C). Three-sequence S–B–S block polymers form networks and show virtually no true viscous flow below T_g(styrene). They are quite viscous in the melt and are therefore prepared to relatively low molecular weight for processability. For these polymers to flow at all, some disruption of the polystyrene domains must occur. Since the domain structure and its disruption by shearing forces are temperature dependent, block polymers generally exhibit a great degree of thermoplasticity. This is illustrated in Fig. 17 with data on a series of 75:25 butadiene copolymers taken at a fixed shear stress. As the degree of nonrandomness increases, so does the slope of the curves. The high melt viscosity of S–B–S block polymers and the lack of temperature-shear rate superposition of the non-Newtonian viscosity encountered in these materials have been confirmed by Holden et al. (63).

The effects of branching in multichain block polymers have been investigated by Kraus and co-workers (64). In polymers of structure [B–S]$_{-x}$, where $x = 2$ (linear), 3, or 4, the length of the terminal butadiene blocks rather than the molecular weight of the entire polymer determines

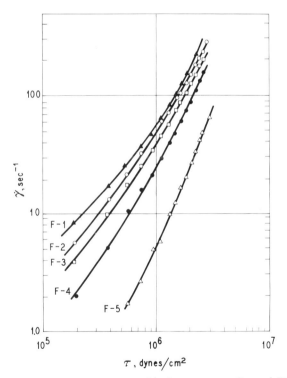

Figure 16 Effect of increasing styrene block content on flow of 75:25 butadiene-styrene copolymers at 106°C (59). Block styrene varies from zero (F–1) to 19.7% (F–5). Reproduced by permission of the *Journal of Applied Polymer Science*.

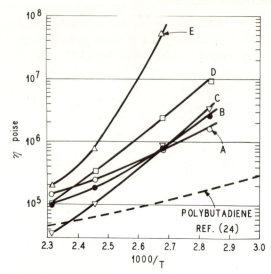

Figure 17 Effect of monomer sequence on the temperature coefficient of viscosity (59). Shear stress $= 5 \times 10^5$ dynes/cm². All copolymers 75:25 butadiene-styrene. *A*, Uniformly (perfectly) random; *B*, random; *C*, B:S/S block (18% block styrene); *D* B/S block; *E*, SBS block. Reproduction by permission of the *Journal of Applied Polymer Science*.

the viscosity. The branch points have no significant effect on the formation of the polystyrene domains (65) and become buried inside them, so that flow is governed primarily by entanglement coupling between butadiene blocks. Similarly, in [S–B]$_{-x}$ polymers the length of the *styrene* blocks (rather than total molecular weight) appears to govern flow, as disruption of the polystyrene domains becomes increasingly difficult with rising molecular weight of the styrene blocks. In these polymers the dynamic viscosity always exceeds the steady flow viscosity when compared at $\omega = \dot{\gamma}$; i.e., equation [15] does not hold. This is at least in part attributable to the disruption of the domain network which cannot be expected to occur to equal extents in steady flow and in small oscillatory deformations.

The effects of deformation rate on the domain structure have been considered by Arnold and Meier (66). From dynamic viscosity measurements on melts the authors conclude that an S–B–S or similar block polymer can exist in three distinct states, depending on the rate of deformation. At rest or at very low deformation rates the molecular network remains intact. At higher rates domains and aggregates of many molecules still occur, but are no longer linked together in a three-dimensional infinite network. Finally, at very high deformation rates the system is thought to behave as an assembly of unaggregated molecules. A rather surprising result of this study is the rather good temperature-frequency superposition of the dynamic viscosity, in view of the obvious thermorheological complexity of the system.

In summary, it is clear that the flow of block polymers is a complex phenomenon. It would be an obvious thought to attempt a theoretical generalization of their flow behavior in terms of the association of styrene blocks. Unfortunately, any such attempt is doomed to failure, for the association is not bifunctional and the resulting branched structures will lead to rheological effects that are not yet predictable by theory.

2.2.9 Polymers with Polar End Groups

Specialty polymers carrying reactive end groups to make them vulcanizable by condensation reactions have achieved some significance as solid propellant binders. Typical of these materials are carboxy terminated polybutadienes with molecular weights of a few thousand, prepared by solution polymerization with dilithium initiators. The flow behavior of these liquid rubbers is not unlike those of similar polymers without polar end groups, except for an enhancement in the viscosity resulting from end group association (61, 67). The apparent degree of association is, however, relatively small; a larger effect on the viscosity would be expected from the normal association equilibrium of carboxylic acids. The results can be explained by assuming that some of the associations are intramolecular (ring closure) so that the molecular weight of the associated species is held down (61). An alternative explanation is that interchange of hydrogen bonds, the rate of which greatly exceeds the shear rate, results in a time averaged degree of association which does not represent the true association equilibrium (68).

2.2.10 Polyisoprene

Synthetic *cis*-polyisoprene differs from the polybutadienes and butadiene–styrene solution polymers by its ease of breakdown by mastication, a property it not unexpectedly shares with the natural product. It is prepared to high molecular weight and usually contains some branching or even gel, either from the polymerization or from the finishing operations. There is no cold flow problem. In normal factory processing, breakdown is accompanied by some recombination and the final rheological properties of compounds are often not too closely related to those of the polymer as originally synthesized. Several basic studies of the flow behavior of raw polyisoprenes exist. Yasuda (69) has shown that the

Newtonian viscosity of narrow distribution polyisoprenes follows the 3.4th power law in molecular weight. On the other hand, Holden (70) reports a 3.95 power dependence, conceivably due to traces of adventitious branching introduced in the handling of the polymers. The data of Holden also show the extended Newtonian flow regime found in other narrow distribution rubbers (Section 2.2.1). Dynamic viscoelastic properties of uncross-linked narrow distribution polyisoprenes have been reported by Vinogradov et al. (54).

Nemoto and associates (17, 20) found the steady state compliance of narrow distribution polyisoprenes to follow equation [11] up to a molecular weight of 50,000, after which J_e became constant. This is four or five times the critical molecular weight for entanglement coupling. The dependence of J_e on diluent concentration was found to be complex (71).

2.2.11 *Summary*

Sufficient data on narrow distribution polymers are now at hand to permit some generalizations on the applicability of the various molecular theories based on the Rouse-Bueche model (14, 15). In general, these theories hold for undiluted polymers up to molecular weights equal to or somewhat exceeding M_c. Thus equation [11] for J_e is applicable to *very roughly* $4M_c$, as is the prediction that star branching lowers η_0 by a factor $g^{3.4}$. Theoretical molecular weight, temperature, and concentration dependences of the non-Newtonian viscosity (equation [9]) likewise appear limited to this range of chain length. It becomes clear that M_c is a very critical parameter in the melt rheology of polymers. For example, the small steady state compliance of *high molecular weight* polybutadiene is readily explained by the fact that M_c is only about one-sixth that of polystyrene. Since J_e is proportional to M up to a value of about $4M_c$ and then becomes independent of molecular weight, M_c governs the compliance at *high* molecular weights. As already discussed, the molecular weight required for the development of abnormally high low shear viscosity in branched polymers also appears to be related to the value of M_c.

2.3 Commercial Polymers

In the preceding we have emphasized the importance of molecular weight distribution and long chain branching on the rheological properties of polymer melts. Although most of the polymers used to illustrate the effects of these variables were specially prepared to emphasize one or the other structural feature, it is nevertheless true that similar polymers prepared under ordinary polymerization conditions still retain most of the characteristics described. Thus, a conventional butyllithium polymerized polybutadiene will, in general, have a somewhat broader molecular weight distribution than the polymers of Section 2.2.1, and it will almost certainly have acquired some long chain branching if dried in a commercial extrusion dryer. Such a rubber will differ significantly in its flow behavior from the ideal structure polymer, especially at low shear rates, but it will still cold flow and lack in melt elasticity.

The fact that many (but not all) solution polymerization systems based on organometallic catalysts produce linear molecules of relatively narrow molecular weight distribution imparts to these polymers processing behavior which differs characteristically from that of emulsion SBR or natural rubber. Considerable effort has been expended to modify the rheological properties of solution polymers either during polymerization or by posttreatments. These are described in Chapters 2–7 in this book. They all operate on the principles of altering the MWD or changing the kind and degree of long chain branching.

2.3.1 Molecular Weight Distribution and Branching in Commercial Polymers

Because of the widespread practice of modifying branching and MWD, commercial polymers exhibit a wide variety of chain architecture (72), and it would be impossible to attempt to describe more than a very few of these in a short exposé. While the MWD can be ascertained with reasonable assurance and rapidity by use of gel permeation chromatography, a complete, general analysis of branching distributions is not yet possible.

The best and most complete method known for characterizing the branch structure of a polymer is to fractionate the polymer by molecular weight and measure the molecular weight and intrinsic viscosity of *each* fraction (73). At constant molecular weight (36)

$$\frac{[\eta]_{br}}{[\eta]_{lin}} = g^{1/2} \qquad [35a]$$

where g is the branching factor already defined (equation [5]). The degree of branching in a given fraction is indicated by the depression of its intrinsic viscosity from the value expected for the linear polymer. Even this extremely laborious method does not furnish a complete description of branching because g is not unique. It expresses the overall effect of branching on the mean square radius of gyration, but gives no information regarding the number and length of the individual branches.

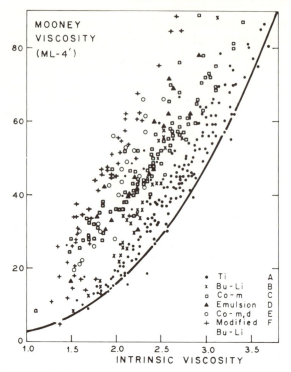

Figure 18 Relationship between Mooney and intrinsic viscosity for polybutadienes (74). ●, high cis, Ti; x, Bu–Li; □, high cis, Co/Al dialkyl monohalide; △, emulsion; ○, high cis, Co/mixed Al dialkyl monohalide and Al alkyl dihalide; +, modified (branched) Bu–Li. Reproduced by permission of *Rubber and Plastics Age*.

A very rough estimate of branching has been proposed by Van der Hoff and associates (74) which takes advantage of the anomalously high melt viscosity of high molecular weight branched polymers. Because of it, the ratio of the intrinsic viscosity to a low shear melt viscosity will be largest for a linear polymer and will decrease with increasing branching. The Mooney viscosity is often sufficient for such a test (Fig. 18). The method can be quite useful, but it has its limitations. As is clear from Fig. 18, the ratio $[\eta]/[ML_4]$ is not constant, but depends on the molecular weight and its distribution. Comparisons should be made only when the MWD is known to be similar for the samples under consideration.

2.3.2 Processing Problems

Common processing problems of the solution rubbers are almost invariably traceable to lack of sufficient long chain branching and/or to narrow molecular weight distribution, leading to insufficient melt elasticity and relatively small zero shear viscosity. The former causes poor banding on roll mills and poor feeding to screw extruders. The outstanding manifestation of the latter is cold flow. In the case of polybutadienes, a contributory factor is the very low glass transition, which means that under conventional processing conditions the polymer is being handled farther above T_g than, for instance, SBR. This tends to further emphasize viscous over elastic deformation during processing. Thus *cis*-polybutadiene will form a tight band on a cold roll mill, but becomes lacey and bags on a hot mill. When the sheet is cut off from a cold mill, it shrinks in the longitudinal direction, but there is little shrinkage when the polymer is cut off from a hot mill. The mill behavior of elastomers has been studied in detail by Tokita and White (75), who interpret the characteristic behavior of polybutadienes as a viscoelastic effect related to their chain architecture. A different explanation has been offered for the "mill transition" of *cis*-polybutadiene by Bianchi and associates (76, 77); the present author favors the interpretation of Tokita and White (75). The mill banding of polybutadiene responds well to broadening of the MWD.

In considering processing problems, it is important to distinguish between characteristics of the raw rubbers and of the fully compounded stocks. The introduction of fillers and plasticizers profoundly affects rheological behavior. Nevertheless, in the case of polybutadienes and solution butadiene–styrene copolymers, which undergo relatively little breakdown during processing, the effects of the original chain architecture persist to a surprising degree in the fully compounded rubbers. Thus, manipulation of the molecular weight distribution and branch structure can produce notable improvements in processing behavior of compounded stocks in spite of the leveling effect of carbon black and extender oils. On the other hand, with polyisoprene, breakdown phenomena are more important and relatively less appears to be gained by control of the original chain structure (see Section 2.2.10).

In one area of processing, linearity and narrow MWD can be advantageous. This is in injection molding, where shorter injection times and lower injection and clamp-down pressures can be realized with solution rubbers.

2.3.3 Extrusion and Injection Molding

Since linear, narrow distribution polymers lack melt elasticity, they exhibit very small die entrance losses during extrusion. In fact, in capillary rheometry the end correction, e, of such polymers (equation [18]) is found to be extremely small (30). The polymers also show relatively little shear

Table 5 Extrusion Rates and Injection Times for Fully Compounded Solution and Emulsion SBR

	Solution Polymer A[a]	SBR 1502	Solution Polymer B[b]	SBR 1708
Mooney viscosity (ML$_4$), raw	56	52	46	55
compounded	57	53	61	58
Extrusion rate (g/min)	128	118	132	116
Injection time (seconds)	5	11	2	5

[a] Solprene 1204.
[b] Solprene 375, oil extended.

sensitivity because of the extended Newtonian flow regime (Fig. 1). Referring now to equation [22] we see that at fixed extrusion conditions and a *comparable viscosity* the flow rate will be larger than for a polymer with relatively little melt elasticity, as both N and e will be smaller. This can explain the higher extrusion rates and faster injection times of solution SBR versus emulsion SBR (Table 5), except at very high shear rates where transport of the stock may occur by plug flow. The solution rubbers in Table 5 match the emulsion rubbers in styrene content, amount and type of extender oil, and Mooney viscosity, but are less branched and have narrower molecular weight distributions (78). Improved extrudability and injection molding characteristics have also been reported for synthetic *cis*-polyisoprenes (79, 80). These too are the result of better control over the initial molecular weight distribution, the effects of which are not completely destroyed in compounding.

3 VISCOELASTIC BEHAVIOR OF VULCANIZATES

In the preceding section we have been concerned with the viscoelasticity of polymer melts as related to their processability, i.e., with the behavior of unvulcanized rubbers, typically some 200°C above their glass transition. We now inquire about the viscoelastic properties of some of these same polymers when cross-linked into networks of practical utility. Vulcanization—the conversion of a substantially linear polymer to an infinite network—clearly produces enormous changes in the rubbery (low frequency, high temperature) region of viscoelastic response. It converts a predominantly viscous material to a highly elastic one. All of this is accomplished with very little change in short-range segmental motion. The degree of cross-linking introduced in normal vulcanization is of the order of one cross-linked unit per sequence length of molecular weight 10,000, which is barely enough to shift the glass transition by a few degrees and which has little effect on viscoelastic behavior in the transition zone between glassy and rubbery behavior. The viscoelastic behavior of networks is thus a function of two sets of factors. The first relates to short-range segmental motion, which depends on the chemical structure of the rubber and determines the temperature range in which rubbery behavior is encountered, but which is fundamentally not different in the unvulcanized and vulcanized polymers. The second is network topology and its effects on long-range motion, which are more nearly independent of the chemical nature of the polymer, but which depend on such physical parameters as the cross-link and entanglement densities, the primary molecular weight, and the number of chain scissions experienced during vulcanization.

3.1 Segmental Friction and Viscoelasticity

Since local segmental mobility is little affected by cross-linking, it is instructive to return briefly to the behavior of unvulcanized polymers. From equation [3] it is obvious that the position of the glass transition temperature is one of the dominant factors affecting segmental friction at any given T; in fact, the quantity $T - T_g$ is of fundamental significance as a specification of corresponding states in polymer viscoelasticity. When compared at corresponding states, or $T - T_g$, the properties of different polymers are not identical, but they are often surprisingly similar. Thus, Kraus and Gruver (61) found that a plot of $\eta_0/Z_w^{3.4}$ against $T - T_g$ for polybutadienes of different vinyl contents (see equations [1] and [3]) yields an essentially universal curve in spite of a very strong dependence of T_g on vinyl content. The parameters in equation [3] are simply not highly structure dependent.

   A most impressive demonstration of this "law of corresponding states" has been given by Sanders and co-workers (81). These authors studied the viscoelastic behavior of polybutadienes of 7 and 91.5% vinyl unsaturation, SBR, natural rubber, and polyisobutylene in the transition zone. Figure 19 shows their loss tangent versus frequency plot, reduced to $T_g + 100°C$. With the striking exception of polyisobutylene, the curves lie quite close together. A very different comparison results at fixed temperature, as shown in Table 6. These results indicate that, at corresponding states, the local segmental mobility of polybutadienes is not very sensitive to microstructure, nor is it greatly affected by the introduction of styrene comonomer. One would therefore expect the random solution copolymers of butadiene and styrene to exhibit segmental friction and mobility

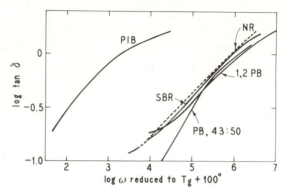

Figure 19 Loss tangents for four rubbers and polyisobutylene (PIB) with frequency scale reduced to $T_g + 100$ for each polymer (81). PB 43:50 is butyllithium polybutadiene; 1,2 PB is 91.5% vinyl polybutadiene. Reproduced by permission of the *Journal of Polymer Science*.

similar to emulsion SBR. Likewise, there is no reason to expect synthetic *cis*-polyisoprene to differ appreciably from natural rubber in this respect. One is thus led to the conclusion that for all rubbers based on butadiene, butadiene–styrene, and isoprene, energy dissipation in the transition zone is, in the zeroth approximation, a function of $T - T_g$ only. This rule agrees quite well with practical experience, and its validity appears to extend part way into the rubbery zone. However, the spectacular failure of polyisobutylene or butyl rubber to follow the same $T - T_g$ correlation is ample proof of its limited nature.

Table 6 Comparison of tan δ for Various Rubbers[a] at 298°K and at $T_g + 100°C$

Polymer	T_g (°K)	log ω at log tan δ = −0.2 298°K	log ω at log tan δ = −0.2 $T_g + 100°C$
Polyisobutylene	205	2.57	2.82
Polybutadiene (91.5% vinyl)	261	2.71	5.61
SBR (23.5% styrene)	210	4.95	5.48
Natural rubber	200	5.41	5.45
Polybutadiene (7% vinyl)	172	6.14	5.55

[a] From Sanders et al. (81).

3.2 Network Structure and Viscoelastic Behavior

In the rubbery zone, the effects of local segmental friction and molecular mobility of vulcanizates are often obscured by differences in network structure. One only needs to compare two polymers (e.g., *cis*-polybutadiene and *cis*-polyisoprene) at different degrees of cross-linking to reverse their hysteresis. Unfortunately, the effects of cross-linking are not simple, but interact strongly with those of the entanglement network existing before vulcanization. There are very few fundamental studies recorded in the literature dealing with the effects of network structure on viscoelastic behavior and these are almost entirely the work of J. D. Ferry and his students.

3.2.1 The Entanglement Network Prior to Vulcanization

The critical chain length for entanglement coupling Z_c of equation [1] is related to the actual spacing of entanglements Z_e in the linear polymer. It has been proposed (82) that $Z_c = Z_e$, but more recent evidence (2, 81) favors $Z_c = 2Z_e$. In addition to the method of evaluation from the break in the viscosity-molecular weight curve (Fig. 1), there are at least two other ways to arrive at estimates of Z_e. These methods usually agree within a factor of about 2.

The viscoelastic functions J, J', G' (defined above) and the relaxation modulus G of *high molecular weight* linear polymers all exhibit a "plateau" region between the transition zone and the rubbery flow regime which is interpreted as a pseudoequilibrium compliance (or modulus) of the entanglement network. The entanglement spacings can then be estimated from the plateau value of the appropriate viscoelastic function by the statistical theory of rubber elasticity. In reality the plateau is more accurately described as a flat region, and some uncertainty exists in locating the entanglement network contribution to the compliance (J_{eN}) or modulus (G_{eN}). Ferry et al. (81, 83) have developed improved methods for locating J_{eN} and G_{eN} from integrations of the loss compliances and loss moduli. Application of the theory of rubberlike elasticity then gives

$$G_{eN} = J_{eN}^{-1} = \frac{j\rho RT}{M_0 Z_e} \qquad [36]$$

where M_0 is the monomer molecular weight, j is the number of chain atoms per monomer unit, and ρ, R, and T have their usual meaning. (Ferry uses the symbol jZ_e for what we, for the sake of consistency in notation, call Z_e.) An alternative estimate of Z_e can be obtained from the minimum in the loss tangent versus frequency curve, applying a theory of Oser and Marvin (81, 84). For details, the reader is referred to the original literature.

Table 7   Entanglement Spacing Estimates (81)

Polymer	$\log J_{eN}$	Kinetic Theory	Oser-Marvin	η_0 vs. M_w
			Z_e from	
Polyisobutylene	−6.40	320	—	270
Polybutadiene				
(91.5% vinyl)	−6.79	132	131	—
SBR (23.5% styrene)	−6.89	186	192	—
Natural rubber	−6.57	360	400	412
Polybutadiene				
(7% vinyl)	−7.18	110	86	208

The entanglement spacing estimates for the five rubbers of Table 6 are reproduced in Table 7. There is fair agreement, although it must be pointed out that Ferry and associates (81) divide the Z_e values obtained by the Oser-Marvin method by an empirical correction factor of 1.48 to produce the agreement shown. Comparison of the Z_e values with $Z_c/2$ from Table 4 also reveals fair agreement. It appears that the entanglement spacing increases in the order polybutadiene < SBR < polyisoprene.

3.2.2   Network Structure of Vulcanized Rubbers

When chemical cross-links are introduced into an entanglement network to produce a typical vulcanizate with substantially zero sol fraction, the resulting structure may be regarded as two interpenetrating networks or as a single network in which the entanglements have been converted into effective cross-links. The argument for the latter point of view stems from the fact that the entanglements will now be unable to disentangle, having become trapped between the fixed chemical cross-links. What is not clear, however, is how the trapped entanglements will affect the physical properties of the vulcanizate and, in particular, under what conditions, if any, do they become physically fully equivalent to the chemical cross-links.

A. M. Bueche (85) was the first to use an equation of the form

$$\nu = 2c - \frac{2\rho}{M_n} + 2\epsilon\left(1 - \frac{\rho}{cM_n}\right) \qquad [37]$$

to express the elastically effective network chain density ν as a function of the cross-link density c, the entanglement density ϵ, and the primary number average molecular weight M_n. The quantity ρ is the polymer mass density. The second term on the right-hand side of equation [37] is the familiar free chain end correction; two free chain ends result from every primary molecule. The third term represents the conversion of entanglements to equivalent cross-links, all entanglements becoming physically effective as c or M_n approaches infinity. This equation has been used with a minor modification by Mullins (86) and by Kraus (87) to describe the effects of primary molecular weight on the network chain density deduced from equilibrium modulus and swelling measurements. The equation correctly expresses the experimental fact that fewer chemical cross-links are required to vulcanize a high molecular weight polymer to the same ν than would be required for a low molecular weight rubber, the free end correction being far too small to account for the difference observed. Application of equation [37] to the data of Moore and Watson (88) on natural rubber yields an entanglement spacing $Z_e \approx 1100$. The author's (87) data on polybutadiene (10% vinyl) yield $Z_e \approx 420$. These are over twice the corresponding values of Table 7. These results seem to suggest that in the examples at hand only about half the total number of entanglements ever become elastically effective at equilibrium. This problem was considered further by Mancke and co-workers (89). These authors point out that equation [37] overestimates the proportion of trapped entanglements T_e by setting it equal to $1 - \rho c/M_n$. This specifies only that two of the strands emanating from an entanglement are tied down. The probability that all four strands terminate in cross-links is $(1 - \rho/cM_n)^2$, so that equation [37] becomes

$$\nu = 2c - \frac{2\rho}{M_n} + 2\epsilon\left(1 - \frac{\rho}{cM_n}\right)^2 \qquad [38]$$

A more complete theory by Langley (90) not only takes into account the necessity of trapping all four network chains radiating from an entanglement but also considers the effects of the initial molecular weight distribution and of chain scissions accompanying vulcanization. For the special case of no chain scission and initially uniform molecular weight, equation [38] gives values negligibly different from Langley's theory for values of $T_e > 0.3$ (89).

If J_{eN} is the compliance level in the frequency range where none of the entanglements are relaxing—as in equation [36]—but is now augmented by the contribution of chemical cross-links, then

$$\frac{J_e}{J_{eN}} = \frac{\nu_{eN}}{\nu} \qquad [39]$$

where J_e is the equilibrium compliance and ν_{eN} is the density of network chains terminated by cross-links or by entanglements, whether or not they are trapped. Obviously

$$\nu_{eN} = 2c\left(1 - \frac{\rho}{cM_n}\right) + 2\epsilon \tag{40}$$

Combining equations [38]–[40] one obtains for the equilibrium shear modulus

$$G_e = J_e^{-1} = J_{eN}^{-1} - J_{eN^\circ}^{-1} + J_{eN^\circ}^{-1}\left(\frac{J_{eN}^{-1} - J_{eN^\circ}^{-1}}{J_{eN}^{-1} - J_{eN^\circ}^{-1} + 2\rho RT/M_n}\right)^2 \tag{41}$$

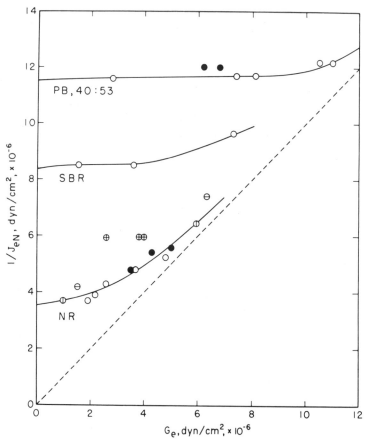

Figure 20 Reciprocal of compliance associated with entanglement network plus cross-links, plotted against equilibrium modulus as suggested by equation [41]. ○, ◐, ⊖, Peroxide vulcanizates; ●, sulfur vulcanizates; ⊕, tetramethyl thiuram disulfide vulcanizates. Dashed line, slope of unity (89). Reproduced by permission of the *Journal of Polymer Science*.

At zero cross-linking $G_e = 0$ and $J_{eN} = J_{eN°}$; at high degrees of cross-linking J_{eN}^{-1} approaches equality with G_e as the quantity in parentheses tends toward unity. At this point, all entanglements are trapped and are equally effective at equilibrium and at high frequencies. A plot of J_{eN}^{-1} versus G_e, reproduced from Mancke et al. (89), is shown in Fig. 20. At high cross-link densities (large G_e) the curves of J_{eN}^{-1} do, indeed, approach G_e. However, it is clear that very high cross-link densities are required in both polybutadiene and SBR to trap all entanglements; $G_e = 1.2 \times 10^7$ dynes/cm^2 corresponds to an effective physical cross-link density of 4.8×10^{-4} moles/cm^3, about three to four times the value of a "good" rubber vulcanizate.

No distinction has been made in this discussion between different kinds of chemical cross-links, although Fig. 20 contains data on various types of vulcanizates. The effects of cross-link type on viscoelastic behavior are not always negligible (91, 92), but there is reason to believe them to be similar in all diene based rubbers with predominantly 1,4 enchainment. Hasa and Van der Hoff (93), studying the elastic behavior of poly(butylacrylate) networks, found no effect of cross-link length up to 16 bonds.

3.2.3 Viscoelastic Behavior of Polybutadiene as a Function of Cross-Linking

Maekawa and co–workers (92) studied the effects of increasing levels of cross-linking in narrow distribution butyllithium initiated polybutadiene of 180,000 and 510,000 primary molecular weights. The first of these polymers was the same sample used in the work of Valentine et al. (48), already referred to in Section 2.2.5. Together, these two papers present the most complete analysis available of the effects of cross-linking on the viscoelasticity of any rubber, in particular a well-characterized, narrow distribution elastomer.

The results of the two studies are combined in Figs. 21 and 22, showing the reduced dynamic storage (J_p') and loss (J_p'') compliances shifted to 25°C. The curves represent increasing degrees of cross-linking, starting with the linear parent polymer (curve 1), followed by a sample cross-linked short of the gel point (curve 2), a sample containing 32% gel (curve 3), and three vulcanizates with (physical) network chain densities $\nu = 0.57 \times 10^{-4}$ (curve 4), 1.85×10^{-4} (curve 5), and 2.68×10^{-4} moles/cm^3 (curve 6).

Examination of the storage compliance data reveals several interesting features. There is a slight shift of the transition zone toward lower frequencies as ν increases. The shift is in the expected direction, since

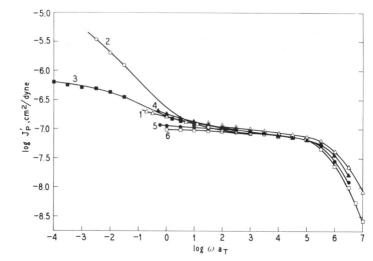

Figure 21 Dynamic storage compliance for butyllithium polybutadiene as a function of cross-linking (48, 92). Degree of cross-linking increases from 1 to 6 (see text). Data reduced to 25°C; subscript p denotes multiplication by $T\rho/T_0\rho_0$.

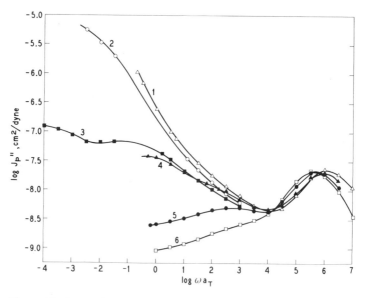

Figure 22 Dynamic loss compliance for samples of Fig. 21 (48, 92).

cross-linking decreases the free volume and raises T_g. The shift is some-what larger than in natural rubber (92). In the rubbery plateau zone ($2 < \log \omega a_T < 5$) cross-linking has very little effect on the storage com-pliance. In this region the level of compliance is governed by the entanglement network (see above). Because of the close entanglement spacing of polybutadiene (Table 7) normal cross-linking does not greatly augment the network chain density in the frequency region where all entanglements are physically effective. In the terminal zone the storage compliance first rises enormously with cross-linking as a result of branch-ing and broadening of the molecular weight distribution (see section 2.2.5), then falls as infinite networks of increasing cross-link density are formed.

The loss compliance (Fig. 22) shows the expected primary dispersion region at $\log \omega a_T = 6$. It is relatively unaffected by cross-linking and is attributed to the entanglement network originally present. By far, the more interesting feature is the secondary, flat low frequency maximum clearly recognizable in all cross-linked samples. This maximum shifts toward lower values of J_p'' and higher frequencies with increasing degrees of cross-linking. It has been observed in every rubber examined by Ferry and co-workers (94–96). Its origin is attributed to relaxation of untrapped entanglements on structures loosely attached to the network. These structures could be individual polymer molecules or several molecules joined by cross-links, but not firmly incorporated into the main network. In spite of the rather complex dependence of the loss compliance on cross-linking and frequency, it is clear that the mechanical losses decrease monotonically with increasing degree of cure.

The effect of the molecular weight before vulcanization (92) is shown in Table 8 at two levels of cross-linking. It is clear that the higher molecular weight rubber is more resilient, a fact well known to rubber technologists.

Table 8 Effect of Primary Molecular Weight on Dynamic Properties[a] of Sulfur Vulcanized Polybutadiene

$M_w/1000$	$\nu \times 10^4$ (moles/cm³)	$\log J_p'$	$\log J_p''$	$\tan \delta$	Resilience[b] (%)
180	1.2	−6.92	−8.13	0.062	68
510	1.2	−6.98	−8.40	0.038	79
180	1.8	−6.94	−8.27	0.047	75
510	1.8	−6.99	−8.53	0.029	83

[a] From Maekawa et al. (92); $\omega = 10$ rads/sec, 25°C.
[b] $R = 100 \times \exp(-2\pi \tan \delta)$.

The improvement is usually attributed to the smaller free end fraction in the network, but the more efficient trapping of entanglements in a polymer of higher molecular weight may well be a contributing factor.

The influence of the primary molecular weight distribution on the viscoelastic properties of vulcanizates has not been investigated in comparable depth. Real effects are to be expected at low degrees of crosslinking, because the trapping of entanglements is sensitive to the MWD (90). In well-vulcanized rubbers, however, the effects are apparently small. Short et al. (97) determined the engineering physical properties of butyllithium polybutadienes blended to different molecular weight distributions, but all of 40 ML_4 Mooney viscosity. In bimodal blends ranging in M_w/M_n from 1.4 to 6.0, no systematic trend in heat generation, determined by Goodrich flexometer, was found. These blends, however, had $M_n \geqq 100,000$ and contained no polymer fractions below 10,000 molecular weight and little below 50,000. In broad distribution polymers prepared by blending many different rubbers to synthesize distributions skewed to both the high and low molecular weight sides of the distribution, a definite trend toward higher hysteresis was noted when the distribution was broadened by adding low molecular fractions progressively. This also lowered the number average molecular weight. Most of the rise in heat buildup occurred when M_n fell below 100,000. These results are in accord with the conclusions presented above; low molecular weight fractions not only increase the number of free chain ends, but are likely to form the type of structures suspected of giving rise to the secondary low frequency loss maximum described by Ferry et al. (48, 92). In a study of well-formed (essentially sol-free) networks prepared from polybutadienes of different molecular weight, distribution, and branch content, Kraus and Rollmann (98) found good correlation of the dynamic storage compliance with the free end fraction, while the loss compliance appeared to be more closely related to the fraction of untrapped entanglements.

3.3 Monomer Sequence in Copolymers

The alkyllithium polymerization employed in the manufacture of solution copolymers of butadiene and styrene is capable of modification to produce rubbers of widely different monomer sequence and distribution (99). The effect of this variable on melt flow behavior was discussed in Section 2.2.8. Its influence on the viscoelastic properties of vulcanizates is also pronounced. Figures 23 and 24 show the storage moduli and loss tangents (at 0.1 Hz) of four copolymers of butadiene and styrene, vulcanized in simple gum formulations, as a function of temperature (100). All four

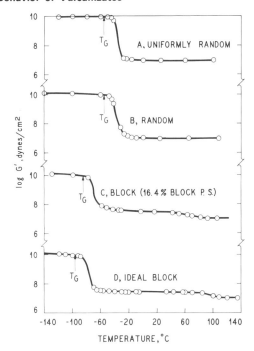

Figure 23 Dynamic storage modulus as function of increasing nonrandomness in 75:25 copolymers of butadiene and styrene (100). Indicated T_g are dilatometric values. Reproduced by permission of the *Journal of Applied Polymer Science.*

polymers contained 25% styrene. In sample A, composition was very nearly independent of conversion, precluding both inter- and intramolecular compositional heterogeneity. For sample B, composition was a function of conversion, but no pure polystyrene block was detected by chemical analysis. Sample C contained 16.4% block styrene in a single block, the remainder of the styrene being dispersed in the butadiene-rich block. Sample D was a two-sequence polymer of compositionally pure butadiene and styrene blocks. The two block polymers clearly display two glass transitions, a result of microphase separation. This effect (63, 101–104) plays a crucial part in the characteristic behavior of block polymers of the ABA type in the un-cross-linked state, to be discussed in detail in Section 4. In vulcanized block polymers, there is little distinction between the AB and ABA (e.g., A = styrene, B = butadiene) block arrangements because conversion of the butadiene blocks to a network produces substantially identical structures (105).

A significant difference is seen in the position and height of the loss maxima of the styrene blocks of polymers C and D in Fig. 24. This is

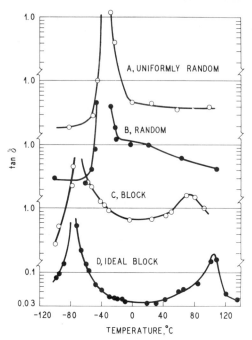

Figure 24 Loss tangents for polymers of Fig. 23 (100). Reproduced by permission of the *Journal of Applied Polymer Science.*

attributed to partial compatibility of the polystyrene domains of polymer C with sections of styrene-rich copolymer immediately preceding the pure polystyrene block. This shifts the maximum toward lower temperature, while at the same time broadening the transition. The effects of breadth in the *intra*molecular composition distribution (and hence also the monomer sequence distribution) persist clearly to polymer B, which has no detectable *long* styrene block. This rubber, like polymer A, exhibits a single dispersion region, but the transition is broadened and the loss tangent shows a definite shoulder. Rollmann and Kraus (60) studied this effect in a series of anionic butadiene–styrene copolymers ranging in breadth of intramolecular composition distribution roughly between polymers A and B. Dynamic viscoelastic properties of the vulcanized rubbers were virtually identical up to the point of incipient appearance of chemically detectable block styrene, even though the raw rubbers exhibited rheological differences clearly attributable to compositional heterogeneity. This result suggests that, as long as the composition distribution does not extend to substantially pure polystyrene, the polymer forms a single phase at vulcanization temperature. Cross-linking

Table 9   Degree of Randomness and Hysteresis

Parameter	Polymer Type (Figs. 23 and 24)	
	A	B
Butadiene/styrene ratio	75 : 25	75 : 25
Mooney viscosity (ML_4)	57	56
300% modulus (kg/cm^2)	128	128
ΔT (°C) (Goodrich flexometer)^a	24.0	28.6
Resilience (%) (Yerzley)^a	78.6	74.6

^{a} For sulfur vulcanizates containing 50 phr HAF carbon black, 10 phr processing oil.

joins together segments of different composition and thus tends to freeze in the single-phase morphology. Even in rubbers polymerized to the point of incipient (long) block styrene formation, which would show considerable blockiness by NMR analysis (due mainly to relatively short styrene sequences), the effects of "underrandomization" on hysteretic behavior in the important rubbery response region are not great. One such example (106), comparing fully compounded vulcanizates, is shown in Table 9.

In vulcanized block polymers the trend toward higher hysteresis is accentuated. When a polymer such as C or D is subjected to repeated flexure at room temperature, the initial hysteresis may be quite low, but the ensuing temperature rise moves the rubber into a region of higher loss. This feedforward mechanism results in increasingly higher energy dissipation and further temperature rise. After the temperature has traversed the loss maximum of the styrene blocks, the network changes from a pseudofilled rubber to an imperfect network with a large (unvulcanizable) free end fraction, another contributory factor to high hysteresis. On the other hand, at room temperature these same rubbers may exhibit good resilience as tan δ is quite small between the two transitions. This somewhat unusual combination of high room temperature resilience and high heat generation on repeated flexure is typical of these rubbers (107).

3.4   Behavior of Commercial Solution Polymers

It has been made clear that the dynamic properties of rubbers are the product of local segmental friction and of long-range motions dependent on the network structure, the former being governed to a considerable extent by the position of the testing temperature relative to T_g, the latter being a function of primary molecular weight, degree of cross-linking, and the "natural" entanglement spacing prevailing in a given polymer. Because of the necessity to control processing behavior, unvulcanized rubbers of essentially similar type may vary considerably in primary MWD and long chain branching. While vulcanization tends to minimize the effects of these differences, they may persist to a minor degree in the final vulcanizates. Two examples of this appear of particular interest.

3.4.1 Effects of Differences in Primary Chain Structure

Hanmer and Railsback (108) compared solution and emulsion SBR in several practical tread formulations. The solution polymers had narrower molecular weight distributions and lower degrees of long chain branching, but slightly higher vinyl content (27 versus 18%). In spite of the latter difference, which would raise T_g by some 4°C, the solution rubbers displayed higher resilience and lower heat generation in all cases. When a polymer is branched prior to vulcanization, it will contribute more than two chain ends per molecule to the final network. The M_n appropriate to the free end correction is thus the hypothetical number average molecular weight (generally unknown) of the polymer with all prevulcanization cross-links cut (98). This further accentuates the differences in *effective* primary molecular weight between the solution rubbers, which are usually free of low molecular weight fractions, and the emulsion polymers, which are not. However, there is nothing inherent in solution processes or stereoregular polymerizations which would make it impossible, at least in principle, to reproduce the branching and MWD of a typical emulsion rubber. If this were done, the differences in dynamic properties would largely disappear.

The same effects of branching and MWD are often seen in comparisons of vulcanized high *cis*-polybutadienes. It has been shown (72, 74) that polybutadienes prepared with cobalt initiators generally have broader molecular weight distributions and are more highly branched than polymers made with iodine based catalysts. The result is slightly higher heat generation in the cobalt based rubbers. This difference is not associated with the steric purity of these polymers and tends to disappear when the primary chain structures are closely matched.

3.4.2 Stereoregularity in Polybutadiene and Polyisoprene

In one of the earliest publications on stereo rubbers the author, together with J. N. Short and V. Thornton, showed the hysteresis of 1,4-polybutadienes to decrease with increasing cis enchainment (109). In view

of the now well-established fact that T_g decreases in the same direction
(72, 100) and the remarkable similarity of the segmental friction of
butadiene based rubbers when compared (see Table 6) at equal $T - T_g$,
such a trend would be expected in comparisons at fixed temperature. In
the correlations of ref. 109, one polymer consistently violated this trend
on the side of lower hysteresis. Although not pointed out in the original
paper, this rubber had a much narrower molecular weight distribution
and virtually no long chain branching. These factors undoubtedly contributed to its lower hysteresis.

The effect of cis content on the properties of polyisoprenes was
investigated by Bruzzone and associates (110, 111) over a relatively
narrow range. (See also Fig. 4 and discussion in Chapter 8.) The authors
compared natural rubber with synthetic polyisoprenes prepared by an
aluminum-titanium catalyst (96% cis) and by a lithium alkyl (92.5% cis).
The glass transition of polyisoprene is not very sensitive to cis content
(112). Accordingly, in measurements well above T_g, little effect of cis
content on segmental motion would have been expected. The results
confirmed this expectation. Rebound resilience was slightly lower for the
synthetic polyisoprenes, but so was the apparent degree of cross-linking
attained. At low degrees of cross-linking, heat buildup was markedly
higher for natural rubber than for the 96% cis-polyisoprene, a difference
that disappeared in the tighter networks. The authors interpreted their
observations as being due to formation of more perfect networks by the
synthetic polyisoprenes at low cross-link densities. In general, cross-link
density, primary molecular weight, and initial long chain branching had
the expected effects on viscoelastic behavior (111). Unfortunately, a
clear-cut characterization of polyisoprene networks is rendered particularly difficult by their complex initial chain structure, breakdown in
processing, and cure reversion phenomena.

Saltman and co-workers (113) have shown that cis-polyisoprene made
by coordination catalysts can be made to match the dynamic properties
of natural rubber exactly. However, rather subtle changes in polymer
structure, which are not reflected in T_g or in the infrared structure
analysis, can evidently cause significant variations, the origin of which is
not clearly understood. The close similarity of the dynamic properties of
natural and synthetic cis-polyisoprenes has been demonstrated repeatedly
(79, 114).

3.4.3 Polymer Blends

Many of the stereo rubbers, in particular the polybutadienes, are used
extensively in blends with either SBR or natural rubber. Dynamic viscoelastic properties have been used to derive information regarding the

compatibility of the polymeric components in such blends. The measurements are usually made at fixed frequency, varying the temperature. In systems that are immiscible, each component polymer gives its characteristic dispersion, almost unchanged in position, just as in the case of block polymers with immiscible monomer sequences (Section 3.3). Compatibility or mutual solubility is rare and results in a single dispersion region intermediate between those of the components on the temperature scale.

Fujimoto and Yoshimiya (115) measured dynamic moduli and loss tangents of blends of a commercial *cis*-polybutadiene with both natural rubber and SBR 1500. Their tan δ curves for the blends with NR are shown in Fig. 25. Virtually complete immiscibility is indicated by the occurrence of separate unshifted loss maxima. In blends with SBR (Fig. 26), the loss maximum shifts with composition, but most of the curves exhibit shoulders suggestive of at least partial incompatibility. Further evidence for a two-phase structure in SBR/BR blends has been obtained by Scott et al. (116). The apparent incompatibility shown by these

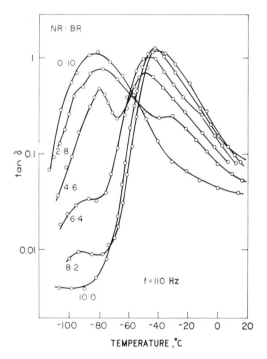

Figure 25 Loss tangents for blends of *cis*-polybutadiene and natural rubber (155). Unfilled sulfur vulcanizates.

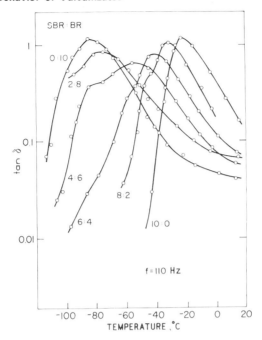

Figure 26 Loss tangents for blends of *cis*-polybutadiene and SBR (115). Unfilled sulfur vulcanizates.

elastomer blends is not a consequence of inadequate mechanical mixing; the same dynamic properties are obtained with solution blended rubbers.

3.4.4 *Viscoelasticity, Friction, and Abrasion*

The low hysteresis of some of the solution rubbers, especially polybutadiene, poses the problem of a potential loss in resistance to skidding in tire tread compounds, since much of the energy dissipated in rubber friction is energy lost in deforming the rubber. At the same time, a low coefficient of friction tends to increase resistance to abrasion. The solution to this problem is compounding with higher amounts of carbon black and extender oil. The additional carbon black increases hysteresis, but tends to maintain the excellent abrasion resistance through higher reinforcement. The additional plasticizer, of course, is needed to control hardness at an acceptable level. The problem of skid and traction has been reviewed by Bevilacqua and Percarpio in an extensive series of papers (117). Details on compounding and tire performance are discussed in other chapters of this book.

4 NETWORK-FORMING BLOCK COPOLYMERS

Block copolymers made up of incompatible monomer sequences can form networks with high rubberlike elasticity without the introduction of chemical cross-links, provided they contain at least two long sequences of a monomer with a higher T_g than the monomer or monomers making up the continuous phase, usually the monomer present in excess (59, 63, 99–105). The outstanding example of this is the three-sequence block polymers with ABA arrangement of the blocks, where A is styrene and B is butadiene or isoprene. The center block can also be a random copolymer of A and B. When phase separation occurs, A blocks from different molecules share in making up domains of A dispersed in a continuum of B, provided that B is present in sufficient excess. Between the glass transitions of A and B, the polymer then consists of rubbery chains (the B blocks) starting and terminating in glassy domains (the aggregated A blocks). This structure resembles a cross-linked, filler reinforced vulcanizate. Above T_g the polymer becomes processable (Section 2.2.8). Because of this characteristic behavior, these polymers have been termed "thermoplastic elastomers" (63). They are made by anionic polymerization techniques, usually utilizing alkyllithium catalysts. Their preparation and basic properties are described more fully in Chapters 4 and 5.

4.1 Structure and Morphology

The fundamental parameters governing the physical behavior of ABA type block polymers are total composition, total molecular weight, and the number, length, and composition of the individual block sequences. These in turn govern the morphology of the domain structure. Clearly, if the normally glassy blocks (polystyrene) total more than 50% in volume, the chances of the polymer being rubbery are strongly diminished, the glassy phase then becoming continuous and load bearing. These materials are no longer typical rubbers and will not be discussed further.

3.1.1 *Morphology of the Domains*

Even in polymers containing less than 50 vol % styrene, there may be appreciable interconnection of the polystyrene domains. These are broken on stretching, giving rise to a strong stress softening effect (105), which is generally reversible only on heating. Fischer and associates (118, 119) have demonstrated this effect with stress birefringence measurements. At low strains the stress optical coefficient (K_σ) is very small;

there is little orientiation giving rise to birefringence, but the network of interconnected polystyrene domains is capable of supporting an appreciable stress. At larger strains K_σ rises as these connections are broken and the rubbery phase becomes increasingly oriented and load bearing. Orientation in the polystyrene domains, if it occurs, contributes little to the birefringence because K_σ of glassy polystyrene is very much smaller than that of the rubbery phase.

Deposition from solvents varying in quality with respect to the two types of block sequences leads to films of different morphologies and physical properties (120–123). There are five types of fundamental domain structures as revealed by electron microscopy, which may be demonstrated by casting, from a common (good) solvent, block polymers of widely varying composition: spherical domains of A in a matrix of B, rodlike domains of A in B, alternating lamellae of A and B, rods of B in A, and spheres of B in A (121). Compression molded samples also exhibit these basic morphologies (102, 123, 124). Orientation and degree of order are strongly dependent on the method of sample preparation, particularly with cylindrical and lamellar morphologies (125). Extruded samples of extremely high degree of order have been prepared from a styrene–butadiene–styrene block polymer (126). These resemble single crystals consisting of hexagonally spaced long rods of polystyrene in the polybutadiene matrix. From electron microscopic examination and small angle x-ray diffraction studies, domain sizes and spacings are known to range from approximately 50 to 1000 Å, depending greatly on polymer composition, molecular weight, and method of sample preparation (102, 127).

4.1.2 Network Characterization

It is scarcely to be expected that the effective chain length between "cross-links" in ABA block polymers (Z_c) would correspond to the entire molecular weight of the center block; trapped entanglements can logically be thought to exist between the glassy domains. The process of trapping of these entanglements might, however, be quite different from what it is in the random cross-linking of a simple rubber. Bishop and Davison (128) estimated effective physical cross-link densities by the equilibrium swelling method. Isooctane, a virtual nonsolvent for polystyrene, was chosen as the swelling agent and the data were analyzed by a theory originally developed for the swelling of vulcanized, filler reinforced rubbers (129). Because there is theoretically no free chain end correction and no chemical cross-linking, the network density is directly convertible to Z_c:

Polymer	Z_c
Styrene–isoprene–styrene	880
Styrene–butadiene–styrene	1140

Both values correspond to a molecular weight between physically effective cross-links of about 16,000, much smaller than the molecular weight of the center blocks, which ranged from about 50,000 to 230,000. Similar results have been obtained by Kawai and associates (121). While these results confirm the presence of entanglements in the rubbery blocks, the agreement with the entanglement spacings (Z_e) of Table 7 is not good, $Z_c \gg Z_e$. This may be the result of the inadequacy of the various theories employed to estimate these parameters, or may reflect a real difference in the state of entanglement.

4.2 Viscoelastic Behavior

The behavior of uncross-linked styrene–butadiene–styrene block polymers in dynamic measurements is similar to that of the block polymer vulcanizates of Figs. 23 and 24, the principal difference being accountable for by the absence of cross-links. Thus tan δ remains relatively high between the two transitions and does not return to a low value above T_g of the polystyrene blocks (100). The upper portions of Fig. 27 illustrate this behavior with data on three block polymers of varying Mooney viscosity (and hence molecular weight), but similar block composition and arrangement: styrene (6)/butadiene–styrene copolymer (88)/styrene (6). The polystyrene maximum shifts toward lower temperature as molecular weight and block length decrease. The lower portion of Fig. 27 shows comparable data for a styrene–butadiene–styrene block polymer of 12.5 : 75 : 12.5 composition. These data also show a flat secondary loss maximum at about 0°C. The presence of a secondary loss maximum between the principal transitions has been found in other polymers of this type. Beecher et al. (104) have shown that this maximum is sensitive to the morphology as determined by the method of sample preparation. Casting films from different solvents can change the height of the maximum, a large secondary maximum being accompanied by a decrease in the height of the polystyrene peak. The authors interpret this behavior as being due to incomplete demixing of the polystyrene and polybutadiene blocks. This is in accord with the observation that the secondary maximum is most pronounced when a film is cast from a good solvent for both monomer

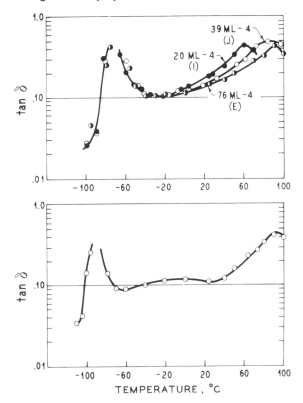

Figure 27 Loss tangents for uncross-linked butadiene-styrene block copolymers (100). Total styrene content 25% in all cases. Upper diagram, 12% block styrene; lower diagram 25% block styrene. Reproduced by permission of the *Journal of Applied Polymer Science.*

sequences. The existence of a mixed interlayer between compositionally pure domains has been proposed by several authors (130–132) and appears to be well founded in theory (133).

Beecher and co-workers (104) also investigated the frequency dependence from 0.001 to 10 Hz of a styrene–isoprene–styrene block polymer of 22% styrene content and $M_w = 130,000$. They were able to temperature shift their data to obtain modulus and tan δ versus frequency master curves for 0°C. The loss maxima found were located approximately at 10^{-4} (polystyrene), 10 (secondary), and 10^5 (polyisoprene) Hz. Their temperature shift function a_T is reproduced in Fig. 28. Up to 0°C the function a_T is reasonably well approximated by the WLF equation (4) with constants applicable to polyisoprene, but deviates from it by an order of magnitude in the flat portion of the curve between 40 and 60°C.

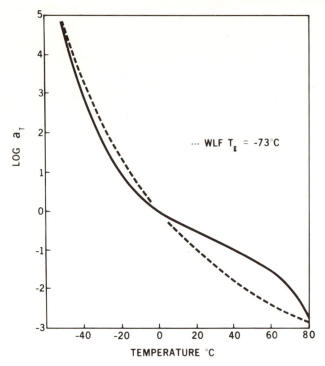

Figure 28 Temperature shift factor for styrene-isoprene-styrene block copolymer (104). Total styrene content 22%. Dashed line is WLF function for polyisoprene. Reproduced by permission of the *Journal of Polymer Science*.

As the glass transition of the polystyrene blocks is approached, a_T becomes very steep. This might be anticipated, since in this region softening of the polystyrene blocks becomes the most temperature sensitive effect. The apparent success of simple time (frequency)-temperature reduction is, however, fortuitous, for when multiple loss mechanisms are involved it is necessary to subject viscoelastic data to a more complex analysis (14, 134). In the example at hand, simple superposition would have been expected only at low temperatures, where the polystyrene domains act as inert filler particles (135). Fesko and Tschoegl (136) have developed a general treatment of time-temperature superposition in thermorheologically complex materials. To apply it to the system under consideration it is necessary to make an assumption concerning the mechanical coupling between phases (parallel or series). Since the type of coupling certainly depends on the morphology—and will in general be more complex than simple parallel or series coupling—shift factors will not be unique for a given polymer. Simple empirical shifts often appear to succeed because of the narrowness of the "experimental window," i.e., the range of times or frequencies accessible to the experimenter. When this is the case, the apparent shift function will, of course, still depend on the morphology. This explains the frequently poor agreement between a_T reported for the same material by different investigators. For example, Smith and Dickie have published an a_T function derived from isochronal stress–strain data on two commercial styrene–butadiene–styrene block polymers containing ca. 30% block styrene (137). Their function, log $a_T = -0.116(T-273)$, drops much more rapidly with temperature between 0 and 60°C (about 7 decades versus 1.5 decades) than occurs in Fig. 28. This is much more than can reasonably be accounted for by differences in polymer composition alone.

Another consequence of fibrillar or lamellar domain morphologies is that block polymers are not only heterogeneous but frequently also anisotropic, and their elastic behavior clearly reflects this. Thus, the "single crystals" of Keller et al. (126, 127) require five independent elastic moduli or compliances to describe their behavior completely (138). One manifestation of domain connectivity in the viscoelastic properties (123) is the appearance of different loss tangents measured in the tensile and shear deformation modes and ratios of tensile to shear moduli exceeding 3: For a homogeneous, isotropic rubber this ratio is always very close to 3 and can never exceed this value.

REFERENCES

1. T. G. Fox and V. R. Allen, *J. Chem. Phys.*, **41,** 344 (1964).

2. G. C. Berry and T. G. Fox, *Adv. Polym. Sci.*, **5,** 261 (1968).

3. H. Vogel, *Phys. Z.*, **22,** 645 (1921).

4. M. L. Williams et al., *J. Am. Chem. Soc.*, **77,** 3701 (1955).

5. J. H. Magill and H. M. Li, *J. Polym. Sci., B,* **11,** 667 (1973).

6. P. J. Flory, *Principles of Polymer Chemistry,* Cornell University Press, Ithaca, New York, 1953, Ch. X.

7. G. V. Vinogradov and A. Ya. Malkin, *J. Polym. Sci., A-2,* **4,** 135 (1966).

8. F. Bueche, *J. Chem. Phys.*, **22,** 1570 (1954).

9. R. A. Stratton, *J. Colloid Sci.*, **22,** 517 (1966).

10. R. A. Stratton, *Macromolecules,* **5,** 304 (1972).

11. F. Bueche and S. W. Harding, *J. Polym. Sci.*, **32,** 177 (1958).

12. G. Kraus and J. T. Gruver, *Trans. Soc. Rheol.*, **13:3,** 315 (1969).

13. S. Middleman, *The Flow of High Polymers,* Wiley-Interscience, New York, 1968, Chs. IV and V.

14. J. D. Ferry, *Viscoelastic Properties of Polymers*, 2nd ed., Wiley, New York, 1970.

15. P. E. Rouse, Jr., *J. Chem. Phys.*, **21**, 1272 (1953).

16. A. V. Tobolsky et al., *J. Chem. Phys.*, **42**, 723 (1965).

17. N. Nemoto et al., *Macromolecules*, **4**, 215 (1971).

18. H. Leaderman et al., *J. Polym. Sci.*, **36**, 233 (1959).

19. W. M. Prest, Jr., *Polym. Prepr., Div. Polym. Chem., ACS*, **10**, No. 1, 137 (1969).

20. N. Nemoto et al., *Macromolecules*, **5**, 531 (1972).

21. W. P. Cox and E. H. Merz, *J. Polym. Sci.*, **28**, 619 (1958).

22. S. Onogi et al., *J. Phys. Chem.*, **68**, 1598 (1964).

23. B. D. Coleman et al., *Viscometric Flow of Non-Newtonian Fluids*, Springer-Verlag, New York, 1966.

24. B. D. Coleman and H. Markovitz, *J. Appl. Phys.*, **35**, 1 (1964).

25. W. W. Graessley and J. S. Prentice, *J. Polym. Sci.*, *A-2*, **6**, 1887 (1968).

26. W. M. Prest et al., *J. Appl. Polym. Sci.*, **14**, 2697 (1970).

27. W. Philippoff and F. H. Gaskins, *Trans Soc. Rheol.*, **2**, 263 (1958).

28. J. Tordella, *Rheol. Acta*, **1**, 216 (1958); *J. Appl. Phys.*, **27**, 454 (1956); *Trans. Soc. Rheol.*, **1**, 203 (1957).

29. J. J. Benbow and P. Lamb, *SPE Trans.*, **3**, 7 (1963).

30. J. T. Gruver and G. Kraus, *J. Polym. Sci.*, *A*, **2**, 797 (1964).

31. H. H. Meyer and W. Ring, *Kaut. Gummi Kunstst.*, **24**, 526 (1971).

32. T. Ree and H. Eyring, *J. Appl. Phys.*, **26**, 793 (1955); **26**, 800 (1955).

33. G. Kraus and J. T. Gruver, *Advances in Polymer Science and Technology*, Soc. Chem. Ind., Monograph No. 26, London 1967, p. 30.

34. R. P. Zelinski and C. F. Wofford, *J. Polym. Sci.*, *A*, **3**, 93 (1965).

35. G. Kraus and J. T. Gruver, *J. Polym. Sci.*, *A*, **3**, 105 (1965).

36. B. H. Zimm and R. W. Kilb, *J. Polym. Sci.*, **37**, 19 (1959).

37. T. Masuda et al., *Macromolecules*, **4**, 763 (1971).

38. R. Suzuki and A. J. Kovacs, Paper presented at 4th I.U.P.A.C. Microsymposium on Macromolecules, Prague, September 1969.

39. J. Pannell, *Polymer*, **13**, 2 (1972).

40. V. C. Long et al., *Polymer*, **5**, 517 (1964).

41. R. P. Chartoff and B. Maxwell, *J. Polym. Sci.*, *A-2*, 455 (1970).

42. J. L. White, *Rubber Chem. Technol.*, **42**, 257 (1969).

43. B. L. Johnson et al., *Proceedings of the International Rubber Conference, Brighton, U.K.*, McLaren, London, 1968, pp. 29–43.

44. G. Ghijsels and H. J. M. A. Mieras, *J. Polym. Sci., Phys. Ed.*, **11**, 1849 (1973).

45. G. Kraus and J. T. Gruver, *J. Appl. Polym. Sci.*, **9**, 739 (1965).

46. G. V. Vinogradov et al., *J. Polym. Sci.*, *A-2*, **8**, 333 (1970).

47. J. K. Hughes, Phillips Petroleum Co., unpublished data.

48. R. H. Valentine et al., *J. Polym. Sci.*, *A-2*, **6**, 479 (1968).

49. R. H. Valentine, Ph.D. Thesis, University of Wisconsin, 1967.

50. D. J. Plazek, unpublished results quoted in ref. 49.

51. G. Kraus, unpublished results.

52. J. L. White and N. Tokita, *J. Appl. Polym. Sci.*, **11**, 321 (1967).

53. G. V. Vinogradov et al., *J. Polym. Sci.*, *A-2*, **9**, 1153 (1971).

54. G. V. Vinogradov et al., *J. Polym. Sci.*, *A-2*, **10**, 1061 (1972).

55. R. F. Boyer, *Rubber Chem. Technol.*, **36**, 1303 (1963).

56. G. Kraus and J. T. Gruver, *Trans. Soc. Rheol.*, **9:2**, 17 (1965).

57. W. W. Graessley et al., *Trans. Soc. Rheol.*, **11:3**, 267 (1967).

58. G. Kraus and J. T. Gruver, *J. Polym. Sci.*, *A-2*, **8**, 305 (1970).

59. G. Kraus and J. T. Gruver, *J. Appl. Polym. Sci.*, **11**, 2121 (1967).

60. G. Kraus and K. W. Rollmann, *Angew. Makromol. Chem.*, **16/17**, 271 (1971).

61. G. Kraus and J. T. Gruver, *Rubber Chem. Technol.*, **42**, 800 (1969).

62. U. Bianchi et al., *Polymer*, **11**, 268 (1970).

63. G. Holden et al., *J. Polym. Sci.*, *C*, **26**, 37 (1969).

64. G. Kraus et al., *J. Polym. Sci.*, *A-2*, **9**, 1839 (1971).

65. C. Price et al., *Polymer*, **13**, 333 (1972).

66. K. R. Arnold and D. J. Meier, *J. Appl. Polym. Sci.*, **14**, 427 (1970).

67. E. A. Collins and W. H. Bauer, *Trans. Soc. Rheol.*, **9:2**, 1 (1965).

68. E. P. Otocka et al., *J. Appl. Phys.*, **40**, 4221 (1969).

69. G. Yasuda, *Nippon Gomu Kyokaishi*, **39**, 194 (1966).

70. G. Holden et al., *J. Appl. Polym. Sci.*, **9**, 2911 (1965).

71. N. Nemoto et al., *Macromolecules*, **5**, 641 (1972).

72. W. S. Bahary et al., *Rubber Chem. Technol.*, **40**, 1529 (1967).

73. G. Kraus and C. J. Stacy, *J. Polym. Sci.*, *A-2*, **10**, 657 (1972).

74. B. M. E. Van der Hoff et al., *Rubber Plast. Age*, **46**, 821 (1965).

75. N. Tokita and J. L. White, *J. Appl. Polym. Sci.*, **10**, 1011 (1966).

76. U. Bianchi and E. Bianchi, *Rubber Chem. Technol.*, **38**, 343 (1965).

77. E. Pedemonte and U. Bianchi, *Rubber Chem. Technol.*, **38**, 347 (1965).

78. H. E. Railsback and G. Kraus, *Kaut. Gummi Kunstst.*, **22**, 497 (1969).

79. J. D. D'Ianni, *Kaut. Gummi Kunstst.*, **19**, 138 (1966).

80. L. H. Krol and P. P. A. Smit, *Kaut. Gummi Kunstst.*, **18**, 589 (1965).

81. J. F. Sanders et al., *J. Polym. Sci.*, *A-2*, **6**, 967 (1968).

82. H. Markovitz et al., *J. Phys. Chem.*, **66**, 1567 (1962).

83. R. G. Mancke and J. D. Ferry, *Trans Soc. Rheol.*, **12:2**, 335 (1968).

84. H. Oser and R. S. Marvin, *J. Res. Nat. Bur. Stand.*, **67B**, 87 (1963).

85. A. M. Bueche, *J. Polym. Sci.*, **19**, 297 (1956).

86. L. Mullins, *J. Appl. Polym. Sci.*, **2**, 1 (1959).

87. G. Kraus, *J. Appl. Polym. Sci.*, **7**, 1257 (1963).

88. C. G. Moore and W. F. Watson, *J. Polym. Sci.*, **19**, 237 (1956).

89. R. G. Mancke et al., *J. Polym. Sci.*, *A-2*, **6**, 1783 (1968).

90. N. R. Langley, *Macromolecules*, **1**, 348 (1968).

91. R. A. Stratton and J. D. Ferry, *Rev. Gén. Caoutch.*, **41**, 635 (1964).

92. E. Maekawa et al., *J. Phys. Chem.*, **69**, 2811 (1965).

93. J. Hasa and B. M. E. Van der Hoff, *J. Polym. Sci.*, *Phys. Ed.*, **11**, 297 (1973).

94. R. A. Dickie and J. D. Ferry, *J. Phys. Chem.*, **70**, 2594 (1966).

95. R. G. Mancke and J. D. Ferry, *Trans. Soc. Rheol.*, **12:2**, 335 (1968).

96. N. R. Langley and J. D. Ferry, *Macromolecules*, **1**, 353 (1968).

97. J. N. Short et al., Paper presented at the International Symposium on Macromolecular Chemistry, Tokyo-Kyoto, 1966.

98. G. Kraus and K. W. Rollmann, *J. Polym. Sci., Symp.*, No. 48, 87 (1974).

99. R. P. Zelinski and C. W. Childers, *Rubber Chem. Technol.*, **41**, 161 (1968).

100. G. Kraus et al., *J. Appl. Polym. Sci.*, **11**, 161 (1968).

101. S. L. Cooper and A. V. Tobolsky, *Textile Res. J.*, **36**, 800 (1966).

102. H. Hendus et al., *Kolloid-Z.*, **216–217**, 110 (1967).

103. R. J. Angelo et al., *Polymer*, **6**, 141 (1965).

104. J. F. Beecher et al., *J. Polym. Sci., C*, **26**, 117 (1969).

105. C. W. Childers and G. Kraus, *Rubber Chem. Technol.*, **40**, 1183 (1967).

106. G. Kraus, unpublished data.

107. J. R. Haws, *Rubber Plast. Age*, **46**, 1144 (1964).

108. R. S. Hanmer and H. E. Railsback, *Rubber Age*, **96**, 73 (1964).

109. G. Kraus et al., *Rubber Plast. Age*, **38**, 880 (1957).

110. M. Bruzzone et al., *Rubber Plast. Age*, **46**, 278 (1965).

111. M. Bruzzone et al., *Fourth International Synthetic Rubber Symposium*, Rubber & Technical Press, London, 1969, pp. 83–87.

112. K. W. Scott et al., *Rubber Plast. Age*, **42**, 175 (1961).

113. W. M. Saltman et al., *Rubber Plast. Age*, **46**, 502 (1965).

114. R. V. Todd, *Rubber World*, **146**, 69 (1962).

115. K. Fujimoto and N. Yoshimiya, *Nippon Gomu Kyokaishi*, **38**, 284 (1965); *Rubber Chem. Technol.*, **41**, 669 (1968).

116. C. E. Scott et al., Paper presented at the 95th ACS Meeting of the Division of Rubber Chemistry, Los Angeles, California, May 1969.

117. E. M. Bevilacqua and E. P. Percarpio, *Rubber Chem. Technol.*, **41**, 832–94 (1968).

118. J. F. Henderson et al., *J. Polym. Sci., C*, **16**, 3121 (1968).

119. E. Fischer and J. F. Henderson, *J. Polym. Sci., C*, **26**, 149 (1969); **30**, 459 (1970).

120. G. L. Wilkes and R. S. Stein, *J. Polym. Sci., A-2*, **7**, 1525 (1969).

121. G. Uchida et al., *J. Polym. Sci., A-2*, **10**, 101 (1972).

122. T. Miyamoto et al., *J. Polym. Sci., A-2*, **8**, 2095 (1970).

123 G. Kraus et al., *J. Polym. Sci., A-2*, **10**, 2061 (1972).

124. M. Matsuo et al., *Polymer*, **9**, 415 (1968).

125. P. R. Lewis and C. Price, *Polymer*, **13**, 20 (1972).

126. J. Dlugosz et al., *Kolloid-Z., Z. Polym.*, **242**, 1125 (1970).

127. A. Keller et al., *Kolloid-Z., Z. Polym.*, **238**, 385 (1970).

128. E. T. Bishop and S. Davison, *J. Polym. Sci., C*, **26**, 59 (1969).

129. G. Kraus, *J. Appl. Polym. Sci.*, **7**, 861 (1963).

130. M. Shen and D. H. Kaelble, *J. Polym. Sci., B*, **8**, 149 (1970).

131. D. F. Leary and M. C. Williams, *J. Polym. Sci., B*, **8**, 335 (1970).

132. M. Girolamo and J. R. Urwin, *Eur. Polym. J.*, **7**, 225 (1971).

133. D. J. Meier, *Polym. Prepr.,* **15** (1), 171 (1974).

134. W. Dannhauser et al., *J. Colloid Sci.,* **13,** 103 (1958).

135. C. K. Lim et al., in N. Platzer, ed., *Multicomponent Polymer Systems,* Adv. Chem. Ser. 99, American Chemical Society, Washington, D.C., 1971, p. 397.

136. D. G. Fesko and N. W. Tschoegl, *J. Polym. Sci.,* C, **35,** 51 (1971).

137. T. L. Smith and R. A. Dickie, *J. Polym. Sci.,* C, **26,** 163 (1969).

138. R. G. C. Arridge and M. J. Folkes, *J. Phys. D: Appl. Phys.,* **5,** 344 (1972).

R. F. Fedors

Jet Propulsion Laboratory.
Pasadena, California

Chapter 12 Uniaxial Rupture of Elastomers

1 INTRODUCTION

The factors that affect or control the values obtained for the the stress-at-break and the strain-at-break for elastomers have been the subject of numerous experimental and theoretical investigations. Early work indicated the great sensitivity of the ultimate properties to the rate and

679

temperature of testing. In addition, the importance of molecular parameters such as the degree of cross-linking was also recognized. For example, at constant test rate and test temperature, increasing degrees of cross-linking lead to decreasing values of the strain-at-break whereas the stress-at-break response passes through a maximum at a relatively low degree of cross-linking. It was also recognized that the basic failure response was closely connected to the viscoelastic nature of the elastomer.

Since the breaking point of an elastomer is the terminus of a stress-strain curve, rupture in a viscoelastic material should not be treated as a separate phenomenon, but rather as part of the more general problem of describing its stress-strain-time properties. A convenient graphical representation for the viscoelastic response is a surface in stress-strain-time space. Rupture is represented on this surface by a boundary or discontinuity, which can be considered to result from the intersection of this surface with another surface representing failure. The projection of the boundary to the stress-strain plane furnishes the relationship between stress-at-break and strain-at-break, and the projections to the remaining two planes furnish the time dependence of rupture. These three projections constitute the basic rupture response of an elastomer. In general, any change in the magnitude of the primary variables such as the degree of cross-linking or polymer structure, for example, will lead to a corresponding change in the nature of the surface and its associated boundary as well.

This chapter is restricted to a description of the effect of variables such as test rate, test temperature, degree of cross-linking, and chemical structure of the polymer chain on the shape and location of both the surface and its boundary in stress-strain-time space. The effect of axiality as well as other test modes such as tear and fatigue will not be discussed.

There are many gaps in our knowledge, not only because comparatively little work has been done on the stereoregular elastomers but also because the polymers reported in the literature have not always been completely characterized as to their basic physical and chemical properties. In order to review the field properly it has been necessary to discuss polymers other than stereoregular in considerable detail. The first part of this chapter is concerned with the behavior of unfilled elastomers, and the second part deals with the corresponding behavior of filled systems.

2 UNFILLED ELASTOMERS

2.1 Stress-Strain Response in the Long Reduced Time Region

For well-vulcanized elastomers, data indicate that over a considerable portion of the reduced time scale, the time dependence and the strain dependence of the stress are factorable (1–3). Hence, the stress σ, based on the undeformed specimen dimensions, can be expressed as

$$\sigma(t, \epsilon) = G(t)f(\epsilon) \qquad [1]$$

where $G(t)$ is a function of the time alone and $f(\epsilon)$ is a function of the strain ϵ. As $\epsilon \to 0$, $f(\epsilon) \to 3\epsilon$ and equation [1] reduces to a time dependent Hooke's law. In addition, for sufficiently long reduced times, $G(t)$ becomes a slowly varying function of time which approaches a limiting value G_e asymptotically as $t \to \infty$.

2.1.1 *Properties of the Time Dependent Function*

Implicit in equation [1] is the assumption that the time dependent function is independent of the strain and is in fact the same function one would obtain at infinitesimal deformations where linear viscoelasticity is applicable. Hence the nonlinear viscoelastic response of an elastomer is apparently a reflection of the nonlinear character of $f(\epsilon)$ (2, 4, 5) rather than a nonlinear viscoelastic character of $G(t)$.

The nature of the time dependent function is related to the type of deformation or test mode considered. In a simple test mode, either the stress or strain is held constant and the other parameter is allowed to vary as a function of time. The ratio $\sigma/f(\epsilon)$, or σ/ϵ for small strains, defines a time dependent function whose nature depends on whether the stress or the strain is the time varying parameter. When ϵ is fixed and σ is permitted to vary, the ratio σ/ϵ defines the tensile stress relaxation modulus, $E(t)$; when σ is fixed and ϵ is permitted to vary, the ratio ϵ/σ defines the tensile creep compliance $D(t)$.

Another convenient test mode is the constant strain rate test where both stress and strain vary with time (the strain varying linearly with time). However, constant strain rate data are more difficult to analyze because stress relaxation occurs continuously throughout the test. In order to separate possible time and strain nonlinearities in this test mode, a procedure developed by Smith may be used, which is based on converting constant strain rate data to isochronal data (constant time) by cross-plotting (3). Values of the stress and strain at fixed times are thereby obtained which are more amenable to analysis. Using such isochronal data, Smith (3) defines a constant strain rate modulus $F(t)$ which is related to the tensile stress relaxation modulus by $E(t) = F(t)[1 + m]$ where m is the slope of a log-log plot of $F(t)$ versus t. Since m is normally a negative number in the range 0 to -0.67, $F(t) \geq E(t)$, and at long reduced times, $F(t) \to E(t)$ since $m \to 0$.

When linear viscoelastic behavior is applicable, all relaxation processes have the same temperature dependence. Hence data obtained at different temperatures can be superposed to yield a master curve that is a continuous function of the reduced time t/a_T. The shift factor a_T, a function of temperature, is commonly obtained from an equation of the WLF type (6), although an Arrhenius type dependence of a_T on temperature has also been observed with the small strain response (7) and the rupture response (8). Furthermore, several studies have shown that a_T is strain independent and hence the shift factors obtained from small strain response are also directly applicable to the finite strain response including rupture (3, 7, 9, 10). It will be assumed throughout that any time dependent quantity is also temperature dependent and the symbol t will be used for both measured time and reduced time.

Since $G(t)$ is apparently the same time dependent function one would measure in the linear region, the results of the many studies on linear viscoelasticity can be used directly. For example, the dependence of $G(t)$ on reduced time can be given a straightforward molecular interpretation in terms of the degree of cross-linking and chain mobility (11–13). In addition, $G(t)$ is readily measured by a number of techniques and the response obtained in one type of test mode (e.g., stress relaxation) can be converted to the response obtainable in another test mode (e.g., creep) using standard methods of linear viscoelasticity (13).

2.1.2 Properties of the Strain Dependent Function

Having briefly discussed $G(t)$, we now consider the function $f(\epsilon)$. In the long reduced time region, $f(\epsilon)$ depends only on the strain. In addition, Halpin has shown that $f(\epsilon)$ is also approximately independent of the test mode in that the same $f(\epsilon)$ can be used to describe the response obtained in creep, stress relaxation, and constant strain rate tests (14).

Recent efforts have been directed toward obtaining explicit forms for $f(\epsilon)$. One such form, which is based on non-Gaussian chain statistics, is (15–17)

$$f(\epsilon) = \frac{n^{1/2}}{3}\left[\mathscr{L}^{-1}\left(\frac{\lambda}{n^{1/2}}\right) - \frac{1}{\lambda^{3/2}}\,\mathscr{L}^{-1}\left(\frac{1}{n^{1/2}\lambda^{1/2}}\right)\right] \qquad [2]$$

where n is a maximum stretch parameter (16) or the number of statistical segments per network chain (17), λ is the stretch or extension ratio ($\lambda \equiv 1 + \epsilon$), and $\mathscr{L}^{-1}$ is the inverse Langevin function defined by the relation $\mathscr{L}^{-1}(x) = y$ and $x = \coth y - 1/y$. It is possible to simplify equation [2] somewhat when the product $n\lambda$ is sufficiently large, say greater than about 25. Under this restriction, which is applicable to most vulcanizates provided that the degree of cross-linking is not excessively high, the second term in brackets can be taken equal to $3/\lambda^2 n^{1/2}$ to a good approximation and equation [2] can be written as (16)

$$f(\epsilon) = \frac{n^{1/2}}{3} \, \mathscr{L}^{-1}\left(\frac{\lambda}{n^{1/2}}\right) - \frac{1}{\lambda^2} \qquad [3]$$

It is interesting to note that when $\lambda \to n^{1/2}$, $f(\epsilon)$ and hence $\sigma \to \infty$; thus for λ approaching $n^{1/2}$, equations [2] and [3] predict a sharp upturn with a maximum extension ratio λ_m equal to $n^{1/2}$. If n is taken as the number of statistical segments per network chain, then n can be expressed in terms of ν_e by

$$n = \frac{\rho g f}{M_0 \nu_e} \qquad [4]$$

where ρ is the density, g is the gel fraction, f is the fraction of effective network chains in the gel fraction, M_0 is the molecular weight of a statistical segment which is a function of the flexibility and chemical structure of the polymer chain, and ν_e is the concentration of effective network chains per unit volume. In order to estimate the value of f one of the several free chain end corrections which have been proposed must be used (18). Since $\lambda_m = n^{1/2}$, the maximum extension ratio can be related to ν_e by

$$\lambda_m = n^{1/2} = \left(\frac{\rho g f}{M_0 \nu_e}\right)^{1/2} \qquad [5]$$

Smith and Dickie (7) consider that for a real material, where nonaffine deformation is assumed to occur, the exponent in equation [5] will in general be greater than $\frac{1}{2}$. As the degree of cross-linking is increased, they envision a decrease in the exponent to a limiting value of $\frac{1}{2}$ at high degrees of cross-linking.

Equations [2] and [3] have been found to be applicable to the stress-strain response of an SBR vulcanizate obtained from different test modes (4) as well as to EPM data obtained under near equilibrium conditions (19). Smith and Frederick found that equation [3] describes equilibrium data, obtained indirectly from constant strain rate data, for butyl, silicone, viton, and SBR vulcanizates (20).

For small λ, equations [2] and [3] reduce to the deformation factor derived from kinetic theory, $\lambda - 1/\lambda^2$. For equal modulus, the stress predicted using equation [2] or [3] will be equal to or greater than that predicted by kinetic theory. Numerous studies have shown that experimental data generally yield values of σ at a given λ which are less than those predicted by kinetic theory and hence lower than values predicted by equations [2] and [3]. For small λ, an equation of the form proposed by Mooney and Rivlin provides a better representation of experimental data (17):

$$f(\epsilon) = \left(\lambda - \frac{1}{\lambda^2}\right)\left(1 + \frac{C_2}{C_1\lambda}\right) \qquad [6]$$

where C_1 and C_2 are parameters. The factor $2C_1$ has been identified with the term $\nu_e RT$ obtained from kinetic theory. This expression is generally found adequate to represent data for small to moderate values of λ.

Bueche and Halpin (21) propose an equation of the form

$$f(\epsilon) = \left(\lambda - \frac{1}{\lambda^2}\right)\left(1 + \frac{C_2}{C_1\lambda}\right)\frac{1}{(1 - \lambda^2/n)} \qquad [7]$$

to represent stress-strain data for both SBR and EPDM. The first two terms are equivalent to equation [6] while the third term is an approximation to the response predicted by equation [3] since

$$\left(\lambda - \frac{1}{\lambda^2}\right)\left(1 - \frac{\lambda^2}{n}\right)^{-1} \approx \frac{n^{1/2}}{3}\,\mathscr{L}^{-1}\left(\frac{\lambda}{n^{1/2}}\right) - \frac{1}{\lambda^2}$$

Essentially the same equation was proposed by Smith and Frederick (20) and was found to be applicable to a sulfur cured butyl vulcanizate. It is given by

$$f(\epsilon) = \left[\frac{n^{1/2}}{3}\,\mathscr{L}^{-1}\left(\frac{\lambda}{n^{1/2}}\right) - \frac{1}{\lambda^2}\right]\left(1 + \frac{C_2}{C_1\lambda}\right) \qquad [8]$$

A somewhat similar relation proposed by Morris has the form (22)

$$f(\epsilon) = \left[f_1(\lambda, n) + \left(\lambda - \frac{1}{\lambda^2}\right)\left(1 + \frac{C_2}{C_1\lambda}\right)\right] \qquad [9]$$

where $f(\lambda, n)$ is a function obtainable only in series form, and is similar to equation [2]. It is based on a theoretical stress-strain response derived by Treloar (23). This derivation does not assume any artificial distribution of chain vectors, such as the three-chain model which leads to equation [2].

Finally, an empirical expression proposed by Martin and co-workers (24) has the form

$$f(\epsilon) = \frac{1}{3}\left(\frac{1}{\lambda} - \frac{1}{\lambda^2}\right)\exp A\left(\lambda - \frac{1}{\lambda}\right) \qquad [10]$$

where A is a parameter having values generally less than unity. Equation [10] has been shown to fit data for several elastomers up to the point of rupture (2, 10). This form for $f(\epsilon)$ is comparable to equations [7], [8], and [9] in that they all predict a sharp upturn at large λ and a response at small λ which lies below that calculated from kinetic theory.

2.2 Stress-Strain Response in the Short Reduced Time Region

At short reduced times, separability of the time and strain dependence as embodied in equation [1] breaks down and it is no longer possible to represent the stress as the product of two functions, one of which depends on time alone and one of which depends of strain alone. Hence a different or modified expression is required to represent the response in this region.

2.2.1 *Properties of the Time and Strain Dependent Functions*

Halpin presented experimental evidence which indicated that the form of equation [1] is preserved provided that one or more of the parameters present in $f(\epsilon)$ is taken to be time dependent (14). Specifically, he showed that if n in equation [3] is taken as a time dependent quantity, then equation [1] can be used to describe the response in the short time region as well. Data for several elastomers, viton, SBR, and poly(ethyl acrylate), indicate that the same form of $f(\epsilon)$, in the region where it too is time dependent, is obtained from creep, stress relaxation, and constant strain rate tests. Furthermore, the form for $f(\epsilon)$ obtained experimentally is consistent with equation [3] provided n is taken as time dependent and proportional to $G(t)^{-1/2}$. Thus, when $G(t)$ becomes essentially time independent as in the long reduced time region, n becomes proportional to $G_e^{-1/2}$ and the behavior described in Section 2.1 is recovered. Thus, the more general stress-strain-time response may be written

$$\sigma = G(t)f(\epsilon, t) \qquad [11]$$

where $G(t)$ is the time dependent modulus obtained from experiments conducted at infinitesimal deformation and $f(\epsilon)$ is the same function appearing in equation [1] except that one or more of the parameters present is taken to be time dependent.

For lightly cross-linked vulcanizates, Halpin observes that $f(\epsilon)$ is time dependent over much of the reduced time scale in contrast to well-vulcanized elastomers in which $f(\epsilon)$ becomes time dependent only at short reduced times (25). He attributes this behavior to an increased average orientation of the network chain occurring in the direction of stretch. Taking this factor into account, Halpin obtains

$$f(\epsilon) = x(t)\left[\frac{n(t)^{1/2}}{3}\,\mathscr{L}^{-1}\left(\frac{\lambda}{n(t)^{1/2}}\right) - \frac{1}{\lambda^2}\right] + [1 - x(t)]p \qquad [12]$$

where

$$p = \frac{1}{\lambda} + \frac{3}{2\lambda(\lambda^3 - 1)} - \frac{3\lambda^2}{2(\lambda^3 - 1)^{3/2}}\arctan(\lambda^3 - 1)^{1/2} \qquad [13]$$

where x is a time dependent parameter having values between 0 and 1. When $x = 1$, equation [12] reduces to equation [3].

Taking n and x as adjustable parameters, Halpin was able to fit the stress-strain response for lightly cross-linked SBR vulcanizates. Values of x ranged from about unity for the most highly cross-linked vulcanizate to about 0.4 for the vulcanizate with the lowest degree of cross-linking. The values of n were found to be proportional to the square root of the reciprocal of the time dependent modulus evaluated at small strains (25).

Smith and Dickie carried out a comprehensive investigation of the stress-strain response of an SBR vulcanizate over a wide range of test temperatures and strain rates (7). To represent the data, they used an equation of the form

$$\frac{\lambda_m^0}{\lambda_m(t)} \frac{\lambda\sigma}{F(t)} = \Omega\left[(\lambda - 1)\frac{\lambda_m^0}{\lambda_m(t)}\right] \tag{14}$$

where λ_m^0 is the maximum extension ratio for an equilibrium stress-strain response, $\lambda_m(t)$ is the time dependent maximum extension ratio, $F(t)$ is the constant strain rate modulus obtained from isochronal data, and Ω is a function whose explicit form is not specified, although it was noted that Ω can be approximately represented at large λ by a relation of the form given by equation [3]. Values of $\lambda_m(t)$ were estimated using equation [3] in the form

$$\frac{3\lambda\sigma}{\lambda_m^2 F(t)} \approx \frac{\lambda}{3\lambda_m} \mathcal{L}^{-1}\left(\frac{\lambda}{\lambda_m}\right) \tag{15}$$

which is valid when $3\lambda\sigma/\lambda_m^2 F(t) \gg 1/\lambda$, i.e., when λ is large. Thus a plot of isochronal data in the form $\log \lambda\sigma$ versus $\log \lambda$ should have the same shape as a plot of $\log (\lambda/3\lambda_m)\mathcal{L}^{-1}(\lambda/\lambda_m)$ versus $\log \lambda$. From the shift distances required for superposition, values of λ_m and $F(t)$ can be obtained. It was noted that the values of $F(t)$ estimated by this procedure were somewhat smaller than those estimated from data obtained at small strains. Values of λ_m obtained by this method, which for the 1 minute data varied from 4.4 at $-45°C$ to 8.1 at $-20°C$, were plotted against both t and $F(t)(273/T)$ on log-log coordinates from which the maximum extension ratio for equilibrium data, λ_m^0, was estimated to be 8.6. Above $-20°C$, λ_m was found to be time independent. In addition, the $\log \lambda_m$ versus $\log F(t)(273/T)$ response was linear, with a slope equal to -0.82, compared to Halpin's results which predict (25) the slope to be -1.0. Using $\lambda_m^0 = 8.6$ and the appropriate λ_m values, 1 minute isochronal data obtained at -20, -35, and $-40°C$ were fitted quite precisely by equation [14].

In summary, the studies of Halpin and Smith indicate that the stress-strain-time response of well-cross-linked elastomers can be represented by four functions: (a) a time dependent modulus; (b) a nonlinear function of the strain which itself becomes time dependent at short reduced times; (c) the time dependent maximum extensibility parameter; and (d) a time-temperature equivalence function such as the WLF or Arrhenius equation. Having briefly discussed the stress-strain-time response, the rupture properties are now considered.

3 UNIAXIAL RUPTURE OF UNFILLED ELASTOMERS

3.1 Effect of Test Conditions on Rupture

In a qualitative way, it has long been recognized that factors such as the test temperature and degree of cross-linking could have a pronounced effect on the values of the stress-at-break σ_b and the strain-at-break ϵ_b for a given elastomer. Villars (26), for example, using a high strain rate tester showed that σ_b for an SBR elastomer increases steadily with increase in strain rate while ϵ_b passes through a maximum. The strain rates were not too reliable, but this was apparently the first demonstration that ϵ_b would pass through a maximum with increasing strain rate. Villars also tested natural rubber and observed qualitatively the same results. Borders and Juve (27) showed that σ_b for elastomers of different chemical structure, measured as a function of temperature at a constant strain rate, could be correlated if σ_b values were compared at equal values of $T - T_g$. Noncrystalline elastomers and elastomers that are subject to strain induced crystallization showed separate responses. Gee demonstrated that, at a given test temperature and test rate, σ_b for natural rubber is strongly dependent on the degree of cross-linking and somewhat dependent on the nature of the curing system employed (28).

In this section, a survey of the rupture of unfilled elastomers is presented. The effect of testing conditions on the observed values of σ_b and ϵ_b is first discussed since an understanding of this aspect of the problem will furnish a basis for attempts to correlate the effects of other factors such as the degree of cross-linking.

According to equation [1], the basic response of an elastomer can be considered in terms of three variables—stress, strain, and time—and a convenient graphical representation is a surface in stress-strain-time space (5), sometimes referred to as a tensile property surface. An example for a gum viton elastomer having a ν_e value of 114×10^{-6} moles/cm^3 is shown in Fig. 1 using logarithmic scales for the coordinate axes. The intersection

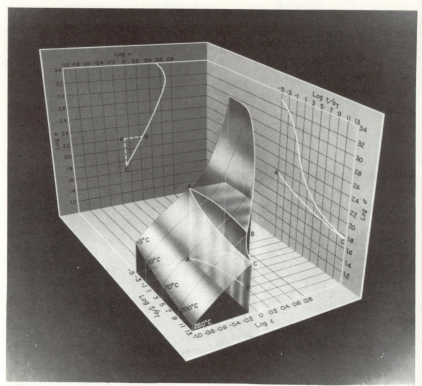

Figure 1 Tensile property surface for a viton gum elastomer. The detailed data for this figure were kindly furnished by Dr. T. L. Smith. Reprinted from Landel and Fedors (5, p. 384), by courtesy of John Wiley & Sons, Inc.

of the surface with a plane parallel to the log σ, log ϵ plane provides the isochronal response given by

$$\sigma(\epsilon, t) = \frac{F(t)}{3} f(\epsilon) \qquad (\text{time} = \text{const}) \qquad [16]$$

Such isochronal curves used to construct Fig. 1 are shown in Fig. 2 for several test temperatures in the range -5 to $230°C$. For long reduced times, where $F(t)$ is only a slowly varying function of time [and $f(\epsilon)$ is independent of time], the isochronal curves will be essentially the same, i.e., the surface will be nearly level.

The intersection of the surface with a plane parallel to the log ϵ, log t plane yields the response at constant σ, i.e., the creep response, given by

$$\sigma(\epsilon, t) = \frac{f[\epsilon(t)]}{3D(t)} \qquad (\sigma = \text{const}) \qquad [17]$$

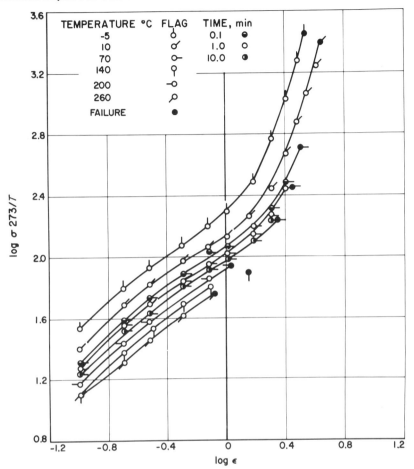

Figure 2 One minute isochronal stress-strain curves for a viton gum elastomer. At 70°C, the 0.1 and 10 minute isochronal curves are also shown for comparison. The filled circles indicate rupture. Data furnished by Dr. T. L. Smith. Reprinted from Landel and Fedors (5, p. 375), by courtesy of John Wiley & Sons, Inc.

One such creep response is shown in Fig. 1 by the curve AB. Curves resulting from cuts made at a number of different σ values will be parallel when projected to the log ϵ, log t plane, for reduced times where $f(\epsilon)$ is time independent.

Finally, intersection of the surface with a plane parallel to the log σ, log t plane provides the response in stress relaxation, which is given by

$$\sigma(\epsilon, t) = \frac{E(t)}{3} f(\epsilon) \qquad (\epsilon = \text{const}) \qquad [18]$$

and depicted in Fig. 1 by the curve AC. As was the case in creep, for time scales for which $f(\epsilon)$ is time independent, multiple cuts through the surface will yield parallel stress relaxation curves when projected to the log σ, log t plane.

Rupture, on this surface, is represented graphically as a discontinuity or boundary, which can be considered to result from the intersection of the tensile property surface with another surface representing failure. In Fig. 1 rupture is represented by the terminating line on the right. If this rupture line is projected to the log σ, log ϵ plane, it represents the failure envelope. If it is projected to the log σ, log t plane, it gives the time dependence of the rupture stress; if projected to the log ϵ, log t plane, the time dependence of the rupture strain results. These projections are also indicated in Fig. 1.

3.1.1 Failure Envelope

In a series of papers, Smith reported the results of his studies on the effects of test temperature and test rate on the rupture behavior of unfilled vulcanizates. He found that a plot of log $\sigma_b(T_0/T)$, where T_0 is an arbitrary reference temperature and T is the test temperature, versus the corresponding value of log ϵ_b, obtained as a function of test rate and test temperature, yielded a characteristic curve which he called the failure envelope (29, 30). A typical failure envelope is shown in Fig. 3 for the same viton elastomer depicted in Fig. 1. As the test temperature is decreased or the strain rate is increased, the points defining the failure envelope move along the curve in a counterclockwise direction and the values of σ_b and ϵ_b both increase. As the temperature is decreased further, ϵ_b reaches a maximum value and then decreases. The values for σ_b, on the other hand, increase continuously. Implicit in the existence of the failure envelope is the applicability of time-temperature superposition; i.e., if a reduced time scale did not exist, the physical property surface, and hence the failure envelope as well, would not be a continuous function. Representing as it does the projection of the terminating line to the log σ, log ϵ plane, the failure envelope is also independent of time and temperature. A change in either of these variables merely moves the point along the curve without affecting its shape or location.

The failure envelope is apparently unique in that it is independent of the test mode. Thus, rupture data obtained in creep (8, 29), and stress relaxation (5, 29, 33), yield the same failure envelope determined at constant strain rate. For the case of a cyclic stress history where specimens were cycled between various load limits and at different temperatures, Smith observed (8) that the failure data tended to fall slightly below the envelope obtained from constant strain rate tests.

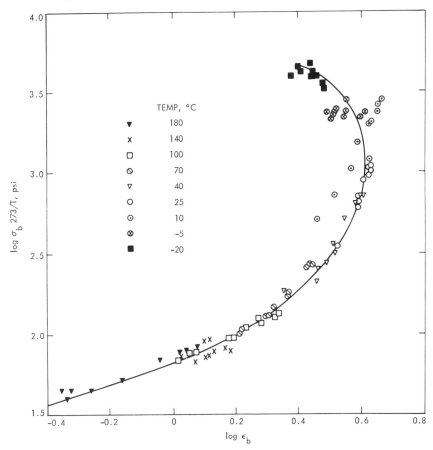

Figure 3 Projection of the rupture points for a viton elastomer to the σ, ϵ plane to generate the failure envelope. Data furnished by Dr. T. L. Smith. Reprinted from Landel and Fedors (5, p. 379), by courtesy of John Wiley & Sons, Inc.

The failure envelope has been shown to be applicable to a wide variety of amorphous, cross-linked elastomers differing in chemical structure (31, 32). Qualitatively, the failure envelopes obtained with different elastomers are similar in shape but differ in location in the log σ, log ϵ plane. However, since the degree of cross-linking usually differs from elastomer to elastomer, it is difficult to separate the relative contributions of the changing chemical structure and the varying degree of cross-linking in determining the location of the failure envelope. Figure 4 shows typical failure envelopes taken from the literature for several elastomers which differ in chemical structure (32). However, in addition to the changes in

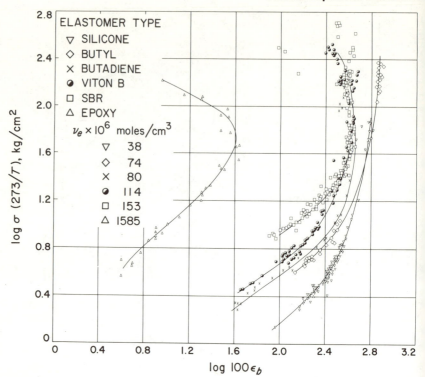

Figure 4 Failure envelopes for six types of rubber each at constant degree of cross-linking.

chemical structure, it will also be noted that ν_e varies by a factor of about 50 through the series.

According to equation [16], for long reduced times, where $F(t)$ is only a slowly varying function of time and where $f(\epsilon)$ is time independent, the tensile property surface will be nearly level and its projection to the log σ, log ϵ plane will be represented by a single curve. Since the terminating line which represents the occurrence of rupture is a boundary of the surface, its projection to the log σ, log ϵ plane will also be a single curve; furthermore, these two projections will be identical. Hence, for long reduced times the failure envelope can be represented by the isochronal stress-strain response of the elastomer (5, 34). The lowest temperature or shortest time scale for which this *can* be true is in the region of the time scale where ϵ_b reaches its maximum value; however, for strongly time dependent elastomers, the identity of the failure envelope and the stress-strain curve will break down at longer times, i.e., temperatures higher than that at which the maximum in ϵ_b is observed.

Thus, the high temperature portion of the failure envelope is nearly the same as the quasi-equilibrium isochronal response. Furthermore, since the time dependence is small, it can also be represented by the constant strain rate stress-strain response itself (5, 34). Hence, if a mathematical description for the stress-strain response is available, the same equation can be used to represent the high temperature portion of the envelope.

Figure 5 shows failure data for several gum elastomers taken from the literature; the full curves represent the response predicted by equation [2]. The values of ν_e and n required for the fit shown are indicated in the figure. It is evident that for these data at least, equation [2] provides a satisfactory fit. In all but one case, curve 6 for butyl, the values of ν_e obtained from independent methods such as from equilibrium swelling correspond closely to the values required for fitting. In addition, these

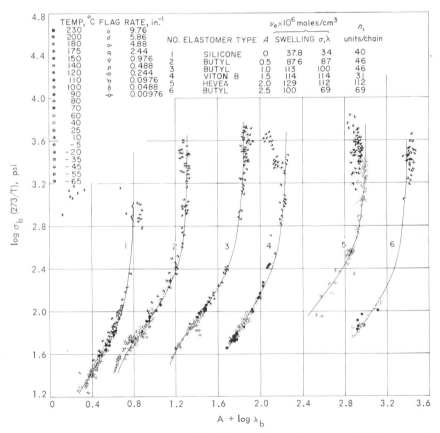

Figure 5 Comparison of equation [2] with experimental rupture data for several elastomers.

results indicate that $F(t)$ and $f(\epsilon)$ can be taken, as a first approximation at least, as time independent over the range of time scale or temperature for which $\epsilon_b \le (\epsilon_b)_m$. This correspondence has an interesting application. It has been shown for an SBR vulcanizate that the failure envelope obtained by testing many specimens at a single test rate and three temperatures is the same as the envelope obtained by testing single specimens over a much more extended range of temperatures and rates (34).

The discussion so far has been limited to amorphous and well-cross-linked vulcanizates. When strain induced crystallization is possible, the failure envelope apparently exists only for those test conditions under which crystallization does not occur. Data for a natural rubber vulcanizate obtained at temperatures between −55 and 120°C and for strain rates between 0.0102 and 10.2 min^{-1} are shown in Fig. 6 (35). For temperatures

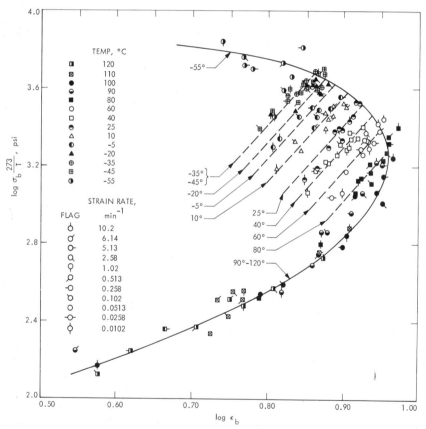

Figure 6 Failure envelope for a natural rubber vulcanizate. Data from Smith (37).

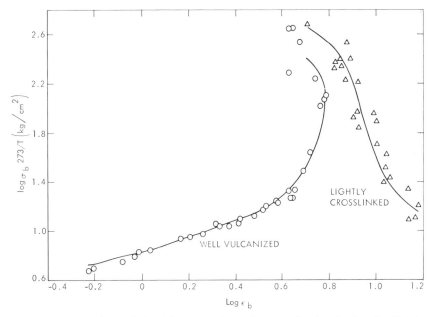

Figure 7 Comparison of the failure envelopes for a well-vulcanized and a lightly vulcanized SBR gum elastomer. Data from Halpin (25).

above about 90°C, the rupture points appear to define a segment of a failure envelope. For temperatures less than 90°C, the data do not. For a given temperature in this region, the failure points appear to define curves as shown by the dashed lines in the figure. As the strain rate is increased at a given temperature, the points move to the right to higher values of σ_b. Smith suggests that each dashed line is actually a segment of a failure envelope for a material that has undergone strain induced crystallization. The presence of crystallites changes the nature of the elastomer which then behaves differently from a material that remains amorphous. For very lightly cross-linked elastomers, a failure envelope exists but the shape is markedly different from that observed with well-vulcanized elastomers as exemplified by the envelope for viton shown in Fig. 3. Some results obtained by Halpin (25) for an SBR elastomer are shown in Fig. 7. The lightly cross-linked elastomer is distinguished by the fact that ϵ_b decreases continuously and monotonically over the entire range of test rate and temperature. The full curves shown in the figure are theoretical predictions of the Bueche-Halpin theory (36).

In this theory, the specimen is considered to be subjected to a tensile creep experiment. The stress-strain response is assumed to be given by equation [17]. During elongation of the specimen, small tears are considered to form and grow. At the crack tip, the stress is increased over the nominal value by a stress concentration factor S. The criterion for break is a critical elongation of the filaments at the crack tip, which is taken to be time and temperature independent. The filament breaking time is assumed to be proportional to the break time for the macroscopic specimen and the stress-at-break is given by the theory as

$$\sigma_b = \frac{K}{3D(t_b/q)} \qquad [19]$$

while the strain-at-break is given by

$$f(\epsilon) = \frac{KD(t_b)}{D(t_b/q)} \qquad [20]$$

where K and q are parameters. Equations [19] and [20] indicate that the creep response in conjunction with the values of the parameters K and q can be used to predict the failure envelope. Reasonable agreement between the theoretical prediction and experimental failure envelope was obtained for several elastomers (36).

3.1.2 *Time Dependence of Rupture*

Working with SBR, Smith (9) showed that the break data obtained at constant temperature at a series of test rates could be shifted along the time axis to yield master curves of log $\sigma_b(T_0/T)$ versus log t_b and log ϵ_b versus log t_b. The shift distances obtained from σ_b, t_b and ϵ_b, t_b data were the same within experimental error and were in agreement with the WLF equation. Hence Smith concluded that ultimate properties vary with temperature because either the internal viscosity or the molecular friction coefficient varies with temperature. The ultimate properties vary with strain rate because the viscous resistance to network deformation increases with rate.

Figure 8 shows the σ_b, t_b response and Fig. 9 shows the ϵ_b, t_b response for the viton elastomer shown in Fig. 1. The time dependence of σ_b and ϵ_b is represented, in terms of the physical property surface, by projections of the surface boundary to the σ, t and ϵ, t plane, respectively. Plots such as those given in Figs. 8 and 9 have been shown to be applicable to a wide variety of elastomers differing in chemical structure; curves of the same general shape but differing in location in the respective planes are

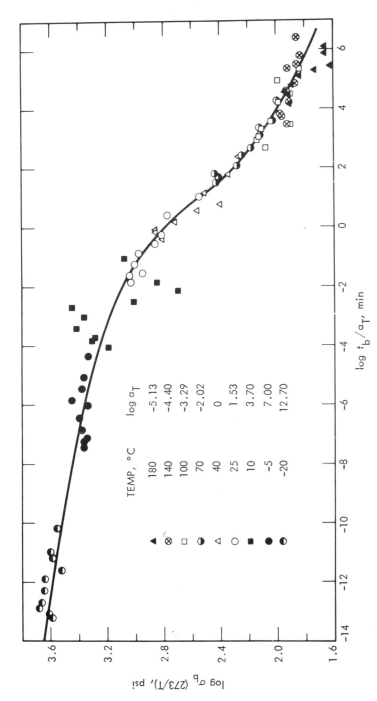

Figure 8 Projection of the rupture points for a viton elastomer to the σ, t plane. Data from Smith (37).

TEMP, °C	log a_T
180	-5.13
140	-4.40
100	-3.29
70	-2.02
40	0
25	1.53
10	3.70
-5	7.00
-20	12.70

log t_b/a_T, min

log σ_b (273/T), psi

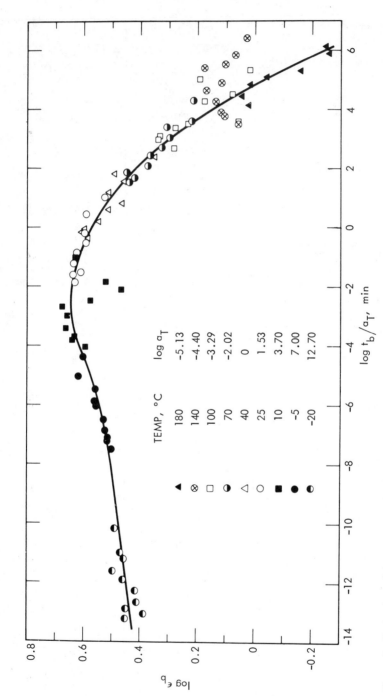

Figure 9 Projection of the rupture points for viton to the ε, t plane. Data from Smith (37).

obtained. However, changes in the chemical structure of the elastomer were not carried out at a constant degree of cross-linking and the effect of chemical structure alone on the response was not previously reported.

The time dependence of the ultimate properties will depend on the test mode. Referring to the physical property surface in Fig. 1, if a viton specimen were brought to point A and held at constant stress (creep), then the response would follow path AB. Rupture is expected to occur when the path AB intersects the boundary, which would occur after about 14 decades of time had elapsed. On the other hand, if the specimen again at point A were held at constant strain (stress relaxation), the response would be path AC. Again the specimen would be expected to break when the path reached the limiting boundary of the surface, which would occur after some 18 decades of time. Thus, starting from a given point on the surface, rupture is expected to occur in a shorter time for creep than for stress relaxation.

3.2 Effect of Degree of Cross-Linking

3.2.1 *Failure Envelope*

For long reduced times, the time dependent function in equation [1] becomes a slowly varying function of time and approaches a limiting value $\nu_e RT$. Under these conditions, equation [1] can be written as

$$\frac{\sigma}{\nu_e} \frac{T_0}{T} = RT_0 f(\epsilon, n) \qquad [21]$$

where the dependence of $f(\epsilon)$ on a parameter, such as the maximum stretch parameter, is shown explicitly, although the dependence of $f(\epsilon)$ on other parameters, such as C_2/C_1, could also be included as well. If $f(\epsilon)$ is represented by equation [2], the dependence on n becomes apparent only for large values of the strain. Hence for small to moderate strains, $f(\epsilon)$ will depend on ϵ alone. In the long time region where the physical property surface is level, the projections of both the surface and the surface boundary to the σ, ϵ plane are very nearly the same, and equation [21] can also be used to represent rupture data. Hence a plot of log $(\sigma_b T_0/\nu_e T)$ versus log ϵ_b should yield a master failure envelope (5, 38), provided the test temperature is sufficiently high (long reduced times). Furthermore, the master envelope should be independent of the precise chemical structure of the polymer. In order to demonstrate this independence, the data in Fig. 4, which represent failure envelopes for six

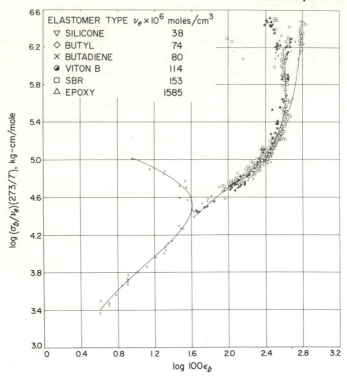

Figure 10 Failure envelopes of Fig. 4 replotted with σ_b reduced to unit network chain concentration.

different elastomers which range in ν_e value from about 40 to 1600×10^{-6} moles/cm^3, have been replotted as suggested by equation [21] and the result is shown in Fig. 10 (32).

In Fig. 11, rupture data obtained at a single rate and temperature are shown for four elastomers which vary both in chemical structure and degree of cross-linking. When these data are reduced to unit ν_e, as suggested by equation [21], the data superpose as shown in Fig. 12 to a single curve to within the experimental uncertainty in the reported ν_e values, especially for the EPDM and SBR elastomers. It will be noted that even rupture points obtained from very lightly cross-linked EPDM and SBR also apparently conform to equation [21], which is unexpected (32). This point is discussed in more detail later.

It might be argued that, since not only ν_e but T_g and type of backbone are being changed, the reduction is not real but only an artifact stemming

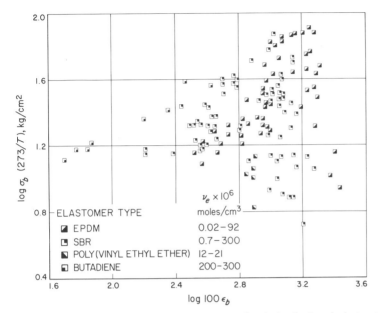

Figure 11 Rupture data for four elastomers varying in both chemical structure and degree of cross-linking.

from the simultaneous change in several variables. Figure 13 shows failure envelopes for a single elastomer, viton A-HV, constructed (32) from rupture data reported by Smith for a series of vulcanizates as a function of the degree of cross-linking (8). Figure 14 shows the reduced failure envelope and it can be seen that the high temperature portions have all merged into a common response curve, independent of ν_e.

The master envelope or common response curve is also independent of statistical variations in the rupture properties. Hence statistical variations in the breaking properties as measured for a sample of a given cross-link density at a given strain rate and temperature, permit the delineation of portions of the failure envelope for that sample. This is illustrated in Fig. 15, where the unfilled squares show portions of the envelope for an SBR rubber as measured at three temperatures, compared to the envelope obtained by varying the strain rate and temperature (the solid line). The filled squares denote the segments formed at 25°C at progressively lower cross-link densities (39). In Fig. 16, these data have been reduced to unit ν_e and compared with the reduced master curve (solid line) obtained from Fig. 12. Again, the reduction principle holds at high temperatures and hence is unaffected by statistical variability (32).

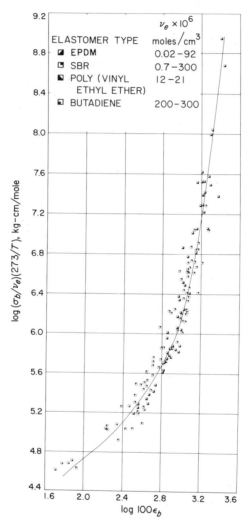

Figure 12 Data from Fig. 11 replotted with σ_b reduced to unit network chain concentration.

The reduced failure envelopes for all the systems discussed so far are shown in Fig. 17. When the breaking stress is normalized to unit ν_e, then for temperatures sufficiently above the glass temperature, the reduced stress is a unique function of the breaking strain, independent of the chemical structure of the chain (epoxy, silicone, butadiene, butyl, fluorocarbon, styrene–butadiene), such that data for some 15 elastomers can be reduced to a single master curve as predicted by equation [21].

However, the independence of the reduced failure envelope on chemical structure is merely a reflection of the fact that the stress-strain response, i.e., $f(\epsilon)$, does not, to a first approximation, depend on chemical structure.

At lower test temperatures, i.e., shorter reduced time scales, the failure data diverge upward from the common response curve. This divergence is caused by two factors: (a) the extension ratio at break is approaching the value of the maximum stretch parameter, and (b) $G(t)$ is becoming more time dependent. The first factor depends on the chemical structure of the polymer in that the maximum stretch parameter, or chain flexibility, depends directly on structure through equations [4] and [5].

Smith has proposed an alternative method for normalizing failure envelopes to take account of variations in ν_e. He considers the rupture data in terms of a plot of $\log \sigma_b \lambda_b (T_0/T)$ versus $\log \epsilon_b$. The ordinate corresponds to the breaking stress based on the deformed cross-sectional

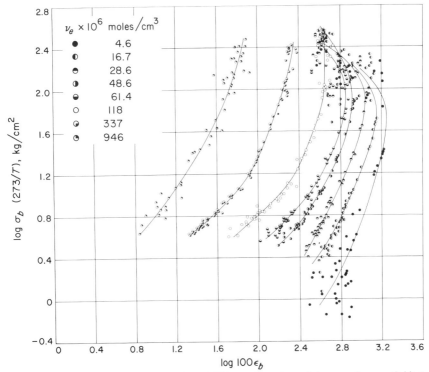

Figure 13 Failure envelopes for viton A-HV as a function of degree of cross-linking. Data from Smith (8).

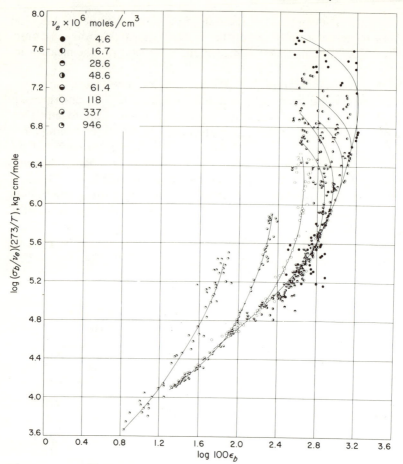

Figure 14 Data of Fig. 13 replotted with σ_b reduced to unit network chain concentration.

area. When plotted in this manner, Smith (31) finds that individual failure envelopes can be superposed by shifting the envelope along the abscissa by a factor equal to log $E_e(T_0/T)$ where E_e is the equilibrium Young's modulus. This procedure superposes the failure envelope up to the region where the maximum in ϵ_b occurs. Figure 18 shows such a plot for five different elastomers, and except for the silicone and natural rubber vulcanizates, Smith's procedure seems to be applicable. However, for the remaining elastomers which do superpose, the modulus data show that they all have very nearly the same degree of cross-linking. Note that the general applicability of this procedure requires that $\sigma_b\lambda_b(T_0/T) =$

$f(E_e\epsilon_b T_0/T)$ which, by implication, requires that this functional relationship describe the equilibrium stress-strain response extending up to strains where $(\epsilon_b)_m$ is observed. Second, it requires that $(\epsilon_b)_m$ be proportional to ν_e^{-1}. It is interesting to note that for small strains, the slope of the failure envelope must be unity since in this region, Hooke's law must be valid; here superposition can be effected by either a horizontal or vertical shift.

Smith also studied the rupture behavior of a series of viton A-HV vulcanizates cross-linked to different extents (40). The response was measured at a series of rates and temperatures. The vulcanizates had moduli, measured at elevated temperatures, which differed by a factor of about 10. The data in the form of plots of log $\sigma_b\lambda_b(T_0/T)$ versus log ϵ_b are shown in Fig. 19. The full curves represent average curves drawn through the individual data points. (Note that these data correspond to the data shown in Figs. 13 and 14.) Failure points on the equilibrium

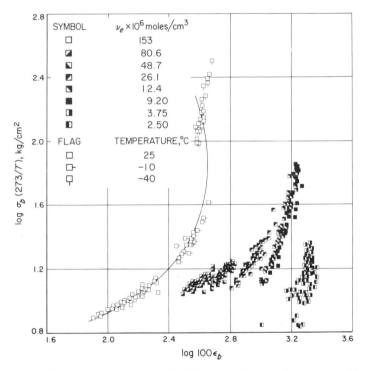

Figure 15 Comparison of portions of a failure envelope (unfilled squares) for an SBR vulcanizate -at different temperatures with the envelope generated by varying the strain rate and temperature (solid line). Filled squares denote segments formed at 25°C at progressively lower network chain concentrations.

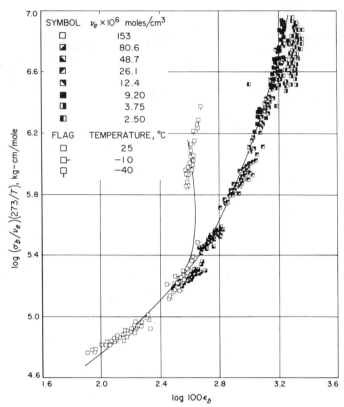

Figure 16   Data of Fig. 15 with σ_b reduced to unit network chain concentration.

stress-strain response, as estimated and depicted in Fig. 19 by the dashed curves, could not be obtained for vulcanizates having low ν_e values even at test temperatures as high as 230°C. Smith attributed this to either the presence of persistent viscous effects or the occurrence of chemical degradation. Figure 20 shows the data of Fig. 19 superposed by a shift along the abscissa. The log $E_e(T_0/T)$ values required for superposition ranged from 1.57 for the lowest degree of cross-linking to 2.50 for the highest degree of cross-linking. The fact that log $E_e(T_0/T)$ values estimated from modulus measurements at high temperature were lower, for the lightly cross-linked vulcanizates, than the values estimated from the shift was attributed to the occurrence of chemical degradation. However, it was subsequently established that degradation did not occur; hence the log $E_e(T_0/T)$ values used to effect superposition must be considered as empirical quantities and are not in general equal to values estimated independently as, for example, from the modulus.

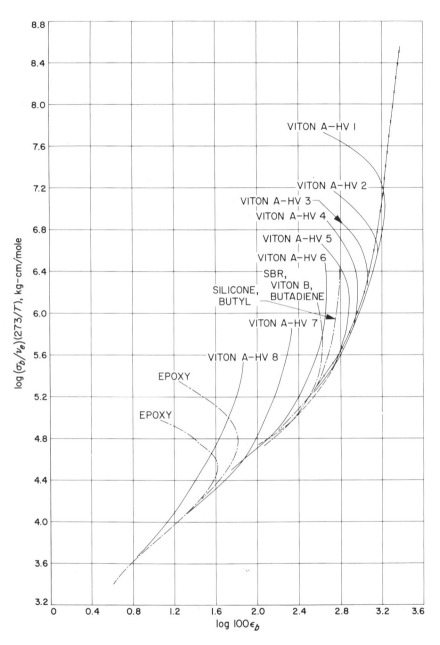

Figure 17 Reduced failure envelope for several elastomers.

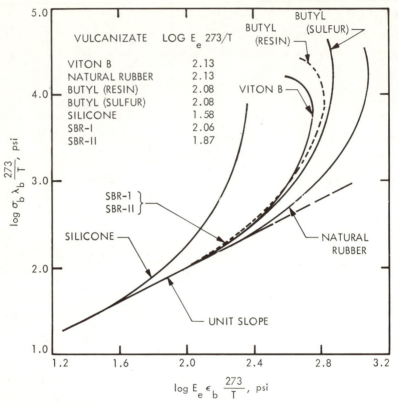

Figure 18 Comparison of failure envelopes for five elastomers plotted as log $\sigma_b \lambda_b (273/T)$ versus log $E_e \epsilon_b (273/T)$. Data from Smith (40).

In a subsequent paper, Smith and Frederick pointed out that these viton data can be represented by an equation of the form

$$\frac{1}{n}\left[\frac{3\sigma_b\lambda_b}{E_e}+\frac{1}{\lambda_b}\right]=f\left(\frac{\lambda_b}{n^{1/2}}\right)$$ [22]

where f is an unspecified function (20). This expression is similar to equation [3] except that the undefined function f replaces the inverse Langevin function. If this equation is applicable, plots of log $\sigma_b\lambda_b(T_0/T)$ versus log λ_b should be superposable by a shift along the abscissa proportional to log $E_e n$, provided $1/\lambda_b$ is small compared to the quantity $3\sigma_b\lambda_b/E_e$. This superposition requires that the product $E_e n(T_0/T)$ be constant for all vulcanizates, i.e., that n is proportional to ν_e^{-1}. Equation [22] was found to be applicable to the viton data but the values of E_e

required for superposition were lower than the values estimated independently from modulus data; hence the values of E_e obtained from equation [22] should also be considered empirical.

Chemical Structure and Extensibility. As pointed out previously, the maximum stretch parameter is related to the degree of cross-linking. Some data on a variety of elastomers are shown in Fig. 21, as a plot of $(\lambda_b)_m$ versus ν'_e, the concentration of effective network chains per unit volume of gel, in log-log coordinates. According to equation [5], for constant f and M_0, the response should be linear with a slope of $-\frac{1}{2}$ and the line shown in the figure has been drawn with this value. As can be seen, the data scatter about this line, which is expected since both f and M_0 are certainly not constant for the elastomers represented. However, it is expected that M_0 will not vary by more than a factor of about 2 or 3 for most elastomers, which is probably why such a relatively good correlation

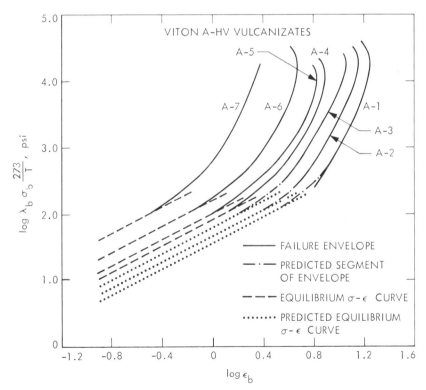

Figure 19 Failure envelopes plotted as true stress versus strain for viton A-HV vulcanizates as a function of degree of cross-linking. Data from Smith (40).

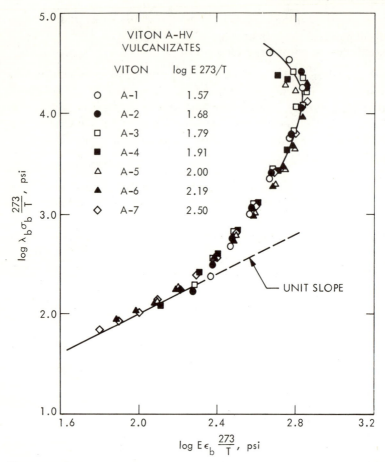

Figure 20 Data of Fig. 19 after superposition by shifting along the abscissa. Data from Smith (40).

is observed. Nevertheless, since the chemical structure of the polymer does vary, the result shown in Fig. 21, while suggestive, is not, however, conclusive.

The viton data of Smith represent a system in which the chemical nature of the chain is fixed and only the degree of cross-linking is varied (40). For these data, Smith found that $(\lambda_b)_m$ is proportional to $\nu_e^{-1/2}$ provided the vulcanizate of the lowest degree of cross-linking is excluded. When the sol fraction is taken into account by plotting $(\lambda_b)_m$ against ν'_e, $(\lambda_b)_m$ is found to be proportional to $(\nu'_e)^{-0.6}$, again provided the vulcanizate with the lowest degree of cross-linking is excluded. In addition to the

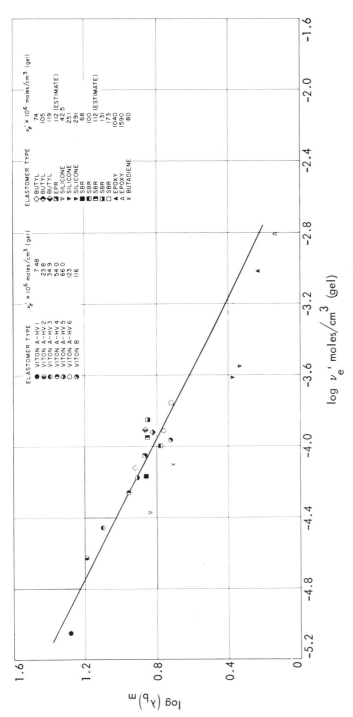

Figure 21 Dependence of $(\lambda_b)_m$ on network chain concentration. Data from Landel and Fedors (38).

sol fraction, equation [5] shows that the fraction of free chain ends should also be taken into account. Using the Flory chain end correction (41), an estimate for the concentration of effective chains per unit volume of gel, ν''_e, is obtained in the form

$$\nu''_e = \frac{\nu_e}{g} + \frac{2\rho g}{M_n} \qquad [23]$$

where M_n is the number average molecular weight of the primary molecules, which was taken to be 2×10^5 g/mole. Using this relationship, Smith finds $(\lambda_b)_m$ proportional to $(\nu''_e)^{-0.75}$ for all vulcanizates including the one with the lowest degree of cross-linking.

Fedors and Landel (42) have also considered the viton data of Smith. An estimate of the molecular weight of the primary molecules as a function of the degree of cross-linking was obtained using sol-gel relationships. It was found that scission accompanies vulcanization, and estimates of the molecular weight in the gel as a function of degree of cross-linking were obtained. They found that the maximum value of the exponent in equation [5] lies in the range 0.78–0.58, depending on the particular form of the free chain end correction employed. The higher maximum estimate (0.78) was obtained using the Flory chain end correction (42), whereas the lower maximum estimate (0.58) was obtained using the Berry chain end correction (43). In obtaining these results, the assumption was made that $\nu_e = \nu_c$. Use of ν_c would reduce the exponent to a value closer to $\frac{1}{2}$. Here ν_c is the concentration of network chains terminated by chemical crosslinks.

Under the assumption that the maximum stretch parameter is related to the number of statistical links per network chain, Bueche et al. (44) calculated values of the ratio $(r_0^2/M)^{1/2}$, where r_0^2 is the mean square chain end-to-end separation and M is the molecular weight, from the experimentally measured $(\lambda_b)_m$, using the following explicit relationship which is derivable from equation [5]:

$$(\lambda_b)_m = n^{1/2} = \frac{l}{M_1} \left(\frac{\rho}{\nu_e} \right)^{1/2} \left(\frac{M}{r_0^2} \right)^{1/2} \qquad [24]$$

where l is the length increment contributed per backbone bond to the fully extended chain ($l = 1.25$ Å for the C–C bond) and M_1 is the molecular weight per backbone atom. The ratio r_0^2/M can be evaluated independently from solution behavior of the unvulcanized polymer. The ratios calculated from $(\lambda_b)_m$ were found to be about 0.7 times as large as the values estimated from solution behavior at the theta temperature, Bueche suggested that the lower value of the ratio obtained from $(\lambda_b)_m$ might be due to nonaffine character of deformations at large strains.

Data have been presented which demonstrate that n estimated from the isochronal stress-strain response and from the failure envelope may not always be identical (20). Thus, for SBR, n estimated from equilibrium stress-strain was 100 whereas n estimated from the failure envelope was 51. However, for butyl, both sulfur cured and resin cured, silicone, natural rubber, and viton B vulcanizates, identical values for n were obtained when both equilibrium and failure data were analyzed using equation [3]. When equation [8] was used, taking the value of the ratio C_2/C_1 equal to 0.394, isochronal stress-strain data for a sulfur cured butyl vulcanizate yielded values of n in the range 47–50, depending on the value used for equilibrium modulus (144–133 psi) (20). Thus, differences in the value of n can be observed depending on the particular explicit form chosen for $f(\epsilon)$.

A comparison was also made between the experimentally observed values of $(\lambda_b)_m$ and estimates obtained from the ratio $(r_0^2/M)^{1/2}$ in conjunction with equation [5] taking both g and f to be unity (20). For butyl and natural rubber, the calculated values of $(\lambda_b)_m$, 6.0 and 8.2, respectively, were found to be lower than values obtained directly from the failure envelope, 7.4 and 10.1, respectively. In contrast, the calculated values for silicone and SBR vulcanizates were higher than obtained experimentally. In addition, Smith noted that somewhat different values for $(\lambda_b)_m$ are obtained if the failure envelope itself is used to estimate $(\lambda_b)_m$ or if the failure envelope is fitted by an equation such as [2] with the assumption that $n^{1/2} = (\lambda_b)_m$.

In another study, Furukawa and co-workers (45) investigated the chain stiffness of both random and alternate copolymers of butadiene and acrylonitrile varying in acrylonitrile content from 27 to about 50 mole %. The molecular weight of the statistical unit was estimated from stress birefringence measurements. Excellent agreement was found between $(\lambda_b)_m$ obtained from failure envelopes and values calculated from equation [5].

Although precise agreement between $(\lambda_b)_m$ obtained from the failure envelope and values calculated from $(r_0^2/M)^{1/2}$ is not always obtained, the results are sufficiently close to serve as a useful guide for estimating the gross effect of chemical structure on failure response.

Table 1 lists data for several polymers grouped arbitrarily into three sets. In set A are listed several hydrocarbon polymers, set B includes the more common elastomers, and set C lists several methacrylate polymers. Included in the table are values (46) of T_g, M_1, $(r_0^2/M)^{1/2}$, l, and also estimates of $(\lambda_b)_m$ for cross-linked polymers having $\rho/v_e' = 10^4$ obtained using equation [24].

According to equation [24], a key molecular parameter is M_1, which

Table 1 Selected Molecular Parameters

Polymer	T_g (°C)	M_1 (g/mole)	$\left(\dfrac{r_0^2}{M}\right)^{1/2}$ (Å)	l (Å)	$(\lambda_b)_m$ [a]
A. Poly(ethylene)	−125	14	0.95	1.25	9.4
Poly(1-butene)	−24	28	0.78	1.25	5.7
Poly(isobutene)	−73	28	0.70	1.25	6.4
Poly(1-pentene)	−40	35	0.80	1.25	4.4
Poly(styrene)	100	52	0.67	1.25	3.6
Poly(2-vinyl naphthalene)	162	77	0.77	1.25	2.1
B. Poly(butadiene) 98% cis	−95	13.5	0.88	1.15	9.7
Poly(butadiene) 98% trans	−83	13.5	0.82	1.84	17
Poly(chloroprene) 85% trans	−45	22	0.72	1.84	11
Poly(isoprene) 100% cis	−73	17	0.85	1.15	8.0
Poly(isoprene) 100% trans	−60	17	0.97	1.84	11
Poly(methylsiloxane)	−123	37	0.67	1.45	6.0
C. Poly(methyl methacrylate)	105	50	0.64	1.25	4.0
Poly(ethyl methacrylate)	65	57	0.56	1.25	4.0
Poly(n-butyl methacrylate)	20	71	0.51	1.25	3.4
Poly(n-hexyl methacrylate)	−5	85	0.53	1.25	2.8
Poly(n-octyl methacrylate)	−20	99	0.50	1.25	2.5
Poly(n-lauryl methacrylate)	−65	127	0.50	1.25	2.0

[a] Calculated from equation [24] assuming $\rho/v_e' = 10^4$ g/mole.

for the examples listed in Table 1 varies from a low value of about 14 for polybutadiene to a high value of 127 for the n-lauryl methacrylate polymer; i.e., an average main chain atom in the methacrylate polymer occupies about 10 times more volume than does an average main chain atom in polybutadiene. In the case of the methacrylate polymer, the long lauryl side chain does not contribute to rubberlike elasticity directly, but it does contribute to the mass or volume of the polymer chain. Considered in this context, M_1 can be related to "network chain efficiency" since the lower the M_1 value, the greater will be the fraction of material present as main chain atoms, and hence a unit volume of polymer will contain a larger number of network chains capable of supporting a load.

The values of $(r_0^2/M)^{1/2}$ do not vary by more than a factor of 2 for any of the polymers listed; furthermore, within a given set, even less variation occurs. Hence the main factor influencing $(\lambda_b)_m$ will be the value of M_1. The final column lists values of $(\lambda_b)_m$ calculated from equation [24]. In general, $(\lambda_b)_m$ is larger the smaller the M_1 value, and this fact contributes to the large $(\lambda_b)_m$ values predicted for the dienes. Surprisingly, the trans configuration has a larger $(\lambda_b)_m$ than the cis and this is a reflection of the fact that l is larger for the trans isomer. However, crystallization occurs in this isomer so that *trans*-polyisoprene has very low extensibility at room temperature. The data would indicate that polyethylene should be a highly extensible elastomer; however, in this case also the structure is so regular and uniform that the material is normally crystalline to a large extent. The data for the methacrylate polymers were included to demonstrate the effect of the increasing side chain length on M_1 and $(\lambda_b)_m$. It is evident that long side chains or bulky side groups are undesirable in elastomers which are intended to exhibit a large strain capability.

3.2.2 *Time Dependence of Rupture*

The projection of a family of surface boundaries to the stress-time or strain-time planes gives rise to a family of curves that represent the time dependence of rupture. In view of the fact that the high temperature segments of failure envelopes could be superposed by normalizing σ_b to unit ν_e, it is of interest to consider whether superposition, over the same range of time scale, could be achieved with the time dependence of rupture as well. In a study of the effects of statistical variability and network chain concentration on the failure envelopes of amorphous gum elastomers, it was noted that changes in ν_e corresponded, qualitatively at least, to changes in the time scale. Thus, it was reported that a decrease in ν_e under constant test conditions, had the same effect on the ultimate properties as if the test rate had been increased or the test temperature had been decreased, and conversely (38). This observation suggests that ν_e might be a suitable parameter for reducing or normalizing the time scale and hence for generating master curves for the time dependence of rupture (47).

Subsequently, Plazek demonstrated that the long time or terminal region of the creep compliance for both natural rubber and SBR gum vulcanizates could be superposed to yield a master curve by using ν_e as a reduction variable on the time scale (48).

The superposed creep compliance response is defined by Plazek in the following way:

$$J_x\left(\frac{t}{a_T a_x}\right) = J_p\left(\frac{t}{a_T a_x}\right)\frac{J_e(R)}{J_e} \qquad [25]$$

where J_x is the universal creep compliance response, J_p is the response for an elastomer with a given ν_e value, $J_e(R)$ and J_e are the compliances at mechanical equilibrium for an arbitrarily chosen reference system and for a system with a given ν_e value, respectively, and a_x is an empirical shift factor that relates ν_e to the time scale of the experiment. For both elastomers, a_x was found to be related to ν_e by means of the equation

$$a_x = \left(\frac{C}{\nu_e}\right)^{15.4} \qquad [26]$$

where C is constant. Plazek found that the elastomers had different C values, which was tentatively ascribed to differences in the monomeric friction coefficients for these two materials. Equation [26] implies that for a given time scale, the greater the ν_e value, the closer the system is to mechanical equilibrium; it also indicates that small changes in ν_e can have an enormous influence on the effective time scale.

It is of interest to ascertain whether the ν_e–time equivalence found by Plazek for small strain creep response is also applicable to the time dependence of rupture (47). It has already been pointed out that the a_T factor, which was first used to relate time and temperature in the small strain region, is also applicable to rupture.

Several studies have been carried out on the rupture behavior of SBR. Smith (9) investigated ring specimens of a sulfur cured SBR gum having a constant value of ν_e (106×10^{-6} moles/cm^3), at strain rates between 0.158×10^{-3} and 0.158 sec^{-1} and at 14 temperatures between -67.8 and $98.3°C$. He found that the rupture data obtained at a given temperature as a function of the strain rate could be shifted horizontally, and the result was a set of master curves relating both σ_b and ϵ_b to the reduced strain rate $(Ra_T)^{-1}$ where R is the strain rate. The magnitude of the horizontal shift was equal to a_T as obtained from the WLF equation.

Baranwal investigated the failure behavior of SBR, in the form of ring specimens, at strain rates between 2.10×10^{-4} and $0.526 \, \text{sec}^{-1}$, and at eight temperatures between -30 and $70°C$ (49). The vulcanizate was sulfur cured and had a ν_e value as estimated from equilibrium swelling of 204×10^{-6} moles/cm^3. In agreement with the findings of Smith, Baranwal observed that the rupture data could be shifted to yield a superposed curve of log σ_b and log ϵ_b as a function of reduced strain rate.

Stress-strain measurements to failure were also carried out by Healy on a sulfur cured SBR gum at strain rates between 2×10^{-3} and $0.2 \, \text{sec}^{-1}$ and at temperatures between -35 and $70°C$ (50). The ν_e value, estimated from equilibrium swelling measurements, is reported as 94×10^{-6} moles/cm^3. Stress-at-break and strain-at-break were plotted against t_b. Master curves were obtained by shifting the isothermal data along the time axis. The a_T factors obtained were in fairly good agreement with the values calculated from the WLF equation, except at the higher temperatures.

Whereas Smith, Baranwal, and Healy carried out their studies with vulcanizates having constant ν_e while the test temperature and test rate were allowed to vary, other studies of the failure behavior of SBR have been carried out at constant rate and constant temperature while ν_e was allowed to vary. For example, Taylor and Darin (51) investigated the effect of chain concentration on the ultimate properties of SBR at $30°C$ employing dumbbell shaped specimens run at constant crosshead speed. The chain concentration was varied by changing the content of the quantitative cross-linking agent decamethylene bis(methylazocarboxylate). Their data indicated that σ_b first increases, then passes through a maximum, and finally decreases as the content of curing agent is increased. Strain-at-break, on the other hand, decreases monotonically under the same conditions. In this study, the concentration of chemically cross-linked units in the network was taken equal to the concentration of curing agent present. No work was carried out to estimate the effective number of network chains, ν_e.

Epstein and Smith (52) carried out a similar study using the same polymer and curing agent. However, instead of relying on the concentration of curing agent as an adequate measure of ν_e, Epstein and Smith estimated ν_e independently, using both swelling and modulus measurements. Their data obtained at room temperature on dumbbell specimens using a constant crosshead speed also demonstrated that σ_b passes through a maximum at a relatively low value of ν_e. In addition, these workers were among the first to report that the σ_b, ν_e and the ϵ_b, ν_e responses were dependent on the conditions of measurement (52).

Fedors and Landel (39) studied the rupture behavior of sulfur cured

SBR vulcanizates as a function of ν_e at fixed temperature (28°C) and rate (4.25 min^{-1}) using ring specimens. In order to arrive at more reliable estimates for σ_b and ϵ_b, a study was made of the statistical variability of rupture ϵ_b, using 46 specimens at each ν_e value, and from these data the most probable values for σ_b and ϵ_b were obtained. In this investigation it was also found that σ_b passes through a maximum at a relatively low value of ν_e, whereas ϵ_b decreases monotonically with ν_e (39).

In using these failure data, the following procedures were employed (47):

1. The data as reported by Smith were reduced to a standard temperature of 263°K while most of the other data were determined at about 298°K. Hence a_T was calculated from the WLF equation and was used to shift Smith's data to a reference temperature of 298°K. His data are reported as log $\sigma_b(263/T)$ versus log $1/Ra_T$ and as log ϵ_b versus log $1/Ra_T$, and these were converted to log $\sigma_b(298/\nu_e T)$ versus log t_b/a_T and log ϵ_b versus log t_b/a_T plots. The value of ν_e used (106×10^{-6} moles/cm^3) was calculated from an equilibrium modulus value obtained at 87.8°C.

2. The data of Baranwal (49) were reduced to a reference temperature of 264°K. The appropriate a_T needed to shift the data to a reference temperature of 298°K was calculated from the WLF equation. The data were reported in graphical form as log $\sigma_b(273/T)$ and log ϵ_b versus log $1/Ra_T$. The reduced strain rate was converted to reduced time-to-break.

3. The data of Healy (50) were reduced to a reference temperature of 273°K. The appropriate a_T needed to shift the data to a reference temperature of 298°K was obtained from a tabulated list of experimentally determined values.

4. The data of Epstein and Smith (52) obtained at a crosshead speed of 2 in./min were used. Plots of log $\sigma_b(298/\nu_e T)$ and log ϵ_b versus log t_b were constructed from the tabulated data. The ν_e values used were those estimated from equilibrium modulus measurements. They also estimated ν_e from equilibrium swelling measurements but had to assume a solvent-polymer interaction coefficient χ_1 in order to calculate ν_e using the Flory-Rehner swelling equation. For this reason, the ν_e values obtained from modulus measurements were considered more reliable. Dumbbell shaped specimens were used, and the strain rate was estimated using the reported crosshead speed and the specimen gage length. In addition, the strain rate was assumed constant even though it is well known that the strain rate experienced by a dumbbell shaped specimen decreases with strain. This effect is negligible for most purposes.

5. For the data of Taylor and Darin (51), plots of log $\sigma_b(298/\nu_e T)$ and log ϵ_b versus log t_b were constructed from the tabulated data. In this

study ν_e values were not independently measured; rather, only the concentration of cross-link agent was reported. Since Epstein and Smith worked on essentially the same kinds of vulcanizates, it was assumed that both sets of data could be characterized by the same ν_e at equal concentration of cross-linking agent. Dumbbell shaped specimens were used, and the strain rate was assumed constant and was estimated from the reported specimen gage length and the crosshead speed.

6. The data of Fedors and Landel were obtained using ring specimens. Plots of log $\sigma_b(298/\nu_e T)$ and log ϵ_b versus log t_b were constructed from tabulated data. Values of ν_e were estimated from equilibrium swelling.

In the initial attempt to effect superposition, a_x was calculated from equation [26] taking $C = 106 \times 10^{-6}$ moles/cm^3; i.e., the data of Smith were chosen as the reference state and the remaining data were plotted as iog $\sigma_b(298/\nu_e T)$ versus log $t_b/a_T a_x$. However, it was determined by trial and error that much better overall superposition could be achieved by taking the exponent in equation [26] as 7.7 rather than 15.4.

The results for the time dependence of σ_b are shown in Fig. 22. The full curve represents the average response. The dashed curve represents the approximate band of scatter for the data of Smith, the dotted curve corresponds to the data of Healy, and the dot-dash curve corresponds to Baranwal's data. These three sets of data were obtained at constant ν_e but varying test rate and temperature. The open triangles represent the data of Taylor and Darin, the open circles the data of Epstein and Smith, and the filled circles the data of Fedors and Landel. These three sets represent data obtained at constant temperature and rate but varying ν_e It can be seen that most of the data fall neatly within the band of scatter characteristic of Smith's data.

Scatter in data such as these is due primarily to statistical variability of rupture and, to a smaller extent, to variability in ν_e from specimen to specimen. When ν_e is kept constant, unique values of σ_b and ϵ_b are not obtained when replicate specimens are tested. Variability of ultimate properties can be represented by a distribution of the double exponential type and the parameters which characterize the distribution have been found to vary with both ν_e and test temperature such that the distribution broadens as either ν_e or test temperature is decreased (5). Hence greater scatter should be expected at short times (low ν_e, low temperature) than at longer times, as is observed. In addition, the shape of the "average" curve obtained with a given set of data depends in part on the number of specimens tested, since this determines the degree of scatter observed, which in turn affects the way in which the average curve is constructed.

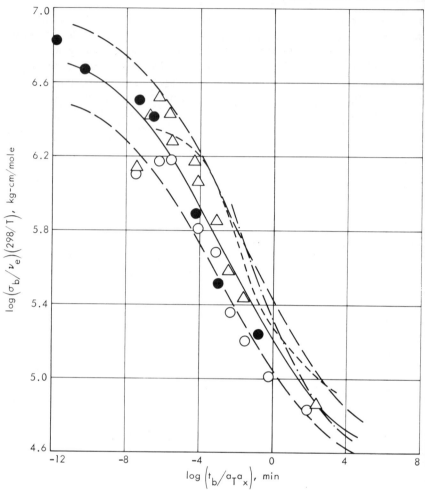

Figure 22 Dependence of reduced stress-at-break on the reduced time-to-break. Data from Fedors and Landel (47).

The slight divergence in curve shape for the various sets of data is not considered significant.

Figure 23 shows the log ϵ_b, log $t_b/a_T a_x$ response employing the same symbols and the same exponent (7.7) used in Fig. 22. At long times, the data superpose to within the band of scatter characteristic of Smith's data. The differences at short times between the data of Fedors on the one hand and of Taylor and Epstein on the other may be due, in part, to the fact that the latter two investigations employed dumbbell shaped specimens rather than rings. Epstein reported that ring specimens provided lower values for both σ_b and ϵ_b than did dumbbell specimens, especially at low ν_e values. The two sets of data obtained with dumbbells are in close agreement but fall above the data obtained with rings. However, as ν_e is increased and hence as the effective time scale is increased, the differences due to the specimen shape decrease and the data converge.

The data of Smith obtained at constant ν_e first increase, pass through a maximum, then finally decrease with decrease in time scale to yield a bell shaped curve. This is also apparent in Baranwal's data, but the time scale at which the maximum occurs is higher than is the case for Smith's data; in addition, the maximum value of ϵ_b is lower for Baranwal's vulcanizate. These effects can be attributed to the variation of n with ν_e. From the discussion in Section 3.2.1 it is evident that as ν_e increases, $(\lambda_b)_m$ must decrease and the time scale at which the maximum occurs must also increase; the ν_e value of Baranwal's vulcanizate is about twice the value

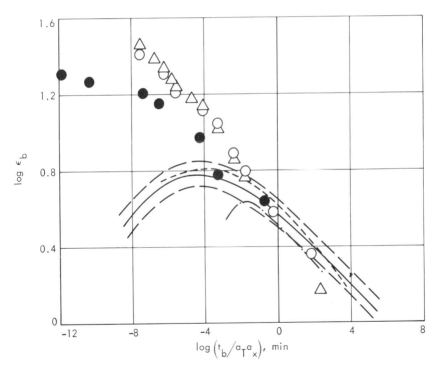

Figure 23 Dependence of strain-at-break on the reduced time-to-break. Data from Fedors and Landel (47).

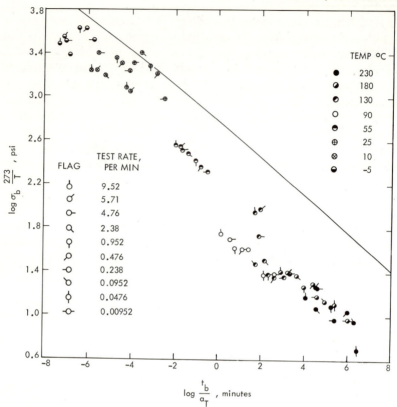

Figure 24 Dependence of stress-at-break on time-to-break for viton A-HV having $\nu_e = 4.6 \times 10^{-6}$ moles/cm^3; $a_T = 1$ at 90°C. Data from Smith (8).

for Smith's sample. It thus appears that there exists a limiting envelope for the log ϵ_b, log t_b/a_Ta_x response which represents the largest ϵ_b that any system can attain. As a result of the $(\lambda_b)_m$, ν_e relationship given by equation [24], however, individual curves representing the response of a system at constant ν_e will peel off the envelope at ϵ_b values which decrease, and at time scales which increase with increase in ν_e.

The results obtained with SBR, though suggestive, need to be extended to other elastomers to determine the general applicability of a ν_e–time shift. The viton A-HV results of Smith (8, 40) represent another system of interest because of the extensive data available.

Tabulated values for both log σ_b and log ϵ_b for the series of viton A-HV vulcanizates appear in ref. 8. Plots of log $\sigma_b(273/T)$ and log ϵ_b, obtained at constant temperature but at different strain rates, were shifted along the log t_b axis until superposition was obtained (53). The values of a_T required for superposition varied slightly but randomly with regard to ν_e; i.e., a_T is apparently independent of ν_e except possibly for the vulcanizate having the lowest degree of cross-linking. The average values of a_T obtained from shifting both σ_b, t_b and ϵ_b, t_b data agreed with the values reported by Smith, who observed an Arrhenius type dependence of a_T on temperature with an apparent activation energy of 31.2 kcal/mole (40). The σ_b, t_b data are shown in Figs. 24–30, while the ϵ_b, t_b data are shown in Figs. 31–37. For the σ_b, t_b data, the response appears to be almost independent of ν_e in that the data points tend to superpose without any shift. This is true for all except the vulcanizate having the

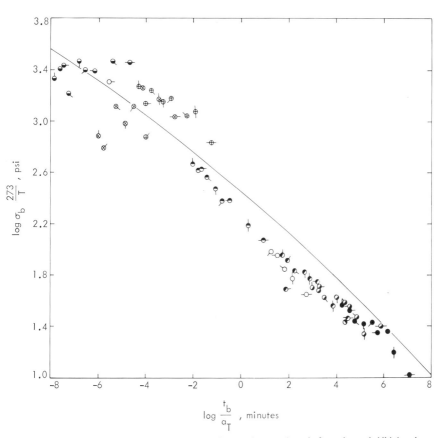

Figure 25 Dependence of stress-at-break on time-to-break for viton A-HV having $\nu_e = 16.7 \times 10^{-6}$ moles/cm³; $a_T = 1$ at 90°C. See Fig. 24 for definition of symbols. Data from Smith (8).

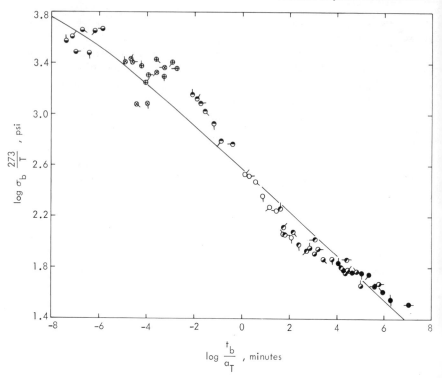

Figure 26 Dependence of stress-at-break on time-to-break for viton A-HV having $\nu_e = 28.6 \times 10^{-6}$ moles/cm³; $a_T = 1$ at 90°C. See Fig. 24 for definition of symbols. Data from Smith (8).

lowest degree of cross-linking (Fig. 24); the response of this vulcanizate tends to be parallel but falls below that observed with the more highly cross-linked vulcanizates. For the ϵ_b, t_b response, on the other hand, the typical bell shaped curves move to longer time scales and lower ϵ_b values as the ν_e value increases.

To test the applicability of the a_x shift, plots of log $\sigma_b(273/\nu_e T)$ versus log t_b were prepared (53). Normalization of σ_b by ν_e is equivalent to a vertical shift in the log σ, log t plane. The response for the vulcanizate with $\nu_e = 48.6 \times 10^{-6}$ moles/cm³, was taken as the reference and the remaining curves were shifted along the time scale until superposition was obtained. The same procedure was carried out for the ϵ_b, t_b data, except that here, no vertical shift preceded the shift along the time scale. The shift distances were found to vary somewhat but the average values of a_x conform to equation [26] except that the exponent is 7.7, the same as found with SBR (47). Hence, for SBR and viton A-HV at least, the a_x

factor appropriate to rupture is given by

$$a_x = \left(\frac{C}{\nu_e}\right)^{7.7} \qquad [27]$$

A master curve was drawn through the average superposed curve and the segment of the master curve appropriate to each set of data is shown in Figs. 24–37 as the full curves. Overall, it appears that a_x as given by equation [27] is applicable to all except the vulcanizate having the lowest degree of cross-linking; in this case, the σ_b, t_b response obtained experimentally falls below that predicted by equation [27]. This vulcanizate represents data, on a plot of σ_b versus ν_e which is characterized by a maximum in σ_b under certain test conditions, that fall to the left of the maximum. In this region, the degree of cross-linking is low and the values of both σ_b and ϵ_b may depend strongly on the fraction of network chains which can undergo deformation, which in turn is a strong function of M_n.

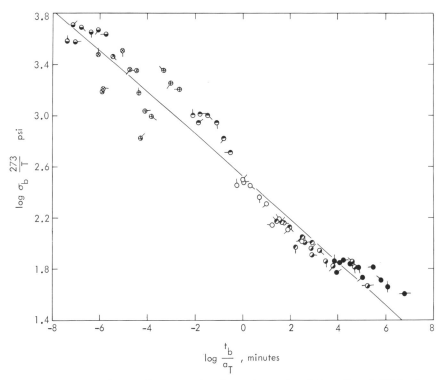

Figure 27 Dependence of stress-at-break on time-to-break for viton A-HV having $\nu_e = 48.6 \times 10^{-6}$ moles/cm³; $a_T = 1$ at 90°C. See Fig. 24 for definition of symbols. Data from Smith (8).

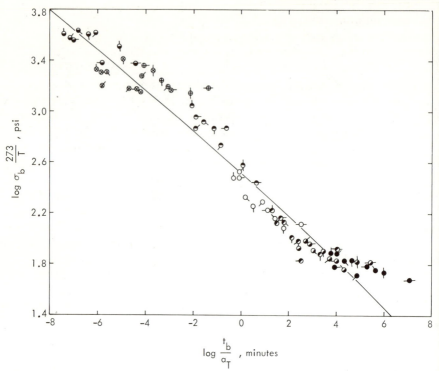

Figure 28 Dependence of stress-at-break on time-to-break for viton A-HV having $\nu_e = 61.4 \times 10^{-6}$ moles/cm³; $a_T = 1$ at 90°C. See Fig. 24 for definition of symbols. Data from Smith (8).

For the remaining vulcanizates, where ν_e varies from 16.7×10^{-6} to 335×10^{-6} moles/cm³, the σ_b, t_b data can be superposed quite well except perhaps in some cases where data obtained at the higher temperatures tend to fall above or to the right of the common response. With the ϵ_b, t_b data, the use of a_x permits superposition of the data up to the region of the maximum in ϵ_b. For the vulcanizate with the highest degree of cross-linking the ϵ_b, t_b response tends to fall slightly below the common response. This may be because it is difficult to obtain reliable values of ϵ_b using ring specimens when the vulcanizate has high modulus and low breaking strains.

The common responses obtained for both the σ_b, t_b and ϵ_b, t_b data are similar in shape to the results obtained with SBR. As a matter of fact, the data for both vulcanizates can be superposed. However, when the same C value is used in equation [27] for both elastomers, an additional horizontal shift, which is probably related to the effect of chemical structure on rupture behavior, is required. More data will be necessary on a variety of elastomers differing in chemical structure in order to evaluate the nature of this shift. It may be mentioned here that a_x as defined by equation [27] has also been shown to be applicable to the rupture properties of a series of fluorosilicone elastomers (53).

It is interesting to note that the common response for the ϵ_b, t_b data turns out to be linear over an extended range of reduced time when plotted as log λ_b versus log t_b. When this is true, λ_b can be expressed as

$$\lambda_b = b\left(\frac{t_b}{a_T a_x}\right)^{-a} \qquad [28]$$

where a is the slope and b is the intercept of the log-log plot. Equation [28] can be rearranged to yield

$$\lambda_b \epsilon_b{}^a = b(a_T R)^a \left(\frac{C}{\nu_e}\right)^{7.7a} \qquad [29]$$

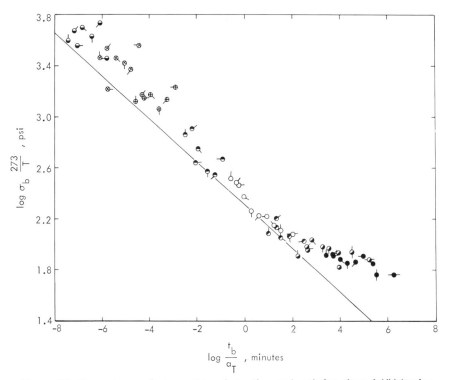

Figure 29 Dependence of stress-at-break on time-to-break for viton A-HV having $\nu_e = 118 \times 10^{-6}$ moles/cm^3; $a_T = 1$ at 90°C. See Fig. 24 for definition of symbols. Data from Smith (8).

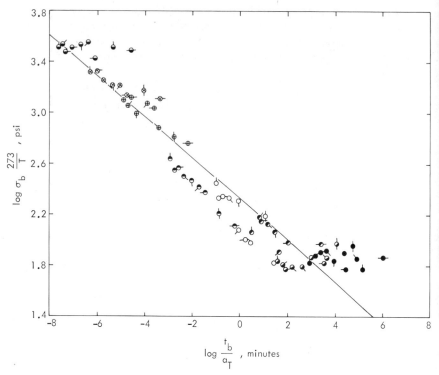

Figure 30 Dependence of stress-at-break on time-to-break for viton A-HV having $\nu_e = 337 \times 10^{-6}$ moles/cm^3; $a_T = 1$ at 90°C. See Fig. 24 for definition of symbols. Data from Smith (8).

and the left-hand side can be written as

$$\lambda_b \epsilon_b{}^a = \lambda_b^{1+a}\left(1 - \frac{1}{\lambda}\right)^a \qquad [30]$$

The factor $(1 - 1/\lambda)^a$ will be very nearly unity especially for large λ and small a and hence we can take $\lambda_b \epsilon_b{}^a = \lambda_b^{1+a}$, and equation [29] becomes

$$\lambda_b = b^{1/(1+a)}(a_T R)^{a/(1+a)}\left(\frac{C}{\nu_e}\right)^{7.7a/(1+a)} \qquad [31]$$

For the viton data, $C = 48.6 \times 10^{-6}$ moles/cm^3, $a = 0.069$, and $b = 6.66$, and hence

$$\lambda_b = \frac{4.11(a_T R)^{0.0648}}{\nu_e^{1/2}} \qquad [32]$$

which predicts that λ_b varies as $\nu_e^{-1/2}$ and furthermore predicts the temperature and strain rate dependence of this response as well. It is also predicted that $\lambda_b \to 0$ as $\nu_e \to \infty$, which indicates that this relationship must break down for large ν_e since λ_b cannot be less than unity. In order to test this relation, Figs. 38–40 show plots of λ_b versus $\nu_e^{-1/2}$ for viton as a function of temperature for three test rates. The points represent the experimental data and the full lines represent the prediction of equation [32]. Except for data representing the vulcanizate having the lowest degree of cross-linking which do not conform to equation [27] and hence to equation [32], and for test temperatures lower than that at which ϵ_b is at its maximum, i.e., temperatures for which equation [28] is no longer applicable, equation [32] appears to provide a useful description of the λ_b, ν_e dependence. A similar linear response of λ_b versus $\nu_e^{-1/2}$ has been observed for SBR data taken from the literature. The slope of the response predicted using equation [31] and the data in Fig. 23 were in excellent agreement with the measured slope (53). A linear dependence

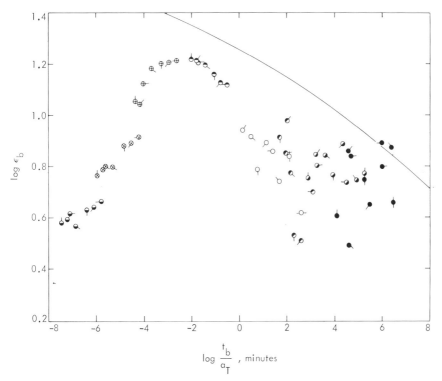

Figure 31 Dependence of strain-at-break on time-to-break for viton A-HV having $\nu_e = 4.6 \times 10^{-6}$ moles/cm³; $a_T = 1$ at 90°C. See Fig. 24 for definition of symbols. Data from Smith (8).

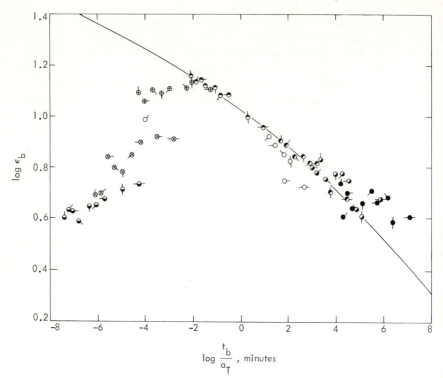

Figure 32 Dependence of strain-at-break on time-to-break for viton A-HV having $\nu_e = 16.7 \times 10^{-6}$ moles/cm³; $a_T = 1$ at 90°C. See Fig. 24 for definition of symbols. Data from Smith (8).

of λ_b on $\nu_e^{-1/2}$ has been reported for EPM (54) and for natural rubber (55).

In the foregoing treatment of the viton data, values of ν_e used for shifting the rupture data were derived from the Mooney-Rivlin parameter $2C_1(1)(T_0/T)$, obtained from 1 minute isochronal stress-strain data (56). Smith observed (56) that $2C_1(1)(T_0/T)$ was independent of temperature for stress-strain data obtained in the temperature range where $\epsilon_b < (\epsilon_b)_m$. At the same time, $2C_2(1)(T_0/T)$ decreases with temperature over the same range, and in addition was found to be strongly dependent on the value of $2C_1(1)(T_0/T)$. It is of interest to determine if the same a_x obtained from the rupture response can be used to superpose $2C_2(1) \times (T_0/T)$ data. In analogy to normalizing σ_b by ν_e, log $(2C_2(1)/2C_1(1))$ is plotted against log $t/a_T a_x$ and shown in Fig. 41 for those data where $2C_1(1)(T_0/T)$ is independent of temperature. The values of a_x were calculated from equation [27]. It can be seen that a single response results when a_x is used as a reduction variable along the time scale. In contrast to the rupture behavior, however, the data obtained with the vulcanizate of the lowest degree of cross-linking are also superposed in such a plot. Figure 41 would indicate that $2C_2(1)$ is predominantly viscoelastic in origin and that small values of this parameter can be obtained either at long times or with vulcanizates having a large ν_e value. The figure indicates, in addition, that there is apparently no tendency for the ratio $2C_2(1)/2C_1(1)$ to approach a constant limiting value.

Smith also reported values of the 1 minute isochronal constant strain rate modulus as a function of temperature and ν_e (8, 56). It was found (53) that values of $F(1)$ obtained as a function of ν_e can be superposed when plotted in the form $\log (F(1)/\nu_e)(T_0/T)$ versus $\log t/a_T a_x$. However, this superposition is again only applicable for the time scale over which $2C_1(1)(T_0/T)$ is constant; this corresponds roughly to the time scale for

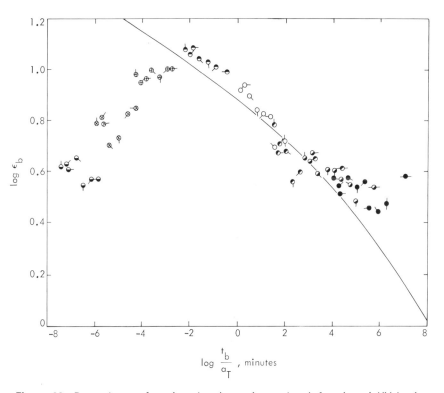

Figure 33 Dependence of strain-at-break on time-to-break for viton A-HV having $\nu_e = 28.6 \times 10^{-6}$ moles/cm^3; $a_T = 1$ at 90°C. See Fig. 24 for definition of symbols. Data from Smith (8).

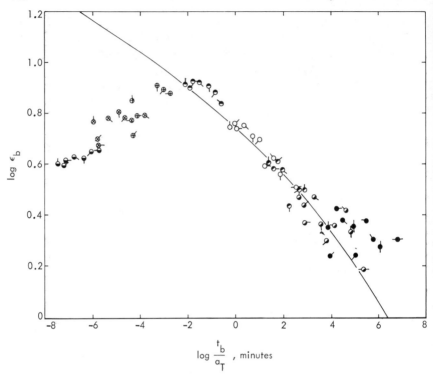

Figure 34 Dependence of strain-at-break on time-to-break for viton A-HV having $\nu_e = 48.6 \times 10^{-6}$moles/cm^3; $a_T = 1$ at 90°C. See. Fig. 24 for definition of symbols. Data from Smith (8).

which ϵ_b is a decreasing function of time. In addition, since $3[2C_1(1) + 2C_2(1)] = F(1)$ a plot of $\log\{[3[2C_1(1) + 2C_2(1)]/\nu_e](T_0/T)\}$ versus $\log t/a_T a_x$ was prepared and found identical, apart from a small vertical shift, to the $\log F(t)T_0/\nu_e T$ versus $\log t/a_T a_x$ response. The required small vertical shift probably means that the Mooney-Rivlin equation departs from linearity at small strains.

According to Figs. 24–30, the reduced stress-at-break can be taken as a universal function of the reduced time-to-break, at least to an approximation representative of scatter in typical data of this type, and hence we can write (47)

$$\frac{\sigma_b}{\nu_e}\frac{T_0}{T} = k\left(\frac{t_b}{a_T a_x}\right)$$ [33]

Likewise, Figs. 31–37 indicate that ϵ_b for strains not too close to $(\epsilon_b)_m$ can also be taken as approximately a universal function of the reduced time-to-break, and hence

$$\epsilon_b = h\left(\frac{t_b}{a_T a_x}\right) \qquad [34]$$

At long reduced times where equation [1] is applicable, the stress-strain response can be written as

$$\frac{\sigma T_0}{\nu_e T} = g(t)f(\epsilon) \qquad [35]$$

where $g(t) \equiv G(t)T_0/\nu_e T$. Since as $t \to \infty$, $G(t) \to G_e$, $g_e = RT_0$.

The rupture point is the terminus of the stress-strain response, and thus equation [35] is also applicable to failure. Hence

$$\frac{\sigma_b T_0}{\nu_e T} = g(t_b)f(\epsilon_b) \qquad [36]$$

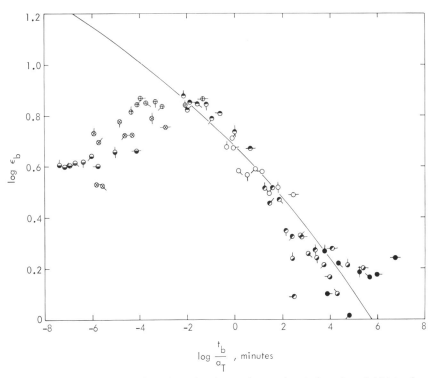

Figure 35 Dependence of strain-at-break on time-to-break for viton A-HV having $\nu_e = 61.4 \times 10^{-6}$ moles/cm^3; $a_T = 1$ at 90°C. See Fig. 24 for definition of symbols. Data from Smith (8).

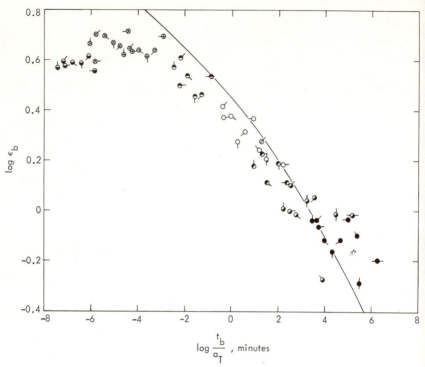

Figure 36 Dependence of strain-at-break on time-to-break for viton A-HV having $\nu_e = 118 \times 10^{-6}$ moles/cm³; $a_T = 1$ at 90°C. See Fig. 24 for definition of symbols. Data from Smith (8).

Substituting equations [33] and [34] for the appropriate quantities in equation [36] yields

$$k\left(\frac{t_b}{a_T a_x}\right) = g(t_b) f\left[h\left(\frac{t_b}{a_T a_x}\right)\right] \qquad [37]$$

To the extent that the strain function f can be assumed to be a universal function, equation [37] implies that g will be a universal function as well. Therefore we can write

$$\frac{\sigma_b T_0}{\nu_e T} = g\left(\frac{t_b}{a_T a_x}\right) f(\epsilon_b) \qquad [38]$$

This expression indicates that a plot of log $\sigma_b T_0 / \nu_e T$ versus log ϵ_b, i.e., the reduced failure envelope, will also be a universal function. It has already been shown that this is indeed the case (5, 32, 38). In addition, the fact that the reduced failure envelope is independent of the chemical structure of an elastomer, again to within an approximation represented by the scatter in such data, indicates that the functions g and f are not very sensitive to chemical structure.

For the case where ϵ_b is in the long time region but close in magnitude to $(\epsilon_b)_m$ the function f becomes dependent on parameters other than ϵ_b. The most important parameter seems to be n, the number of statistical units per network chain. For this more general case, equation [35] can be written

$$\frac{\sigma_b T_0}{\nu_e T} = g(t_b)f(\epsilon_b, n) \qquad [39]$$

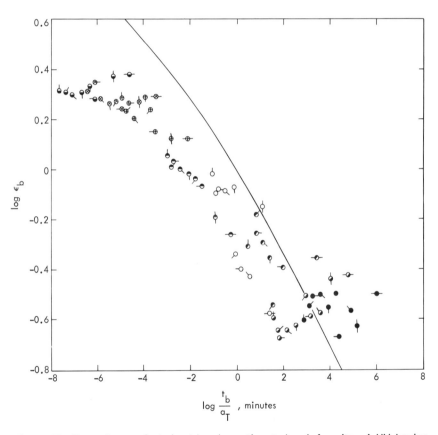

Figure 37 Dependence of strain-at-break on time-to-break for viton A-HV having $\nu_e = 337 \times 10^{-6}$ moles/cm^3; $a_T = 1$ at 90°C. See Fig. 24 for definition of symbols. Data from Smith (8).

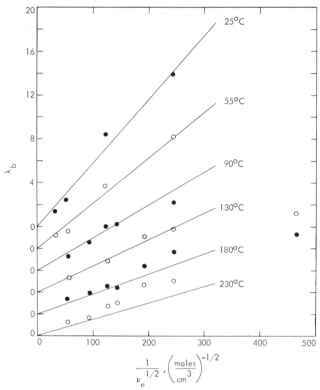

Figure 38 Temperature dependence of λ_b on $\nu_e^{-1/2}$ for viton A-HV obtained at a strain rate of 0.00952 min^{-1}. Data from Smith (8).

However, Figs. 24–30 indicate that equation [33] is still valid, and hence

$$k\left(\frac{t_b}{a_T a_x}\right) = g(t_b)f(\epsilon_b, n) \qquad [40]$$

Thus, for values near $(\epsilon_b)_m$, ϵ_b is no longer a universal function of the reduced time-to-break, since f depends on n as well as on ϵ_b. In this region, instead of a universal curve, a plot of ϵ_b against reduced time-to-break should yield a family of curves each one of which is related to the value of the parameter n. However, to the extent that equation [2] or [3] can represent experimental data, a plot of $f(\epsilon_b, n)$ rather than ϵ_b versus $t_b/a_T a_x$ should lead to a universal curve, i.e.,

$$f(\epsilon_b, n) = r\left(\frac{t_b}{a_T a_x}\right) \qquad [41]$$

Assuming this to be the case, equation [40] becomes

$$k\left(\frac{t_b}{a_T a_x}\right) = g(t_b) r\left(\frac{t_b}{a_T a_x}\right) \tag{42}$$

or $g(t_b)$ must again be universal. Hence

$$\frac{\sigma_b T_0}{\nu_e T} = g\left(\frac{t_b}{a_T a_x}\right) f(\epsilon_b, n) \tag{43}$$

Now, a plot of log $\sigma_b T_0/\nu_e T$ versus log ϵ_b, i.e., the reduced failure

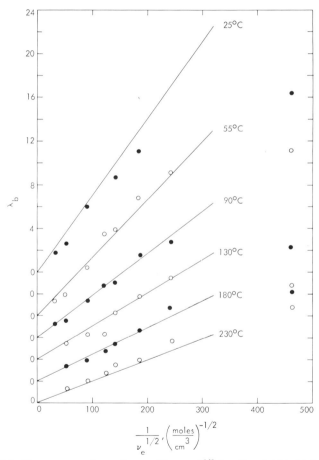

Figure 39 Temperature dependence of λ_b on $\nu_e^{-1/2}$ for viton A-HV obtained at a strain rate of 0.238 min^{-1}. Data from Smith (8).

envelope, will consist of a family of curves each one of which is charac-terized by its particular n value. This family will converge to a single master curve for those conditions under which $f(\epsilon, n) \rightarrow f(\epsilon)$, i.e., when ϵ_b is small and/or n is large.

The preceding discussion has indicated that for SBR and viton, the effect of ν_e on σ_b, t_b and ϵ_b, t_b is simply to shift the response along the time scale without, however, changing the shape of the response. This is true only over a limited region of the time scale, specifically the time scale over which ϵ_b is a decreasing function of time. In addition, for viton the same a_x shift is applicable to the small strain behavior as well.

These results imply that the shape of the physical property surface for strains up to the maximum in ϵ_b is independent of the degree of

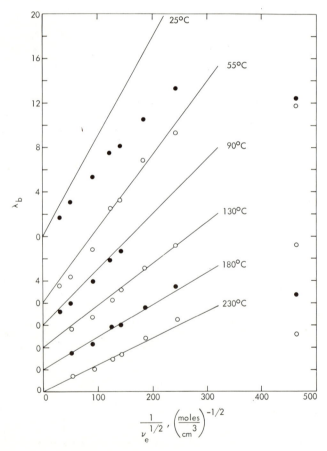

Figure 40 Temperature dependence of λ_b on $\nu_e^{-1/2}$ for viton A-HV obtained at a strain rate of 9.52 min⁻¹. Data from Smith (8).

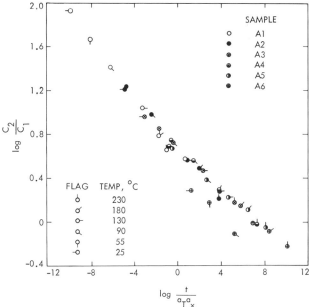

Figure 41 Dependence of the ratio C_2/C_1 on reduced time for viton A-HV; $a_T = 1$ at 90°C and $a_x = 1$ at $C_1 = 10$ psi. The samples had the following ν_e values (moles $\times 10^{-6}$/cm^3): A1 = 4.6, A2 = 16.7, A3 = 28.6, A4 = 48.6, A5 = 61.4, and A6 = 118. Data from Smith (56).

cross-linking provided that ν_e is not too small. Changes in ν_e merely serve to shift the response along the time scale by the factor a_x. To the extent that the same dependence of a_x on ν_e, equation [27], is applicable to other elastomers, the shape of the tensile physical property surface in the long time region will be unique and surfaces representing several elastomers will be superposable by a translation along the time scale. In addition, ν_e can be used as a convenient variable to expand the experimental time scale, i.e., to obtain data for times or temperatures that may not be accessible experimentally. For example, if significant degradation occurs at an elevated temperature for an elastomer having a given ν_e value, then equivalent information in the absence of degradation can be obtained by studying the behavior at lower temperatures for specimens having higher ν_e values; these data can then be shifted along the time axis to yield the desired information.

3.3 Breaking Energy of Rubber

3.3.1 Energy Density and Energy Loss at Break

In a recent study, Grosch and associates (57) found that the energy required to break a specimen, U_B, is related to the hysteresis loss at

break, H_B, by
$$U_B = KH_B^{2/3} \qquad [44]$$

When both U_B and H_B are expressed in joules per cubic centimeter, K has the value 4.1. Alternatively, equation [44] can be written as

$$U_B = K^3 h_B{}^2 \qquad [45]$$

where h_B is the hysteresis loss ratio at break, and is defined as H_B/U_B. Since K^3 is the value of U_B when $h_B = 1$, which corresponds to the limiting case where all the applied work invested in deforming the specimen is dissipated, equation [45] can also be written as (59)

$$U_B = U_{B(\max)} h_B{}^2 \qquad [46]$$

The value of U_B is obtained from the area under the stress-strain response while H_B is taken as the area between the extension and retraction curves, until just before break occurs, both obtained on fresh specimens and tested under the same conditions used to evaluate U_B.

Equation [44] was found to hold for several rubbers such as natural rubber, isomerized natural rubber, SBR, butyl, cis-1,4-polybutadiene, and acrylonitrile-butadiene when tested at temperatures between -50 and $160°C$. In addition, the equation was also found to be applicable to natural rubber and SBR vulcanizates in the swollen state as well as to specimens of these two elastomers which had been aged for various periods of time at elevated temperatures (57).

In subsequent studies (58, 59), this earlier work was extended to an investigation of the effect of temperature, at a single strain rate, on the values of U_B and H_B for the same elastomers used by Grosch (57). It was pointed out that equation [44] furnishes a sensitive criterion for failure in that a different U, H response is obtained for strains less than ϵ_b (59). Figure 42 shows data for an acrylonitrile-butadiene rubber. The U, H response for strains less than ϵ_b and for temperatures in the range $50–140°C$ is shown as the dashed lines and filled points while the U_B, H_B response observed at rupture is depicted by the full line and unfilled circles (59). It is evident that U is everywhere greater than H. Rupture occurs when $\epsilon \to \epsilon_b$ and the U, H response (dashed lines) intersects the U_B, H_B response (full line).

For the acrylonitrile-butadiene rubber, the rupture data do not follow equation [44] when the test temperature approaches the vulcanization temperature. For SBR and cis-1,4-polybutadiene polymers, this expression was applicable over the whole range of test temperatures employed. However, for natural rubber, agreement was obtained only for test temperatures less than $60°C$ or greater than $140°C$. This was attributed to the occurrence of strain induced crystallization (59).

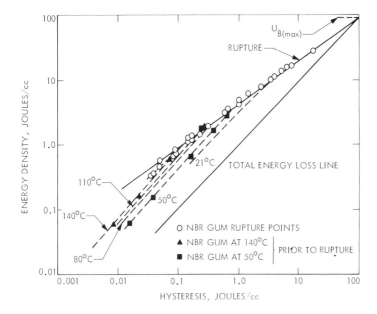

Figure 42 Dependence of energy density on hysteresis for an acrylonitrile-butadiene gum vulcanizate. Data from Harwood and Payne (59).

Although the form of equation [44] or [46] was found applicable in this more extensive study, contrary to the results of Grosch (57), it was found that the value of K^3 varies with the rubber tested. Values of the parameter K^3 are listed in Table 2. The second column contains the values reported in ref. 58 while the third column lists the corresponding values reported in ref. 59. It may be seen that the parameter differs appreciably in the case of acrylonitrile-butadiene and butyl rubber. Although no

Table 2 Selected Constants for Various Rubbers (58, 59)

Rubber	K^3 (J/cm³) (58)	K^3 (J/cm³) (59)	$U_{B(max)}$	$\dfrac{\Delta H}{R}$ (J-°C/cm³)
SBR	61	60	61	5.7×10^2
Acrylonitrile-butadiene	207	100	97	8.3×10^2
cis-1,4-Polybutadiene	88	91	88	—
Butyl	36	12.2	49	8.3×10^2
Isomerized natural and natural rubber	124	119	124	5.5×10^2

information was provided on the dependence, if any, of K^3 on the degree of cross-linking, the different values reported for these two rubbers may reflect differences in the degree of cross-linking of the vulcanizates used in the two separate studies (58, 59).

Using an equation of Chang's (60) it can easily be shown that $K = k v_e^{1/3}$, where k is a constant, at least for long reduced times or for high test temperatures (53) i.e., K depends implicitly on v_e.

3.3.2 Effect of Temperature on Energy Density at Break

For the noncrystallizable elastomers, it was found that the dependence of the energy required to break a specimen, U_B, on temperature was expressible by means of an equation of the form

$$U_B = U_{B(\max)} \exp\left[\frac{\Delta H}{R}\left(\frac{1}{T} - \frac{1}{T_g}\right)\right] \qquad [47]$$

where ΔH is an apparent activation energy (58, 59). This expression predicts $U_B \to U_{B(\max)}$ as $T \to T_g$. Values of $U_{B(\max)}$ and $\Delta H/R$ are included in Table 2. It was pointed out that the magnitudes of $U_{B(\max)}$ rank in the same order as the cohesive energy densities of the elastomers. For the crystallizable natural rubber and cis-1,4-polybutadiene, no simple relationship between U_B and T was observed, even though equation [44] was found applicable to the latter elastomer over the whole temperature range studied (59).

Considering the response at high temperatures, where $U_B \gg H_B$, an attempt was made to calculate a minimum value for the energy density at break, $U_{B(\min)}$, using two methods, one based on considerations of the critical stress required to extend a hole and the other based on consideration of tearing energy. Both methods yielded estimates for $U_{B(\min)}$ which were roughly the same as the values obtained experimentally at high temperatures. Assuming the existence of a lower limit for the energy density, equation [47] is rewritten as (59)

$$U_B = U_D + U_{B(\min)} = U_{B(\max)} \exp\left[\frac{\Delta H}{R}\left(\frac{1}{T} - \frac{1}{T_g}\right)\right] \qquad [48]$$

where U_D is defined as the energy of deformation which becomes negligible compared to $U_{B(\min)}$ at temperatures greater than about 200°C above T_g.

3.3.3 Effect of Swelling on Energy Density at Break

When obtained from data at a single test temperature and strain rate, the energy density at break, normalized to unit volume of the dry rubber and denoted by U_{Bd}, was generally strongly dependent on the degree of swelling (58) when bromobenzene was used as the swelling agent. For SBR, acrylonitrile-butadiene, and natural rubber, the U_{Bd}, v_r response is linear at small v_r such that $U_{Bd} \to 0$ as $v_r \to 0$, but increases sharply and in a nonlinear fashion as $v_r \to 1$. Here, v_r is the volume fraction of rubber in the swollen vulcanizate. The value of v_r at which deviations from linearity occur and the magnitude of the deviation depend on the type of rubber. For cis-1,4-polybutadiene, however, the U_{Bd}, v_r response was linear over the whole range of v_r. Except for natural rubber, when plotted in log-log coordinates, the U_{Bd}, v_r response for the different elastomers could be superposed by a horizontal shift along the log v_r axis; the magnitude of the shift was found to be a linear function of $T - T_g$. The failure of natural rubber to superpose in this manner was attributed to the occurrence of crystallization. For the superposed response, there is apparently no limiting minimum value of U_{Bd} at high degrees of swelling (58).

In addition to studying the U_{Bd}, v_r response, the dependence of σ_b and λ_b on v_r was also reported. For high degrees of swelling, the extension ratio at break based on the swollen dimensions, λ_{Bs}, is a linear function of $v_r^{1/3}$ such that $\lambda_{Bs} \to 0$ as $v_r^{1/3} \to 0$. For cis-1,4-polybutadiene, the constant of proportionality is identical to λ_{Bd}, the extension ratio at break based on the dry dimensions. For the remaining three elastomers, the constant is less than λ_{Bd}; however, for all four elastomers, the constant had the same numerical value and this was attributed to the fact that the vulcanizates had the same degree of cure (61). It is evident that this linear relationship must break down at low v_r since the minimum value of λ_{Bs} is unity. No attempt was made to superpose the λ_{Bs}, v_r response, although the similar shapes of the curves obtained with the elastomers would indicate that superposition is possible. When the λ_{Bs}, v_r data were replotted in log-log coordinates, superposition was obtained by shifting horizontally, but the magnitudes of the shifts were much smaller than those required to superpose the U_{Bd}, v_r response (53).

It was also demonstrated that σ_b, based on the unswollen, undeformed cross section obtained as a function of v_r, could be superposed when plotted in log-log coordinates by shifting horizontally along the v_r axis. This was found to be the case for all the elastomers except natural rubber. The shift factors required were the same as those used to superpose the U_{Bd}, v_r response (61).

These data indicate that U_{Bd} and λ_{Bs} decrease to limiting values of zero and unity, respectively, while σ_b decreases to a constant limiting value greater than zero as the degree of swelling increases. In addition, the U_{Bd}, v_r and σ_b, v_r response for noncrystallizing elastomers can be superposed by a horizontal shift along the log v_r axis. The magnitude of the shift is independent of v_r and is a linear function of $T - T_g$. For the λ_{Bs}, v_r response, superposition can also be obtained but the values of the shift factors are lower (53). It is interesting to note that shift factors which account for the effect of swelling on the time scale of an experiment have been proposed and used to superpose stress relaxation data obtained at small strain (62, 63). Hence it may be possible to convert data such as the σ_b, v_r response to an equivalent σ_b, t_b response.

4 MODULUS OF FILLED ELASTOMERS

4.1 Small Strain Behavior

It is well known that the presence of inclusions in a matrix generally gives rise to a change in the stiffness or modulus. When the inclusions have a higher modulus than the matrix, the modulus of the composite is greater than that of the matrix; when the inclusions have a lower modulus, the reverse is true. Several attempts have been made, by way of theoretical analysis, to develop analytical expressions which permit the estimation of the elastic modulus of a composite from a knowledge of the properties of the individual components. These attempts have followed two separate but related approaches. On the one hand, the relative viscosity of a suspension as a function of concentration of suspended particles is calculated from hydrodynamic theory. Generally, it is assumed that the suspension is Newtonian and further that slow flow prevails so that inertial terms can be neglected, which results in linear rather than nonlinear differential equations. The basic equations of hydrodynamics and of elasticity can be interconnected if the components of velocity and Newtonian viscosity in the former are replaced by the components of displacement and modulus in the latter. Hence, expressions relating relative viscosity, η_r, of a suspension to the volume fraction of filler, ϕ, can be used directly to calculate the dependence of the relative modulus on ϕ by simply substituting relative modulus for relative viscosity. Solutions based on hydrodynamic theory generally assume constant volume and this one-to-one correspondence between relative viscosity and relative modulus is strictly valid only when Poisson's ratio is $\frac{1}{2}$.

On the other hand, the dependence of the relative modulus on ϕ has been treated directly in terms of elasticity theory. Generally, these calculations assume that the composite is isotropic, macroscopically homogeneous, and Hookean. The results obtained from application of elasticity theory are limited to small deformations. However, the elastic solutions do not assume constant volume, and as will be indicated later, volume changes are explicitly taken into account by the appearance of Poisson's ratio.

Both the hydrodynamic and elastic solutions are limited, generally, to small particle concentrations in that the two-phase system is considered so dilute that particle-particle interaction can be neglected. In addition, with a few exceptions, the particles are assumed to be spherical in shape.

A third approach is based on what may be termed empirical or semiempirical curve fitting. This approach recognizes that the solution, by either hydrodynamic or elastic theory, for the case of finite ϕ, which is the range of great practical importance, is probably hopelessly complex or impossible. One reason for this is that in the range of finite concentration, where particle-particle interaction occurs, the response of the composite depends not only on the concentration of inclusions but also on their statistical and geometrical distribution within the composite. This combination of statistics and geometry poses a difficult problem which has yet to be solved.

In the following sections we review briefly, rather than exhaustively, some of the expressions obtained using hydrodynamic theory which relate η_r to ϕ. Following this, the expressions obtained from elastic theory are presented. Finally, some of the more useful empirical descriptions are given.

4.1.1 Equations Based on Hydrodynamic Theory

Einstein (64) was apparently the first to derive from the equations of hydrodynamics, a relationship for the shear viscosity of a very dilute suspension of rigid particles. In the derivation, the assumptions were made that (a) the suspension is so dilute that particle-particle interaction can be neglected; (b) the motion of the suspending liquid is slow; (c) the particles are rigid and spherical; (d) the liquid wets the particles and no slippage occurs; and (e) the volume is constant. The final result is given by

$$\eta_r = 1 + 2.5\phi \qquad [49]$$

These assumptions demonstrate that this expression is an idealized limiting law. According to equation [49], the presence of a rigid spherical filler increases the relative viscosity. In addition, η_r depends only on the volume fraction of filler; it is independent of both the particle size and the particle size distribution.

When slippage occurs, the relationship becomes (65)

$$\eta_r = 1 + \phi \qquad [50]$$

A comparison between equations [49] and [50] demonstrates that the nature of the boundary conditions assumed to exist at the particle surface plays an important role in the results that are obtained.

For rigid spherical particles that do not interact with the solvent or stick together, the coefficient of ϕ (2.5) seems to be well established. For rigid asymmetric particles, the coefficient of ϕ is greater than 2.5; the value of the coefficient depends on the details of the particle shape as well as on whether Brownian motion occurs. Hence for asymmetric particles, there is a difference between the behavior of microscopic particles (Brownian motion) and macroscopic particles (no Brownian motion).

When the inclusions are spherical voids for which surface tension effects can be ignored, Guth and Mark obtain for the relative viscosity (66)

$$\eta_r = 1 - \phi \qquad [51]$$

Thus the relative viscosity decreases as the void content increases. As with the rigid spherical filler, the viscosity is independent of the size and size distribution of the voids.

For the intermediate case where the viscosity of the dispersed phase is comparable to that of the medium, Taylor derived the following equation (67):

$$\eta_r = 1 + 2.5\left(\frac{2\eta_1 + 5\eta_2}{5\eta_1 + 5\eta_2}\right)\phi \qquad [52]$$

where η_1 and η_2 are the viscosities of the medium and dispersed phase, respectively. Equation [52] indicates that the viscosity of a suspension is increased by the presence of liquid particles. Taylor's result assumes that the dispersed phase is composed of spherical particles that remain spherical during flow and further that no slippage occurs. It is interesting to note that when $\eta_2 \to \infty$, i.e., the dispersed particles are rigid compared to the medium, equation [52] reduces to the Einstein result. On the other hand, when $\eta_2 = 0$, which corresponds to the case of slippage, equation [50] is obtained.

Several attempts have been made to extend the hydrodynamic treatment beyond the linear range. For finite concentration, the relative viscosity can be expressed as a power series in ϕ:

$$\eta_r = 1 + \alpha_1\phi + \alpha_2\phi^2 + \alpha_3\phi^3 + \cdots \qquad [53]$$

For rigid spheres, α_1 is the Einstein coefficient 2.5 and the extension to finite concentrations involves the estimation of the numerical coefficients, $\alpha_2, \alpha_3, \ldots$. For example, Guth and Gold (68) estimate the value of α_2 from a consideration of the effects of doublets on flow and obtain

$\alpha_2 = 14.1$. Simha (69), taking into account additional factors, finds $\alpha_2 = 12.6$, whereas Vand (70) obtained the value $\alpha_2 = 7.35$, which however, becomes 10.05 when corrections for the true lifetime of a doublet are made (71). Both higher and lower values for α_2 have been proposed but experimental evidence on the viscosity of suspensions of spherical particles suggests that values of α_2 in the range of about 10–14 are probably correct. Little work has been carried out on the evaluation of the higher coefficients.

For rigid rodlike particles, Guth (72) derived the equation

$$\eta_r = 1 + 0.67f\phi + 1.62f^2\phi^2 + \cdots, \qquad f \gg 1 \qquad [54]$$

where f is a shape factor equal to the length/diameter ratio.

Robinson (73) carried out a series of experiments on the viscosity of suspensions of closely sized glass beads in several media. He considers that the total volume of liquid medium present is not the appropriate measure for the true liquid volume because some of the liquid will be immobilized within clusters or particle aggregates. Based on this assumption, the effective volume fraction of filler ϕ_{eff} in a suspension is taken as

$$\phi_{\text{eff}} = \frac{V_p}{V_T - V_p/\phi_m} = \frac{\phi}{1 - \phi/\phi_m} \qquad [55]$$

where V_p is the volume of dispersed phase, V_T is the total volume of the suspension, and ϕ_m is the maximum volume fraction of filler to which the particles can pack. The effective volume occupied by the aggregates is given by the ratio ϕ/ϕ_m. If in equation [49], the effective volume fraction of filler as given in equation [55] is used instead of ϕ itself, then

$$\eta_r = 1 + \frac{2.5\phi}{1 - \phi/\phi_m} \qquad [56]$$

As $\phi \to 0$, equation [49] is recovered while for $\phi \to \phi_m$, $\eta_r \to \infty$. Hence, this expression predicts that the viscosity will become infinite when the volume fraction of particles becomes equal to the maximum volume fraction to which the particle can pack. Robinson estimated ϕ_m independently from sedimentation experiments and good accord was found with the values of ϕ_m estimated from viscosity measurements. However, the numerical constant was observed to vary from 3 to 5, whereas equation [56] predicts a value of 2.5.

Another approach for obtaining an η_r, ϕ relation applicable to finite ϕ has been developed independently by Vand (74), Brinkman (75), and Roscoe (76) and is based on a simple method of estimating the dependence of one variable at finite values of another if the result at infinite dilution is known. Starting with equation [49], which represents the dependence of η_r on ϕ at infinite dilution, and assuming that the suspension behaves hydrodynamically as a continuous and homogeneous medium with respect to the addition of another spherical particle, the following expression is easily derived:

$$\eta_r = (1 - \phi)^{-2.5} \qquad [57]$$

Justification for the assumption that a suspension can be considered as a continuous and homogeneous medium with respect to the addition of another particle is supplied by the study of Fidleris and Whitmore (77). They observed that the resistance encountered by a falling sphere in an aqueous suspension of nonsettling spheres is unaffected by the size of the suspended spheres. This is true even when the falling sphere is the same size as the suspended spheres. This result indicates that the approach of Vand, Brinkman, and Roscoe is reasonable.

Roscoe made the important point, as did Robinson, that for high ϕ values, where aggregation of the suspended spherical particles would be expected to occur, a fraction of the liquid medium will be effectively entrapped in the interstitial space within the aggregate; this effect therefore reduces the fraction of free liquid and hence increases the effective concentration of particles. Assuming the most dense regular packing of spheres, face centered cubic packing with $\phi_m = 0.74$, Roscoe defines the effective volume fraction of filler to be

$$\phi_{\text{eff}} = \frac{\phi}{\phi_m} \qquad [58]$$

which differs from Robinson's definition (equation [55]). Combining equation [58] with equation [57] yields (76)

$$\eta_r = \left(\frac{\phi}{\phi_m - \phi}\right)^{2.5} \qquad [59]$$

Roscoe demonstrated that the data of Ward and Whitmore (78) on the η_r, ϕ behavior of closely sized poly(methyl methacrylate) spheres could be represented by this expression. Landel (79) generalized equation [59] by assuming ϕ_m to be a parameter whose value depends on the packing characteristics of the particular material used. Using values of ϕ_m estimated from sedimentation experiments, Landel was able to fit η_r, ϕ data on several suspensions containing particles having a range of measured ϕ_m from 0.31 to 0.85.

It is interesting to note that equation [59] does not reduce to the Einstein equation in the limit of infinite dilution, and hence it must break down as $\phi \to 0$. In addition, it may be mentioned that it is apparently not permissible to substitute the value of the effective volume fraction of filler directly into equation [57]. On going through the derivation with ϕ_{eff} defined by equation [58], the following expression is deduced (53):

$$\eta_r = (1 - \phi)^{-2.5/\phi_m} \tag{60}$$

in which ϕ_m enters as an exponent.

Another widely used expression was derived by Mooney (80). Assuming that the particles are monodisperse spheres, Mooney considers that a suspension containing a volume fraction $\phi = \phi_1 + \phi_2$ is made by two successive additions. First, ϕ_1 spheres are added to the solvent and this increases the viscosity from η_0 to η_1. The second addition, ϕ_2, is made to the suspension containing ϕ_1, which increases the viscosity from η_1 to η_2. Mooney takes for the effective volume fractions

$$\phi_{1\text{eff}} = \frac{\phi_1}{1 - k\phi_2} \quad \text{and} \quad \phi_{2\text{eff}} = \frac{\phi_2}{1 - k\phi_1} \tag{61}$$

The term $k\phi_2$ is the fraction of liquid medium which has been effectively tied up by the particles from the second addition, and similarly for the term $k\phi_1$. Here, k is called the crowding factor and is in effect a measure of the actual or effective volume occupied by the particles, i.e., $k = 1/\phi_m$. According to equation [61], only mutual crowding of the particles is considered. If self-crowding were included, the denominators in the equation would be $1 - k(\phi_1 + \phi_2) = 1 - k\phi$. Functional analysis provides a relationship of the form

$$\ln \eta_r = \frac{2.5\phi}{1 - k\phi} \tag{62}$$

which reduces to the Einstein result when $\phi \to 0$.

Based on considerations of regular packing of equal sized spheres, Mooney takes as the lower bound of packing the simple cubic array characterized by $\phi_m = 0.52$. The upper bound is taken as face centered cubic with $\phi_m = 0.74$. Hence the limits of k are assumed to be $1.35 \le k \le 1.93$.

Mooney used experimental data published by Vand (70), Eilers (81), and Robinson (73) to test equation [62]. Good fits were obtained and the ϕ_m values required for fit were 0.70, 1.33, and 1.20, respectively. For any system $\phi_m \le 1$, and values of ϕ_m greater than unity were attributed to polydispersity of the suspended particles.

For the case of polydisperse particles, Mooney obtained the general expression

$$\ln \eta_r = 2.5 \left[\sum_{i=1}^{n} \frac{\phi_i}{\sum_{j=1}^{n} \lambda_{ji}\phi_i} \right] \tag{63}$$

where λ_{ji} are the crowding factors appropriate to components j and i which in general are expected to be functions of the index. In order to use this expression, the functional form of λ must be known; in general this information is not available and as a first approximation, Mooney, by an expansion of equation [63] to the second power of ϕ, obtained

$$\ln \eta_r = \frac{2.5\phi}{1 - \lambda\phi} \qquad [64]$$

which has the same form as equation [62]. The value of λ is expected to be greater than k.

It may be pointed out that if the viscosity of a suspension is independent of the order of addition of the particles, and if self- as well as mutual crowding is considered, then equation [62] can be written as

$$\ln \eta_r = \frac{2.5\phi}{1 - \phi/\phi_m} \qquad [65]$$

which is valid for both mono- and polydisperse particles (53). Here, ϕ_m is the appropriate maximum packing fraction for the particles.

Using a cell model, Simha considers the existence of aggregates in which surrounding particles effectively shield a central particle from interaction with particles other than its nearest neighbors (69). He obtains for the relative viscosity

$$\eta_r = 1 + 2.5\phi\left[\frac{4(1 - y^7)}{4(1 + y^{10}) - 25y^3(1 + y^4) + 42y^5}\right] \qquad [66]$$

where $y = \phi^{1/3}/(f - \phi^{1/3})$. When $y \to 0$, this expression reduces to the Einstein equation; on the other hand, when $y \to 1$, $\eta_r \to \infty$ and this will only occur when $f = 2\phi^{1/3}$. This condition implies that when $\eta \to \infty$, $\phi \to \phi_m$, and hence $f = 2\phi_m^{1/3}$. This also corresponds to the maximum value that f can attain. Simha considers f to be a slowly increasing function of ϕ in the range $1 < f < 2$ with the upper limit given by $f = 2\phi_m^{1/3}$.

On expansion of equation [66], Simha obtains for small ϕ

$$\eta_r = 1 + 2.5\phi\left[1 + \frac{25\phi}{4f^3} + \cdots\right] \qquad [67]$$

whereas for high ϕ

$$\eta_r = 1 + \frac{54}{5f^3}\left[\frac{\phi^2}{(1 - \phi/\phi_m)^3}\right] \qquad [68]$$

Thus far, the equations that have been discussed were derived from the application of hydrodynamic theory. For finite ϕ where hydrodynamic theory alone is inadequate, several empirical equations have been proposed. One such equation has been described by Eilers (81, 82). He found that the η_r, ϕ response of suspensions containing hard polydisperse spherical bitumen particles could be described by an equation proposed by Van Dijck, which has the form

$$\eta_r = \left(1 + \frac{1.25\phi}{1 - \phi/\phi_m}\right)^2 \qquad [69]$$

In this equation, the Einstein result is obtained when $\phi \to 0$; on the other hand, when $\phi \to \phi_m$, $\eta_r \to \infty$.

Based on considerations of regular packing of equal sized spheres, ϕ_m was set equal to 0.74, the value characteristic of face centered cubic packing. It was shown that the data could be well fitted for ϕ values up to about 0.4. However, a good fit over the entire range of ϕ, $0 \le \phi \le 0.71$, was obtained using $\phi_m = 0.78$, which is slightly higher than the value expected for the dense regular packing. Eilers concluded that this higher value of ϕ_m was due to the polydispersity of the bitumen spheres.

Maron and Belner (83) found that equation [69] could be used to describe their experimental data on SBR latex for ϕ values up to about 0.56. The value of ϕ_m required for fit varied from 0.68 to 0.72, depending on the shear stress range used. When allowance was made for the layer of soap present on the surface of the latex particles, the value of the numerical coefficient was 1.25. Robinson (84) also observed that equation [69] is applicable to glass bead suspensions.

Since the parameter ϕ_m appears in several of the equations relating η_r to ϕ, it is appropriate to discuss briefly several of the factors that are known to have an effect on the ϕ_m value of a system.

4.1.2 *Factors That Affect ϕ_m*

The conditions that lead to the attainment of the most dense packing of particles are of interest in the fields of ceramics, concrete, solid propellants, asphalts, and powder metallurgy and in disciplines such as soil mechanics and geology, as well as in rubber reinforcement.

Two general kinds of packing may be distinguished: systematic and random. In systematic or regular packing, equisized spheres are considered to pack on a regular lattice. Table 3 describes some of the characteristics of several stable systematic packings obtainable using spheres of diameter d (85). As may be seen, packing on a simple cubic lattice results in the most open system, whereas packing on a face centered cubic lattice produces the densest system. Between these two

Table 3 Characteristics of Stable Systematic Packing

Name	Coordination Number	Unit cell,[a] Volume	Voids,[a] Volume	ϕ_m
Cubic	6	1.00	0.48	0.52
Tetragonal	8	0.87	0.34	0.60
Rhombohedral	10	0.75	0.23	0.70
Face centered cubic	12	0.71	0.18	0.74

[a] Actual volume of unit cell or of voids is this value multiplied by the sphere diameter cubed.

extremes, there are an infinite number of possible systematic arrangements having packing fractions in the range $0.52 \leq \phi_m \leq 0.74$. Random packing occurs when the placement of the particles is controlled by random variables. An example of this type of process is the sedimentation of equisized spheres from an initially uniform dispersion. Studies of random packing of monodisperse spheres of various materials such as glass and metal show that for large particle size, where surface energy effects can be ignored, ϕ_m falls within a surprisingly narrow and well-defined range.

Experimentally, ϕ_m for a randomly packed bed is determined as the ratio of the volume of the particles to the volume they occupy. In this kind of experiment, it is known that the measured ϕ_m value depends on the container size (i.e., on the number of particles present) and shape, and on the presence of mechanical vibration either during or after the packing of the particles. Thus, Scott (86), working with 2×10^4 uniform steel balls $\frac{1}{8}$ in. in diameter, and using a variety of container types, found that without mechanical vibration, a loose random packing is obtained characterized by $\phi_m = 0.60$; when mechanical vibration is employed, however, dense random packing occurs and $\phi_m = 0.63$ is obtained. These ϕ_m values were essentially independent of container geometry. Susskind et al. (87) working with up to 1.5×10^5 uniform glass beads 0.118 in. in diameter and with $\frac{1}{8}$ in. steel balls, also with a variety of container geometries, obtained $\phi_m = 0.63$ for the loose random packing and $\phi_m = 0.65$ for the dense random packing.

Scott and Kilgour (88), continuing earlier work, measured ϕ_m using steel spheres in both air and paraffin oil, and polished Plexiglas spheres and unpolished nylon spheres in air. The data were analyzed using the equation

$$\phi_m(R, H) = \phi_m(\infty) + \frac{A}{R} + \frac{B}{H}$$
[70]

Table 4 Densities of Random Packed Beds (88)

Material	ϕ_m (Loose)	ϕ_m (Dense)
Steel in air	0.608	0.638
Steel in oil	0.611	0.636
Plexiglas	0.605	0.636
Nylon	0.575	0.629

where $\phi_m(R, H)$ is the apparent maximum packing for a container of radius R and for a packed bed of height H, $\phi_m(\infty)$ is the maximum packing fraction for infinite volume, and A and B are constants. For dense random packing of steel spheres, they obtained $\phi_m(\infty) = 0.6366 \pm 0.0005$, $A = 0.01285$ cm, and $B = 0.04603$ cm. The results for both loose and dense random packing for the various types of particles studied are shown in Table 4. It may be seen that with the exception of the nylon spheres, remarkably consistent values of ϕ_m were obtained. The lower values for the nylon spheres were attributed to surface roughness. It is interesting to note that ϕ_m for dense random packing is approximately the same as ϕ_m for a systematic packing consisting of a $1:1$ mixture of simple cubic and face centered cubic. Such a mixture has $\phi_m = 0.632$.

For large size uniform spheres, the upper limit of ϕ_m for dense random packing is about 0.637. As the particle size decreases to small values, the ratio of the surface area to the volume increases and effects that depend on surface energy come into play. Thus as the particle size decreases, the tendency for particles to stick and agglomerate increases.

Figure 43 shows some data on the dependence of ϕ_m on particle size for two materials, calcium sulfate (anhydrite) (89) and aluminum spheres (90). For the calcium sulfate, ϕ_m was calculated from measurements of the apparent density as a function of particle size, while for the aluminum, ϕ_m was obtained from sedimentation experiments. It may be seen that at large particle size, ϕ_m is constant and has a value of about 0.63 for the aluminum and about 0.59 for the calcium sulfate. The lower value for this latter material indicates that the particles are asymmetric. For smaller particle size, the response enters a transition zone wherein ϕ_m decreases with decrease in particle size. For very small particle size it is probable that a third region exists where ϕ_m again becomes almost independent of particle size. The shape of the ϕ_m particle size response as well as its location, can be expected to depend on factors such as the particle shape, particle size distribution, and nature of the particle surface.

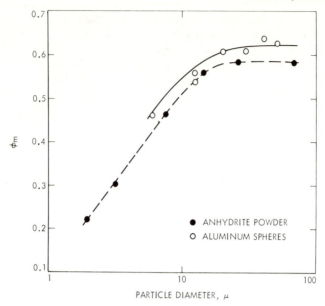

Figure 43 Dependence of ϕ_m on particle size for calcium sulfate (89) and aluminum (90).

In the transition zone, the dependence of ϕ_m on particle size is determined primarily by particle-particle interaction, and for particles that have the same size and shape, ϕ_m decreases with increase in surface energy of the particles. When the packed bed is formed in the presence of a liquid medium, the surface energy of the particles can influence ϕ_m either by particle-particle or by particle-medium interactions. In an interesting study, de Waele and Mardles observe that ϕ_m for TiO$_2$ is smaller in xylene than in the more polar solvent n-butanol (91). However, the sediment formed in n-butanol does not increase in volume when it is permeated by xylene. These results imply that ϕ_m is controlled in large measure by particle-particle interaction rather than by particle-medium interaction even when a liquid medium is present. However, under certain conditions, the nature of the surface forces between particles can be profoundly influenced by the presence of adsorbed molecules. For example, ϕ_m for small size aluminum spheres in air is 0.45. When the surface of the particles was coated with 0.05 wt % of asolectin, ϕ_m increased to 0.64 (79, 90). Corresponding increases in ϕ_m for particles coated with asolectin were also observed in several liquids. Harkins and Gans (92) observed that ϕ_m for TiO$_2$ in water-free benzene is increased by a factor of about 3 when enough oleic acid is added to form an approximately unimolecular layer on the TiO$_2$ surface. Thus, Moser finds that ϕ_m for aluminum spheres in air is 0.44. The ϕ_m value measured on the same material in low molecular weight poly(oxypropylene) is 0.55 and this same ϕ_m value is obtained when the particles are first coated by 0.4 wt % poly(oxypropylene) and ϕ_m again determined in air (90).

Effect of Polydispersity. Furnas considered the important problem of estimating the upper bound to the maximum density of a packed bed formed by blending uniform particles of different size (93, 94). He considered the volume of a packed bed formed when the particles of largest size are permitted to pack first. If particles of sufficiently small size are now added, it is evident that these could be placed in the voids associated with the large particles. The net effect will be an increase in the number of particles present without any overall increase in volume of the packed bed. In principle this process can be repeated by adding a third, fourth, etc., component of sufficiently small size such that the particle will fit into the void space associated with the next larger particle. Thus, the blending of particles of appropriate size will, in general, lead to an increase in the packing density of the bed. Furthermore, it is also evident that at some ratio of large to small particle size as well as at some fraction of the large to small component, the packing density will reach a maximum.

Using the method of Furnas, a general expression can be derived for the volume fraction of particles in a packed bed of maximum density produced by blending n components, each of uniform size, but not all of which necessarily have the same void volume. If the n components are arranged in order of decreasing particle size such that the component of largest size carries the subscript 1 and the component of smallest size carries the subscript n, then the void volumes available to each size component may be estimated in succession. Hence the volume fraction of particles at maximum density, ϕ_m^*, in a packed bed prepared by blending n components of appropriate size, can be shown to be

$$\phi_m^* = \phi_{m1} + (1 - \phi_{m1})\phi_{m2} + (1 - \phi_{m1})(1 - \phi_{m2})\phi_{m3} + \cdots$$
$$+ (1 - \phi_{m1})(1 - \phi_{m2}) \cdots (1 - \phi_{m(n-1)})\phi_{mn} \qquad [71]$$

When $\phi_{m1} = \phi_{m2} = \cdots = \phi_{mn}$, this expression is equivalent to the equation of Furnas (94).

Because ϕ_m^* does not depend explicitly on particle size, only a knowledge of the individual ϕ_{mi} values is necessary to estimate ϕ_m^*. However, there is an implicit dependence on size because ϕ_{mi} itself is a function of particle size. Since each factor on the right-hand side of equation [71] is less than unity, it is evident that the relative contribution of each term, representing contributions from components of successively smaller size, decreases very rapidly, and the major contributions to ϕ_m^* are provided by the first few terms.

The expected composition of the mixture at which the maximum density occurs can also be calculated from equation [71]. If each term is divided by ϕ_m^*, there is obtained

$$\frac{\phi_{m1}}{\phi_m^*} + \frac{(1-\phi_{m1})\phi_{m2}}{\phi_m^*} + \frac{(1-\phi_{m1})(1-\phi_{m2})\phi_{m3}}{\phi_m^*} + \cdots$$

$$\frac{(1-\phi_{m1})(1-\phi_{m2})\cdots(1-\phi_{m(n-1)})\phi_{mn}}{\phi_m^*} = 1 \qquad [72]$$

The first term represents the volume fraction of component 1 at maximum density, X_1^*, the second, the volume fraction of component 2, X_2^*, etc.

When $\phi_{m1} = \phi_{m2} = \cdots = \phi_{mn} = 0.63$, the ϕ_m^* values as well as the composition $X_1^*, \ldots, X_n^*$ predicted using equations [71] and [72] are listed in Table 5. It is apparent from these calculations that the increase in packing obtained by the addition of other components rapidly diminishes as the number of components increases beyond three.

For the case of a two-component blend, it is also possible to calculate the upper bound on ϕ_m of the mixture as a function of composition (93). Consider a packed bed formed by the largest component and characterized by a packing fraction ϕ_{m1}. The volume of the large component at any composition X, where X is the true volume fraction of the large component, is given by

$$V_1 = \left(\frac{X}{1-X}\right)V_2 \qquad [73]$$

where V_1 and V_2 are the true volumes occupied by the large and small

Table 5　Calculation of Packed Bed Characteristics

Number of Components	ϕ_m^*	X_1^*	X_2^*	X_3^*	X_4^*	X_5^*
1	0.63	1.0	—	—	—	—
2	0.86	0.73	0.27	—	—	—
3	0.95	0.67	0.25	0.09	—	—
4	0.98	0.64	0.24	0.09	0.04	—
5	0.99	0.64	0.23	0.08	0.03	0.01

components, respectively. The total volume occupied by V_1 is $V_{T1} = V_1/\phi_{m1}$, and

$$V_{T1} = \left(\frac{X}{1-X}\right)\frac{V_2}{\phi_{m1}} \qquad [74]$$

If it is assumed that the addition of the small component does not increase the volume of the bed, then

$$\phi_m = \frac{V_1 + V_2}{[X/(1-X)](V_2/\phi_{m1})} = \frac{\phi_{m1}}{X} \qquad [75]$$

When $X = 1$, which corresponds to all component 1, this expression reduces to $\phi_m = \phi_{m1}$. Equation [75] is valid only over the range $X^* \leq X \leq 1$. Now consider a packed bed consisting only of component 2. If it is assumed that the addition of component 1 increases the volume of the packed bed in direct proportion to the true volume of component 1, then by arguments similar to those leading to equation [75], there is obtained

$$\phi_m = \frac{\phi_{m2}}{1 - X(1 - \phi_{m2})} \qquad [76]$$

which is valid in the range $0 \leq X \leq X^*$. Equations [75] and [76] provide the upper bound for ϕ_m as a function of X.

4.1.3 Equations Based on Elasticity Theory

An early application of elasticity theory to the calculation of the modulus of a composite was carried out by Smallwood (95). He considered the case of a rigid spherical particle embedded in an elastic medium and calculated the displacements and stresses in the vicinity of the inclusion when the composite is deformed. The assumption is made that ϕ is sufficiently small so that interaction between particles can be neglected. For a medium with a Poisson's ratio of $\frac{1}{2}$, the relative shear modulus is given by

$$G_r = 1 + 2.5\phi \qquad [77]$$

which is identical to the Einstein equation for the viscosity of a suspension. The derivation assumes that the medium is isotropic and obeys Hooke's law.

Hashin (96) also considered the problem of a dispersion of spherical elastic particles in an elastic medium. His result, valid for small ϕ, is

$$G_r = \frac{(7-5\nu_1)G_1 + (8-10\nu_1)G_2 - 15(1-\nu_1)(G_1-G_2)\phi}{(7-5\nu_1)G_1 + (8-10\nu_1)G_2} \qquad [78]$$

Here the subscripts 1 and 2 refer to the matrix and inclusion, respectively, and ν is Poisson's ratio. This expression specifically takes into account the elastic nature of the inclusion through the presence of the quantity G_2, and as such is useful when the particles are deformable. Equation [78] was also derived independently by Eshelby (97). When $G_2 \gg G_1$ and when $\nu_1 = \frac{1}{2}$, this expression reduces to Smallwood's equation.

It is also applicable to the case where $G_2 \ll G_1$, i.e., where the inclusion is a spherical hole, in which case

$$G_r = 1 - \frac{15(1-\nu_1)}{7-5\nu_1} \phi \qquad [79]$$

This result was also obtained earlier by Mackenzie (98). For the case of an elastomeric matrix where $\nu_1 = \frac{1}{2}$, equation [79] becomes

$$G_r = 1 - \frac{5}{3} \phi \qquad [80]$$

This expression assumes that the bubbles deform during deformation. In practice however, both the gas pressure and surface tension would tend to maintain a spherical shape.

Oldroyd (99) considered the case of dispersions of small viscous inclusions on the mechanical properties of an elastic solid for ϕ so small that interaction can be ignored. His result for the analogous case of elastic inclusions in an elastic medium can be written as

$$G_r = \frac{(7-5\nu_1)G_1 + (8-10\nu_1)G_2 - (7-5\nu_1)(G_1-G_2)\phi}{(7-5\nu_1)G_1 + (8-10\nu_1)G_2 + (8-10\nu_1)(G_1-G_2)\phi} \qquad [81]$$

It is interesting to note that when the inclusion is rigid compared to the matrix, $G_2 \gg G_1$, and for $\nu_1 = \frac{1}{2}$ this expression becomes

$$G_r = 1 + \frac{2.5\phi}{1-\phi} \qquad [82]$$

which for small ϕ is equivalent to Smallwood's result. It can easily be shown that equation [78] can be obtained from equation [81] by expanding the denominator and neglecting terms involving the square and higher powers of ϕ.

Kerner (100), Uemura and Takayanagi (101), and Okano (102) have also independently derived equation [81].

An approximate theory was developed by van der Poel who represented the real material by a hypothetical composite consisting of three concentric spheres (103). The inner sphere represents the filler, the

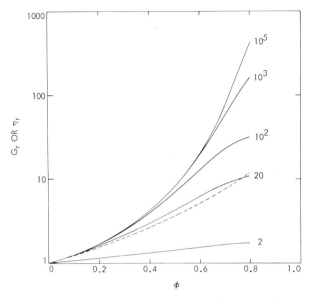

Figure 44 Dependence of the relative modulus on the volume fraction of dispersed phase according to the model of van der Poel. The number associated with each curve represents the value of the ratio G_2/G_1. The dashed curve represents the behavior predicted by equation [82].

intermediate sphere represents the binder, and the outer sphere represents the suspension. The resulting solution for the shear modulus cannot be expressed in closed form; however, the modulus of the composite was shown to depend on four parameters: ϕ, Poisson's ratio of the matrix and filler, and the ratio of the shear modulus of the filler and binder. Van der Poel demonstrated that his theoretical results were in good agreement with experimentally determined relative Young's modulus data for several composites. The form of van der Poel's results is shown in Fig. 44 for several values of the ratio G_2/G_1, assuming as a typical value $\nu_1 = \frac{1}{2}$. It is evident that a wide range of behavior can be accommodated by van der Poel's theory. Also shown for comparison as the dashed curve is the response predicted by equation [82].

Instead of trying to develop an exact solution for the modulus of a composite, Paul used energy theorems of elasticity to find the upper and lower bounds on the elastic modulus for a two-phase system containing arbitrary ϕ (104). Employing a constant stress field to establish the lower bound and a constant strain field to establish the upper bound, Paul obtains

$$\frac{1}{(1-\phi)+\phi(G_1/G_2)} \le G_r \le (1-\phi)+\phi\frac{G_2}{G_1} \qquad [83]$$

For the case of rigid inclusions in an elastomer, where typically the ratio G_2/G_1 is of the order of 10^4, this relation becomes

$$\frac{1}{1-\phi} \le G_r \le 10^4 \phi + (1-\phi) \qquad [84]$$

The lower bound for small ϕ does not reduce to the Smallwood expression, and in addition the upper bound is very large. Thus the bounds given by this relationship are too wide to give useful information for typical filled elastomeric systems.

Hashin and Shtrikman also addressed themselves to this problem. The inclusions are assumed to be spherical and the composite isotropic and quasi homogeneous (105). The results obtained are independent of the statistical distribution of the particles within the matrix. The lower bound for the relative shear modulus, G_r, is identical to equation [81] derived by Oldroyd. For the upper bound, $\bar{G}_r$, the result is

$$\bar{G}_r = \frac{G_2}{G_1} \left[\frac{15(1-\nu_2)(G_1/G_2) - (7 - 5\nu_2)(G_1/G_2 - 1)\phi}{15(1-\nu_2) + (8 - 10\nu_2)(G_1/G_2 - 1)\phi} \right] \qquad [85]$$

In common with Paul's results, these bounds are too wide to provide useful information for a system such as an elastomer containing rigid inclusions. However, as the properties of the components become more similar, the bounds tend to converge. In a later paper, Hashin (106) showed that for small ϕ, the upper and lower bounds (equations [81] and [85]) converge to a common expression which is identical to equation [78].

Wu attempted to take into account the effect of the geometric distribution of inclusions within the matrix by the introduction of a scaling parameter defined by application of the energy principle (107). This approach simplifies the problem since the detailed distribution of stress and strain need not be explicitly calculated. The result for the relative Young's modulus is

$$\frac{1}{E_r} = \left[1 - \frac{(1 - E_1/E_2)^2}{\lambda(1 + (E_1/E_2)(1-\phi)/\phi)} \right](1-\phi) + \phi \frac{E_1}{E_2} \qquad [86]$$

where λ, which is taken as independent of ϕ, is a parameter that characterizes the distribution of inclusions. Wu points out that when $\lambda = 1$, equation [86] closely approximates Paul's upper bound, whereas when $\lambda \rightarrow \infty$, the expression reduces to lower bound. In a later paper, Wu considered the problem of inclusion shape and obtained explicit results for dish, needle, and spherical shaped inclusions (108). For the spherical inclusion, Wu's result is identical to equation [81]. His results also indicate that dish shaped inclusions are more effective in increasing the modulus than are needle shaped inclusions.

From the foregoing discussion, it is apparent that the application of rigorous hydrodynamic and elastic theory to the dependence of the relative viscosity or relative modulus on ϕ leads to equations which are valid only for small ϕ.

For the finite concentration range, a rigorous solution has not yet been presented, and the difficulty of obtaining such a solution lies in the fact that at finite ϕ, the properties of the composite depend not only on ϕ but also on the geometric distribution of the particles as well. In an effort to circumvent these difficulties, lower and upper bounds to the relative modulus have been obtained; these bounds are independent of the geometric distribution of the inclusions. However, the bounds are only useful when the properties of the matrix and inclusion are similar. Another approach replaces the real material by a model for which solutions can be obtained; this is the approach taken by van der Poel. Moreover, none of the work has taken explicitly into account the fact that there is an upper limit to volume fraction of particles which can be added to a matrix. This might be accomplished by using existing equations and defining an effective ϕ as has been done by Robinson and Roscoe. However, this approach may lead to expressions that do not reduce to Einstein's or Smallwood's equation in the limit $\phi \to 0$.

Another possible approach that can be considered is to use the more applicable of the many empirical expressions that have been proposed. Although not based on rigorous theory, the Eilers-van Dijck equation, [69], seems to be applicable to a wide variety of experimental data, as will now be demonstrated.

4.1.4 Applicability of the Eilers-van Dijck Equation

Relative Viscosity. Saunders (109) measured the flow behavior of monodisperse polystyrene latices having ϕ values up to 0.25. Particle diameters ranged from 990 to 8710 Å. He showed that his data could be described by equation [62], provided corrections for the absorbed layer of emulsifier (ca. 9 Å thick) were taken into account. The observed values of the crowding factor ranged from a low of 1.118 for the large diameter latex to a high of 1.357 for the small diameter latex. These crowding factors, which correspond to ϕ_m values of 0.894 and 0.737, respectively, are rather large for monodisperse spheres.

The data of Saunders have also been fitted to the Eilers-van Dijck equation (53). Table 6 lists a portion of the original η_r, ϕ data as well as values of η_r calculated from equation [69]. The values of

Table 6   Viscosity-Concentration Data for Monodisperse Polystyrene Latices (109)

Diameter (Å)	ϕ	η_r Measured	η_r Calculated	k^a	ϕ_m Apparent	ϕ_m Corrected
990	0.02394	1.068	1.068	1.345	0.506	0.545
	0.09576	1.339	1.33			
	0.1915	2.009	2.00			
1480	0.02422	1.067	1.067	1.32	0.544	0.575
	0.09688	1.331	1.34			
	0.1938	1.945	1.95			
2490	0.02372	1.064	1.065	1.295	0.572	0.592
	0.09487	1.315	1.32			
	0.1897	1.886	1.87			
3420	0.04695	1.135	1.13	1.285	0.578	0.598
	0.1409	1.538	1.53			
	0.2348	2.271	2.28			
4240	0.02332	1.063	1.062	1.280	0.585	0.599
	0.09328	1.302	1.30			
	0.1866	1.827	1.82			
8710	0.04823	1.137	1.14	1.27	0.589	0.598
	0.1447	1.543	1.54			
	0.2411	2.329	2.31			

[a] Apparent k.

the coefficient of ϕ (listed as the k value) and of ϕ_m used in the calculation for each latex are also listed. First of all, it will be noted that the calculated values of η_r are in good agreement with the measured values. Second, the values of k increase and those of ϕ_m decrease as the particle size of the latex decreases; furthermore, in all cases the k values are greater than the value 1.25 which appears in equation [69]. However, these k and ϕ_m values do not take into account the absorbed layer of emulsifier present on the particle surface. If the thickness of the layer is denoted by Δ, the relationship between the apparent concentration ϕ_0 and the true concentration ϕ_T for small Δ is

$$\phi_T = \phi_0\left(1 + \frac{6\Delta}{D}\right), \qquad \Delta \ll 1 \qquad [87]$$

where D is the particle diameter. Using this relationship, the Eilers-van

Dijck equation can be written as

$$\eta_r = \left[1 + \frac{1.25(1 + 6\Delta/D)\phi_0}{1 - (1 + 6\Delta/D)\phi_0/\phi_m}\right]^2 \tag{88}$$

If the apparent value of k is plotted versus the reciprocal of the particle diameter, Δ as well as the true k value can be estimated. It was observed that the response was linear over the whole range of particle size; the intercept for infinitely large particles yields a k value of 1.26 which is very close to the predicted value of 1.25. The slope yielded a value of Δ equal to 14.4 Å. The corrected values of ϕ_m are shown in the last column of Table 6. For particle size greater than 2490 Å, the ϕ_m values are essentially independent of particle size and equal to about 0.60, which lies within the range expected for random packing. For smaller particle size, ϕ_m decreases as expected. The lowest value obtained is 0.545 for a particle size of 990 Å.

Another interesting study was carried out by Gillespie (110). He measured the flow behavior of dialyzed monodisperse polystyrene latices as a function of particle size. The initial experiments resulted in values for the Einstein coefficient much greater than 2.5, which was attributed to the presence of permanent aggregates.

The degree of permanent aggregation was varied by subjecting the latices to ultrasonic irradiation for varying periods of time. The percentage of unaggregated spheres ranged from about 62 to 97%. Gillespie showed that his data could be fitted to a modified Vand-Roscoe-Brinkman equation of the form

$$\eta_r = \left(\frac{1}{1 - s\phi}\right)^{k_E} \tag{89}$$

where $k_E = 5\alpha/2s$, and α is the ratio of the hydrodynamic volume of a particle or aggregate to its volume in the dry state and s is the ratio of the increase in volume of the suspension when one particle or aggregate is added, to the volume of the particle or aggregate in the dry state. Plots of $\log \eta_r$ versus $\log (1/1 - \phi)$ were linear with slopes which depended on the extent of irradiation. From the slope, values of α were obtained (assuming $s = 1$) which were then related to the average number of spherical particles per aggregate.

It is also possible to fit Gillespie's data to the Eilers-van Dijck equation (53). The form given by equation [69] assumes that the particles are nonaggregated; when aggregation is present, however, the volume fraction occurring in the numerator is not ϕ but rather the ratio ϕ/ϕ_m. Making this substitution, the Eilers-van Dijck equation for aggregated particles may be written as

$$\eta_r = \left(1 + \frac{1.25\phi/\phi_m}{1 - \phi/\phi_m}\right)^2 \qquad [90]$$

or

$$\eta_r = \left(1 + \frac{k_a\phi}{1 - \phi/\phi_m}\right)^2 \qquad [91]$$

which has the same form as the original expression but a variable coefficient k_a instead of the constant 1.25. Table 7 compares the measured η_r of two of Gillespie's samples with values calculated using equation [91]. In most cases, the agreement is satisfactory. Also included are the values of k_a and ϕ_m used in this calculation and for others of Gillespie's samples. These ϕ_m values are minimum estimates since no account was taken of a possible absorbed layer of emulsifier on the particle surface. An examination of the k_a values shows that they are all greater than 1.25. According to equations [90] and [91], the product $\phi_m k_a$ should equal 1.25 and, as may be seen in Table 7, this is approximately the case.

Table 7   Viscosity-Concentration Data for Uniform Particle Size Polystyrene Latex (Diameter 1.3×10^4 Å) (110)

	Sample A			Sample B	
ϕ	η_r (Measured)	η_r (Calculated)	ϕ	η_r (Measured)	η_r (Calculated)
0.046	1.40	1.27	0.051	1.41	1.41
0.110	2.40	2.35	0.115	2.38	2.32
0.128	3.04	2.79	0.130	2.67	2.64
0.173	4.46	4.47	0.174	4.30	4.00
0.233	9.36	10.25	0.233	8.10	8.20

Sample	ϕ_m	k_a	$k_a\phi_m$
A	0.365	3.40	1.24
B	0.390	3.20	1.25
C	0.400	3.14	1.25
D	0.420	3.00	1.26
E	0.400	3.15	1.26
F	0.430	2.91	1.25

The values of ϕ_m required for fit to equation [91] range from 0.365 to 0.430 which are much smaller than $\phi_m = 0.63$ characteristic of random packing of large spheres. However, as pointed out previously, aggregates pack less efficiently than do single particles. Thus, if the primary particles pack to form aggregates having maximum packing ϕ_m, and if the aggregates themselves pack with a maximum packing characterized by ϕ'_m, then the maximum packing for the system is simply the product $\phi_m \times \phi'_m$. If we take $\phi_m = \phi'_m = 0.60$, then the packing fraction for the system of aggregates is 0.36.

In another study, Thomas (111) conducted a survey of the extensive published literature on the η_r, ϕ dependence for spherical particles such as glass beads and latices. When plotted as log η_r versus ϕ, the data show a wide divergence, especially at the higher ϕ values. Thomas attempted, where possible, to extrapolate the data so as to minimize non-Newtonian flow behavior and inertial effects. Treatment of suitable data yielded an η_r, ϕ response in which the divergence at large ϕ was much reduced.

Thomas fitted the data by means of the empirical equation

$$\eta_r = 1 + 2.5\phi + 10.05\phi^2 + A \exp(B\phi) \qquad [92]$$

where A and B are parameters having the values 0.00273 and 16.6, respectively. The first three terms on the right-hand side correspond to the Vand equation. It should be noted that this expression does not reduce to Einstein's equation in the limit $\phi \to 0$. As good a fit was also obtained using the Eilers-van Dijck equation when ϕ_m is taken equal to 0.65 (53).

Robinson studied the flow behavior of suspensions of three sizes of glass spheres, 3–4 μ, 4–10 μ, and 10–30 μ, in aqueous glucose of two different densities and in mineral oil (84, 112). He also measured the sedimentation volumes from which ϕ_m can be estimated. It was found that these data could be fitted by the Eilers-van Dijck expression using for ϕ_m the values estimated from Robinson's reported sedimentation volumes (53).

Thus far it has been shown that equation [69] is applicable to the dependence of η_r on ϕ; the suitability of this expression to the dependence of the relative modulus on ϕ will now be considered.

Relative Modulus. Payne carried out an extensive series of measurements on the dynamic behavior of filled elastomers as a function of amplitude at a constant frequency of 0.1 Hz (113). He observed that for very small strains, <0.001%, the dynamic modulus is independent of the strain. In this region, the strain independent dynamic shear modulus is roughly analogous to the limiting shear modulus at zero strain that would be calculated from a static experiment. As the strain amplitude increases, a transition zone is reached, and a rather marked decrease of the modulus is observed. At still higher strain amplitudes, greater than about 10%, the modulus again levels off and becomes independent of the amplitude. The difference between the limiting small strain modulus G_0 and the limiting large strain modulus G_∞ is a function of the type and concentration of filler; this difference is essentially zero for a gum vulcanizate (113).

Data of Voet and Cook (114) indicate that the decrease in modulus is a reversible process. For example, when the modulus has been measured at increasing strain amplitudes and the test specimen is allowed to rest for about 20 minutes, a markedly lower curve is obtained when the modulus is remeasured at decreasing strain amplitudes. However, the original response under increasing strain amplitudes is reproduced when the test specimen is permitted to rest for 2–4 days; this indicates that the processes leading to decreasing modulus or softening are reversible. They also observed that high structure blacks show a faster rate and more complete recovery for short rest periods than do low structure blacks. In addition, for high structure blacks the difference $G_0 - G_\infty$ is greater than for low structure blacks. Voet and Cook consider that this reversible change in modulus is caused by the reversible breakdown of transient or loosely associated carbon black aggregates and not to the breakdown of the permanently fused aggregates. The nature of the surface of the carbon black does not seem to play a large role in softening since graphitized channel black showed essentially the same behavior as the nongraphitized black (114). Voet and Cook consider that transient structures are destroyed by deformation at high amplitudes and reform at lower amplitudes or when the system is at rest. At all strain amplitudes they envision the existence of a dynamic equilibrium between breakdown and reformation of transient structure which provides a distinct mechanism for the absorption and release of energy. An alternative explanation which can account for these facts is based on the idea that reversible dewetting or a separation of the rubber from the filler surface occurs; the extent of dewetting is primarily dependent on the extent of deformation imposed (53).

Payne measured G_0 for natural rubber containing nonreinforcing fillers such as glass beads, a calcium carbonate, and a large size carbon black, P-33 (115). His data are shown in Fig. 45. Shown as the full curve is the response predicted by equation [69] with ϕ_m assumed to have the value 0.63.

Payne also reported numerical values of both G_0 and G_∞ as a function of ϕ for HAF filled natural rubber vulcanizates. These data are plotted in Fig. 46 where G_{0r}, the relative limiting small strain modulus, is shown as

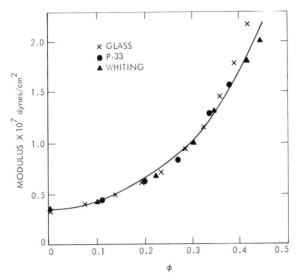

Figure 45 Dependence of modulus on volume fraction of filler for natural rubber vulcanizates. The full curve is the prediction of equation [69] when $\phi_m = 0.63$. Data from Payne (115).

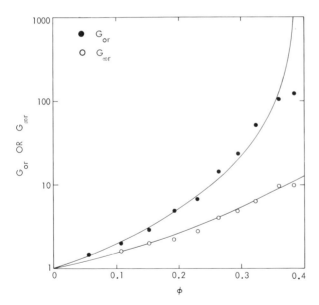

Figure 46 Effect of volume fraction of HAF black on the relative limiting small strain shear modulus G_{or} and on the limiting large strain shear modulus $G_{\infty r}$ of natural rubber. The curves represent the fit of equation [91]. Data from Payne (115).

the filled circles and $G_{\infty r}$, the relative limiting large strain modulus, is shown as the open circles. Since it is well known that HAF is a high structure black that consists almost entirely of aggregates, the form of the Eilers-van Dijck equation which had been shown to be applicable to the aggregated latices and given by equation [91] was compared to these HAF data (53). The response predicted by this equation is shown in the figure as the full curves. For the G_{0r} data the values required for the fit shown are $k_a = 3.1$ and $\phi_m = 0.40$; the product $k_a \phi_m$ equals 1.24, which is close to the expected value of 1.25. For the $G_{\infty r}$ data $k_a = 2.07$ and $\phi_m = 0.60$, and the product of these two factors is 1.25. It is also possible to fit the HAF data obtained in the transition zone where the modulus decreases with increasing strain amplitude. The ϕ_m values increase uniformly from 0.40 for the limiting small strain response, to 0.41 when strain amplitude is 0.01, to 0.48 when strain amplitude is 0.1, and finally to 0.60 for the limiting large strain response. An increase in ϕ_m with strain implies that the packing of the HAF aggregates is becoming more efficient, which in turn implies that the asymmetry of the particles is decreasing, i.e., the structure is breaking down although the occurrence of dewetting will also provide a basis to explain these facts. To show that the ϕ_m value required for fitting the G_{0r} data is reasonable, estimates of this parameter can be obtained from dibutyl phthalate absorption or from specific volume measurements. Using the results for HAF published by Dollinger et al. (116), one can calculate from dibutyl phthalate absorption the value $\phi_m = 0.36$ and from specific volume measurements the value $\phi_m = 0.43$. Since the mechanical work applied to the carbon black in these two tests is probably much less than that involved in incorporating the black into an elastomer, these estimates of ϕ_m should correspond to the value appropriate to G_{0r} data. The value of ϕ_m required for fitting the G_{0r} data thus falls within the range estimated from absorption and specific volume measurements.

Payne also studied the effect of conditions of mixing on both G_0 and G_∞. Specifically, he studied butyl rubber filled with HAF which had been (a) mixed normally or (b) heat treated (117). These data are shown in Fig. 47. As can be seen, heat treatment drastically reduces G_{0r} but has a relatively small effect on $G_{\infty r}$. For the normally compounded materials, there appears to be a transition or break in both G_{0r} and $G_{\infty r}$ at a ϕ value of about 0.23. For ϕ greater than this value, the relative moduli tend to converge so that when ϕ is about 0.29, there is apparently no effect of mixing procedure on modulus. For ϕ values up to 0.23, G_{0r} data for the normally compounded vulcanizates can be fitted to equation [91] if $\phi_m = 0.28$. Such a low value indicates very large asymmetry or poor dispersion. The $G_{\infty r}$ data for the normally compounded material can also

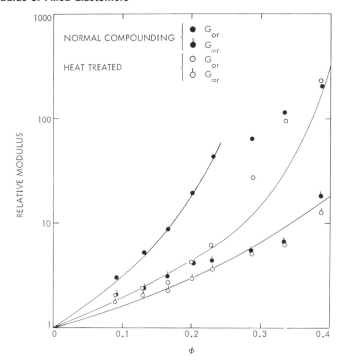

Figure 47 Effect of volume fraction of HAF black on G_{0r} and $G_{\infty r}$ for butyl rubber. The curves represent the fit of equation [91]. Data from Payne (117).

be fit over the same range of ϕ if $\phi_m = 0.43$. For the heat treated materials, G_{0r} data require $\phi_m = 0.43$ while the $G_{\infty r}$ data require $\phi_m = 0.55$. These latter two values are close to the values obtained with the natural rubber HAF system. When ϕ is less than about 0.20, $G_{\infty r}$ for the normally compounded material is essentially the same as G_{0r} for the heat treated samples. Furthermore, when ϕ is greater than about 0.20, $G_{\infty r}$ for the normally compounded samples becomes essentially the same as $G_{\infty r}$ for the heat treated material.

One further point is worth noting. Payne showed that the dynamic modulus of a carbon black suspension is essentially independent of whether the matrix is an elastomer or a low viscosity paraffin oil (118).

4.2 Large Strain Behavior

It has been shown that the dependence of the relative modulus on ϕ at small strains can be represented by equation [69] or [90]. Also of interest is the effect of ϕ on the stress-strain response at large strains. For a filled system, the simplest assumption one can make is that, analogous to equation [1] for gum elastomers, the effects of ϕ, time, and strain are factorable and hence the stress can be expressed as the product of three separate functions:

$$\sigma = G(t)H(\phi)f(\epsilon) \tag{93}$$

where $G(t)$ and $f(\epsilon)$ are identical to the corresponding functions appearing in equation [1] and obtainable from measurements carried out on an unfilled elastomer, and $H(\phi)$ is a function of ϕ alone. Based on the successful application of equations [69] and [90] to modulus, ϕ data obtained at small strains, these expressions may be tentatively taken as explicit forms for $H(\phi)$. Equation [93] predicts that isochronal stress-strain curves for filled vulcanizates as well as the corresponding gum will be parallel on a log-log plot of σ versus ϵ. Furthermore, these curves are superposable by a vertical shift, and the magnitude of the shift is log $H(\phi)$.

Smith studied the uniaxial stress-strain response for butyl rubber filled with HAF as function of strain rate and temperature (8). Using data obtained at several strain rates, 1 minute isochronal stress-strain curves were constructed for strains in the range 10–100%. Essentially linear plots of unit slope were obtained when log $\sigma\lambda$ was plotted against log ϵ. From such data obtained at several temperatures, the 1 minute constant strain rate modulus $F(1)$, was evaluated and plots of log $F(1)(273/T)$ versus temperature were constructed. Taking the unfilled elastomer as a reference, Smith shifted the data for the filled vulcanizate both vertically and horizontally to effect superposition. The vertical shift distance corresponds to log $F_r(1)$ where $F_r(1)$ is the relative 1 minute modulus, while the horizontal shift distance was denoted by T_r. Thus, the vertical shift corresponds to the effect of filler on the modulus, while the horizontal shift corresponds to the effect of filler on the time (temperature) scale. From the T_r values obtained, it is evident that the presence of filler causes the relaxation times to increase. A possible mechanism by which filler particles effectively produce an increase in the relaxation times has been advanced by Halpin and Bueche (119).

Smith found that the dependence of $F_r(1)$ on ϕ could be represented by equation [53] with $\alpha_1 = 2.5$ and $\alpha_2 = 14.1$. Values of $F_r(1)$ and T_r are shown in Table 8. Also listed are values of $F_r(1)$ calculated from equation [90] using ϕ_m values of 0.28 and 0.43, which correspond to the values required to fit the G_{0r} data of Payne for butyl-HAF composites, normally mixed and heat treated, respectively.

It is seen that the experimentally observed values of $F_r(1)$ are smaller than either of the two calculated values. If it is assumed that this

Table 8   HAF Carbon Black in Butyl Rubber

ϕ	$F_r(1)^a$	$F_r(1)^b$	$F_r(1)^c$	T_r (°C)
0	1.00	1.0	1.0	0
0.028	1.08	1.30	1.18	9
0.059	1.08	1.78	1.43	24
0.101	1.53	3.02	1.91	24
0.144	1.68	5.40	2.66	29
0.220	2.37	31.1	5.36	42

[a] Estimated from experimental data (8).
[b] Calculated from equation [90] with $\phi_m = 0.28$.
[c] Calculated from equation [90] with $\phi_m = 0.43$.

difference is due to a decrease in the degree of cross-linking of the matrix, then the data in Table 8 indicate that the degree of cross-linking decreases with increase in ϕ. Furthermore, if the a_x factor is applicable to filled systems, then the observed increase in the relaxation time with increase in filler can be attributed simply to the progressive decrease in the degree of cross-linking.

Mullins and Tobin measured the stress-strain response of dicumyl peroxide cured natural rubber containing varying amounts of MT and HAF carbon black (120). To treat their data, they introduced the concept of a strain amplification factor. In this approach, the strain experienced by the matrix ϵ_1 in a composite is considered to be greater than the overall or measured strain ϵ. An effective extension ratio Λ is defined by

$$\Lambda = 1 + E_r\epsilon \qquad [94]$$

where E_r is the relative Young's modulus of the filled elastomer. In order to test the concept of the strain amplification factor, values of Young's modulus were derived from the linear stress-strain response obtained at small strains; the response was reported linear for strains up to 2%. When the stress-strain data were plotted in Mooney-Rivlin coordinates as $(\sigma/2)(\lambda - 1/\lambda^2)^{-1}$ versus $1/\lambda$, separate and distinct curves were obtained for each ϕ value, as shown in Fig. 48 for the HAF system. However, if Λ is used instead of λ as the appropriate deformation variable, then the data should be plotted in the form $(\sigma/2)(\Lambda - 1/\Lambda^2)^{-1}$ versus $1/\Lambda$. The results for HAF plotted in this manner are shown in Fig. 49. It is evident that the use of the strain amplification factor tends to superpose the data at small strains. This superposition is expected since in the limit of zero strain $\Lambda \to 1$ and $(\sigma/2)(\Lambda - 1/\Lambda^2)^{-1} \to E_e/6$, where E_e is the equilibrium value of Young's modulus for the unfilled elastomer. Hence the limiting ordinate

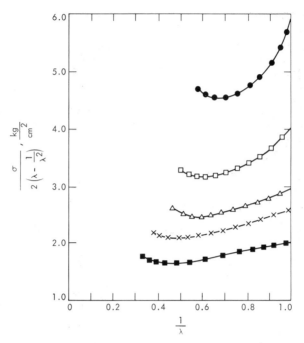

Figure 48 Stress-strain response of natural rubber filled with HAF black. The volume fractions of HAF are as follows: ■, 0; ×, 0.0383; △, 0.0611; □, 0.0954; ●, 0.144. Data from Mullins and Tobin (120).

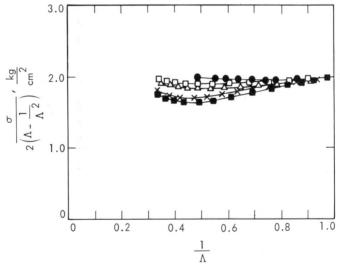

Figure 49 Data of Fig. 48 replotted using an effective extension ratio defined by equation [94]. Symbols have the same meaning as for Fig. 48. Data from Mullins and Tobin (120).

value at $\Lambda = 1$ is $E_e/6$, which is independent of ϕ. However, at larger strains the curves tend to diverge. Similar results were obtained with the MT system.

Mullins and Tobin tested the validity of the strain amplification factor by assuming an explicit form for the stress-strain response, i.e., the Mooney-Rivlin form. As will now be shown, this can be accomplished without the necessity of specifying the form of the stress-strain response. The existence of a strain amplification factor implies that equation [1] is applicable to filled elastomers provided the measured strain is replaced by the effective strain, which is $E_r\epsilon$. Hence, for long times or for temperatures well above T_g, equation [1] can be written as

$$\sigma = E_e f(E_r\epsilon) \tag{95}$$

According to this expression a log-log plot of σ versus ϵ should yield a family of parallel curves which can be superposed to the response of the unfilled elastomer by a horizontal shift along the log ϵ axis. The magnitude of the shift is then equal to log E_r. Thus, using this procedure, no assumption need be made concerning the explicit form of $f(\epsilon)$. The data of Mullins and Tobin have been treated in this manner and are shown after the horizontal shift has been applied in Figs. 50 and 51 for the MT and HAF systems, respectively. As may be seen, satisfactory superposition has resulted over the entire range of ϵ from about 10 to 250%. From the magnitude of the horizontal shift, values of E_r at each filler concentration have been estimated and these are listed in Table 9. A comparison of the estimated values with those obtained by direct measurement of Young's modulus indicates good agreement for the MT data, but a small discrepancy in the case of HAF. For the MT data, the E_r values employed by Mullins and Tobin in their calculation of Λ are also listed, and it is seen that these values differ slightly from those calculated from Young's modulus.

In view of the success of the strain amplification factor in correlating the stress-strain response for filled natural rubber, it is of interest to compare the same data for fit to equation [93]. This expression predicts that the stress-strain response plotted as log σ versus log ϵ will yield a family of parallel curves which can be superposed to that of the gum vulcanizate by a vertical shift. The magnitude of the vertical shift is equal to log $H(\phi)$ or log E_r. The data of Mullins and Tobin (120), shifted in this manner, are shown in Figs. 52 and 53 for the MT and HAF systems, respectively. The values of E_r estimated from the vertical shift are listed in Table 9. It may be seen that values estimated in this manner are perhaps in better overall agreement with values calculated directly from

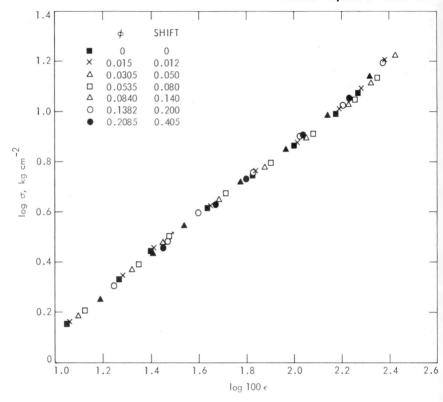

Figure 50 Stress-strain data for natural rubber filled with MT black superposed by horizontal shifts. Data from Mullins and Tobin (120).

Young's modulus than those estimated from horizontal shifting. A comparison of E_r obtained from the horizontal shift and from the vertical shift shows that the latter method of evaluation leads to lower E_r values. This is primarily a reflection of the fact that shifting is done with curves that have slopes which are less than unity. Also listed are values of E_r calculated from the Eilers-van Dijck equation. For MT black, equation [69] was used and ϕ_m was taken as 0.58, while for HAF, equation [90] was used with ϕ_m taken equal to 0.40, which, incidentally, is the same value obtained for HAF from the limiting small strain dynamic modulus G_{0r}. It may be seen that E_r values calculated using the Eilers-van Dijck equation are in good agreement with the values estimated from the vertical shift. As in the case of HAF, independent estimates of ϕ_m for MT black may be obtained from measurements of dibutyl phthalate absorption and apparent specific volume, which yield the values 0.56 and 0.70, respectively. Thus, as noted previously for HAF, the value of ϕ_m required for fitting modulus data lies in the range estimated from absorption and specific volume measurements. The stress-strain response for both MT and HAF can apparently be described by either equation [93] or [95] (using different values of E_r) over a wide range of strain, i.e., from about 10 to about 250%. In order to demonstrate which representation is more applicable, small strain data in the linear Hookean range are also required.

The uniaxial stress-strain response of dicumyl peroxide cured SBR vulcanizates containing MT black has been studied under conditions of constant strain rate over a wide temperature range (53). The stress-strain response at 25°C and a rate of 4.25 min^{-1} for various values of ϕ were shifted both horizontally (Fig. 54) and vertically (Fig. 55), and the filled symbols in the figures denote the breaking points. The values of the relative modulus estimated from the shifts are listed in Table 10. Also listed are the relative modulus values calculated from equation [69] using for ϕ_m the value 0.58 obtained from fitting the natural rubber-MT data of

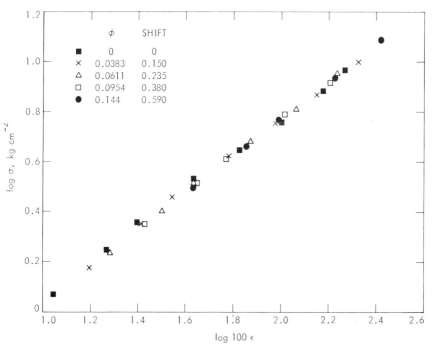

Figure 51 Stress-strain data for natural rubber filled with HAF black superposed by horizontal shifts. Data from Mullins and Tobin (120).

Table 9   MT Carbon Black Vulcanizates

ϕ	E_r^a	E_r^b	E_r^c	E_r^d	E_r^f
0	1.00	1.00	1.00	1.00	1.00
0.015	1.09	1.03	1.04	1.04	1.04
0.0305	1.13	1.12	1.09	1.08	1.09
0.0535	1.24	1.20	1.14	1.15	1.17
0.0840	1.27	1.38	1.29	1.26	1.31
0.1382	1.50	1.59	1.43	1.51	1.55
0.2082	2.21	2.54	2.04	1.98	2.13

HAF Carbon Black Vulcanizates

ϕ	E_r^a	E_r^b	E_r^c	E_r^e
0	1.00	1.00	1.00	1.00
0.0383	1.30	1.41	1.26	1.28
0.0611	1.50	1.72	1.42	1.50
0.0954	2.02	2.40	1.82	1.93
0.144	3.12	3.90	2.57	2.88

[a] Calculated from the reported Young's modulus (120).
[b] Calculated from horizontal shift, equation [95].
[c] Calculated from vertical shift, equation [93].
[d] Calculated from equation [69] with $\phi_m = 0.58$.
[e] Calculated from equation [90] with $\phi_m = 0.40$.
[f] Values of E_r employed by Mullins and Tobin to calculate Λ.

Mullins and Tobin. It is evident from the figures that either procedure will superpose the stress-strain response for strain less than about 200%. Of the two shifting procedures, the method of Mullins and Tobin yields relative modulus values which for ϕ up to about 0.25 are in good agreement with the predictions of equation [69]. However, the relative modulus estimated from the vertical shift is in particularly poor agreement with prediction. If the results obtained from equation [69] are assumed to be correct, then the lower values of E_r obtained experimentally presumably are attributable to the interference of the black with reactions leading to cross-linking. It is known that the cross-linking of SBR by dicumyl peroxide proceeds with very high efficiency such that 12.5 moles of cross-links are formed per mole of peroxide decomposed. For natural rubber, the efficiency is essentially unity (121). Hence SBR is

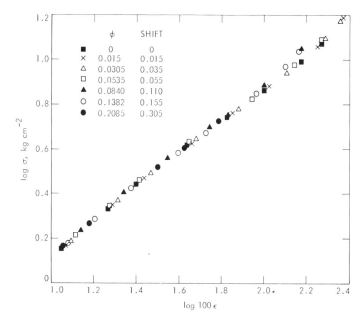

Figure 52 Stress-strain data for natural rubber filled with MT black superposed by vertical shifts. Data from Mullins and Tobin (120).

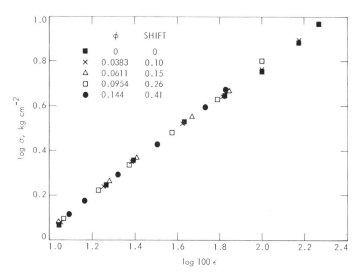

Figure 53 Stress-strain data for natural rubber filled with HAF black superposed by vertical shifts. Data from Mullins and Tobin (120).

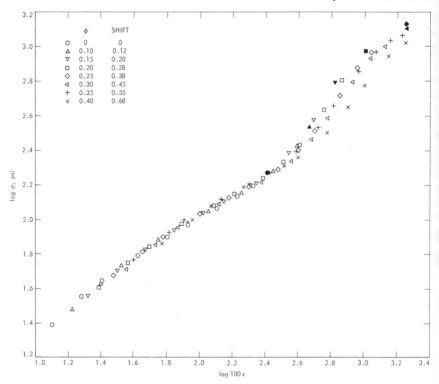

Figure 54 Stress-strain data for SBR filled with MT black superposed by horizontal shifts.

more sensitive to nonproductive side reactions that waste small amounts of the peroxide.

Thus far, the behavior of systems in which the matrix is apparently firmly bound to the filler particles has been discussed, and for contrast, it is of interest to consider a case in which this is not so. Fedors and Landel (122) studied the stress-strain response of SBR filled with glass beads with ϕ ranging from 0 to about 0.34. The glass filler was composed of spherical particles with diameters in the range 20–60 μ.

Each composite was characterized for the degree of cross-linking by determining both the extent of equilibrium swelling, using several solvents, and the stress-strain response in uniaxial compression on specimens swollen to equilibrium. The techniques employed in the characterization work as well as the detailed findings may be found in the original publications (123, 124). In summary, it was observed that the determination of the extent of equilibrium swelling does not, in general, provide

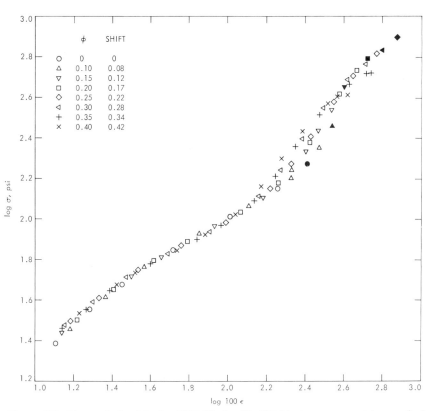

Figure 55 Stress-strain data for SBR filled with MT black superposed by vertical shifts.

Table 10 MT Carbon Black in SBR

ϕ	E_r^a	E_r^b	E_r^c
0	1.00	1.00	1.00
0.10	1.20	1.32	1.33
0.15	1.32	1.59	1.57
0.20	1.48	1.91	1.91
0.25	1.66	2.37	2.40
0.30	1.88	2.82	3.16
0.35	2.16	3.55	4.43
0.40	2.60	4.79	6.79

[a] Estimated from vertical shift.
[b] Estimated from horizontal shift.
[c] Calculated from equation [69] with $\phi_m = 0.58$.

correct estimates of ν_e for the SBR-glass bead system. Even though ν_e of the binder was constant, the extent of equilibrium swelling in five solvents decreased as ϕ increased.

This finding is contrary to the proposal of Bills and Salcedo (125) that, for composites where the filler-binder interface is disrupted by the solvent in the swelling process, the equilibrium extent of swelling will be independent of filler level. If the values of ν_r were used directly as proposed by Bills, then the equilibrium swelling data would provide erroneous values of ν_e. The Bills treatment is based on the assumption that none of the filler particles are in contact with each other in the original unswollen composite; i.e., each filler particle is isolated from every other filler particle. It is difficult to see how this assumption of nonagglomeration can remain valid for high loadings.

The behavior of swollen composites in uniaxial compression tests was also studied (124). It was observed that the response obtained in only a single solvent is not in general sufficient to define a unique ν_e value. When several solvents are employed, a range of apparent ν_e values is obtained such that the apparent ν_e increases with decrease in solvent power, i.e., as the extent of equilibrium swelling decreases. However, it was possible to extrapolate the data to obtain a unique value for ν_e. Moreover, it was demonstrated that the composites have essentially the same ν_e (110×10^{-6} moles/cm^3) independent of ϕ.

Typical stress-strain data are shown in Fig. 56 for one strain rate and three temperatures for the binder and three composites. The open circles represent the break point. Inspection of the stress-strain curves indicates that at small strains the stress required to reach a given strain increases as ϕ is increased. At the higher strains, the curves for the filled materials flatten out and approach the stress-strain response of the unfilled gum. As the break point is approached, the stress-strain curves for the filled systems are essentially indistinguishable from that of the unfilled binder. This behavior was observed for all the test rates and test temperatures investigated. The data in Fig. 56 represent the response of a filled system in which the only variable parameter is ϕ.

One way to account for the fact that the stress-strain response for the composites converges to that of the gum binder, is to assume that dewetting takes place. As the strain is increased, the binder-filler interface is disrupted with the concomitant formation of voids. This effect has two immediate consequences: (a) it reduces the surface area of filler remaining in contact with the binder, which reduces the effective value of ϕ, and (b) it produces a volume dilatation which should be proportional to ϕ. The first effect indicates that ϕ should be replaced by an effective value ϕ_{eff}, which is a function of strain such that when ϵ is zero, ϕ_{eff} is

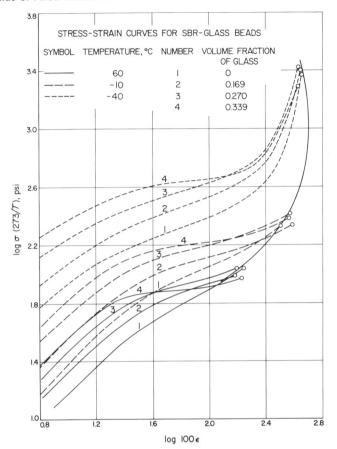

Figure 56 Stress-strain behavior of SBR-glass bead composites. The strain rate is $4.25\,\text{min}^{-1}$.

equal to the initial filler volume before dewetting has occurred, ϕ_0, and as ϵ increases, ϕ_{eff} decreases monotonically to zero.

Based on a simple picture of the dewetting process in which it is assumed that conical cavities or voids develop around the dewetted particle, a first approximation theory was developed in an attempt to account for both the stress-strain response and the volume dilatation response for composites (122). Briefly, it was shown that the effective volume fraction can be written as

$$\phi_{\text{eff}} = \frac{\phi_0}{1 + k\epsilon_{\text{eff}}} \qquad [96]$$

where k is a parameter related to the strength of the filler-binder interface and to the binder modulus. The effective strain in the binder ϵ_{eff} is taken from Bueche (126) to be

$$\epsilon_{eff} = \frac{\epsilon}{1 - \phi_0^{1/3}} \qquad [97]$$

where ϵ is the overall strain on the specimen. It was also shown that the volume dilatation $\Delta V / V_0$ depends on strain according to

$$\frac{\Delta V}{V_0} = \frac{\phi_0}{2} \frac{k^2 \epsilon_{eff}^2}{1 + k \epsilon_{eff}} \qquad [98]$$

In deriving equations [96] and [98], it was assumed that the filler-binder interaction is essentially zero and hence that all particles begin to dewet simultaneously as soon as the specimen is subjected to strain.

In order to test the form of equation [98], the volume dilatation-strain response was measured by the hydrostatic weighing technique using water as the confining fluid. Measurements were carried out at 23 and 0°C and were continued until the specimens broke. Since this technique is necessarily time-consuming, requiring about 4 hours to carry out a run, the results correspond to stress-strain experiments carried out at very low (as well as nonuniform) strain rates (122). The results are shown in Fig. 57 for data obtained at 23°C. Comparable data were obtained at 0°C. The circles represent the experimental data obtained and the full curves represent the predictions of equation [98] using $k = 0.52$. For the 0°C data, $k = 0.56$. It is evident that the fit is reasonable for strains up to about 100%. Deviations from prediction at high strains are expected and result from a breakdown in the simple picture assumed for the dewetting process. The fact that the values of k do not depend strongly on temperature lends support to the idea that for the SBR-glass bead system the energy of the filler-binder bond is very small.

In an attempt to represent the stress-strain response, it was assumed that equation [93] could be used for dewetting systems and that $H(\phi)$ is given by equation [69] with ϕ_{eff} replacing ϕ. This is equivalent to writing

$$\log \sigma = \log G(t) + 2 \log \left(1 + \frac{1.25 \phi_{eff}}{1 - \phi_{eff}/\phi_m} \right) + \log f(\epsilon) \qquad [99]$$

In this expression, the terms $\log G(t) + \log f(\epsilon)$ represent the time-strain response of the unfilled material; the response for the composite is obtained by including the additive term

$$2 \log \left(1 + \frac{1.25 \phi_{eff}}{1 - \phi_{eff}/\phi_m} \right)$$

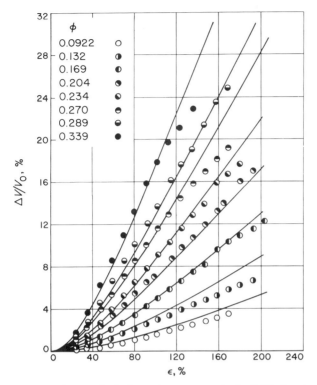

Figure 57 Dependence of relative volume on strain for SBR-glass bead composites at 23°C.

which is a function of the strain and takes into account dewetting of the filler particles. For glass beads in mineral oil, a system similar to the SBR-glass bead composites, ϕ_m was found (79) to be 0.63. Using $k = 0.52$ to relate ϕ_{eff} to ϕ_0 by means of equations [96] and [97] and with $\phi_m = 0.63$, estimates for the stress-strain response for the composites can be obtained. Typical results are shown in Figs. 58 and 59, where the predicted stress-strain response is shown as the dashed curve, which is to be compared with the experimentally observed response denoted by the unfilled circles. It is evident that the model provides a satisfactory representation of the stress-strain response in the low and intermediate strain region. At high strains, however, higher stresses are predicted at a given strain than observed experimentally. Further, the strains at which the divergence becomes apparent decrease with increase in ϕ. This was attributed to a breakdown in the simple dewetting model.

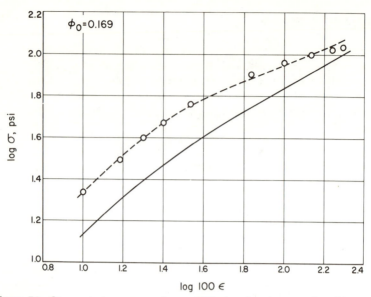

Figure 58 Stress-strain response for an SBR-glass bead composite with $\phi = 0.169$ at 25°C and a strain rate of 4.25 min^{-1}. The circles denote the experimental data, the dashed curve is the predicted response, and the full curve is the response of the unfilled gum.

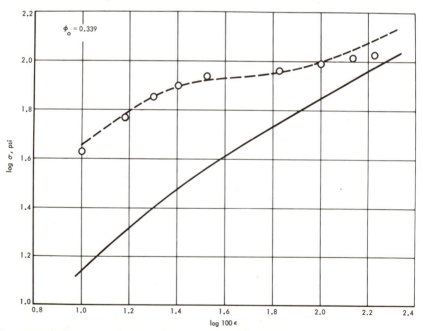

Figure 59 Stress-strain response for an SBR-glass bead composite with $\phi = 0.339$ at 25°C and a strain rate of 4.25 min^{-1}. Same symbols as Fig. 58.

784

Experiments were also carried out to ascertain the effects on subsequent response of various prior stress-strain histories (122). In experiments conducted at 25°C specimens of each composite were strained to about 170% at a rate of 0.85 min^{-1}; they were then immediately unloaded by reversing the direction of the crosshead travel. This cyclic procedure was repeated several times. For the most highly filled composite, the maximum strain was 165% rather than 170%. Typical results are shown in Figs. 60–62. The unfilled symbols represent the stress-strain response on the loading portion of each cycle, and the filled symbols denote the corresponding behavior during unloading. The data indicate that for the unfilled material, the stress-strain response on both loading and unloading for the first cycle is slightly higher than the response obtained on subsequent cycles. For the composites, however, the initial loading curve is unique whereas the initial unloading and all subsequent load-unload curves differ only slightly. Furthermore, these latter curves essentially reproduce the stress-strain response shown by the unfilled

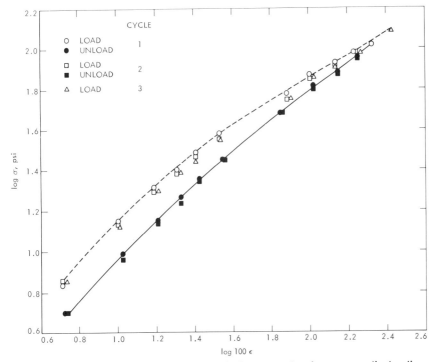

Figure 60 Stress-strain response of an SBR gum vulcanizate to cyclic loading-unloading tests carried out at 25°C and a strain rate of 0.85 min^{-1}.

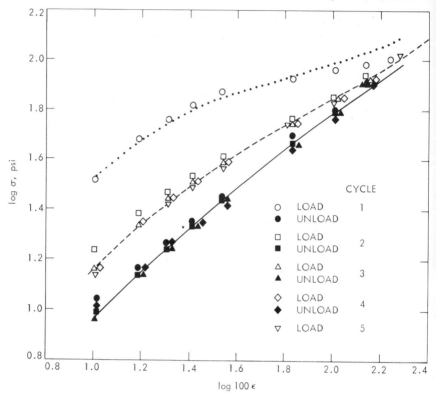

Figure 61 Stress-strain response of an SBR-glass bead composite with $\phi = 0.169$ to cyclic loading-unloading tests carried out at 25°C and a strain rate of 0.85 min⁻¹. The dotted curve is the predicted response.

vulcanizates, except in the neighborhood of the maximum strain used in the experiment. For comparison, the dashed curve and the solid curve in this set of figures represent the response of the gum vulcanizate. It is also found that if the specimens are permitted to rest for a few days, complete or near complete rewetting takes place, and the original first cycle stress-strain response can be reproduced.

5 UNIAXIAL RUPTURE OF FILLED ELASTOMERS

5.1 The Effect of Filler on the Failure Envelope

It is well known that the presence of fillers such as carbon black in noncrystallizable elastomers produces marked improvement in properties such as tensile strength, ultimate elongation, tear resistance, and abrasion

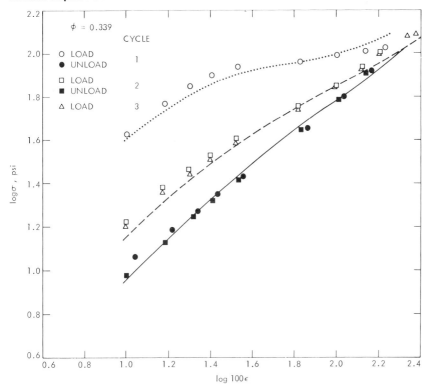

Figure 62 Stress-strain response of an SBR-glass bead composite with $\phi = 0.339$ to cyclic loading-unloading tests carried out at 25°C and a strain rate of 0.85 min^{-1}. The dotted curve is the predicted response.

resistance. However, the magnitude of the enhancement in such properties can be expected to be influenced by effects such as deformation induced dewetting. Two limiting or ideal cases can be considered: (*a*) systems in which total filler dewetting occurs at rupture and (*b*) systems in which essentially no dewetting occurs at rupture. In this section, a survey of the rupture of filled elastomers is presented in terms of the failure envelope and time dependence of rupture. It will be convenient to discuss dewetting and nondewetting systems separately.

5.1.1 Dewetting Systems

In analogy with the behavior of unfilled elastomers, filled elastomers can also be characterized by failure envelopes. Figure 63 shows failure envelopes in the form log $\sigma_b(308/T)$ versus log λ_b for the SBR-glass bead

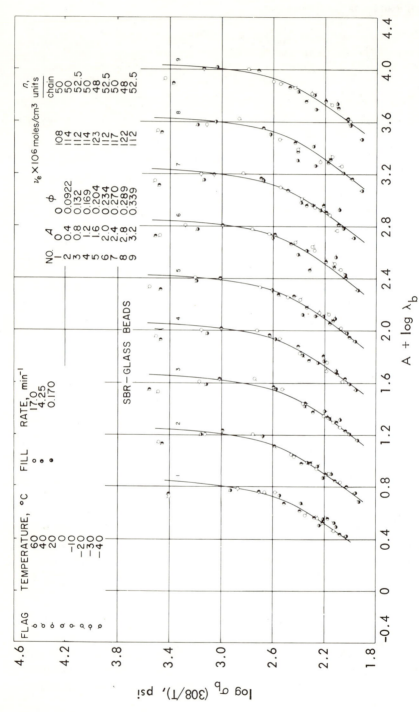

Figure 63 Failure envelopes for SBR-glass bead composites as a function of ϕ. The full curves represent the fit of the data to equation [2].

system as a function of ϕ (123). The individual sets of data, representing particular ϕ values, have been shifted along the abscissa for clarity. An inspection of the data will show that at a particular rate and temperature of testing, both σ_b and ϵ_b are essentially independent of ϕ, implying that for constant ν_e rupture is determined by the properties of the binder alone, the presence of the glass bead filler having no appreciable influence. This was attributed to the occurrence of almost total dewetting for strains close to ϵ_b over the experimental time and temperature range studied. This contention is supported by the fact, already noted, that when cycled to strains near ϵ_b, the first retraction as well as subsequent extension-retraction cycles of the filled materials were virtually indistinguishable from the corresponding behavior of the unfilled elastomer.

The full curves shown represent the fit of the data to the form of $f(\epsilon)$ given by equation [2]. The values of ν_e and n required are listed in the figure, and it can be seen that both values are constant and independent of ϕ. Hence, for the SBR-glass bead system at least, the failure envelope for the filled composites is the same as that of the gum elastomer. Moreover, the dependence of the envelope on parameters such as ν_e can be readily predicted.

5.1.2 *Nondewetting Systems*

Elastomers containing carbon black fillers, especially the reinforcing types, are generally considered to have strong elastomer-filler adhesion. Judged on the basis of volume change with deformation, which is very small, such systems can be considered to be nondewetting especially in comparison to the SBR-glass bead composites. Halpin and Bueche and Smith presented data on carbon black filled SBR (119), EPDM (119), and butyl (8) which demonstrated that the failure envelope is also applicable to these systems. It was found that the shape of the failure envelope does not change markedly as ϕ is increased; however, the location depends on ϕ in that the envelope moves upward in the log $\sigma_b T_0/T$, log ϵ_b plane with increase in ϕ. Halpin and Bueche further demonstrated that the failure envelope for a filled material (SBR–HAF with $\phi = 0.07$) is independent of the test mode since rupture in stress relaxation and in creep yielded essentially the same failure envelope as obtained with constant strain rate tests (119).

Figures 64–71 show the failure envelopes for an SBR elastomer containing from 0 to 0.40 volume fraction of a medium thermal carbon black, plotted in the form log $\sigma_b T_0/T$ versus log λ_b (53). As was observed with the failure envelope of a gum elastomer, the rupture points move counterclockwise along the envelope as the test temperature is decreased or

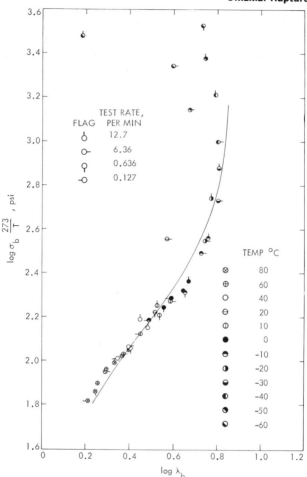

Figure 64 Failure envelope for an SBR gum vulcanizate.

the test rate is increased. Also, the envelopes are characterized by a maximum in λ_b (or ϵ_b). For the gum elastomer, the maximum in ϵ_b occurs at a test temperature of about 233°K. As the carbon black content increases, the test temperature at which the maximum occurs also increases such that for the vulcanizate with $\phi = 0.40$, $(\epsilon_b)_m$ is observed at test temperatures in the neighborhood of 273°K. Considering the entire set of data from $\phi = 0$ to $\phi = 0.40$, it may be observed that the shape of the envelope is essentially independent of ϕ; however, as ϕ increases, the envelopes are shifted progressively upward. Also shown as the full curves is the fit of the data to the $f(\epsilon)$ function given by equation [2]. The values of ν_e and n required for fit arc listed in Table 11. Also shown are values of the relative chain concentration, $\nu_{e,r}$, for the various ϕ estimated from the fit of the rupture data to equation [2] and also as estimated from superposing the high temperature portions of the failure envelopes. These values are in good accord with the relative modulus evaluated from the superposition of the stress-strain curves (see Table 10). Also listed are values of n which generally increase with increase in ϕ; this behavior is, no doubt, a reflection of the decrease in ν_e of the binder with increase in ϕ.

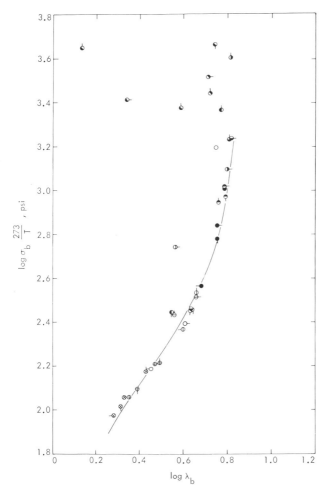

Figure 65 Failure envelope for SBR containing MT black with $\phi = 0.10$. The full curve represents the fit of the data to equation [2].

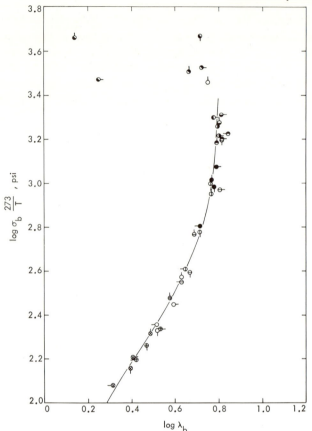

Figure 66   Failure envelope for SBR containing MT black with $\phi = 0.15$. The full curve represents the fit of the data to equation [2].

Smith (8) studied the rupture behavior of a butyl elastomer containing HAF black with ϕ varying from 0 to 0.22. His results are in agreement with the data on the SBR–MT system in that the failure envelopes do not change shape appreciably when filler is added; the major effect seems to be a shift of the envelope progressively upward as the filler content is increased. However, Smith observed that the failure envelopes consisted of two groups of points, one group representing data obtained at high temperature (40–100°C) and the second group representing data obtained at low temperature (25 to −20°C). This segregation of the data points was previously also noted for a gum silicone and a gum sulfur cured butyl vulcanizate. From an analysis of the time and temperature dependence of the data, Smith concluded that this effect is caused by the occurrence of strain induced crystallization.

Smith also studied the effect of prior cycling on the rupture behavior of a butyl vulcanizate containing 22 vol % of HAF. It was found that prior cycling, i.e., cycling a specimen three times between 0 and some selected load limit at a crosshead speed of 10 in./min following which the specimen was extended to rupture at a series of crosshead speeds, had no effect on the observed rupture response when testing was carried out at high temperatures (70–100°C). However, at low test temperatures (below 70°C) the rupture data obtained by cycling were significantly different from the behavior observed using the normal test procedure. When

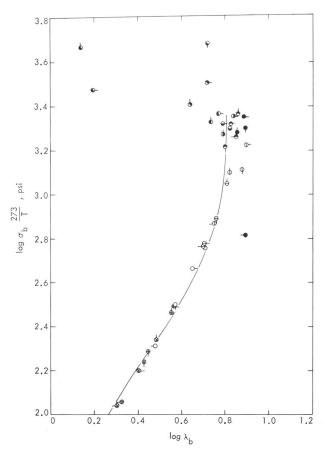

Figure 67 Failure envelope for SBR containing MT black with $\phi = 0.20$. The full curve represents the fit of the data to equation [2].

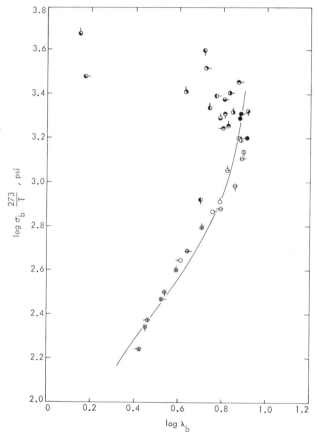

Figure 68 Failure envelope for SBR containing MT black with $\phi = 0.25$. The full curve represents the fit of the data to equation [2].

cycled prior to rupture, the values of σ_b are almost independent of test rate, but a marked increase is observed in ϵ_b. This unusual behavior was attributed to the occurrence of strain induced crystallization during cycling (8).

5.2 Time Dependence of Rupture

5.2.1 *Dewetting Systems*

In comparison with gum elastomers, relatively little work has been reported on the time dependence of rupture for filled systems. For systems that dewet extensively, such as the SBR-glass bead system, the time dependence of rupture should be the same as that observed with the unfilled gum. Although the rupture data were not obtained at a sufficient number of strain rates to establish the superposed σ_b, t_b and ϵ_b, t_b responses, an examination of the rupture data in Fig. 63 shows that for a particular strain rate and temperature, the values of σ_b and λ_b are essentially independent of ϕ and hence, by implication, the σ_b, t_b and ϵ_b, t_b responses are also independent of ϕ. Investigations of the effect of noninteracting fillers on the stress relaxation modulus at small strains showed that the time dependence per se was essentially independent of filler (127–129).

5.2.2 Nondewetting Systems

For the SBR–MT system discussed earlier, data were not obtained at a sufficient number of test rates to enable time-temperature superposition

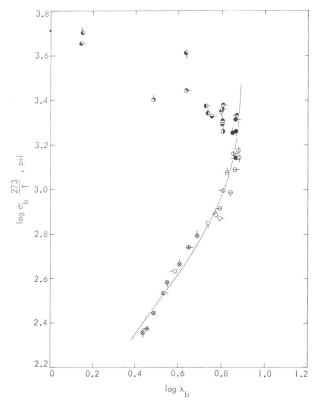

Figure 69 Failure envelope for SBR containing MT black with $\phi = 0.30$. The full curve represents the fit of the data to equation [2].

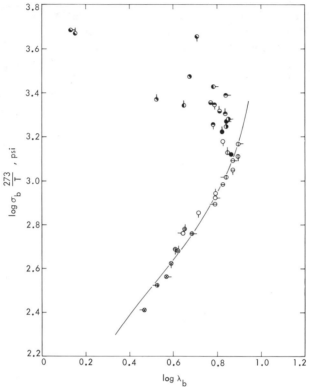

Figure 70 Failure envelope for SBR containing MT black with $\phi = 0.35$. The full curve represents the fit of the data to equation [2].

to be employed. For the butyl-HAF vulcanizates (8), the occurrence of crystallization at test temperatures of 40°C and lower precludes obtaining the reduced σ_b, t_b and ϵ_b, t_b responses except for the two higher test temperatures, namely 70 and 100°C.

The rupture behavior of both SBR and EPDM filled with HAF carbon black and having ϕ values of 0, 0.07, and 0.14 was studied by Halpin and Bueche (119). In this work, they were primarily concerned with the fit of experimental data to the theory which they developed, and in general good agreement was obtained. The values of q and K in equations [19] and [20] increased, somewhat, with an increase in ϕ. With the SBR system, for example, at $\phi = 0$, $K = 3.16$ and $q = 10^{7.5}$, while at $\phi = 0.15$, $K = 3.55$ and $q = 10^{8}$. Similar small variations were also obtained with the EPDM system.

It was also observed that the time-temperature shift factors obtained using the rupture data could be fitted to an Arrhenius type dependence,

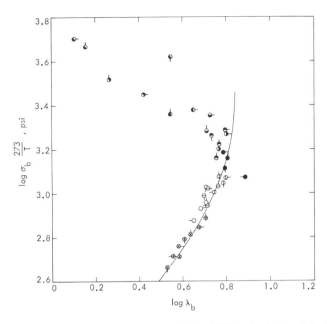

Figure 71 Failure envelope for SBR containing MT black with $\phi = 0.40$. The full curve represents the fit of the data to equation [2].

Table 11 MT Carbon Black in SBR

ϕ	$\nu_e \times 10^6$ (moles/cm^3)^a	$\nu_{e,r}{}^a$	$\nu_{e,r}{}^b$	n (units/chain)^a
0	124	1.00	1.00	58
0.10	143	1.15	1.20	51
0.15	161	1.30	1.35	43
0.20	183	1.48	1.41	47
0.25	210	1.69	1.66	80
0.30	250	2.02	1.86	83
0.35	271	2.18	2.05	100
0.40	320	2.58	2.57	80

^{a} Values obtained from fit of rupture data to equation [2].
^{b} Values obtained from reduced failure envelope.

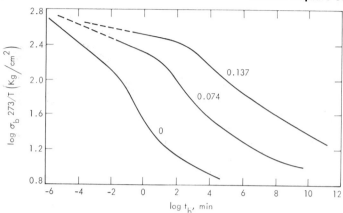

Figure 72 Dependence of σ_b on t_b for SBR containing HAF black. Data of Halpin and Bueche (119).

and were the same, within the rather wide experimental scatter, as the values obtained from superposition of creep and stress relaxation data (119). However, the activation energy progressively increased with increase in ϕ. For SBR, $\Delta H = 26$ kcal/mole for $\phi = 0$, while $\Delta H = 41$ kcal/mole for $\phi = 0.14$. The time dependence of σ_b for the SBR–HAF system is shown in Fig. 72. It may be observed that at short reduced times, i.e., low test temperatures, the response shows a weak dependence on ϕ and the curves tend to converge as the test temperature approaches T_g. At long reduced times, the shape of the response is roughly independent of ϕ so that the curves representing systems having different ϕ values are nearly parallel. It is probable that such differences in shape that may exist are related, in large measure, to the different temperature dependence of a_T, i.e., to a variation of ΔH with ϕ.

The similar shapes of the long time response suggest the possibility of superposing the individual curves by a horizontal shift along the time axis analogous to the a_x shift factor used for the SBR and viton gum elastomers. The data for the SBR–MT system demonstrate that the same vertical shift factors used to superpose the stress–strain response also superpose the high temperature portions of the failure envelopes; this shift factor is the modulus. This suggests that normalization of σ_b by the modulus, whether unfilled or filled, will lead to σ_b/E, t_b and ϵ_b, t_b curves whose long time segment may be superposable by a horizontal shift along the t_b axis. The magnitude of the shift for a filled material could depend on ϕ as well as on ν_e.

Such a superposition has been achieved (53) using the butyl data reported by Smith (8). In order to avoid complications due to crystallization, only data obtained at the two highest temperatures, 70 and 100°C, could be used. Values of log $\sigma_b T_0/T$ and log ϵ_b were plotted as a function of log t_b for each temperature. The isothermal response was shifted along the time scale until superposition was achieved. This procedure yields a_T. The same a_T was found to be applicable to both σ_b and ϵ_b data and furthermore was independent of ϕ. After dividing σ_b by the appropriate $F(1)$ value, the log $\sigma_b/F(1)$, log t_b response for the various ϕ values was then shifted along the time scale, taking the unfilled elastomer as the reference. It was observed that superposition could be obtained; furthermore, the same shift factors were found to be applicable to both $\sigma_b/F(1)$ and ϵ_b. The magnitude of the shift increased progressively with increase in ϕ and was qualitatively similar to what would be expected in an unfilled system if ν_e had decreased. If it is assumed that $F_r(1)$ for these butyl vulcanizates is given by equation [90] with $\phi_m = 0.28$, and the differences between the calculated $F_r(1)$ and the values estimated experimentally are due to a decrease in ν_e of the binder, then the a_x factors calculated from equation [27] are equal to the shift factors required for superposition. This is true for ϕ values up to 0.101.

5.3 Breaking Energy of Filled Elastomers

Harwood and Payne have shown that equation [44], relating energy density at break to the hysteresis loss at break for a gum elastomer, is also applicable to elastomers that contain filler (59). For example, data for an acrylonitrile-butadiene rubber containing HAF carbon black with ϕ varying from 0 to 0.31, yield a linear relationship with a slope of $\frac{2}{3}$ except at low temperatures where the data tended to converge. As the filler level increased, the response, while still linear and parallel to that of the gum, was displaced progressively downward; this fact demonstrates that for a given value of U_B, the hysteresis loss at break increases with increase in ϕ. It was suggested that this additional energy loss is caused by the fact that the actual deformation of the binder in the filled system is greater than the overall or measured deformation. It was shown also that the log U_B, log H_B response could be superposed by shifting along the H_B axis. The magnitude of the shift, X, increased with increase in ϕ and it was assumed in fact that $X = E_r$. The dependence of E_r on ϕ was assumed to be given by equation [53] with $\alpha_1 = 2.5$ and $\alpha_2 = 14.1$. However, it was noted that the value of the shift factor as given by this expression tended to overcorrect some of the data, i.e., shift the data too far. Actually, it is doubtful whether this expression can describe the modulus dependence on ϕ for an active black such as HAF over any extended range of ϕ.

Mullins and Tobin found for the same black, but in natural rubber, that this equation was not applicable (120). However, they were able to fit their data using equation [54] with $f = 6.5$, which predicts a higher modulus at a given ϕ value than does equation [53]. Hence, the use of a more applicable expression for the modulus, ϕ response would apparently introduce an even greater overcorrection.

Harwood and Payne conclude that for the filled acrylonitrile-butadiene system, the data for both gum and filled vulcanizates can be described by

$$U_B = \left(\frac{H_B}{X}\right)^{2/3} \tag{100}$$

where $X = 1$ for the gum vulcanizate. By plotting U_B as a function of ϵ_b, curves similar to failure envelopes were obtained. Applying the X shift to the ϵ_b axis, the high temperature portions of such envelopes were found to superpose and the data for temperatures below that at which $(\epsilon_b)_m$ is observed could be described by

$$U_B = A(\epsilon_b X)^2 \tag{101}$$

where A is a constant. Eliminating X between [100] and [101] yields

$$U_B = B(H_B \epsilon_b)^{1/2} \tag{102}$$

where B is a constant. It was shown that this equation was applicable to all the data for the acrylonitrile-butadiene system.

The effect of swelling on the rupture behavior of filled elastomers was studied by Grosch (57). He observed that σ_b, based on unit volume of dry rubber, decreases with decrease in v_r. This is similar to the behavior observed with gum elastomers. Below a certain value of v_r which decreases with ϕ, σ_b apparently tends toward a limiting value which increases progressively with increase in ϕ. The σ_b, v_r response for a given elastomer could be superposed by shifting both vertically and horizontally; the magnitude of the vertical shift was found to be equal to the modulus, while the magnitude of the horizontal shift increased with ϕ. For a gum rubber, the horizontal shift factors were shown to be directly proportional to $T - T_g$. However, for carbon black filled rubber, the shift factors were too large to be accounted for by the $T - T_g$ factor alone; this was attributed to the presence of shells of rubber surrounding the filler particles and containing a locally higher concentration of network chains. Both the vertical and horizontal shift depend in magnitude on the type of rubber and type of filler.

The dependence of U_B, referred to unit volume of dry rubber, on v_r was also investigated, and it was found that all curves for a given elastomer and representing different ϕ values were similar in shape. A master curve could be obtained by applying both a vertical and horizontal shift. For SBR and *cis*-1,4-polybutadiene vulcanizates, the same horizontal shift used to superpose the σ_b, v_r response was also found to be applicable to the U_B, v_r data. On the other hand, the vertical shifts for SBR obtained from the U_B, v_r response were significantly lower than the values obtained from the σ_b, v_r response. This discrepancy in the case of SBR was ascribed to differences in the shape of the stress-strain curves.

REFERENCES

1. E. Guth et al., *J. Appl. Phys.*, **17**, 347 (1946).

2. R. F. Landel and P. J. Stedry, *J. Appl. Phys.*, **31**, 1885 (1960).

3. T. L. Smith, *Trans. Soc. Rheol.*, **6**, 61 (1962).

4. J. C. Halpin, *Polym. Lett.*, **2**, 969 (1964).

5. R. F. Landel and R. F. Fedors, "Rupture of Amorphous, Unfilled Polymers," in B. Rosen, ed., *Fracture Processes in Polymeric Solids, Phenomena and Theory*, Wiley-Interscience, New York, 1964.

6. M. L. Williams et al., *J. Am. Chem. Soc.*, **77**, 3701 (1955).

7. T. L. Smith and R. A. Dickie, *J. Polym. Sci.*, A-2, **7**, 635 (1969).

8. T. L. Smith, *Technical Documentary Report No. ASD-TDR-63-430*, Wright-Patterson Air Force Base, Ohio, May 1963.

9. T. L. Smith, *J. Polym. Sci.*, **20**, 89 (1956).

10. J. A. C. Harwood and A. Schallamach, *J. Appl. Polym. Sci.*, **11**, 1835 (1967).

11. F. Bueche, *Physical Properties of Polymers*, Wiley-Interscience, New York, 1962.

12. A. V. Tobolsky, *Properties and Structures of Polymers*, Wiley, New York, 1960.

13. J. D. Ferry, *Viscoelastic Properties of Polymers*, Wiley, New York, 1961.

14. J. C. Halpin, *J. Appl. Phys.*, **36**, 2475 (1965).

15. W. Kuhn and F. Grün, *Kolloid-Z.*, **101**, 248 (1942).

16. H. M. James and E. Guth, *J. Chem. Phys.*, **11**, 455 (1943).

17. L. R. G. Treloar, *The Physics of Rubber Elasticity*, Oxford Univ. Press, London, 1958.

18. R. F. Fedors and R. F. Landel, *Space Programs Summary, 37-36*, Vol. IV, Jet Propulsion Laboratory, Pasadena, California, December 1965, p. 137.

19. G. Natta et al., *J. Polym. Sci.*, A-1, **1**, 3569 (1962).

20. T. L. Smith and J. E. Frederick, *J. Appl. Phys.*, **36**, 2996 (1965).

21. F. Bueche and J. C. Halpin, *J. Appl. Phys.*, **35**, 36 (1964).

22. M. C. Morris, *J. Appl. Polym. Sci.*, **8**, 545 (1964).

23. L. R. G. Treloar, *Trans. Faraday Soc.*, **50**, 881 (1954).

24. G. M. Martin et al., *Trans. Inst. Rubber Ind.*, **32**, 189 (1956).

25. J. C. Halpin, *J. Polymer. Sci.*, **C16**, 1037 (1967).

26. D. S. Villars, *J. Appl. Phys.*, **21**, 565 (1950).

27. A. M. Borders and R. D. Juve, *Ind. Eng. Chem.*, **38**, 1066 (1946).

28. G. Gee, *J. Polym. Sci.*, **2**, 451 (1947).

29. T. L. Smith, *J. Polym. Sci.*, **A1,** 3597 (1963).

30. T. L. Smith, *ASTM Special Technical Bulletin No. 325*, 1962, p. 60.

31. T. L. Smith, *J. Appl. Phys.*, **35,** 27 (1964).

32. R. F. Landel and R. F. Fedors, *Rubber Chem. Technol.*, **40,** 1049 (1967).

33. J. C. Halpin, *J. Appl. Phys.*, **35,** 3133 (1964).

34. R. F. Landel and R. F. Fedors, *J. Polym. Sci.*, **31,** 539 (1963).

35. T. L. Smith, *Proc. R. Soc.*, **282A,** 102 (1964).

36. F. Bueche and J. C. Halpin, *J. Appl. Phys.*, **35,** 36 (1964).

37. T. L. Smith, *Technical Documentary Report No. ASD-TDR*-62-572, Wright-Patterson Air Force Base, Ohio, June 1962.

38. R. F. Landel and R. F. Fedors, in E. H. Lee, ed., *Proceedings of the Fourth International Congress of Rheology*, Part 2, Wiley-Interscience, New York, 1965, p. 525.

39. R. F. Fedors and R. F. Landel, *Trans. Soc. Rheol.*, **9,** 195 (1965).

40. T. L. Smith, in E. H. Lee, ed., *Proceedings of the Fourth International Congress on Rheology*, Part 2, Wiley-Interscience, New York, 1965, p. 525.

41. P. J. Flory, *Chem. Rev.*, **35,** 51 (1964).

42. R. F. Fedors and R. F. Landel, *J. Appl. Polymer Sci.*, **19,** 2709 (1975).

43. J. P. Berry and W. F. Watson, *J. Polym. Sci.*, **18,** 201 (1955).

44. F. Bueche et al., *Polym. Lett.*, **3,** 399 (1965).

45. J. Furukawa et al., *Polym. Lett.*, **8,** 25 (1970).

46. J. Brandrup and E. H. Immergut, eds., *Polymer Handbook*, Wiley-Interscience, New York, 1966.

47. R. F. Fedors and R. F. Landel, *Space Programs Summary, 37-58*, Vol. III, Jet Propulsion Laboratory, Pasadena, California, August 1969, p. 180; *J. Polym. Sci., Polymer Physics*, **13,** 419 (1975).

48. D. J. Plazek, *J. Polym. Sci.*, **4,** 745 (1966).

49. K. C. Baranwal, Ph.D. Thesis, University of Akron, Ohio, 1967.

50. J. C. Healy, Ph.D. Thesis, University of Akron, Ohio, 1967.

51. G. R. Taylor and S. R. Darin, *J. Polym. Sci.*, **17,** 511 (1955).

52. L. M. Epstein and R. P. Smith, *Trans. Soc. Rheol.*, **9,** 195 (1965).

53. R. F. Fedors, unpublished results.

54. E. DiGiulio et al., *Rubber Chem. Technol.*, **39,** 726 (1966).

55. H. W. Greensmith et al., in L. Bateman, ed., *The Chemistry and Physics of Rubberlike Substances*, MacLaren, London, 1963, p. 255.

56. T. L. Smith, *J. Polym. Sci.*, **16C,** 841 (1967).

57. K. A. Grosch et al., *Nature*, **212,** 497 (1966).

58. K. A. Grosch et al., Physical Basis of Yield and Fracture: Conference Proceedings, Institute of Physics and Physical Society Conference Series No. 1, 1966, p. 144.

59. J. A. C. Harwood and A. R. Payne, *J. Appl. Polym. Sci.*, **12,** 889 (1968).

60. F. S. C. Chang, *J. Appl. Polym. Sci.*, **8,** 37 (1964).

61. K. A. Grosch, *J. Appl. Polym. Sci.*, **12,** 915 (1968).

62. H. Fujito and A. Kishimoto, *J. Polym. Sci.*, **28,** 547 (1958).

63. J. D. Ferry and R. A. Stratton, *Kolloid-Z.*, **171,** 107 (1960).

64. A. Einstein, *Ann. Phys.* **17,** 549 (1906).

65. R. Eisenschitz, *Phys. Z.*, **34,** 411 (1935).

66. E. Guth and H. Mark, *Ergeln. Exakt. Naturwiss.* **12,** 115 (1933).

67. G. I. Taylor, *Proc. R. Soc. A*, **138,** 41 (1932).

68. E. Guth and O. Gold, *Phys. Rev.*, **53,** 322 (1938).

69. R. Simha, *J. Appl. Phys.*, **23,** 1020 (1952).

70. W. Vand, *J. Phys. Colloid Chem.*, **52,** 277 (1952).

71. R. S. J. Manley and S. G. Mason, *Can. J. Chem.*, **33,** 763 (1955).

72. E. Guth, Proceedings of the Second Rubber Technology Conference, London, 1948, p. 353.

73. J. V. Robinson, *J. Phys. Colloid Chem.*, **53,** 1042 (1949).

74. W. Vand, *Nature*, **155,** 364 (1945).

75. H. C. Brinkman, *J. Chem. Phys.*, **20,** 571 (1952).

76. R. Roscoe, *Brit. J. Appl. Phys.*, **3,** 267 (1952).

77. V. Fidleris and R. L. Whitmore, *Rheol. Acta*, **1,** 38 (1958).

78. S. G. Ward and R. L. Whitmore, *Brit. J. Appl. Phys.*, **1,** 286 (1950).

79. R. F. Landel et al., in E. H. Lee, ed., *Proceedings of the Fourth International Congress on Rheology, Brown University, Rhode Island, August 26–30, 1963* Part 2, Wiley-Interscience, New York, 1965, p. 663.

80. M. Mooney, *J. Colloid Sci.*, **6,** 162 (1951).

81. H. Eilers, *Kolloid-Z.*, **97,** 313 (1941).

82. H. Eilers, *Kolloid-Z.*, **102,** 154 (1943).

83. S. H. Maron and R. J. Belner, *J. Colloid Sci.*, **10,** 523 (1953).

84. J. V. Robinson, *Trans. Soc. Rheol.*, **1,** 15 (1957).

85. L. C. Graton and H. J. Fraser, *J. Geol.*, **43,** 785 (1935).

86. G. D. Scott, *Nature*, **188,** 908 (1960).

87. H. Susskind et al., *Brookhaven National Laboratory Report*, BNL 50022, T-441, June 1966.

88. G. D. Scott and D. M. Kilgour, *Brit. J. Appl. Phys. (J. Phys. D)*, **2,** 863 (1969).

89. P. S. Roller, *Ind. Eng. Chem.*, **22,** 1206 (1930).

90. B. G. Moser et al., *Space Programs Summary, 37-34*, Vol. IV, Jet Propulsion Laboratory, Pasadena, California, August 1965, p. 133.

91. A. de Waele and E. W. J. Mardles, Proceedings of the International Rheology Congress, 1st Congress Scheveningen, Part II, 1949, p. 166.

92. W. D. Harkins and D. M. Gans, *J. Phys. Chem.*, **36,** 86 (1932).

93. C. C. Furnas, *U.S. Bureau of Mines, Rep. Invest.* 2894, October 1928.

94. C. C. Furnas, *Ind. Eng. Chem.*, **23,** 1052 (1931).

95. H. M. Smallwood, *J. Appl. Phys.*, **15,** 758 (1944).

96. Z. Hashin, *Bull. Res. Council Israel*, **5c,** 46 (1955).

97. J. D. Eshelby, *Proc. R. Soc. London*, **241A,** 376 (1957).

98. J. K. Mackenzie, *Proc. Phys. Soc. London*, **B63,** 2 (1950).

99. J. G. Oldroyd, in R. Grammel, ed., *Deformation and Flow of Solids, Colloquium Madrid, September 26–30, 1955,* Springer-Verlag, Berlin, 1956, p. 304.

100. E. H. Kerner, *Proc. Phys. Soc.,* **1956,** 698, 808.

101. S. Uemura and M. Takayanagi, *J. Appl. Polym. Sci.,* **10,** 113 (1966).

102. K. Okano, *Rept. Prog. Polym. Phys. Japan,* **3,** 69 (1960).

103. C. van der Poel, *Rheol. Acta,* **1,** 196 (1958).

104. B. Paul, *Trans. Am. Inst. Mining Metall. Pet. Eng.,* **1960,** 218, 36.

105. Z. Hashin and S. Shtrikman, *Franklin Inst. J.,* **271,** 336 (1961).

106. Z. Hashin, *J. Appl. Mech.,* **29E,** 143 (1962).

107. T. T. Wu, *J. Appl. Mech.,* **1965,** 211.

108. T. T. Wu, *Int. J. Solid Struct.,* **2,** 1 (1966).

109. F. L. Saunders, *J. Colloid Sci.,* **16,** 13 (1961).

110. T. Gillespie, *J. Colloid Sci.,* **18,** 32 (1963 .

111. D. G. Thomas, *J. Colloid Sci.,* **20,** 267 (1965).

112. J. V. Robinson, *J. Phys. Colloid Chem.,* **55,** 455 (1951).

113. A. R. Payne, *J. Appl. Polym. Sci.,* **6,** 57 (1962).

114. A. Voet and R. Cook, *Rubber Chem. Technol.,* **40,** 1364 (1967).

115. A. R. Payne, *J. Appl. Polym. Sci.,* **6,** 368 (1962).

116. R. E. Dollinger et al., *Rubber Chem. Technol.,* **40,** 1311 (1967).

117. A. R. Payne, in G. Kraus, ed., *Reinforcement of Elastomers,* Wiley-Interscience, New York, 1965, p. 72.

118. A. R. Payne, *J. Colloid Sci.,* **19,** 744 (1964).

119. J. C. Halpin and F. Bueche, *J. Appl. Phys.,* **35,** 3142 (1964).

120. L. Mullins and N. R. Tobin, *J. Appl. Polym. Sci.,* **9,** 2993 (1965).

121. L. D. Loan, *J. Appl. Polym. Sci.,* **7,** 2259 (1963).

122. R. F. Fedors and R. F. Landel, *Space Programs Summary, 37-41,* Vol. IV, Jet Propulsion Laboratory, Pasadena, California, October 1966, p. 97.

123. R. F. Fedors and R. F. Landel, *Space Programs Summary, 37-40,* Vol. IV, Jet Propulsion Laboratory, Pasadena, California, August 1966, p. 80.

124. R. F. Fedors and R. F. Landel, *Space Programs Summary, 37-43,* Vol. IV, Jet Propulsion Laboratory, Pasadena, California, February 1967, p. 177.

125. K. Bills and F. S. Salcedo, *J. Appl. Phys.,* **32,** 2364 (1961).

126. F. Bueche, *J. Appl. Polym. Sci.,* **4,** 107 (1960).

127. R. F. Landel, *Trans. Soc. Rheol.,* **2,** 53 (1958).

128. R. F. Landel, Paper presented before the Division of High Polymer Physics, American Physical Society, Boston, Massachusetts, 1959.

129. C. W. Van der Wal et al., *J. Appl. Polym. Sci.,* **9,** 2143 (1965).

K. A. Grosch

Uniroyal European Tire Development Center
Aachen, Germany
formerly with
The Malaysian Rubber Producers' Research Association, Hertford,
Great Britian

Chapter 13   Some Physical Aspects of Elastomers Used in Tires

1   INTRODUCTION

It is appropriate that a book on stereoregular rubbers should also contain
a chapter on tires and their properties because more than 60% of the
world's rubber production is used in tires. It was also the unprecedented

growth of the car industry that stimulated the growth of the rubber industry.

Before World War II all tires were made of natural rubber (NR), but by the end of the war the first synthetic rubber, styrene–butadiene (SBR), had become established in the United States as a second general purpose rubber. Its use in certain parts of tires became rapidly established in the rest of the world also. Although this process was undoubtedly hastened by economic considerations, this could not have occurred if SBR did not also offer some decisive technical advantages over the natural product.

Today SBR has been joined by a whole series of synthetic elastomers that enable the tire designer to select an elastomer according to its best technical and economic merit. Of these latest additions to general tire rubbers the stereoregular rubbers culminating in *cis*-polybutadiene (BR) and *cis*-polyisoprene (IR), the synthetic "natural" rubber, made the most significant contributions.

Although many different compounds may be used in building a tire, most of the rubber is found in three regions: the tread, the carcass, and the sidewalls. The compounds in the different sections undergo different physical processes during the service life of the tire and hence have to meet different physical requirements. For example, a tread should primarily give good road holding and show a low rate of wear; the carcass must not fail because of overheating under high speed running of the tire; and the sidewalls should stand up to repeated flexing without cracking and show adequate resistance to cutting, oxygen and ozone aging, and curb chafing.

These requirements are common to all tires, but different applications may, in addition, demand one property to be particularly outstanding. Thus a car tire is often driven to the limit of its road holding capacity, and even a marginal increase in this property is important to the safety of the driver. In contrast, for large truck tires, a low heat buildup in the carcass is paramount in order to achieve the required durability. This necessarily means a different type of tread compound than that which would be used in passenger car tires.

Today, not only are tires no longer made of a single elastomer, but even the compounds in the different sections of the tires consist of blends of two or even three different rubbers in order to obtain maximum technical and economic advantage. The selection of materials and their compounding for the use in tires is still largely a matter of trial and error, since the stresses and strains in tires at any point and the response of a particular material to them are too complex to be predictable or completely understood. In addition, increasing sophistication of tire design also makes continuous new demands on tire compounds.

Until recently virtually all tires were of the "bias type" construction. Rubber coated plies made of equally spaced rayon or nylon cords were placed on top of each other in such a way that there was a constant angle between cords of adjacent layers, the so-called crown angle of the tire. Several plies together formed the carcass. The only criteria open to the designer were the type of carcass material, the number of plies, the choice of crown angle, and the shape of the mold. This is considerably changed at present with the advent of the radial ply tire and the bias belted tire.

In the radial ply tire the carcass cords run virtually at a 90° crown angle from bead to bead. In order to lend stiffness to the structure a belt is placed around the circumference of the tire between carcass and tread. This consists of several layers of cord with a relatively small crown angle, usually between 20 and 30°, depending on the type of cord. Although the original commercially available radial ply tires were made with steel cord for the belt, for a short time rayon was widely used. In most recent years, however, steel is again becoming dominant. The design has undergone such a revolution that there is as yet no end of change in sight. While in Europe the radial ply construction is beginning to dominate the scene, in the United States a halfway construction became widespread: the bias belted tire in which the carcass plies, with a conventional crown angle, are supplemented by a belt under the tread with a smaller crown angle than that of the carcass but usually larger than is commonly used in radial ply tires. Glass fiber is most commonly used as the belt material for bias belted tires. It is obvious that such fundamental changes in design also made specific new demands on compounds, such as adhesion to steel cord, or glass fiber, and green strength of the carcass coating.

This chapter attempts to give a brief survey of the various physical processes, as far as they can be defined, which rubber undergoes in tire service. These are invariably only met in part by any one type of rubber, but the existence of the wide range of elastomers allows the tire designer through selection and blending to achieve a balance of properties which makes the tire the versatile engineering product it has become today.

2 THE ROAD HOLDING OF TIRES

2.1 The Forces on a Tire

It is the prime function of the tire as the link between self-propelled vehicle and road to permit the driver accurate steering control of the vehicle at all practical speeds. That this is possible at all is due to the high elasticity of the tire and the high friction between rubber and road. Figure 1 gives a view of the contact area of a small model wheel (*a*) when rolling

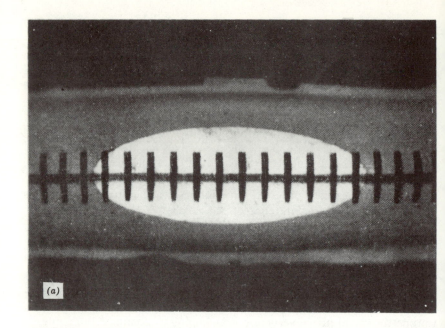

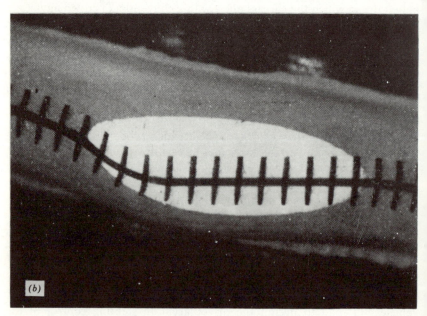

Figure 1 View of the contact region of a model wheel (a) when rolling freely, (b) when running under slip, photographed through a glass plate (1).

freely and (b) when traveling at an angle to the direction of forward motion (1). In the second case it is seen that the rubber initially adheres to the track surface, distorting the periphery of the wheel out of its original plane along the forward traveling direction, until at some point in the contact area the elastic forces of the wheel are large enough to overcome the friction between track and rubber and pull the distorted section back into its original position.

The sum of the elastic forces in the wheel adds up to an external force, referred to as *side force*, acting normally to its plane and thus making it possible to steer. When traveling along a circular path (as in driving around a corner) the combined side forces of the four tires balance the centrifugal force which tends to move the car out of its intended path. The larger the centrifugal force, the more the tire has to be distorted to produce sufficient side force in order to balance the centrifugal force.

For a wheel model in the form of a thin disk, in which neighboring elements can deflect independently of each other, the center line of the wheel in the contact area will distort as shown in diagrammatic form in Fig. 2, and the side force is then given by

$$S = \int_0^a My\,dx \qquad [1]$$

where M is a stiffness constant of the model. In a real tire the distorted region is not confined to the contact area but reaches beyond it on either side so that equation [1] is only obeyed approximately. However, Figure 2 displays all the essential features of a tire which are of concern here.

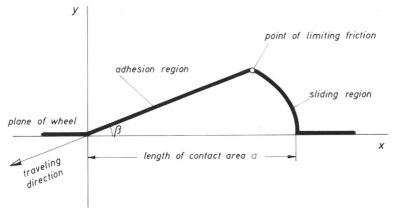

Figure 2 Deflection of a wheel model in the contact area slipping under angle β in the direction of forward motion. (1)

The point at which the tire begins to slide back is determined by the limiting friction coefficient between road and tire surface, the local ground pressure, and the lateral stiffness of the tire. As the angle β between the forward traveling direction and the plane of the wheel, referred to as the slip angle, is increased, the point of initial sliding moves toward the front of the contact area until eventually sliding occurs over the whole contact area. As this point is approached the side force tends toward a limiting value given by

$$S_1 = \mu L \qquad [2]$$

where μ is the coefficient of friction and L is the wheel load (1). The direction is then no longer perpendicular to the plane of the wheel but acts opposite to the direction in which the car traveled when this condition was reached, and continues to act opposite to any direction in which the car might subsequently move. All steering control is therefore lost at this point (2).

Figure 3 shows typical side force-slip angle curves on a dry surface. Similar curves have been extensively reported in the literature and the references quoted are only a small selection (2–5). At small slip angles,

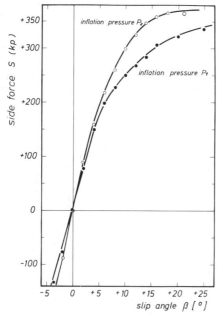

Figure 3 The side force S that a tire generates when running at different slip angles β. The two curves were obtained at two different inflation pressures. The side force slip angle characteristic depends also on normal load, construction, and coefficient of friction between tire and surface.

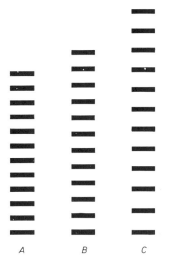

A   B   C

Figure 4 A rubber toothed wheel when (A) transmitting a driving torque, (B) rolling freely, (C) transmitting a braking torque (1)

the side force increases linearly with the slip angle. At large slip angles the side force tends toward the limiting value μL, which is virtually reached between 15 and 25° slip angle.

For reasons of simplicity the case of steering was considered where the forces act perpendicularly to the plane of the tire. A similar situation exists when a car is accelerated or braked. The forces on the car have to be supported by forces in the road acting across the tire and in this case in its plane. Figure 4 shows a model wheel when freely rolling (B), when braked (C), and when accelerated (A), in each case for one revolution (i.e., for the same number of teeth) (1). Clearly, the wheel travels different distances in the three cases. With an accelerating torque the teeth of the wheel are pressed closer together on entering the contact area. They remain in this state until at the rear of the contact area the contact pressure has dropped to a value such that the frictional force is smaller than the elastic forces in the wheel and sliding takes place to restore the wheel to equilibrium; i.e., it goes into tension to balance the compression at the front of the wheel. Exactly the reverse is the case when the wheel is braked during rolling. In this case the wheel is extended on entering the contact area and compressed on leaving it. In all these cases there is a region of static friction and a sliding region, the proportion of the latter to the former depending on the size of the forces to be transmitted, on the nature of the surface, the tire construction, the pressure distribution in the contact region, and finally on the friction coefficient between tire and road surface.

2.2   The Coefficient of Friction of Elastomers

2.2.1   *Its Temperature and Speed Dependence*

In defining a friction coefficient μ it is assumed that the classical relation holds, namely that the tangential force F to move a piece of rubber under a normal load L is proportional to that load. This is approximately true for the type of surface and compound and the range of loads which occur in tire applications (6), although it no longer holds on smooth surfaces under very high pressures (7).

For solids such as metals and wood, the friction coefficient depends remarkably little on the type of material, the surface finish, or the sliding speed and temperature (8). In contrast, the friction coefficient of rubber varies considerably with the roughness of the surface on which it slides and, above all, it depends critically on temperature and sliding speed.

In discussing the temperature and speed dependence of the coefficient of friction it is convenient to start with speeds at which the temperature

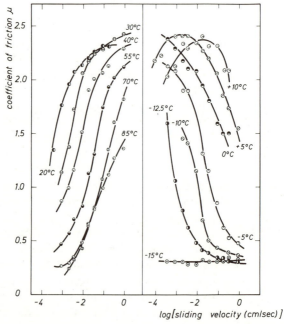

Figure 5 Coefficient of friction of unfilled acrylonitrile-butadiene rubber (NBR) on pinhead glass at various temperatures as a function of sliding speed (9).

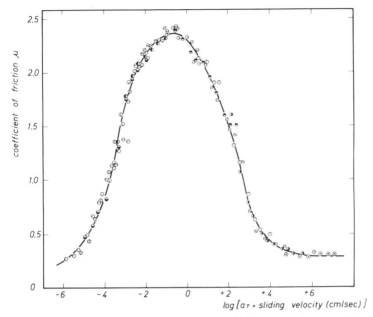

$log [a_T \times sliding\ velocity\ (cm/sec)]$

Figure 6 Master curve for coefficient of friction of NBR gum rubber at 20°C, derived from the graphs of Fig. 5 (9).

rise due to frictional heat generation can be ignored so that temperature and speed can be treated as independent of each other. Figure 5 shows the friction coefficient of a pure gum nitrile (NBR) rubber as a function of speed at various temperatures when sliding on a smooth glass surface (9). The speeds were varied between 10^{-3} and about 1 cm/sec. Inspection of the curves shows that by a shift along the speed axis these curves can be assembled to a single master curve (Fig. 6) and a shift function whereby the master curve represents the friction coefficient as a function of speed at a convenient reference temperature multiplied by a temperature dependent factor a_T, its temperature dependence being determined by the shift function (10).

This type of transformation is always possible whenever a relaxation process which itself is temperature dependent is at work (11). In rubber physics it was first employed successfully to describe the frequency dependence of the modulus (12, 13) but later also such diverse properties as the rate dependence of the tensile strength (14) and even the rate and temperature dependence of sliding abrasion (15, 16) (see also Section 3.1.2). In all cases and for all polymers the shift function has a similar

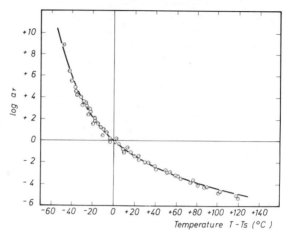

Figure 7 The shift function, log a_T, of four different gum rubbers, obtained from friction data on dry glass tracks, reduced to a standard reference temperature $T_s = T_g + 50$. The solid line represents the WLF equation (9, 10). ⊖, NBR; ⊖, SBR; ⊙, butyl; ⊙, NR.

shape and can itself be transformed into a universal function if the reference temperature chosen for the master curve lies at least 50°C above the glass transition temperature T_g as first demonstrated by Williams, Landel, and Ferry (the WLF equation) (10). This is shown in Fig. 7 for shift functions obtained from friction data on dry glass tracks for four different gum rubbers with the solid line representing the WLF equation.

These results demonstrate clearly that the friction coefficient is closely linked to the viscoelastic properties of the rubber; this becomes even more apparent if master curves of different gum rubbers on glass are compared with each other. Figure 8 shows the master curves of the four gum rubbers whose shift functions were mentioned above, and one for a BR gum rubber also on a glass track. All curves are similarly shaped, displaying a single, well-defined maximum. They rise at low transformed speeds (log $a_T v$ where v is the speed) from very low values of about 0.2 to more than 10 times this value, and fall again to about 0.2 at very high, transformed speeds. The major differences in the frictional behavior between the different rubbers arise then from the position of the curves on the transformed speed axis, conveniently represented by the speed v_m at which the friction coefficient is a maximum. The NBR has its maximum at a lower speed than the SBR, and the friction maximum of BR occurs at a speed and temperature (transformed speed) that were outside the experimental range. An exactly similar behavior is observed with the position of the master curves for the real and imaginary parts of the dynamic moduli of these rubbers and tan δ, the loss factor, all plotted as a function of the transformed frequency. Comparing the speed v_m of maximum friction with the frequency f_m of maximum loss modulus (the imaginary part of the modulus) yields the relation

$$\frac{v_m}{f_m} = \lambda \qquad [3]$$

where λ is the same constant of 6×10^{-7} cm for all rubbers that have been examined. This is a distance of molecular dimensions and has been termed the jump distance (17).

Various models of rubber friction have been developed, notably by Schallamach (18, 19), Bartenev and El'Kin (20), and Bulgin et al. (21). The simplest model, by Bulgin, envisages small molecular complexes that make adhesional contact with the glass and under sliding deform until a constant limiting adhesional force is reached. At this point the contact is

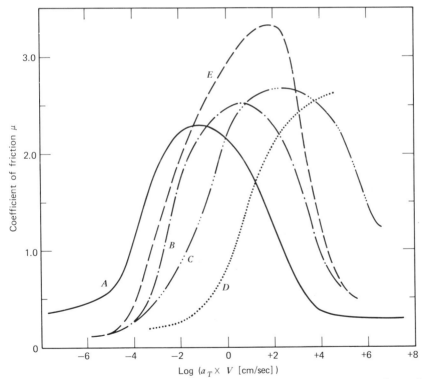

Figure 8 Master curves of the friction coefficient of five gum rubbers on glass, all reduced to a reference temperature of 20°C. A, NBR; B, SBR; C, NR; D, BR; E, butyl.

broken and the molecular complexes contract to repeat the process. Because of its viscoelastic nature, energy is lost in this molecular stress-strain cycle, the amount of which depends on the temperature and frequency in the same way as the bulk properties. Schallamach's theory is more sophisticated in that he assumes that the number of contact sites available are temperature and speed dependent; i.e., adhesion itself is a relaxation process.

On rough surfaces a further mechanism of friction comes into play. The experimental data can still be transformed into master curves and the universal Ferry shift function, but the shape of the master curve is different, as shown in Fig. 9 which compares the master curves of an NBR gum rubber on two surfaces, glass and an abrasive carborundum paper. At the speed at which, on the smooth surface, the friction reached its maximum value, a hump is visible for the friction curve on the rough surface. But instead of falling off at still higher speeds, the friction coefficient rises further, reaching a very sharp maximum at a very high transformed speed. In this case, a constant relation exists between the transformed frequency of the maximum loss factor tan δ and the speed at maximum friction on the abrasive surface.

The proportionality constant is of the same order of magnitude as the roughness of the surface, measured in terms of the spacing of the abrasive particles. It appears that the rubber is deformed around the particles undergoing a stress cycle as it passes over the roughness of the surface, again losing energy because of its viscoelastic nature. This process is

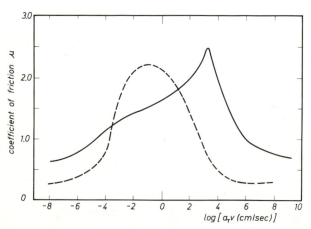

Figure 9 Master curve for the friction coefficient of a gum NBR obtained on a silicon carbide (grade 180) paper track (full line). The dashed line shows the curve obtained on a smooth glass track, both transformed to the same reference temperature.

therefore termed deformation friction; the first theory attempting to describe it quantitatively was formulated by Greenwood and Tabor (22). On a perfectly lubricated rough surface it would be expected to be the only friction mechanism operating, the rubber being deformed by the asperities of the track with a constant strain amplitude acting normally to the sliding direction. However, a simple calculation shows that this contribution to the friction is small and could certainly not explain the very high friction maximum obtained in the experiments described above. Moreover, in the region of low transformed speeds, where both adhesional friction and deformation losses in the rubber are small, the friction on the rough track is three times higher than on the smooth one. It appears that deformation losses contribute to rubber friction because of the presence of adhesion. Even small adhesional forces along the slopes of the asperities can produce large tangential stresses in the rubber. It is also quite feasible that, under tangential stresses in the rubber, elastically stored energy is lost in the frictional process. This is obviously the case for the abraded particles which are detached from the rubber during the friction process. Moreover, as the deformed volume involved in an abrasion process is much larger than the volume of the abraded particles, additional elastically stored energy can be lost if that energy cannot be transferred back to the track. This would be the case if the distortion in the rubber at the rear of the contact area was so large that contact between rubber and asperity would be lost. The elastic energy would then presumably be dissipated as heat in a damped vibration. Except for perfectly lubricated systems, therefore, adhesional friction also contributes to the deformation friction.

2.2.2 The Effect of Carbon Filler

Gum rubbers, used above to demonstrate simply and clearly the elastomer effects on the friction of rubber, play no part in tires because they are too soft. Tire compounds invariably contain a substantial amount of carbon black filler. The inclusion of black does not affect the transformation characteristics of the rubber but it considerably changes the shape of the master curve, as shown in Fig. 10. Master curves were obtained using the same gum NBR discussed above, but filled with 20 parts of HAF black per 100 parts of rubber (phr) and with 50 phr, respectively. Three types of surface were used: glass, abrasive paper, and the same paper dusted with a fine layer of magnesium oxide powder. This was done in order to suppress the rubber adhesion effect and isolate the deformation friction as far as possible.

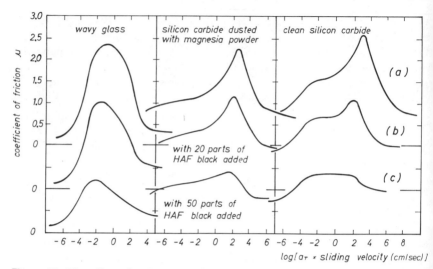

Figure 10 The effect of carbon black filler on the shape of the coefficient of friction master curves of an NBR compound (a) as a gum rubber, (b) filled with 20 parts of HAF black, and (c) filled with 50 parts of HAF black. Each curve was obtained on three surfaces: glass, clean silicon carbide, silicon carbide dusted with magnesia powder.

    In all cases the position of the curves on the transformed speed axis is essentially maintained, except that on glass, a very small shift toward lower speeds is noticeable as the carbon black content is increased. However, the maximum value that can be achieved is progressively reduced with increasing amounts of black. The same holds for the deformation friction maximum on the dusted surface. The application of magnesium oxide to the track eliminates the hump in the master curve of the gum rubber but leaves the deformation peak unchanged. On the addition of 20 parts of black to the rubber the peak is substantially reduced, and this reduction is even more pronounced when 50 phr of black are present (15).

    On the clean abrasive the adhesion hump and the deformation peak are reduced to about the same values by the inclusion of 50 phr of HAF black so that the master curve displays a broad plateau, which starts approximately at the speed at which the adhesion friction has its maximum and ends at the speed of maximum deformation friction, the friction being lower on either side of it. This feature is common to all rubbers usable in tire tread compounds, with the exception of butyl rubber, as shown in Fig. 11. Because of its exceptional viscoelastic behavior, the adhesion peak and deformation peak of the friction master curve occur at almost the same transformed speed. Even when loaded

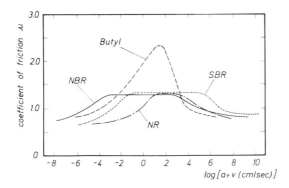

Figure 11   Master curves of the friction coefficient on a clean silicon carbide track for various rubbers, all filled with 50 parts of HAF black.

with 50 phr of HAF black, butyl still shows a pronounced single peak at that speed.

2.2.3   *Friction on Lubricated Tracks*

Measurements similar to those described above were also carried out on lubricated tracks, although confined in this case to carborundum abrasive with distilled water as lubricant. Keeping the experimental speeds again confined to values below 1 cm/sec it was possible to transform the experimental curves into master curves, as demonstrated in Fig. 12, for the gum NBR. The master curve obtained in this way displays the same features as that obtained on the dry clean abrasive track, only the maximum of the deformation peak is much reduced. But, again, the adhesion "hump" is noticeable. And just as the hump could be suppressed by coating the dry track with magnesium oxide powder, so could it be eliminated on the lubricated track by adding a detergent to the water lubricant, as shown in Fig. 13. Clearly, adhesion between rubber and track still takes place on water lubricated tracks. Indeed, the interaction between rubber and track can be strong enough for abrasion to occur, although invariably on a much reduced scale. It is interesting to note that the shift function followed more closely the Ferry transform if account was taken of the temperature dependence of the viscosity of the water lubricant by transforming the speed at a constant water viscosity (23).

    When detergent was added, only the hump disappeared, otherwise the absolute value of the friction coefficient was not affected to any significant extent. Since these values are much larger than can be accounted for by normal deformation losses in a perfectly lubricated system, tangential

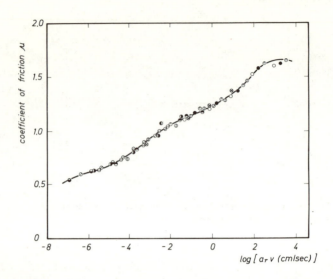

Figure 12 Master curve of the friction coefficient of a gum NBR on a carborundum stone track of 180 grade, lubricated with distilled water. The reference temperature was 20°C.

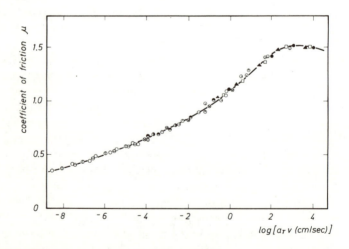

Figure 13 Master curve of the friction coefficient of a gum NBR on a carborundum stone track lubricated with distilled water to which 10 parts of detergent have been added. Same conditions as Fig. 12 except for the added detergent.

stresses must contribute to the friction coefficient. If it is assumed that the detergent formed a layer on the surface of the rubber, thereby suppressing the molecular adhesion contribution of the rubber, it also follows that the layer itself must be able to transmit considerable force from the rubber to the track surface. Even in lubricated friction, therefore, adhesion between rubber and track, possibly across a thin liquid layer, appears to play a considerable role.

Since carborundum tracks have sharp asperities with small radii of curvature at their tips, real contact pressures at the tips of the asperities are inevitably large so that the film thickness is invariably small.

To study lubricated friction between rubber and single asperities of well-defined radii of curvature of larger magnitude, the author used steel spheres sliding over thick rubber tracks, lubricated with mixtures of water and glycerol. The hydrodynamic effects are then well defined and their effect on friction can be determined fairly readily. Figure 14 shows results with a natural rubber gum compound carried out by sliding a steel ball over a thick rubber plate lubricated with a 50:50 mixture of water and glycerol. Speed and temperature range were similar to those employed in the previously described experiments. At high temperatures the curves rise with increasing speed and in this region they can be transformed, again following the Ferry transform as shown in Fig. 15. However, they reach a maximum value at much lower speed than they did on a dry surface. Beyond the maximum value the Ferry transform no longer holds. The maximum value appears not only to shift toward lower speeds at lower temperature, as would be expected from the Ferry transform, but

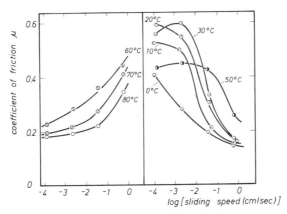

Figure 14 Coefficient of friction of a natural rubber gum compound at various speeds and temperatures obtained by sliding a $\frac{1}{4}$ in. steel ball over a rubber track lubricated with a mixture of 50 parts of water and 50 parts of glycerol. Load 20 N (23).

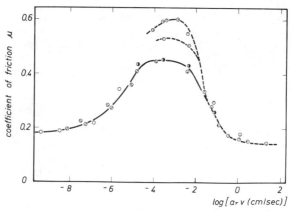

Figure 15   The data of Fig. 14, transformed into a master curve. The shift function a_T follows the WLF equation only for points on the slow speed side of the maximum.

also to increase in magnitude. It is at present not certain whether this is a real effect or is due to lack of reproducibility in the experiments. To afford a rational and easy interpretation, the data have been transformed whereby the part of the curve for which the Ferry transform was applicable has been drawn as a full line, the remainder as dotted lines. These, too, have been transformed as far as the data would allow it, with much smaller shift factors than would have been expected from the WLF equation.

Beyond the maximum the friction drops rapidly with increasing speed. This drop in friction with increasing speed on lubricated tracks has been discussed in the literature by Cohen and Tabor (24) and Walters (25) and is attributed by them to hydrodynamic effects. A very instructive paper on the nature of these effects has been published by Roberts and Tabor (26).

Examination of different types of rubber shows that the transformed speed at which the maximum is reached depends on the type of rubber (see Fig. 16) and occurs in the same order as on dry surfaces, although at much lower transformed speeds. Addition of carbon black shifted the maximum toward higher speeds (Fig. 17) or lower temperatures (transformed speed). Increasing the viscosity of the lubricant by varying the ratio of water to glycerol shifted the maximum toward lower speeds, as seen in Fig. 18.

These findings are most easily explained if it is assumed that rubber adhesion is still effective across a thin liquid film (boundary lubrication), but that it rapidly decreases once the film has reached a critical thickness (hydrodynamic lubrication). Classical lubrication theory states that the liquid film thickness between a surface and a slider increases with the

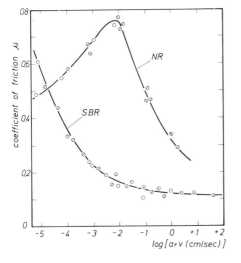

Figure 16 Master curves of the coefficient of friction of an NR and an SBR gum compound, respectively, obtained by sliding a steel ball over a rubber track. Experimental details as for Fig. 14.

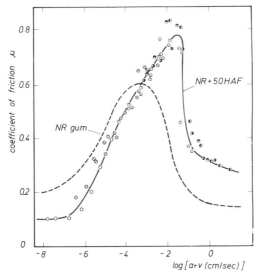

Figure 17 Comparison of master curves of the coefficient of friction for an NR gum compound and a similar compound filled with 50 pph of HAF black. Experimental details as for Fig. 14.

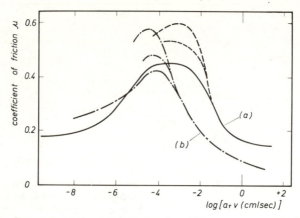

Figure 18 Comparison of master curves of an NR gum compound with two types of lubricant: (*a*) a 50:50 water-glycol mixture, (*b*) a 90:10 glycol-water mixture. Other experimental details as for Fig. 14.

product of liquid viscosity and sliding speed. Lowering the viscosity of the liquid should shift the maximum of the friction coefficient toward higher speeds.

The film thickness is also inversely proportional to the mean contact pressure. An increase in hardness by adding carbon black to the rubber would decrease the contact area and increase the contact pressure, moving the friction maximum toward higher speeds. A reduction in ball diameter should have the same effect. This, too, is observed. The absolute values of the friction coefficient are larger for a smaller indentor at the same normal load. This was first reported by Sabey (27) and discussed by Greenwood and Tabor (22) who explain the phenomenon in terms of deformation losses.

However, again the coefficients of friction in the present experiments are too large to be explained by normal deformation losses alone; tangential stresses must be present to contribute to them, and they can only occur if adhesion between track and rubber takes place. It appears from the limited data available that the critical film thickness at which boundary lubrication becomes hydrodynamic depends on the viscoelastic state of the rubber. For different types of rubber it appears to be a function of the difference between experimental temperature and glass transition temperature.

Roberts (28) has recently described some very instructive experiments on the nature of adhesion between rubber and glass in the presence of pure water and water containing a wetting agent, respectively. He shows that pure water does not truly wet the contact area but forms small globules so that distinct dry regions remain. In the presence of a wetting agent, dry regions can no longer be established, the polar nature of the wetting agent molecules preventing any real contact.

2.2.4 *Friction at High Speeds*

Sliding two solid bodies over each other at any but very low speeds generates frictional heat to a sufficient degree to raise the temperature at the interface and in the adjacent region significantly. Figure 19 shows temperature rise measurements in the interface and the coefficient of friction between natural rubber and copper as a function of the sliding speed as obtained by Schallamach (29). The temperature rise was shown to be proportional to the square root of the velocity, a finding that is in agreement with the expected temperature rise at the surface over which a heat source of constant power passes with a velocity v.

The coefficient of friction (shown in Fig. 20) rises at small velocities but soon reaches a maximum value and then falls again. It also depends on the normal load; the higher the load, the lower is its final value. Since the speed and temperature were known for all the friction coefficients, a transformation into a master curve, using the WLF equation described above, was possible, as shown in Fig. 21 (30). Now all values indeed lie on the rising branch of a single master curve. This also means that the different coefficients of friction obtained for different loads were not due to a nonlinear dependence of the frictional force on load but were due to the fact that at the same speed, the interfacial temperature depended on

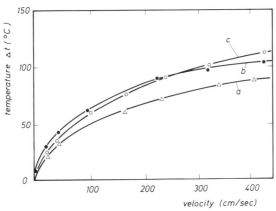

Figure 19 Temperature rise in the interface between a natural rubber track and a metal slider at various sliding speeds. Curves *a*, *b*, and *c* were obtained at three different loads (29).

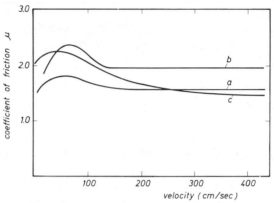

Figure 20 Coefficient of friction of a natural rubber tread compound at high sliding speeds and various loads. Curves *a*, *b*, and *c* as for Fig. 19 (29).

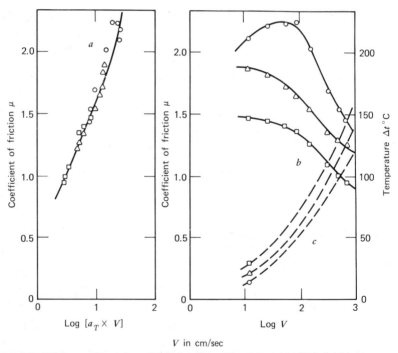

V in cm/sec

Figure 21 WLF transformation of the data from Figs. 19 and 20 (30). (*a*) Master curve reduced to 20°C, (*b*) coefficient of friction versus speed, (*c*) temperature rise versus speed. The sets of three curves are for the three loadings.

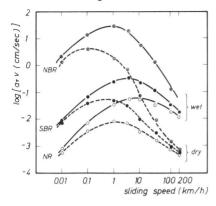

Figure 22 Log a_Tv as a function of the sliding velocity for friction experiments at high sliding speeds.

the load. The maximum value that Schallamach obtained is seen to have nothing to do with the maximum value of the master curve described in Section 2.2.1, but is due to the fact that speed and temperature have opposing effects on the friction coefficient. Their combined effect is easily demonstrated if log a_Tv is plotted as a function of the sliding speed, as shown in Fig. 22. Values of a_T were obtained using actual measured temperatures in the interface and the WLF equation. It is seen that log a_Tv varies remarkably little with speed and, like the friction coefficient, actually begins to fall with increasing speed. The log a_Tv values that occur in practice in skidding on roads therefore only cover a narrow band of the possible range.

Recently temperature measurements were carried out on wet surfaces (31). At low speeds the temperature rises with increasing speed while at higher speeds it passes through a maximum. Hydrodynamic lubrication prevents contact between rubber and the thermocouple slider. The a_Tv values obtained in this way are also shown in Fig. 22, and demonstrate that on wet roads also the range of a_Tv values obtainable in practice is strictly limited.

Skidding on icy roads is as important a consideration as under dry and wet conditions. In this case the temperature cannot exceed 0°C since the frictional heat is not likely to be able to supply the latent heat needed to melt all the ice, so that on ice a different portion of the master curve comes into play than on either wet or dry tracks.

The portions of the master curves which are likely to be of importance in tire skids on dry, wet, and icy road surfaces are marked in Fig. 23 for three rubbers NBR, SBR, and NR. Although this is only an academic exercise, because the master curves are those for gum rubbers on glass and would certainly have lower absolute values if they had been obtained

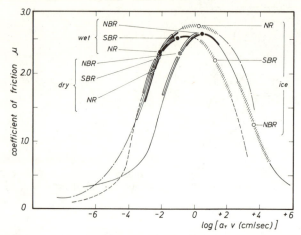

Figure 23 Master curves for three gum rubbers on dry glass tracks with the positions marked which are of importance in tire skids on roads.

with filled compounds on road surfaces, in particular on wet or icy roads, it is nevertheless instructive. It explains rationally two important findings which are generally observed on the skid performance of tires (as discussed further in Sections 2.3.3 and 2.3.6), namely (*a*) on wet roads elastomers are generally *ranked* according to their glass transition temperature, and (*b*) on icy roads the ranking is completely reversed. In Fig.

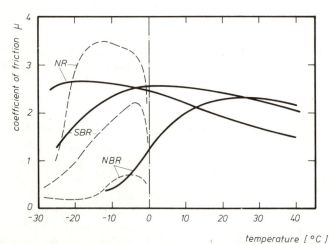

Figure 24 Comparison of friction coefficients obtained on ice at various temperatures and a sliding speed of 5 cm/sec with master curves (solid lines) obtained on glass transformed to a temperature scale (33).

23 the three master curves were in each case referred to their standard reference temperature T_s, which lies about 50° above the glass transition temperature T_g. The curves then take up approximately the same position on the log $a_T v$ axis.

The fact that the portions of the master curves obtainable on ice differ in no way from those obtained on glass has recently been demonstrated by Southern and Walker (32) and Gnörich and Grosch (33) independently. Even the absolute values are in close agreement if care is taken to keep the icy surface absolutely free from contamination (see Fig. 24).

2.3 The Skid Resistance of Tires

2.3.1 *The Braking Coefficient-Time Curve in Locked Wheel Skidding*

Two methods are commonly employed to measure the skid coefficient during locked wheel braking. The first is to lock the front wheel brakes of a car (34), driven at constant speed, with the engine disconnected during the actual braking experiment, and then measure its deceleration. Care has to be taken of dynamic effects such as transfer of load and tilt of the car. The first influences the reaction forces between tire and road, the second the accelerometer reading. The rolling resistance of the car, apart from the contribution of the front wheels, also enters the calculation.

In the second type of experiment, a car pulls a trailer on which the test tires are mounted (35). In this case the trailer wheels are locked, again with the engine of the car disconnected, after the car has reached a prescribed speed. The towbar pull between car and trailer is measured. This measurement also contains the rolling resistance of the car, which can, however, be determined in a separate experiment.

Figure 25 shows a typical deceleration-time curve of a locked wheel braking experiment (34, 36). It is seen that the curve rises from zero to a maximum and then falls to a lower steady value. A slotted disk fitted to the wheel also shows that the instant at which the wheel is completely blocked occurs just beyond the point at which the friction reaches its maximum value. As the wheel is braked the slip rises at first proportionally with the braking force. This part of the curve corresponds to the rising part in the side force-slip angle curves of Fig. 3. While the theory predicts a limiting value F_1 of the braking force,

$$F_1 = \mu L \qquad [4]$$

in practice the curve passes through a maximum and has a final sliding value somewhat lower than the maximum, the exact ratio of sliding value

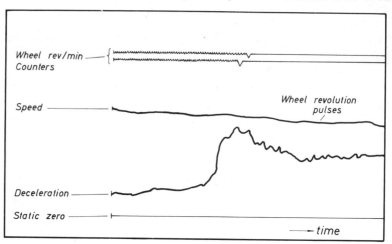

Figure 25 Deceleration-time curve obtained in a locked wheel braking experiment, when braking the front wheels with the rear wheels disconnected (36).

to maximum value depending on the experimental conditions. Two causes contribute to this effect, which are not allowed for in the theory.

First, the temperature rises in the interface with increasing slip. In this sense the braking force-slip curves are identical to the friction coefficient speed curves described by Schallamach for friction at high speeds (see the preceding section), if it is remembered that the slip speed, which determines the coefficient of friction, is defined as the difference between the circumferential velocity of the wheel and its forward traveling velocity, both being referred to its axle (i.e., when the wheel is rolling freely the slip velocity is zero and when the wheel is finally locked it has risen to the forward velocity). At small slip velocities the temperature rise is small and the coefficient of friction increases with speed. Very soon, however, this is overcompensated by the temperature rise, resulting in a decrease in the friction coefficient.

Second, on wet surfaces, hydrodynamic effects intervene in the friction process, tending to reduce the available frictional forces. These too become more effective as the slip velocity increases. A liquid film forms between the rubber and the road surface asperities, the thickness of which depends on the shape of the surface asperities, the hardness of the rubber compound, and the sliding velocity, as described in the lubricated friction experiments with spherical sliders (Section 2.2.3). With increasing velocity, the friction coefficient reaches a maximum value which is no longer determined by the viscoelastic properties of the rubber but by the thickness of the lubricating film.

In addition inertia effects come into play. The water has to be accelerated in order to be expelled from the interface. At moderate velocities this effect is usually much smaller than the viscosity effects but can become important on flooded surfaces or at very high speeds.

From these considerations, it is expected that the braking coefficient of tires on wet tracks would depend not only on the compound but also on the nature of the track surface, the tread pattern of the tire, and the degree of wetness of the track surface. A detailed study of this is, however, outside the scope of this chapter. In the present context it is important to see how far compound effects become apparent in the braking of tires and how far they depend on such secondary variables as mentioned above, i.e., speed, track surface, tread pattern, and state of road wetness.

2.3.2 Peak and Sliding Braking Coefficients

For practical purposes the braking coefficient-time curve can be described with sufficient accuracy by two coefficients, the peak value μ_p and the sliding value μ_s (see discussion of Fig. 25).

Figure 26 shows the peak and the sliding coefficient of smooth, straight ribbed, and patterned ("siped") tires of the same construction and the same tread composition on two different surfaces, quartzite and fine asphalt (34). The same rate of water was applied to both by means of spray bars. Since quartzite has a sharp open textured surface, from which the water can drain easily, only a thin film of water is present at any time. On the other hand, fine asphalt has a close textured surface from which the water can only run off along the sides, so that a much thicker film forms on the surface for the same rate of water application. On quartzite, the smooth tire actually outperformed the straight ribbed and the patterned tire, and all three tires had only a very small speed dependence. In addition, the ratio of peak to slide changed very little with increasing speed. This was not so on the fine asphalt. The friction coefficient of the smooth tire dropped very rapidly with increasing speed; indeed, the sliding coefficient reached values of below 0.1 at quite moderate speeds. With the straight ribbed and the patterned tires an essentially linear decrease with increasing velocity was observed over the whole experimental speed range, but again the negative friction speed coefficient was larger on fine asphalt than on quartzite. Finally, the speed dependence of the sliding value was in general not the same as that of the peak value, so that the ratio of the peak to slide coefficient also changed with speed. This was more pronounced on asphalt than on quartzite. All these effects can be explained, at least qualitatively, in terms of hydrodynamic effects (2,

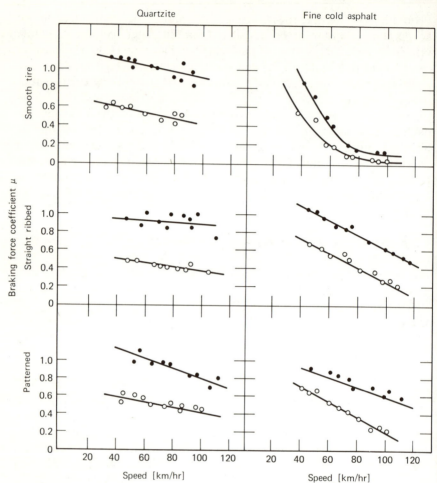

Figure 26 Peak and sliding coefficients of smooth, straight ribbed, and patterned tires of the same construction and tread composition on two wet surfaces at different approach speeds (34). ● peak ○ slide.

24–26). In the present context it suffices to show that the available friction on a wet road surface can depend considerably on the pattern of the tire profile, the speed, and the type of road surface on which the tire is braked.

2.3.3 The Influence of Compounds

Figures 27 and 28 compare peak and sliding coefficients of tires of similar construction and tread profile but different tread compositions on two

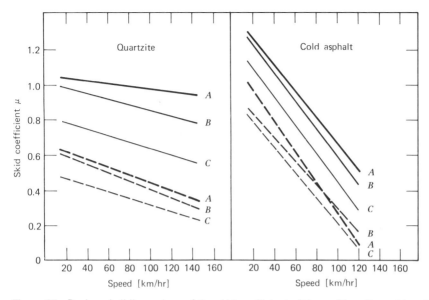

Figure 27 Peak and sliding values of the skid coefficient of tires with patterned tread on two different surfaces as a function of vehicle speed (34). A, High styrene-SBR (H-SBR); *B*, SBR; *C*, NR. Peak–full lines; slide–dashed lines.

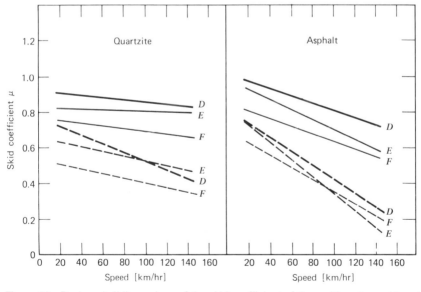

Figure 28 Peak and sliding values of the skid coefficient of tires with patterned tread on two different surfaces as a function of vehicle speed (36). *D*, OE-SBR; *E*, OE-NR; *F*, NR. Peak—full lines; slide—dashed lines.

different surfaces at various speeds. The experimental points have been omitted for the sake of clarity, and the straight lines were computed from the data using the least squares method. The tread compositions consisted in two cases of NR, in the others of oil extended SBR and high styrene-BR (H-SBR), respectively (34, 36). At low speeds the ranking of the compounds in relation to NR is the same on both surfaces, and this remains so up to speeds of about 100 km/hr. Above that speed occasional reversals in ranking can be observed. On quartzite OE-SBR reverses with OE-NR and on asphalt OE-NR reverses with NR and H-SBR with SBR. The question arises as to what extent such reversals with increasing speed are significant and what their likely cause is, and, more generally, how far skid ratings of tread compounds are dependent on the testing conditions.

Skid experiments are difficult to repeat exactly because the coefficients are subject to a large number of influences usually outside the control of the experimenter, obvious examples being the ambient temperature and exact conditions of the road surface. Statistical methods therefore form a valuable tool to ascertain the significance of the results. Maycock and Grosch (36) have tested six compounds at 10 different speeds on four different surfaces, measuring in each case peak and sliding values of the skid coefficients. Analysis of covariance (a combined regression and variance analysis) allows the assessment of the significance of the mean skid coefficients and their speed dependence for the various factors involved (surface, compound, peak, and slide). It was of particular interest to discover the extent to which interactions (i.e., the dependence of one variable upon another) such as compound rating upon speed, compound rating upon surface, and compound rating upon peak or slide measurements, were statistically significant.

Table 1 shows the ratings at three different speeds of those four of the six compounds which differed significantly in their composition. The skid

Table 1   Average Skid Ratings of Four Tread Compounds at Three Different Vehicle Speeds[a]

Compound	32 km/hr	80 km/hr	128 km/hr
OE-SBR	125	124	123
OE-NR 67/33[b]	117	113	109
OE-NR 55/45[b]	128	128	128
NR	100	100	100

[a] These ratings are averaged peak and slide measurements on four surfaces (36).
[b] Oil content, %

ratings are averaged over the four surfaces and the peak and sliding measurements. It is seen that the ratings are independent of speed.

The speed dependence of ratings is defined as

$$R = \frac{\mu_1}{\mu_2} \times 100 = \frac{\mu_{01} + b_1 v}{\mu_{02} + b_2 v} \times 100 \qquad [5]$$

where μ is the skid coefficient, μ_0 is the skid coefficient at zero speed, b is the speed regression coefficient, and v is the speed. The subscripts refer to compounds 1 and 2, respectively. Expansion leads to the relation

$$R = \frac{\mu_{01}}{\mu_{02}} \left[1 + \left(\frac{b_1}{\mu_{01}} - \frac{b_2}{\mu_{02}} \right) v \right] \times 100 \qquad [6]$$

Although the speed regression coefficients b can be very large and differ greatly between surfaces, their differences between compounds on any one surface are usually so small that they are statistically not significant. Hence a very large number of readings would be required to establish a significance, if any really existed. Theoretically such a difference between the speed dependence of compounds could arise from differences between the temperature conditions in the interface and their effect on the skid coefficients (see Section 2.2.4). In general the rubber with a high glass transition temperature and hence a high mean braking coefficient would also be expected to have a higher speed dependence. This can be deduced, for instance, from the slopes of log $a_T v$ versus v curves. It is also verified by the fact that the reversals in ranking observed with increasing speed in the present experiment, although statistically not significant, are in agreement with the prediction just given (see also Section 2.3.4).

Table 2 gives the average ratings at 80 km/hr of the four compounds

Table 2 Average Skid Ratings Showing (a) Effect of Peak and Slide, Measurements, (b) Surface Effects

	Skid Rating					
	(a) Averaged over Four Surfaces		(b) Averaged over Peak and Slide Measurements			
Compound	Peak	Slide	Asphalt	Quartzite	Gravel	Concrete
OE-SBR	127	120	123	129	130	115
OE-NR 67/33[a]	114	111	109	121	117	105
OE-NR 55/45[a]	128	128	119	139	141	116
NR	100	100	100	100	100	100

[a] Oil content, %

(*a*) derived from all peak and sliding coefficients, irrespective of the surface, and (*b*) derived from the skid coefficients determined on the four surfaces, but averaged over the peak and sliding values. It is seen that in all cases the ranking is maintained. Even average ratings derived from the peak and slide values are closely similar on such widely different surfaces as quartzite and gravel, although quartzite is a very sharp surface, and gravel a smooth pebble surface. Both, however, are well drained. On flooded surfaces such as asphalt or concrete, differences between compound ratings are in general smaller, although here too rankings are maintained.

Because statistics deals only with experimental data and not with functional relationships, it cannot prove a universal truth. The present argument, that ranking of compounds in order of their skid resistance is independent of surface and speed, is therefore limited to the data presented. However, since they include the extreme in practical tire compounding and surface roughness, respectively, it may be concluded that this is also likely to be the case for compound variations such as blending of rubbers or carbon black variation, if based on the currently available rubbers.

Measurement of peak and sliding coefficients forms, therefore, a useful and relatively simple means of assessing compound contribution to the skid resistance of tires under all driving conditions.

2.3.4 *The Effect of Oil Extension on the Skid Rating*

Apart from the viscoelastic properties of the "pure" elastomer, the skid coefficients are also influenced by the replacement of part of the rubber by a high viscosity mineral oil (oil extension), as seen by curves E and F

Table 3 Effect of Oil Extension on the Average Skid Ratings of Tread Compounds

	Skid Rating	
Compound	From ref. 36	From Pendulum Skid Tester Measurements
NR	100	
OE-NR 67/33[a]	113	
OE-NR 55/45[a]	128	
SBR		100
OE-SBR 67/33[a]	125	112

[a] Oil content, %

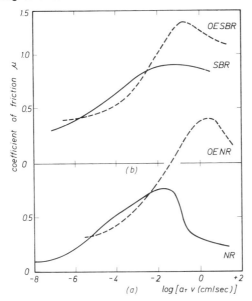

Figure 29 Master curves of the friction coefficient on lubricated carborundum tracks comparing (*a*) OE-NR with NR and (*b*) OE-SBR with SBR.

in Fig. 28. In general, oil extension increases the skid coefficient compared with the unextended tread material. Table 3 shows average skid ratings for SBR and OE-SBR, NR and OE-NR.

Figure 29 shows master curves of the friction coefficient on a carborundum track lubricated with a $50:50$ glycol-water mixture, comparing an OE-NR with an NR and an OE-SBR with an SBR compound. It is seen that the oil extension raises the friction coefficient mainly in the region around the maximum. At low log $a_T v$ values the friction coefficient decreases more rapidly than that of the unextended rubber, and even falls below that of the unextended rubber. The practical log $a_T v$ range extends approximately from -2 to about $+1$, so that an improvement in the skid resistance of tires would be expected. It is also apparent that with increasing speed (decreasing $a_T v$ values) the coefficient falls more rapidly for the oil extended rubber than for the nonextended ones, in qualitative agreement with the slightly steeper speed dependence of these compounds in the skid experiment discussed above (Section 2.3.3).

2.3.5 The Pendulum Skid Tester

Although friction measurements as described in Section 2.2 form a valuable basis for a fundamental understanding of the mechanism of

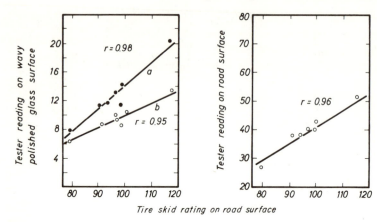

Figure 30 Correlation between skid rating of tires and skid coefficients obtained with the pendulum skid tester (38). Curves *a* and *b* represent two different types of glass surface. Tire skid rating on road surface at 32 km/hr.

rubber friction, it is clearly desirable also to have a simple laboratory test to screen tread compounds as to their skid properties before conducting much more expensive tire tests. The pendulum skid tester, developed by the British Road Research Laboratory (37), has found wide acceptance in the tire industry as such an instrument. Bevilacqua and Percapio (38) have examined the friction coefficients obtainable with the tester on a great number of actual road surfaces and some easily reproducible laboratory surfaces, such as sand blasted glass and carborundum stone. Although some surfaces were very much more discriminating than others, they found in general a good correlation between readings obtained on road surfaces and laboratory surfaces. In addition, Fig. 30 shows correlations which Bevilacqua and Percapio obtained between skid tester readings and tire skid ratings of different compounds. Grosch and Maycock (36) also found excellent correlation between their skid measurements on four road surfaces and the pendulum tester on concrete slabs. Giles et al. (39) demonstrated that the skid tester measurements correlated well with tire measurements if different surfaces were tested with the same compound. Used carefully the tester can therefore form a convenient screening method to select promising tread compounds. The reason for the good correlation is again because log $a_T v$ values are similar to those obtained in skid experiments with cars. The sliding speed in the pendulum tester is about 11 km/hr. This is lower than sliding speeds obtained with cars, but since the temperature rises are also lower, the log $a_T v$ values again fall within the range obtained in full-scale skid experiments.

2.3.6   *The Skid Resistance of Tread Compounds on Ice*

As already indicated in Section 2.2.6, the skid rating between SBR and NR based tread compounds is reversed on icy surfaces. This is shown in Fig. 31, where the skid resistance measured with the RRL pendulum skid tester is shown as a function of the track temperature. The sample temperature in these experiments was kept constant at 20°C. At temperatures above 0°C the track is wet and the SBR compound clearly has the higher skid coefficient; at 0°C the coefficient of friction drops sharply; and below 0°C ice has formed on the track and the natural rubber is clearly superior to the SBR compound.

This behavior is also observed in tire skid experiments on ice. French and Patton (40) have published a graph very similar to that shown in Fig. 31 from tire skid measurements. Extensive skid trials on ice by Grosch et al. (35, 41) gave the average skid ratings as shown in Table 4. Reproducibility is strongly affected by uncontrollable factors such as polishing of the track after repeated skids and rapidly changing weather conditions, and discrepancies in the ratings in Table 4 are probably due to this. Nevertheless it is clear that NR and OE-NR based compounds are superior to SBR based ones. It may be surprising that blending SBR with BR has little effect in improving SBR in relation to NR. However, this too can be deduced from the master curves and the log $a_T v$ range which is operative on ice at about 0°C. In this range BR has not yet reached its maximum value. BR becomes beneficial to skidding only at much lower temperatures.

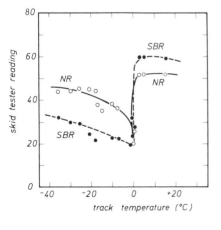

Figure 31 The skid coefficient, determined with the pendulum skid tester, as a function of the track temperature on a wet and icy track. Both rubbers contain 50 phr HAF black.

Table 4 Effect of Compound Variations on the Skid Rating on Ice

| | Skid Ratings | | | | | | | |
| | Lake Ice (−2°C) | | Road Ice, First Trial (−2 to −6°C) | | Road Ice, Second Trial (−6°C) | | Breakaway on Closed Circle | |
	Nonstudded	100 Studs	Nonstudded	100 Studs	Nonstudded	100 Studs	Nonstudded	Studded
NR	119	119	—	—	99	157	124	131
OE-NR	118	119	142	149	109	173	111	120
OE-SBR/BR	101	137	100	131	103	123	92	111
Commercial tire (60 SBR = 40 BR)	100	123[a]	—	—	100	155[a]	100	127[a]

[a] This tire contained 133 studs.

3 THE WEAR OF TIRES

3.1 Abrasion Mechanisms

3.1.1 *The Pressure Dependence of Abrasion*

The wear of tires and of rotating test pieces on certain laboratory abraders depends not only on the intrinsic abrasion resistance of the material (tread compound), but in equal measure on their overall mechanical properties. This is clearly demonstrated by the difference in wear between conventional and radial ply tires with identical tread compounds. The theory on the wear of slipping wheels by Schallamach and Turner (1) attempts to take account of such effects (see Section 3.2). One basic assumption of the theory is that the intrinsic abrasion loss, obtained in ordinary sliding between rubber and a track surface (as distinct from a partially slipping wheel), is proportional to the energy dissipation that occurs during the sliding process. It is now known that this is generally not the case (42). For a test piece consisting essentially of a square block of rubber sliding over the abrading surface at various normal loads, the energy dissipation per unit distance is numerically equal to the frictional force. If this is proportional to the normal load, which is usually the case in the range of pressures commonly met with in the use of tires, it would follow that the volume abrasion per unit sliding distance should also be proportional to the normal load. Figure 32 shows the abrasion loss per unit sliding distance of natural rubber and SBR on a tarmac road surface, a silicon carbide Akron abrader disk, and silicon carbide paper as a function of the normal pressure, both plotted on a logarithmic scale (43). The abrasion loss is shown as

$$h = \frac{V}{A} \qquad [7]$$

where V is the volume loss per unit distance and A is the apparent contact area of the test piece, so that the dimensions of the test piece have been eliminated from the graphs. The dependence on the pressure p can be written as

$$\frac{h}{h_1} = \left(\frac{p}{p_1}\right)^n \qquad [8]$$

where h_1 is the loss at pressure p_1. The power index n depends on the compound and the type of surface on which it slides, as shown in Table 5.

The power index appears to increase with decreasing abrasive power of the track. Comparing different rubbers it is generally larger for SBR and

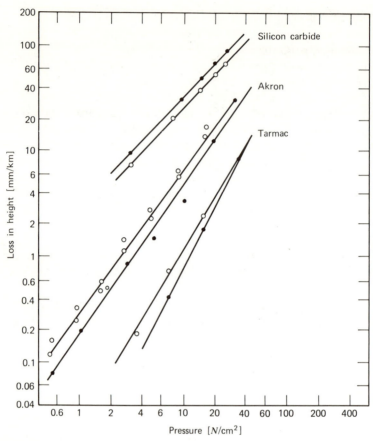

Figure 32 Sliding abrasion loss of two different rubbers on three types of surface as a function of the contact pressure (43). ○, NR; ●, SBR.

Table 5 Values of n in $h/h_1 = (p/p_1)^n$ on Different Surfaces and for Four Different Rubber Compounds (43)

Type of Surface	Abrasion Loss h(mm/km) of NR at 10 N/cm² Pressure	Power Index n			
		BR	NR	SBR	NBR
Carborundum paper 180	26	1.00	1.03	1.03	1.02
Carborundum disk	6.6	1.21	1.36	1.43	1.51
Tarmac road surface	1.1	2.25	1.57	1.96	1.90

NBR than for NR. For BR it is very large on the road surface and small on the carborundum disk. On the carborundum abrasive paper on which the abrasion is very severe, as shown by the abrasion loss for NR in Table 5, n becomes 1 for all rubbers. Ratner and Klitenik (42) correlate the power index n obtained in abrasion experiments on a metal gauze with the cohesive energy density of the rubber. Power indices obtained on the metal gauze are much larger than on the types of track considered in Fig. 32, presumably because abrasion losses are very much smaller. Ratner proposes that the value of n is a measure of the molecular interaction between track and rubber surface. Increasing polarity in the rubber is supposed to increase the power index n, and swelling of the rubber with a liquid decreases it.

Reznikovskii (44) relates the power index n to the index b in his fatigue life relation

$$N = \left(\frac{\sigma_0}{\sigma}\right)^b \qquad [9]$$

where σ_0 is the tensile strength of the material, σ is the stress amplitude at which a rubber specimen is cycled to break, and N is the number of cycles to failure.

Although these arguments are only qualitative (attempts to propose quantitative theories have in the opinion of the author met with only limited success) they indicate that the nonlinear load dependence of abrasion of rubbers on different tracks is intimately connected with the nature of the abrasion mechanism. Moreover, since differences in power indices between rubbers depend on the type of surface, a severity effect of compound ratings in tire wear is to be expected.

3.1.2 The Abrasion Mechanism on Sharp Abrasives (Silicon Carbide Paper)

Sliding rubber over a silicon carbide abrasive paper track produces a very high abrasion volume loss per unit load and sliding distance as shown above (Fig. 32; Table 5), and it appears almost the only surface on which the abrasion loss is proportional to the energy dissipation. An investigation into the temperature and sliding speed dependence of the abrasion loss per unit energy dissipation (referred to as the abradability) has shown that the WLF transformation can be applied. This is shown in Fig. 33 in which the master curve for the abradability is shown for an NBR gum compound (15, 16). At low speeds the abradability decreases with increasing sliding speed and reaches a minimum beyond which an increase in the abradability is observed. This coincides with a change in the

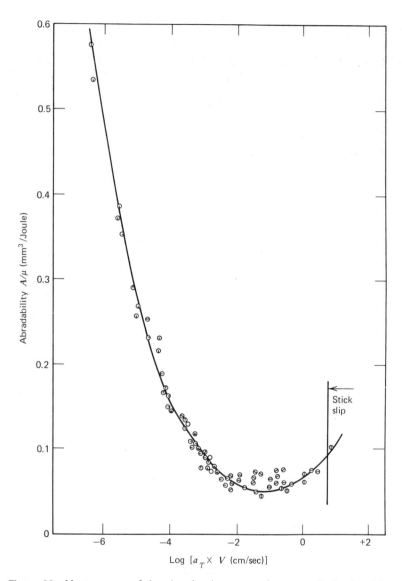

Figure 33 Master curve of the abrasion loss per unit energy dissipation (abradability) for a gum NBR compound on silicon carbide (16). ⊙, 97°C; ⊖, 85°C; ⊕, 75°C; ⊖, 65°C; ⊘, 55°C; ⊘, 45°C; ⊖, 20°C; ⊕, 10°C.

appearance of the abraded surface. While at low speeds the surface shows the tendency to produce abrasion patterns of the type first described by Schallamach (45), at high transformed speeds scoring marks in the direction of the sliding speed are clearly visible, indicating that the rubber has become brittle to the abrasion attack.

Similarly shaped curves are obtained if the reciprocal of the energy density at break, U_B, in a rupture experiment is plotted as a function of the transformed extension rate.

A comparison of the two curves then yields on the transformed speed axis an equivalence between rate of extension and sliding speed, so that the abrasion loss per unit energy dissipated

$$\left[\frac{V}{F}\right]_{\text{sliding speed}} = \left[\frac{\text{const}}{U_B}\right]_{\text{extension rate}} \qquad [10]$$

where V is the abrasion volume loss per unit distance and F is the frictional force. The subscripts indicate constant sliding speed and extension rate.

The speed and extension rate to achieve best superposition of the experimental curves showed that 1 cm/sec in the abrasion experiment corresponded approximately to a strain rate of 10^2 sec^{-1} so that the dimensions of the particles of the test piece being strained to break would be approximately 10^{-2} cm in length, which in turn is about the dimension of the contact points of the abrasive paper.

In this case of extremely severe abrasion, the mechanism appears therefore to be one of detachment of small particles through mechanical rupture. The abrasive particles strain a small region around them to near breaking point (see Fig. 34). After a few passes a cut opens and eventually a particle is detached. Such a mechanism was first suggested by Schallamach (46, 47) and demonstrated by scratching rubber with a needle. Bulgin and Walters (48) propose a theory in which asperities, approximated to cones, plow grooves out of the rubber surface. Both mechanisms lead to a linear load dependence of abrasion, as is indeed observed on this surface. A comparison between the actual energy dissipated to remove a volume V from the rubber and the one calculated from the maximum strain energy which can be stored in that volume shows that this process is still extremely inefficient. Over 10^3 times more energy is dissipated in the abrasion process than the energy to break which can be stored in the abraded volume. Either each particle executes approximately 1000 cycles to near breaking point before it is detached or energy is lost in other ways, of which sliding friction would be the most obvious.

Figure 34 Deformation of rubber around a pointer pressed into the rubber and then dragged along it, simulating the deformation due to a single abrasive asperity (46).

3.1.3 The Abrasion Mechanism on Mild Abrasives

On mild abrasives also the abrasion mechanism is probably essentially one of mechanical failure but at lower strain amplitudes, so that the cut growth characteristics determine the abrasion behavior of different compounds. Although clear-cut quantitative relations between abrasion and cut growth rate do not yet exist, there are a number of qualitative similarities in the two processes which deserve comparison:

1.   The rate of growth of a cut in a test piece that is being cycled at a given strain amplitude increases in general with the power β of the tearing energy (49, 50), which in turn is proportional to the stored elastic energy density (see also Section 4). Hence the abrasion loss would be expected to be proportional to a power function of the energy dissipation rather than simply proportional to it (see Section 3.1.1).

2.   It is known that the power index β of the cut growth characteristic for the synthetic butadiene rubbers is in general larger than that for natural rubber (50). This is also the case for the power index of the

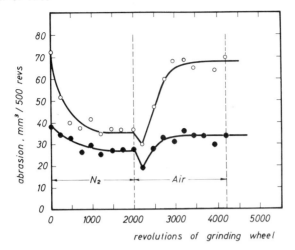

Figure 35 Abrasion of natural rubber in nitrogen and air, respectively (52). ○, Without antioxidant; ●, with antioxidant.

abrasion loss-pressure dependence, although differences between the power indices of different rubbers are smaller for the abrasion process than they are for the cut growth characteristic.

3. Cut growth behavior is influenced by the presence of oxygen (50). The rate of growth in air is faster than in nitrogen, and it is slowed down by the presence of antioxidants in the rubber. In the same way abrasion on mild abrasives is larger in air than in nitrogen (51, 52), as shown in Fig. 35, despite the fact that matters are complicated by contamination of the abrasive through the oxidized abraded particles, which tends to reduce the abraded volume loss. Figure 35 also shows that the presence of antioxidants strongly influences the rate of abrasion on these types of surface, reducing the volume loss in relation to unprotected compounds. This is not the case on very sharp abrasives such as carborundum paper.

3.1.4 The Speed Dependence of Abrasion on Mild Abrasives

The author has carried out some experiments with standard Akron test pieces at 15° slip at different speeds but constant load (pressure) on carborundum stone of the type used on the Standard Akron Angle Abrader. Curves of the abrasion coefficient (volume loss per unit distance and load) as a function of the track speed are shown in Fig. 36 for the natural rubber and SBR samples (53).

For the SBR the abrasion coefficient decreases with increasing speed in the low speed range, while the abrasion coefficient of NR is almost

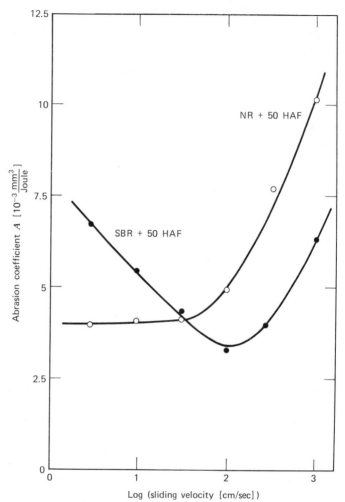

Figure 36 Abrasion coefficient (volume loss per unit length and unit load) on an Akron type abrader disk as a function of the track speed.

unaffected by changes in speed. The abrasion mechanism in this range is probably similar to that described earlier, when discussing the speed-temperature transformation for sliding abrasion on silicon carbide paper; only the number of cycles required to detach a volume of rubber is much larger. With increasing speed, and hence increasing rate of extension, the resistance to tearing increases for the SBR and hence the abrasion loss falls. Natural rubber crystallizes on extension (54) and hence has a reinforcing mechanism that is not rate dependent. Therefore, its abrasion rate is not greatly dependent on speed. At very high rates of extension there is not sufficient time for crystals to form, and this mechanism no longer operates. Natural rubber is then correspondingly weak.

At higher speeds both rubbers show a marked increase in the rate of abrasion. Unlike the increase of the abrasion rate observed with increasing speed discussed in Section 3.1.2, in the present experiments it is probably due to a rise in the interfacial temperature. This rise may be sufficient to cause oxidation to proceed more rapidly than would be the case at lower temperatures. Local temperature rises may be large enough to cause thermal degradation of the rubber. As the mean temperature in the interface rises with increasing sliding speed, the likelihood of finding a spot of sufficiently high temperature for the rubber to degrade thermally is proportionally larger.

It is well known that natural rubber begins to degrade at lower temperatures than SBR or BR (55). This would explain the increase in the abrasion rate at lower speeds for NR than for SBR.

A very low abrasion rate observed in similar measurements for BR is partly due to its lower friction coefficient; at the same speed the mean interfacial temperature is lower than for the other two rubbers. In addition, BR has the highest degradation temperature of the three rubbers.

3.1.5 *Secondary Abrasion Processes*

In addition to the primary processes discussed above, secondary processes can often contribute to increase the abraded volume beyond that expected from the primary processes. Well known are the Schallamach abrasion patterns (45), which seem to constitute a departure from randomness of the abrasion process. Instead of strained regions acting independently, strains are sufficiently large to interact with neighboring regions. Abrasion patterns then form, i.e., ridges with a fairly regular spacing at right angles to the abrasion direction. The abrasion volume loss is invariably larger when these patterns are present than when they are suppressed under otherwise similar conditions.

Abrasion pattern formation is always more pronounced on smooth surfaces than on sharp ones and with soft compounds than with hard ones. It can take extreme forms with gum rubbers on polished glass, as recently discussed by Reznikovskii and Brodskii (56). On such surfaces the abrasion loss of soft compounds can suddenly become very severe, with the formation of very large abrasion patterns. However, their importance in tire wear seems limited to accelerated tests. The appearance of distinct abrasion patterns during the normal running of a car usually signals the presence of a misaligned wheel.

The temperature dependence of the abrasion volume loss also appears to depend strongly on the nature of the abrasion process. While in the abrasion on sharp abrasives (Section 3.1.2), abrasion increased with the surrounding temperature by approximately 0.2–0.5%/°C in the range 20–70°C, Viehmann (57) obtains values ranging between 4–14%/°C, depending on the compound. His experiments were carried out at small slips on a grinding stone, which is, of course, much less sharp than abrasive paper. While on sharp surfaces the temperature dependence is related to the temperature dependence of the reciprocal of the energy density at break, the temperature dependence on the Viehmann machine is not so clear. Whether it constitutes thermal degradation, oxidation, or the temperature dependence of low stress fatigue must be left to future research. Its importance to tire wear will become clear in the next section.

3.2 The Wear of Slipping Wheels

It was stated in Section 2 that slip occurs in part of the contact area of the tire whenever a force is transmitted from the car to the road, and the magnitude of the slip depends on the magnitude of the transmitted force. This results inevitably in tire wear. Schallamach and Turner in their theory of the wear of slipping wheels (1) assume that the volumes loss per unit distance traveled is proportional to the energy dissipated during the slipping process. A tire running under a force F (side force, braking or accelerating torque) produces a wear per unit distance, W:

$$W = \gamma F s \qquad [11]$$

where γ is a constant, the volume loss per unit energy dissipated, often referred to as the abradability of the material, and s is the slip. For a wheel running under a small slip angle Θ the side force F_s is given by

$$F_s = K\Theta \qquad [12]$$

and as s itself is approximately proportional to Θ, the wear becomes

$$W = \gamma K \Theta^2 \qquad [13]$$

where K is the stiffness of the tire and a measure of its resistance to angular distortion.

Equation [13] predicts that the volume abrasion per unit distance increases with the square of the slip angle. It would follow from this that

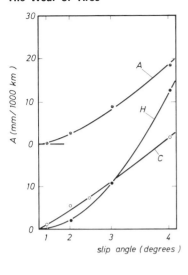

Figure 37 Tire wear A of tread compounds as a function of the square of the slip angle on a two-wheeled trailer (58) A=SBR+150 HAF; H=NR+150 HAF; C=150 NR/50 BR+150 HAF blend.

the relative rating of two compounds

$$\frac{W_1}{W_2} = \frac{\gamma_1}{\gamma_2} \frac{K_1}{K_2} \qquad [14]$$

is independent of the slip angle. Figure 37 shows the wear of a number of different tread compounds, all retreaded on tires of the same carcass construction and hence of the same stiffness K, as a function of the square of the slip angle (58). The experiments were carried out with a two-wheeled trailer, the wheels of which could be set at the required slip angle to the traveling direction. Mounting tires with the same compounds and the same carcass simultaneously on the two axle positions ensured that both tires were running at the same slip angle. It is seen that the curves are not straight lines, as would have been expected from the theory, but the wear always rises more sharply with slip angle than predicted. Moreover, the degree of the deviation from the straight line varies with the compound examined. It is largest for natural rubber (H), smaller for SBR (A), and smallest for the NR–BR blended compound (C). Possible reasons for this deviation are discussed below.

3.3 The Wear Rating of Tread Compounds

3.3.1 *The Severity Dependence*

The nonlinear pressure dependence of sliding abrasion (see Section 3.1.1) also influences the wear dependence of slipping wheels on the slip angle.

Schallamach and Grosch (43) have modified the original Schallamach and Turner theory on the wear of slipping wheels to take account of the nonlinear pressure dependence of abrasion. The basic relation then becomes

$$\frac{W}{W_1} = \left(\frac{Fs}{F_1 s_1}\right)^n$$

[15]

where W is the volume loss per unit distance, W_1 is the volume loss at side force F_1 (and slip s_1), F is the side force, s is the slip, and n is the power index defined in Section 3.1.1. The theory is beyond the scope of this chapter but its main conclusions are that the relative wear rating of two compounds cured to tires of identical construction depends weakly on the normal load, the type of road surface on which the rating is measured, and the slip angle. As load, road surface, and slip angle decide the severity of the abrasion loss, it follows that the relative wear rating is severity dependent, in agreement with practical experience. Thus Davison et al. (59) found that the rating of two tread compounds differed under otherwise constant conditions, when the experiment was carried out on different types of road surface. Bulgin and Walters (48), examining the slip angle dependence of tire wear, proposed the equation

$$W_1 = K[P\Theta^2 + Q\Theta^{3.5}]$$

[16]

where K, P, and Q depend on the compound.

Schallamach's modified theory states that the slip angle dependence of tire wear in crabwalk is proportional to

$$W = \text{const } \Theta^{2+(n-1)}$$

[17]

under otherwise constant conditions where n is again the power index for the pressure dependence of the sliding abrasion. From the n values for NR and SBR given in Table 5 for a tarmac road surface it follows that the rate of wear of SBR in relation to NR:

$$\frac{W_{\text{SBR}}}{W_{\text{NR}}} = \text{const } \Theta^{1.92-1.63} = \text{const } \Theta^{0.29}$$

[18]

would increase with increasing slip angle, and the wear rating, the inverse ratio of the wear rate, would correspondingly decrease. But this is contrary to the findings given in Fig. 37 which shows that NR becomes worse in relation to SBR as the slip angle is increased. Another effect appears therefore to intervene to affect the slip angle dependence of the tire wear results shown in Fig. 37.

3.3.2 The Temperature Dependence of Tire Wear

Figure 38 shows the volume loss per unit distance of an SBR and a natural rubber compound as a function of the ambient temperature. The experiments were carried out with a trailer as described above at a constant slip angle Θ, constant speed, and on the same type of road surface, but during a dry period in which the ambient temperature rose steadily. The wear of both compounds increased as the temperature increased. However, it did so more for the natural rubber than for the SBR tread stock. From the straight line graphs, temperature coefficients can be derived which are about 3%/°C for the NR but only 1.5%/°C for the SBR stock. Such temperature dependence of wear of tread compounds has been reported in the literature (60–62). It can lead to complete reversals in the ranking of the wear rating, as indeed is apparent from Fig. 38, and has led to the proposal that the employment of different rubbers to their best advantage from a wear point of view may be geographically conditioned: natural rubber and, in the future, polyisoprene based compounds being employed essentially in the colder regions, and SBR or SBR/BR blends in the warmer regions of the world. Although no direct evidence is known to the author of the abrasion temperature dependence of BR, indirect evidence suggests that it should have only a small abrasion temperature coefficient. One would therefore expect that in blends with them it would reduce the temperature dependence of natural rubber or even of SBR.

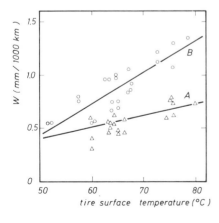

Figure 38 Tire wear W of an SBR and an NR tread compound at constant slip and road surface conditions as a function of the ambient temperature (58). Curve A ($\triangle$), SBR+50 HAF; curve B ($\bigcirc$), NR+ 50 HAF.

With rising ambient temperature, road surface and tire surface temperature rise simultaneously so that the tire itself is in a different temperature state. It is presumably the tire surface temperature that is responsible for the change in wear resistance.

3.3.3 The Dependence of Wear on the Wetness of the Road Surface

Figure 39 shows the wear of tires with NR and SBR tread compounds on the same road surface when dry and when wet under otherwise constant conditions of load and slip angle and with almost the same ambient temperature conditions. In the range between these two extreme conditions, stretches of wet and dry road and partly wet road were encountered to varying degrees. To quantify these conditions, the percentage of wetness was defined as the ratio of wet to total distance covered. It is seen that on the dry road surface the SBR compound showed the higher resistance to wear while on the wet road surface the reverse was the case. Similar reversals have been reported by Westlinning (63). Several explanations of this phenomenon are possible. The author proposed that the tire surface temperature was drastically reduced by the water on the road (58) and although the wetness affected the absolute rate of wear, the reversal in ranking was essentially due to the different temperature-wear coefficients of these two rubbers. However, other investigators see in this

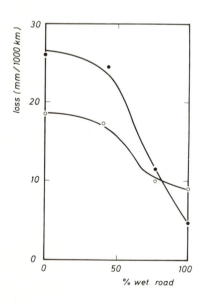

Figure 39 Tire wear of an NR and an SBR tread compound on wet and dry roads under otherwise similar testing conditions (58). ○, SBR; ●, NR.

behavior a shift in the importance of the two different abrasion mechanisms (48), fatigue and abrasive wear, and this remains a possibility, although not currently open to direct experimental verification.

3.3.4 The Influence of the Tire Surface Temperature on the Slip Angle Dependence of the Wear Rating

Tires running at different slip angles dissipate different amounts of energy, part of which appears as heat. This heat raises the temperature of the tire, initially in a surface layer, and, after the tire has run for a sufficiently long time, throughout the tire structure. Figure 40 shows the equilibrium temperature rise of a natural rubber and an SBR tire after the tire has been run at some angle for a sufficiently long time. The temperature was determined by stopping the vehicle (a two-wheeled trailer described in detail in ref. 58) and placing a temperature probe on the surface. The linear increase in temperature with slip angle has been explained by Schallamach (64), who also attempted to relate this measurement to the actual tire surface temperature as occurs in the contact region during sliding. He finds that these two temperatures must be proportional to each other, although the latter is not amenable to direct measurement. In the sliding region of the contact area the tire temperature reaches very high values while cooling takes place in the adhesion region and outside the contact area.

Assuming that the basic relation of wear of slipping wheels follows the original equation of Schallamach and Turner, i.e., neglecting the severity

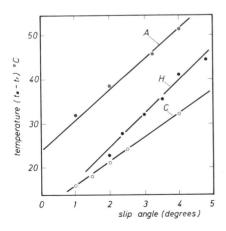

Figure 40 Temperature rise on the tire surface of three different tread compounds as a function of the slip angle in a controlled slip experiment with a trailer (58). $A = SBR + 50 \, HAF$; $H = NR + 50 \, HAF$; $C = 50 \, NR/50 \, BR + 50 \, HAF$ blend.

Table 6   Wear-Temperature Coefficients of NR and SBR Tread Compounds[a]

Compound	Temperature Coefficient $\alpha(\%/°C)$
NR + 50 HAF (Avon)	4.25
NR + 50 HAF (Michelin)	2.92
NR + 50 HAF (retread)	4.25
NR + 50 HAF (MRPRA retread)	4.02
SBR	3.09
50 NR=50 BR blend (retread)	1.68

[a] Deduced from the deviations of the "square law" (wear is proportional to the square of the slip angle) (58).

effect due to the nonlinear load dependence, but allowing for the temperature rise that takes place with an increase in slip angle, i.e.,

$$W = K\Theta^2[1 + \alpha\Delta t] \tag{19}$$

it is possible to calculate the temperature coefficient α and the constant K. Table 6 shows these constants for the natural rubber, SBR, and a natural rubber/BR blend (58). It is immediately obvious that they are very close in magnitude to those found independently by direct measurements, as discussed in Section 3.3.2.

Indeed, if the relative wear rating of NR versus SBR obtained in trailer measurements over a wide range of differing slip angle and weather conditions is plotted as a function of the tire surface temperature, a single function results, independent of the testing conditions, as seen in Fig. 41.

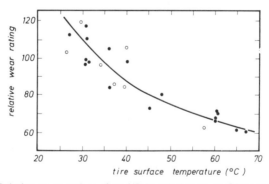

Figure 41   Relative wear rating of an NR in relation to an SBR tread compound as a function of the tire surface temperature (65). ●, Trailer results; ○, test car results.

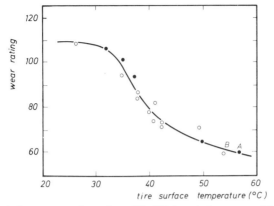

Figure 42   Relative wear rating of an OE=NR in relation to an OE-SBR compound (65), ●, Trailer results; ○, test car data. For an explanation of points marked *A* and *B* see text.

This function can also be extended to wear trials with cars where the slip values are no longer under the control of the experimenter (65), as seen by the points represented by open circles. These were obtained under different weather (i.e., wet and dry road) and driving conditions. This is particularly impressive for the two points marked *A* and *B* in Fig. 42, which shows a similar curve for the wear rating of OE-NR in relation to ⎺ ᏚᏴᏴ compound. The rating *A* was obtained with the trailer with a constant slip angle of 2° and a speed of 30 mph, while point *B* was obtained in a car trial during motorway driving on a mainly straight road. The absolute loss rate differed by more than a factor of 20, the only common parameter being the same tire surface temperature. These measurements have recently been extended to include light heavy service tires. Trailer and measurements under service conditions again showed a similar relation between the wear rating of SBR and NR. However, the temperature at which the two rubbers had the same rating was higher (53).

Undoubtedly there are severity effects that influence the wear rating of tread compounds. However, it appears that temperature plays such an important role that it can reverse the ranking of compounds in a way that could not have been foreseen from severity effects such as surface and pressure dependence of the abrasion rate.

Accepting that service conditions as far as they affect the relative rating of compounds are primarily described by the tire surface temperature, average relative wear ratings of tread compounds can be estimated from a wear rating-tire surface temperature relation and a knowledge of the average tire surface temperature to be expected in service. The rating temperature relation can be obtained fairly quickly with a trailer running at various slip angles. The average tire surface temperature in service would depend on the geographical situation and on the habits of the driver (65).

3.3.5 *The Velocity Dependence of Tire Wear Due to Cornering*

Driving a car around a corner of a given radius at various speeds produces wear that increases very rapidly with the speed, as shown in Fig. 43. This increase must not be confused with the increase in abrasion with increasing speed discussed in Section 3.1.4.

When a car is driven around a corner the centrifugal forces have to be balanced by the side forces of the tires; i.e., the tires must run at a slip angle to the forward velocity of the vehicle. Assuming for simplicity's sake that all four wheels of the vehicle coincide with the center of mass of the car, the centrifugal force F_c is then balanced by the side force of the tires

$$F_c = \frac{mv^2}{R} = \text{const } \Theta \tag{20}$$

where m is the mass, v is the forward velocity, R is the radius of curvature, and Θ is the slip angle of the tires. But as the abrasion loss is to

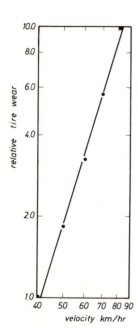

Figure 43 Tire wear as a function of vehicle speed on the Monthlery racing track (66).

a first approximation proportional to the square of the slip angle, the abrasion loss A can be written

$$A = \text{const} \frac{v^4}{R^2} \qquad [21]$$

The very steep dependence of the abrasion loss on speed is due to purely mechanical reasons; it is derived directly from the behavior of slipping wheels and does not stem from any changes in the abrasion mechanism or the material properties of the tread compounds. Temperature considerations and thermal effects stemming from them would tend to increase the abrasion loss still further. It will be noticed that the slope in Fig. 43 is less than 4. This is because the drivers were unable to stick precisely to the same radius of curvature at all speeds. At high speeds they tended to "cut" the corners a little.

3.4   Conclusions on Abrasion and Tire Wear

It is clear from what has been said about abrasion and tire wear that the vision that has sustained research in this subject for so long of finding a single simple laboratory test to determine or at least to estimate the likely wear resistance of a particular compound in service or, more modestly still, to determine its rating in relation to a known compound is not likely to become fact. There are too many unknowns in the abrasion process which to a varying and almost unpredictable degree influence tire wear in service.

Nevertheless, the service testing of the wear resistance of tread compounds is a very expensive item in the development of new compounds. An intelligent screening can be achieved quickly and efficiently by using trailers, which allow a measure of control over the slip conditions, thereby eliminating driver influence. Because wear is accelerated, results are obtained quickly under reasonably constant weather conditions. But even here, wrong conclusions are possible because accelerated tire wear also brings about an increased temperature at the tire surface, which, as has been shown, influences the wear rating of tread compounds.

4   FATIGUE FAILURES OF TIRE COMPOUNDS

4.1   Introduction

A tire must not fail in any other way but through wear. For a passenger car tire a reasonable tire life may be anywhere from 40,000 to 65,000 km. During this life each point on the tire passes through the contact area about $2-3 \times 10^7$ times, i.e., it has to sustain that number of stress-strain cycles of an amplitude which depends on its position on the tire, the inflation pressure, and the driving habit of the user. No cracks should appear and no pieces should break off from the pattern ribs during this time.

Two regions in which cracks might appear prematurely are the grooves of the tread pattern and the sidewall. Groove and sidewall cracking are essentially low strain fatigue failures, while a third phenomenon, rib tearing and profile chunking, is a high strain fatigue failure limited mainly to heavy service tires or passenger car tires when driven under extremely severe conditions, such as during rallies.

A third region occurs at construction discontinuities, such as ply turnups, bead reinforcing structures, or at the edge of breakers in the case of radial ply tires. Since these are regions inside the tire a systematic investigation of the precise nature of failures is very complex, particularly because in-service failure is due not only to the repeated stressing but also to the temperature buildup that occurs.

4.2 Tearing Energy and Cut Growth

Considerable evidence has been gathered since Rivlin and Thomas first reported their concept of critical tearing energy (49), that virtually all dynamic mechanical failure phenomena of elastomers are related to a fundamental material property, namely their cut growth characteristics (50) as described by

$$\frac{\Delta c}{\Delta n} = f(T) \tag{22}$$

where $\Delta c/\Delta n$ is the rate of growth of a cut c when the number of deformation cycles n has increased by Δn, T is the tearing energy; and f is a characteristic function of the material. A typical curve for NR showing the relationship in equation [22] is given in Fig. 6 of Chapter 9. The discussion in Section 5 of Chapter 9 should be of interest since it treats this same subject from a chemical viewpoint.

Extensive experimental evidence (50) shows that in general $\Delta c/\Delta n$ takes the forms

$$\frac{\Delta c}{\Delta n} = A(T - T_0) \qquad \text{for small values of} \quad T \tag{23}$$

and

$$\frac{\Delta c}{\Delta n} = BT^\beta \qquad \text{for larger values of} \quad T \tag{24}$$

A, T_0, B, and β determine the cut growth behavior of a material for the whole range of tearing energies from the limiting value T_0, below which no mechanical tearing takes place, to T_c, the critical tearing energy at which catastrophic tearing occurs.

The value of T depends on the shape of the test piece (49, 67) and the strain energy density. It is defined basically as

$$T = -\left[\frac{dE}{dA}\right]_L \qquad [25]$$

where E is the elastic energy stored in the test piece and A is the new surface area created as the cut proceeds. The constant L denotes that during the process of crack propagation the test piece dimensions should remain constant, a condition which in practice is not always fulfilled.

From this definition, usable forms for a number of experimental test pieces can be derived. The two that are important in the present context are (a) the tensile strip with a cut length c as shown in Fig. 44a for which

$$T = 2KWc \qquad [26]$$

where W is the stored strain energy density for the strain amplitude of the deformation cycle, and is easily obtained from the stress-strain curves as

$$W = \int \sigma \, d\epsilon \qquad [27]$$

K is a numerical constant that depends a little on the strain but is approximately 2, σ is the stress, and ϵ is the strain imposed on the test piece; (b) the pure shear test piece (Fig. 44b) for which

$$T = Wh_0 \qquad [28]$$

where h_0 is the free length of the test piece.

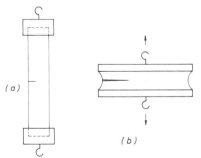

(a)

(b)

Figure 44 Two test pieces for which tearing energies can be calculated from easily measurable parameters: (a) tensile strip; (b) pure shear test piece.

For a long cut the relation [26] for the tensile strip approaches relation [28]. Thus for the same W, the tearing energies are the same for both test pieces when

$$2Kc = h_0 \qquad [29]$$

i.e., when

$$c \simeq \tfrac{1}{4}h_0 \qquad [30]$$

since $K \simeq 2$. When the cut reaches a length equal to or larger than a quarter of the free length of the test piece, it would be expected that the tearing energy would become independent of the cut length c.

4.3 Groove Cracking in Tires

4.3.1 *Tearing Energies in the Grooves of Tires*

It is generally not possible to calculate accurately the tearing energy operating in a tire, because a tire is a complex structure in which the assumption of a uniform strain field is not applicable. Lake (68, 69) has attempted to calculate tearing energies in the grooves of heavy service tires by assuming that the grooves act like pure shear test pieces. He obtained an estimate of the stored energy density by inserting cuts of various lengths and depths into the grooves and determining their width. It is clear from Fig. 45 that the width w of a cut in a pure shear test piece is related to the strain by

$$w = \epsilon h_0 \qquad [31]$$

provided the cut is so long that the region near the center of the cut is

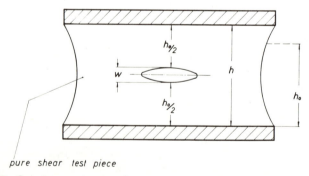

pure shear test piece

Figure 45 Opening w of a cut of length c in a pure shear test piece under a strain $(h - h_0)/h_0$, where h is the strained and h_0 the unstrained width of the test piece.

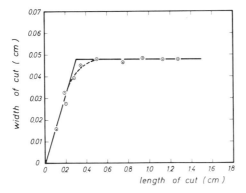

Figure 46 The opening w of a cut in a pure shear test piece as a function of its length.

strain free. For shorter cuts the width w of the cut is proportional to its length c:

$$w = 2\epsilon c \qquad [32]$$

Therefore, plotting the width w as a function of the cut length c produces a graph as shown in Fig. 46 with an initial slope of 2ϵ, according to equation [32], and a final ordinate ϵh_0, according to equation [31]. This enables the determination of both the strain and the free length h_0. For a tire groove the latter is approximately equal to the groove width.

The author has used this method to determine semistatically the strains in grooves of passenger car tires under load. Figure 47 shows the strains in the four grooves as a function of the periphery angle, with zero at the center of the contact area. Inside the contact patch the grooves are in compression and no measurements can be taken with this method. At points well away from the contact region the grooves are in tension, with strains reaching approximately 5%. This is due to the inflation pressure. Just outside the contact region, however, the strain rises to almost three times this value.

Assuming that the groove acts like a pure shear test piece, the tearing energy T can be calculated from equation [28] to be

$$T = \tfrac{2}{3}E\epsilon^2 h_0 \qquad [33]$$

where W, the stored energy density, has been replaced by E, the modulus of the material, and ϵ, the maximum of the strain cycle. The equivalent groove width h_0 is determined from graphs such as that shown in Fig. 46 and discussed above. It must be remembered that this value of the tearing energy is only reached when the cut already has a length of at least

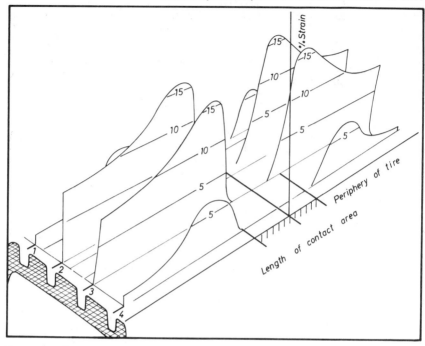

Figure 47 The strains in the grooves of a loaded passenger tire as a function of the periphery angle with zero strain at the center of the contact patch.

one-quarter of the free width of the groove (see equation [31]), i.e., at least 2 mm in most cases for passenger tires. For smaller cuts the tearing energy would be smaller and would increase with the cut length until the limiting value above had been reached.

4.3.2 *The Rate of Crack Propagation*

Figure 48 shows the growth of single cracks in the outside and center grooves of cross ply passenger car tires as a function of the running distance, since the latter is proportional to the number of deformation cycles. These curves were obtained by running a car on the expressway in Northern Italy at a constant speed of 130 km/hr under very hot but constant dry weather conditions with an average ambient temperature of 30°C. A total of eight different compounds were tested, all mounted on similar cross ply carcasses. In each case cracks grew linearly with traveling distance in the range over which measurements were taken.

Such linear growth would be expected if the tearing energy does not depend on the crack dimensions, and this would be expected after the crack has reached approximately 2 mm in length, as pointed out above. In fact, if the tearing energy concept is correct, the slopes of the crack length-distance curves can be calculated from laboratory measurements.

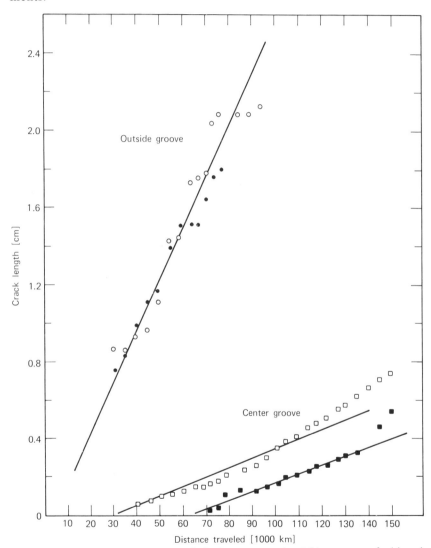

Figure 48 The growth of single cracks in the center and outside grooves of a bias ply car tire during a motorway trial.

From equation [23] the slope b becomes

$$b = \frac{\Delta c}{\Delta s} = 500 \times \frac{\Delta c}{\Delta n} = 2A(T - T_0) \times 500 \qquad [34]$$

where s is the distance covered in kilometers; the factor 500 denotes the fact that 500 revolutions were made by the wheels of the car to cover 1 km; and A and T_0 are the cut growth constants at small tearing energies. The factor 2 arises because cracks grow simultaneously in both directions.

Table 7 shows average slopes b for five different compounds as they occurred in the grooves of the tires that were used in the trial described above. For two of the compounds laboratory cut growth measurements were carried out. To simulate the road conditions as far as possible, these measurements were made at 80°C under nitrogen and air and at a frequency of 30 Hz. Measurements in two atmospheres were included because the oxidation effect on small test pieces was judged to be much more severe than would be the case with relatively bulky tires.

The laboratory measurements are seen to rate the two compounds in the same way as the road trials, although in absolute measure the agreement is only moderate. In particular, it would have been expected from strain measurements that the rate of crack growth would have been larger in the center grooves than in the outside ones, while in fact the opposite was the case. This is probably because the strains in the grooves

Table 7   Rates of Crack Growth in Tire Grooves During a Motorway Groove Cracking Trial

	Rate of Groove Cracking, b (cm/1000 km)					
	Road Trial		From Laboratory and Static Strain Measurements			
			Outside Groove		Center Groove	
Compound	Outside Groove	Center Groove	Air	N_2	Air	N_2
100 NR[a]	0.27	0.052	0.112	0.106	0.25	0.23
90 NR/10 BR	0.19	No cracks				
80 NR/20 BR	0.17	No cracks				
70 NR/30 BR	0.16	No cracks	0.075	0.068	0.19	0.17
80 SBR/20 BR	0.30	No cracks				

[a] All compounds were extended with 50 phr oil and filled with 82.5 phr HAF carbon black.

of the tire were of necessity determined semistatically, which can undoubtedly differ substantially from dynamic measurements.

4.3.3 Crack Initiation

The preceding analysis holds only for cracks that have already reached a length of about 2 mm. Below 2 mm the tearing energy depends on the crack dimensions and increases with the crack length. If the crack length is so small that the tearing energy is smaller than the fatigue limit T_0, cracks cannot grow by mechanical means, although they can still grow under ozone attack. Since cracks have to reach a critical initial length:

$$c_i = \frac{T_0}{2KW} \qquad [35]$$

where the symbols are the same as in equation [26], from which equation [35] has been derived, the interval before crack growth begins to become critical depends on the fatigue limit T_0 of the compound. Figure 49 shows the average crack length as a function of distance covered for five compounds differing in their polymer content. Because results tend to scatter, average crack lengths for a large number of single cracks were calculated by means of regression analysis. The straight lines shown were fitted to the data by the least squares method.

As the BR content in the basic NR compound is increased the cracking rate decreases slightly (see also Table 7), but more important, the onset of cracking is delayed. The SBR/BR blend included in the tests required the largest distance before cracks appeared, but once there, they grew faster than in any of the other compounds. Table 8 lists the fatigue limits T_0 for two of the compounds obtained in air and nitrogen at 80°C and 30 Hz. The presence of BR has increased the fatigue limit by about 50%. It is known that SBR has a higher fatigue limit than NR (50) so that the SBR/BR blend would be expected to have the longest initiation interval. The faster growth rate after cracking has started would also be expected for the SBR/BR blend because it is known that the cut growth characteristics A, B, and β (equations [23] and [24]) are larger for SBR than for NR (50). The effect of T_0 is also felt after the initial crack length c_i has been reached, because T_0 significantly influences the effective tearing energy $(T - T_0)$ as long as T is of the same order of magnitude as T_0. Thus, under the strain conditions that prevail in cross ply passenger tires, it would take approximately 1900 km further running distance for a cut to grow from 1 mm length to 2 mm for the 70 NR/30 BR blend than for the basic NR compound.

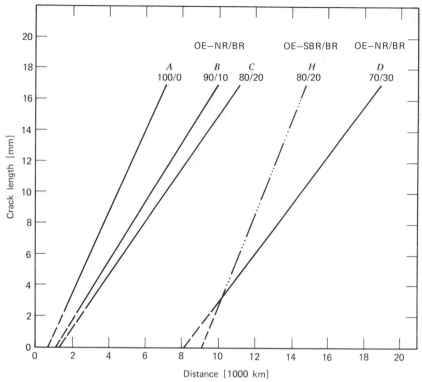

Figure 49 The average crack length as a function of distance run for five different polymer blends on cross ply tires.

The smallest cut growth length from which cracks could grow by the mechanooxidative process for which the cut growth characteristics apply can be estimated from equation [35]. With the values for the energy density W in the grooves of cross ply passenger tires this would lead to c_i values of approximately 0.05–0.08 cm. Since these are very large, either considerable ozone attack would have to exist for the tire to reach such

Table 8 Limiting Tearing Energies T_0 (N/cm) (Fatigue Limit) for Two of the Compounds of Table 7

	100 NR	70 NR/30 BR
Air	0.4	0.6
Nitrogen	0.5	0.7

flaw sizes or initiation by inflicting flaws on the tire surface by external means would have to be stipulated. Lake and Thomas (69), in investigating groove cracking on bus tires, suggested that this could come about through stone cutting. In the experiments described here this was almost certainly not the case. Instead, since cracks always grew from sites that were well defined by the tread profile, it is possible that "flaws" in the rubber were created through local overheating or local stress concentrations.

4.3.4 *Conclusion on Groove Cracking*

Although the application of cut growth theory to groove cracking is only semiquantitative, it allows a rational explanation of a number of empirical rules regarding the prevention of groove cracking (70):

1. It is known that the presence of BR, particularly in NR compounds, increases the cracking resistance of the compound. It is shown that this is due to the higher fatigue limit of BR. Even a small increase is sufficient to delay the onset of cracking significantly.

2. The tearing energy operating in the groove is determined by the stress-strain characteristics of the cords in the carcass, which bear the load due to inflation pressure and the forces on the tire. The lower the strain in the cords for a given load, the lower are the tearing energies in the rubber compound surrounding them. Nylon, which, in use, creeps more than rayon, tends to increase the strains in the grooves of the compounds. Without post-inflation, which takes out most of the growth of nylon directly after curing, nylon would create major problems with groove cracking. Steel, on the other hand, would probably eliminate the phenomenon altogether. Radial ply tires, with the very stiff belt underneath the tread, also have much lower strains in the grooves, and groove cracking has never been a problem with these tires.

3. Since the strain in the rubber is determined by the cords, the tearing energy in the rubber can be reduced by lowering the modulus. Thus, too high a black loading encourages groove cracking. Oil extension, on the other hand, would by itself not be expected to have an adverse effect on groove cracking unless its softening effect would be overcompensated by excessive black loading.

4. Driving habits can clearly influence the occurrence of groove cracks. Fast long-distance driving on expressways in a warm climate would encourage its formation because of high temperature conditions and mild wear. Continuous use under severe wear conditions would shorten the life of a tire so much that groove cracking would not become a problem.

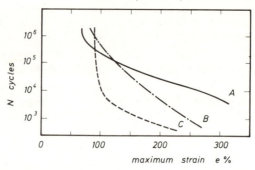

Figure 50 Fatigue life of different elastomers as a function of the strain amplitude. (50). *A*, NR gum; *B*, SBR gum; *C*, BR gum.

4.4   Rib Tearing and Chunking

Unlike groove cracking, rib tearing and chunking occur only under severe driving conditions. Rib tearing is mainly confined to heavy service vehicles, driving across large obstacles such as pavement stones, when the whole load is momentarily transferred to a single rib of the tire. Chunking is similar in that pieces are torn out of the profile when driving under severe conditions of speed and slip. It is beyond present knowledge to calculate tearing energies operating during this process; it is obvious, however, that this is a large strain fatigue phenomenon. Figure 50 shows the fatigue lives of a number of compounds based on different elastomers as a function of strain. At low strain amplitudes both BR and SBR have fatigue lives superior to that of NR; at high strains, however, the reverse

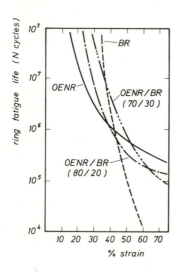

Figure 51 Fatigue life of NR, BR, and two NR/BR blends as a function of strain.

is the case. Natural rubber would therefore be the preferred elastomer if rib tearing has to be contended with.

Blending can produce a synergism which gives acceptable behavior over the whole strain region. NR/BR blends are good examples. At blend ratios up to 70 NR/30 BR, the BR improves the low strain fatigue properties of the NR and hence its groove cracking resistance, with only a small loss in the high strain fatigue resistance, as seen in Fig. 51. Hence NR/BR blends are used today in heavy service tires to protect the tire against groove cracking failure without reducing the rib tearing resistance unduly.

4.5 Sidewall Failures

4.5.1 *The Strains in the Sidewalls of Tires*

Sidewalls are subject to failure through fatigue cracking because, like the grooves, they are continuously subjected to a dynamic strain cycle. In addition, cracking through ozone attack may play a role.

It is not possible to calculate tearing energies for sidewalls because their strain pattern cannot be easily compared with those of simple laboratory test pieces. Strains in sidewalls vary with the distance r from the axle and the rotational angle ϕ measured from the center of the contact patch. Moreover, not only the magnitude but also the direction of the principal strain is dependent on r and ϕ. To determine the principal strains ϵ_1 and ϵ_2 (the extensive strain and the compressive strain, respectively) and their directions, three measurements are required (71).

Dynamic strains as they occur in the sidewall, in the radial, tangential, and 45° directions, are usually measured with strain transformers in the shape of soft U-springs that have strain gages attached. Figure 52 shows the principal strains ϵ_1 in the sidewall of a radial ply tire along a constant radius. The principal strains ϵ_1 are largest before entering the contact region and on leaving it. On entering, their direction is approximately 45° to the radial direction, counting counterclockwise as positive, but is changed to approximately $-45°$ on leaving. Hence a rotation in the direction of approximately 90° occurs as a point passes through the contact region. At a point opposite the center of the contact region the principal extensive strain is in the radial direction. Unlike grooves, strains in sidewalls do not, in general, relax completely but change only in magnitude and direction. This too complicates any calculation of tearing energies. Strains at different radii around the radial ply tire follow approximately the same pattern as far as their direction is concerned, but differ in magnitude.

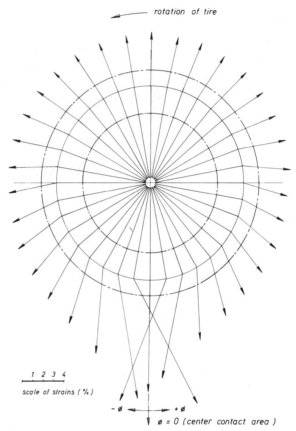

Figure 52 Principal strains ϵ_1 along a constant radius r in the sidewalls of a radial ply car tire as a function of the rotational angle ϕ.

In cross ply tires, strains are somewhat smaller. The largest strains occur at the center of the contact area in the tangential direction, while at all other points the strains are again approximately in the radial direction, as shown in Fig. 53. Again the strain cycle is nonrelaxing for each point as the tire rotates.

4.5.2 Crack Propagation in Sidewalls

In order to study the effect of strains and their direction on the crack growth rate in sidewalls of tires, cuts of equal length and depth were inserted at various points and directions in the sidewalls of radial and cross ply tires. The tires were then run under constant load, inflation pressure, and speed on an outdoor wheel, and the growth of the cracks was

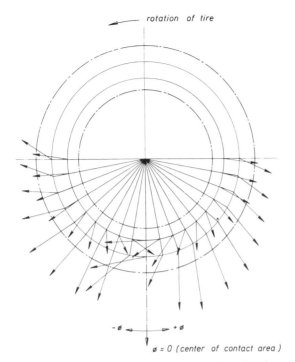

Figure 53 Principal strains ϵ_1 at various radii in the sidewalls of a cross ply car tire as a function of the rotational angle ϕ.

measured at regular intervals. Figure 54 shows the direction of the cuts that were originally inserted and the direction of the cracks that grew from the cuts. It is seen that the cracks always grow in a direction normal to that of the maximum principal strains, irrespective of the direction in which cuts were inserted into the sidewalls. For radial ply tires the principal maximum strains lie both $+45$ and $-45°$ to the radial direction, so that cracks would be expected to grow normally to either one of these directions, which is always the case.

For cross ply tires, the maximum principal strain is in the tangential direction, so that the growth of cracks would be expected to be in the radial direction. This was again the case, also irrespective of the direction in which cuts were inserted into the tire.

The growth of single cracks as a function of the number of radial ply tire revolutions, i.e., the number of strain cycles, is shown in Fig. 55 for cuts inserted at different points along a constant radius, all in the same direction. Individual cuts, although nominally under the same strain condition because they were cut in the same direction and at the same

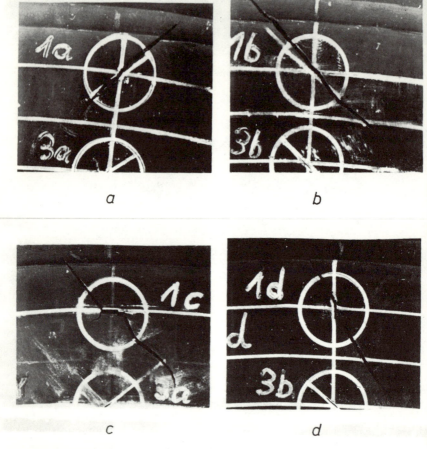

Figure 54 Direction of cuts inserted into sidewalls of radial ply tires and the direction of cracks growing from them.

radius from the center, show considerable differences in their rate of growth as shown by the different curves in Fig. 55. Initially the crack rate increases with the number of strain cycles, but after the cuts have reached a length of between 10 and 20 mm the crack rate becomes constant.

Applying, for brevity's sake, the concept of tearing energy and the analogy of laboratory test pieces for which relations for tearing energies are known (Section 4.2), this behavior can be described satisfactorily if it is assumed that

$$\frac{\Delta c}{\Delta n} = 2A(T - T_0) \qquad [36]$$

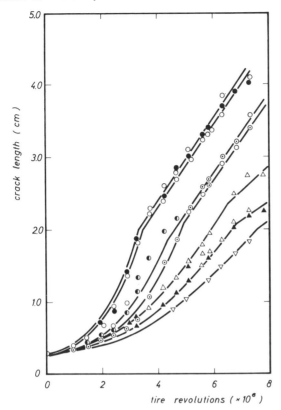

Figure 55 Growth of single cracks in the sidewall of a radial ply tire as a function of the number of revolutions. Different lines represent the growth of different cuts all under identical experimental conditions, indicating the statistical nature of cut growth phenomena.

where $T = 4Wc$ for small crack lengths c (below 10–20 mm) and $T = Wh_0$ for large crack lengths (above 20 mm), A is a crack growth constant, and h_0 is the equivalent width of a pure shear test piece which the sidewall would represent. The stored energy density W is calculated from modulus and strain measurements.

Values of A and h_0 obtained from average cut growth rates on points along two different radii but all in the direction of the maximum principal strains are shown in the first two rows of Table 9. The stored energy density was calculated using the maximum strains and ignoring the fact that these strain cycles were nonrelaxing. As this is known to affect the rate of crack growth, cracking rates were also measured in the laboratory simulating the amplitude of the strain cycle as it occurred in the sidewalls, ignoring, however, its changes in direction. For comparison a growth rate constant obtained in the laboratory with pure shear test pieces has been shown in the last column of Table 9. Considering the very crude assumptions that were made, the agreement between tire and laboratory measurements is remarkably good.

The crack growth rate from cuts inserted in the radial or tangential direction instead of the direction of principal strain is slower at first but then becomes similar to the rate of cracks starting from cuts in the directions of principal strains in that it increases with the cut length c for small cut lengths and becomes independent of the cut length once the cuts have reached a length of about 1 cm. The initially slower rates are presumably due to the fact that the effective cut length normal to the principal strain direction is reduced so that an effectively smaller tearing energy is operating. Using as effective crack length the projection at a direction normal to the direction of the maximum principal strain, cracking rate constants A and equivalent widths h_0 can also be calculated. They are shown in the last two rows of Table 9 for tangential and radial cuts along a constant radius. It is interesting that the h_0 values deduced from cut growth measurements are at least the right order of magnitude to have some geometric association. They are of the same order in magnitude as the width of the sidewall.

Table 9 Cut Growth Constants A and Equivalent Pure Shear Test Piece Width $h_0{}^a$

	ϵ_{max} (%)	ϵ_{min} (%)	A (10^{-7} cm^2/N)	h_0 (cm)	A from Laboratory Measurements (10^{-7} cm^2/N)
1	13.0	5.0	0.25	2.9	0.36
2	11.2	5.0	0.30	2.7	
Point 1 initial cut in radial direction	13.0	5.0	0.20	7.9	
Point 1 initial cut in tangential direction	13.0	5.0	0.21	7.8	

^{a} Determined in the sidewall of a radial ply tire according to $\Delta c/\Delta n = 2A(T - T_0)$ and in the laboratory on pure shear test pieces.

Unlike groove cracking or mechanical sidewall cracking, ozone cracking is not bound to a dynamic stress-strain cycle (a tire has to run under load before groove cracking will ever commence). It requires only a static minimum tearing energy (72, 73). If this is present, for instance because the inflation pressure produces sufficiently large strains in the sidewall, cracking will begin. Cracks increase in length with time, independent of the number of revolutions the tire executes. Their rate of increase depends on the ozone concentration and the type of polymer. The nature of ozone attack and methods of protection against ozone initiated degradation are considered in greater detail in Chapter 9, Sections 4 and 5.

5 CONCLUSION

There is no single polymer that can fulfill all the requirements demanded in the various parts of a tire. Because of the large number of special polymers available, many different compounds can be employed in one tire, each selected to best suit the demands of the particular area in which it is used. Today compounds consist almost invariably of blends of polymers. Thus, the high groove cracking resistance of BR and the high rib tearing resistance of NR can be combined by blending to produce through a synergism a compound that is acceptable in both properties.

In passenger car tread compounds a balance must be achieved between skid resistance and wear. Here, physical properties naturally favor the use of SBR in warm climates i.e., under conditions in which friction on wet roads must combine with good wear resistance. The use of BR improves the wear resistance still further, but lowers the skid resistance. Oil extension can compensate for this, so that oil extended SBR/BR blends have become established as the best wear and skid resistant tread compounds for summer passenger tires. For winter tire use, on the other hand, natural rubber or polyisoprene has the optimum physical properties of high skid resistance on ice and a high wear resistance at low ambient and tire surface temperatures.

This chapter has dealt only with the properties of the vulcanized tire in use. There has not been enough room to discuss behavior of carcass compounds in tires. High adhesion to the cord material and a low heat buildup are their main requirements. These are best met by blends of NR with SBR and a low reinforcing black. The more stringent the requirement for low heat buildup, the higher is the content of NR.

Because of the wide choice of elastomers available, tires of virtually optimal properties in every respect can be offered to the customer. Ever increasing demands of safety, speed capability, and economy require us to keep abreast with new developments in tire materials and construction.

REFERENCES

1. A. Schallamach and D. M. Turner, *Wear*, **3**, 1 (1960).
2. A. Schallamach, "Skid Resistance and Directional Control," in S. K. Clark, ed., *The Mechanics of the Pneumatic Tire*, Nat. Bur. Stand. Monograph 122, U.S. Gov. Printing Office, Washington, D.C., 1971.
3. V. E. Gough, *Automob. Eng.*, **44**, 137 (1954).
4. B. Förster, *Dtsch. Kraftf. Forsch. Zwischenbericht*, **22** (1938).
5. T. J. P. Joy and D. C. Hartley, *Tyre Characteristics as Applicable to Vehicle Stability Problems*, Institute of Mechanical Engineers, London, Proceedings of the Automobile Division, 1954.
6. A. Schallamach, *Wear*, **1**, 384 (1958); *Rubber Chem. Technol.*, **31**, 982 (1958).
7. P. Thirion, *Rev. Gén. Caoutch.*, **23**, 101 (1946).
8. F. B. Bowden and D. Tabor, *Friction and Lubrication of Solids*, Oxford University Press, London, 1954.
9. K. A. Grosch, *Proc. R. Soc.*, **A274**, 21 (1963).
10. M. L. Williams et al., *J. Amer. Chem. Soc.*, **77**, 3701 (1955).
11. K. W. Wagner, *Elektrotech. Z.*, **36**, 135 (1915).
12. H. Leaderman, *Elastic and Creep Properties of Filamentous Materials and Other High Polymers*, The Textile Foundation, Washington, 1943.
13. J. D. Ferry, *Viscoelastic Properties of Polymers*, Wiley, New York, 1961.
14. T. L. Smith, *J. Polym. Sci.*, **32**, 99 (1958).
15. K. A. Grosch, Ph.D. Thesis, University of London, 1963.
16. K. A. Grosch and A. Schallamach, *Trans. Rubber Ind.*, **41**, T80 (1965); also *Rubber Chem. Technol.*, **39**, 287 (1966).
17. G. M. Bartenev, *Dokl. Akad. Nauk SSSR*, **96** (6), 1161 (1954) (DSIR translation).
18. A. Schallamach, *Proc. Phys. Soc.*, **66**, 386 (1953).
19. A. Schallamach, *Wear*, **6**, 375 (1963); *Rubber Chem. Technol.*, **39**, 320 (1966).
20. G. M. Bartenev and A. J. El'Kin, *Wear*, **8**, 8 (1965).
21. D. Bulgin et al., Proceedings of the 4th International Rubber Technological Conference, London, 1962, p. 173.
22. F. A. Greenwood and D. Tabor, *Proc. Phys. soc.*, **71**, 989 (1958).
23. K. A. Grosch, to be published.
24. S. C. Cohen and D. Tabor, *Proc. R. Soc.*, **A291**, 186 (1966).
25. M. H. Walters, in W. H. Edwards, ed., *Proceedings of the D. Mat. (Av.), Ministry of Technology Conference on Friction and Wear in Tires*, H.M.S.O., London, 1969, p.29.
26. A. D. Roberts and D. Tabor, *Wear*, **11**, 163 (1968).
27. B. Sabey, *Proc. Phys. Soc.*, **71**, 979 (1958); *Rubber Chem. Technol.*, **33**, 119 (1960).
28. A. D. Roberts, *Proceeding of the Symposium on the Physics of Tire Traction*, Plenum Press, New York, 1974, p. 179.
29. A. Schallamach, *Trans. Inst. Rubber Ind.*, **32**, 142 (1956).
30. A. Schallamach, in L. Bateman, ed., The Chemistry and Physics of Rubberlike Substances, MacLaren, London, 1963, Ch. 13.

31. K. A. Grosch, *Proceedings of the Symposium on the Physics of Tire Traction,* Plenum Press, New York, 1974, p. 143.

32. E. Southern and R. W. Walker, *Nature Phys. Sci.,* **237** (78), 142 (1972).

33. W. Gnörich and K. A. Grosch, *J. Inst. Rubber Ind.,* **6,** 5, 192 (1972).

34. G. Maycock, *Proc. Inst. Mech. Eng.,* **180,** 122 (1965–66).

35. K. A. Grosch, *Rubber Age,* **99** (2), 63 (1967).

36. K. A. Grosch and G. Maycock, *Trans. Inst. Rubber Ind.,* **42,** T280 (1966).

37. British Road Research Laboratory, *Instructions for Using the Portable Skid Resistance Tester,* Department of Scientific and Industrial Research, Road Research Note No. 27, H. M. Stationery Office, London, 1960.

38. E. M. Bevilacqua and E. P. Percapio, *Rubber Chem. Technol.,* **41,** 832 (1968).

39. C. G. Giles et al., *Development and Performance of the Portable Skid Resistance Tester,* Department of Scientific and Industrial Research, Road Research Technical Paper No. 66, H. M. Stationery Office, London, 1964; *Rubber Chem. Technol.,* **38,** 840B (1965).

40. T. French and R. G. Patton, Proceedings of the 4th International Rubber Technological conference, London, 1962, p. 196.

41. K. A. Grosch et al., Proceedings of the International Rubber Conference, Moscow, 1969.

42. S. B. Ratner and G. S. Klitenik, *Zav. Lab.,* **25,** 1375 (1959); D. James, ed., *Abrasion of Rubber,* MacLaren, London, and Palmerston Publishing Co., New York, 1967, p. 64.

43. K. A. Grosch and A. Schallamach, *Kaut. Gummi Kunstst.,* **22,** 288 (1969).

44. M. M. Reznikovskii, *Sov. Rubber Technol.,* **9,** 32 (1960); D. James, ed., *Abrasion of Rubber,* MacLaren, London, and Palmerton Publishing Co., New York, 1967, p. 119.

45. A. Schallamach, *Trans. Inst. Rubber Ind.,* **28,** 256 (1952).

46. A. Schallamach, *J. Polym. Sci.,* **9,** 385, (1952).

47. A. Schallamach, *Proc. Phys. Soc.* **B67,** 883 (1954).

48. D. Bulgin and M. H. Walters, Fifth International Rubber Conference, Brighton, 1967.

49. R. S. Rivlin and A. G. Thomas, *J. Polym. Sci.,* **10,** 291 (1953).

50. G. Lake and P. B. Lindley, *Rubber J.,* **146,** 24 (1964).

51. G. I. Brodskii et al., *Sov. Rubber Technol.,* **19,** No. 8, 22 (1960).

52. A. Schallamach, *J. Appl. Polym. Sci.,* **12,** 281 (1968).

53. K. A. Grosch, in W. H. Edwards, ed., *Proceedings of the D. Mat. (Av.), Ministry of Technology Conference on Friction and Wear in Tires,* H.M.S.O., London, 1969, p. 69.

54. K. A. Grosch and L. Mullins, *Rev. Gén. Caoutch.,* **39,** 1781 (1962).

55. S. L. Madorsky, *Thermal Degradation of Organic Polymers,* Wiley-Interscience, New York, 1964.

56. M. M. Reznikovskii and G. I. Brodskii, in D. James, ed., *Abrasion of Rubber,* MacLaren, London, and Palmerton Publishing Co., New York, 1967, p. 14.

57. W. Viehmann, *Kaut. Gummi,* **8,** WT 227 (1955).

58. K. A. Grosch and A. Schallamach, *Wear,* **4,** 356 (1961).

59. S. Davidson et al., *Rubber World,* **151,** 83 (1965).

60. J. Mandel et al., *Ind. Eng. Chem.,* **43,** 2901 (1951).

61. H. C. J. deDecker et al., Proceedings of the 3rd Rubber Technology Conference, London, 1954, p. 749.

62. C. Prat, *Rev. Gén. Caoutch.*, **32,** 991 (1955).

63. H. Westlinning, *Kaut. Gummi,* **9,** WT 273 (1956).

64. A. Schallamach, *J. Inst. Rubber Ind.*, **1,** 40 (1967).

65. K. A. Grosch, *J. Inst. Rubber Ind.*, **1,** 35 (1967); J. A. Brydson, ed., *Developments with Natural Rubber*, MacLaren, London, 1967, p. 113.

66. H. Geesink and C. Prat, *Rev. Gén. Caoutch.*, **33,** 973 (1956).

67. A. G. Thomas, *J. Appl. Polym. Sci.*, **3,** 168 (1960).

68. G. J. Lake, Proceedings of the Conference of Yield and Fracture of Materials, Cambridge, 1970, p. 5.311.

69. G. J. Lake and A. G. Thomas, in press.

70. R. Snyder, Gordon Research Conference on Elastomers, New London, New Hampshire, 1964.

71. W. F. Kern, *Kaut. Gummi Kunstst.*, **13,** No. 3 (1960); W. F. et al., *Kaut. Gummi Kunstst.*, **16,** No. 11 (1963).

72. A. N. Gent and M. Braden, *J. Appl. Polym. Sci.*, **3,** 90 (1960).

73. M. Braden and A. N. Gent, *J. Appl. Polym. Sci.*, **6,** 449 (1962).

Index